Electron Microscopy and Analysis Group Conference 2009

(EMAG 2009)

AA002768

Journal of Physics: Conference Series Volume 241

Sheffield, United Kingdom
9-11 September 2009

Editors:

Richard Baker
Richard Brydson
Stephen Donnelly
Ian MacLaren
David McComb
Günter Möbus
Peter Nellist
Dogan Ozkaya
Thomas Walther

ISBN: 978-1-61738-807-1
ISSN: 1742-6588

Printed from e-media with permission by:

Curran Associates, Inc.
57 Morehouse Lane
Red Hook, NY 12571

Some format issues inherent in the e-media version may also appear in this print version.

Printed by Curran Associates, Inc. (2010)

For permission requests, please contact the Institute of Physics
at the address below.

Institute of Physics
Dirac House, Temple Back
Bristol BS1 6BE UK

Phone: 44 1 17 929 7481
Fax: 44 1 17 920 0979

techtracking@iop.org

Additional copies of this publication are available from:

Curran Associates, Inc.
57 Morehouse Lane
Red Hook, NY 12571 USA
Phone: 845-758-0400
Fax: 845-758-2634
Email: curran@proceedings.com
Web: www.proceedings.com

TABLE OF CONTENTS

PLENARY PRESENTATIONS

ADVANCES IN IMAGING AND SPECTROSCOPY TECHNIQUES

ELECTRON MICROSCOPY IN THE LIFE AND EARTH SCIENCES

FUNCTIONAL MATERIALS

HIGH RESOLUTION IMAGING AND SPECTROSCOPY OF MATERIALS

IN-SITU TEM

MODELLING AND QUANTIFICATION

NANOFABRICATION AND SEM

NANOMATERIALS

STRUCTURAL MATERIALS

PREFACE

The latest biennial conference of the Electron Microscopy and Analysis Group (EMAG) of the Institute of Physics was held at the University of Sheffield on 9–11 September, 2009. In addition, the Advanced School associated with the conference was run at the University of Sheffield on 8 September. It was particularly pleasing to return to Sheffield after ten years, the successful and memorable EMAG 99 having been held here too. The subject areas covered at EMAG 2009 were advanced electron microscopy techniques; investigating structure-property relationships in advanced materials; nanophysics and nanotechnology.

The EMAG 2009 conference attracted 172 delegates while the Advanced School had a full complement of eighteen attendees. Three plenary lectures were given to the whole conference and invited contributions were presented within the theme of each of nine parallel sessions. There were 54 contributed oral presentations within these parallel Sessions and a further 89 poster presentations. All authors were invited to contribute a paper to this Proceedings volume and 108 papers are presented here. I thank all who presented at EMAG 2009 and those who provided a paper for this Proceedings. Each paper was peer reviewed by two reviewers and I also want to thank those colleagues who helped with this essential task. In this volume, the plenary papers are presented first followed by all papers presented in each themed session. These sessions are ordered alphabetically. Within each Session, the invited presentations are presented first, followed by oral and poster contributions together.

Another activity of EMAG which is directed primarily at less experienced scientists is the Advanced School. This year, this was on Nanofabrication and Nanomanipulation and I want to thank Guenter Moebus and his colleagues at th University of Sheffield for putting on such an excellent Advanced School.

The EMAG series of conferences are well-known not only for the academic conference but also for the major trade exhibition which runs in parallel. This requires a great deal of additional planning and effort on the part of the conference department of the IoP but particularly by representatives of the exhibiting companies. This year there were 29 exhibitors, several of whom brought major items of equipment to demonstrate to delegates. Here I want to thank all the exhibitors, Jill Cowlard and Nicola Deedman of CEM and Pete Lander of JEOL for their efforts in making the trade exhibition such a success.

Finally, sincere thanks to the other members of the EMAG committee, especially Pete Nellist for his work on the scientific programme, and Guenter Moebus, Thomas Walther and colleagues for their invaluable work 'on the ground' at Sheffield, and also to Claire Garland and Lisa Cornwell at IoP for all their hard work and for keeping the academics under control!

Richard Baker
University of St Andrews
EMAG Chair and EMAG 2009 Proceedings Editor

Session Editors

Richard Brydson
Stephen Donnelly
Ian MacLaren
David McComb
Günter Möbus
Peter Nellist
Dogan Ozkaya
Thomas Walther

All papers published in this volume of *Journal of Physics: Conference Series* have been peer reviewed through processes administered by the proceedings Editors. Reviews were conducted by expert referees to the professional and scientific standards expected of a proceedings journal published by IOP Publishing.

Journal of Physics: Conference Series **241** (2010) 012001 doi:10.1088/1742-6596/241/1/012001

The nanoworld through aberration corrected lenses

David Cockayne

Department of Materials, University of Oxford, Parks Road, Oxford OX13PH, UK

david.cockayne@materials.ox.ac.uk

Abstract The advent of aberration correction for electron microscope lenses has produced a significant advance in the improvement of resolution in microscopy. This improvement, while significant in itself, promises to have its most profound impact in materials science when it delivers quantitative information to challenge models and modellers. This capability for an electron microscope-modelling synergy to deliver useful results at the atomic level is not yet firmly established. For this reason, one of the major challenges for electron microscopy in materials science over this decade is to study systems where electron microscopists and modellers can work collaboratively. This paper explores some examples.

1. Microscopy and modelling

The history of microscopy has many important milestones, but certainly the most significant of the past decade is the realization of aberration corrected electron lenses. Unlike the step change in resolution which electron lenses brought, the advantages of aberration correction are being more slowly realized and even more slowly brought to bear upon solving materials problems.

Discussions about aberration correction should involve far more than its effect upon resolution, but that is a reasonable place to start. The improvement in achievable resolution, however defined, while significant in itself, promises to have its most profound impact in materials science when it delivers quantitative information (e.g. atom positions, occupancy, local strain) to challenge models and modelers. This is because reliable predictive modeling is a powerful tool for developing new materials, revealing how a material might organize itself and respond to its environment – stress, temperature, local inhomogeneities. The reliability of modeling at the atomic scale is often in question because of the many parameters used (e.g. cell size, potentials, code), and it is this which can be tested against precise and accurate atomic-level experimental data (e.g. images) or high energy resolution EELS. Confirmation of modeling capability gives confidence in predictive modeling. For this reason, a major focus and one of the major challenges for electron microscopy in materials science over this decade should be to study systems where electron microscopists and modelers can work collaboratively, but where the free parameters (i.e. excuses for disagreement) are few.

Journal of Physics: Conference Series **241** (2010) 012001 doi:10.1088/1742-6596/241/1/012001

Success in this collaboration will require an objectivity about the interpretation and reporting of results which has, unfortunately, been somewhat lacking in our field in recent years. The pressures that the culture of "impact factors" brings sometimes result in the objective reporting of uncertainty, essential to the effective comparison between modeling and experiment, being lost. As an example, progress requires a clear reporting of experimental errors and their basis of determination, and, on the modeling side, a thorough understanding of the limitations of the code used. The increasing use of short Letters, with no follow-up substantive paper, contributes to this problem; however the opportunities for depositing the tedious but essential details on archived sites, or in methodology sections or supplements as an appendix to a paper, offer a solution and should continue to be explored.

Resolution and visibility

Rayleigh's criterion is always a starting point for considering resolution, even though it was established for a topic far from the consideration of electron microscope images – the distinguishability of two stars [1] – which accounts for the frequent reappraisal of a suitable definition for our subject. Many use *ad hoc* definitions to suit particular circumstances, but O'Keefe [2] has argued that the determination of resolution in atomic resolution electron microscopy should be based upon the distinguishability of atoms or atoms columns that exist in the sample, and by this definition resolution beyond the Rayleigh criterion has been achieved (e.g. [3]). Provided that "distinguishability" is well-defined, this is an objective approach. However "distinguishability of atoms or atoms columns" is not the same as "separation of atom peaks in the image" [2] unless one is certain that peak positions accurately reflect atom positions. Unfortunately the traffic-stopping outcome if this assumption *cannot* be made sometimes allows the traffic to flow where it should not. The precise determination of atomic positions in strained systems or studies of defect structures which depend upon a direct relationship between image intensity and projected potential are areas where this comment applies in particular. The question should be, what accuracy and precision is necessary to solve a particular problem? In this context, chasing resolution for its own sake is a sterile activity.

The ability to not only reduce but also choose particular values of aberrations has opened up the possibility of tuning lenses (TEM) and detectors (STEM) for particular purposes – both to improve resolution and to improve contrast. With the availability of computer codes (e.g. multislice codes) to simulate both STEM and TEM images, simulated images from model structures can be compared with the image, or with a set of images, to differentiate between candidate models, even for relatively complicated materials systems. This can be done even when there is no obvious relationship between the image detail and the model, even though then many parameters have to be determined (e.g. sample thickness, orientation, defocus, lens aberrations). Of particular interest, however, are those situations in which the image is directly interpretable in terms of structure. For TEM, we consider image formation as a two step process – formation of the exit wave function $\psi(r)$ and transfer of the exit wave function to the image, involving the aberration function $\chi(u) \equiv 0.25C_s\lambda^3u^4 + 0.5\ \Delta f\lambda u^2$, with the parameters having their usual meanings. For phase objects (i.e. $\psi(r)\ \psi^*(r)$ = constant), visibility of any detail in the image depends upon using $\chi(u)$ to shift the phase of some Fourier components at frequency u. If the sample is sufficiently thin to be a weak phase object and if the phase is proportional to the projected potential, then a uniform phase shift for all u (except u=0) results in an image wave function in which the contrast is linearly proportional to the projected potential. In HRTEM the value of defocus Δf is generally chosen to make the aberration function $\chi(u) \simeq \pi/2$ over as large a range Δu of u as possible, as in phase contrast light microscopy using a phase plate. Despite the attractive simplicity of this argument, the majority of samples, even those studied by HRTEM, are *not* weak phase objects. Then the transfer function reports the phase shift at frequency u caused by aberrations and defocus. For some purposes, (e.g. C nanotube imaging) uniform transfer of frequencies might be optimum, but, since there is often a complex relationship between the exit wave and the object, the ability to differentially phase shift particular frequencies could be useful. This is nowhere better illustrated than in the technique of through-focal series reconstruction, in which a set of images, each with particular frequencies enhanced, is used to reconstruct the complex exit wave function.

Maximum resolution is not the only aim of HRTEM – visibility of high resolution detail is also important. The recent work of Jia et al [4] discusses the image intensity

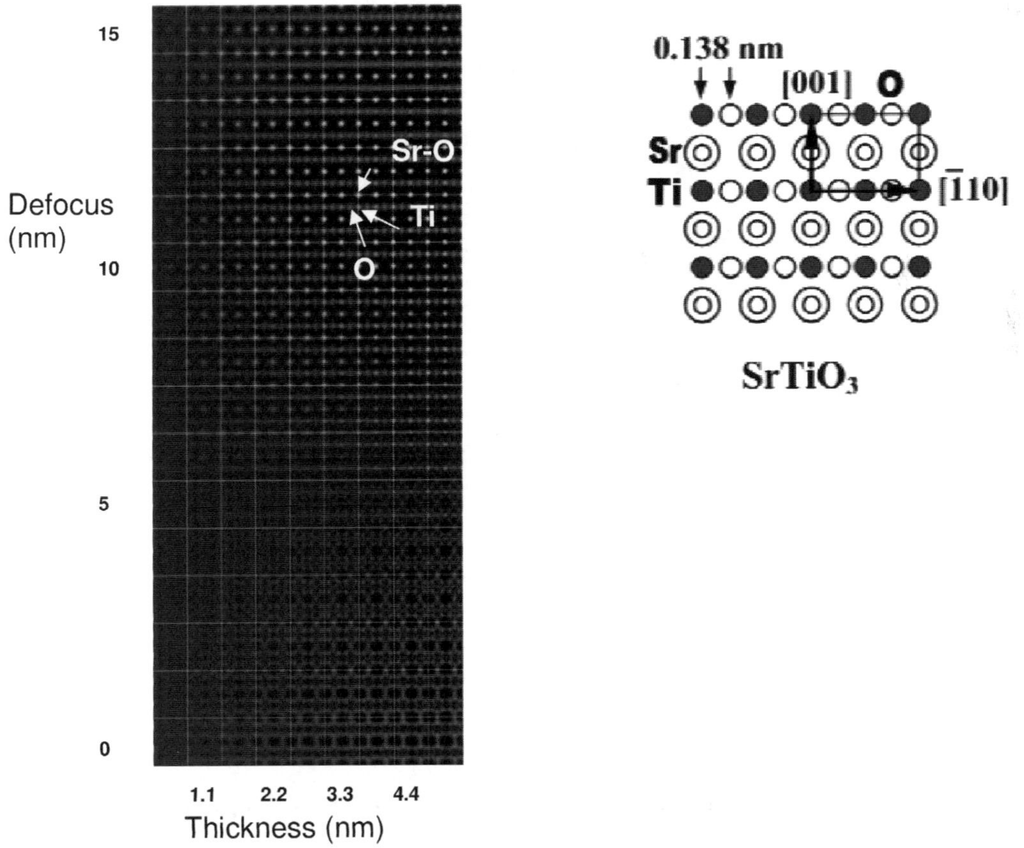

Figure 1 Defocus-thickness map for [110] SrTiO3 for C_s = - 40 micron (200kV) (courtesy S-J Shih)

from a weak phase object in the case where $\chi(u) \simeq -\pi/2$ over a wide range of u, achieved with negative C_s and defocus positive. The important outcome was that atomic columns of low-Z atoms showed high relative contrast using this technique, as seen in Figure 1 for [110] SrTiO3 for values of C_s (- 40 micron) and defocus (10 nm) (rather than C_s positive and defocus negative) set to give $\chi(u) \simeq -\pi/2$ for the JEOL aberration corrected MCO2200. It is seen that the map is a good representation of the projected structure up to 5.5 nm, after which the Sr columns fade.

Jia et al [4] emphasized that such an analysis is only valid for very thin samples, but their full dynamical simulations supported the idea that negative C_s combined with positive defocus can be used to enhance contrast from light atom columns. Other authors have discussed the opportunities offered for enhancing contrast and/or resolution by using negative C_s. Figure 2 shows the thickness defocus map for [001] SrTiO3 for a small negative $C_s = -2\ \mu$ (a value close to that proposed by Chang et al [5] for optimising the contrast transfer function for the JEOL 2200MCO).

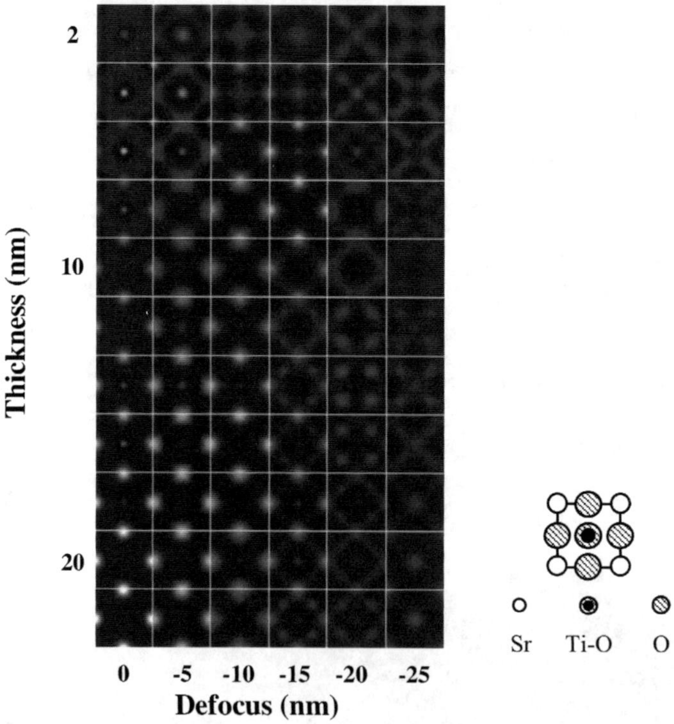

Figure 2 Defocus-thickness map of [001] $SrTiO_3$ for $C_s = -2\mu$ (courtesy S-J Shih)

It is seen that for the thicknesses of ~19.5 nm (well beyond the validity of the weak phase object approximation) and defocus near zero, only O columns appear bright. These are the parameters chosen by Shih [6] to successfully study the atomic structure of interfaces in $SrTiO_3$.

Urban et al [7] commented upon the observed advantage of the negative C_s technique for *thicker samples* in terms of the sample being a mixture of a phase object and an amplitude object. For the amplitude object, where the exit wave amplitude (at appropriate thicknesses) is maximum (or minimum) at the atomic columns (through channeling), then perfect transfer ($\chi(u) = 0$) is optimum. So, they argued, for the sample of intermediate thickness, optimum contrast is achieved for some C_s and defocus intermediate between those appropriate for the weak phase object (-40 μ and 10 nm respectively in their microscope) and the thick sample (zero C_s and zero defocus).

Aberration correction also provides new opportunities for improved contrast and resolution in STEM, by allowing use of larger beam convergence and detector angles. For example recently Okunishi et al [8] reported improved imaging of light elements in STEM using an annular detector positioned within the bright field cone.

Interfaces

Planar interfaces provide possibilities for collaboration between modeling and high resolution electron microscopy, of the kind discussed above, particularly if the crystallography of the material on either side of the interface is well-characterised.

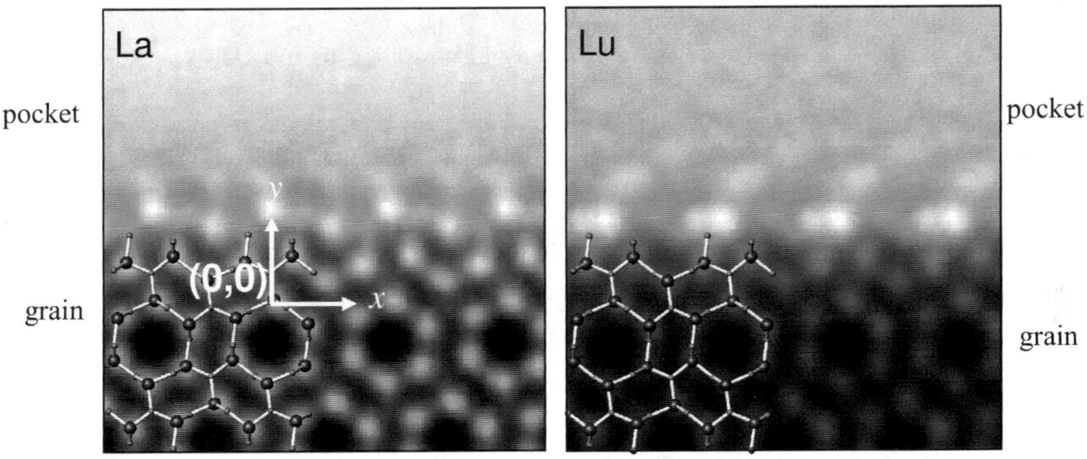

Figure 3 High resolution HAADF images (periodically averaged parallel to the interface) of rare-earth dopants on {100} faces of Si3N4 grain boundaries [9]

Position	*x* (nm)	*y* (nm)	*z* (fractional coordinate)	Position	*x* (nm)	*y* (nm)	*z* (fractional coordinate)
La(1)	0.35 ± 0.02	0.50 ± 0.02	0.25	**Lu(1)**	0.52 ± 0.02	0.54 ± 0.02	0.25
La(2)	0.72 ± 0.03	0.61 ± 0.02	0.25	**Lu(2)**	0.69 ± 0.02	0.55 ± 0.02	0.25
La(3)	0.71 ± 0.04	0.80 ± 0.03	0.75	**Lu(3)**	0.81 ± 0.05	0.81 ± 0.04	0.75
La(4)	0.34 ± 0.02	0.84 ± 0.04	0.75				

Table 1 Coordinates of dopant atom sites on interfaces in Figure 3 (the errors shown are standard deviations; the coordinate system is marked in Figure 3).

As an example, Figure 3 shows an interface between a Si_3N_4 grain and a glassy intergranular pocket (admittedly, the glassy phase was not well-characterised – see below). The question to be answered initially was whether dopant atoms (in this case Lu and La) segregated to the boundary, and if so, whether to particular crystallographic sites. As can be seen from the figure, La and Lu do segregate to specific sites (different for the two dopants). By imaging in two directions in the boundary plane normal to each other, Winkelman et al [9] measured the 3-D coordinates of the dopant sites, as shown in Table 1. They paid particular attention to the error estimates, to facilitate comparison with modeling. Early attempts to predict these observed positions through modeling were only partially successful, the problems being attributed to how to account for the interatomic potentials due to the glassy intergranular film abutting the boundary and surrounding the dopant atoms. Recently Painter et al [10] have reported new modeling results in better agreement with the experimental observations.

Grain boundaries in $SrTiO_3$ have proved to be a fertile ground for refining structures by comparing experimental and model structures. Dudeck et al [11, 12] have used through focal series reconstruction to investigate the structure of the $\Sigma3$ {112} GB, for which there were two competing models of indistinguishable energy. A detailed quantitative comparison of experiment and theory (DFT) of the GB led to a refined structural model of the GB plane [12].Shih et al [6, 13] made a direct comparison of individual experimental images and images simulated from model structures to solve a problem concerning the role of planar {100} GB surfaces in abnormal grain growth in $SrTiO_3$. In this material, many abnormal grains, important to grain growth during sintering, have {100} surfaces which are observed to be "atomistically planar" (e.g. Figure 4). HREM imaging using the conditions suggested by Figure 2 to enhance the contrast on O columns allowed a detailed study to be made of the atomic structure of these surfaces [12]. Comparison of experimental and simulated images led to the conclusion that the {100} surfaces have significant Ti vacancies, which could play a role in explaining the unexpected mobility [14] of these faces.

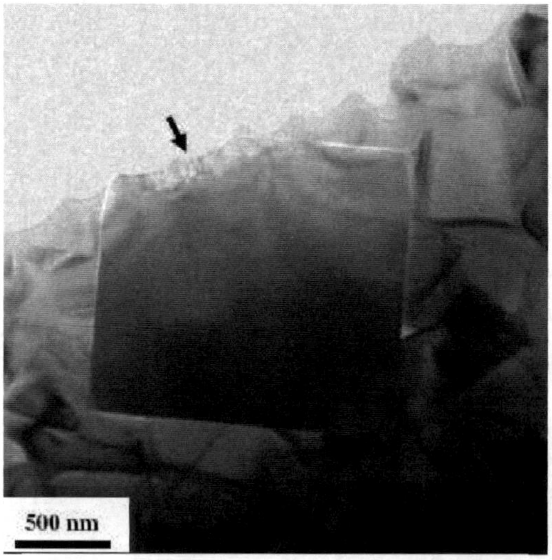

Figure 4 Typical microstructure of SrTiO3 showing the atomically flat {100} surfaces of an abnormal grain (arrowed) surrounded by matrix grains [6,12].

Nanoparticles

Nanoparticles, and in particular catalytic nanoparticles, are a fertile ground for collaboration between modeling and imaging. For single element particles (e.g. Au, Pt), there is the question of structure, particularly surface structure, and of strain. In the case of alloys, the distribution of atomic species within, and especially at the surface of, the particle is of primary interest. A study of elemental distribution in GeSi quantum dots [15] demonstrated the influence of strain and surface energies on such distributions, and warned of ignoring strain or inhomogeneous distribution in analyzing images from nanoparticles. Recent studies of elemental distributions in nickel-doped and yttrium-doped ceria particles [16] have shown layering to be determined mainly by diffusion of the precursor ions, which in turn can be related to their hydration energy. These results make it clear that any analysis requiring the interpretation of images as though they show the projected potential needs to be approached cautiously, and especially for images taken along zone axes. The greatest advances in understanding and then controlling nanoparticle structures will be achieved when tomographic or confocal imaging [17] give reliable 3D structures at the atomic level.

Selected area diffraction

The contribution of high resolution microscopy to understanding the structure of amorphous materials has been relatively small, despite the importance of amorphous phases in controlling the growth and properties of many technologically important materials. The reason is clear – the difficulty of interpreting images of amorphous material at or near the atomic level. Fluctuation electron microscopy appears to be sensitive to medium range order [18], even though it cannot be said to give a measure of it. Short range order can now be studied to high precision and accuracy using electron diffraction combined with RMC refinement of models based on DFT [e.g. 19, 20]. Attempts to investigate inhomogeneities in amorphous structure at the nanoscale were hindered by the dispersion of the selected area in the sample plane caused by spherical aberration (an aberrated objective lens causes the selected area diffraction intensity at scattering angle α to arise from an area of the sample displaced by $C_s\alpha^3$). This obstacle is overcome in an aberration corrected system, and has proved to be of importance in the study of amorphous nanovolumes by RDF electron diffraction analysis, where diffraction data is collected to high scattering angles [21].

Conclusion

The aberration corrected high resolution electron microscope is a tool for investigating the structure of materials to below the 1 A level. It is however a tool which can as easily mislead as it can clarify. Because it gives results in pictorial form, they can be seductively misleading, especially to the biased eye. To use the instrument confidently, it is essential to have a thorough understanding of diffraction and imaging theory, the basis of which is clearly laid down in the literature of the past 50 years. Research into this literature provides a means for avoiding many of the pitfalls of interpretation which await the unwary.

At the same time, properly used, the high resolution microscope gives quantitative data about structure suitable for linking into, and challenging, atomistic models. In combination, high resolution microscopy and modeling can reveal the structure of the nanoworld at an unprecedented level of detail.

References

[1] Rayleigh, Lord (a.k.a. Strutt, J. W.), 1874 *Phil. Mag.* **47** 81
[2] O'Keefe M A, 2008 *Ultramicroscopy* **108** 196
[3] O'Keefe M A, Nelson E C, Wang Y C and Thust A, 2001 *Phil Mag B* **81** 1861
[4] Jia C-L, Lentzen M and Urban K, 2004 *Microsc. Microanal.* **10** 174
[5] Chang L Y, Kirkland A I and Titchmarsh J M, 2006 *Ultramicroscopy* **106** 301
[6] Shih S-J, 2009 *Nanometre-scaled structural studies of strontium titanate by electron microscopy and microanalysis* (D.Phil. thesis) (Oxford University)
[7] Urban K W, Jia C-L, Houben L, Lentzen M, Mi S-B and Tillmann K 2009 Phil Trans R Soc A **367** 3735
[8] Okunishi E, Ishikawa I, sawada H, Hosokawa F, Hori M and Kondo Y 2009 *Microsc Microanal.* **15 (Suppl 2)** 164
[9] Winkelman B, Dwyer C, Hudson T S, Nguyen-Manh D, Döblinger M., Satet L, Hoffmann M J and Cockayne, D.J.H. 2005 *Appl. Phys. Lett.* **87/061911** 1
[10] Painter G S, Averill F W, Becher P F, Shibata N, van Benthem K, and Pennycook S J 2008 *Phys. Rev. B* **78** 214206
[11] Dudeck K J and Cockayne D J H 2009 *(this volume)*
[12] Dudeck K J, Cockayne D , Benedek N A and Finnis M W *Phys Rev* (submitted)
[13] Shih S-J and Cockayne D J H (to be published)
[14] Bäurer M, Weygand D, Gumbsch P and Hoffmann M J 2009. *Scripta Mater.* **61,** 584
[15] Lang C, Cockayne D J H and Nguyen-Manh D 2005 Phys. *Rev. B.* **72/15** 155328
[16] Shih S-J, Li G, Cockayne D J H and Borisenko K B 2009 *Scripta Materialia* **61** 832
[17] Nellist P D, Behan G, Kirkland A I and Hetherington C J D, 2006 *Appl. Phys. Lett.* **89** 124105
[18] Voyles P M, Gibson J M and Treacy M M J, 2000 *J. Electron Microsc* **49(2)** 259
[19] Li G, Borisenko K B, Chen Y, Nguyen-Manh D, Ma E and Cockayne D J H, 2009 *Acta Materialia* **57** 804.
[20] Li G, Borisenko K B and Cockayne D J H 2009 *(this volume)*
[21] Cockayne D J H, Chen Y, Li G and Borisenko K 2009 *(this volume)*

Journal of Physics: Conference Series **241** (2010) 012002 doi:10.1088/1742-6596/241/1/012002

Protons, Ions, Electrons and the Future of the SEM

David C Joy

University of Tennessee, Knoxville, TN 37996-0840,
Center for Nanophase Materials Science, Oak Ridge National Laboratory,
Oak Ridge, TN 37831-6064, USA

E-mail: djoy@utk.edu

Abstract. The Scanning Electron Microscope (SEM) is the most widely used high performance microscope in all fields of science but it is now reaching the theoretical limits of its performance. While advances in sources, optics, and detectors, can result in some improvement in performance the ultimate resolution is limited by fundamental physical constraints. One potential alternative is a scanning microscopes utilizing light ions such as H+ and He+ . Such an instrument shares all of the benefits of the conventional SEM but is free from the constraints encountered when using electrons and could significantly extend the scope and success of scanning microscopy.

1.Introduction

The scanning electron microscope (SEM) was the first practical imaging device that did not rely on visible light. It was conceived in Germany and the USA in the 1930s and 1940s, finally becoming a practical instrument through the work Prof. Oatley's group in Cambridge [1] in 1951. Now, nearly 60 years later, present day SEMs are still recognisably direct descendants of that original design and they have become the most widely used high performance tools for imaging and microanalysis with more than 50,000+ instruments in operation world wide producing literally millions of images per day for every area of science and technology. However, even with the benefits of six decades of continuous development, the SEM is now struggling to maintain the level of performance required for current applications in nanoscience, and the worlds of biology, and soft materials. It is therefore necessary to ask if the SEM still has a future and if the current level of SEM performance can be sufficiently enhanced so as to satisfy ever more demanding specifications, or whether alternative technologies must now be considered as replacements.

2. How good is the SEM performance now ?

It is regrettable that quantitative data on the imaging performance of SEMs now in operation is difficult to obtain and is little better than anecdotal in quality. This is because of the unwillingness of both manufacturers and users to accept the consequences and potential disappointments of employing techniques for determining resolution that are relevant, reliable and consistent with those used in other areas of microscopy such as for the TEM, or advanced optical systems. The situation is compounded by the fact that the performance of an SEM depends strongly on the nature of the sample. As a consequence this has led to the development of test specimens - such as gold particles on carbon -

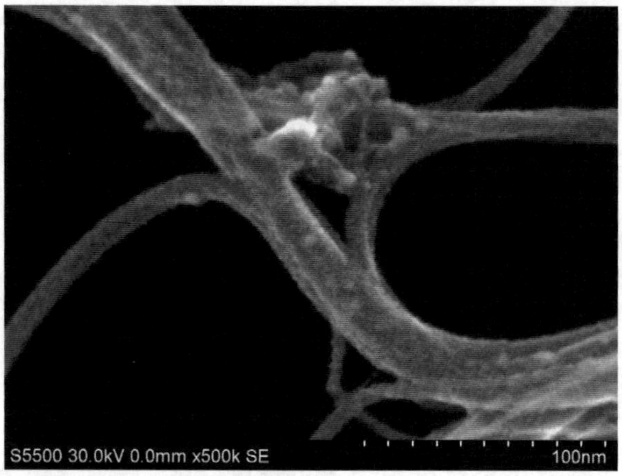

Figure 1. SE image of multi-wall carbon nano-tube imaged at 30keV demonstrating resolution and surface detail. Field of view 250nm. Image courtesy of Bill Roth (HHTA)

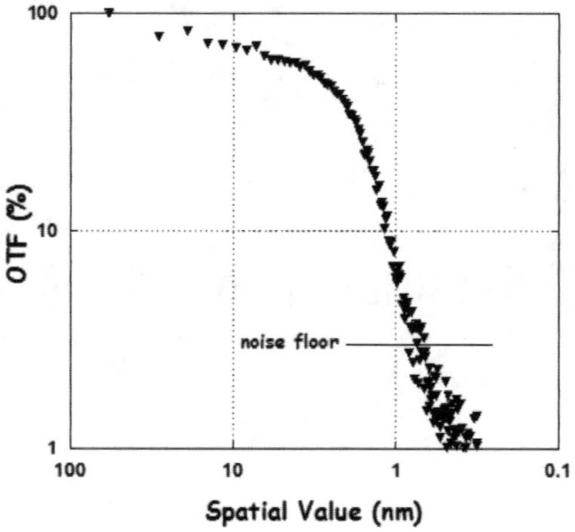

Figure 2 . Experimental Optical Transfer from a current production high performance field emission SEM at 5keV confirming sub-nm information transfer at noise limit

whose only merit has been to offer unrealistically high contrast as compared to 'real' samples. As a result the data represented as 'measuring' SEM performance may be optimistic and provide little guidance as to either the real capabilities, or the operational deficiencies, of these instruments. The resolution of current production, top grade, SEMs is from visual observation certainly of the order of a nanometer (figure 1), and established methods such as the super-position diffractograms or Optical Transfer Function (OTF) analysis of the instrument [2] confirm that such machines can indeed retrieve some image information at the sub-nanometer level (figure 2). In fact several high performance SEMs have been shown to be capable of displaying bright field lattice fringes from carbon (0.32nm spacing) at 30keV [3] when operated in STEM mode (figure 3). Fringe imaging is not, of course, a measure of the probe size but does confirm that the high voltage supply, lens controller, and mechanical stability of the column, are of a quality comparable with those in a TEM. At lower beam energies (3keV or less) chromatic aberration and the reduction in source brightness become dominant factors causing a rapid fall in contrast and signal to noise although information retrieval down to 2-5nm is still achieved. While these levels of performance are an advance on those of a decade or more ago they remain marginal for the routine sub-nanometer scale imaging of most specimens, and are hardly competitive with scanned probe (AFM) tools or, far less, the TEM or STEM so significant enhancements in instrumentation must be sought to close this gap.

3. Can SEM performance be improved?

The imaging performance of an SEM is determined by the probe size at the sample - which is in turn set by the electron optics parameters - and by signal to noise - set by the brightness of the electron source and by the detectors that collect the signal. One important step that has already taken to improve SEM performance has been the replacement of tungsten hairpin thermionic sources with either cold field, or Schottky, emitters either of which offer a significant enhancement in source brightness giving higher currents into the final probe. This, in turn, enhances signal to noise and raises the highest spatial frequency at which image information is transferred to the observer. High brightness sources are also essential for the low voltage (i.e. sub-3keV) operation which is such an important part of contemporary SEM usage.

Figure 3. 30keV bright field STEM image of CNT showing 0.32nm lattice fringes. Horizontal field of view 30nm. Image Courtesy of Hitachi High Technology

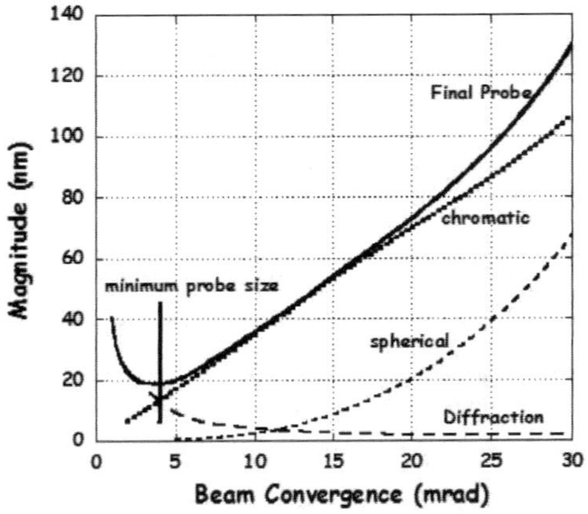

Figure 4. Contributions to final probe size from diffraction, spherical, and chromatic aberrations as a function of the beam convergence (m.rad) at 5keV

The detector in the SEM is the first step in the signal chain and is the place where the signal to noise is poorest, so detector efficiency has a major influence on the quality of SEM imaging. In addition a detector should be selective – collecting a maximum of the signal of interest and a minimum of unwanted contributions. Unfortunately it is probably true to say that SEM detectors are never designed, rather they are copied from one instrument to the next and if they generate an image then they are regarded as fully acceptable. Because of this neglect it is probable that there are still opportunities to improve image signal to noise, contrast, and information content by studying, designing and building with the same care given to other components in the instrument. In summary, although much could and should be done to enhance detector performance, there are no novel candidate sources whose brightness could exceed that of either cold field emission guns or Schottky emitters so further improvements in performance must be achieved from the electron-optics.

Three factors - chromatic aberration, spherical aberration, and diffraction limiting - determine the probe diameter and profile (figure 4). The dominant contributions are diffraction limiting and chromatic aberration and because these move in opposite senses as a function of the beam convergence α then there is an optimum value of α which gives the highest current into the smallest probe. Typically this optimum α is a few milliradians in size, a value which severely limits the incident beam current (which is proportional to α^2) although providing a useful depth of field in the image (proportional to $1/\alpha$). The route taken by high performance STEMs and TEMs to overcome the constraints of these limitations is to 'correct' the spherical (Cs) and chromatic (Cc) aberrations associated with the probe forming, or objective, lens. Reducing Cs and Cc magnitudes from millimeters to micrometers shifts the optimum beam convergence from a few mrads. to 30 or 40mrads, making it possible to generate sub-angstrom probes containing 20-30x more current than for the equivalent uncorrected system. The same procedure can be applied to the SEM [4], and prototype aberration corrected SEMs (ACSEMs) have been successfully demonstrated by several companies (for example FEI Inc. and JEOL Inc.) although only one such instrument – from Holon – seems to have made its way into production.

In fact there are several reasons why a full fledged ACSEM is unlikely to become the SEM of the future. First, the increase in beam convergence α made possible by aberration correction reduces the imaging depth of field at high magnification settings to a nanometer or less – which makes the ACSEM of little use for real world samples. Second, optimization of aberration correctors is achieved by a complex, computer controlled, iterative, procedure. For TEMs and STEMs this is guided by the analysis of the Ronchigrams generated from ultra-thin transparent metal foils, but for an SEM imaging a bulk sample there is no equivalent of a Ronchigram and so the correctors must be adjusted by observing the real space image. This is slow, unreliable, hard to optimize, and easily confused by effects such as sample charging and contamination.

Two limited forms of aberration correction for SEMs are, despite these issues, presently in use . One, the retarding field mode – is in fact the most widely used form of aberration correction for any electron beam system. In this technique , the incident beam of energy E_B is abruptly decelerated to some lower final energy E_F over a distance L in the region between the probe forming lens and the sample surface. In this case the effective aberration coefficients of the combined optical system are then [5] effectively become

$$Cs = -Cc = L.\ E_F/E_B \qquad (1)$$

and so are independent of the actual lens parameters. For example, a 5keV beam decelerated to 500eV over a distance of 1mm creates effective aberration coefficients in the micrometer range. Significant enhancements in resolution are achievable with minimal effort and consequently this approach has become the standard for the semiconductor industry's "Critical Dimension-SEMs" which perform nanometer scale metrology of devices. However the retardation mode is difficult to engineer if high landing energies are required, and does not work well for samples which are small in size, or have prominent topography. Alternatively , the electron beam can be 'mono-chromated" to reduce its energy spread (ΔE) and hence the associated chromatic aberration error ($\sim \alpha.\Delta E/E$) to enhance performance. The "Uni Colore" system from FEI Inc. employs a dispersive lens structure built into the Schottky emitter source which is paired with a slit aperture to eliminate the lowest energy electrons from the beam and so reduce the energy spread from about ~0.7eV to below 0.3eV - a value comparable with a cold field emission source. The improvement in resolution achieved at low energies is significant and requires no elaborate alignment or optimization but this enhancement comes at the expense of total beam current and is of less value at high energies (10keV and above).

4. What happens next?

Within the energy range 0.5-30keV over which the SEM is most commonly used, and even allowing for the more widespread adoption of the kinds of improvements discussed above, there is little scope for substantial gains in SEM performance. Firstly, electrons have a wavelength which at low energies is a significant fraction of a nanometer in size making diffraction limiting an impassable barrier. Increasing the beam energy minimizes this problem but also leads to a rapid rise in beam penetration. Thus better resolution is achieved - but only at the expense of losing surface specific detail in favour of deep sub-surface information carried by backscattered electrons. While the enhanced brightness of the electron source at higher energies puts more current into the focused probe this gain is offset by the reduction in signal intensity which results from the reduction in secondary electron yield that occurs. There is, thus, neither any instrumental development nor optimal of operational conditions which can sufficiently enhance the performance to maintain the competitive position of the SEM. Comparable alternative technologies must therefore be devised and exploited instead.

5. Ion imaging – the new scanning microscopy?

A promising option is to consider the use of ion, rather than electron, beams. As first pointed out by Levi-Setti more than thirty years ago [6], ions are much more massive than electrons and so their wavelength λ is at least 50 to 100 x smaller than that of an electron of the same energy. Since scanning beam systems are always ultimately limited in performance by diffraction, reducing λ by such a large factor makes a substantial improvement in resolution a definite possibility. A further consequence of the large mass of an ion relative to an electron is that, at the same energy, its velocity is much smaller. For example, for He + ions (mass 7297x that of an electron) their velocity v_{He} compared to that of an electron v_{e-} of the same energy is (under non-relativistic conditions)

$$v_{He} = v_{e-} * 0.012 \qquad (2)$$

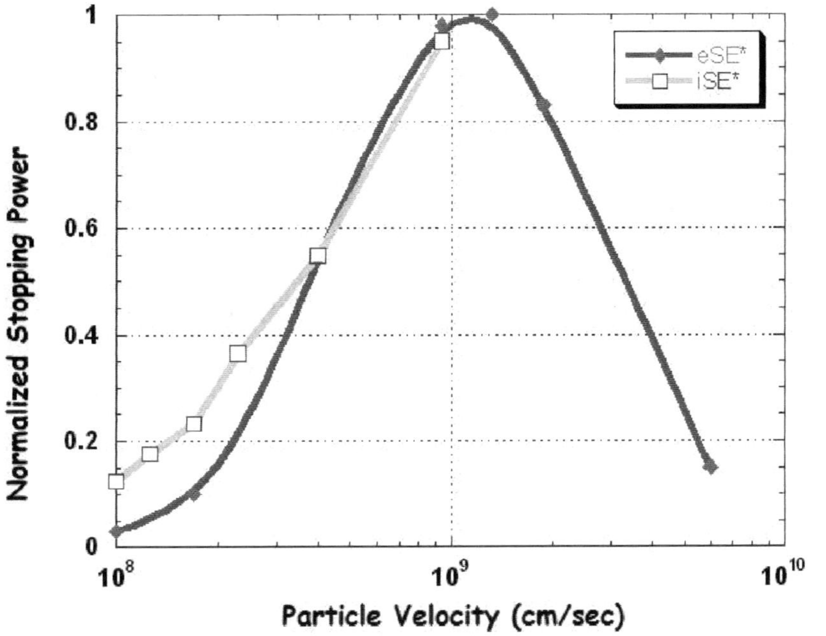

Figure 5. The variation of electronic stopping power for electrons (eSE) and ions (iSE) plotted as a function of their velocity. The data is normalized because the maximum SP values for ions and electrons differs by an order of magnitude

This difference in velocities has important implications. When a charged particle enters a material the rate at which its energy is lost depends on the 'stopping power' of the material which in turn determines not only the range of the particle but also the profile of the energy deposition along the trajectory and, consequently, the magnitude and depth distribution of the various interactions and signal emissions that occur. Thompson [7] showed that stopping power is a function of particle velocity not energy and if, as shown in figure (5) the stopping power of electrons and of ions is plotted as a function of their velocity then the normalized curves resemble each other closely although their absolute magnitude may differ significantly. For both projectiles the maximum in the stopping power occurs at a velocity of about $6x10^8$ cm/sec which for electrons correspond to an energy of about 200eV but to about 700keV for a He+ ion. Electron microscopes thus operate on the right hand side of the profile where the stopping power falls with increasing energy – and consequently as the energy is increased the range increases rapidly, most of the energy is deposited at the end of the range i.e. deep in the sample, and secondary electron and other emissions fall as the energy rises. It is this chain of events which leads to the problem of 'optimizing' the SEM discussed above. For ions the situation is quite different. A proton or a He+ ion microscope operating at 30-50keV is on the left hand side of the profile where stopping power is rising with energy. An increase in beam voltage therefore results in only a relatively modest increase in range, the majority of signal generation occurs at the surface where the energy and hence the stopping power is highest, and the overall SE yield rises. Because of this behaviour for ion beams all parameters of interest - from source brightness and wavelength to beam range and signal production – are simultaneously optimized as the energy is increased.

6. The Scanning Ion Microscope (SIM)

To exploit the potential of its greatly reduced wavelength a scanning ion microscope (SIM), like any SEM, also requires a high performance beam source and quality lenses. As shown by Levi-Setti [7] protons can be generated in a standard Butler-Crewe cold field emission gun by the expedient of replacing the usual high vacuum with a low pressure of hydrogen. Helium ions, in the ZeissSMT 'Orion' instrument [8], are generated from a field ionization source developed from that devised by Muller in 1951 for the field ion microscope. In both cases the ion source can be operated at modest vacuum levels, and exhibits both a brightness which is comparable with a field emission electron gun and a nanometer source size. Ion optical lenses are usually of the simple 'Einzel' electrostatic type but offer competitive aberration characteristics All of the components required for competitive ion microscope performance are therefore already available. Ion induced secondary electron (iSE) emission is - as in an SEM – the imaging mechanism of the most immediate interest. The iSE yield (iSE/ion) is a factor of 5 to 10 time higher than the electron yield (SE/electron) from the same material and rises with beam energy only reaching a maximum for protons and He+ at energies in the range 400-700keV. Consequently the signal to noise of a SIM is superior to that of an SEM for comparable beam currents. For protons and helium ions the effective beam range R (nm) in a material of density ρ (gm/cc) at energy E(keV) is well described [9] by the relation

$$R_{ion} = 80.(E^{0.73}/\rho) \quad (3)$$

which compares with the Kanaya-Okayama [10] estimate of

$$R_{electron} = 80.(E^{1.66}/\rho) \quad (4)$$

for electrons. The ratio of electron to ion beam range is therefore of the order of magnitude 'E' (i.e. 30x more at 30keV) confirming that the iSE image will always contain significantly more surface information than the electron signal at all energies. The iSE image is further enhanced by differences in the origin of the SE between electrons and ions. For electron irradiation typical 50% of the total SE signal is generated by exiting backscattered electrons, the resultant 'SE2' contribution being of poor

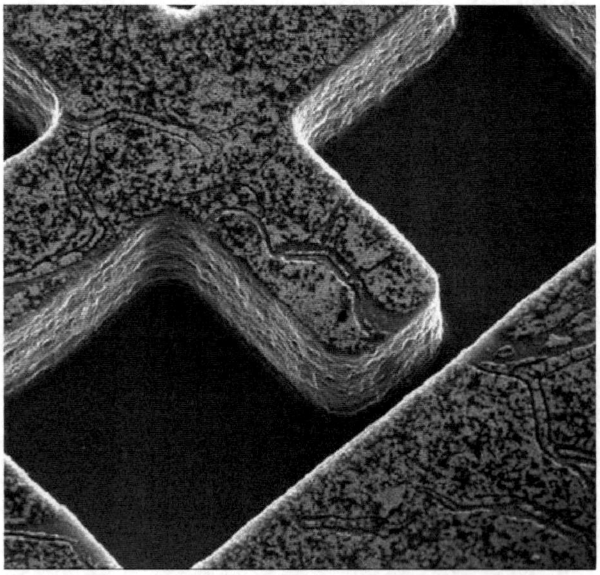

Figure 6. Helium ion beam secondary electron image of an aluminum registration mark. Note the clarity of surface topography, the visibility of surface contamination, and the large depth of field. Beam energy 35keV, horizontal field of view 4.5 micrometers. Image courtesy Dr. L.Scipioni, ZeissSMT

resolution as compared to the SE1 contribution generated by the incident beam. In the case of ions while the ratio iSE2/iSE1 is comparable to the value for electrons at low energies this falls steadily with increasing energy both because of the reduction in the ion backscatter yield and because the lower stopping power of the backscattered ions reduces their iSE generation efficiency. Consequently at higher energies the iSE signal tends towards being pure SE1 in character which further improves the contrast of the high resolution image. In addition (figure 6) ion induced SE images display high sensitivity to changes in surface chemistry, high contrast from channelling interactions with the crystallography of a sample, and an impressive depth of field which is the result of the small convergence angle (less than 1 milli-rad) typically chosen.

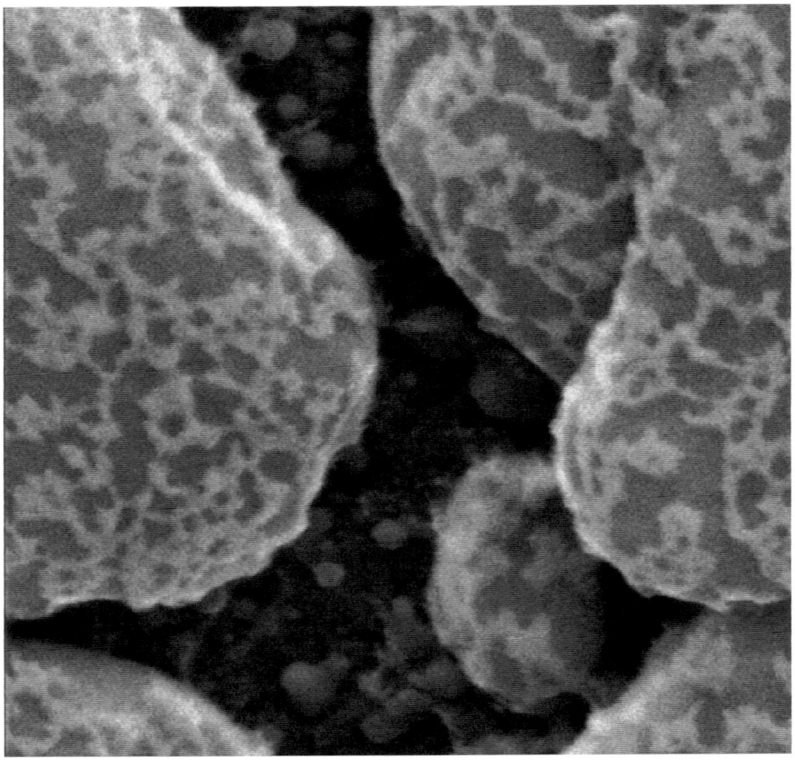

Figure 7. He+ ion secondary electron image of gold on carbon. The beam energy was 35keV and the field of view is 200nm in width. Image courtesy of Dr.L. Scipioni, ZeissSMT

Figure (7) shows an example of a He+ ion SE image of a gold on carbon specimen recorded at 35keV The material contrast between the gold and carbon is strong and clearly reveals detail that is no evident in electron induced SE images of the same material. Diffractogram and contrast transfer function analysis of such images demonstrates the information limit to be well below 1nm, and ultimately to be limited by signal to noise rather than by probe size or mechanical instability. The beam range, and secondary electron yields, from both proton and helium beams are quite similar at the same energy and both have the virtue that, because of their low mass and high velocity, their nuclear stopping power component is small and falls with increasing energy so they cause little sputter damage from specimens. Scanning ion microscopes using argon or gallium beams would also offer many valuable benefits. However, because such ions are massive compared to H+ and He+ high beam energies (>1MeV) are necessary to generate useful secondary electrons yields comparable to those produced by lighter ions. Heavy ion beams are also more prone to exhibit multiple ionization states which increase the chromatic spread of the beam and so degrade the probe size, and to producing significant sample damage as the result of their substantial nuclear interactions.

Scanning ion microscopes do also have some operational limitations. Even for H+ and He+ beams the probe current may need to be restricted to the 1-10pA range in order to minimize damage, to nanoscale objects, and to limit the energy broadening of the beam which could occur as the result of the Boersch effect [11]. Sample charging which unlike the electron situation is always positive in polarity, and which is substantially higher in intensity than that encountered with electrons, is also to be expected on all poorly conducting and insulating samples unless some form of charge compensation such as a low energy electron flood gun is applied. Finally, X-ray microanalysis by light ion beams is not generally possible except at beam energies in excess of 1MeV [12] although energy spectroscopy of the Rutherford Back Scattered ion (RBS) signal might provide a method for surface analysis with extremely high depth resolution.

7. Conclusions

The SEM has established itself as a uniquely powerful and versatile tool for imaging and analysis, but the performance of the instrument is now constrained by limits set by the wavelength of the electron, and the problem that there is no one beam energy which permits all of the parameters of performance to be simultaneously optimized. A scanning microscope which employs a beam of light ions combines the convenience and familiarity of the SEM with advantages that come from the shorter wavelength of ions and their higher stopping power. This in turn leads to restricted beam ranges, enhanced surface sensitivity, and higher secondary electron yields. Contrast effects associated with sample crystallography, and with surface chemistry, are an order of magnitude stronger than for an electron beam. Increasing the energy of the ion beam to 100 or 200keV from the current 30 to 50keV levels would increase source brightness by a factor of 10x, reduce the ion wavelength by a factor of 3x, and increase secondary yields by a factor of 3-4x while only increasing beam range by a factor of 3-4x. Such an instrument would have the potential to achieve sub-angstrom resolution in SE mode -without aberration correction of any type - under such conditions, and would position Scanning Microscopy to remain the universal imaging technique of choice.

Acknowledgments

The author is grateful to Drs. Ranjan Ramachandra, Brendan Griffin, Lucille Giannuzzi, Andras Vladar, David Bell, Dale Newbury, Joe Michael, Scott Wight, and John Villarubbia for valuable discussions and to Clarke Fenner and Drs. John Notte , Bill Ward, and Larry Scipioni of Zeiss SMT for their generous help in supplying information on the HIM.. This work was partially supported by the SRC, contract manager Dr. Dan Herr.

References

[1] Oatley, C W 1972 *The SEM: Part 1 The instrument* (Cambridge: University Press)
[2] Joy D C 2002 *J. of Microscopy* **208**, 24-34
[3] Ngo V, Hernandez M, Roth W, and Joy D 2007 *Microscopy Today* **15**, 12-16
[4] Joy D C 2008 Chapter 3 in *Biological Low-Voltage Scanning Electron Microscopy,* ed H Schatten and J Pawley, (New York: Springer)
[5] Mullerova I and Frank L 1993 *SCANNING* **15**, 193-201
[6] Levi-Setti R 1974 *Proc. SEM 1974* Part 1 ed O Johari, (Chicago:IITRI), 125- 130
[7] Thompson J J 1912 Phil. Mag. **28**, 449-457
[8] For details see US patent # 7368727, May 6th 2008
[9] Ramachandra R, Griffin B, Joy DC 2009 *Ultramicroscopy* **109**, 748-757
[10] Kanaya K, Okayama S 1976 *J.Phys.D. Appl.Phys.* **5**, 43-7
[11] Kruit P, Jansen G H 1997 *Handbook of Charged Particle Optics* ed J Orloff, (NY:CRC)
[12] Joy D C, Meyer H M, Bolorizadeh M, Lin Y, Newbury D E 2007 *Scanning* **29**, 1

Journal of Physics: Conference Series **241** (2010) 012003 doi:10.1088/1742-6596/241/1/012003

High resolution transmission imaging without lenses

J M Rodenburg[1,2], A C Hurst[1] and A Maiden[1,2]

[1]Department of Electronic and Electrical Engineering, University of Sheffield, Sir Frederick Mappin Building, Mappin Street, Sheffield, S1 3JD, UK.
[2]Phase Focus Ltd, The Kroto Innovation Centre, University of Sheffield, North Campus, Broad Lane, Sheffield S3 7HQ, UK.

Email: J.M.Rodenburg@shef.ac.uk

Abstract. The whole history of transmission imaging has been dominated by the lens, whether used in visible-light optics, electron optics or X-ray optics. Lenses can be thought of as a very efficient method of processing a wave front scattered from an object into an image of that object. An alternative approach is to undertake this image-formation process using a computational technique. The crudest scattering experiment is to simply record the intensity of a diffraction pattern. Recent progress in so-called diffractive imaging has shown that it is possible to recover the phase of a scattered wavefield from its diffraction pattern alone, as long as the object (or the illumination on the object) is of finite extent. In this paper we present results from a very efficient phase retrieval method which can image infinitely large fields of view. It may have important applications in improving resolution in electron microscopy, or at least allowing low specification microscopes to achieve resolution comparable to state-of-the-art machines.

1. Ptychographic iterative phase retrieval

Consider the optical arrangement shown in Figure 1a. A coherent source of radiation illuminates an aperture. Some distance downstream of the aperture there is an object of interest, which can be shifted laterally with respect to the illumination. A long way downstream of this is a detector that can measure the intensity of the Fraunhofer diffraction pattern. We have demonstrated that it is possible to obtain a complex representation of the object (that is, an object that can be regarded as a transmission function which introduces a phase change and an absorption term into the incident wave-front), even with such a simple experimental set-up, in principle only limited in resolution by the wavelength of the radiation used [1]. The only requirement on the dimensions of the experiment is that the diameter of the illuminating wavefield is sufficiently small so that the Nyquist sampling condition is satisfied in the diffraction plane. If the object was completely disordered, then this would imply that the speckle pattern so produced must be sampled at an angular resolution of half the speckle size.

To form an image from diffraction data alone, we are faced with what may at first appear to be an intractable problem. To use only this intensity data, we must have some means of recording or computing the *phase* of the scattered radiation. Only then will we be able to back propagate the detected wave to the plane of the object, thus creating an image of it. The great strength of a lens is that it preserves phase, retarding the progress of the wave-fronts (their phase) in a pattern that reverses the curvature of the wave-front so that any point in the image is brought to a focus in the image plane. The history of the so-called phase problem (solving for phase from intensity alone) is long. Suffice it to say that it has been known since about the early 1980s that if we have a prior knowledge that our object conforms to certain constraints, such as having finite size, then it is possible to construct an iterative procedure to converge upon the phase of every pixel of a diffraction pattern. A combination of the principle of ptychography (wherein multiple diffraction patterns are recorded for many relative positions of the object and illumination [2]) and iterative phase retrieval, leads to an algorithm (which we call PIE), that can solve for phase with incredible efficiency. Indeed, it turns out if only these relative displacements are known accurately, both the object *and* the illumination function can be solved for [3]: a variation of PIE (ePIE) can do this in real time, as data is collected [4].

Why should we undertake microscopy in such a round-about way? In the context of short (atomic-sized) wavelength radiation (X-rays or high energy electrons) it is difficult to make lenses of large numerical aperture. For example, in electron microscopy, the diffraction pattern lying in the back focal plane of the objective lens (i.e. the selected area diffraction pattern) displays significant intensity up to very high angles of scatter. In contrast, the diffractogram is much narrower. The interferometer in diffraction is the atoms within the object itself, whereas conventional imaging relies on the interference of beams that have propagated through macroscopic distances (millimeters/centimeters).

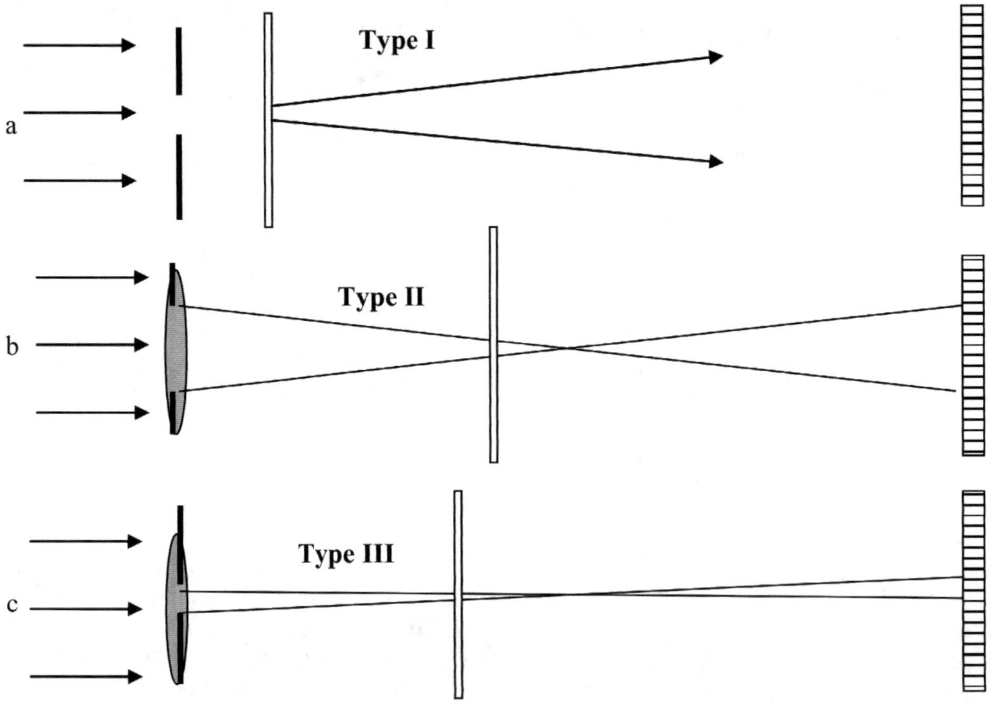

Figure 1: Three configurations for ptychography and lens-assisted pytchography, with coherent radiation incident from the left. a) A simple aperture, a specimen downstream of this, and a detector in the far-field. b) A convergent probe of high numerical aperture, defocused at the specimen plane to create a patch of illumination. c) A narrow-beam convergent probe, with the specimen substantially out of the Gaussian focal plane to create a large patch of illumination: this is the closest analogy to (a), making the most use of high-angle diffracted data.

2. Visible light ptychography

We start by considering visible light ptychography. Unlike X-ray and electron sources, laser light is to a very good approximation totally coherent. The scale of the required accuracy of the relative displacement between the specimen and the illuminating beam is also relatively large (of the order of

0.5mμ) and can be easily achieved by commercially-available optical microscope specimen stages. Furthermore, the use of a simple aperture arrangement (as in Figure 1a) is practicable. If the detector is a CCD camera with 1kx1k pixels, then to achieve sub-micron resolution while still satisfying the Nyquist sampling condition in the diffraction plane, the aperture size can be as large as 0.5mm. This means that scanning over a large field of view can be achieved in a relatively short time. Just like a conventional lens, the resolution of the final reconstruction is determined by the effective numerical aperture of the scattered wave presented to the detector, whether it is a lens or a CCD camera. In practice, a flat detector limits the total solid angle of the scattered radiation that can be processed, especially if we require it to be far enough downstream of the object so that it lies in the Fraunhofer diffraction plane. There is no requirement in the PIE or ePIE algorithms on the nature of the propagator between the object and the detector plane. To obtain high resolution with a flat detector we can place it in the Fresnel regime a short distance downstream of the object, thus increasing the solid angle it subtends at the specimen plane. This angle can also be expanded by employing a large, poor quality lens downstream of the object, which can bring the Fraunhofer diffraction pattern to a focus in its back focal plane. Aberrations in such a lens manifest themselves only as geometrical distortions in the diffraction pattern, and so can be easily removed computationally.

Figure 2 shows a typical image that we can obtain using the ePIE algorithm. Its quality is as good as what we would expect from a lens, although of course we have the added benefit of obtaining both phase and modulus of the specimen transmission function. Resolution in this micrograph is limited by the solid angle of the detector, as described above, at about 2μm.

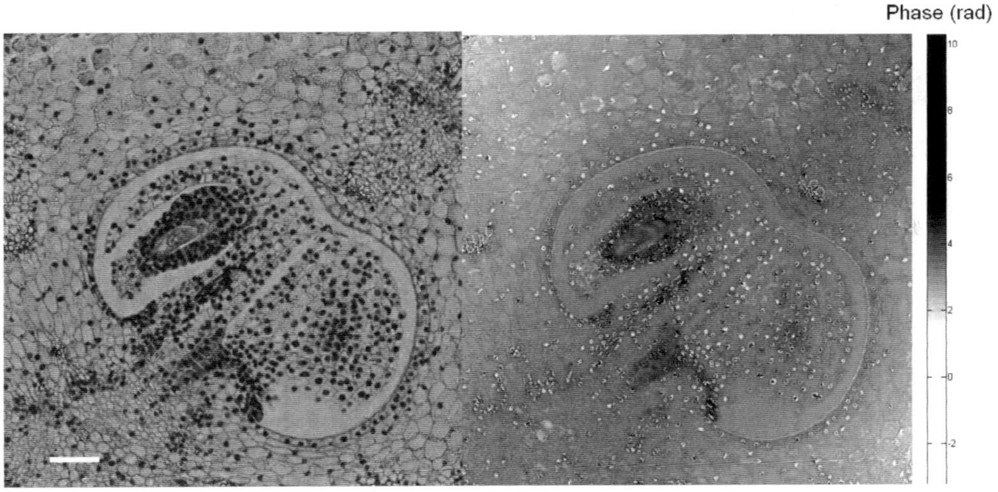

Figure 2: Visible light ptychograph of lily eggs, reconstructed using ePIE. The scale bar is 100μm. Note that the dynamic range of the phase has been considerably expanded.

Some principal benefits of visible-light ptychography are:

1) *Very large working distance*. Because there is no need to have an objective lens close up against the specimen, reasonable resolution can be obtained with the detector mounted centimeters away from the object. This is particularly useful in the context of imaging through enclosed containers, such as Petri dishes, say for the purpose of imaging the growth cycle of live cells.

2) *Post collection focus swim-through*: because the entire wave-field is reconstructed in modulus and phase, it is possible to produce images at any defocus required, after the data collection. This is particularly useful in the context of focusing in on biological features after undertaking a fast throughput experiment where it may not be practicable to focus accurately on the object of interest in real-time.

Journal of Physics: Conference Series **241** (2010) 012003 doi:10.1088/1742-6596/241/1/012003

3) *Absolute phase measurement*: Unlike Zernike contrast which distorts the image of the edges of objects, the absolute phase can be used to quantitatively measure the optical thickness of objects over a wide field of view. This in turn leads to the ability to perform accurate partitioning of the image, say for the purpose of automatically counting cells or similar features. The technique can also measure very large phase changes (many multiples of 2π) say for characterization of, for example, contact lenses.

3. Hard X-ray ptychography

The first demonstration of iterative ptychography at short wavelengths was at the Swiss Light Source using 8keV X-rays to image a zone plate [5] (see Figure 3). The type I configuration (Figure 1a) was used with an aperture size of 5μm. The first reconstructions obtained (Figure 3, top – from [5]) used an estimate of the illumination function calculated by propagating a wave-field coming from the aperture to the object. The distance of propagation could only be measured approximately, and was about 1.1mm. Because the quality of a PIE reconstruction hinges on how well the illumination function is known, different propagation distances were investigated in order to minimize a least-squares error metric between the calculated and measured diffraction pattern intensities. If we use ePIE (i.e. solving for both the object function and the illumination function simultaneously) – see centre row of Figure 3 – artefacts in the reconstruction are significantly reduced, especially in the modulus reconstruction. The phase reconstruction is also smoother, although at the cost of slightly reduced resolution. The illumination function calculated this way is also somewhat smoother than that estimated assuming the aperture is illuminated by perfectly coherent waves. In trying to find a general solution for both functions simultaneously, it would seem that the ePIE algorithm settles on a smoothed average that has least inconsistency.

4. Electron ptychography

Although iterative ptychography works spectacularly well for photons, the implementation of electron ptychography, at least at the time of writing, has proved to be much more difficult: we have so far not obtained any high-resolution publishable results. The simplest optical geometry (shown in Figure 1a) is impractical for electrons. Manufacturing an aperture which is both small enough at its centre and opaque enough around its perimeter is extraordinarily challenging. To obtain sub-Ångstrom resolution using a 1kx1k detector, the aperture diameter must be of the order of 100Ångstoms. To avoid transmission of substantial intensity outside the aperture, the thickness of the aperture blade would have to be of the order of microns, leading to an aspect ratio of about 1,000. It may be possible to use a much thinner blade if we were to filter out electrons which have passed outside the aperture and consequently lost energy, but this would still be a very difficult structure to make.

The only practicable optical set up is to use a focusing lens to form the illumination function on the specimen. In other words, we still require a lens, albeit one that can have relatively poor intrinsic resolution. If this lens is strong and produces a beam with a high angle of convergence, we classify it as the type II configuration (Figure 1b). On the contrary, the type III configuration (Figure 1c) uses a small angle of convergence. It should be emphasized the PIE or ePIE algorithms are indifferent to form of illumination – it can be soft-edged, consist of curved wavefronts, or have any range of solid angle of incident k-vectors. However, the form of the illumination can be optimized with respect to counting statistics, the coherence of the source and other experimental constraints such as drift rates and specimen damage. Moreover, the best arrangement will depend on the existing specification of the microscope upon which we intend to implement the method.

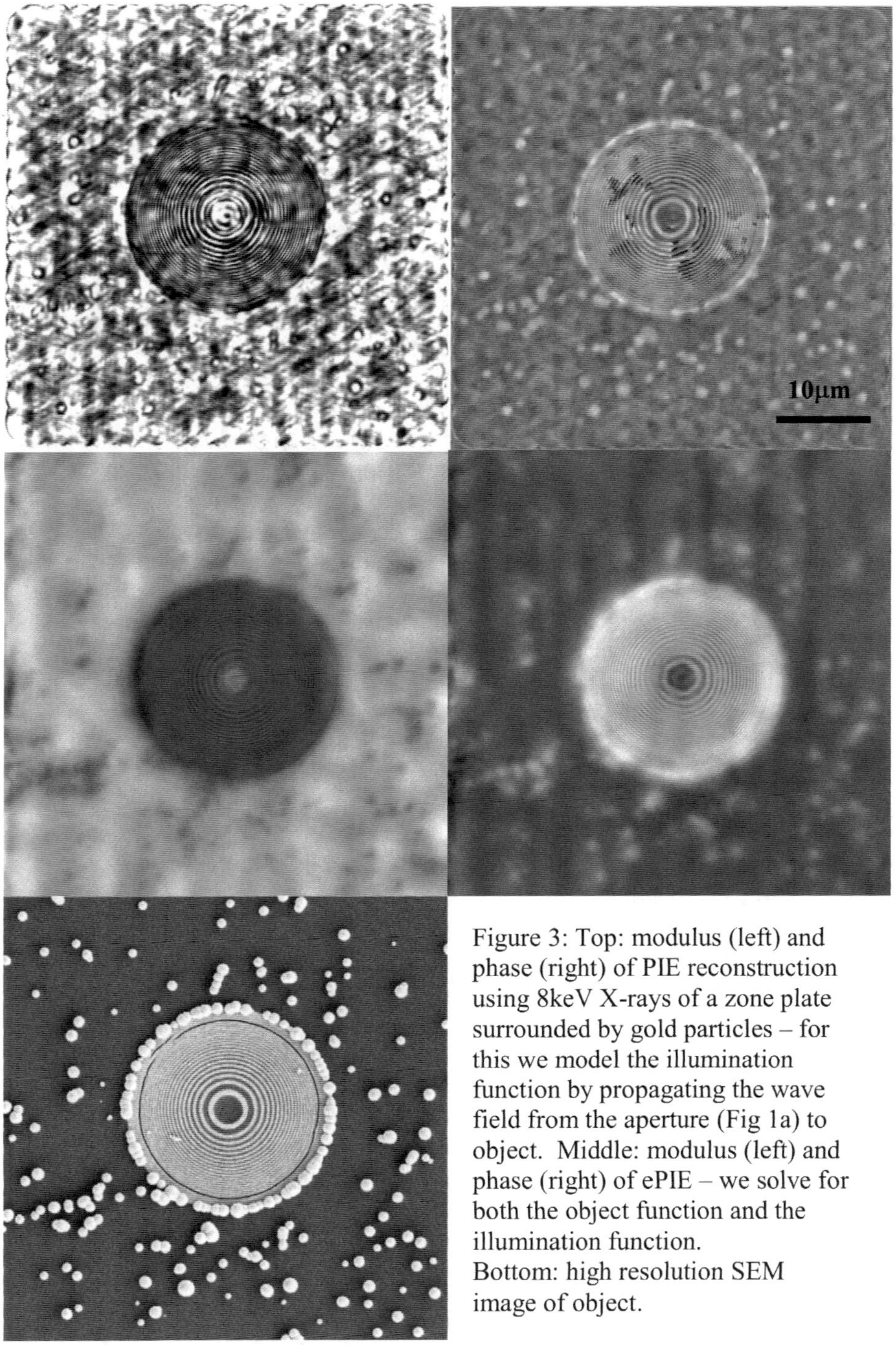

Figure 3: Top: modulus (left) and phase (right) of PIE reconstruction using 8keV X-rays of a zone plate surrounded by gold particles – for this we model the illumination function by propagating the wave field from the aperture (Fig 1a) to object. Middle: modulus (left) and phase (right) of ePIE – we solve for both the object function and the illumination function.
Bottom: high resolution SEM image of object.

Some relative advantages/disadvantages of these different configurations are as follows:

Type II: Counting statistics within the central disc (which we call the 'bright-field' disc) are much higher: provided the specimen is relatively weakly scattering, any diffracted amplitude that gets scattered to within the central disc interferes with a pseudo-reference beam (the unscattered wave). If the object consists of large areas of empty space, this configuration is essentially equivalent to Gabor holography. If we write the amplitude of the unscattered wave as A_r and the amplitude of the scattered wave as A_s, then the intensity measured is $\left|A_r^{\,2}\right| + A_r A_s^* + A_r^* A_s + \left|A_s\right|^2$, where the cross terms are much larger than the intensity of the scattered term. Related to this is the fact is the further advantage that if we only aim to process the bright-field disc, the intensities we need to measure all have a similar dynamic range. To record diffracted data outside the central disc (the dark-field data) requires a detector with a high dynamic range, or at least a detector with two regions (inside and outside the disc) which have different gains. Ptychographical information (that is, relative phase information between scattered diffraction orders) is also very strongly expressed because each point in the diffraction pattern of the object is convolved with a function which is the diameter of the (large) central disc. The most obvious disadvantage is that if it is our intention to improve the resolution of the microscope via ptychography, this configuration requires us to start with a relatively good (or at least, stable) lens. The demands on the coherence of the beam are also high because the back-focal plane of the illumination-forming lens must be smaller than the coherence width at that plane. However, if we can handle the dynamic range issue of the dark-field intensity, the gains in resolution could be several factors better than is currently achievable, even on aberration-corrected machines.

Type III: On the contrary, the type III configuration can use a poor lens. If we process the data scattered outside the central disc up to large scattering angles, then we really can greatly improve upon the intrinsic resolution of the lens by many factors. In the limit of near-parallel illumination, we can place a stop over the (very small) central disc, losing only very low resolution components in the image, and increase the gain of the detector so that all the dark-field data can be collected within a similar dynamic range. This configuration promises to provide a method of upgrading a microscope with rather modest resolution to one that has very high resolution.

So, what are obstacles to high-resolution electron ptychography? At the time of writing, the most powerful algorithm we have developed to data (ePIE) can solve for both the object function and the illumination, but only if the positions of their relative shifts are known accurately. Even on microscopes that can generate very precise scans for STEM images, we have found that when large movements (step sizes) are made, the position of the beam is not accurately reproducible. This is exacerbated by specimen drift, although we are developing strategies to retrospectively calculate the probe illumination positions (see Hurst et al., this volume). Furthermore, ePIE relies on the multiplicative approximation – that is that the exit wave is the product of the illumination function and the specimen transmission function. The influence of thickness and dynamical effects are not serious if the illumination function is known [6], but they do limit the usefulness of ePIE. However, the most challenging problem we face at present is that when we examine in detail the dark-field diffraction (where the high-resolution data resides), we find that the intensity distribution is often not stable. This could be due to a number of factors: instability in the probe; instability or damage in the specimen; instability in the source, or the column itself; or the migration of contamination over the specimen surface. These factors do not seriously impact on conventional imaging, but must be addressed carefully before electron ptychography will deliver images similar in quality to those we can now routinely obtain with photons.

References

[1] Rodenburg J M, Hurst A C and Cullis A G 2007 *Ultramicroscopy* **107** 227
[2] Rodenburg J M *Advances in Imaging and Electron Physics* 2008 **150** 87
[3] Thibault P, Dierolf M, Menzel A, Bunk O, David C, Pfeiffer F 2008 *Science* **321** 379
[4] Maiden A M and Rodenburg J M 2009 *Ultramicroscopy* in press
[5] Rodenburg J M, Hurst A C, Cullis A G, Dobson B R, Pfeiffer F, Bunk O, David C, Jefimovs K and Johnson I 2007 PRL **98** art no. 034801
[6] Cheng Lui, Walther T and Rodenburg J M 2009 *Ultramicroscopy* in press

Journal of Physics: Conference Series **241** (2010) 012004 doi:10.1088/1742-6596/241/1/012004

Probe position recovery for ptychographical imaging

A C Hurst, T B Edo, T Walther, F Sweeney and J M Rodenburg
Department of Electronic and Electrical Engineering,
University of Sheffield, Sheffield S1 3JD, United Kingdom

E-mail: J.M.Rodenburg@sheffield.ac.uk

Abstract. Ptychographical iterative phase retrieval is a promising new transmission imaging technique which uses a set of measured intensities from the consecutive illumination of overlapping regions of the specimen to form an image of the transmission function in phase and amplitude. Although the technique has been shown to work very effectively in both the light-optical and X-ray domains, electron ptychography poses significant difficulties, not least of which is the uncertainty in probe position due to drift and other instabilities. We demonstrate three methods for deriving the relative positions of the illumination spot on the specimen a-posteriori.

1. Introduction

Ptychographical imaging is a promising new technique with important advantages at short radiation wavelengths when compared to conventional transmission imaging, where the performance of the lens and the coherence of the radiation source can limit the available resolution. A computational method called PIE[1,2] is used to recover the phases of a set of diffraction patterns each taken from a different but overlapping area of the specimen, and has been demonstrated using both visible laser light[3] and hard X-rays[4].

Knowledge of the relative probe position associated with each diffraction pattern is a requirement of the PIE algorithm, and in both cited experimental demonstrations this was accurately controlled. In the case of high energy electrons both the specimen and the beam are subject to degrees of instability which can be significant over time scales required to record a set of diffraction patterns.

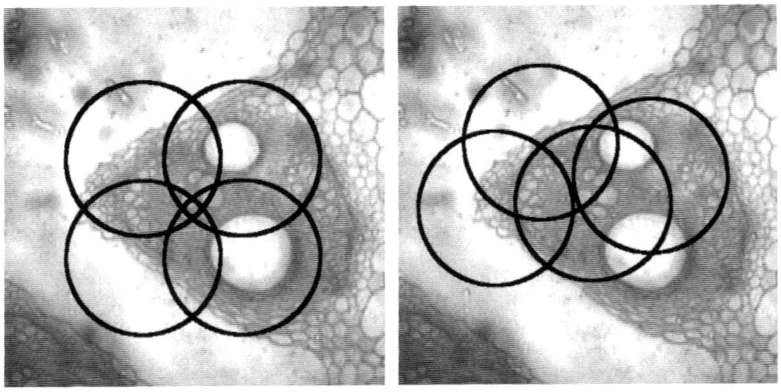

Figure 1. Illustrating how the intended position of the probe at four points in the PIE data collection process(left) can differ significantly from the experimental reality (right) in the presence of random specimen and beam drift.

In this paper we investigate three methods for deriving the relative probe positions from experimental data. It is assumed that if we can find the correct relative position between any given pair of neighbouring probes, the process can be repeated to derive the relative positions of all probes.

2. Optimisation of the diffraction-plane SSE

The convergence of PIE can be measured using the sum-squared error (SSE) in the diffraction plane (defined in [2]), which is calculated at each iteration of the algorithm using the difference between each measured diffraction pattern intensity and the detector-plane intensity from the current estimate of the specimen.

Using a simulated data set comprising just two diffraction patterns, the PIE algorithm was run for a limited number of iterations and the resulting SSE recorded. The process was repeated over a range of probe positions and thus a map of the SSE versus the relative probe position was obtained (figure 2, left) in which the global minimum lay at the correct probe position.

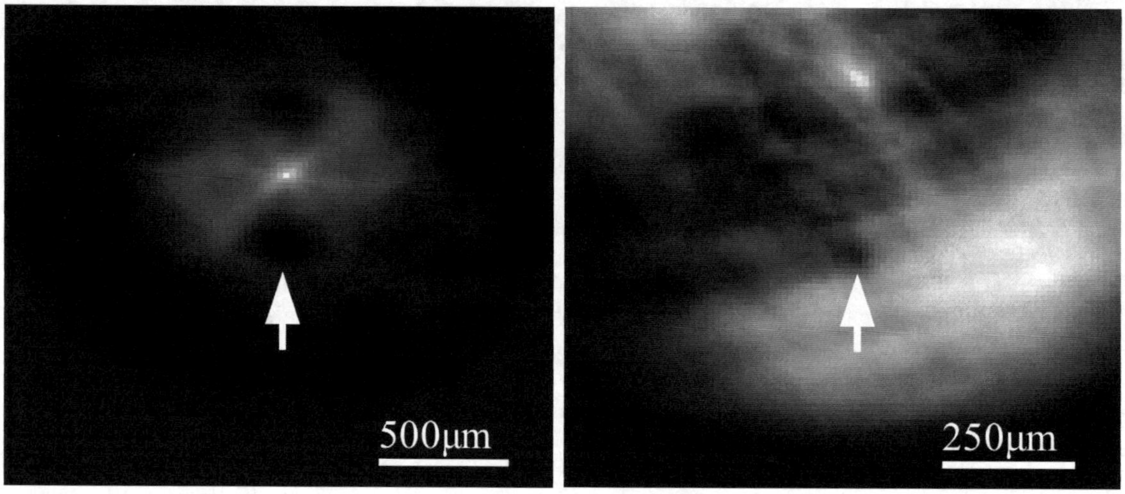

Figure 2. Map of SSE versus probe position using simulated data (left) and experimental data (right). The correct probe position (arrowed, both) is in agreement with the global minimum when using perfect simulated data, while contamination and noise in the experimental data make the SSE unreliable.

The process was repeated using experimental data from a successful light-optical reconstruction [3]. In the resulting SSE map (figure 2, right) the global minimum does not lie at the correct probe position, and there are no features which distinguish the local minimum at the correct probe position from any of the many others. This is probably due to the effect of noise and contamination in the data and errors in the probe model, which are known to make the SSE metric less reliable.

Not only is the calculation of the SSE map computationally expensive, requiring a complete run of the PIE algorithm for every point in the map, but under experimental conditions the SSE cannot be relied upon to locate the correct offset between a pair of probe positions.

3. Cross correlation of shadow images

The central disc in electron diffraction patterns formed using a focused probe often contains a shadow image (Ronchigram) which indicates the relative positions of the specimen in the beam. Using cross-correlation the shadow images in two electron diffraction patterns (figure 3) were used to find the optimum alignment of the specimen.

First a value around the mean intensity in the central disc was subtracted from the diffraction patterns and all negative values truncated to zero to reduce the dominant effect of the bright field disc (figure 4, left). Next the 2D cross-correlation was performed, resulting in a map with a clear maximum at the point where the specimen is aligned optimally.

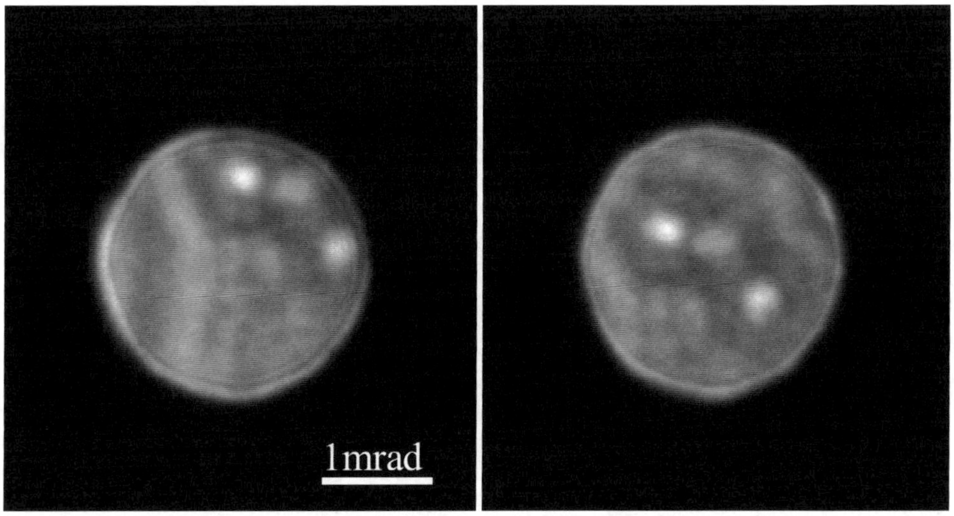

Figure 3. Two electron diffraction patterns from a sample of gold particles on holey carbon film, displaying features which can be aligned using cross-correlation to yield the probe positions.

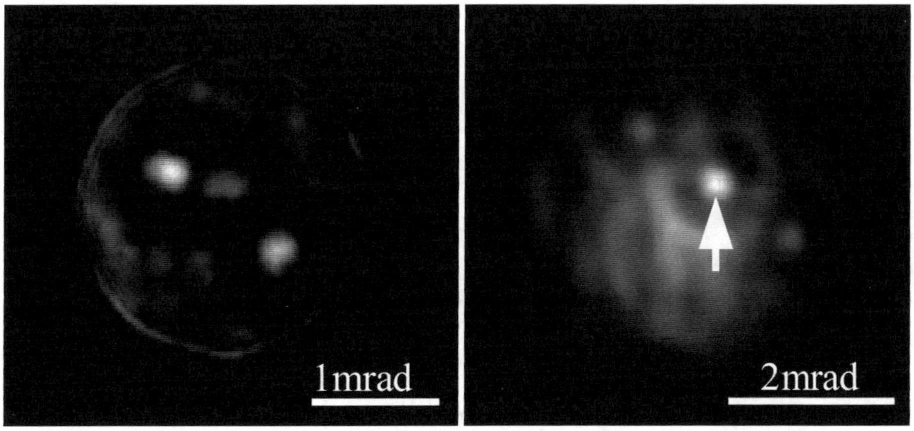

Figure 4. The truncated data (left) is used to produce a cross-correlation (right) wherein the maximum value (arrowed) indicates the required relative position of the probes.

This technique is very effective where the shadow image contains the appropriate features but the values obtained relate to the diffraction plane, and require subsequent scaling to account for the different sampling in the plane of the reconstruction. This scaling factor can be derived in the process of modelling the probe.

4. Cross correlation of Gabor hologram

If an accurate estimate of the experimental probe is available then under certain conditions a Gabor hologram can be obtained from any single diffraction pattern in the data set. The holographic reconstructions obtained from adjacent probe positions (figure 5) may be cross-correlated (figure 6) as with the bright field disc data above so as to detect the correct alignment of features in the specimen. The advantage in this case is that the result obtained is already in the plane of the specimen and does not require any further transformation.

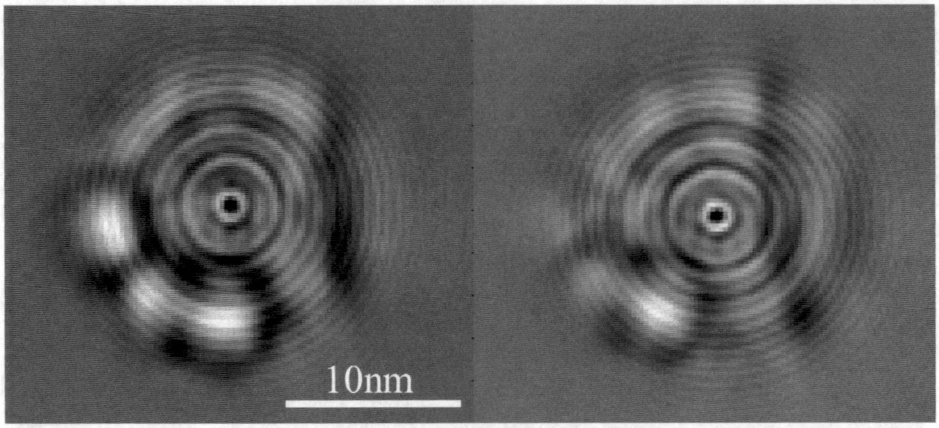

Figure 5. After running one iteration of PIE (equivalent to forming the Gabor hologram in real space for each diffraction pattern) features are recovered on the scale of the reconstruction which when cross correlated produce a maximum at the correct probe position. The observed residual phase from the probe (concentric ring pattern above) gives rise to a single spike in the centre of the cross-correlation which is not physically significant.

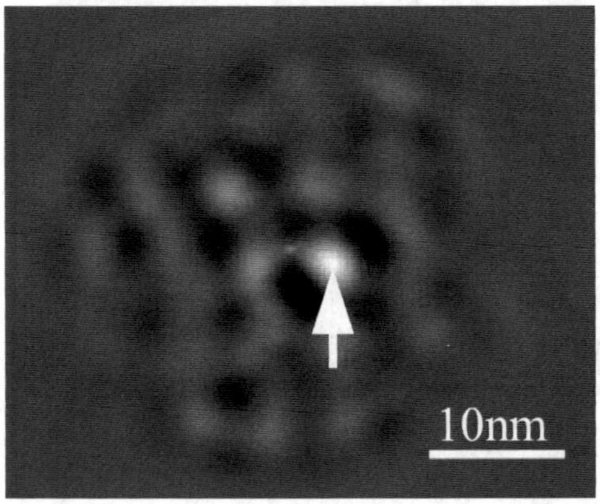

Figure 6. The cross correlation between Gabor holograms produces a strong peak (arrowed) at the correct probe position.

5. Conclusions

The implementation of PIE is significantly more demanding in the field of high resolution electron imaging than in either light-optical or X-ray microscopy in part due to the uncertainty in the probe position caused by drift. The SSE is too sensitive to the quality of the data to determine probe positions over a long range but may be useful in fine tuning a good first estimate produced by either of the other two techniques discussed, since the SSE map features a small local minimum at the right position. The cross-correlation techniques can both succeed in producing just such a good first estimate.

References

[1] Faulkner HML and Rodenburg JM 2004 *Phys. Rev. Lett.* **93** 023903/1
[2] Rodenburg JM and Faulkner HML 2005 *Appl. Phys. Lett.* **85** 4795
[3] Rodenburg JM, Hurst AC and Cullis AG 2007 *Ultramicroscopy* **107** 227-231
[4] Rodenburg JM, Hurst AC, Cullis AG, Dobson BR, Pfeiffer F, Bunk O, David C, Jefimovs K and Johnson I 2007 *Phys. Rev. Lett.* **98** 034801
[5] Fienup JR 1982 *Appl. Opt.* **21** 2758

Electron Microscopy and Analysis Group Conference 2009 (EMAG 2009) IOP Publishing
Journal of Physics: Conference Series **241** (2010) 012005 doi:10.1088/1742-6596/241/1/012005

A phase plate for transmission electron microscopy using the Aharonov-Bohm effect

C J Edgcombe

TFM Group, Department of Physics, JJ Thomson Avenue, Cambridge CB3 0HE, UK

E-mail: cje1@cam.ac.uk

Abstract. The Aharonov-Bohm effect provides a way to change the phase difference between scattered and direct electron waves in the transmission electron microscope by 90 degrees, and so to convert phase variation into amplitude information. In a thin ring of magnetic material a continuous loop of azimuthal flux provides the state of lowest magnetic energy (the vortex state). Permanently-magnetized rings can maintain this flux pattern in the presence of a strong axial field such as that of a microscope objective lens. The radial width required for such a ring is of the order of 50 nm and so the radial fraction of the scattered electrons that is intercepted can be very small.

1. Introduction

In cryo- and other transmission electron microscopy of biological specimens, there is a need to minimise the electron dose by increasing the image contrast for given beam current. It is known that this can be achieved by use of a phase plate, designed to change by 90 degrees the difference between the phases accumulated by scattered and direct electron waves in their transit from the specimen to the detector. We propose to achieve the required phase change by use of the Aharonov-Bohm effect, produced by a ring of magnetic flux set up in a thin magnetic ring inserted in the plane of the objective aperture. The ring dimensions are chosen so that scattered electrons pass outside the ring while direct electrons pass through it.

2. The Aharonov-Bohm effect

When an electron passes through a region in which the vector potential **A** is not zero, the phase of its wave function changes by $\delta\theta = (e/\hbar)\int A.dr$ along its path [1,2]. If parts of a coherent electron wave can arrive at a single point on the detector by traversing more than one path, each part i accumulates a phase change $\delta\theta_i$ along its own path. The change in relative phase due to the different values of $\delta\theta$ for two paths is then

$$\Delta\theta = \delta\theta_2 - \delta\theta_1 = \frac{e}{\hbar}\oint A \cdot dr = \frac{e}{\hbar}\int B \cdot dS = \frac{e}{\hbar}\Phi$$

where $(-e)$ is the electronic charge and Φ is the flux enclosed between the two paths (figure 1). This effect has been demonstrated by electron holography [3], when it is necessary that the wave parts travelling by the different paths should arrive with similar amplitudes.

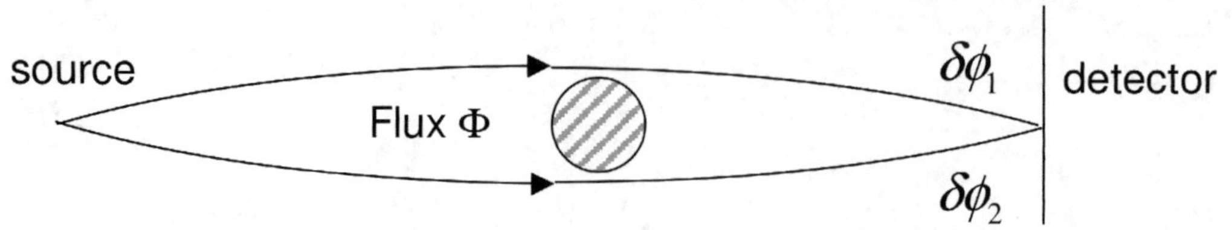

Figure 1. The basic Aharonov-Bohm effect

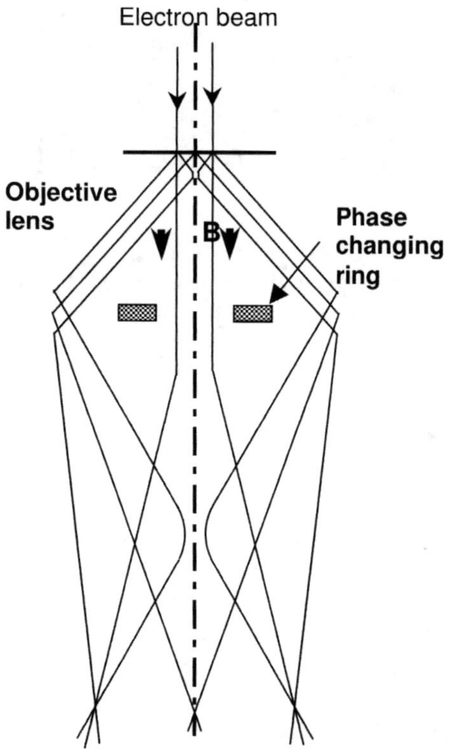

Figure 2. Schematic of TEM with scattered trajectories and ring in back focal plane.

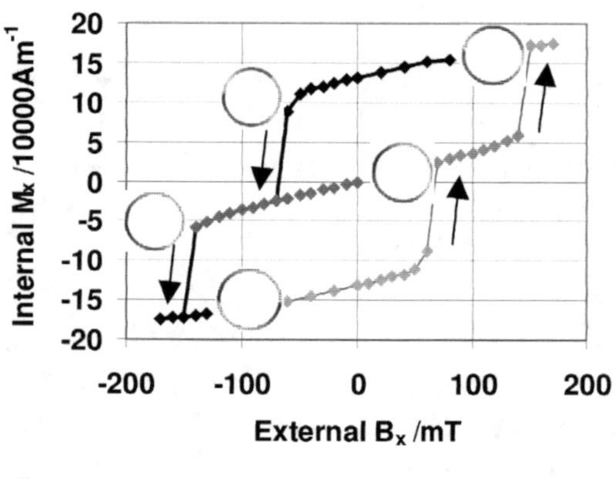

Figure 3 (colour online) Simulation of M_x vs B_x for a cobalt ring. For ring colours see text.

3. Use of thin magnetized ring

For application in a transmission electron microscope (TEM), we seek to produce a phase shift between the electrons that are scattered and those with zero net scattering, and this phase shift should be the same at all angles in azimuth around the axis. Thus we require a ring of magnetic flux directed azimuthally around the axis, and located where the direct and the scattered electrons are spatially separated. The most obvious place to locate it is in the region of the back focal plane, making use of the standard objective aperture holder (figure 2).

The azimuthal (or ϕ-directed) magnetic flux needed is small by macroscopic standards – about 10^{-15} Wb for a phase difference of $\pi/2$. However, to be useful in biological imaging this azimuthal flux must be maintained in the presence of axial magnetic flux density of the order of 1T from the objective lens of the microscope. These two requirements appear to conflict since, in a macroscopic ring of a soft material such as permalloy, an axial field of this magnitude turns all spins to near-axial directions and randomizes the directions of the remaining transverse flux. However, JAC Bland suggested in 2006 [4] that the desired behaviour can be obtained in thin rings of material with greater coercivity. For such a ring, a continuous azimuthal loop of flux (the 'vortex' state) is the state of lowest magnetic energy, and has zero accompanying external field. If the width and thickness of the ring are sufficiently small, the ring may be induced into this state by suitable application of in-plane fields. The existence of the vortex state in rings of sub-micrometric widths has been found by experiment [5] and by simulation [6] as shown in figure 3. In this figure, the colours (visible online) in the rings show the local direction of magnetic field: red corresponds to +x; mid-green, +y; cyan, −x; mid-blue, −y; with continuous gradation between these. The results differ from those of macroscopic structures in that plots of in-

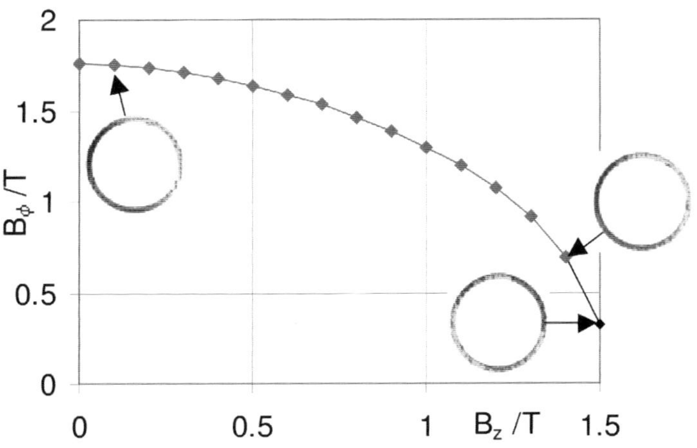

Figure 4. (colour online) Simulation of B_ϕ vs B_z for the ring of figure 3.

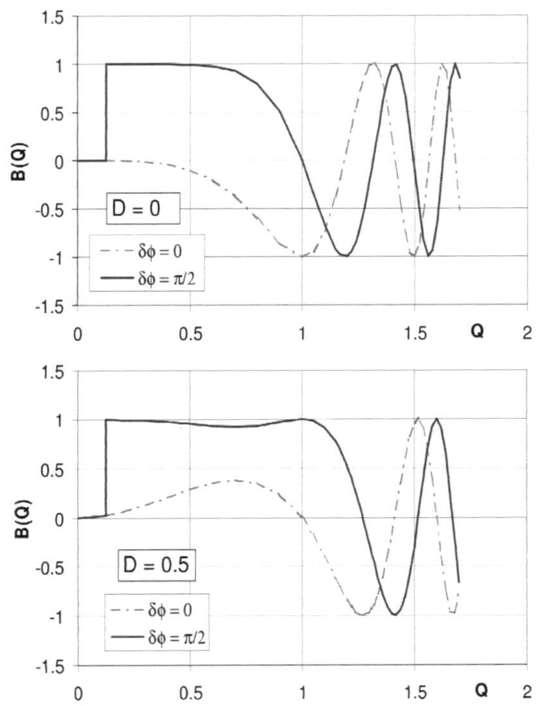

Figure 5. (colour online) Contrast transfer function as function of spatial frequency, with and without π/2 phase plate. Here Q is the reduced spatial frequency, $q(C_s/\lambda)^{1/4}$, where q is the spatial frequency, C_s is the third-order spherical aberration coefficient and λ is the electron wavelength; B(Q) is half the contrast transfer function B(q) as defined by Reimer [7] and D is the reduced defocus, given by (physical defocus) / $(C_s\lambda)^{1/2}$.

plane magnetisation M_x against in-plane applied flux density B_x show steps indicating that the vortex state is maintained over a substantial central range of B_x. The specific values at which switching occurs depend on the dimensions of the ring, but it appears that a range of the order of ± 0.1 T may be achievable.

If the axial thickness of the ring is less than its radial width, a further beneficial effect occurs. The spins resist alignment by an external field perpendicular to the plane of the film, unless that field is very strong. The planar surfaces of the ring thus have a confining effect (shape anisotropy). If the ring consists of a material of sufficiently high coercivity, the vortex state of azimuthal flux can persist in applied axial flux density of the order of 1T. Further simulation (figure 4) suggests that a ring of thickness of the order of 20 nm, as required for a phase plate, might tolerate an axial field of more than 1T while maintaining the vortex of flux. Thus it appears that, in suitable material, a ring of flux can be maintained even in the presence of an axial field comparable with that of typical objective lenses. There is some variation of azimuthal flux density with applied axial field. However, the geometry of conversion of phase variation to amplitude information suggests that some variation in the phase difference from the nominal value of π/2 may be tolerable.

The minimum width needed for such a ring appears to be of the order of 50 nanometres. Thus there will be some interception of the scattered electrons over this range of radius, at the axial position of the objective aperture. However, this range of interception is very much smaller than those of some other forms of phase plate that have been suggested, such as an electrostatic lens. Hence the contrast transfer function can be maintained near its maximum, down to much lower spatial frequencies than are usable at present (figure 5).

In order to define suitable dimensions for the ring, details will be needed for the objective lens with which it is to be used. To check that the phase shift obtained is that intended, measurement will be needed; holography appears to provide the most suitable method.

4. Conclusion

Thin, permanently-magnetized rings have properties that are convenient for use as phase plates in TEMs. Detailed design will require some knowledge of the instrument into which each is to be fitted. It may be possible to design them to fit existing TEMs as well as new ones, with or without aberration correction.

References

[1] Ehrenberg W and Siday R E 1949 *Proc Phys Soc* (London) B **62** 8-21
[2] Aharonov Y and Bohm D 1959 *Physical Review* **115** 485-91
[3] Tonomura A, Osakabe N, Matsuda T, Kawasaki T, Endo J, Yano S and Yamada H 1986 *Phys. Rev. Lett.* **56** 792-795
[4] Bland J A C 2006 oral communication
[5] Kläui M, Vaz C A F, Heyderman L J, Rüdiger U and Bland J A C 2005 *J Magnetism & Magnetic Materials* **290–291** 61-67
[6] Donahue M J and Porter D G 1999 *OOMMF User's Guide* Version 1.0, Interagency Report NISTIR 6376, National Institute of Standards and Technology (Gaithersburg, MD)
[7] Reimer L 1997 *Transmission Electron Microscopy* (Berlin: Springer) section 6.4.1

Electron Microscopy and Analysis Group Conference 2009 (EMAG 2009) IOP Publishing
Journal of Physics: Conference Series **241** (2010) 012006 doi:10.1088/1742-6596/241/1/012006

The technique of RDF of nanovolumes using electron diffraction

David Cockayne, Yixin Chen, Guoqiang Li and Konstantin Borisenko

Department of Materials, University of Oxford, Parks Road, Oxford OX13PH, UK

david.cockayne@materials.ox.ac.uk

Abstract. Amorphous materials are not, as the Greek etymology suggests, "without form". Indeed, as with crystalline materials, their structure determines their properties. The overwhelming number of structural studies of amorphous materials use the diffracted intensity to give the radial distribution function, which can be used as data against which to test, or refine, structural models. Neutron and X-ray diffraction examples abound. However, neutrons and X-rays are of no use when small volumes are involved. This paper reports the current state of RDF analysis of nanovolumes using electron diffraction.

1. The RDF technique

The structure of amorphous materials is widely investigated by using the direct mathematical relationship between $I(q)$, the scattered intensity as a function of the scattering vector q, and the radial distribution function $J(r)$. The relationship assumes single scattering. Comparing $J(r)$ (or its associated parameters $\phi(q)$ (the reduced intensity) or $G(r)$ (the reduced density function)) from experiment with the equivalent functions derived from modeled structures gives a test of the model structure, at least in terms of the first order correlation function. The model can be refined against the experimental data using e.g. Reverse Monte Carlo (RMC) or other techniques, and constraints can be applied during the refinement.

This procedure has been widely employed using both neutron and X-ray scattering. It is also used in electron diffraction from gases, where the density of the gas can be such as to ensure single scattering. This paper reviews the current state of the technique using electron diffraction for studying amorphous solids.

2. The electron RDF technique

Grigson [1] and Gragzyk and Moss [2] were the first to use electron $I(q)$ to give $J(r)$ from thin samples of solid materials. Difficulties with maintaining calibration and collecting $I(q)$ to sufficiently high q resulted in there being no further progress until Cockayne and McKenzie [3] realized that advances in electron optics and data collection overcame some of these difficulties. The key issues in obtaining reliable data are the following.

2.1. Calibration of q

Since $G(r)$ is a Fourier transform of a function of $I(q)$ (see equation below), inaccuracy in the calibration of q leads directly to errors in the scale of r (bond lengths). Consequently a major part of

RDF analysis with electrons is determining and maintaining a stable calibration of q (camera length) throughout an experiment. The optimum approach is to use an internal crystalline standard; using external standards requires a thorough appreciation and minimization of changes in camera length which can result from experimental variables such as sample height and lens hysteresis. With carefully developed protocols, an accuracy of $\pm$ 0.5% can be achieved.

To obtain precision in $G(r)$, $I(q)$ needs to be collected to large q. We routinely use CCDs with 2048 ×2048 pixels (or even larger), which, operating at a small camera length, allows collection to q = 220 nm^{-1}. Transmission electron microscopes are not optimized for linearity over such a wide range of q, but calibration of the camera length as a function of q can minimize problems.

2.2 *Sample thickness*

The conversion of $I(q)$ to $G(r)$ assumes single scattering, easily achieved with neutrons and X-rays, but more difficult to achieve with electrons. There are two tests available to determine whether a sample is sufficiently thin. Firstly, multi-slice calculations using a large super-cell of a model amorphous structure can give $I(q)$ as a function of thickness. Secondly, the goodness of fit of the atomic scattering curve $Nf^2(q)$ (N a fitting parameter) to $I(q)$ throughout the range of q can be tested (the fit deteriorates with multiple scattering) (See Figure 1). Thirdly, if the ratio of elastic to inelastic mean free paths is known, the latter can be determined from EELS. Finally there is the possibility of recovering single scattering $I(q)$ from multiple scattering data Chen et al (to be published).

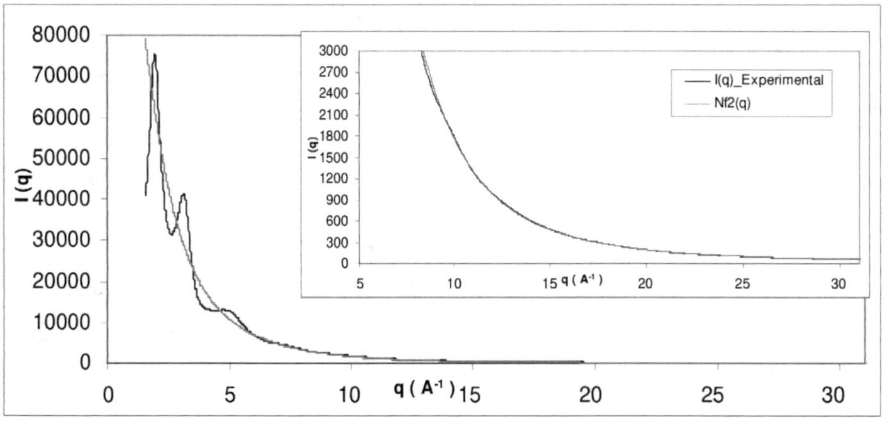

Figure 1. Comparison of single scattering $I(q)$ with no correlations and experimental $I(q)$ from GeSbTe. The inset is the enlargement at high q up to 25 $Å^{-1}$.

2.3 *Inelastic scattering*

The conversion of $I(q)$ to $G(r)$ assumes elastic scattering. In early studies, this was thought to be a major problem, and elastic data was collected using energy filters (e.g. [1], [4]). However our recent studies show that, provided one is within the single scattering regime of thickness, the major difference between filtered and unfiltered data arises from dispersion within the energy filter i.e. drift of the energy selecting aperture on the energy loss spectrum with q. Figure 2 demonstrates this effect. When this problem is overcome, we have observed little advantage in using energy filtered data. TDS could in principle be important, but its effects can be reduced by cooling the sample.

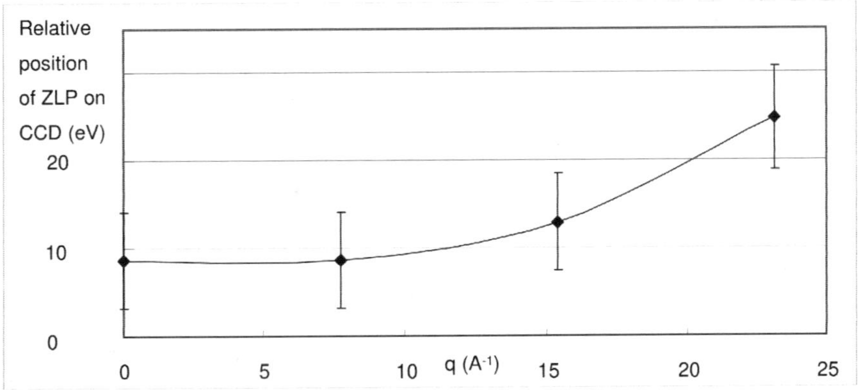

Figure 2. Displacement of zero loss peak (ZLP) relative to energy selecting slit as a function of q (JEOL 2200 MCO with Omega filter).

3. Electron RDF from nanovolumes

To obtain the RDF from a local volume of amorphous material, three methods are possible:-

3.1 Small probe

Modern TEM/STEMs can produce a focused electron beam as small as 0.1nm in width. Such a narrow beam results in a diffraction pattern without sufficient averaging for RDF analysis. A probe of width 1nm can be usefully used, provided it can be made sufficiently parallel not to convolute $I(q)$ with the angular probe function. McBride et al [5] have discussed the possibility of deconvoluting such data, but this is difficult to do routinely. Establishing a protocol which will maintain a calibrated camera length requires care to achieve in practice.

3.2 Selected area diffraction

It is well known that the diffraction data at the scattering angle q arises from the local sample at position C_sq^3. If one is investigating the RDF from local regions, this can cause a difficulty. The use of an aberration corrected TEM removes the problem. After that, C_c will be the limiting aberration.

3.3 Exit wave reconstruction

Koch and Garofalini [6] showed that a small local region of the exit wave, reconstructed from a through-focal or through-tilt series, could be used to give the selected area diffraction pattern without the effects of lens aberrations as in 3.2. The shortcoming of this method is that the resultant range of q over which $I(q)$ is obtained is limited to the range of data faithfully passed by the restoration process. This is well below q = 220 nm^{-1} achievable by straightforward diffraction, and so limits the achievable precision in $G(r)$.

4. Converting $I(q)$ to $G(r)$

$I(q)$ can be investigated as a function of scattering direction, but generally improved signal to noise is achieved by azimuthally averaging the data. First the centre of the diffraction pattern must be found. Various techniques are available, including a Digital Micrograph plug-in involving drawing circles through points of equal intensity. Tests show that the centre can be determined to a precision of 1 pixel. Averaging data azimuthally (taking into account pixels necessarily masked because of the beam stop) leads to $I(q)$ with good signal to noise even at the highest q. For improved signal to noise at high q, diffraction patterns with different exposure times can be overlapped, or repeated patterns can be accumulated.

. Figure 1 shows the output from a program which compares the experimental *I*(*q*) with the scattering from an amorphous alloy with no correlations, using the parameterised forms of the atomic scattering factors $f_i(q)$ [e.g. [7]], and which accepts N and c_i (the number of atoms and the atomic concentrations of different atomic species) as variables. The c_i are generally known for a given sample, or can be determined by EDX or EELS, and N can be chosen to match the two curves at a selected q. It is the difference between these two curves which carries the information about local order, and which, when Fourier transformed, gives *G*(*r*). In this way, *G*(*r*) can be obtained from *I*(*q*) on line at the microscope.

The expression which relates *G*(*r*) to *I*(*q*) is

$$G(r) = 4\int_{o}^{\infty}\phi(q)\sin(qr)dq \quad \text{where} \quad \phi(q) = \left[\frac{I(q) - N\langle f_i(q)^2\rangle}{N\langle f_i(q)\rangle^2}\right]$$

is the reduced intensity function. $\phi(q)$ can be damped by an expression such as $\exp(-bq^2)$, where b is the damping factor, to reduce the effects of limited experimental data range and noise at high q.

5. Using *G*(*r*) to refine models of amorphous structures

There is an extensive literature on the interpretation of *G*(*r*) from X-rays and neutrons in terms of nearest neighbour distances and coordination numbers, and this is directly transferable to electron RDF. There is, however, no technique for experimentally obtaining partial RDFs with electrons, as there is with neutrons and X-rays. For single element materials, nearest neighbour distances can be read directly from the peak positions in *G*(*r*), we find to an accuracy of 0.01A. For alloys, the interpretation of *G*(*r*) is not necessarily straightforward. *G*(*r*) (or $\phi(q)$) can be compared with the equivalent expression derived from a model structure, and the model can be refined, with or without constraints, through a reverse Monte Carlo (RMC) procedure which follows techniques used with X-rays and neutrons (see e.g. [8]). A typical protocol is shown in Figure 3.

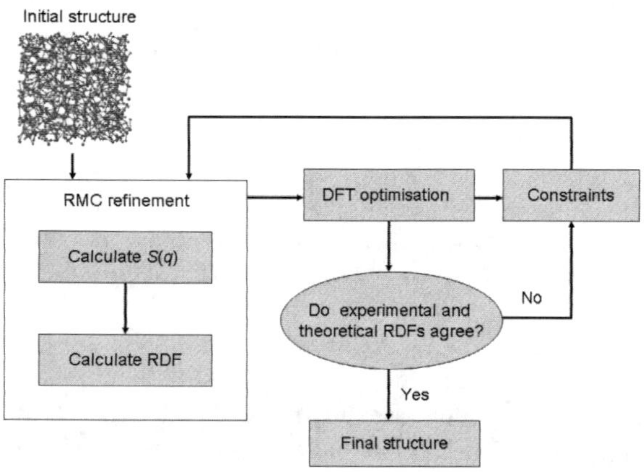

Figure 3. Protocol (flow diagram) for model refinement

References

[1] Grigson C W B, 1962 *J. Electron. Control* **12** 209
[2] Graczyk J F and Moss S C 1969 *Rev. Sc. Instrum.* **40** 424
[3] Cockayne D J H and McKenzie D R 1988 *Acta Cryst.A* **44** 870
[4] Hirata A, Hirotsu Y, Ohkubo T, Hanada T, Bengus VZ. 2006 *Phys Rev B* **74** 214206
[5] McBride W, Cockayne D J H and Nguyen- Manh D 2003 *Ultramicroscopy* **96** 191
[6] Koch C T and Garofalini S H 2006 *Ultramicroscopy* **106** 383
[7] Kirkland E J 1998 *Advanced Computing in electron Microscopy (New York: Plenum Press)*
[8] McGreevy R L 2001 *J. Phys. Condens. Matt.* **13** R877

Journal of Physics: Conference Series **241** (2010) 012007 doi:10.1088/1742-6596/241/1/012007

Obstacles and optimisation in weak-beam dark-field tomography of defects

Sharp J H*, **Barnard J S**, **Midgley P A**

Department of Materials Science and Metallurgy, University of Cambridge, Pembroke Street, Cambridge, CB2 3QZ, UK

E-mail: *jhd28@cantab.net

Abstract. Weak-beam dark field tomography of dislocations has so far been done in a heuristic way with little theoretical analysis of how to optimise the technique. In this paper a more quantitative analysis of the limits of WBDF tomography is summarised. It can be shown that the object reconstructed from WBDF images by a conventional tomography algorithm is the locus of dislocation images laid down as central slices at the appropriate tilt angles, as would be intuitively expected. By comparison with projections of the displacement field, a reconstruction from dark-field images gives a better indication of the dislocation's path than a projection of the displacement field itself. Reconstruction is performed on a simulated tilt series of stacking fault images; the resulting structure follows the correct plane but is a 3D array of rods, not a flat plane. This is produced by the 'movement' of dynamical fringes in the stacking fault images as the sample is tilted and is an example of WBDF images being suboptimal for tomography. Finally, the influence of misalignment between the tilt axis and systematic row is modelled, and is shown to degrade the reconstruction, especially in the out-of-plane direction. This misalignment changes the excitation error with tilt, which introduces another variable to the reconstruction process and breaks the requirement that the images be a straight-ray projection of a monotonic function of the required parameter.

1. Introduction

Tomography is the reconstruction of a 3–dimensional object or structure from a tilt series of 2–D projections. In electron microscopy, tomography has mainly been used for reconstructing the shapes of samples such as particles or nanotubes, using mass-thickness contrast, for example [1]. Diffraction contrast, such as that in dark-field images of defects, has been considered too complex for use in tomography. Using weak-beam dark field (WBDF) defect images for tomography was first attempted in 2005 and was surprisingly successful [2]. To do this, the tilt axis of the tilt-rotate holder is set up parallel to the systematic row that will be used for the weak-beam condition, and the beam tilted along the systematic row to achieve the weak-beam condition. The sample is then tilted around the eucentric axis by up to 70° in each direction and a tilt series of images are recorded; ideally the weak-beam condition is kept constant during tilt. This experiment was originally done to test the feasibility of the concept; the next stage, in progress, is to consider the theoretical basis of this technique and determine its limits.

This paper summarises the results of some simulated experiments to investigate the effects of small and large changes in sample orientation while setting up tomography. On the large scale, for reconstructing stacking faults, the movement of dynamical fringes as the sample is tilted

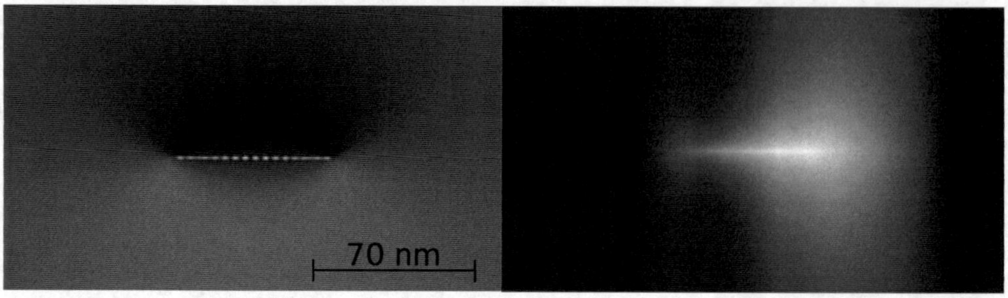

Figure 1. Projections through crystal of Al, showing screw dislocation running left to right from top to bottom of 100 nm thick film. Left: simulated kinematical dark-field image. Right: projection of magnitude of in-plane component of displacement field in same orientation. No background added to either calculation.

has an important impact on the reconstruction; an orientation must be found to minimise this movement. On the small scale, the effect of a small misorientation between the tilt axis and the **g** used for imaging is investigated; as expected, misalignment is found to impair reconstruction quality, especially perpendicular to the film.

2. Tomography theory

When image intensity can be represented as an integral of the property of interest along straight lines through the sample [3], the Fourier slice theorem relates the object $f(x, y)$ to this parallel projection through the sample $P_\theta(t)$ along ray t according to Equation 1. The Fourier transform of the object's projection at angle θ to the x axis, gives a central slice of the Fourier transform of the object taken at an angle θ to the x-frequency axis [4]. In this case, the object can be reconstructed by backprojection tomography. The process of backprojection is that of extending the values in the Fourier-transformed projection over the stripe of Fourier space that corresponds to the projection's tilt angle [3]. When all the projections are backprojected across the same space, a flawed version of the object is reproduced.

$$\int_{-\infty}^{\infty} P_\theta(t) \exp\left[-2\pi i w t\right] dt = \int_{-\infty}^{\infty}\int_{-\infty}^{\infty} f(x,y) \exp\left[-2\pi i w(x\cos\theta + y\sin\theta)\right] dxdy \quad (1)$$

A WBDF dislocation image is not, however, a straightforward projection; in the kinematical approximation, the integral is not over the displacement field $\mathbf{R}$ itself, but over $\exp(-2\pi i \mathbf{g}.\mathbf{R})\exp(-2\pi i s z)$, which is then squared in calculating intensity. By substituting the dark-field image intensity instead of the simple straight-ray projection in the Fourier slice theorem, it becomes a statement that the Fourier transform of an object taken at a tilt angle θ is equivalent to the Fourier transform of the intensity of the dark field image set at a tilt of θ. We will not get a reconstruction of $|\mathbf{R}|$ from a tilt series of WBDF images; reconstruction will yield this image intensity 'object'.

Figure 1 compares a projection of the magnitude of the dislocation displacement field, and a dark-field image calculated using R. Schäublin's CUFOUR program [5]. The parallel-sided Al crystal is oriented with [111] beam direction pointing directly down; the screw dislocation intersects the top of the 100 nm thick foil on the left and runs down with line direction [110] and Burgers vector $\mathbf{b} = \frac{1}{2}[110]$. The image is calculated using the systematic row $n(20\bar{2})$ in $\mathbf{g}(3.5\mathbf{g})$ weak-beam condition, with image taken from $\mathbf{g} = 20\bar{2}$. The dark-field image shows the familiar narrow line, dotted with oscillations from Bloch wave interference. The projection of the displacement field gives an image that becomes more diffuse as the dislocation approaches the bottom surface, becoming a low intensity region around a high-intensity core. Reconstructed,

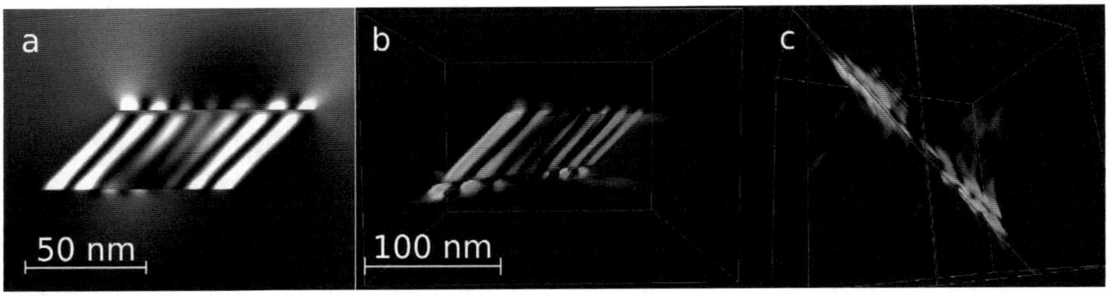

Figure 2. (a) Zero tilt image from simulated stacking fault tilt series; intersects top surface on left of image, bottom surface on right. (b) Same view of reconstruction, with perspective. (c) View of reconstruction along $[\bar{2}11]$, along which the (111) fault plane should appear flat, but is reproduced as an array of rod-like structures because of the movement of fringes as the crystal is tilted.

this will obscure the dislocation line. The dark-field image gives a *better* tilt series image than a simple projection of the displacement field, if the purpose of the study is to reconstruct the path of the dislocation core.

3. Stacking faults

When WBDF tomography is attempted on defects containing stacking faults, the reconstruction quality is found to depend on the choice of tilt axis/imaging reflection. This may be due to the difficulty of tilt series alignment with fringes present, or a basic property of stacking fault contrast. Two tilt series of $\mathbf{g}(3\mathbf{g})$ weak-beam images were simulated using CUFOUR, changing the beam direction to achieve tilt. This simulated tilt series starts with perfect alignment, because the defect is placed at the same point relative to the centre of the image every time. The defect modelled was in Al, on a stacking fault with displacement $\frac{1}{3}(111)$ bounded by partial dislocations running along $[01\bar{1}]$ from top to bottom of the 100 nm thick crystal, with Burgers vectors $\frac{1}{6}[\bar{2}11]$ and $\frac{1}{6}[\bar{1}2\bar{1}]$ according to the Thompson tetrahedron convention [6]. A partial dislocation separation of 50 nm was used, which is larger than would be found in nature but allowed the fringes to be more easily observed. Images were simulated from 45° to -45° for $\mathbf{g} = 200$ and reconstruction was carried out using conventional SIRT, 30 iterations, using FEI's Inspect3D^{TM}. Adaptations were made so that the image became appropriately longer and shorter as the sample was 'tilted', as for an experimental tomography series.

The tilt axis was along the projected width of the stacking fault, so the depth range of the stacking fault changed and the fringes changed position along the length as the tilt angle changed — they appeared to 'move' along the stacking fault through the tilt series. The simple projection object that would give this behaviour is an array of rods, and this is what is reconstructed; this is apparent when viewed parallel to the $[\bar{2}11]$ direction, from which a (111) stacking fault should appear as a flat plane. Unlike for the case of a dislocation, this shows that WBDF contrast is detrimental to tomography of defects containing stacking faults, and another technique such as STEM LAADF imaging would be better [7].

4. Tilt axis misalignment

To test the effect of misalignment between the tilt axis and the systematic row, three tilt series of weak-beam images were modelled using CUFOUR, applying different amounts of misalignment. A mixed dislocation was used, $\mathbf{b} = \frac{1}{2}[1\bar{1}0]$ along $[0\bar{1}1]$ in an Al crystal of foil normal [001]; it passed through the crystal at 45° to the surfaces. A tilt range of -45° to $+45^o$ was used for $\mathbf{g} = 200$ and reconstruction done using SIRT 30 iterations as before. The measure of tilt

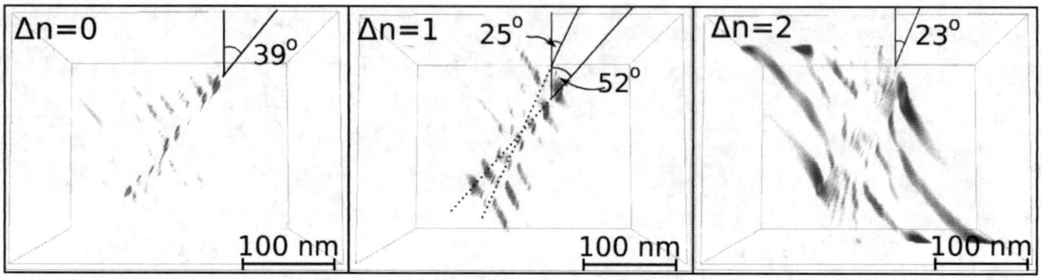

Figure 3. Reconstructions of misaligned weak-beam dark-field tilt series of a single dislocation inclined at 45° to the foil plane. Left: no misalignment. Centre: $\Delta n = 1$ (see text). Right: $\Delta n = 2$. The dislocation's apparent midline is rotated by misalignment, and its shape and thickness change. This rotation has been seen in threading dislocations in experimental reconstructions [2].

axis misalignment Δn for weak-beam condition $\mathbf{g}(n\mathbf{g})$ is the difference in n between the two ends of the tilt series. Simulations were done for $\Delta n = 1$, $\Delta n = 2$ and perfectly aligned, for zero tilt image conditions $\mathbf{g}/3.5\mathbf{g}$ (as used previously for experimental WBDF tomography). As the diffraction condition is changed, the excitation errors change also, making the images more inconsistent.

The results, shown in Figure 3, show that misalignment of the tilt axis has a profound effect on the reconstruction. The perfectly aligned tilt series produces a dislocation of the correct shape, though with dynamical fringes; as the misalignment increases, the intensity deviates more in shape, and the inclination of the dislocation is no longer accurately found. It is therefore important to carefully align the tilt axis along the systematic row before taking a tilt series.

5. Conclusions

Weak-beam dark field tomography of defects is described by an adapted Fourier slice theorem and reconstructs an 'object' corresponding to the dark field tilt series image intensity, not the displacement field of the defect; this is a better function for showing the path of a dislocation core. Simulations show WBDF stacking fault images are less suitable, giving a reconstruction that is an array of rods, not a flat planar fault. Misalignment of the tilt axis with the systematic row causes changes in excitation error that break the projection requirement, giving a dislocation reconstruction of impaired resolution and changed direction.

6. Acknowledgements

We wish to thank Robin Schäublin for contributing the CUFOUR program, and JHS acknowledges funding through an EPSRC Doctoral Training Account.

7. References

[1] Weyland M and Midgley P 2003 *Ultramicroscopy* **96** 413
[2] Barnard J S, Sharp J, Tong J R and Midgley P A 2006 *Philosophical Magazine* **86** 4901–4922
[3] Slaney M and Kak A 1988 *Principles of computerized tomographic imaging* (New York: IEEE Press)
[4] Kak A 1985 *Array Signal Processing* ed Young T and Fu K (Englewood Cliffs, New Jersey: Prentice–Hall)
[5] Schäublin R and Stadelmann P 1993 *Materials Science and Engineering* **A 164** 373–378
[6] Hirth J P and Lothe J 1982 *c* (New York: Wiley)
[7] Sharp J H, Barnard J S, Kaneko K, Higashida K and Midgley P A 2008 *J. Phys. Conf. Ser. EMAG 07: Characterisation, manipulation and fabrication on the nanoscale* vol 126 (Journal of Physics) p 012013

Hybrid Tomography for structural and chemical 3D imaging on the nanoscale.

G Möbus, Z Saghi, W Guan, T Gnanavel, X Xu, Y Peng

Dept Engineering Materials, University of Sheffield, Sheffield S1 3JD, UK.

g.moebus@sheffield.ac.uk

Abstract. Electron tomograms can be reconstructed from a variety of projection imaging techniques in TEM, some of which are sensitive to the surface or the bulk, to the structure or chemistry, or operate under different resolution regimes. We introduce a concept of combining two acquisition series using two different complementary modes to form a single hybrid tomogram. As the main application, composite nanoobjects are presented for which a low-resolution 3D chemical mapping mode for the object interior can be combined with a high-resolution surface sensitive mode to precisely define the object exterior shape.

1. Introduction

Electron tomography in materials science is currently the most powerful method of 3D reconstruction of nanostructures. Using projection image tilt series in the TEM, a level of resolution and optional chemical sensitivity can be achieved which is unmatched by any X-ray techniques. Recently, the family of TEM tomography methods has grown rapidly to now include: bright field (BF), dark field (DF) TEM, ADF-STEM, EFTEM, holography, EELS, and EDX [1, 2].

2. The concept of hybrid tomography

In order to be able to exploit the complementary benefits of the different electron tomography modes, applied to the same sample, we outline a strategy of combination-tomograms. At first, we divide the diversity of imaging modes into the following classes or pairs of opposites:

(i) *surface* sensitive versus *bulk* sensitive tomography. Here *surface* refers to the particle shape (or habit in crystallography); surface tomography is the discipline that ideally returns a 3D function with value 1 along all segments (voxels) of the surface and values 0 both inside and outside the surface (or equivalent: value 1 inside and value 0 outside the material). Here *bulk* refers to the 3D interior density function (e.g. atomic number density) which is determined by bulk tomography, and which has either floating number values or a set of gray levels more than binary.

(ii) *structural* versus *chemical* tomography. This classification separates the aim of reconstructing a 3D morphology (both surface and inner microstructure) from reconstructing a 3D chemical map. A core-shell nanostructure with a crystalline core of element A and an amorphous shell of element B would be detected by both modes. Structural modes would include some sensitivity to atomic number (Z-contrast or scattering contrast), but would integrate all elements in projection in case of a composite, while chemical imaging modes in this sense image uniquely one chemical element (or a molecule/unit cell in case of more advanced chemical fingerprinting modes).

(iii) *geometric* versus *computed* tomography (GT vs. CT). This purely mathematical classification separates modes that measure metrological parameters (e.g. diameters or supports of objects) from methods that use full projections by Radon transform. Its applicability is closely related to the class (i) above, as CT would reveal the 3D bulk density, while GT is often ideal for 3D surface tracking. GT subdivides into many disciplines, of which we have introduced shape-from-silhouette [3, 4] using a binarised projection, and shape-from-contour [4] using an edge-enhanced projection. These techniques are principally applicable for objects of a minimum convexity, subject to exceptions [4].

(iv) *low-resolution* vs *high-resolution* tomography: a hybrid combination of this type would only be meaningful in conjunction with (ii), as the low-resolution information must be valuable (e.g. chemical) and complementary to the high-resolution information (e.g. atomic resolution imaging).

(v.a) *thin* specimen versus *thick* specimen tomography, and closely related:

(v.b) *linear* versus *non-linear* tomography:

Different TEM modes (or different settings within one mode) can be used to record per tilt angle a pair of images, one optimized for best definition of the thin areas and one for most reliable projection of the (difficult) thick areas. A hybrid reconstruction requires segmentation and/or superposition depending on the orientation of thin/thick areas relative to the rotation axis. This type of combination of two tomograms was perhaps the first hybrid CT application in the field of x-ray tomography of materials [5].

(vi) *reconstructive* versus *destructive* tomography.We add an instructive further combination of two modes, the first one "reconstructive" comprising all tomography modes considered so far, whether CT or GT, in the sense that a line integral (a ray of light or an "x-ray" along a line **l**) produces either the density projection (case of CT) of $P_{CT} = \int \rho(x,y,z)\, \mathbf{dl}$ or (case of GT) a binary response ($P = 1$ if at least one voxel >0, $P = 0$ if all voxels $= 0$, i.e. outside the object O). To introduce a destructive tomography mode (DT) within the context of particle beam based nanofabrication we have to set the response to P_{DT} $(x, y, x_0, y_0) = I(t) * P_{CT}$, with $x_0,y_0 = x_0(t,\Theta),y_0\,(t,\Theta)$ indicating the beam position during a processing sequence $\mathbf{r}_0(t)$, as a function of time and tilt angle. The drilling power intensity I(t) is assumed binary and delivers P=0 for $x=x_0$, $y=y_0$, and I>0, while $P=P_{CT}$ for either $x\neq x0$, $y\neq y0$ for any I, or for I=0 for any x,y . The integration along the beam paths has to be understood as iterated over the tilt angular range, such that each new destructive exposure operates on the previous object $O_{N-1} = \rho\,(\Theta_{N-1}; x,y,z)$, with N counting the nth tilt angle Θ_N or the nth destructive modification rather than using the original object for N=0. The combination of voxels at different Θ_N then follows a logical NOR. This latter mode of operation has been shown topologically identical to the GT mode of (iii) [6].

3. Applications

We illustrate the concept of hybrid tomography with highlights of recent work, reinterpreted within the concept of section 2, and extended by a specifically selected application on core-shell particles.

3.1. Hybrid Spectrum Imaging (SI) tomography combined with ADF-STEM.

A primary example of hybrid TEM is the definition of a spectrum image region of interest (linescan or 2D scan) on an ADF STEM image, whether for EDX mapping or EELS-SI. The STEM image provides sharp details on the object shape and size, while the analytical signal in the SI adds internal chemistry. The latter is not only blurred due to delocalization of inelastic processes, but mostly limited in resolution to the spot increment in the scan, typically at least one order of magnitude worse than STEM resolution. In an earlier demonstration of combining a tomographic EDX line scan with an ADF-tomogram we added a sharp contour to a low-resolution multi-element RGB-tomogram [7].

Here we present the combination of an ADF-tomogram with a tomographic set of EELS-SI linescans of a glass-nanocomposite with a CeO_2 particle (Fig 1a), with a tilt series over 120° in 10° increment (JEM 2010F). Suppression of star artefacts, noise and improved contour-resolution result from the multiplication of a binarised ADF-reconstruction (Fig 1b) with the Ce-M-edge slice (the 3D chemical map) of the SI tomogram (Fig 1c), see [8] for details. The dominating artefacts in 1c are due to the large tilt increment and missing wedge, especially the false bright "corners" of the glass, while the poor particle definition is due to the coarse SI step size.

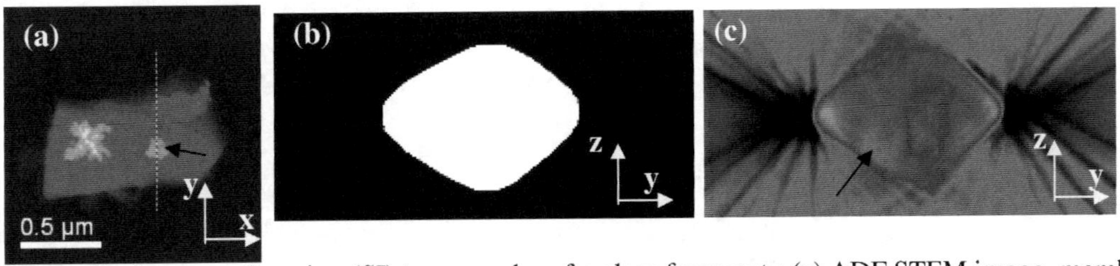

Figure 1. Spectrum Imaging (SI) tomography of a glass fragment ; (a) ADF STEM image, member of ADF-tilt series; (b) binarised contour of cross-sectional ADF-reconstruction along dashed line of (a); (c) EELS-linescan reconstruction at around 900 eV loss (Ce-M-edge) with one precipitate arrowed in (a) and (c). The hybrid tomogram results from multiplication or application of logical OR.

3.2. Dual-Energy EFTEM of core-shell objects.

Core-shell nanoparticles (or their agglomerates) are another promising field for hybrid tomography, as illustrated by Al particles surrounded by amorphous oxide shell and embedded in carbon. Reconstruction (from a 120° tilt series in 5° increment) of the crystalline core via CT from the Al-plasmon at 15eV loss is combined with a projection at 20-25eV plasmon loss, including both amorphous Al_2O_3 and carbon. The resulting tomograms can be superimposed into a multi-segmented combined tomogram.

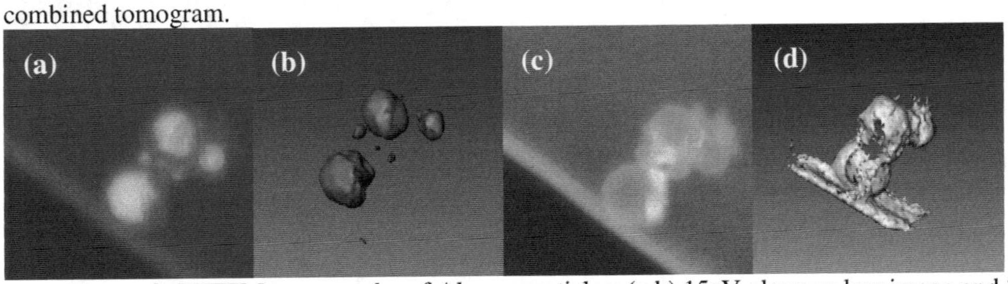

Figure 2. EFTEM tomography of Al nanoparticles: (a,b) 15eV plasmon loss image and 3D reconstruction; (c,d) 20-25eV plasmon loss image and 3D reconstruction (JEM 2010F)

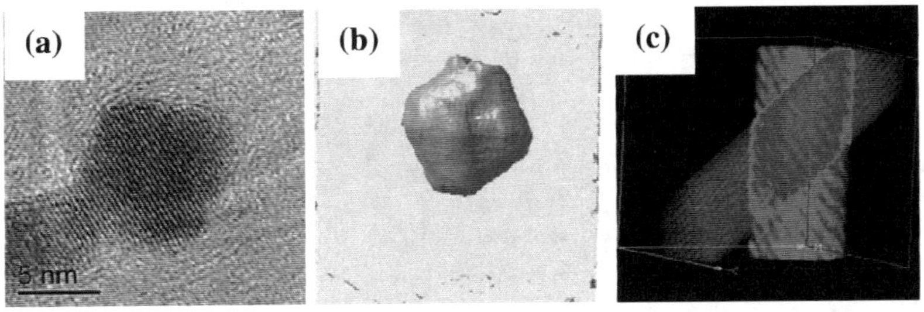

Figure 3. (a) example of a HREM projection of a CeO_2 nanocrystal from a tilt series (JEM3010); (b) geometric tomography reconstructed particle shape from all tilt members; (c) superposition of two backprojected zone axis patterns, revealing crystallography but not the correct particle shape.

3.3. Hybrid HREM tomography

Tomographic reconstruction of 2 or 3 zone axis patterns from a nanoparticle can result in a basic solution of unit cell geometry according to the principle of lattice-fringe based "goniometry" [9]. However, the information about the extent of the nanocrystal (its shape) is lost as the fringes are backprojected to infinity. A multiplication with a complementary geometric tomogram from a full fine-increment tilt series can therefore deliver particle morphology in both external shape and internal crystal structure (if single-grained) [10]. The second tomogram could either be from low-resolution BF-TEM or ADF-STEM modes or from a binarized extract of the support of the particle from a full HREM tilt series. The latter is demonstrated for a CeO_2 nanoparticle in Fig 3 using a 130° tilt series with 10° increment [11].

3.4. Tomographic nanofabrication (destructive combined with reconstructive tomography)

A nickel nanotip is cut and drilled by a FEG-beam at 200kV in three viewing directions to generate a perforated binary structure of homogeneous Ni metal (Fig 4). A subsequent tomographic reconstruction from binarised bright-field images then exactly inverses the fabrication process and leads to a 3D model of the as-fabricated nanostructure. During the fabrication process, extrusions of metal are partially observed, see [6,12] for more details.

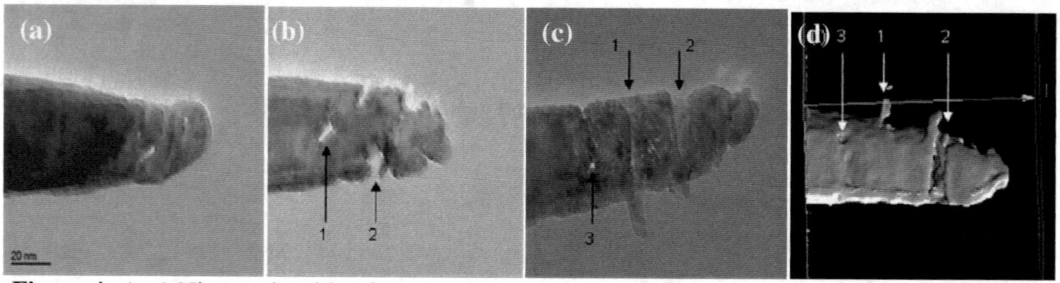

Figure 4. (a-c) Ni-nanotip with drilling events at -45°, 0°, +45° tilt direction; (d) 3D reconstruction from a full geometric tomography tilt series from -45° to 45° in 5° increments with numbered arrows indicating corresponding holes (JEM 2010F).

4. Conclusions The examples of combining two tomograms acquired with different TEM modes or different data registration, serve to demonstrate that the resulting multi-mode tomogram can exceed in its information content any of the single-mode tomograms. Hybrid tomography is a promising research field in early stage of development with many more opportunities to be explored.

Acknowledgements
The work has been funded by EPSRC, UK, through its Basic Technology Programme, GR/S85689/01. We thank Qinetiq, UK, for providing part of the nanoparticles.

References

[1] Weyland M, Midgley P A 2004 *Materials Today* **7** 32-40
[2] Möbus G, Inkson B J 2007 *Materials Today* **10** 18
[3] Xu X, Saghi Z, Gay R, Möbus G 2007 *Nanotechnology* **18** 225501
[4] Saghi Z, Xu X and Möbus G 2008 *J Microsc.* **232** p186.
[5] Tam KC 1987, *J nondestructive evaluation*, **6** 189.
[6] Saghi Z, Gnanavel T, Peng Y, Inkson B J, Cullis A G, Gibbs M R, Möbus G 2008 *Appl Phys Lett* **93** 153102
[7] Saghi Z, Xu X, Peng Y, Inkson B J, Möbus G 2007 *Appl Phys Lett* **91** 251906
[8] Saghi Z, Xu X, Möbus G 2008 Proceed. 14th EMC, Aachen, **1** 427
[9] Qin W, Fraundorf P 2003 *Ultramicroscopy* **94** 245
[10] Saghi Z, Xu X, Möbus G 2009 *J.Appl.Phys* **106** 024304
[11] Saghi Z, Gnanavel T, Xu X, Möbus G 2009 *Mater.Res.Soc.Symp.Proc.* **1184** HH02-03
[12] Gnanavel T 2010 *J Phys Conf Ser, Proceed. EMAG2009, this volume*

Journal of Physics: Conference Series **241** (2010) 012009 doi:10.1088/1742-6596/241/1/012009

Imaging electron detectors for low-voltage TEM

G Moldovan and A I Kirkland

Department of Materials, University of Oxford, Parks Road, Oxford, OX1 3PH, UK

Email: grigore.moldovan@materials.ox.ac.uk

Abstract. There is an increasing effort for the development of transmission electron microscopes operating at accelerating voltages of less that 100 kV, down to 20 kV. This work aims to clarify if the technology of conventional indirect scintillator-CCD cameras is suitable to record images formed with electrons of such low energy and how its performance compare to that of novel direct silicon imaging detectors. The performance of these imaging detectors is discussed in terms of modulation transfer function and detective quantum efficiency. It is demonstrated that whilst the performance of conventional scintillator-CCDs improves as the electron energy is dropped, it then peaks at 70 keV and then drops again. Optimum imaging performance at even lower energies is expected for the novel directly exposed detectors, which are capable of near 100% detective quantum efficiency at very low electron energies.

1. Introduction

Degradation of radiation-sensitive materials under the electron beam of high-voltage transmission electron microscopes (TEM) imposes significant limitations to the characterization of new soft materials such as carbon-based structures. The principal approach to overcome this problem is to lower the voltage below the radiation damage threshold, thus preventing any degradation and allowing for high dose imaging.

However, as the voltage of the microscope is reduced, conventional imaging cameras are required to work outside their design parameters and may not have sufficient modulation transfer function (MTF) and detective quantum efficiency (DQE). Further, several detector technologies can be used for detection of electrons in this energy range, of which indirect scintillator-charge coupled device (SCCD), direct monolithic active pixel sensors (MAPS) and direct double-sided Si strip detectors (DSSD) will be considered here. Since experimentation in this voltage range is not yet available, simulations must be employed to calculate the expected performance of these detectors. The methodology used for simulations will be described first, followed by the resultant imaging metrics and selected model images that illustrate the expected performance of each detector.

2. Method

A simplified structural model was built for each detector, which was then used to simulate the trajectories of beam electrons and distribution of charge induced in each pixel for each microscope voltage. This was then used to calculate MTF, DQE and a low-dose model image. Simulations of induced charge use Joy's model, which consists of straight electron trajectories between discrete elastic Rutherford scattering events and continuous energy loss along these trajectories [1]. Accurate relativistic Rutherford and Bethe formulae were used, as well as the generation of fast secondary electrons [2].

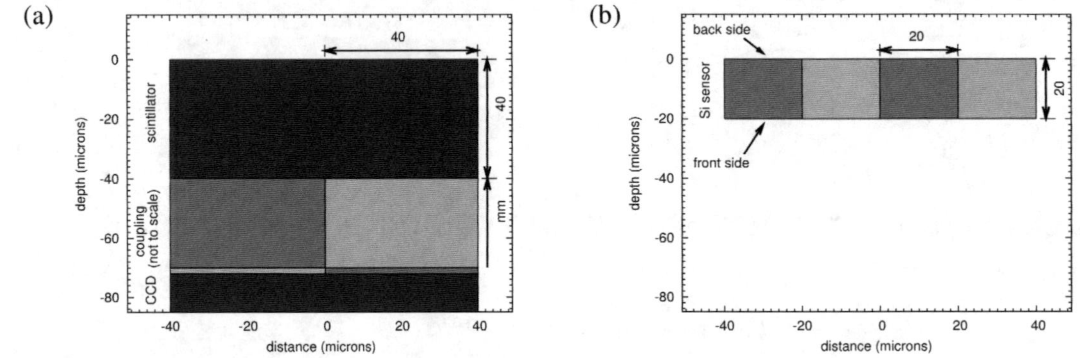

Figure 1: Diagrams of SCCD (a), MAPS and DSSD (b) models used for Monte Carlo simulations.

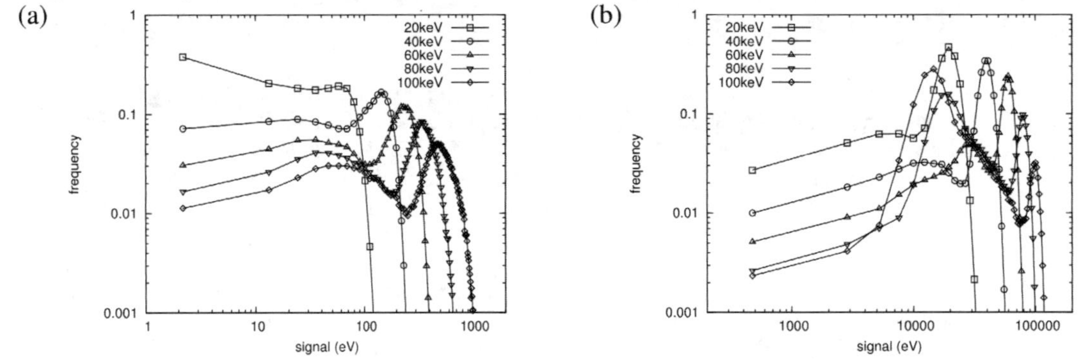

Figure 2: Distribution of signal generated in the SCCD (a), MAPS and DSSD (b).

The structural model used to simulate the performance of indirect SCCD consists of 40 μm thick scintillator with 70% volume content of ZnCdS phosphor grains in a SiO_2Z_2 glass support, coated with 1 μm of Al and mounted on a 2 mm thick glass support (Fig. 1a). The optical fiber bunch was considered to have an entrance aperture semi-angle of 60 degrees, an optical focus at 41 μm and an optical point spread function of 10 μm at normal incidence. To illustrate the importance of the mode of operation, the same structural model was used for the MAPS and DSSD detectors (Fig 1b). This consisted of back-illuminated fully-depleted 20 μm thin Si where the read-out layers were approximated as one front-side continuous 2 μm thick Al coating for MAPS and double-sided 0.2 μm thick continuous Al coatings for DSSD. The electronic noise was set to 25 well-electrons RMS for SCCD and MAPS, which is somewhat better than commercial SCCDs. Sensitivity of DSSD was set to 3,600 eV. A pixel size and pitch of 20 μm was considered for all models.

3. Results and Discussion

SCCD and MAPS operate as integrating detectors, where the signal recorded corresponds to the energy deposited in each pixel during the exposure time, whilst DSSD operate as a counting detector, where the signal recorded corresponds to the number of electrons detected in each pixel during a set time interval. Because of this difference in operation the two types of detectors will be discussed separately.

3.1. Integrating detectors

For the case of 100 keV electrons, both the 40μm thick scintillator and 20μm thick Si are relatively transparent, with some electrons penetrating through the sensor and depositing only a fraction of their total energy in the sensitive volume. Their corresponding signal distribution has two peaks corresponding to full and partial energy loss (Fig. 2a-b). For the case of 20 keV electrons, the sensors

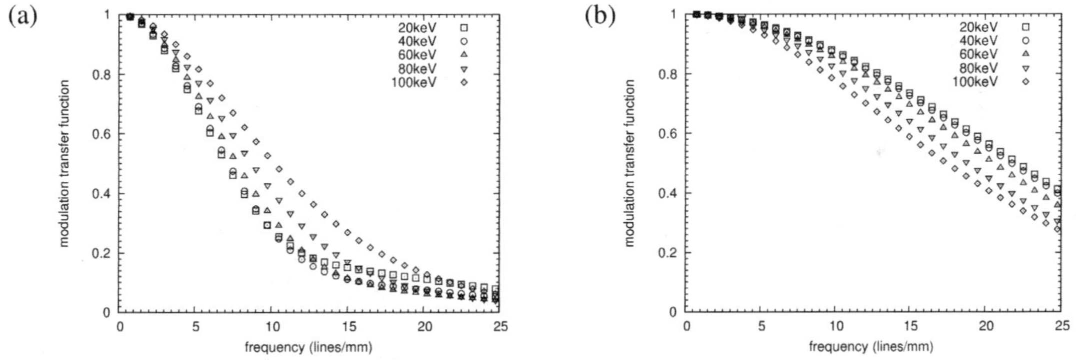

Figure 3: Modulation transfer function as a function of energy for SCCD (a) and MAPS (b).

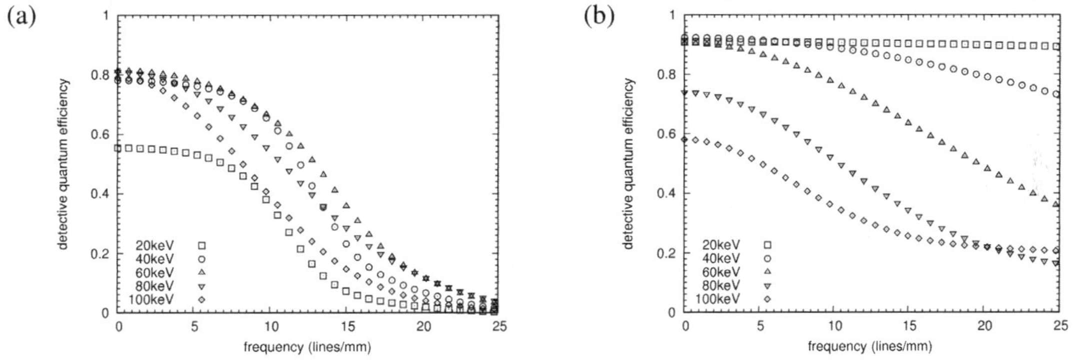

Figure 4: Detective quantum efficiency as a function of energy for SCCD (a) and MAPS (b).

are opaque for both phosphor and Si sensors; however the 1 μm thick Al coverage of the SCCD absorbs an important fraction of the electron energy. Overall, significantly more energy is deposited in the MAPS sensing volume as compared with SCCD, suggesting that a much higher signal/noise ratio is expected for direct detectors operating in this energy range.

As is the case for high-voltage TEM, the SCCD suffers not only from significant lateral displacement and back-scattering of beam electrons in the scintillator, but also from internal reflection and poor collection of photons from the scintillator. Whilst the modulation transfer function of SCCD does improve for 100 kV compared with higher voltages (Fig. 3a), the MTF of MAPS is close to ideal behaviour throughout the entire energy range (Fig. 3b). This is attributed to the MAPS 20μm thin sensor that limits lateral spreading and the lack of support that prevents backscattering.

Detective quantum efficiency of SCCD at 100 kV is somewhat better than that of high-voltage operation, showing a DQE of ~80% for frequencies below 5 lines/mm, which then drops to less than 5% at Nyquist (Fig. 4a). As the voltage is dropped to 60 kV, a much improved DQE is observed due to a smaller frequency of electrons with partial energy loss and therefore lower scattering noise. However, as the voltage is dropped further, the DQE decreases because of the absorption of electrons in the Al coating (Fig. 2a). For the case of high-voltage, the much improved MTF of the MAPS detector results in an increased DQE for frequencies above 15 lines/mm with a Nyquist value of ~20%. As the voltage is dropped to 20 kV, much improved DQE is obtained due to near-ideal MTF and full electron absorption. MAPS DQE is limited by the fraction of electrons that are backscattered as they enter the detector and thus do not deposit sufficient energy to be detected (Fig. 2b).

Whilst MTF and DQE characterize the performance of these detectors under high-dose conditions, simulated images are better suited to illustrate operation under low-dose conditions. Low-dose images have been simulated using a 256 x 256 pixels model consisting of a dark clock in a white background (Fig. 5). A total electron budget of 1,000,000 electrons was used, giving an average dose of 15.3

(a) (b)

Figure 5: Simulated images for SCCD (a) and MAPS (b) for a voltage of 20 kV.

electrons / pixel. The poorer MTF of the SCCD can be observed as blurring of small features and the lower DQE can be observed as stronger noise in the bright background.

3.2. Counting detector

Given the similar structure with the MAPS detector, the DSSD has the same signal distribution (Fig 2b). However, for each electron arriving at the sensor, the energy deposited in each pixel is compared with a threshold set by the sensitivity of the readout electronics and a counter incremented with one if signal is higher than threshold. To remove lateral scattering, a further condition was added before the counter is incremented, that no neighboring pixel holds signal above threshold. This results in ideal MTF limited only by spatial sampling, i.e. pixel size.

Following these two conditions for detection, there are only two cases where a beam electron is not detected (a) if the electron deposits signal below threshold and (b) if the electron deposits energy above threshold in several neighboring pixels. For this voltage range the second condition predominates, resulting in a low efficiency of 42% for the case of 100 kV and high efficiency of 88% for the case of 20 kV.

4. Conclusions

Performance of SCCD, MAPS and DSSD detectors has been calculated in the 20-100 kV voltage range using Monte Carlo simulations of the electron-sensor interaction. At the low end of this voltage range, ideal MTF and a DQE of ~90% have been obtained for the MAPS and DSSD detectors. At the high end of this voltage range it is only the DSSD that retains ideal MTF due to its counting mode, however with a much reduced detective efficiency of 42%. SCCD and MAPS show similarly poor DQE at 100 kV, suggesting that further structural optimisation such as reduced thickness or increased pixel size must be used to obtain higher counting efficiency.

Acknowledgements

This work was supported under EPSRC grant EP/C009509/1 and Framework 6 program for an Integrated Infrastructure Initiative 026019 ESTEEM.

References

[1] Joy D C 1995 *Monte Carlo Modelling for Electron Microscopy and Microanalysis* (Oxford: University Press).

[2] Meyer R R and Kirkland A, *Microscopy Research and Technique*, **49**, 269-280, 2000.

Journal of Physics: Conference Series **241** (2010) 012010 doi:10.1088/1742-6596/241/1/012010

Smart Acquisition EELS

K Sader[1,2], B Schaffer[1,3], G Vaughan[1], P Wang[1,4] A L Bleloch[1], R Brydson[2], A Brown[2]

[1] SuperSTEM, J Block, Daresbury Laboratory, Warrington, Cheshire, WA4 4AD
[2] Institute for Materials Research, University of Leeds, LS2 9JT
[3] Department of Physics and Astronomy, University of Glasgow
[4] now at Department of Materials, University of Oxford, OX5 1PF

E-mail: kasim.sader@gmail.com

Abstract.
Electron energy loss (EEL) spectroscopy and high angle annular dark field (HAADF) imaging in aberration-corrected electron microscopes are powerful techniques to determine the chemical composition and structure of materials at atomic resolution. We have implemented Smart Acquisition, a flexible system of scanning transmission electron microsocpy (STEM) beam position control and EELS collection, on two aberration-corrected dedicated cold field emission gun (FEG) STEMs located at SuperSTEM, Daresbury Laboratory. This allows the collection of EEL spectra from spatially defined areas with a much lower electron dose possible than existing techniques such as spectrum imaging.

1. Introduction

The STEM is the ideal instrument for combined imaging and analysis because of the potential for multiple signals to be collected simultaneously. For example it is possible to image single atomic columns with the high angle annular dark field signal and simultaneously record EEL spectra to give atomic resolution chemical information [1]. However, this is only possible for radiation insensitive materials because of the large electron doses that are required: for radiation sensitive materials this goal is far off.

It is well known that to attain atomic resolution information from frozen hydrated biological macromolecules, only about $10e^-/\text{Å}^2$ can be used [2]. Such a small electron dose requires extensive averaging to increase the signal to noise ratio. The principle of averaging can be applied to EEL spectra, preventing damage by spreading the dose over a large area. However, the systematic noise in charge coulpled devices (CCDs) prevents the summation of individual spectra collected from each pixel of a spectrum image taken with low electron doses or at high edge energies, as is also the case with extremely low dose TEM images [3].

Our currently implemented solution is Smart Acquisition. This involves scanning a STEM probe over similar areas of interest while collecting a single spectrum, therefore avoiding readout noise for low electron counts and the time overhead of multiple readouts. This concept is not entirely new, with similar ideas proposed by John Hunt [5; 6]. However, with the successful implementation of aberration correction and improved stage stability, atomically resolved areas can now be defined and analysed.

We have used Smart Acquisition to fabricate nanostructures in aluminium trifluoride AlF_3 and subsequently collect EEL spectra from spatially defined positions. The use of high current density STEM probes of 10^7A/m^2 is a promising method of producing nanoscale structures in metal halides and oxides [7]. In particular AlF_3 is of interest because its extreme beam sensitivity allows films in the hundred nm range to easily be cut through. A demonstration of the application of AlF_3 hole drilling has been the production of electron optical phase plates [8]. The sensitivity to beam irradiation has also made it difficult to define the exposure mechanism[9], though there have been extensive studies [10].

2. Methods

Images were collected on a 5th order aberration corrected Nion UltraSTEM situated at Daresbury Laboratory, UK, with a third generation Nion C_3/C_5 corrector with 12 quadrupoles and 3 combined quadrupole-octupoles [11]. This uses a cold field emission filament operating at 100kV and is equipped with a Gatan Enfina electron energy loss spectrometer giving a total spectral resolution of 0.4eV, CCD camera, bright field photo-multiplier tube (PMT), medium and high angle annular dark field PMT detectors.

Digital Micrograph (DM, Gatan) scripts have been written to utilize existing beam control commands to control the position of the electron probe on the Nion UltraSTEM, utilising the Digiscan II scaning unit. Shapes to be scanned can be defined in a HAADF image from any combination of regions of interest (ROIs) in DM, or from binary image masks created by thresholding the image itself. This allows the averaging of the EEL signal over defined features.

The beam current was measured by diverting the electron probe onto the drift tube of the Enfina spectrometer connected to an ampmeter. Under standard conditions and after flashing, the beam current was measured to be 100 pA. The beam current was reduced to about 5pA by decreasing the extraction voltage. For a probe size of 1.3Å full width half maximum (FWHM), this gives a current density of $3.8\text{x}10^8$A/m^2. For a 0.8ms dwell time this gives an electron dose of $3\text{x}10^3$ C/m^2 (7000e$^-$/Å^2).

AlF_3 (anhydrous, 99%, Alpha Aesar) was evaporated in a molybdenum boat (modified to prevent the spitting of the powder due to adsorbed water [10]) onto Quantifoil 7/2 square mesh grids covered with 2nm carbon film. The film thickness was monitored using a quartz crystal oscillator and the final thickness measurement was 26.5nm.

3. Results and Discussion

Five positions were exposed to the ~1.3Å FWHM STEM probe for 0.6s each to drill holes in the AlF_3 (HAADF image following exposure - Figure 1a). Seven sets of pixel positions of circles of radii 5.6nm to 0.7nm around the centre of each hole were generated in DM. The Smart Acquisition script was used to move the beam pixel by pixel (with Digiscan) around the set of points for each radius while using an independent command to collect a single EEL spectrum for each set (Figure 1b).

The EEL spectra are shown in Figure 2. The increasing shoulder of the plasmon peak at 15eV (Al bulk plasmon) suggests that the STEM probe irradiation did not remove all the material, but instead concentrated Al metal.

The threshold current density for AlF_3 is 10^7A/m^2, an order of magnitude below the current density estimated for the probe. It is therefore unlikely that metalisation occurred due to a low probe current, but instead due to a constant drilling rate [10] due to the low electron dose of $2.3\text{x}10^4$C/m^2, which is also know to cause metalisation [7].

The other mechanism of hole formation often observed is an initially slow drilling rate, followed by sudden mass loss. This is explained by Berger [12] as being caused by bubbles of fluorine/oxygen, which suddenly burst. It was proposed by Macaulay [10] that the two mechanisms of hole formation- a constant drilling rate or sudden mass loss are due to either the diffusion of oxygen/fluorine molecules to the surface and loss to the vacuum or trapping inside

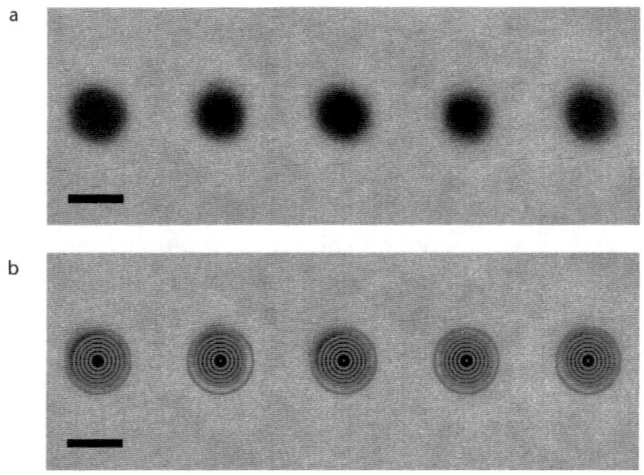

Figure 1. Spatially defined positions of EEL spectra of exposed AlF_3 thin films. A) HAADF image of points exposed to STEM probe for 0.6s. B) Sets of positions (red circles) from which EEL spectra were collected from the exposed AlF_3 films- from outer to inner radii: 5.6nm, 4.2nm, 3.5nm, 2.8nm, 2.1nm, 1.4nm, 0.7nm. Scale bar is 10nm.

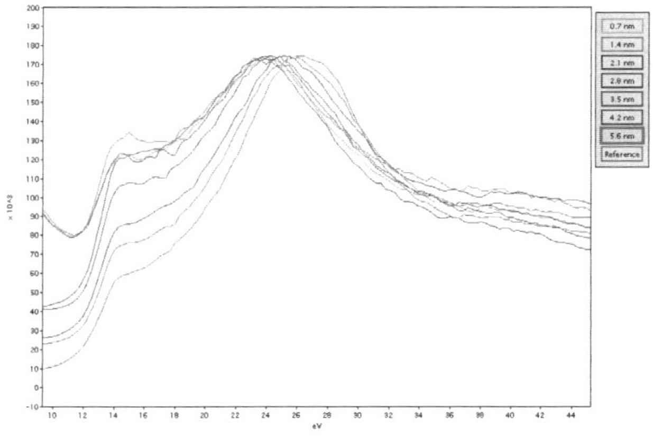

Figure 2. Low loss EEL spectra from each radius. The spectra were aligned vertically to the maximum of the large bulk plasmon signal. This clearly shows a growing shoulder at 15eV- the bulk plasmon of Al metal.

the bulk. Hole drilling by bubble formation was observed with higher doses but holes were larger and less regular.

One of the major problems in our experiments was the inability to control blanking of the electron beam. The movement of the probe was not exactly synchronized with the EELS acquisitions in this application, and therefore an area without specimen (vacuum) was required as a safe position at which the beam could start and end. This did allow the full exposure of the sample to be collected in the EEL spectrum (minimal time taken to move to safe point). However, this was a limiting factor because only areas near vacuum could be studied, and in the loss low region of the spectra, the tails of the zero loss peak had significant intensity. Scripting control of beam blanking has now been implemented on the Nion UltraSTEM. The solution for the first microscope at SuperSTEM, a VG HB501 fitted with second generation 3rd order aberration correctors, was to script control of the printer port to modulate a 5V transistor-

transistor logic (TTL) signal. This is used to directly control the pre-specimen beam blanking and has been incorporated into the Smart Acquisition scripts.

A second issue has been further reducing the electron dose. The Smart Acquisition script takes a list of pixel positions, and requires 0.8ms to move from one position to another. This is largely a communications issue, with 0.8ms being the time required for the computer to communicate with the Digiscan unit over a firewire connection. Further increases of the speed of beam movement could be possible by directly interfacing with the scan unit, as done by Hunt [5], but this may be at the expense of greater inaccuracies in beam position due to the use of electromagnetic scan coils. Further reduction in the extraction voltage appears possible, with an apparent change only in astigmatism, which is easily adjusted in bright field images of amorphous carbon. Another option would be to use the beam blanker to stroboscopically illuminate the sample by fast blanking of the beam during the movement of the beam. A signal generator is already in use at SuperSTEM to attenuate the zero loss peak during the tuning of the Enfina spectrometer and acquisition of loss low spectra, without noticeable drift. The settling time of the blanking plates is in the tens of nanoseconds [13], so a periodic blanking on the order of tens of microseconds could significantly reduce the dose more than 10-fold.

In summary, a flexible system for the recording of low dose EEL spectra, Smart Acquisition, has been developed. EEL spectra can be collected from areas defined at high resolution. We have demonstrated the technique by studying hole drilling in AlF_3 thin films. We have now implemented pre-specimen blanking solutions on both microscopes at SuperSTEM, and we are now working on decreasing the minimum dose to extend the range of beam sensitive materials that we can image and analyse.

4. Acknowledgments

The authors are grateful to the Engineering and Physical Sciences Research Council (EPSRC) for continued funding to the UK SuperSTEM facility at Daresbury laboratory.

References

[1] Muller D A, Kourkoutis L F, Murfitt M, Song J H, Hwang H Y, Silcox J, Dellby N and Krivanek O L 2008 *Science* **319** 1073–1076

[2] Henderson R 1995 *Quarterly Reviews of Biophysics* **28** 171–193

[3] McMullan G, Cattermole D, Chen S, Henderson R, Llopart X, Summerfield L, Tlustos L and Faruqi A R 2007 *Ultramicroscopy* **107** 401–413

[4] McMullan G and Faruqi A R 2008 *Nuclear Instruments and Methods in Physics Research Section A: Accelerators, Spectrometers, Detectors and Associated Equipment* **591** 129–133

[5] Hunt J A, Leapman R and Williams D B 1993 *Microbeam Analysis* vol 2 pp S272–S273

[6] Hunt J A, Disko M M, Behal S K and Leapman R D 1995 *Ultramicroscopy* **58** 55–64

[7] Kratschmer E and Isaacson M 1986 *Journal of Vacuum Science and Technology B: Microelectronics and Nanometer Structures* **4** 361–364

[8] Ito Y, Bleloch A L and Brown L M 1998 *Nature* **394** 49–52

[9] Chen G S, Boothroyd C B and Humphreys C J 1996 *Applied Physics Letters* **69** 170–172

[10] Macaulay J M 1989 *The Production of Nanometre Structure in Inorganic Materials by Electron Beam of High Current Density* Ph.D. thesis University of Cambridge

[11] Krivanek O L, Corbin G J, Dellby N, Elston B F, Keyse R J, Murfitt M F, Own C S, Szilagyi Z S and Woodruff J W 2008 *Ultramicroscopy* **108** 179–195

[12] Berger S D, Salisbury I G, Milne R H, Imeson D and Humphreys C J 1987 *Philosophical Magazine B* **55** 341 – 358

[13] Craven A J, Wilson J A and Nicholson W A P 2002 *Ultramicroscopy* **92** 165–180

Application of 80-200 kV Aberration Corrected Dedicated STEM With Cold FEG

M Konno [1], Y Suzuki[1], H Inada[2] and K Nakamura[2]
[1]Naka application Center, Naka Division, Nanotechnology Products Business Group, Hitachi High-Technologies, 11-1, Ishikawa-cho, Hitachinaka-shi, Ibaraki-ken, 312-0057, Japan
[2]Advanced Microscope System Design 2nd Department, Naka Division, Nanotechnology Products Business Group, Hitachi High-Technologies, 882, Ichige, Hitachinaka-shi, Ibaraki-ken, 312-8504, Japan

E-mail: konno-mitsuru@naka.hitachi-hitec.com

Abstract. We have developed new STEM instrumentation with a cold field emission source (Hitachi HD-2700) in order to perform structural characterization and elemental mapping at the atomic level. The instrument utilises the CEOS GmbH (Germany, managing director: Dr. Max Haider) aberration corrector. The accelerating voltage range is between 80 kV and 200 kV. The cold field emission source proves to be the ideal emitter for analytical transmission electron microscopes due to its high brightness, high current density and small energy spread. In this study, we have examined low accelerating voltage conditions for obtaining high image contrast and high performance elemental analysis (in which FWHM of zero loss peaks are 0.3 eV for acquisition time of 1 second and 0.34 eV for acquisition time of 40 second by accelerating voltage of 80 kV, respectively). We have observed high contrast bright field STEM images of graphene carbon at an accelerating voltage of 80 kV, in which lattice fringes can be clearly seen.

1. Introduction

Electron energy loss spectroscopy (EELS) is a powerful technique for qualitative and quantitative elemental analysis. The spectrum shape also contains information on the chemical bonding state, commonly referred to as energy loss near edge structure (ELNES). Recently, the demands for ELNES analysis have increased as material scientists investigate novel materials and devices on the nanoscale. For this reason cold field emission guns [1] or monochromators [2] have been utilized in transmission electron microscopy (TEM) and scanning transmission electron microscopy (STEM) and the achievable energy resolution has reached around 0.2 eV. Monochromated electron sources, however, have a few weaknesses, such as low probe current and enlarged probe size. In comparison, a cold field emitter has high current density while maintaining a small energy spread, without the drawbacks described above, and therefore maybe the ideal source for an analytical TEM or STEM. Furthermore, spherical aberration (Cs) correctors are now commercially available devices. It is indeed possible to make up for some of the weaknesses of a monochromator with the advantages of a Cs corrector in STEM (with the

1 konno-mitsuru@naka.hitachi-hitec.com

Cs corrector improving probe current and reducing probe size) but it is clearly preferable have an initial source with higher current and lower probe size. Additionally, according to optical principles, the contrast of the elastically scattered electron in aberration corrected STEM is less than that of STEM without Cs corrector. This clearly has an effect on imaging of light elemental materials such as nm-sized carbon particles and some polymers and it is necessary to determine methods to improve the contrast such as lower voltage operation. In this study, we have investigated low accelerating voltage conditions for obtaining high performance imaging and EELS using a STEM instrument equipped with a with Cold FE gun and Cs corrector.

2. Experiment

In order to respond to such demands, we have developed a dedicated STEM instrument with a newly-designed ultrahigh vacuum Cold FE gun (Hitachi HD-2700), in which the CEOS GmbH (Germany, Managing Director: Dr. Max Haider) Cs corrector has been utilized and optimized [3]. The convergence angle of the aberration corrected STEM is around 40mrad[4] at 200kV and 25mrad at 80kV, and the accelerating voltage range is between 80 kV and 200 kV. Source brightness are 1×10^9 $A/cm^2\cdot Sr$ at 200 kV, 6×10^8 $A/cm^2\cdot Sr$ at 120 kV and 4×10^8 $A/cm^2\cdot Sr$ at 80 kV. The instrument is equipped with an EELS spectrometer (Gatan, ENFINA, 16 bit 100×1340 pixel CCD) and a CCD camera (Gatan, ORIUS, 14 bit 2.6k×2.6k) for the capture of diffraction and Ronchigram patterns. Additionally, the column unit is fitted with a magnetic sheilding cover (which showed significant improvement in EELS stability due to the shielding of electromagnetic interference). Here, we have measured image resolution, energy spread and beam stability using EELS at accelerating voltages of 80 kV, 120 kV and 200 kV. We have also compared the EELS spectrum of graphene carbon, known for its susceptibility to radiation damage, at accelerating voltages 200 kV and 80 kV.

3. Results and discussion

Figure.1 shows high resolution Annular Dark Field (ADF) STEM images observed along the Si[110] zone axis at accelerating voltages of 80 kV(a), 120 kV(c) and 200 kV(e), and the Fourier transform (b, d, f) of each ADF STEM image. Not surprisingly the 200 kV condition demonstrates the best result, but at 80 kV it is still possible to slightly-separate the silicon dumbbell structure. The Fourier transform image shows the 004 reflection spot of the silicon dumbbell and the 2-24 reflection spot, corresponding to 0.111 nm at 80 kV. The result at 80 kV reveals the potential for atomic visualization in carbon with minimal radiation beam damage. The Fourier transform image at 120 kV shows the 3-33 and 2-24 reflection spot, corresponding to 0.105 nm. In the 200 kV condition sub-angstrom resolution in the Fourier transform image is detected, with the 4-44 reflection spot (corresponding to 0.078 nm) and 3-35 reflection spot (corresponding to 0.083 nm). This result expands the possibility for structural characterization at the atomic level.

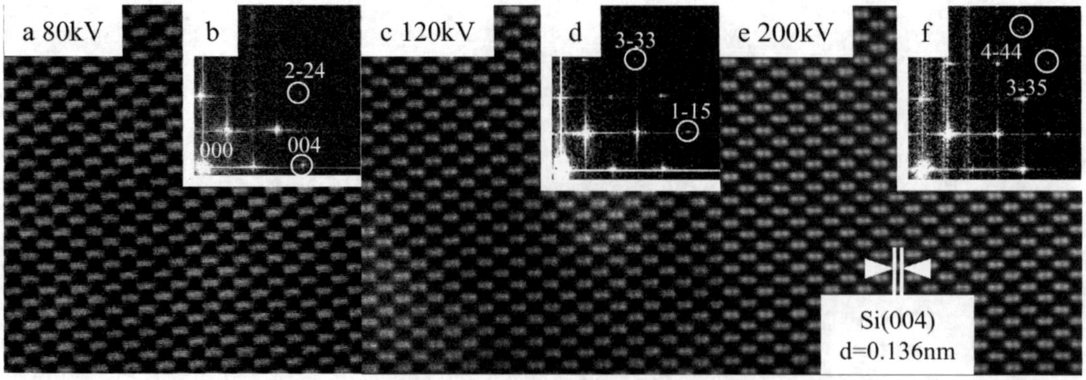

Figure 1. ADF-STEM images observed along the Si[110] zone axis with accelerating voltage of 80 kV, 120 kV and 200 kV, and the corresponding Fourier transform of each image.

Next, we observed an EELS energy time trace of the zero loss peak under the normal analysis condition. The beam current was measured, using a Faraday cup in the specimen holder, to be 29 pA in a 0.1 nm diameter probe at 200 kV, 27 pA in a 0.2 nm diameter probe at 120kV and 5 pA in a 0.2 nm diameter probe at 80 kV. Figure 2 shows the EELS energy time trace for zero loss peak at 200kV(a), 120kV(b) and 80kV(c) for 40 seconds duration. The time trace images give 800 EELS spectra, each of which is acquired with an exposure time of 50 ms (total exposure time is 40 seconds). The minimal fluctuation of the bright line, which is the zero loss peak, demonstrates the stability of the beam energy. Table 1 indicates the FWHM of fluctuation and stability by statistical processing (Stability = FWHM of fluctuation / Beam energy) of the zero loss peak in each line of Figure.2. The FWHM of fluctuation and stability is 146 meV, 0.73 ppm at 200 kV, 133 meV, 1.11 ppm at 120kV and 121 meV, 1.51 ppm at 80 kV, respectively. The FWHM of zero loss peaks are 384 meV at 200 kV, 320 meV at 120 kV and 300 meV at 80 kV for an acquisition time of 1 second, which shows the inherent performance of the Cold FE gun (These spectra was smoothed)(Figure 3). Here, the 200 kV condition has the best result for stability. However, 80 kV shows the best result for FWHM.

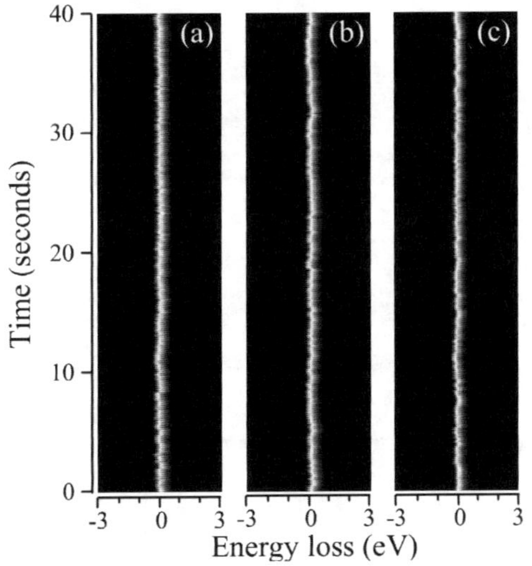

Figure 2. EELS energy time trace of zero loss spectra at 200kV(a), 120kV(b) and 80kV(c) for 40 seconds duration.

Table 1. Fluctuation of zero loss peak and stability of beam energy.

Beam energy (keV)	Fluctuation (meV)	Stability (ppm)
80	121	1.15
120	133	1.11
200	146	0.73

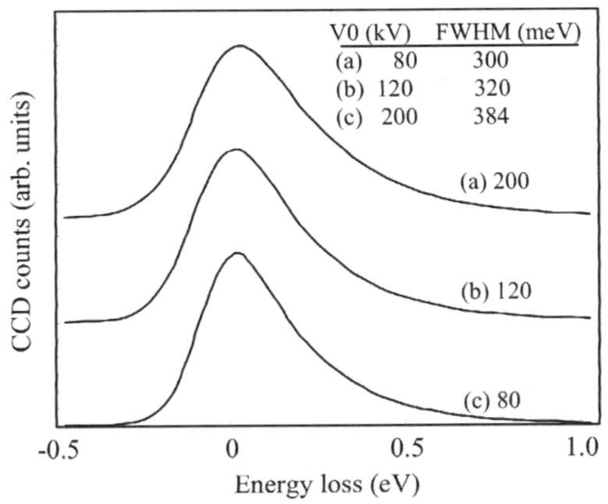

Figure 3. The FWHM of zero loss peaks at 200kV (a), 120kV (b) and 80kV (c) for 1s duration.

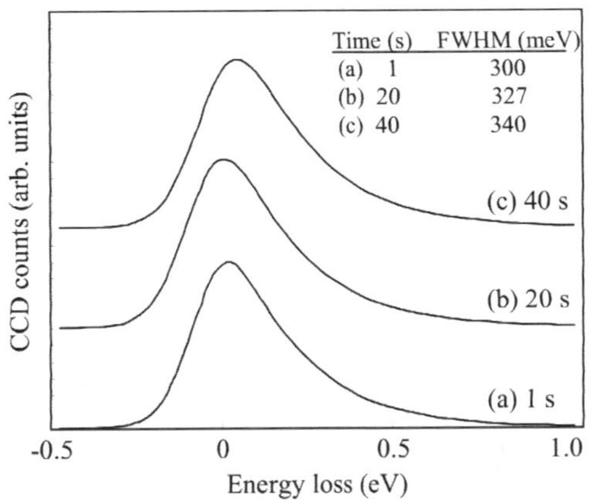

Figure 4. The FWHM of zero loss peaks at 80kV for 1s (a), 20s (b) and 40s (c) duration.

Next, we have measured the FWHM over longer acquisition times. Figure 4 shows the FWHM of the zero loss peak at 80 kV for 1 second (a), 20 second (b) and 40 second (c) duration (These spectra was smoothed). FWHM of the zero loss peaks are 0.30 eV for acquisition time of 1 second, 0.327 eV for 20 second and 0.340 eV for 40 second, respectively. The result reveals the high stability and low energy spread performance of HD-2700 system. Figure 5 shows the Bright Field (BF) STEM image of graphene carbon sheet at an accelerating voltage of 80 kV, in which lattice fringes can be clearly seen. The atom to atom space of graphene carbon is precisely 0.142 nm. This result emphasizes that high resolution imaging and analysis are possible with minimal damage through the use of aberration-corrected lower-energy operation. Figure.6 shows the EELS spectrum of graphene carbon at accelerating voltages of 80 kV(a) and 200 kV(b). The spectrum at 80 kV clearly shows the characteristic five peaks showing bonding information. In comparison, the spectrum at 200 kV is rounded and shows a low peak to background ratio, which is attributed to radiation damage by the 200 kV electron beam.

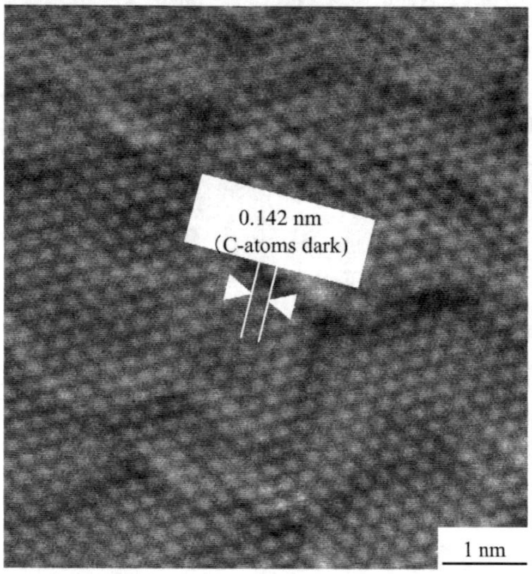

Figure 5. BF-STEM image of graphene carbon sheet.

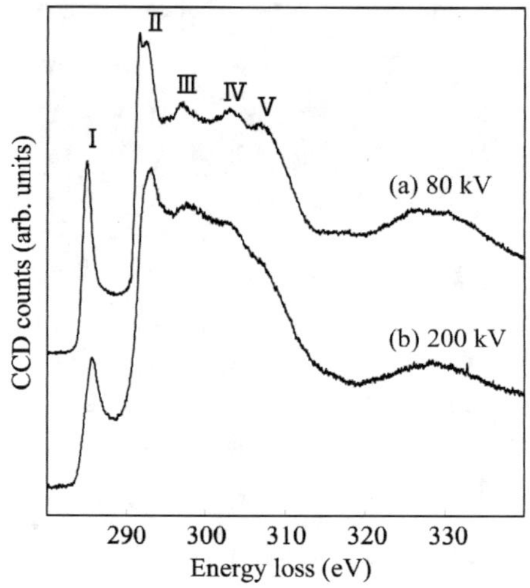

Figure 6. Comparison of EELS spectra accelerating voltage of 80 kV (a) and 200 kV (b).

4. Summary

We have shown high-resolution ADF STEM images and EELS data of graphene carbon using aberration corrected dedicated STEM. Sub-angstrom information, corresponding to 0.078 nm at 200 kV and a narrow energy spread of 0.3 eV in the FWHM of the zero loss peak at 80 kV is obtained under routine analysis conditions. We have also shown stability improvements in the instrumentation which will be a key parameter for chemical bonding state analysis in EELS at the atomic level.

References

[1] Kimoto K, Ishizuka K, Asaka T, Nagai T and Matsui Y 2005 *Micron* 36 465-469
[2] Kimoto K, Kothleitner G, Grogger W, Matsui Y and Hofer Fet al 2005 *Micron* 36 185-189
[3] Nakamura K, Konno M, Yaguchi T, Kamino T, Terauchi D, Inada H, Tanaka H, Taniguchi Y and Isakozawa S 2006 *proc. 16th International Microscopy Congress* 633
[4] Inada H, Zhu Y, Wall J, Volkov V, Nakamura K, Konno M, Jarausch K and Twesten D 2008 *Microsc. Microanal.* 14 1374.
[5] R. F. Egerton, Electron Energy-Loss Spectroscopy in the Electron Microscope , *Plenum, New York*, 1996.
[6] Suenaga K, Sandre E, Colliex C, Pickard C, Kataura H and Iijima S 2001 *PHYSICAL REVIEW B, VOLUME* 63,165408

Electron Microscopy and Analysis Group Conference 2009 (EMAG 2009) IOP Publishing
Journal of Physics: Conference Series **241** (2010) 012012 doi:10.1088/1742-6596/241/1/012012

Experimental Setup for Energy-Filtered Scanning Confocal Electron Microscopy (EFSCEM) in a Double Aberration-Corrected Transmission Electron Microscope

P Wang, G Behan, A I Kirkland and P D Nellist
Department of Materials, University of Oxford, Parks Road, Oxford OX1 3PH, UK

E-mail: peng.wang@materials.ox.ac.uk

Abstract. Scanning confocal electron microscopy (SCEM) is a new imaging mode in electron microscopy. Spherical aberration corrected electron microscope instruments fitted with two aberration correctors can be used in this mode which provides improved depth resolution and selectivity compared to optical sectioning in a conventional scanning transmission geometry. In this article, we consider a confocal optical configuration for SCEM using inelastically scattered electrons. We lay out the necessary steps for achieving this new operational mode in a double aberration-corrected instrument with uncorrected chromatic aberration and present preliminary experimental results in such mode.

1. Introduction

With the development of spherical aberration correctors for both transmission electron microscopy (TEM) and scanning transmission electron microscopy (STEM), the spatial resolution in 2D images has been dramatically improved [1]. The correction of spherical aberration allows a larger probe forming aperture to be used, which not only increases the lateral resolution, but also reduces the depth of field to a few nanometres [2]. This offers the feasibility of depth sectioning in similar fashion to confocal scanning optical microscopy (CSOM) [3]. Spherical aberration corrected TEM/STEM instruments fitted with two aberration correctors have been shown to be capable of working in this confocal mode, named as scanning confocal electron microscopy (SCEM) [4]. The theoretical calculation of the contrast mechanisms showed inelastic SCEM or energy-filtered SCEM (EFSCEM) gives much better depth selectivity and three-dimensional image contrast than elastic SCEM [5]. In particular the calculation showed that the 3D transfer function of EFSCEM does not have a 'missing cone', so that even laterally extended objects can be located in depth. Nellist et al. [6] have demonstrated a confocal optical configuration for elastic SCEM on Oxford-JEOL 2200MCO. The instrument is also fitted with an in-column Ω energy filter, allowing the possibility of EFSCEM experiments. However, chromatic aberration changes the focal length of a lens by $\Delta z = \Delta E_{loss} C_C / E_0$ for the inelastic electrons with an energy-loss of ΔE_{loss} relative to the elastically scattered electrons where E_0 is the accelerating voltage. Although spherical aberration correctors are becoming widespread because of their commercial availability, there is currently only one instrument fitted with a prototype chromatic aberration corrector because it is technologically much harder [7].

In this paper, we will present the 3D optical transfer function (OTF) of EFSCEM and list the steps necessary for achieving confocal electron trajectories suitable for EFSCEM in an instrument with uncorrected chromatic aberration. This procedure allows the confocal point to be regained for the energy loss of interest. Furthermore, we will show preliminary results of EFSCEM from an extended object, an amorphous carbon film, as a proof of principle of the depth discrimination mechanism of the EFSCEM on an electron microscope.

2. OTF of Energy filtered SCEM

For a completely incoherent scattering object in the completely confocal situation with a point detector, the incoherent SCEM image intensity [6] can be written as the three-dimensional convolution

$$I = \psi' \otimes PSF_{confocal} = \psi' \otimes [|P_1|^2 |P_2|^2] \quad (2.1)$$

where ψ' is a local inelastic scattering potential and $PSF_{confocal}$ is the inelastic confocal point spread function (PSF), a product of $|P_1|^2$ and $|P_2|^2$ (the PSFs of the illumination-side or imaging-side optics, respectively). $PSF_{confocal}$ is only considered in an optical system using an infinitesimal energy-selecting window. In practice, a finite energy-selecting window ΔE_{window} is normally implemented. Typical energy filtered electron microscope images are recorded using electrons with energy losses that range over an energy window of several eV to ensure a high enough signal [8]. Therefore a defocus spread introduced to the imaging-side optics is given by a top-hat function S with width of $\Delta E_{window} C_C / E_0$. So the PSF with a finite energy-selecting window is thus given by

$$PSF_{confocal} = |P_1|^2 (|P_2|^2 \underset{z}{\otimes} S) \quad (2.2)$$

where the convolution is along the z-direction.

The optical transfer function (OTF) is the 3D Fourier transform of the PSF. Figure 1 shows the OTFs for STEM a) and EF-SCEM with infinitesimal b) and 5eV c) energy-selecting window, respectively. With both SCEM modes the depth resolution is improved over STEM. The OTF for STEM has a missing cone region where no spatial frequencies can be transferred, whereas there are no such missing regions in the EF-SCEM, even with a 5eV energy window. This is important because laterally extended objects have many components at low radial spatial frequencies.

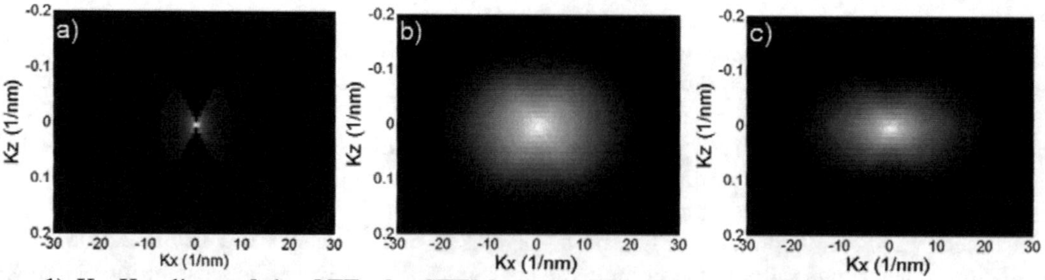

Figure 1) Kx-Kz slices of the OTFs for STEM **a)** and EFSCEM with infinitesimal **b)** and 5 eV **c)** energy-selecting window, respectively. The parameters for each calculation are for a double aberration-corrected instrument with an accelerating voltage of 200kV and a semi-angle of convergence of 22mrad.

3. Optical configuration setup for EFSCEM

Next we outline methods for establishing a confocal trajectory for energy filtered electrons with aberration-corrected illumination and imaging optics [9] in a single 200 keV instrument: the Oxford-

JEOL 2200MCO with an in-column Ω energy filter [10] . The chromatic aberration coefficients in the pre- and post-specimen optics are Cc_1 1.4mm and Cc_2 1.43mm [11]. Hexapole-based correctors [12] also contribute a positive value to *Cc*, and hence the overall value of *Cc* is increased over that of an uncorrected lens.

Establishing confocal trajectories for elastic electrons requires the simultaneous tuning of both correctors about a mutual optical axis. Both correctors need to be aligned so that the upper column optics and lower one are confocal at the point A in figure 2 a). The details of this procedure can be found in Ref [6]. A typical elastic probe image formed through vacuum is shown as in figure 3 a). To regain the confocal point for the inelastic electrons (with a desired energy loss of ΔE_{loss}), we keep the optical alignment for the elastic electrons intact and simply increase the accelerating voltage E_0 by a factor k times the energy-loss of interest ΔE_{loss} where $k = Cc_2/(Cc_1 + Cc_2)$. The pre-specimen optics, due to its chromatic aberration Cc_1, now focuses the electrons of the second incidence energy onto the point C, distant from a by $k\Delta E_{loss} Cc_1 / E_o$ shown in figure 2 c). After the energy-loss event, the electrons have an energy of $E_o + (k-1)\Delta E_{loss}$ and the formula for k given above ensures that the electrons are accurately focussed by the post-specimen optics onto the selecting collector aperture. This optical configuration is kept fixed and the specimen brought down to the confocal point C using z-stage in figure 2 c). Finally, the post-specimen optics are switched into the spectrum mode and the energy-selecting slit with the size of ΔE_{window} is centred at the energy loss ΔE_{loss} of interest using an in-column Ω energy filter. On returning the post-specimen optics to imaging mode, the entire microscope is reconfigured into a confocal condition for the inelastic electrons with energy loss of ΔE_{loss}.

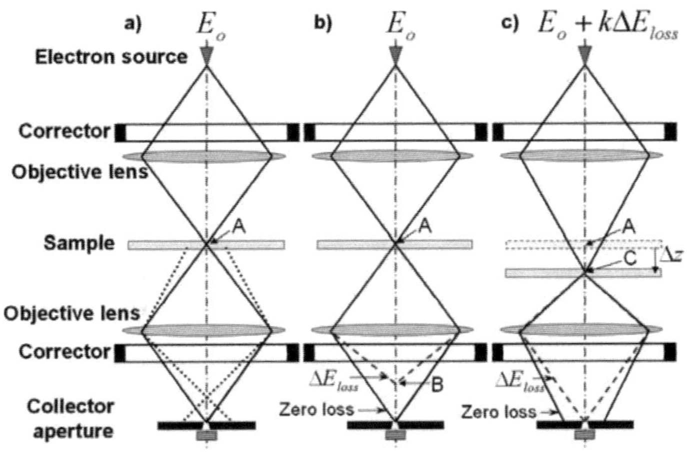

Figure 2 Schematic diagram of the confocal trajectories for elastic SCEM **a)**, which shows the beams (····) away from the confocal point A are rejected by the collector aperture, and **b)**, which shows the post-specimen optics, due to its chromatic aberration focuses the beams (----) with the energy loss of interest ΔE_{loss} are focused at the point B. The diagrams **c)** shows an increment of $k\Delta E_{loss}$ applied to the accelerating voltage E_0 can regain A new confocal point C for the beams (----) with the energy loss of interest ΔE_{loss}.

Figure 3 b) and c) shows the inelastic probe image observed in the collector aperture plane formed though an amorphous carbon thin film recorded on a Gatan CCD with 5 eV energy selecting window placed at 290 eV when the specimen sits at the confocal point and at 164 nm away from it, respectively. It is clear to see that the signal largely drops off at the centre of the probe image when the specimen moves away from the confocal point. This experimentally demonstrates that the scattering from points away from the confocal point become out-of-focus at the collection plane. By placing a pinhole at the collection plane, it is feasible to reject or detect less strongly the out-focus scattering than the in-focus scattering. With a high resolution stage-scanning or beam scan-descan system, we can collect the intensity passing through the pinhole while the specimen is moved with respect to the confocal point. As a result, a 2D elemental map can be acquired with a transverse section at any depth in a thick specimen and a 3D elemental distribution accordingly can be formed by stacking a series of the 2D maps. EFSCEM has the potential to be a powerful 3D chemical analysis technique at a nanometre-resolution, an entirely new form of electron microscopy.

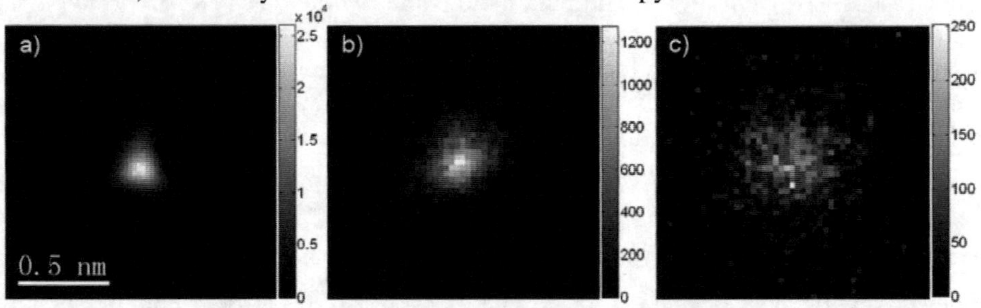

Figure 3 a) shows experimental images of the probe in the elastic confocal mode formed through vacuum and **b)** through a carbon film in the EFSCEM mode with 5eV energy filtering window placed at the carbon K-edge, when the film is just at the confocal point and **c)** a point of 164nm away from it, respectively.

4. Conclusion

We showed that EFSCEM with a 5eV energy-selecting window has no missing cone region in its OTF. We devised a technique whereby the instrument is aligned for the inelastic electrons via increasing the accelerating voltage by a factor k for the energy-loss of interest. We demonstrated experimentally the depth discrimination mechanism in the EFSCEM mode on an electron microscope.

Acknowledgements

P. Wang would like to thank Dr. Andrew Bleloch for his open discussion on EFSCEM project. G. Behan would like to thank Intel Ireland Ltd for financial support. P. Wang, A.I. Kirkland and P.D. Nellist would like to acknowledge the financial support of the Leverhulme Trust, the EPSRC and the Department of Materials at the University of Oxford.

References

[1] Nellist P D et al. 2004 *Science* **305** 1741.
[2] van Benthem K et al. 2005 *Applied Physics Letters* **87** 034104.
[3] Gu M 1996 *Principles of Three Dimensional Imaging in Confocal Microscopes* (World Scientific).
[4] Frigo S P, Levine Z H, and Zaluzec N J 2002 *Applied Physics Letters* **81** 2112.
[5] D'Alfonso A J, Cosgriff E C, Findlay S D, Behan G, Kirkland A I, Nellist P D, and Allen L J 2008 *Ultramicroscopy* **108** 1567.
[6] Nellist P D, Behan G, Kirkland A I, and Hetherington C J D 2006 *Applied Physics Letters* **89** 124105.
[7] Haider M, Muller H, Uhlemann S, Zach J, Loebau U, and Hoeschen R 2008 *Ultramicroscopy* **108** 167.
[8] Rose H 2005 *Ultramicroscopy* **103** 1.
[9] Haider M, Uhlemann S, Schwan E, Rose H, Kabius B, and Urban K 1998 *Nature* **392** 768.
[10] Hutchison J L, Titchmarsh J M, Cockayne D J H, Doole R C, Hetherington C J D, Kirkland A I, and Sawada H 2005 *Ultramicroscopy* **103** 7.
[11] Chang L Y, Kirkland A I, and Titchmarsh J M, 2006 *Ultramicroscopy* **106** 301.
[12] Haider M, Braunshausen G, and Schwan E, 1995 *Optik (Jena)* **99** 167.

Finding phase information in the darkness

S.J.Haigh[1], A.I. Kirkland[1]

[1.] Department of Materials, University of Oxford, Parks Road, OX1 3PH, UK.

sarah.haigh@materials.ox.ac.uk

Abstract. High resolution local structure information is revealed when the complex exit wavefunction is restored from a series of transmission electron microscope (TEM) images. Aberration corrected image data sets allow high spatial frequency information to be restored but this is often at the expense of lower spatial frequencies. In this paper we present a new approach to exit wavefunction restoration using a novel five image focal series with non-uniform focal steps. This data set extends the spatial frequency range over which information can be successfully restored. We present simulation studies comparing this approach to a conventional focal series and demonstrate the benefit of including low spatial frequencies in the restoration. We propose that this approach will be important for the study of large unit cell materials or those with long range ordering.

1. Introduction

High resolution Transmission Electron Microscopy (TEM) is an established technique for atomic scale imaging of materials. With the recent development of electron optical elements capable of direct aberration correction, current instruments are able to routinely record "aberration corrected" images of inorganic materials with a resolution better than 0.1nm [1-3]. Fig. 1 compares the contrast transfer function (CTF) for the JEOL 2200MCO (S)TEM [4] at the Scherzer defocus condition, $f_{Sch} = -\sqrt{(C_s \lambda)}$, with and without 3rd order spherical aberration correction. The CTF is defined here as E(**k**) sin (χ(**k**)), where **k** is the spatial frequency, E(**k**) the envelope function resulting from spatial and temporal coherence effects and χ(**k**) is the phase change due to the wave aberration function assuming all aberrations above 3rd order are negligible. Adjustment of the 3rd order spherical aberration to 10 μm [5] clearly eliminates oscillations in the CTF at high spatial frequencies as shown in figure 1. In Fig. 1(a) and (b) the interpretable limit of 10% transfer is improved from 0.22 nm to 0.14 nm when the spherical aberration is corrected. However, a disadvantage of aberration corrected imaging is that the transfer at low spatial frequencies is significantly reduced. At the Scherzer defocus condition the lower limit of 10% transfer, for an aberration corrected microscope, is 0.63 nm whereas for a conventional C_s the limit is 1.74 nm. This does not affect imaging of small unit cell crystalline materials although it reduces the contrast of long range features such as specimen bending and contamination. However, for large unit cell structures such as macromolecules this loss of long range information is critical.

For imaging large organic structures, a large defocus (>200 nm) is often used in order to improve the transfer of low spatial frequencies but this has the disadvantage of introducing rapid oscillations to the CTF at high spatial frequencies as shown in Fig. 1c. These phase reversals can be partially removed using 'CTF correction' [6,7] however, the accuracy of defocus determination limits interpretability for resolvable distances smaller than 0.5 nm. Slightly better resolution is achievable by 'single particle reconstruction'[8] but requires averaging many thousands of images.

Journal of Physics: Conference Series **241** (2010) 012013 doi:10.1088/1742-6596/241/1/012013

2. Extending the low spatial frequency transfer in exit wavefunction restoration.

Restoring the exit wavefunction from a series of TEM images with differing defocus values reveals local structural information in the phase of the exit wavefunction. Due to the need for multiple images this technique is not generally suitable for radiation sensitive samples. However, for materials where stability is related to the total electron dose, exit wavefunction restoration using a Wiener restoring filter optimally suppresses noise by a factor of √N where N is the number of images in the data set [9]. Thus for an equivalent total electron dose, the noise in the restored exit wavefunction will be reduced in comparison to a single image.

A Wiener filter approach to exit wavefunction restoration [9] requires a minimum of 3 images, although for inorganic materials a typical experimental data set consists of 10-20 images centered about the Scherzer defocus with a small uniform focal increment [10,11]. For radiation sensitive materials the optimal number of images is determined by experimental considerations, such as the minimum detectable signal in a single image, the noise characteristics of the detector used and the minimum reproducible exposure time, as well as by the specimen's total dose tolerance. In practice a five image data set has been found to be an experimentally feasible compromise for beam sensitive specimens.

A five image dataset with a focal step of 5 nm centered about the Scherzer defocus is limited to a long range (low spatial frequency) resolution of ~1nm. In this paper we propose using an alternative five image dataset for exit wave restoration with non-uniform focal increments. The defoci of the novel data set were chosen as -5 nm, -10 nm, -15 nm, -54 nm and -200 nm (for the microscope used). These focal increments do not compromise the transfer of high spatial frequencies but the inclusion of images at larger defoci optimizes transfer of long range information, extending the lower 10% transfer limit to 4 nm. The overlap of the consecutive CTF's has been optimised such that the transfer does

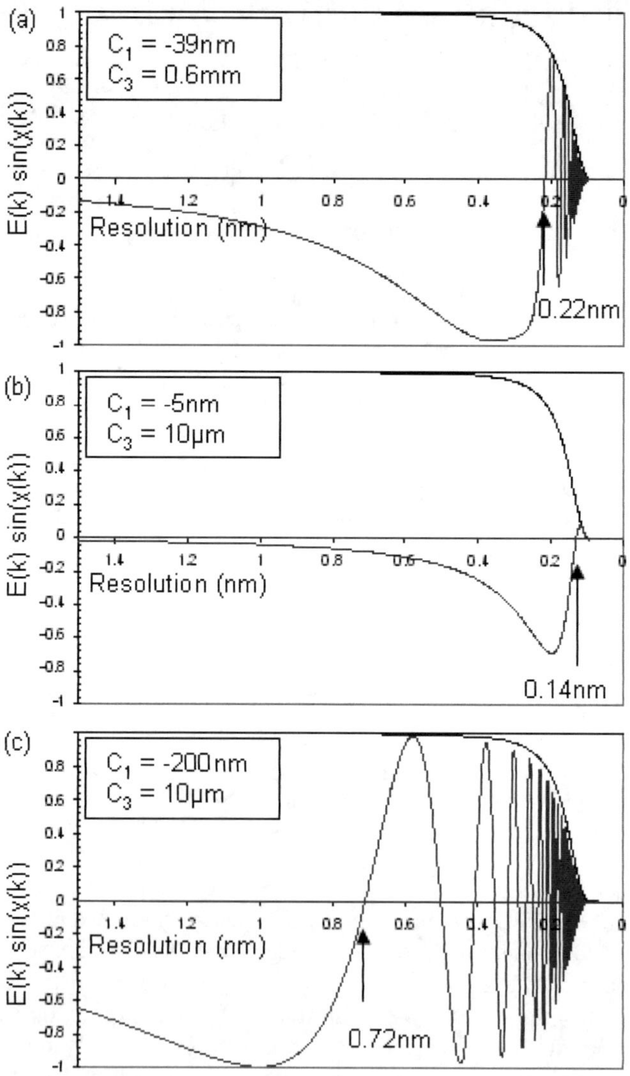

Fig. 1. Contrast transfer functions for the JEOL 2200MCO (S)TEM (200kV) with a focal spread of 3.8 nm at different values of defocus (C_1) and spherical aberration (C_3). For uncorrected C_3 the Scherzer defocus condition (a) transfers phase information without contrast reversals for resolutions between 1.74 nm and 0.22 nm. The equivalent range for a corrected value of C_3 at the Scherzer defocus (b) lies between 0.63 nm and 0.14 nm and at a large underfocus (200nm) (c) lies between 4 nm and 0.72 nm

not fall below 60% of the maximum for any spatial frequency (Fig. 2).

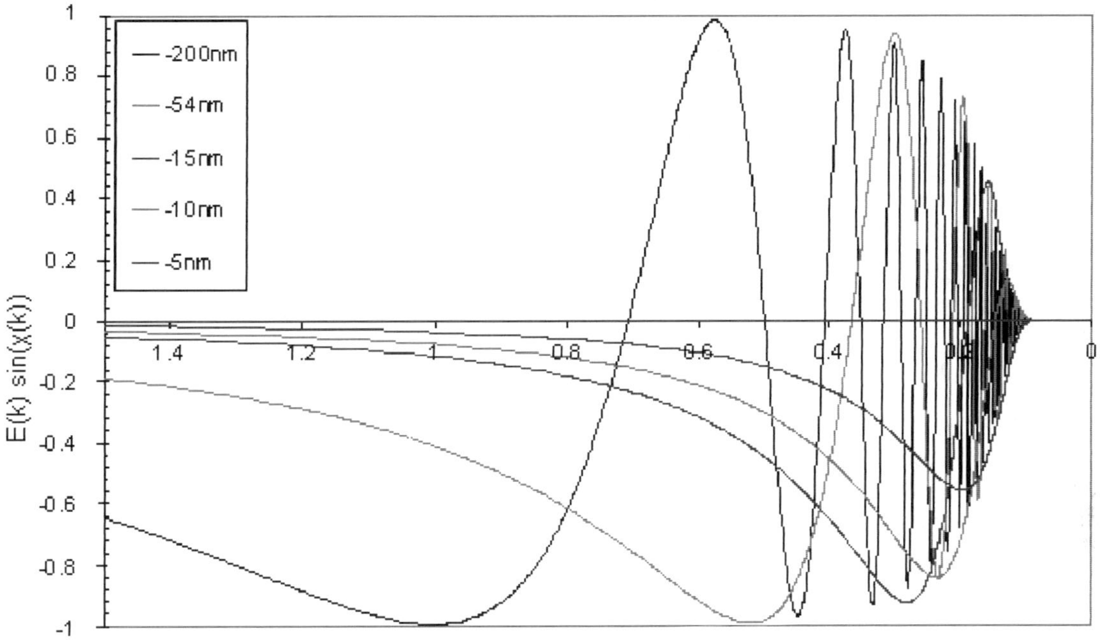

Fig. 2. Comparison of CTF's for the JEOL 2200MCO (S)TEM with a focal spread of 3.8 nm and a third order spherical aberration of 10μm at various defoci as marked in the legend. The real space distance (1/k) in nm is indicated on the horizontal axis.

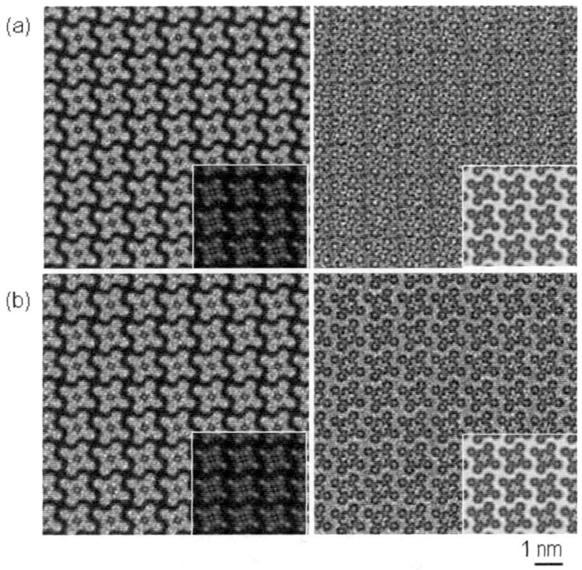

Fig. 3. Modulus (left) and phase (right) of the exit wavefunctions restored from five simulated images of an 18.5 nm thick sample of copper phthalocyanine with an electron dose of 100 electrons $Å^{-2}$. The restoration in (a) used a standard focal series of equally spaced defocus values and (b) used a focal series with unequal focal steps to extend the spatial frequency range of the restoration. The original exit wavefunction used to simulate the input TEM images is shown inset.

3. Simulation studies.

To evaluate this new focal series geometry for low spatial frequency transfer image simulations have been performed using the multislice method [12] for the α-copper phthalocyanine (CuPc) structure [13]. Specimen thicknesses of 3.7 nm, 8 nm, 18.5 nm and 37 nm were investigated at various electron doses appropriate to the imaging of beam sensitive materials (10, 25 and 100 electrons $Å^{-2}$). These simulated images were then used as input to the exit wavefunction restoration procedure. The exit wave recovered from the extended focal series geometry (defoci of -5nm, -10nm,-15nm, -54nm and -200nm) was subsequently compared to that recovered from the traditional focal series (defoci of -5nm, -10nm,-15nm, -20nm and -25nm). The comparison for a sample thickness of 18.5nm and a moderate electron dose (100 electrons $Å^{-2}$) is shown in Fig. 3. Using the new extended focal series approach the phase of the restored exit wavefunction closely resembles that of the original exit wavefunction, whereas that restored from the standard focal series approach is clearly missing low spatial frequency information.

4. Conclusions

In this paper we have presented a new geometry for exit wavefunction restoration which improves the transfer of low spatial frequencies to the exit wavefunction. This approach extends the spatial frequency range for a five image data set to between 4 nm and 0.09 nm (compared to between 1nm and 0.09 nm for a standard focal series). The benefit of this approach for a large unit cell test material has been demonstrated using simulations.

References

[1] Rose, H. H. Historical aspects of aberration correction, *J. Electron Microscopy*, 58, 77-85, (2009).
[2] Dahmen, U. et al., Background status and future of the Transmission Electron Aberration-corrected Microscope Project. *Phil. Trans. R. Soc. A* 367, 3795-3808 (2009)
[3] Haigh, Sarah J. and Sawada, Hidetaka and Kirkland, Angus I. Atomic Structure Imaging Beyond Conventional Resolution Limits in the Transmission Electron Microscope, Physical Review Letters, 103, 126101-4, (2009).
[4] Hutchison, J. L. Titchmarsh, J. M, Cockayne, D. J. H. Doole, R. C. Hetherington, C. J. D. Kirkland, A. I. and Sawada, H. A versatile double aberration-corrected, energy filtered HREM/STEM for materials science, *Ultramicroscopy*, 103, 1, 7-15, (2005).
[5] Chang, L. Y. Kirkland, A. I. and Titchmarsh J.M. On the importance of 5th order spherical aberration for a fully corrected electron microscope, Ultramicroscopy, 106, 301-306 (2006)
[6] Mindell, J.A. and N. Grigorieff, Accurate determination of local defocus and specimen tilt in electron microscopy. *J. Structural Biology*, 142(3) 334 (2003).
[7] Wade, R.H., A Brief Look at Imaging and Contrast Transfer. *Ultramicroscopy*, 46, 145-156 (1992)
[8] Zhang, X., et al., Near-atomic resolution using electron cryomicroscopy and single-particle reconstruction. *Proc. Natl Acad Sci.*, 105, 1867-1872. (2008).
[9] Saxton, W. O. Accurate Atom Positions From Focal And Tilted Beam Series Of High-Resolution Electron-Micrographs, *Scanning Microscopy*, 213-224 (1988)
[10] Saxton, W. O. Accurate Alignment Of Sets Of Images. *J Microsc. Oxford* 174, 61-68 (1994)
[11] Kirkland, A. I. and Meyer, R. R. "Indirect" high-resolution transmission electron microscopy: Aberration measurement and wavefunction reconstruction, *Microsc. Microanal.* 10, 401-413, (2004)
[12] Kirkland, E.J. Advanced Computing in Electron Microscopy, Plenum Press, New York, 1998
[13] Hoshino, A. Takenaka Y. and Miyaji H. Redetermination of the crystal structure of α-copper phthalocyanine grown on KCl, *Acta Cryst.* **B59**, 393-403 (2003).

Application of lattice strain analysis of semiconductor device by nano-beam diffraction using the 300 kV Cold-FE TEM

T Sato[1], **H Matsumoto**[1], **K Nakano**[1], **M Konno**[1], **M Fukui**[1], **I Nagaoki**[2], **Y Taniguchi**[2]
[1]Hitachi High-Technologies Corp., 11-1 Ishikawa-cho, Hitachinaka, Ibaraki, 312-0057 Japan
[2]Higachi High-Technologies Corp., 882 Ichige, Hitachinaka, Ibaraki, 312-8504 Japan

E-mail: sato-takeshi-3@naka.hitachi-hitec.com

Abstract. The lattice strain of the semiconductor device was measured by the nano-beam diffraction method using a 300 kV TEM (HF-3300) equipped with a Cold-FE gun. The various samples thickness prepared by the Focused Ion Beam system. For the sample thickness measurement with accuracy, the mean free path of inelastic scattering (λ) was determined from the correlation between the $In(I_t/I_0)$ value from the Electron Energy Loss Spectrum and the thickness of the pillar-shaped sample measured by the TEM image. As a result, the rate of the lattice strain in device increased with changing sample thickness. In the sample thickness of 494 nm, the lattice strain has concentrated on the neighbourhood Si substrate by the NBD.

1. Introduction

The evaluation of lattice strain is important to control the electrical characteristic of semiconductor devices. Generally, the method of measuring the lattice strain in the devices by the electron beam has mainly two methods. One is the Nano-Beam Diffraction (NBD[1-4]), and another one is Convergent Beam Electron Diffraction (CBED[5-6]) method, it is necessary to tilt the sample from the crystal plane, and there is a problem that the area of an effect lattice strain measurement is greatly limited. On the other hand, the comparison accuracy of the NBD is low in the CBED method. However, the NBD is effective for the application and the analysis because analysis method is easy, and a possibility to be able to automate. In this paper, we show the result of measurement the lattice strain of the devices using the NBD by the HF-3300 TEM.

2. Experimental procedure

TEM images and the NBD patterns of the samples were taken by the HF-3300 TEM, equipped with Cold-FE gun and the charge-coupled device (Gatan, Ultrascan1000 2048x2048 pixels). A post-column energy filter (Gatan, GIF Tridium 863) with the HF-3300 was used on the Electron Energy Loss Spectroscopy (EELS) measurement. For the sample thickness (t) measurement with accuracy, we obtained the mean free path of inelastic scattering (λ) from TEM image and EELS measurement. In general, we have to measure the intensity under the zero-loss peak (I_0) and ratio this to the total intensity (I_t). The relative intensity of the zero loss and the total intensity, is given by

[1] Corresponding author Tel.: +81-29-354-1970; fax: +81-29-354-1971. Email address: sato-takeshi-3@naka.hitachi-hitec.com (Takeshi Sato)

$$t = \lambda In\left(\frac{I_t}{I_0}\right)$$

where λ is mean free path of inelastic scattering. Therefore, all we need is to determine λ for the sample. Figure 1 shows a schematic illustration of the sample thickness measurement by the TEM image and the obtaining of the $In(I_t/I_0)$ value by the EELS method in TEM column. First, we fabricated the pillar-shaped sample using a Focused Ion Beam (FIB) system with the FIB micro-sampling mechanism[7] and the three-dimensional (3D) sample holder with the needle stub[8]. Next, the sample thickness was measured by TEM image, as shown in Fig. 1(a), and then, the rotated needle stub, the $In(I_t/I_0)$ value was obtained from the EELS, as shown in Fig. 1(b). At this time, the direction of the electron beam to the pillar-shaped sample was decided from the diffraction pattern of Si substrate. In this experiment, this working procedure was repeated at nine times. The result was shown in Fig. 2. The value of λ was determined as 197±13 nm. On the other hand, we get this value from a parameterization based on many experimental measurements[9]. The expression is

$$\lambda = \frac{106FE_0}{\left\{E_m In\left(\frac{2\beta E_0}{E_m}\right)\right\}}$$

where λ is in nm, E_0 in keV, β in mrad, F is a relativistic correction factor, and E_m is the average energy loss in eV which, for a material of average atomic number Z, is given by

$$E_m = 7.6Z^{0.36}$$

The relativistic factor (F) is given

$$F = \frac{\left\{1+\frac{E_0}{1022}\right\}}{\left\{1+\left(\frac{E_0}{511}\right)^2\right\}}$$

From these formulas for computation, the λ value of Si is 193 nm. Therefore, the value estimated from novel method was the almost equal to the theoretical value.

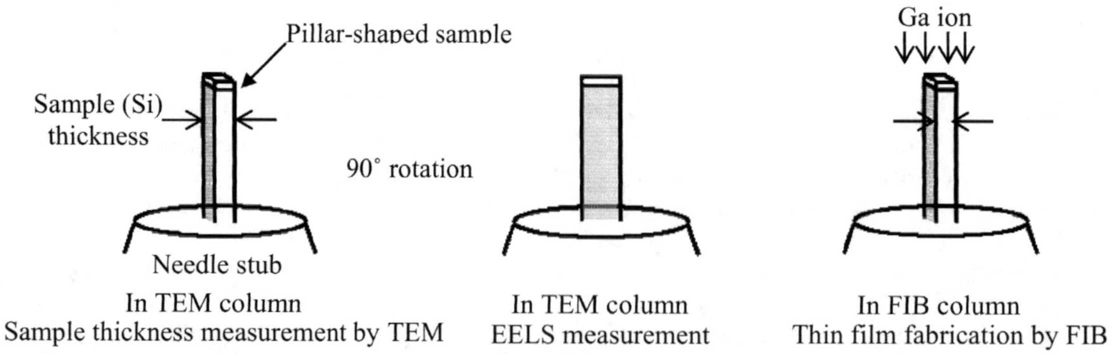

Figure 1. A schematic illustration of the novel method of the sample thickness measurement by the TEM image and the obtaining of the $In(I_t/I_0)$ value by the EELS method.

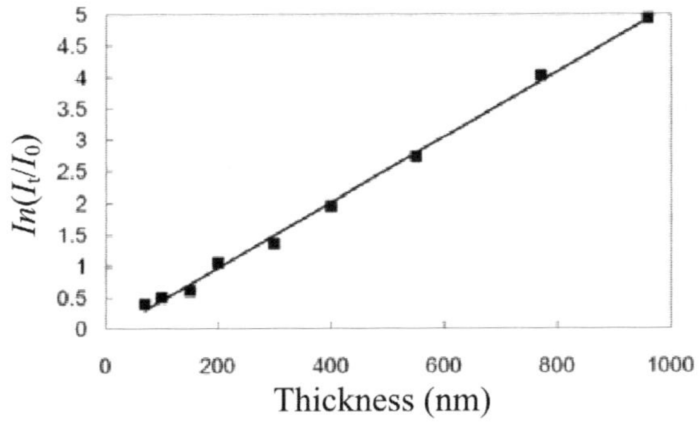

Figure 2. Correlation between the sample thickness and the $In(I_t/I_0)$ value obtained from EELS. Value of λ determined as 197±13 nm. I_0: zero-loss intensity, I_t: total intensity

3. Experimental results and discussion

The four different thick samples of the P-MOS devices were prepared by the FIB. The relativistic correction factor of each sample was obtained from the EELS measurement, and the sample thickness of each sample was estimated to be 97 nm, 207 nm, 275 nm and 494 nm from the value of λ. Figure 3 shows the TEM image of the P-MOS structure by the FIB. Figure 3(a) shows the magnified image of a rectangle region in Fig. 3(a). Dotted circles indicate the NBD measurement position. We measured the NBD pattern on the position of the ten points from the gate, as indicated circles in Fig. 3(b). Figure 4 shows the measurement conditions of the NBD. Figure 4(a) shows the TEM image of focused probe on Si(110) plane. Lattice image of the Si(111) can be clearly observed, as shown in Fig. 4(a). In this experiment, we used that the probe size with the NBD is the 10 nm with the intensity profile, as shown in Fig. 4(b). Figure 4(c) shows the NBD pattern on this probe condition.

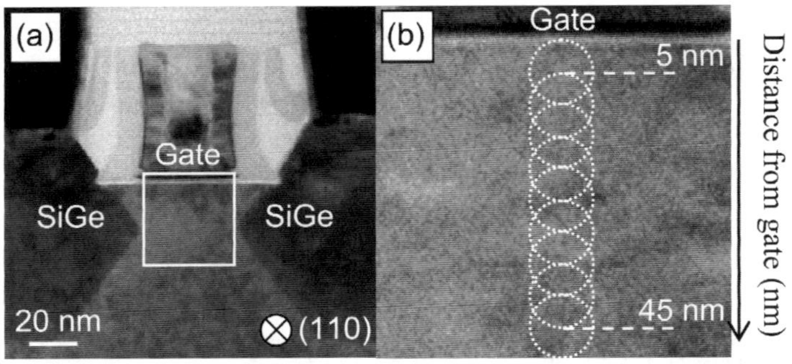

Figure 3. TEM images of P-MOS structure and the indicated the measurement position of the NBD.

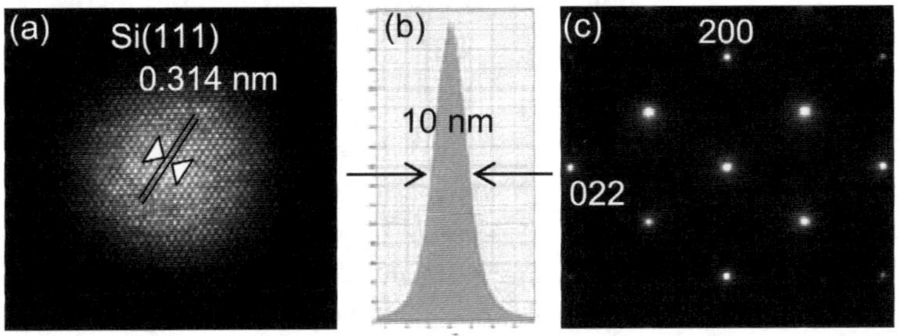

Figure 4. Measurement condition of the NBD. (a) TEM image of focused probe on Si(110) plane. (b) Intensity profile of the probe. (c) The NBD pattern on these conditions.

The distance between Si(022) diffraction spots, i.e., lattice strain of the horizontal direction against the gate was measured using the diffraction analysis function on the Hitachi EMIP software. The electron diffraction pattern of the reference was assumed the diffraction pattern of a place left from the gate by

500 nm. Moreover, it was confirmed that the ratio of the distance of Si(022) (d = 0.192 nm) and Si(200) (d = 0.271 nm) became 1.41 in the electron diffraction pattern of the reference using the diffraction analysis function. Figure 5 shows the change in the strain rate by the each sample thickness in the P-MOS. The strain rate decreases corresponding in the sample thickness, as shown in Fig. 5. The degree of lattice strain is affected by the sample thickness, in which thinner samples have weaker lattice strain rate on the 97 nm and the 207 nm thickness, as shown in Fig. 5. In the sample thickness of 494 nm, the lattice strain was concentrated on the neighbourhood Si substrate, as shown in the circle dotted line in the Fig. 5. These results may be showed the reflection of the stress strain in the bulky state. The accuracy of measurement between diffraction spots using the diffraction analysis function will be confirmed clearly in the future.

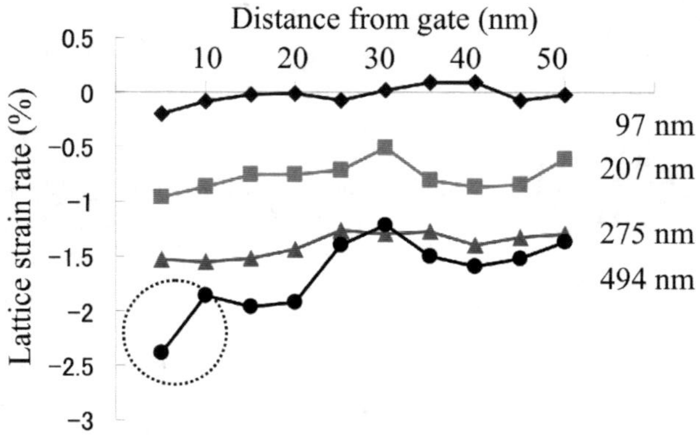

Figure 5. The result of the lattice strain rate of the different thickness at several points in the P-MOS devices.

4. Summary

We measured lattice strain on the P-MOS by the NBD using 300 kV Cold-FE TEM. From the EELS of the pillar-shaped sample, the mean free path of inelastic scattering (λ) was determined as 197±13 nm, by the FIB system and the 3D holder. The rate of the lattice strain in the P-MOS increased with changing sample thickness. In the sample thickness of 494 nm, the lattice strain has concentrated on the neighbourhood Si substrate.

References

[1] Kouguchi M, Nakamura K and Umemura K 2001 *Ext. Abst. of the SSDM.* 198
[2] Kouguchi M, Nakamura K and Umemura K 2001 *Proc. Electrochem. Soc.* 326
[3] Shimizu A, Hachimine K, Ohki N, Ohta H, Kouguchi M, Nonaka Y, Sato H and Oostuka F 2001 *IEDM. Tech. Dig.* 433
[4] Huang J and Kim M J 2006 *Appl. Phys. Lett.* **89** 063114
[5] Kimoto K and Tanaka M 1993 *Jpn. J. Appl. Phys.* **32** 211
[6] Tsuda K and Tanaka M 1999 *Acta. Cryst.* **A55** 939
[7] Ohnishi T, Koike H, Ishitani T, Tomimatsu S, Umemura K and Kamino T 1999 *Int. Symp. For Testing and Failure Analysis.* 449
[8] Kamino T, Yaguchi T, Konno M Ohnishi T and Ishitani T 2004 *J. Electron. Microsc.* **53(6)** 589
[9] Malis T, Cheng S C and Egerton R F 1988 *J. Electron Microsc. Technol.* **8** 193

An Integrated Multiple Silicon Drift Detector System for Transmission Electron Microscopes

H.S. von Harrach[1], P. Dona[1], B. Freitag[1], H. Soltau[2], A. Niculae[2], M. Rohde[3]

[1]FEI Company, Achtseweg Noord 5, 5600 KA Eindhoven, The Netherlands

[2]PN Sensor GmbH, Otto-Hahn-Ring 6, D-81739 Munich, Germany

[3]Bruker AXS Microanalysis GmbH, Schwarzschildstr. 12, 12489 Berlin, Germany

E-mail: sebastian.von.harrach@fei.com

Abstract. A new EDX system, consisting of multiple SDDs has been developed for an FEI 200kV TEM/STEM in which the SDDs, designed by PN Sensor with a total collection angle approaching 1 srad has been obtained by placing 4 SDD's symmetrically around the electron beam axis in the objective lens chamber. The massive increase in solid angle of collection compared to previous designs in S/TEMs leads to a huge reduction in the time for EDX mapping. First results from the detector are reported.

1. Introduction

Silicon Drift Detectors (SDD) [1] are rapidly replacing Si(Li) detectors for EDX microanalysis in SEM, but have yet to have an impact in the S/TEM world. Main reason for this difference is the low X-ray count rate created by thin S/TEM samples [2] compared to the bulk samples in SEM. These low count rates make EDX mapping a very slow process in S/TEM. However, the introduction of probe correctors and more recently the introduction of higher brightness electron sources [3] has led to significantly increased beam currents in small electron probes and, potentially, to higher EDX count rates. Since a key advantage of the SDD is the high count rate capability [1], the throughput improvement compared to the Si(Li) detectors will be considerable in these new instruments. In addition, since many materials are easily damaged by fast electrons it is essential to improve the collection efficiency of the X-rays generated, in order to reduce the dose on the sample.

2. EDX detector design

For the above reasons a new EDX system (patents pending), consisting of multiple SDDs has been developed for an FEI 200kV TEM/STEM in which four SDDs, designed by PN Sensor with a total active area of 120 mm^2, are placed symmetrically around the electron beam axis in the objective lens chamber. The design is shown schematically in figure 1.

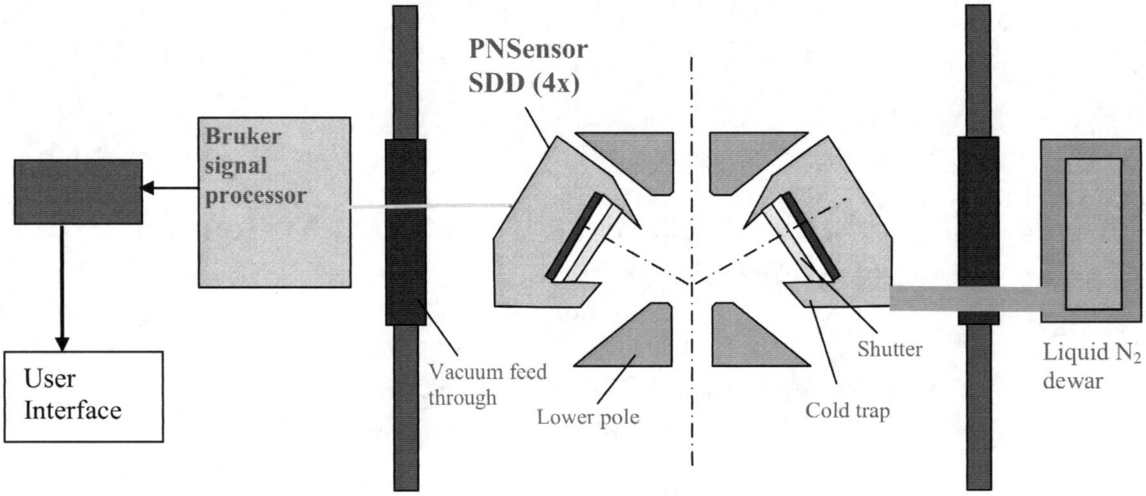

Figure 1. Schematic of Super-X detector

The detectors are integrated into the cold-trap for best performance. The windowless design allows for a better sensitivity for light-element detection than conventional thin-window detectors. Except during analysis, shutters are closed in front of the detectors to prevent contamination. The signals from the SDDs are processed by an ultra fast four-channel pulse processor produced by Bruker AXS, with pixel dwell times down to 10 μs and output count rates up to 60 kcps per channel.

3. Performance

The detector has been characterised using a nickel oxide standard sample of 50 nm thickness on a Mo grid.

Figure 2 shows a spectrum of the NiOx sample at 200kV. The spectrum shows a high peak-to-background ratio, exceeding a Fiori number of 5000, and the absence of spurious peaks. The Ni-L and O-K peaks are 1.3 and 2.5 times higher, respectively, than for a Si (Li) thin window detector.

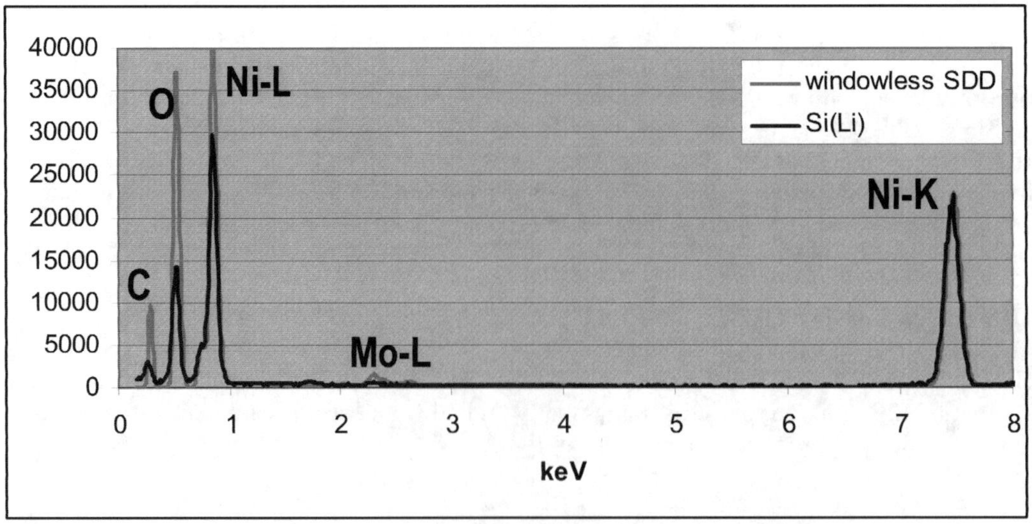

Figure 2. Super-X and Si(Li) detector spectra of NiOx specimen on carbon film and Mo grid at 200kV under identical conditions; spectra normalised to Ni-Kα peak

The input and output count rates were measured for the detectors at a nominal shaping time of 1μs. Fig.3 shows output count rates up to 220 kcps. The dead time is below 60% up to 200 kcps output count rate. The energy resolution rises from 132 to 135 eV over this count rate range.

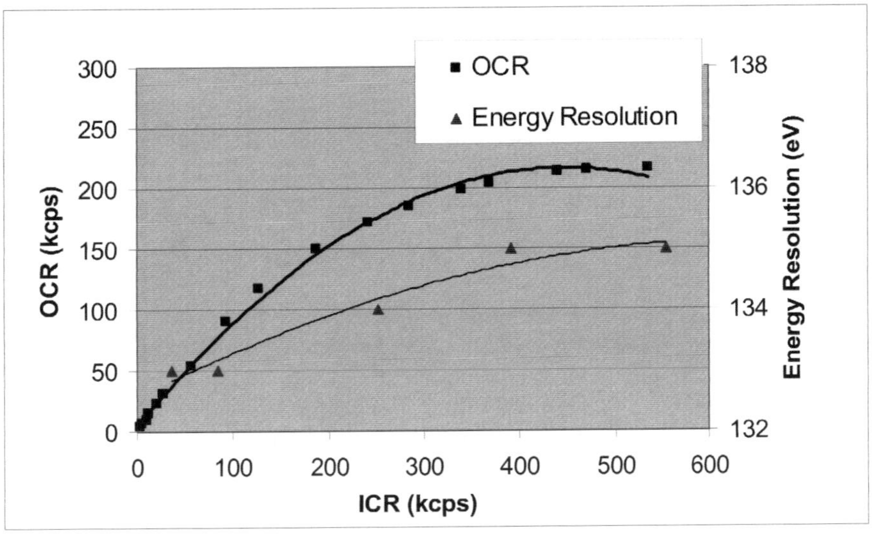

Figure 3. Output count rate (OCR) and energy resolution (RH scale in eV) vs. Input count rate (ICR)

4. Fast mapping

The mapping speed is illustrated in figure 4. A 300 x 50 pixel map was acquired on a Bi2212 superconductor specimen at 200kV. Using a pixel dwell time of 50μs per pixel, the map was acquired in 1.5 minutes and shows clearly the Bi, Sr and Ca lattice planes. This acquisition time is estimated to be 30 times faster than with conventional EDX detectors. Note the missing planes of Ca atoms indicated by arrows in Fig.4. At these defects the BiO spacing is reduced (see model of unit cell in Fig.4), but the spacing of SrO planes is unchanged at about 0.8 nm.

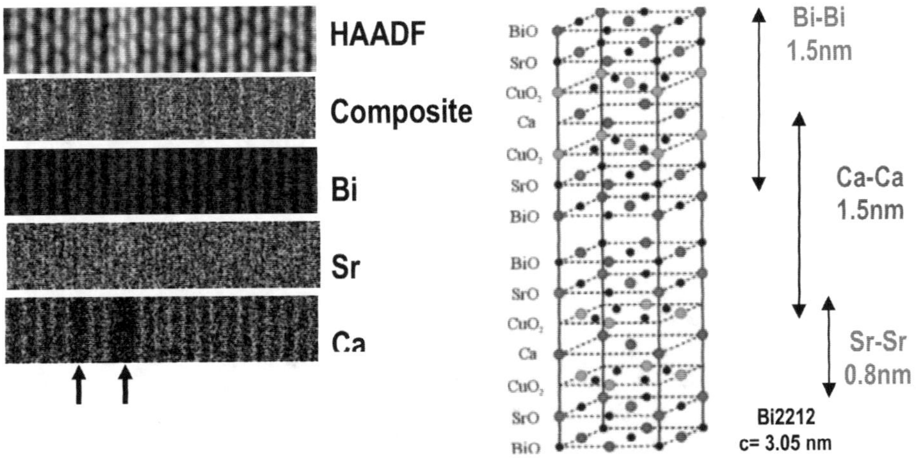

Figure 4. Elemental EDX maps (300 x 50 pixels) of Bi2212 superconductor specimen and structure of unit cell (RHS). Pixel dwell time=50 us, 7 frames. Total mapping time = 94 sec.

5. Conclusions

The SuperX detector design results in a massive increase in solid angle of collection to 0.9 sr compared to 0.1 – 0.3 sr of previous designs in S/TEMs, and hence, a huge increase in count rate for a given beam current. When combined with a high-brightness electron beam a large reduction in the time for EDX mapping from hours to minutes is achieved. Sub-nanometer resolution is achieved in less than 2 minutes' acquisition of X-ray spectrum images.

References

[1] Terborg R. and Rohde M. 2004 *Micosc. Micoanal.* **10** (Suppl 2) 942
[2] Williams D.B., Goldstein J.I. and Fiori C.E. 1986 Principles of X-Ray Energy-Dispersive Spectrometry in the AEM *Principles of Analytical Electron Microscopy*, eds D.C. Joy, A.D. Romig Jr., J.I. Goldstein (New York: Plenum) chapter 4 pp 123-154
[3] Freitag B., Knippels G., Kujawa S., Tiemeijer P., van der Stam M., Kisielowski C., Denes P., Minor A. and Dahmen U. 2008 *Micosc. Micoanal.* **14** (Suppl 2) 55

An improved approach to quantitative X-ray microanalysis in (S)TEM: thickness dependent *k*-factors

T Walther

Department of Electronic & Electrical Engineering, University of Sheffield, Sir Frederick Mappin Building, Mappin Street, Sheffield S1 3JD, UK

t.walther@sheffield.ac.uk

Abstract.
Quantifying energy-dispersive X-ray (EDX) spectra acquired in a transmission electron microscope can produce inconsistent and unreliable results even if correct values for specimen thickness and density are being used for calculating an appropriate absorption and fluorescence correction. This can be due to various factors, such as incorrect *k*-factors, faulty absorption corrections due to specimen roughness, specimen buckling or wrong estimates of the detector's take-off angle, complicated fluorescence effects not accounted for in the software etc. For a thorough investigation of the InGaAs system, wedge shaped specimens were cleaved from GaAs and InAs wafers and studied by EDX spectroscopy in a JEOL2010F instrument, with thicknesses ranging from a few nanometres up to several microns. The thickness dependence of the relative intensities of X-ray lines that remains after nominal absorption and fluorescence correction were taken into account could be used to perform linear least-squares fits, thereby eliminating to a first order the apparent remaining thickness dependence of the *k*-factors. The resulting fits demonstrate this correction improves the quantification of the chemical concentrations of indium and gallium in InGaAs samples of known thickness to 1at%.

1. Introduction

X-ray microanalysis of bulk materials in scanning electron microscopy (SEM) can be quantified by taking into account the different fluorescence yields of the elements, the absorption and fluorescence effects of the various X-ray lines produced, the transmission of the detector window and the detector geometry. As the target is bulk-like, the complete interaction volume between electrons and X-rays needs to be taken into account, either by modelling or by parameter fitting [1]. Quantification of elemental concentrations depends on the knowledge of the relative ionisation cross-sections of the corresponding elements (*Z*-effect), modelled already in 1930 by Bethe [2] and later refined by Duncumb and Reed [3], and proper corrections for absorption (*A*) [4] and for fluorescence (*F*) [5], which can be combined for a given specimen thickness and density in the so-called *ZAF*-correction. More recently, full self-consistent spectrum modelling has become available which allows the user to model X-ray generation as a function of the product of mass and thickness and then integrate the result over the specimen thickness, e.g. with the *PROZA96* [6] program.

X-ray microanalysis of thin specimens in transmission electron microscopy (TEM), on the other hand, has the advantages of higher spatial resolution and the possibility of using a thin film approximation in the form of so-called *k*-factors that are simply experimentally determined sensitivity factors which are the product of the relative fluorescence yields of the various elements and the detector sensitivity [7, 8]. For thin layer structures at different depths within a thicker specimen, the top-bottom effect should be taken into account as well, although this problem may be solved by recording two spectra at different take-off angles [9]. For medium-thick specimens in the range 10nm<t<1000nm or low take-off angles, absorption and fluorescence must be taken into account and should in principle yield consistent results for homogeneous material analysed at different thicknesses (although, not necessarily the right result!).

Testing this on our JEOL 2010F electron microscope operated at 197kV with Oxford Instruments Si:Li detector and ISIS 300 software the consistency of the thickness correction is not perfect, as the results obtained for a homogeneous specimen do slightly depend on the specimen thickness being used despite absorption and fluorescence having been taken into account. For a more thorough investigation, wedge shaped specimens were cleaved from GaAs and InAs wafer material and studied by energy-dispersive X-ray spectroscopy. In order to fit into the 2mm gap in the URP pole piece of our JEOL 2010F, the specimens were glued at 45° orientation with silver epoxy dag onto copper grids specially cut out so that the wedges stuck out symmetrically on both sides. The {110} cleavage facets formed 90° wedge specimens that could be investigated near <100> zone axes orientations. In this geometry the local thickness is given by twice the distance from the specimen edge [10]. The thickness ranges investigated covered several nm up to several μm. The remaining thickness dependence of the relative intensities of X-ray lines could be used to perform linear least-squares fitting, thereby eliminating to a first order the apparent remaining thickness dependence of the k-factors.

2. Experimental

Figures 1 and 2 plot the apparent concentration ratios of elements in GaAs (Fig. 1, left) and InAs (Fig. 2, right) specimens, based on peak integration of various X-ray lines using the corresponding k-factors of the ISIS 300 software and assuming either a constant thickness of T=100nm (a typical thickness of TEM specimens; red squares) or modelling the local thickness correctly as twice the distance from the specimen edge (blue triangles). As peak fitting and built-in k-factors were unavailable in the software above 20kV energy, for the In K-lines at 24-27kV the literature value of $k_{InK,Si}$=7.99 was used which, however, had a massive error bar of ±1.8 (i.e. 23% relative error) attached to it [13]. Also, intensity ratios were calculated in this instance from the peak heights rather than by peak integration and background subtraction, using a program outside the ISIS software. This curve is shown by green circles at the top of Figure 2.

Had the k-factors of the elements been correct, then all extrapolated curves should have intersected the vertical axes for zero thickness at unity to reproduce the correct stoichiometry of the compounds. Had the corrections for absorption and fluorescence worked properly, then all blue curves for the true crystal thicknesses should have produced constant values of zero slope, independent of the actual thickness. Obviously, both k-factors and absorption/fluorescence corrections are not very reliable.

Results from linear fitting the curves in Figures 1 and 2 are reported in Tables 1 and 2. It can be seen that for GaAs the remaining thickness dependence is linear to good approximation for all three blue curves, while for InAs there seem to be non-linearities at small thicknesses, which could be artefacts due to improper cleavage and hence wrongly estimated thicknesses near the crystal edges. While for GaAs the k-factors were wrong by 19±1% relative for Ga_K/As_K, and much more for Ga_K/As_L and Ga_L/As_L, the k-factors for InAs were not out by so far. However, only after least-squares fitting could the relative uncertainties for all k-factors be brought down to 3±1% and that of the slopes down to $(1.7\pm1.5)\times10^{-4}nm^{-1}$. For GaAs, it is advisable to use Ga_K/As_K and/or Ga_L/As_L the k-factors of which can be corrected to 2% accuracy, and the remaining slopes are small enough to change this by only ~1% if the thickness is known to 100nm accuracy. For InAs, on the other hand, it is best to use the mixed In_L/As_K ratio, with a similar result. The typical and maximum relative errors would be ~2% and ~3%, respectively, for the Ga/As and In/As ratios; hence, for ternary $In_xGa_{1-x}As$ it should be possible to determine the In concentration x in the group-III sub-lattice to better than ±1at% (Δx=±0.01) using cross-checks based on the above methods (i.e. determining the Ga/As and the In/As ratio independently and then checking the overlap region).

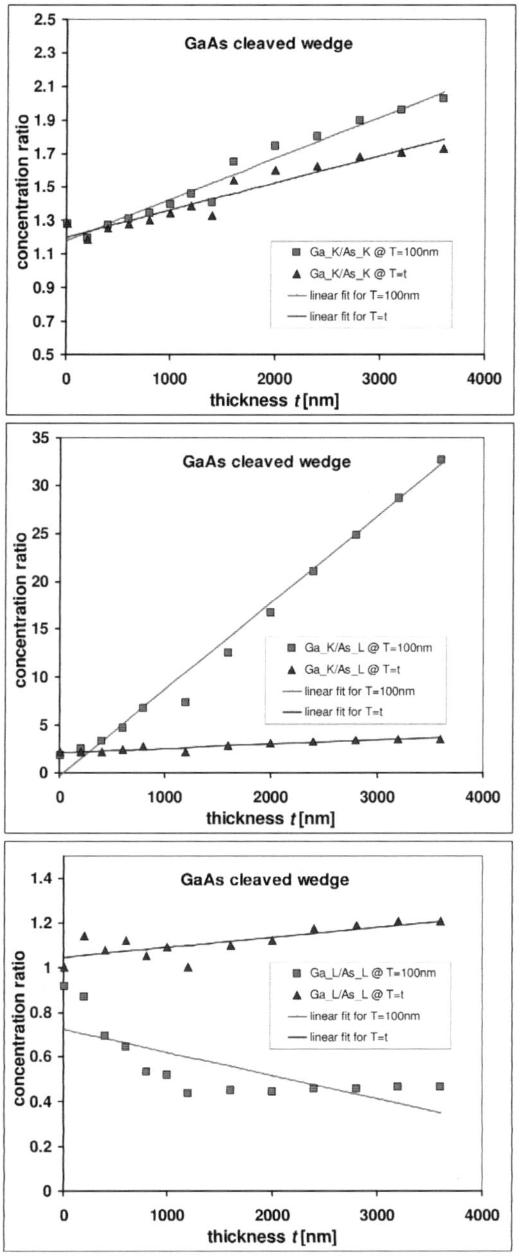

Figure 1. Experimental result of measuring the Ga_K/As_K ratio (above), the Ga_K/As_L ratio (middle) and the Ga_L/As_L ratio (below) for cleaved GaAs using the LINK ISIS300 software, assuming a thickness of T=100nm throughout (red squares) or the true thickness (blue triangles).

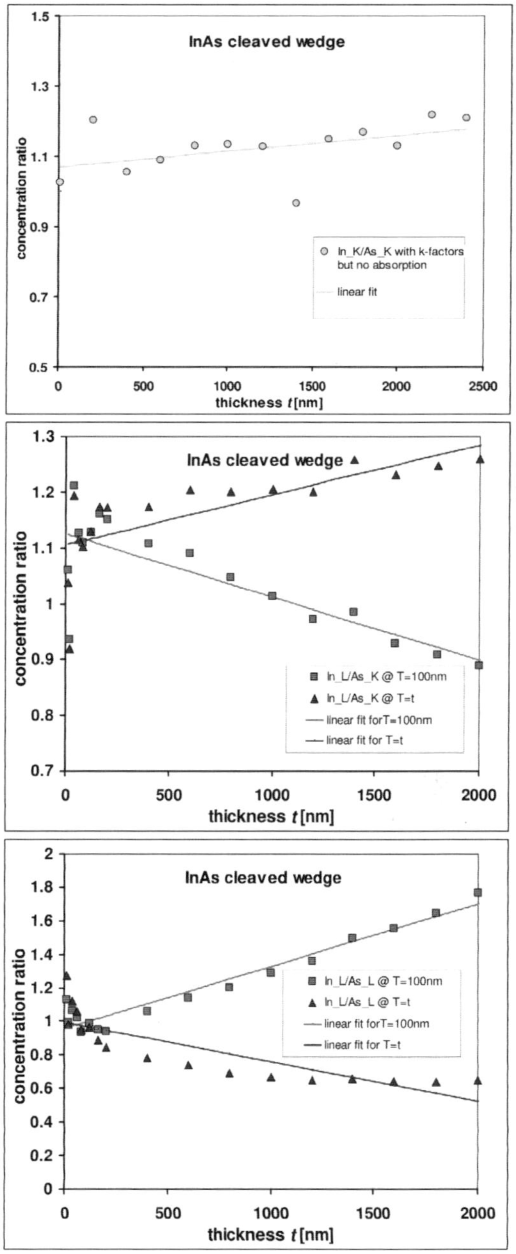

Figure 2. Experimental result of measuring the In_K/As_K ratio (above), the In_L/As_K ratio (middle) and the In_L/As_L ratio (below) for cleaved InAs using the LINK ISIS300 software, assuming either a thickness of T=100nm throughout (red squares) or the true thickness (blue triangles). For the In K-line no absorption was taken into account (green circles).

Table 1. Result of linear regression fitting to the GaAs data shown in Figure 1

X-ray lines	modelled T	linear corr. coefficient	vertical axis intercept	slope of fit [10^{-6}/nm]
Ga_K / As_K	T=100nm	+0.9798	1.182±0.027	245±14
	T=t	+0.9607	1.202±0.025	162±14
Ga_K / As_L	T=100nm	+0.9932	−0.315±0.645	8861±329
	T=t	+0.9352	2.117±0.100	426±51
Ga_L / As_L	T=100nm	−0.7389	0.723±0.054	−103±28
	T=t	+0.7656	1.046±0.022	45±11

Table 2. Result of linear regression fitting to the InAs data shown in Figure 2

X-ray lines	modelled T	linear corr. coefficient	vertical axis intercept	slope of fit [10^{-6}/nm]
In_K / As_K	T=0	+0.4621	1.071±0.036	44±26
In_L / As_K	T=100nm	−0.7989	1.125±0.021	−113±22
	T=t	+0.7136	1.106±0.022	89±23
In_L / As_L	T=100nm	+0.9679	0.957±0.024	372±25
	T=t	−0.8271	0.993±0.039	−234±41

3. Summary

It has been shown that absorption/fluorescence corrections applying commercial software to experimental energy-dispersive X-ray microanalysis spectra can yield relative errors above 10%. This can be improved by making systematic measurements of relative X-ray intensities as a function of known sample thickness and fitting the remaining thickness dependence to first order. For $In_xGa_{1-x}As$ this means the indium concentration thus calculated is accurate to better than 1at%.

References

[1] Goldstein JI, Newbury DE, Echlin P, Joy DC, Lyman CE, Lifshin E, Sawyer L and Michael JR 2003 *Scanning Electron Microscopy and X-ray Microanalysis*, New York: Springer, 3rd ed.
[2] Bethe HA 1930 *Ann. Phys.* **397**, 325
[3] Duncumb P and Reed SJB 1968 Quantitative Electron Probe Microanalysis *NBS Spec. Publ.* (vol 298) ed KFJ Heinrich, p. 133
[4] Philibert J 1963 *Proc. Int. Symp. X-ray Optics and X-ray Microanalysis (Academic Press, New York, USA)* ed HH Pattee, VE Cosslett and E Engström, p. 379
[5] Reed SJB 1965 *Brit. J. Appl. Phys.* **16**, 13
[6] Bastin GF, Dijkstra JM and Heijligers HJM 1998 *X-ray Spectrom.* **27**, 3
[7] Goldstein JI, Costley JL, Lorimer GW and Reed SJB 1977 Quantitative X-ray Analysis in the Electron Microscope *Scanning Electr. Microsc.: Proc. Workshop Anal. Electron Microsc. (IIT Research Institute, Chicago, IL, USA, March 1977,* vol 1*)* ed O Johari, p 315
[8] Williams DB and Carter CB 1996 Transmission *Electron Microscopy IV– Spectrometry*, New York: Springer
[9] Walther T 2008 *J. Phys.: Conf. Ser.* **126**, 012090
[10] Buffat PA, Stadelmann P, Garnière JD, Martin D and Reinhart FK 1987 Proc. 5th Microsc. Semicond. Mater. Conf., Oxford, UK, *Inst. Phys. Conf. Ser.* **87**, 207
[11] Sheridan PJ 1989 *J. Electr. Microsc. Tech.* **11**, 41

Journal of Physics: Conference Series **241** (2010) 012017 doi:10.1088/1742-6596/241/1/012017

Quantification of carbon contamination under electron beam irradiation in a scanning transmission electron microscope and its suppression by plasma cleaning

AJV Griffiths and T Walther

Department of Electronic & Electrical Engineering, University of Sheffield, Sir Frederick Mappin Building, Mappin Street, Sheffield S1 3JD, UK

t.walther@sheffield.ac.uk

Abstract. We have measured the build-up of carbon surface contamination as a function of time and irradiated area size for various specimens in a JEOL 2010F (scanning) transmission electron microscope, employing both t/λ mapping with our Gatan imaging filter and recording changes in annular dark-field image intensity. It is shown that the total number of carbon atoms deposited per time for a given beam intensity is roughly constant at room temperature for as-received specimens while it is significantly lower for plasma cleaned specimens. This explains why contamination is generally only an issue at the highest magnifications where the contamination regions become smaller and the carbon layers correspondingly thicker. A Fischione plasma cleaner was then used to remove these carbon layers, and the rate of carbon removal has been determined for contamination spots produced in stationary spot mode as well as for extended regions scanned for a minute so that optimal cleaning times can be chosen.

1. Introduction

The quantitative analysis of scanning transmission electron microscopy data, images as well as spectra, is often hampered by surface contamination during electron irradiation. The gradual build-up of carbon layers on the specimen will reduce the image contrast and increase the signal from C K-edge energy losses and corresponding X-rays, thereby causing artefacts in spectroscopy.

Contamination is caused by the cracking of hydrocarbons present both in the vacuum chamber and on the specimen and their subsequent surface diffusion to the site being irradiated [1,2]. Although our JEOL2010F field-emission gun (scanning) transmission electron microscope (FEG-(S)TEM) is equipped with a completely dry vacuum pumping system that involves only scroll, turbo molecular drag and ion getter pumps but no rotary or oil diffusion pumps, carbon contamination is still an issue when imaging or scanning small regions at high magnifications. For this study, the microscope has been tested just before a bake-out of its column was due to improve vacuum levels. We have calculated quantitatively the build-up of carbon surface contamination as a function of time and irradiated area size for different specimens and holders. A Fischione Model 1020 high-frequency argon/oxygen plasma cleaner was then used to remove these carbon layers, and the rate of carbon removal has been determined so that optimal cleaning times can be determined.

2. Experimental

Samples of C, InAs and InGaAs were investigated in the JEOL 2010F FEG-(S)TEM in scan or spot mode at nominal magnifications between 60kX and 4000kX (4MX), using an electron beam ~0.5nm in diameter and with ~0.1nA current (nominal spot size M, 100-130μA emission, 197kV). The electron beam was either held stationary (spot mode) or raster scanned over square regions (scan mode). In order to facilitate quantitative evaluation, usually several frames of 1min exposures were taken at successively smaller magnifications so that the regions scanned at e.g. 4MX times were then imaged at 1MX times, 250kX and 60kX, revealing a superposition of the corresponding carbon contamination layers, almost square in size if drift was small, in the final image.

In high-angle annular dark field (ADF) imaging, the signal detected depends on the product of the square of the average atomic number of the atoms ('Z-contrast') and the number of atoms (i.e. the thickness) [3], and carbon contamination deposited on any of the free surfaces of a sample will show up as an increase in intensity and a reduction of image contrast from the underlying original specimen material. Figure 1 depicts such a typical ADF image taken at 120kX of an area scanned before at higher magnifications, where the contamination windows produced by the previous scans are clearly visible. From such an image we can measure the sizes of the contaminated areas, as in Figures 1 and 2. Figure 3 shows that at liquid nitrogen temperature carbon is etched away instead of being deposited, for the same sample as before. The inner collection angle for ADF imaging in all cases was ~40mrad.

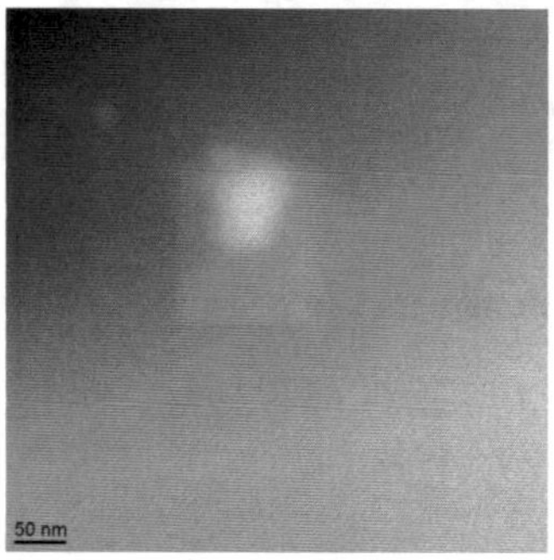

Figure 1. ADF-STEM at 120kX magnification of contamination windows on InAs produced during prior scanning for 1min each at magnifications of 2MX and 500kX. Room temperature

Figure 2. ADF-STEM at 120kX magnification of contamination windows on C test specimen produced during prior scanning for 1min each at magnifications of 1MX, 500kX and 250kX. Room temperature

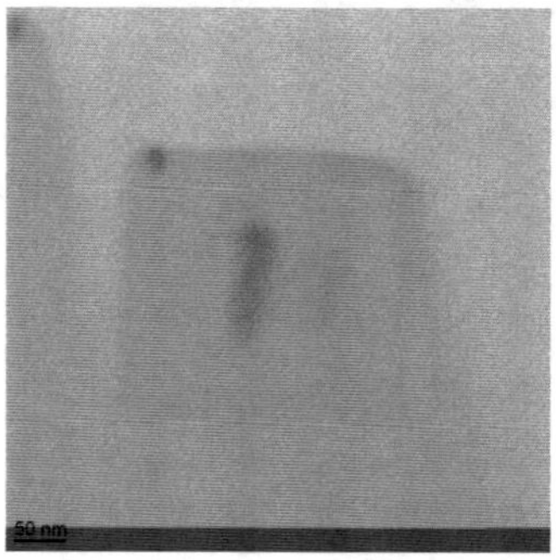

Figure 3. ADF STEM at 120kX magnification of contamination windows on C test specimen produced during prior scanning for 1min each at magnifications of 1MX, 500kX and 250kX. $T = -173$ °C (100K)

3. Results and Discusssion

Figure 4 depicts t/λ maps from an area of InAs with carbon contamination areas and contamination spots after several periods of plasma cleaning at standard conditions (25% O_2, 75% Ar mixture, 13.56MHz, no shield). The gradual disappearance of the carbon contamination can be clearly monitored. λ denotes the inelastic mean free path of the material, which is a measure of the degree of inelastic scattering. As scattering probabilities add up, for several materials on top of each other the mean free paths are additive, i.e. for a stack of i different materials $t/\lambda = \Sigma\, t_i/\lambda_i$ [4]. Absolute thicknesses of carbon overlayers on InAs can thus be determined if the inelastic mean free paths of both materials are known ($\lambda_C \approx 130$nm and $\lambda_{InAs} \approx 84 \pm 13$nm for U=197kV, β=40mrad [4]) and if the InAs thickness is measured. The latter can be determined indirectly by ADF imaging of the clean InAs region at the beginning of the experiment, assuming the intensity scales linearly with the product of thickness, t, and mean atomic number, Z, to the power of some exponent ε: $I_{ADF} - I_0 = ctZ^{\varepsilon}$, where I_0 is the dark current signal offset (background in vacuum without specimen or beam off) and c=const. The proportionality constant can be determined for given ε. For an inner collection angle β=40mrad we used $\varepsilon \approx 1.6$, based on earlier experiments studying ADF image intensities quantitatively as a function of collection angle [5].

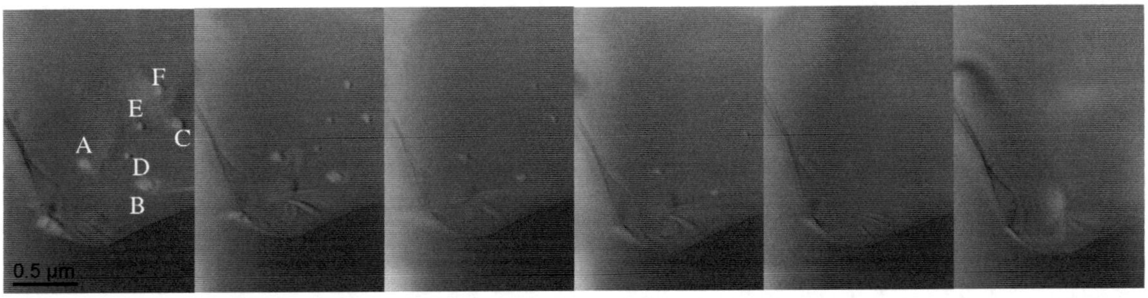

Figure 4. Relative thickness maps obtained of the same InAs area after successive plasma cleaning. The leftmost image shows the condition at the start (*t*=0), the series from second left to right shows maps after 1, 2, 4, 6 and 8 minutes of cleaning. The labels indicate the positions of the contamination windows produced in scan mode (A-C) or pillars (D-F) in spot mode, respectively.

The effectiveness of the plasma cleaner is demonstrated in Figure 5, which plots the carbon thickness, *d*, thus determined on top of the InAs samples as a function of time, *t*. The relationship is clearly exponential and can be fitted by functions of the form $d=d_0 \exp(-\lambda t)$, with time constants of λ^{-1}~2min for flat contamination windows produced during scanning and ~4min for contamination pillars produced in stationary spot mode. We explain this by the difference in aspect ratios: contamination spots have the form of pillars and thus offer a smaller surface for the plasma to etch away the carbon, compared to flat carbon films that have a larger free surface. Our findings are in agreement with previous studies on the same plasma cleaner that indicated 2min [6] to 5min [7] would suffice to remove surface C contamination almost completely.

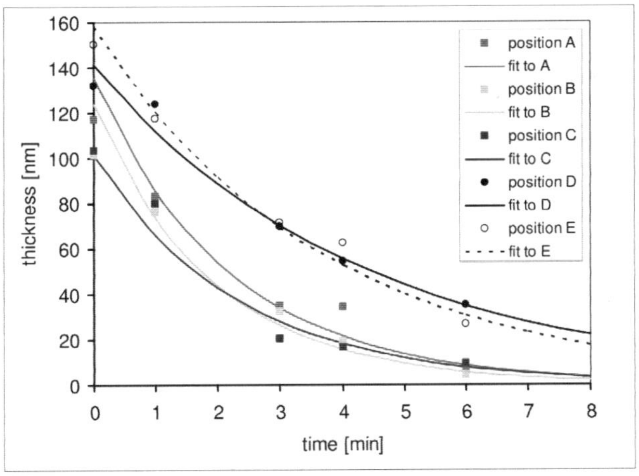

Figure 5. Plot of thickness of remaining carbon deposit as a function of plasma cleaning time. Square symbols (A, B, C) are for flat contamination windows produced in scan mode, round symbols (D, E) represent contamination pillars obtained in spot mode.

position	t_0 [nm]	λ [1/min]	*R*
carbon films (scan mode)			
A	135.3	0.4596	–0.9665
B	123.8	0.5178	–0.9843
C	101.2	0.4290	–0.9796
average	120.1	0.47±0.05	–0.9768
carbon pillars (spot mode)			
D	141.4	0.2323	–0.9938
E	157.9	0.2741	–0.9863
average	149.7	0.25±0.03	–0.9901

Table 1. Statistical evaluation of contamination data from Figure 5. t_0= initial thickness ; λ=inverse time constant for exponential fit; *R*=regression coefficient for exponential fit

The volume of carbon contamination thus determined from the product of lateral extension along *x* and *y* directions and thickness can be converted into atomic numbers, given a molar volume of 5.29cm^3. The number of carbon atoms deposited while keeping the beam scanning for 1min at a given magnification is plotted in Figure 6 and indicates, for uncleaned InAs or InGaAs at room temperature, values around 1-3×10^8 C atoms for both specimens. For a beam current of 0.1nA this implies that ~200 electrons are needed for depositing a single carbon atom. This observation may be explained by the total charge of the electron beam being the relevant parameter for cracking of hydrocarbon bonds. Thus, scanning *n*-times smaller areas at *n*-times higher magnifications will lead to n^2 times thicker carbon layers, which explains the importance of the effect only at highest magnifications. Plasma cleaning the sample for 8min reduces contamination by a factor of ~20, in agreement with [7], i.e. ~4000 electrons would deposit a single carbon atom. An amorphous C film cooled by liquid nitrogen to ~100K, however, is gradually etched away, as reported in [8], and it is interesting to note the etching rate decreases with magnification, as shown in the right part of Figure 6. This confirms the rôle of the free surfaces from where carbon atoms are lost by knock-on processes.

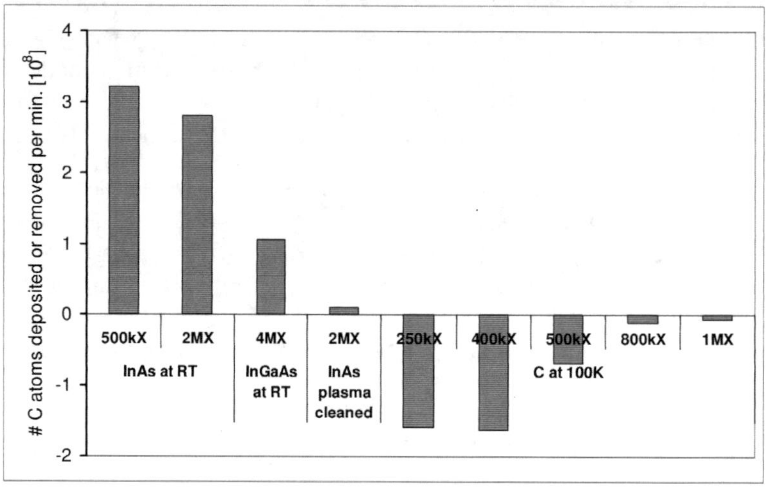

Figure 6. Plot of total number of carbon atoms deposited (positive sign) or etched away (negative sign) by electron irradiation in our JEOL 2010F during 1min scan mode with medium sport size at various conditions (RT=room temperature). Note the only small difference between different specimens (InAs vs. InGaAs), but the large influence of magnification, prior plasma cleaning (for the InAs sample) and specimen temperature.

4. Summary

It has been shown that scanning a small electron beam in FEG-STEM experiments leads to surface carbon contamination, the rate (volume of carbon per electron dose) of which is almost constant. In our JEOL2010F microscope, just before bake-out, ~200 electrons at 197kV yield one C atom of contamination for a normal argon ion milled sample, while plasma cleaning the sample for 8min reduces contamination by a factor ~20. For a given specimen cleanliness, a constant contamination rate explains why carbon contamination is only an issue at high magnifications, as the area scanned is then very small and the contamination film correspondingly thicker. Plasma cleaning reduces the amount of carbon contamination on the surface exponentially as a function of plasma cleaning time, with a time constant for the decay of the remaining carbon thickness of 2-4 minutes.

References

[1] Ennos AE 1953 *Brit. J. Appl. Phys.* **4** 101
[2] Ennos AE 1954 *Brit. J. Appl. Phys.* **5** 27
[3] Walther T and Humphreys CJ 1997 *Inst. Phys. Conf. Ser.* **153** 303
[4] Egerton RF 1996 *Electron Energy-Loss Spectroscopy in the Electron Microscope* (New York: Plenum, 2nd ed.)
[5] Walther T 2006 *J. Microsc.* **221** 137
[6] Isabell TC, Fischione PE 1998 *Mat. Res. Soc. Symp. Proc.* **523** 31
[7] Isabell TC, Fischione PE, O'Keefe C, Guruz MU and Dravid VP 1999 *Microsc. Microanal.* **5** 126
[8] Egerton RF and Rossouw CJ 1976 *J. Phys. D: Appl. Phys.* **9** 659

Effects of functionalization on the targeting site of carbon nanotubes inside cells

Alexandra E. Porter[1*] James S. Bendall[2], Mhairi Gass[3] Karin Muller[4], Jeremy Skepper[4], Paul Midgley[5] and Mark Welland[2]

[1]Dept. Materials, Imperial College London, South Kensington, London SW7 2AZ UK, [2] UK SuperSTEM, Daresbury Laboratory, Daresbury, Cheshire WA4 4AD, UK, [3]The Nanoscience Centre, University of Cambridge, 11 J. J. Thompson Avenue, Cambridge CB3 OFF, [4]Multiimaging Centre, Dept. of PDN (Physiology, Development and Neuroscience, Anatomy Building, University of Cambridge, Downing Street, Cambridge CB2 3DY, UK, [5]Dept. of Materials Science and Metallurgy, University of Cambridge, Pembroke Street, Cambridge CB2 3QZ, UK

a.porter@imperial.ac.uk

Abstract. Functionalized single-walled carbon nanotubes (SWNTs) are currently being investigated for a variety of applications, including contrast agents for medical imaging[1]. However before they can be used commercially it is necessary to assess whether they enter cells, the site they target within the cell and whether they cause any cytotoxicity. Here we characterize uptake of unlabelled, acid-treated, COO^- functionalized SWNTs by human monocyte derived macrophage cells using both low-loss energy loss spectroscopy and compare our findings to previous work on unpurified SWNTs. The acid-treated SWNTs were less aggregated within cells than unpurified SWNTs. Acid treatment was found to affect the distribution of intracellular SWNTs. Bundles, and also individual acid treated SWNTs, were found frequently inside lysosomes, cytoplasm and also inserting into the plasma membrane whereas unpurified non-functionalised SWNTs entered lysosomes and occasionally the nucleus.

1. Introduction

Single walled carbon nanotubes (SWNTs) are currently being explored in medical applications due to their unique structure and properties which make them ideal candidates for drug delivery systems [1]. To be suitable for medical applications, it is necessary first to water solubilise the SWNTs so they are presented to the cell in individual form and second to purify the SWNTs to minimise any cytotoxicity arising from the catalyst nanoparticles. One simple technique to solubilise SWNTs is to reflux and dissolve them via a multi-step acid treatment using both hydrochloric and nitric acid [2]. Nitric acid treatment will alter the surface properties of the SWNTs by generating defects which break the end caps apart and introducing hydrophilic carboxylic acid groups at the tube ends and possibly, at defects on the sidewalls [3]. Therefore it is important to assess the uptake and cytotoxicity of acid-treated SWNTs as they are the precursors for the synthesis of a number of other functionalizations of SWNTs.

Direct imaging of unlabelled, individual SWNTs by electron microscopy is very challenging, because it is difficult to distinguish carbon based nanotubes from carbon-rich organelles due to similarities in their chemical composition and dimension. Visualizing SWNTs inside cells will help us understand how SWNTs enter cells, where they migrate to, and their fate after uptake. Previously we exposed human monocyte-macrophages to unpurified HiPco SWNTs and demonstrated that a combination of low-loss energy-filtered transmission electron microscopy (EFTEM) combined with electron energy loss (EEL) spectrum imaging—a method that yields characteristic energy-loss information—enables clear differentiation of SWNTs in the cell [4]. These techniques gave a good image contrast of unstained sections that cannot be achieved using conventional imaging techniques. These images confirmed that unpurified SWNTs localise primarily as bundles within lysosomes, but were also seen to enter the cytoplasm and localize within the cell nucleus. They were seen to cause cell mortality in a dose dependent manner. Visualizing acid treated COO^- functionalised SWNTs within cells will be particularly challenging as the SWNTs will be more disperse than pristine untreated SWNTs.

The goal of this study will be to characterize and visualise acid-treated water soluble SWNTs alone and within human monocyte derived macrophage cells (HMMs). HMMs were chosen as a model as this cell type comprises the first line of defence against foreign organisms or particles and are central to the body's immune responses. Specifically we aim to establish the sites they target within the cell and the uptake mechanism, relating the difference in uptake to changes in the chemistry of the SWNT walls.

2. Materials and Methods

2.1. Materials

SWNTs were purchased from CNT@Rice (HiPco as-prepared). The nanotubes underwent a purification procedure to remove any unwanted non-tubular carbon species and catalyst particles. The nanotubes were initially refluxed in concentrated nitric acid (70%) for 3 hours, filtered through a PTFE membrane with a pore size of 200nm, washed and dried and heated in air for 40 minutes at 380°C. These steps were effective at releasing the catalyst particles from their carbon shells and oxidising them. The sample was then refluxed for 6 hours in concentrated HCl (33%) after which they were filtered through a 200nm pore PTFE membrane, washed and dried. TEM characterisation of the sample showed that the purification procedure was effective at removing the catalyst particles and amorphous carbon material, leaving the nanotubes themselves intact. Raman spectroscopy confirmed that the acid treated SWNTs had a greater number of functional groups on the carbon walls than non-treated SWNT.

2.2. Transmission electron microscopy

TEM studies of SWNTs nanotubes in cells were performed at a concentration of 5 μg/ml for 4 days in accordance with previous work [4]. Purified SWNT were non-cytotoxic after 4 days exposures to HMMs at concentrations ranging from 0-20μg/ml. To establish the mode of cellular uptake, cells were exposed to acid treated SWNTs at both 4°C and 37 °C for 4 hours.

High-resolution imaging, high angle annular dark field scanning transmission electron microscopy (HAADF-STEM) and EELS were performed at the UK SuperSTEM laboratory on a 100kV VG HB501 dedicated STEM fitted with a Nion second generation spherical aberration corrector and a Gatan Enfina electron-energy loss spectrometer. The convergence semiangle of the electron probe was 24 mrad for both imaging and spectroscopy. The collection semiangle for the EELS was 4 mrad and 70 – 210 mrad for HAADF imaging. SuperSTEM enables both BF and HAADF images to be acquired simultaneously. Low-loss EELS spectrum images were acquired at every pixel over a used-defined area with an energy dispersion of 0.1eV per channel, covering an energy range of ~0eV to ~100eV.

3. Results and Discussion

High magnification lattice imaging confirmed the graphitic structure of the carbon nanotubes had been preserved by acid treatment and that the iron catalyst had been removed. To confirm the intracellular localisation of individual SWNTs, EEL spectra were acquired from HMMs exposed to SWNTs using unstained cell sections [2]. Medium resolution HAADF-STEM appeared to show SWNTs bundles within the cell (Fig. 1a). Low loss EEL spectrum images were taken from the boxed-out region in Fig. 1c.

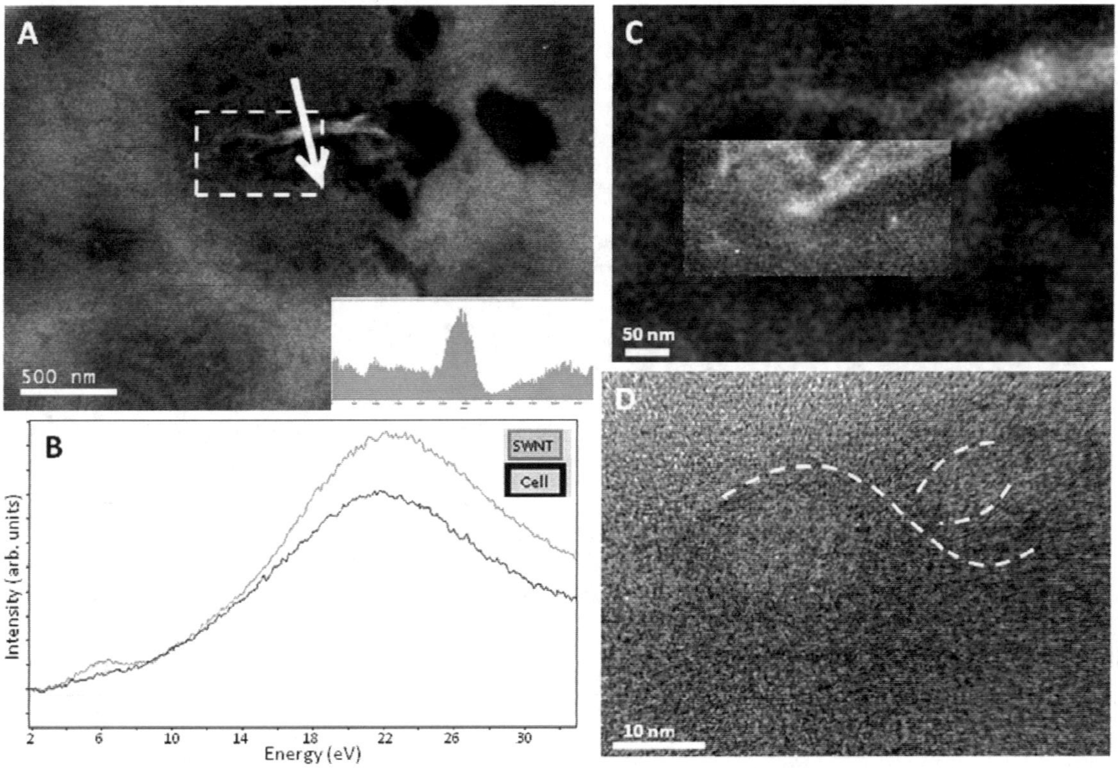

Figure 1 (a) HAADF-STEM image of SWNT bundles within the cytoplasm of an unstained cell after 4 days exposure. The white box shows area analyzed by low-loss EELS spectroscopy. The inset shows an intensity profile taken along the direction of the arrow, the increase in intensity at the location of the SWNTs is evident. Low-loss EELS spectra for SWNTs in the cell (green/light) and the cell (grey) are shown in (b). The π to π^* transition at ~6eV is seen clearly only in the spectrum taken from the SWNT bundle, it also exhibits a higher bulk plasmon energy. c) Map showing the change in the plasmon peak position. Range from 17.5eV (blue/ dark) to 21.5eV (white). Contrast due to SWNTs can be seen that is not identified in conventional imaging. d) High resolution lattice image of intracellular SWNTs. Some of the SWNTs have been indicated with a white dotted line as the contrast is weak.

A low-loss EELS spectra for SWNTs in the cell and the cell are shown in Fig. 1b.The map of plasmon energy was created by fitting a Gaussian to the (σ+π) bulk plasmon peak [5] in the aligned spectrum image using Digital Micrograph. The energy of the peak was extracted from the Gaussian fit. The position of the plasmon maximum ranged from 17.5eV from the cell to 21.5eV from the intracellular SWNTs. Hence, by mapping the peak position (Fig. 1c), enhanced contrast due to SWNTs was possible. The observed energy shift is small compared to the individual values because the nanotubes are within the cellular structure resulting in a plasmon containing information from both the SWNTs and cell. Although the measured shift is small, the stability of the cold-FEG tip combined with the high energy dispersion means that it is still possible to extract such information. The resolution of such a fit has previously been studied by Sanchez et al [6] using an Enfina spectrometer on a VG 601 STEM where the standard deviation of the plasmon energy from 800 spectra (GaAs material) was 7meV. As the same spectrometer is used in this work, and the Cs corrected VG 501 has a very high stability it is concluded that a shift in plasmon energy in the order of 1eV can confidently be observed. High resolution lattice imaging also confirmed the presence of SWNTs within the cell (Fig. 1d).

The mechanisms by which SWNTs enter cells are not fully understood. Confocal microscopy studies and computer simulations have indicated that SWNTs could either traverse the cellular membrane via endocytosis [7] or insert into and diffuse through the lipid bilayer [8]. Low loss EEL spectrum imaging confirmed that SWNTs interacted with, and possibly inserted into, the plasma membrane of macrophage cells (Fig. 2).

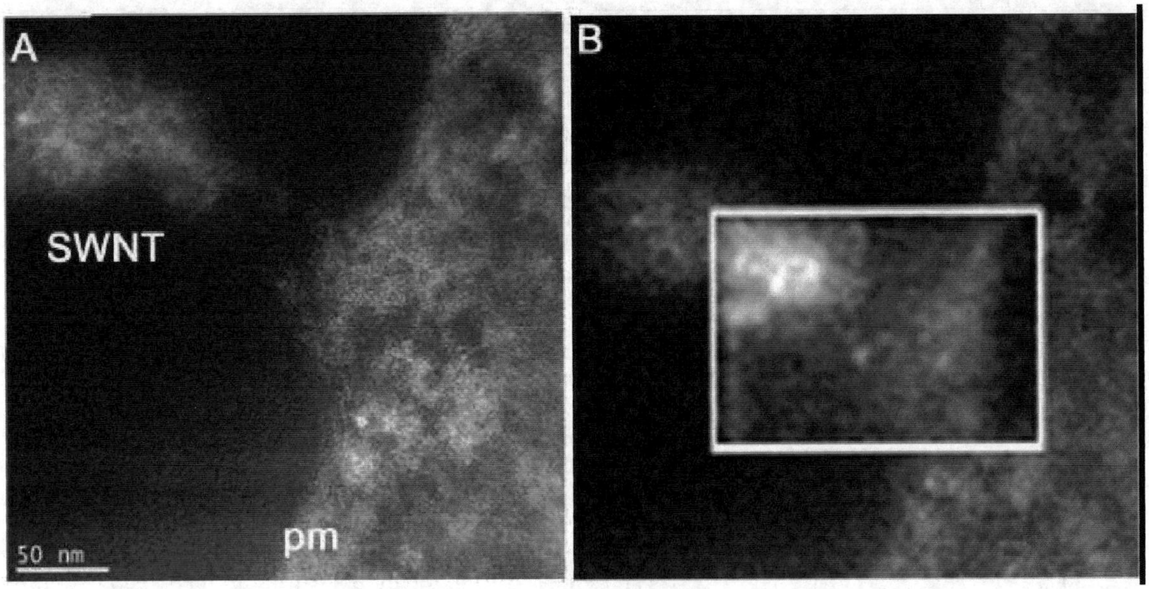

Figure 2 Bundles of SWNTs fusing with the plasma membrane of a cell in unstained cell sections after 4 days exposure. (a) A bundle of SWNTs fusing with the plasma membrane of the cell causing membrane disruption (HAADF STEM, stained section). (b) Plasmon HAADF image of SWNT fusing the plasma membrane of unstained cell. Range from 17.5eV (blue/dark) to 21.5eV (white).

Imaging intracellular SWNTs after 4 hours at 4°C was found to be very challenging. One simple method to enhance contrast from SWNTs inside cells is to fill them with heavy metal halides as described in ref.9. This method completely encapsulates the $GdCl_3$ inside the SWNTs, does not introduce functional groups onto the surface and also removes toxicity issues [10, 11]. Therefore cells were also exposed to $GdCl_3$ filled SWNTs (Gd-SWNTs [12]. After 4 hours at 37°C, individual Gd-SWNTs inserted into the plasma membrane of the cell (Fig. 3a,b). Notably, the Gd-SWNTs entered the cell parallel to the plasma membrane and then entered the cytoplasm with a cork-screw morphology (Fig. 3b). No evidence of cell penetration by Gd-SWNTs was observed at 4°C, suggesting that cellular uptake of the Gd-SWNTs was only active and not passive.

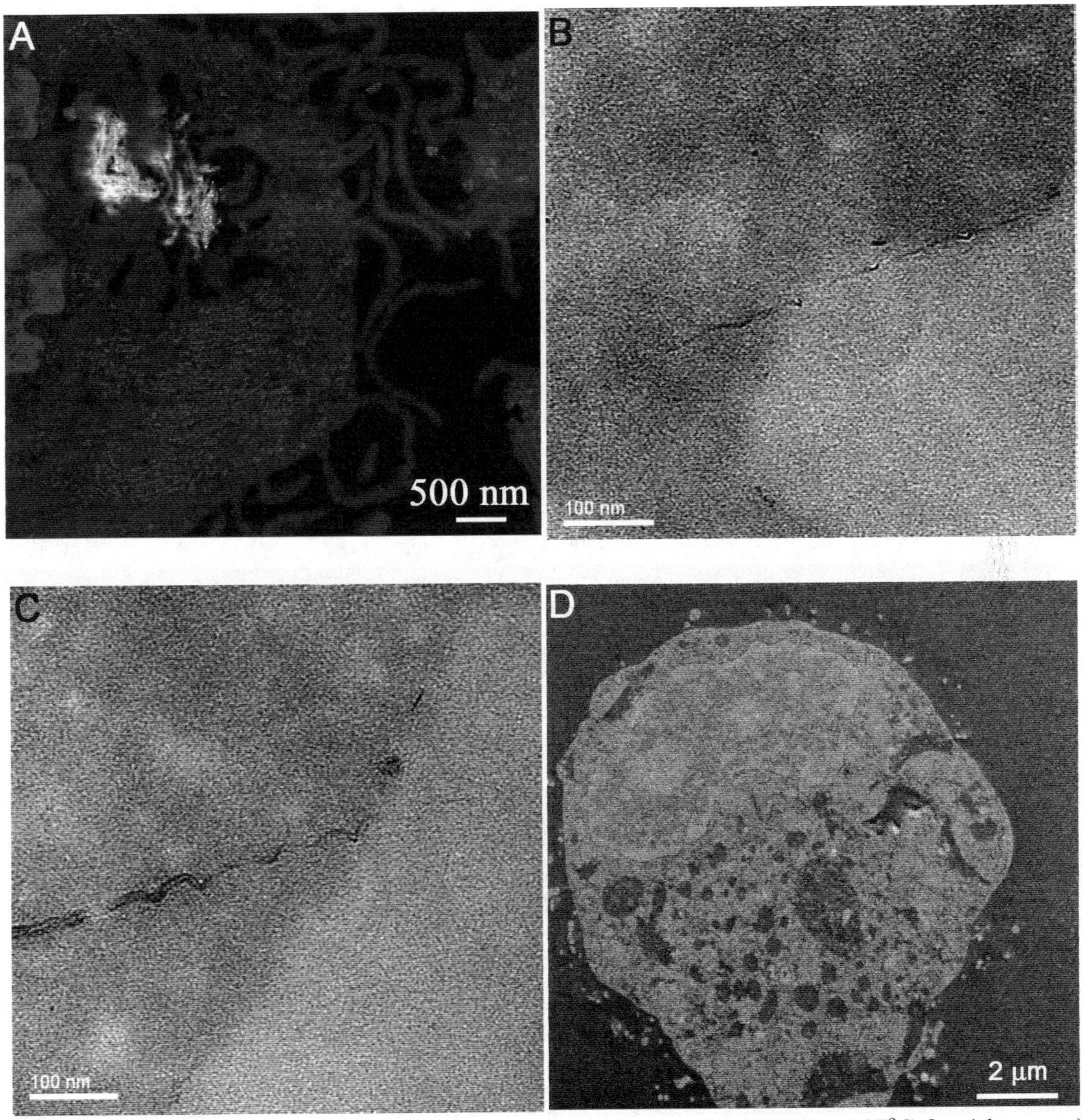

Figure 3 Human monocyte derived macrophages exposed to Gd-SWNTs at 37°C for 4 hours. a) Individual Gd-SWNTs inserting into the plasma membrane of unstained cells. b) Individual Gd-SWNTs inserting parallel to the plasma membrane of an unstained macrophage cell then entering the cytoplasm with a "cork-screw" morphology.

In summary, we were able to directly visualise acid treated SWNTs entering different intracellular sites than untreated SWNTs [4], where negatively charged SWNTs localise more frequently within the cytoplasm of the cell. SWNTs entered both cell lysosomes and also inserted into the plasma membrane. This platform will be used in future investigations to guide studies to assess uptake of more medically relevant classes of functionalized SWNTs being developed as vehicles for the delivery of therapeutic and diagnostic agents.

4. References

[1] Kam, NW, O'Connell, M, Wisdom, JA, Dai, H. *Carbon nanotubes as multifunctional biological transporters and near-infrared agents for selective cancer cell destruction.* Proc. Natl. Acad. Sci. USA 2005. **102**(33): p. 11600-5.

[2] Kajiura, H, Tsutsui, S, Huang, H, Murakami, Y. *High-Quality Single-Walled Carbon Nanotubes from Arc-Produced Soot.* Chem. Phys. Lett. 2002. **364**: p. 586-592.

[3] Banerjee, S, Hemraj-Benny, T, Wong, S. Covalent Surface Chemistry of Single-walled carbon Nanotubes. *Adv. Mater.* 2005. **75**: p.17-29.

[4] Porter, AE, Gass, M; Muller K, Skepper JN; Midgley PA, Welland M. *Direct Imaging of Single-Walled Carbon Nanotubes in Cells.* Nat. Nano. 2007. **2**: p. 713-7.

[5] Saito, Y., Shinoara, H, Ohshita, A. *Bulk Plasmons in Solid C_{60}.* Japanese Journal of Applied Physics. 1991. **30:** p. 1068-1070.

[6] Sanchez, AM, Beanland, R, Gass, MH, Papworth, AJ, Goodhew, PJ, Hopkinson M. *Mapping quantum dop-in-well structures on the nanoscale using the plasmon peak in electron energy loss spectra*. Physical review B 2005. **72** (7): p. 075339.

[7] Kostarelos, K, Lacerda, L, Pastorin, G, Wu, W, Wieckowsk, S, Luangsivilay, J, Godefroy, S, Pantarotto, D, Briand, JP, Muller, S, Prato, M, Bianco, A. *Cellular uptake of functionalised carbon nanotbues is independent of functional group and cell type. Nat. Nano.* 2007. **2**: p.108-113.

[8] Lopez, CF, Nielsen, SO, Moore PB, Klein ML. *Understanding nature's design for a nanosyringe. Pro. Natl Acad Sci* USA. 2004. **101(13)**: p. 4431-4.

[9] Brown, G. *et al. High yield incorporation and washing properties of halides incorporated into single walled carbon nanotubes.* Appl. Phys. A 2003 **76**: p. 457–462.

[10] Horak, D, Cervinka, M, and Puza, V. *Radio-opaque poly(2-hydroxyethyl methacrylate) particles containing silver iodide complexes tested on cell culture.* Biomaterials 1998. **19**: p. 1303–1307.

[11] Sloan, J, Kirkland, AI, Hutchinson, JL and Green, MLH. *Aspects of crystal growth within carbon nanotubes.* C. R. Phys. 2003 **4**: p. 1063–1074.

[12] Miyawaki J et al. *Biodistribution and Ultrastructural Localization of Single-Walled Carbon Nanohorns Determined In Vivo with Embedded Gd2O3 Labels.* ACS Nano, 2009 **3**(6): p. 1399–1406.

Microscopy at the life sciences / physical sciences interface

Paul D Brown, [1]Hannah K Edwards & *Mike W Fay

Department of Mechanical, Materials and Manufacturing Engineering;
Materials, Mechanics & Structures Research Division, Faculty of Engineering;
*Nottingham Nanotechnology & Nanoscience Centre;
University of Nottingham, University Park, Nottingham, NG7 2RD, UK.

paul.brown@nottingham.ac.uk

Abstract. Procedures for investigating biomedical implant surfaces and biological cell / surface interactions in plan-view and cross-sectional geometries are described. Ultramicrotomy, FIBSEM and cryogenic-FIBSEM techniques, as developed for the controlled cross-sectioning of human osteoblasts on Ti substrates for TEM investigation, are compared.

1. Introduction

The four basic design considerations for load bearing biomedical implants are biocompatibility, corrosion resistance, tribological performance and the matching of mechanical properties. Such design considerations have led to many exciting advances in ceramic, polymeric or composite-based biomaterials in recent years, although it is recognised that the vast majority of biomedical implants are still metallic-based, covering a wide range of sites and time dependencies in the body. In particular, Ti-based alloys are made interesting due to their beneficial osteo-conductive properties, with the development of native passivating oxide or hydroxide surfaces that continue to grow within the body, helping to provide protection against metal ion release [1]. Such alloys may be used within orthopaedic devices such as hip and knee joints and bone fixation plates, along with heart valves, maxillofacial implants and dental implants, and even though there is a huge industry dedicated to the application of biomedical implants, better understanding of how living organisms interact with foreign material is still required, in order to progress the development of biomaterials with appropriately designed surfaces.

The process of introducing a surface-modified, foreign material into the body often seeks to promote controlled cell differentiation and proliferation, with the aim of stimulating rapid tissue repair. However, it is recognised that body tissue combines many functions, *e.g.* the partitioning of biochemical reactions, the transmission of mechanical loads and neurological signals, and the provision of energy sources, in addition to incorporating mechanisms for functional repair. The cellular level, in particular, introduces complicated associations of proteins, nucleic acids, polysaccarides and lipids that adopt 3D conformations according to the local chemical environment. In this context, the initial stage of bio-integration is concerned with the sequential process of protein layer adsorption, biomolecule absorption, cell interaction and tissue integration. However, close

[1] Now at Lockheed Martin, Ampthill, Bedford, UK

a) b) c)
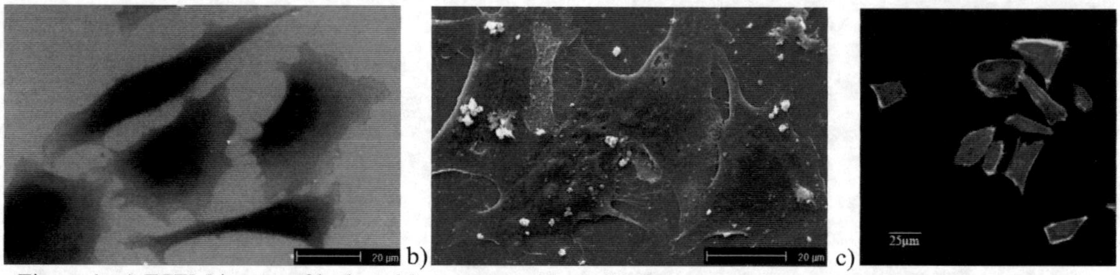

Figure 1: a) ESEM image of hydrated human osteoblasts (HObs) cultured on a TiO_2 anatase surface and b) the equivalent sample after dehydration and imaging in high-vacuum SEM. The samples were fixed after 6 hours of incubation. c) CLSM image of HObs cultured on a TiO_2 anatase surface fixed with gluteraldehyde and labelled with FITC conjugated phalloidin and propidium iodide after 6 hours of incubation, revealing details of the actin cytoskeleton.

apposition of tissue to an implant followed by a certain degree of fibrous encapsulation may lead to inflammation, tumour formation and possible tissue necrosis. Thus, improved understanding is also required of the combined effects of biodegradation and toxicology, in the context of biomaterial design and construction, along with clinical aspects of surgery and sterility, to make best sense of the biological response to an implant. The challenge is to tailor the implant with appropriate levels of bioactivity (or bioinertness) depending on the intended application.

By way of example, research on the controlled modification of Ti-based alloy surfaces has shown clear differences in the biological cell response to surface chemistry [2]. The surface modification process of shot peening (grit blasting), presently favoured by the bioengineering industry, can lead to the amorphisation of surfaces, in turn offering better corrosion resistance behaviour [3]. Further, simple heat treatments in air can be effective in reducing metal ion release, *e.g.* reducing platelet activation or the oxidative stress of endothelial cells [4]. Other approaches, such as highly polished surfaces, can have advantages in terms of sterilisation but may lead to other problems such as higher initial corrosion behaviour. Adding hard ceramic layers to prevent metal ion release is also an option.

In this context, there is need for an appropriate combination of complementary techniques to characterise (i) the interaction of biological cells with implant materials (in planar-view geometry), in terms of cell morphology, motility and viability (complementing genetic marker and cell assay-based investigations); (ii) the surface modified implant material in terms of morphology, crystal structure and chemistry in advance of introduction to the biological environment and *in vitro* (or *in vivo*) cell trials; and ideally (iii) the interfaces between biological cells and biomedical implant materials (in cross-sectional geometry), on the molecular scale. Here, we comment briefly on these various approaches to inform the design and development of biomaterials and biomedical coatings.

2. Overview of characterisation techniques

2.1 Planar view investigation of cells on surfaces

The combined techniques of environmental scanning electron microscopy (ESEM), high-vacuum SEM and confocal laser scanning microscopy (CLSM) provide valuable information on cell morphology and the internal actin distribution. ESEM investigation, in particular, when performed in environmental (wet) mode, allows the morphology of hydrated cells (Figure 1a) to be assessed in relation to the local chemistry of a material surface without the extraneous effects of the dehydration process (Figure 1b). The operation of the ESEM naturally creates a conducting pathway in the vicinity of an insulating material under examination due to exploitation of the 'poor vacuum' ambient and this obviates any concerns about charging effects during sample observation and analysis [5]. For the purpose of CLSM, fixed cells may be labelled using antibodies against vinculin to visualize focal adhesions, and FITC tagged phalloidin to visualize F-actin filaments, thereby giving an indication of the organisation of the cytoskeleton (Figure 1c).

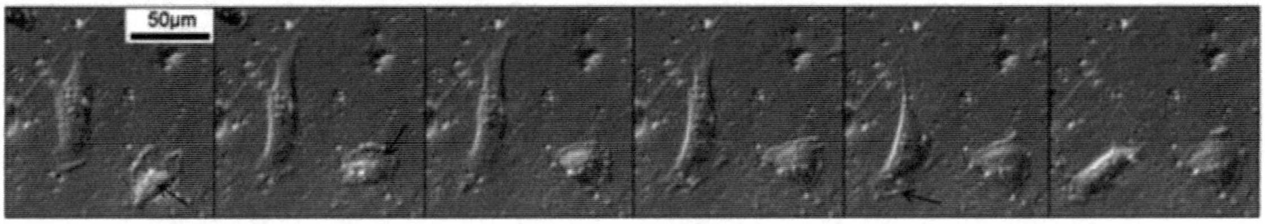

0 min 20 min 40 min 60 min 80 min 100 min

Figure 2: Series of time lapse DIC images showing the motion of two HObs on cp-Ti. The arrows label a main cell body (0 min), a pseudopod (20 min) and ruffles / blebs (80 min).

Conversely, reflectance differential interference contrast (DIC) microscopy may be used for time lapse microscopy of cell motility on opaque surfaces (Figure 2), recognising that migration is essential for the colonisation of biomaterial scaffolds in tissue engineering and of orthopaedic implants. From images of viable cells crawling on a biomaterial surface, two types of sustained motion are generally observed: i) a steady gliding movement where the various crawling processes take place simultaneously and ii) a two-step movement where a slow lamellipod extension occurs, followed by adhesion and then translocation, detachment and retraction, in accordance with the cell crawling models of Abercrombe [6] and Alt and Dembo [7].

These combined planar-view techniques are ideally suited to the investigation of cell-surface associations, given that the process of cell migration over surfaces and through tissue is fundamental to many biological phenomena. Indeed, this approach has been used to inform the development of a mathematical model of cell crawling activity [8].

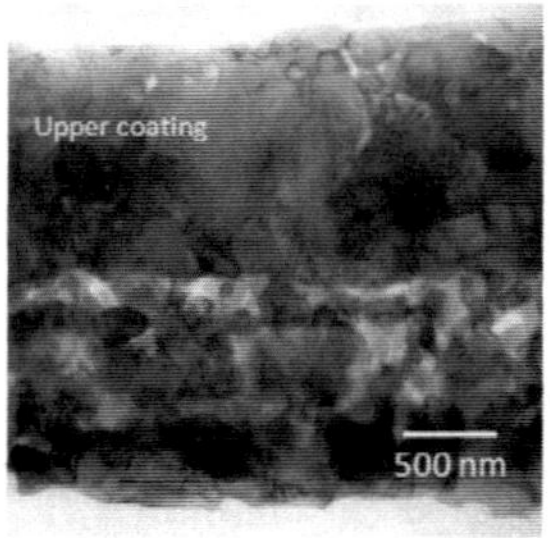

3)

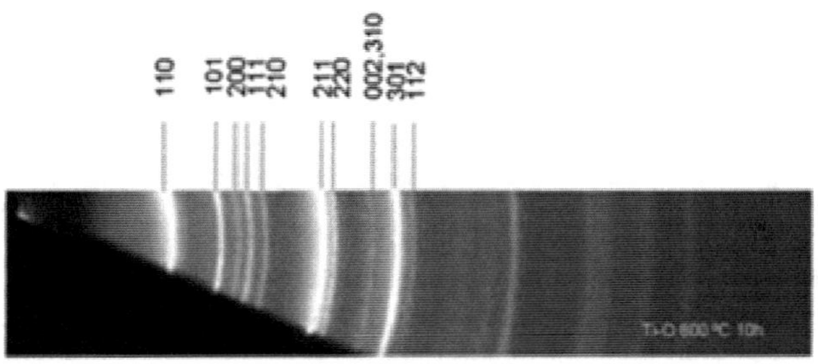

4)

Figure 3: Bright field diffraction contrast TEM image from a cross-sectioned sample of oxidised Ti, annealed at 900 °C, showing the development of the TiO_2 rutile phase and revealing a buried interface consistent with the line of the original Ti surface.

Figure 4: Section of a RHEED pattern taken from a Ti sample annealed in air for 10 h at 600 °C, showing well defined polycrystallinity. Diffraction spacings corresponding to the TiO_2 rutile phase are indicated.

2.2 Characterisation of biomaterial surfaces and surface modified implants

Ideally, the biomedical implant surface should be fully characterised in terms of morphology, crystal structure and chemistry, in advance of cell seeding and *in vitro* cell trials. Hence, it is recognised that SEM, atomic force microscopy (AFM) and profilometry techniques are valuable for providing insight into surface morphology and roughness, whilst reflection high energy electron diffraction (RHEED) and X-ray photoelectron spectroscopy (XPS) are ideal for the assessment of the surface crystal structure and chemistry, respectively, complementing X-ray diffractometry (XRD) and energy dispersive X-ray (EDX) analysis in the SEM that provide for more 'bulk' information on the sample structure and chemistry. Given that the surface structure of the biomedical implant is very often graded, *e.g.* as a consequence of the surface modification procedures applied during processing, the technique of cross-sectional transmission electron microscopy (TEM) potentially provides valuable insight into the nature of the surface microstructure being presented to the biological environment. For example, Figure 3 shows the TiO_2 rutile scale developed at the surface of a mechanically polished

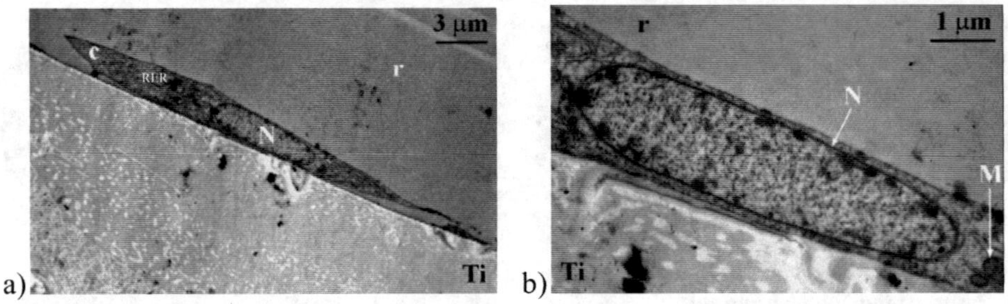

Figure 5: (a,b) Bright field TEM images of a HOb/Ti-foil sample prepared by ultramicrotomy containing a cell (c), Ti foil (Ti) and resin (r). The cell ultrastructure (nucleus (N), mitochondria (M) and rough endoplasmic reticulum (RER)) is rendered visible by staining with osmium tetroxide and uranyl acetate.

Ti sample following annealing at 900°C. The limitation, as ever, is the time consuming nature of the conventional sample preparation procedures of sequential mechanical polishing and low angle Ar ion milling, for cross-sectional TEM investigation. Accordingly, attention is drawn to the little used technique of RHEED that can be used to characterise rapidly the near surface crystal structure of a biomaterial. The RHEED technique is sensitive to the top few nanometres and hence is found to be very useful for the rapid assessment of surface modified titanium, allowing for a 'process-map' for thermal oxidation to be constructed, in advance of more detailed TEM investigation. For example, the RHEED pattern of Figure 4 confirms the development of very fine grained TiO_2 rutile phase at the surface of a Ti sample annealed in air for 10 h at 600°C.

2.3 Cross-sectional TEM investigation of biological cells on metallic implants

Extending this investigation process to the appraisal of the biological cell / metallic implant interface constitutes a significant technological challenge, given the very different mechanical properties of the material either side of the interface. For the purpose of cross-sectional TEM investigation, the challenge is to retain representative material from the interfacial region, free of artefacts, for the purpose of detailed analytical investigation. In this context, the techniques of ultramicrotomy, and conventional and cryogenic focused ion beam scanning electron microscopy (FIBSEM) have been developed for the controlled cross-sectioning of model system implant metallic and ceramic templates, exposed *in-vitro* to human osteoblasts (HObs) [9].

Cells cultured on Ti-foil substrates were fixed, dehydrated, stained and embedded in Araldite-Epon resin for the purpose of sectioning using a Reichert Jung Ultracut microtome. Conversely, two different sample geometries were employed for the sectioning of cell/Ti disc or foil samples using an FEI Quanta 200 3D FIBSEM, featuring a Quorum Technologies PP2000T cryo-transfer unit, an Omniprobe micromanipulator and an Oxford Instruments INCA energy dispersive X-ray (EDX) analysis system. Conventional FIBSEM site-specific 'lift-out' procedures were applied to fixed, dehydrated, stained and embedded cell/Ti samples, whilst 'H-bar' procedures were used for the cryogenic FIBSEM preparation and investigation of fixed, stained and hydrated cell/Ti-foil 'edge-on' samples. The details of the experimental procedures used are given by Edwards *et al* [9]. In particular, the use of a cryogenic-transfer system allows hydrated biological specimens to be rapidly frozen in slush liquid nitrogen, thereby stabilising the specimen for subsequent imaging and milling at low temperature [10]. Electron transparent specimens were examined using an FEI Tecnai 12 Biotwin TEM (100 and 120 kV) and a Jeol 2100F field emission gun TEM (100 and 200 kV) for which a Gatan liquid nitrogen cooled sample holder was employed for the support of both FIBSEM and cryo-FIBSEM sectioned specimens.

Figures 5a,b present bright field TEM images of an ultramicrotomed HOb on a Ti-foil substrate, viewed in cross-section. The cell ultrastructure is clearly visible, with high contrast, and the cell morphology has been well maintained. Mitochondria (M), rough endoplasmic reticulum (RER) and cell nucleus (N), may be discerned within the cell.

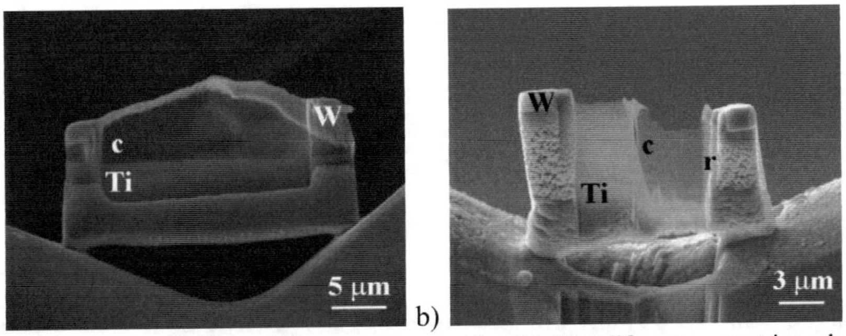

Figure 6: Secondary electron images of the two site-specific FIBSEM lift-out geometries adopted. (a) A 'top-down' HOb/Ti lift-out section (cultured for 5 days) containing cell (c), Ti and W; and (b) a 'side-on' HOb/Ti lift-out section (cultured for 21 days) containing cell (c), resin (r), Ti and W, both attached to TEM support grids.

Figure 6a shows a thinned area of a HOb attached to a Ti substrate during the preparation of a 'top down' lift-out section, whilst Figure 6b shows a HOb/Ti specimen attached to a TEM support grid following 'side-on' lift-out. The 'top down' geometry favours the lift-out of a single cell, whilst the 'side-on' lift-out approach produces a section containing resin as well as cells, favouring the preparation of specimens containing multilayered cells.

7)

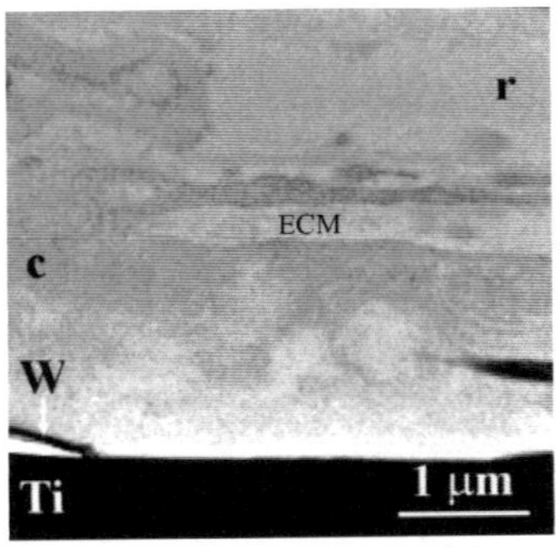

8)

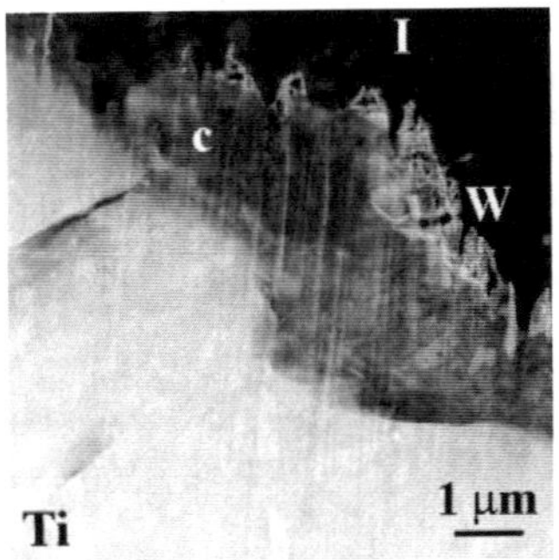

Figure 7: Bright field TEM image of embedded multilayered HObs (cultured for 21 days) on a Ti-foil prepared through FIBSEM 'side-on' lift-out, showing cells (c), collagenous extra-cellular matrix (ECM), resin (r), Ti and artefact W at a point of delamination between the cell and Ti.

Figure 8: HAADF image of a HOb/Ti electron transparent section produced through cryo-FIBSEM milling showing the morphology of the cell (c) with respect to the Ti-foil, limited detail within the cell and milling artefacts across the specimen and illustrating the build up of ice crystals formed during transfer of the specimen from FIBSEM to TEM (I).

Figure 7 presents a bright field TEM image of a 'side-on' lift-out section containing HObs cultured for 21 days on a Ti-foil. In this instance, interpretable contrast attributable to the cell ultrastructure and collagenous extra cellular matrix (ECM) can be discerned, and the region of the cell/Ti interface has been preserved and although the cell appears to exhibit intermittent contact with the Ti substrate.

Figure 8 presents an HAADF image a HOb/Ti-foil specimen (cultured for 21 days) produced through cryo-FIBSEM milling in the H-bar configuration. A pronounced curtaining effect was found to be more strongly associated with hydrated specimens prepared through using cryo-FIBSEM procedures as compared with conventional FIBSEM techniques applied to dehydrated specimens, being attributed to the non-uniformity of the gas injection system (GIS) deposited W coating at cryogenic temperatures. By reducing the GIS temperature and carrying out metal deposition in the absence of the ion beam [10], superior coatings with reduced porosity were achieved.

By way of general overview, it was found that conventional ultramicrotomy was capable of producing cross-sections with a large field of view and high contrast, with the sectioning of cell/Ti-foil specimens embedded in resin, but significant problems, *e.g.* with cell detachment, using this technique were found to restrict its general applicability. Conventional FIBSEM 'top-down' and 'edge-on' lift-out procedures were found to be effective for the preparation of uniform sections of fixed and dehydrated cell/Ti specimens, but the control of cell staining was found to be an issue. Cryogenic-FIBSEM procedures used with an 'H-bar' sample geometry enabled hydrated cell/Ti-foil specimens to be sectioned, but issues remain over ion beam induced artefacts. This was due in part to practical problems associated with the ion beam deposition of protective W coatings under cryogenic conditions, leading to the 'curtaining' effect. Further, refined procedures for the complete inhibition of ice formation on such specimens during cryo-transfer are also required.

3. Summary

Procedures for the characterisation of biological cells on biomedical implant materials have been described. In particular, ultramicrotomy, FIBSEM and cryogenic FIBSEM techniques, as developed for the controlled cross-sectioning of human osteoblasts on titanium substrates for TEM investigation, have been compared. Each method was found to be viable, enabling the initial response of biological cells on hard metallic implants to be investigated, but technical challenges still remain for the refinement of these techniques. It is considered that the fundamental understanding gained through the TEM assessment of such sectioned biological cell / engineering alloy specimens will have a significant effect on the design and development of future biomaterials and biocompatible coatings.

References

[1] Long M and Rack HJ (1998) Biomaterials **19** 1621
[2] Ryhanen J, Niemi E, Serlo W, Niemela E, Sandvik P, Pernu H & Salo T (1997) J. Biomed. Mat. Res. **35** 451
[3] Plant, SD, Grant DM and Leach L (2005) Biomaterials **26** 5359
[4] Armitage DA, Parker TL and Grant DM (2003) J. Biomed. Mat. Res. **66A** 129
[5] McKinlay KJ, Scotchford CA, Oliver J, King J, Grant DM, Wood JV and Brown PD (2004) J. Biomedical Materials Research **69A** 359
[6] Abercrombie M (1980) Proc. R. Soc. Lond. (B) **207** 129
[7] Alt W and Dembo M (1999) Mathematical Biosciences **156** 207
[8] Oliver JM, King JR, McKinlay KJ, Brown PD, Grant DM, Scotchford CA and Wood JV (2005) Mathematical Medicine and Biology **22** 53
[9] Edwards HK, Fay MW, Anderson SI, Scotchford CA, Grant DM & Brown PD (2009) J. Microscopy **234** 16
[10] Hayles M, Stokes D, Phifer D and Findlay K (2007) J. Microscopy **226** 263.

Acknowledgements

The author would like to acknowledge the EPSRC for funding under grants GR/M74603/01, GR/R04744/01 and EP/E015379/1. With thanks to Karen McKinlay, James Oliver, John King and John Wood for the cell/surface motility, ESEM and modelling work; Spyridoula Marlafeka, Yisong Han, Robin Walton and Tianfeng Tao, for the RHEED and cross-sectional TEM investigations of surface modified Ti; Marie Smith and Denise Christie and Sue Anderson for the preparation of ultramicrotomy sections; David Grant and Colin Scotchford for many technical discussions; and Martin Roe, George Anderson, Nigel Neate, Nikki Weston, Sam Coe, Julie Thornhill and Keith Dinsdale in the Department of M3 for technical support.

Imaging Alzheimer's disease-related protein aggregates in human cells using a selenium label

E K McGuire[1], M Motskin[2], T P J Knowles[3], C M Dobson[4], D W McComb[1] and A E Porter[1]

[1] Dept. of Materials, Imperial College, Exhibition Rd, London SW7 2AZ, UK
[2] Dept. of Anatomy, University of Cambridge, Downing St, Cambridge CB2 3DY, UK
[3] Nanoscience Centre, University of Cambridge, JJ Thomson Ave, Cambridge, CB3 0FF, UK
[4] Dept. of Chemistry, University of Cambridge, Lensfield Rd, Cambridge CB2 1EW, UK

Email: e.mcguire07@imperial.ac.uk

Abstract. The aberrant folding and subsequent aggregation of proteins and peptides is associated with a range of pathological conditions from the systemic amyloidoses to neurodegenerative diseases such as Alzheimer's and Parkinson's diseases. While this link is well established there is a lack of understanding of the exact role protein aggregates play in disease pathogenesis. Part of the reason for this is that it has proved extremely challenging to characterise the localisation and structure of amyloid fibrils within the cellular environment due to a lack of contrast between the carbon rich protein aggregates and the carbon rich cell. We report a novel method for visualising Alzheimer's disease-related amyloid fibrils inside human cells without the use of invasive or unreliable stains or tags. The naturally occurring sulfur atom in the amyloid-β peptide is replaced with a selenium atom, a heavier element in the same group of the periodic table of elements. Using high angle annular dark field (HAADF) in a scanning transmission electron microscopy (STEM) the selenium-labelled aggregates can be identified within the cellular environment.

The aberrant folding and subsequent aggregation of proteins and peptides into insoluble plaques known as amyloid fibrils[1, 2] is associated with a number of diseases including the systemic amyloidoses, Alzheimer's and Parkinson's diseases and late onset diabetes[1, 3, 4]. Although recent research has revealed increasingly detailed molecular events associated with the formation of such aggregates[5, 6], the localisation of these structures within the cell and the detailed mechanisms that characterise their involvement in disease states are crucial questions which largely remain to be addressed[7]. The challenges in providing answers to such questions originate from difficulties in imaging carbon containing polypeptide aggregates within the carbon rich cell[8]. Previous strategies for overcoming this lack of contrast have involved the use of potentially intrusive tags or of heavy element stains that have an extremely low level of specificity.

In this paper an alternative approach for increasing selectively the contrast of intracellular peptide aggregates by labelling the fibrils in such a way as to increase their scattering power relative to the cellular environment is described. The selenium analogue of the sulfur-containing amino acid methionine, which possesses a larger electron scattering cross section than its natural counterpart but is chemically essentially identical to it has been incorporated into the peptide. The samples were imaged using high angle annular dark field (HAADF) imaging in the scanning transmission electron microscope (STEM). This technique is highly sensitive to local variations in the atomic number within the sample because of the Z^n-dependency of Rutherford scattering [9].

The aggregation and toxicity of a peptide corresponding to residues 25-36 human amyloid-β (Aβ) protein, the main constituent of the plaques found in the brains of Alzheimer's patients[2], and from an analogous peptide in which the methionine residue in position 35 is replaced by its selenium-labelled analogue. The protein was prepared by dissolution in trifluoroacetic acid followed by lyophilisation and dissolution in 1,1,1,3,3,3-hexafluoro-2-propanol. The protein was then diluted in phosphate buffered saline at pH7.4 with 100mM NaCl at a protein concentration of 100μM at 37°C. The aggregation was examined using thioflavin-T fluorescence and transmission electron microscopy (TEM); branched, globular aggregates were formed after three hours and longs thin amyloid fibrils were formed after about 12 hours. Human monocyte derived macrophage cells (HMMs) were exposed to the aggregates at a concentration of 10μM for varying lengths of time from 2 to 24 hours before fixation, according to a protocol published previously[10]. Samples were then ultramicrotomed into thin sections (40 – 250nm) and mounted on 300 mesh bare copper grids for examination by HAADF-STEM. No formvar or carbon support films were used in order to minimise the total sample thickness. HAADF-STEM experiments were performed on an FEI Titan microscope operating at 300kV with a convergence angle of 14mrad and a probe size of approximately 0.24nm.

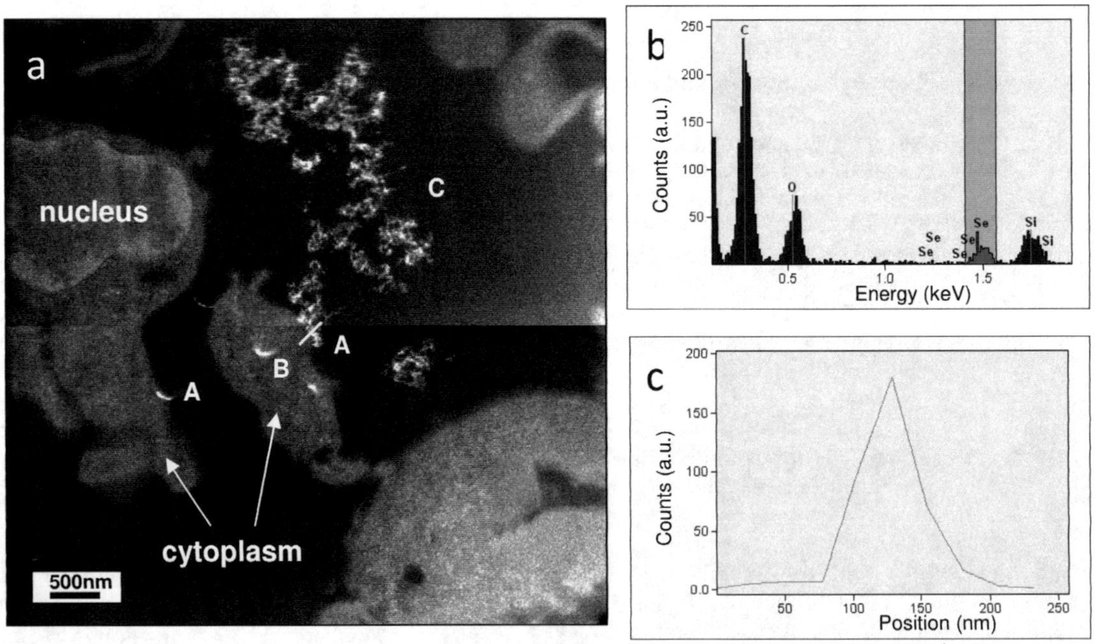

Figure 1: (a) HAADF-STEM image of HMMs exposed to selenium-labelled fibrils showing fibrils associated with the cell membrane, A, in the cell cytoplasm, B, and in the extracellular space, C. The line indicates the fibril across which an EDX line spectrum image (SI) was recorded. (b) The EDX spectrum at the point of highest concentration of selenium and highest HAADF signal on the line SI. (c) Selenium profile extracted from SI.

Figure 1a shows a HAADF-STEM image recorded with inner (θ_{in}) and outer (θ_{out}) HAADF collection angles of 12.8mrad and 63.7mrad respectively, showing selenium-labelled fibrils associated with the cell membrane, A, in the cell cytoplasm, B, and in the extracellular space, C. To confirm that the observed increased scattering is due to the selenium atoms incorporated into the fibrils an energy dispersive X-ray (EDX) spectrum image was recorded across the fibril image (figure 1c). Figure 1b shows the EDX spectrum at the HAADF peak and clearly shows the presence of selenium. The total counts for an energy window around the selenium peak across the spectrum image is shown in figure 1c and shows that the highest concentration of selenium corresponds with the highest intensity of electrons falling on the HAADF detector showing that the increased intensity at the position of the fibrils is indeed due to the presence of selenium.

For biological samples mass-thickness is the "conventional" source of contrast when imaging in the electron microscope. If θ_{in} is small then the HAADF image will be dominated by mass-thickness contrast (figure 2a). As θ_{in} is increased the contribution to the HAADF image from electrons that have undergone Rutherford scattering is increased while the contribution from mass-thickness contrast is reduced. At high θ_{in} the HAADF image is dominated by electrons that have undergone Rutherford scattering (figure 2b); this means that the main source of contrast in the image is due to local variations in atomic number. The low collection angles used when recording figure 2a allow cell organelles such as the nucleus to be identified, due to variations in sample density, as well as the selenium-labelled fibrils. The higher collection angles used when recording figure 2b mean that cell organelles are not visible whereas the selenium-labelled fibrils can be identified. This method allows the fibrils to be identified within the cellular environment without the use of any other staining thanks to the incorporated selenium atom.

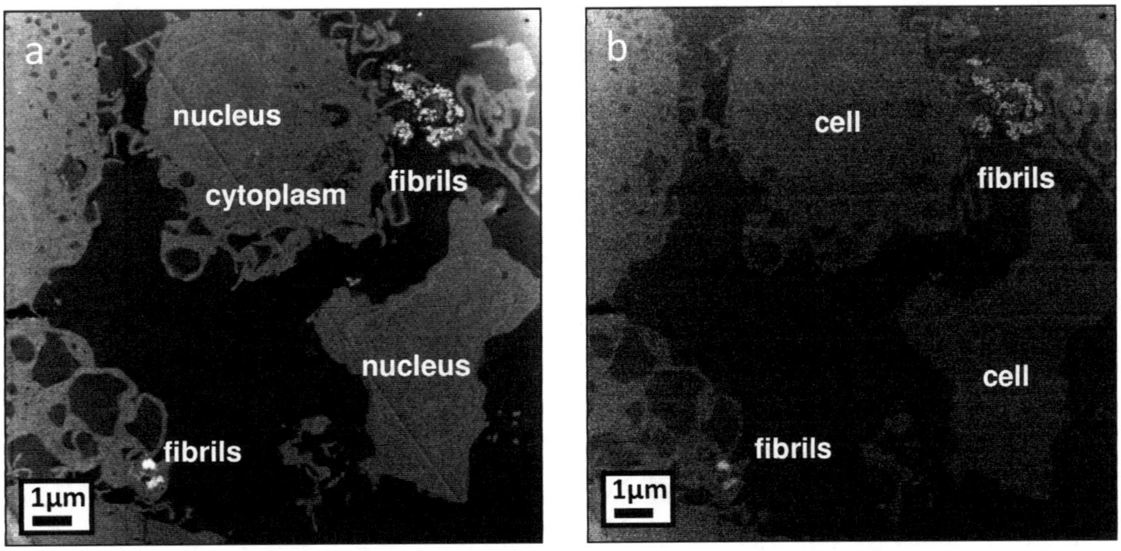

Figure 2: (a) HAADF-STEM image of HMMs exposed to selenium-labelled fibrils recorded with θ_{in} 8.3mrad and θ_{out} 41.5mrad. (b) HAADF-STEM image of the same area of the sample recorded with θ_{in} 32.5mrad and θ_{out} 161.0mrad. Both images were recorded using a probe size of approximately 0.24nm.

3-D electron tomography was performed on selected sections of cells exposed to the Se-peptide in order to visualise the distribution of protein aggregates in three dimensions. HAADF-STEM tomographic datasets were acquired over a tilt range of -36° to +42° using a step size of 2°. For these STEM-tomography measurements, we used samples of 250 nm thickness in order to increase volume information. Three-dimensional reconstruction was carried out using the simultaneous iterative reconstruction technique[11, 12] using EMISPEC 3-D reconstruction software; reconstructed images were visualised in slices using AmiraTM 3D visualization software. Figure 3a shows a HAADF-STEM image of a selenium-labelled aggregate which has entered the cell cytoplasm from the extracellular medium. Figure 3b shows a Voltex 3D reconstruction of the boxed area in figure 3a showing that the aggregates are inside the cell cytoplasm.

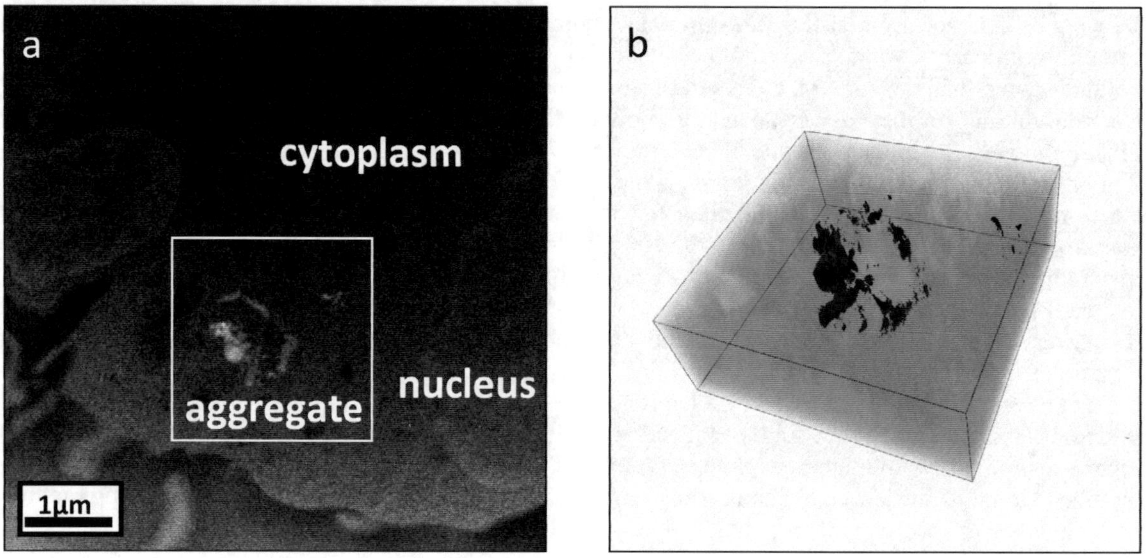

Figure 3: (a) HAADF-STEM image of a selenium-labelled aggregate in the cell cytoplasm. (b) Voltex 3D reconstruction of a HAADF-STEM tomography series taken of the boxed region in (a).

In conclusion, the results of this study show that the use of peptides in which the natural sulfur atoms are replaced with selenium enables the observation of amyloid fibrils within individual cells and cell compartments. This approach therefore represents a powerful tool for future investigations of the molecular processes that underlie the normal and aberrant effects of protein self-assembly and misassembly in living systems.

References

1. Chiti, F. and C.M. Dobson, Annual Review of Biochemistry, 2006. **75**: p. 333-366.
2. Dobson, C.M., Nature, 2003. **426**(6968): p. 884-890.
3. Pepys, M.B. *Pathogenesis, diagnosis and treatment of systemic amyloidosis*. in *Discussion Meeting on Protein Misfolding and Disease*. 2000. London, England.
4. Kelly, J.W., Nature Structural Biology, 2002. **9**(5): p. 323-325.
5. Wasmer, C., et al., Science, 2008. **319**(5869): p. 1523-1526.
6. Collins, S.R., et al., Plos Biology, 2004. **2**(10): p. 1582-1590.
7. Lansbury, P.T. and H.A. Lashuel, Nature, 2006. **443**(7113): p. 774-779.
8. Jensen, G.J., Science, 2009. **323**(5913): p. 472-473.
9. Eckman, C.B., et al., Human Molecular Genetics, 1997. **6**(12): p. 2087-2089.
10. Porter, A.E., et al., Acta Biomaterialia, 2006. **2**(4): p. 409-419.
11. Kak, A.C. and M. Slaney, *Principles of Computerized Tomographic Imaging*. 1988: IEEE Press.
12. Gilbert, P., J Theor Biol, 1972. **36**(1): p. 105-17.

The application of ESEM to biological samples

J E McGregor & A M Donald
Cavendish Laboratory, University of Cambridge, J. J. Thomson Avenue, Cambridge CB3 OHE, UK

E-mail: jem60@cam.ac.uk

Abstract. The Environmental Scanning Electron Microscope (ESEM) differs from a conventional SEM in that a differential pumping system maintains a pressure of gas (typically H_2O) in thespecimen chamber whilst the gun remains at high vacuum. Ionizing collisions between electrons and these gas molecules create positive ions which drift down onto the sample neutralising specimen charge. It is therefore possible to image insulating samples without the need for metallic coating. The presence of water vapour in the chamber also means that a high relative humidity can be maintained and samples can be imaged in a hydrated state without the need for dehydration and fixation. These features suggest that ESEM could be well suited to imaging biological samples undergoing natural biological processes. We present a proof of principle study on the closure of stomatal pores in *Tradescantia andersonia* leaf tissue. An imaging protocol is developed and the advantages and limitations of this technique are discussed. Images of *Vicia faba* leaf tissue are also presented. Challenges include minimising beam damage and reconciling the need for an adequate physiological temperature and a low gas pressure favourable for imaging, with the thermodynamic constraints on achieving a high relative humidity.

1. Introduction

ESEM was first proposed by Danilatos in 1979 [1] and offers the ability to image insulating samples at high relative humidity, features which are not available with conventional SEM. These abilities suggest it may have substantial capacity for high resolution imaging of biological specimens. Donald [2] discusses the ESEM technique and its advantages in more detail. Our work focuses on optimizing the ESEM for biological use and exploring its applications in that field. Specifically, we demonstrate the possibility of using ESEM to image live cells undergoing dynamic morphological changes.

The system under study here is the closure of stomatal pores in leaf epidermal tissue. Each pore is formed by a pair of guard cells which can swell, change shape and open the pore - thereby controlling gas exchange with the atmosphere. A more detailed discussion of stomatal biology can be found in Taiz & Zeiger [3]. In nature stomatal movements occur in response to a range of environmental cues including changes in atmospheric humidity. Varying the temperature of the sample in the microscope chamber allows control of relative humidity with a view to inducing stomatal closure.

ESEM imaging of plant tissue in it's native state was first explored by Danilatos [4] in the early days of the instrument. By optimizing the microscope to view stomatal movements in living *Tradescantia andersonia* (Spiderwort) leaf tissue, we explore the possibility of using ESEM to image dynamic natural processes in living tissue

2. Materials and methods

2.1 Plant Material

Young *Tradescantia andersonia* and *Vicia faba* (broad bean) plants were grown in commercial potting compost (John Innes no. 2) in 25cm plastic pots. A light intensity of 100-300 $\mu molm^{-2}s^{-1}$, dependent on leaf height, (measured using an Apogee quantum sensor) was generated by a 600W Sunmaster metal halide growth lamp. The plants were entrained to a 16h photoperiod and were well watered. Temperatures ranged between approximately 15°C and 25°C with relative humidities in the range 35-55%. Plants were transferred to a perspex box with the same temperature range but higher humidity several hours before imaging to ensure the stomatal pores were open wide. The relative humidity levels in this box varied between 40 and 65% during the day and a maximum of 90% during the night. Experiments took place within a 10am to 4pm time window. Leaf tips were cut from the plants using a scalpel and transferred to the microscope chamber within a minute and a half.

2.2 ESEM setup and protocol

The sample was placed, lower epidermis up, on an adhesive carbon tab fixed to a 3.5 cm x 3 cm copper plate in thermal contact with a water cooled Peltier chip in the sample chamber. This allowed temperature control of the whole sample and meant that complete leaves or leaf tips could be accommodated, minimising the number of potentially dehydrating cut edges. All experiments were conducted using an FEI XL30 FEG ESEM with the Secondary electron signal collected using a needle detector, developed by Toth and Baker [6]. The accelerating voltage was 10kV, 7kV or 5kV. A final water vapour pressure of 7.3 Torr (unless otherwise stated) was achieved using a custom pumpdown procedure developed by Cameron and Donald [7] minimizing sample evaporation. Samples were typically maintained at 7 °C throughout pumpdown unless otherwise stated. The relative humidity at the sample surface was estimated by expressing the actual pressure of water vapour in the sample chamber as a percentage of the theoretical saturated vapour pressure of water at the temperature of the Peltier stage. Saturated vapour pressure increases exponentially with increasing sample temperature, thus it is possible to lower the relative humidity by raising the sample temperature but leaving the gas pressure unchanged. The time elapsing between cutting the leaf tissue and capturing the first image was minimized such that the stomatal pores were maximally open and the tissue was as fresh as possible; this preparation and pumpdown time typically ranged from four and a half to ten minutes. After an initial image was taken, the temperature of the Peltier stage was raised to lower the relative humidity and trigger the closure. Images were taken at five or ten minute intervals, between which the beam was blanked.

3. Results and discussions

3.1 Imaging the closure of stomatal pores

Following the general protocol outlined in section 2.2, stomatal closure could be observed reproducibly occurring in the ESEM chamber. The sequence presented in figure 1 is closure in response to a reduction in humidity from 97 to 91%. In this case the process occurred on a time scale of 20 minutes. Single imaging scans were taken at approximately ten minute intervals with a 10kV beam, spotsize 3. The beam was blanked between scans to minimize beam damage. Tripling the magnitude of humidity decrease halved the closure time, suggesting that we are observing a natural living process rather than the consequences of apoptosis. If, as a control, no step in humidity was introduced closure occurred on a time scale of 45-50 minutes, consistent with the expected response of an excised leaf tip in dark, cold conditions. If the sample had been introduced into the high vacuum conditions of a conventional SEM, the absence of any water water vapour in the air would also have triggered closure however the rapid dehydration of the tissue would severely and irreversibly damage the guard cells; impairing normal function. One would also anticipate imaging difficulties due to charging in high high vacuum conditions.

3.2 Limitations

The images of *vicia faba* presented in figure 2 illustrate the consequences of imaging using unsuitable environmental conditions. In this case the relative humidity was 89% and should therefore have triggered a rapid closure response. Instead the stomata in image B appear open yet the epidermal cells appear flaccid. This appearance was not limited to the irradiated region indicating that it was not a beam damage effect. We suggest that this unusual behaviour was an unexpected consequence of the tissue growth conditions which resulted in tissue adapted to high humidities. As a consequence, the tissue did not respond well to the conditions of the microscope chamber. Non closing stomata have previously been reported in tissue grown at high humidity [5].

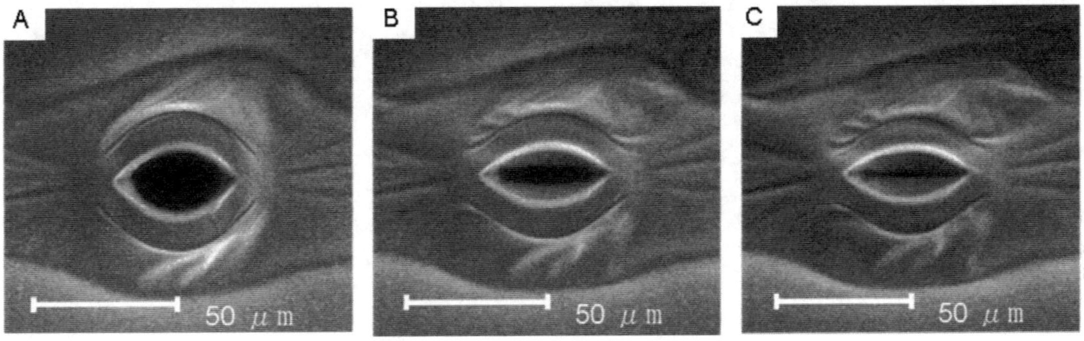

Figure 1: A sequence of three ESEM images following the closure of a single stomatal pore in response to the reduction in chamber relative humidity from 97% to 91%. **(A)** 12 minutes, **(B)** 25 and **(C)** 32 minutes after cutting. Accelerating voltage 10kV, beam current 0.09nA. Water vapour pressure 7.3 Torr, temperature 7°C m in image (A), subsequent pressure 7.2 Torr, temperature 8 °C , in (B) and (C).

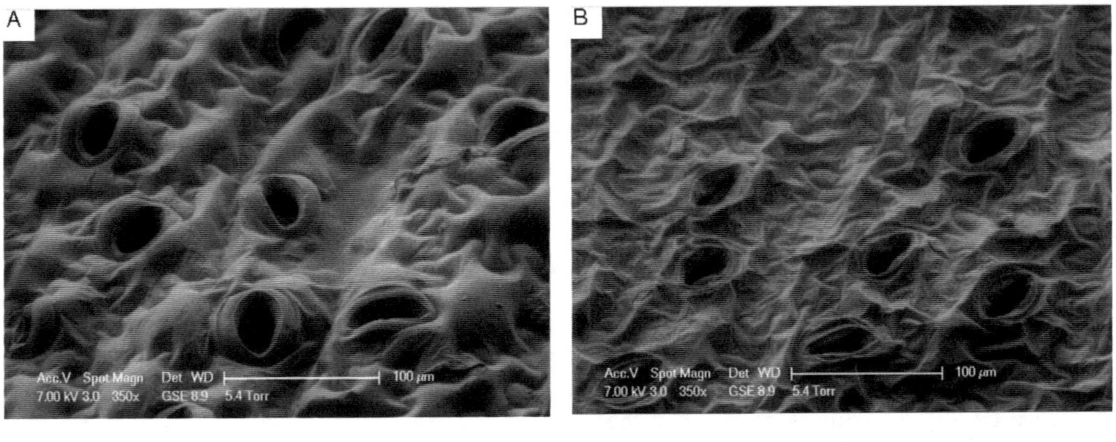

Figure 2: Images of the lower epidermis of *vicia faba* grown at high humidity (described in section 2.1) Imaged using a 7kV beam, FEI spotsize 3. After pumpdown, the sample maintained at 4°C, 5.4 Torr, 89% relative humidity throughout imaging. Image B is taken 5 minutes after image A. The sample is tilted to 45 degrees, revealing more of the three dimensional structure.

Figure 3 shows the changes to tissue character induced by the action of the beam. Beam damage is distinct from environmental damage, being localised to the region being scanned. The tissue in figure 3B has been continuously scanned for 35 minutes with a 10kV beam. This epidermal cells appear deflated and the surface wrinkled. The damage caused by continuous scanning at 10kV necessitated the single imaging scans and beam blanking adopted in the protocol generating the sequence in figure 1, recent work has, however shown that continuous scanning at 5kV is less damaging to the surface structure.

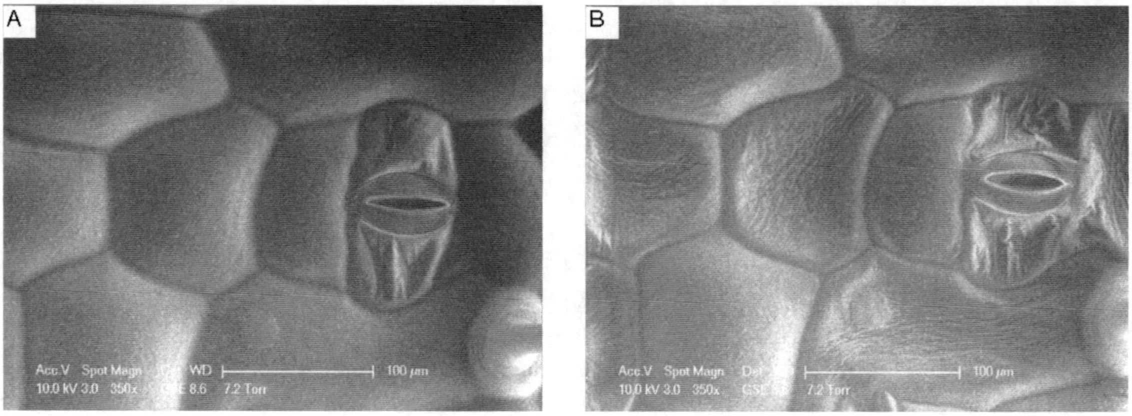

Figure 3: Two images of the epidermis of Tradescantia andersonia imaged at 10kV, FEI spotsize 3, 7.2 Torr, 7ºC Image A is the initial image taken mins after excision. Image B is the same region as B but taken following 35 minutes of continuous scanning. The cells in B appear sunken and the surface is wrinkled.

4. Conclusions

This work indicates that it is possible to observe dynamic biological processes occurring in plant tissue. Stomatal closure can be observed inside the ESEM chamber using the protocol presented in section 2. We observe that progress toward continuous imaging may be possible by imaging at 5kV and stress the importance of adapting plant tissue for the imaging environment.

Acknowledgments

The authors wish to acknowledge Prof. Howard Griffiths and Dr Alex Webb, Cambridge University, for useful discussions on aspects of plant science and Dr Peter Franks, University of Sheffield for practical assistance. We also wish to thank Trevor Fairhead for his invaluable help with the XL30 ESEM. This work was funded by the EPSRC.

References

[1] Danilatos G D, Robinson V N E, 1979 *Scanning* **2** 72-82.
[2] Donald A M 2003 *Nature Materials* **2** 511
[3] Taiz L, Zeiger E, Plant Physiology 2002 3rd Edn Sinauer
[4] Danilatos G D, 1980 *J. Microsc.* **121** 235
[5] Toth M, Baker F S 2004 *Microscopy and Microanalysis.* **2** (suppl. 2) 1062
[6] Cameron R E, Donald A M, 1994 *J. Microsc.* **173 (3)** 227
[7] Nejad A R, Harbison J, van Meeteren U 2006 *J. Exp. Bot* **57** 3669

Journal of Physics: Conference Series **241** (2010) 012022 doi:10.1088/1742-6596/241/1/012022

Microscopical analysis of synovial fluid wear debris from failing CoCr hip prostheses

M B Ward[1], A P Brown[1], A Cox[2], A Curry[3] and J Denton[4]

[1]Institute for Materials Research, University of Leeds, LS2 9JT

[2]Centre for Analytical Sciences, University of Sheffield, S10 2TN

[3]Manchester Royal Infirmary, Manchester, M13 9WL

[4]Department of Laboratory Medicine, University of Manchester, M13 9PL

m.b.ward@leeds.ac.uk

Abstract. Metal on metal hip joint prostheses are now commonly implanted in patients with hip problems. Although hip replacements largely go ahead problem free, some complications can arise such as infection immediately after surgery and aseptic necrosis caused by vascular complications due to surgery. A recent observation that has been made at Manchester is that some Cobalt Chromium (CoCr) implants are causing chronic pain, with the source being as yet unidentified. This form of replacement failure is independent of surgeon or hospital and so some underlying body/implant interface process is thought to be the problem. When the synovial fluid from a failed joint is examined particles of metal (wear debris) can be found. Transmission Electron Microscopy (TEM) has been used to look at fixed and sectioned samples of the synovial fluid and this has identified fine (< 100 nm) metal and metal oxide particles within the fluid. TEM EDX and Electron Energy Loss Spectroscopy (EELS) have been employed to examine the composition of the particles, showing them to be chromium rich. This gives rise to concern that the failure mechanism may be associated with the debris.

1. Introduction

The materials used in commercially available hip prostheses have undergone significant developments since their integration into routine surgery. When a hip implant is necessary, the replacement should mimic the natural predecessor as closely as possible. The choice of material is important as it should have little or no impact on the biological environment it is used in. If the material impacts on the health of the patient, maintenance or even replacement of the part may need to be carried out, causing them stress and discomfort. Metal on polymer prostheses have been shown to have high wear rates [1], thus generating significant amounts of wear debris which can lead to osteolysis [2]. Ceramic on ceramic prostheses are also available, but these are prone to damage due to the materials brittle nature [3]. In more recent years, metal on metal hip replacements produced using cobalt-chromium (CoCr) alloys with extremely low wear rates have become popular. It has been shown that wear rates are low compared to metal on polymer replacements [1], and gross wear debris is not as significant in these types of prostheses and so osteolysis is a less common problem.

However, a recent observation at Manchester in metal on metal CoCr hip prostheses is failure through an as yet unidentified mechanism. The primary symptom of this mode of failure is chronic pain in the patient, and in cases it can lead to replacement of the synovial fluid or even of the hip itself. It has already been established that fine wear debris particles are still produced in this type of prostheses [4], and the presence of such fine particles in the body and their impact on health is a matter of concern. In addition, the presence of chromium deserves some attention, as in certain chemical states the element can have an impact on biological systems [5]. This report describes an investigation into the nature of these wear particles using Transmission Electron Microscopy (TEM). Parallel Electron Energy Loss Spectroscopy (EELS) has proved a useful technique, giving information on the chemical nature as well as the composition of the wear particles.

2. Experimental techniques

TEM specimens were prepared by fixing the synovial fluid with glutaraldehyde before dehydrating, embedding in resin and cutting microtome sections. These sections were then lifted onto formvar coated copper grids (Agar Scientific) suitable for TEM. Prior to examination in the microscope, specimens were carbon coated with a < 5 nm layer to minimise charging. Samples used to measure standard Cr $L_{2,3}$-edges for different oxidation states were prepared using Cr(II)F_2, Cr(III)Cl_3, Cr(III)$_2O_3$ and Cr(IV)O_2 powders (Sigma-Aldrich). These were crushed using a pestle and mortar in acetone, and then the mixture was drop cast onto a holey carbon coated copper grid (Agar Scientific). Onto the same grids, TiO_2 (anatase) powder (Sigma-Aldrich) prepared in the same way was drop cast, so each of the grids contained a mixture of both the chromium based powders and TiO_2 (the O EELS K-edge from TiO_2 was used to calibrate the spectra).

Samples were examined using a Philips CM200 FEGTEM operated at 197kV, fitted with an Oxford Instruments UTW EDX detector and a Gatan GIF 200 imaging filter. Spectra were collected onto a 1024 pixel CCD camera, at a dispersion of 0.1 eV per pixel. An approximate energy resolution of 0.8 eV was recorded. For EELS, the microscope was operated in diffraction mode (image coupled) with a collection semi-angle of 6 mrads and a convergence semi-angle of approximately 1 mrad. Energy calibrations were carried out on the standard samples using the distinctive O K-edge for TiO_2

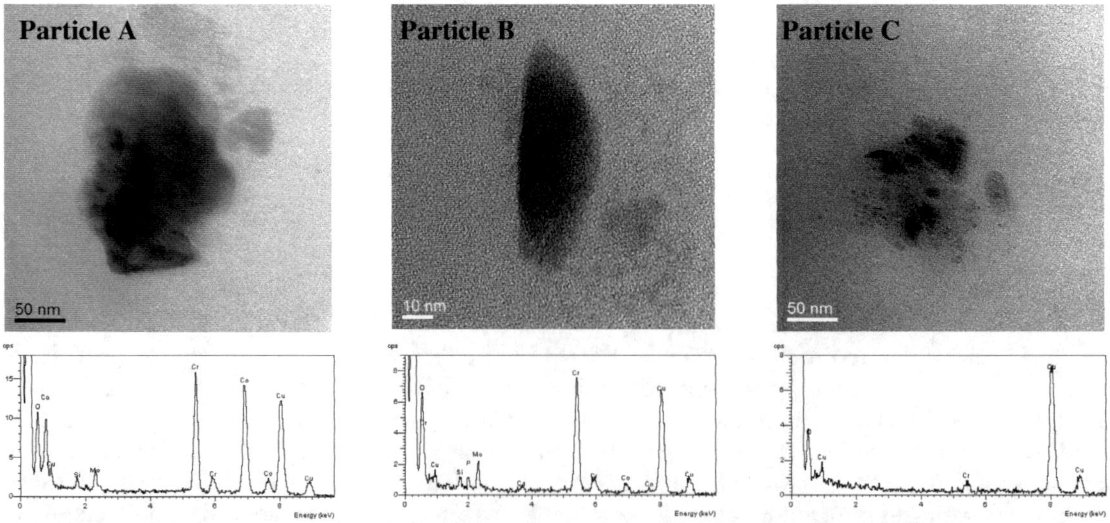

Figure 1 Examples of three typical particles found within the synovial fluid. Below the images are associated EDX spectra collected for the particles. Particle A has a similar composition to the bulk material (cobalt, chromium and molybdenum) and is crystalline, but particle B is richer in chromium and molybdenum and the particles shown in C appear to be composed predominantly of chromium and oxygen. Particles B and C were both amorphous.

(anatase) [6]. Energy calibrations were carried out on the sectioned samples using the zero loss peak and relying on the accuracy of the applied drift tube voltage.

3. Results and Discussion

3.1. Morphology and composition of wear debris particles

Sections of the synovial fluid showed a distribution of fine particles of varying size, morphology and structure. TEM Energy Dispersive X-ray (EDX) analysis of the particles showed the larger particles to be closer to the bulk composition of the metal hip component (cobalt, chromium, molybdenum) and crystalline, whereas the smaller particles were more commonly chromium rich and amorphous. It may be that the top surface of the joint is chromium rich and amorphous, and this is why the smaller particles that were identified were of this composition/structure. Figure 1 shows three different particles that were typical of the sample along with EDX spectra collected from them. Particle C is actually composed of many particles, ranging in size from 50 nm down to < 5 nm. That they are clustered together is of interest, they may have been generated together, or may have become clustered while inside the fluid.

3.2. Cr standards used for EELS

The standards were chosen to give Cr $L_{2,3}$-edges for Cr(II), Cr(III) and Cr(IV). A comparison of the edges (figure 2) shows an increasing energy loss value (measured from the L_3 peak) with increasing oxidation state. As well as changes in position, a distinctive change in shape of the L_3 peak can be observed in the spectra. The Cr(II) L_3 peak is intensity-weighted to the lower energy loss side, whereas the Cr(IV) L_3 peak is weighted to the higher energy loss side.

3.3. EELS from wear debris particles

The chromium $L_{2,3}$-edge was collected for the three particles shown in figure 1. The low loss spectrum was also collected from each for calibration. A comparison of the spectra can be seen in figure 3.

The positions of the L_3 peaks in each of the spectra lie between 575.4 and 576.1 eV energy loss. The oxidation state that these edges most closely resembles (with respect to energy loss) is Cr(II). However, the shape of the L_3 peak better matches that of Cr(III). It is probable that the particles studied are composed of Cr(III), as the shape of the Cr(II) L_3 peak is quite distinctive (and not evident here). An exception may be the spectrum for particle C. Although it is very noisy, there is some evidence for weighting of the L_3 peak to the high energy loss side. This would match Cr(IV) from figure 2, and may suggest that the particle is composed wholly or partially of Cr(IV). This would agree with the difference in composition measured by EDX. In this instance, peak shape rather than position is probably a better method for giving a qualitative indication of oxidation state, as the calibration method used differed from that of the standards. Using an internal calibration (as was done

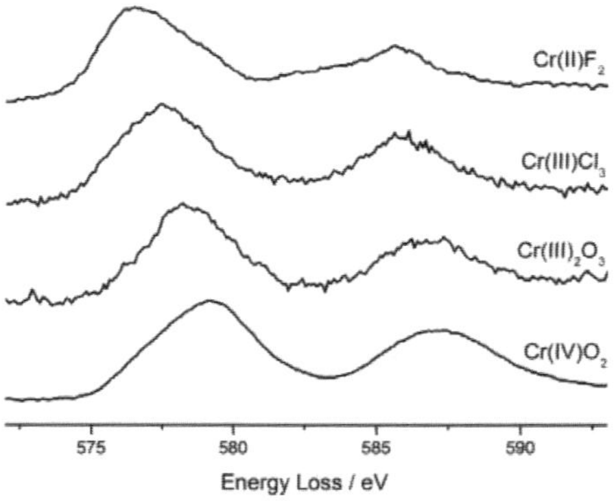

Material	L_3 peak position / eV
Cr(II)F_2	576.0
Cr(III)Cl_3	576.9
Cr(III)$_2O_3$	577.8
Cr(IV)O_2	578.6

Figure 2 EEL spectra collected from standard samples. An increase in energy loss of the edge can be observed with increasing oxidation state, along with a distinctive change in shape of the L_3 peak.

with the standards, using the TiO_2 O *K*-edge) is a more accurate method of calibration than using the zero loss peak. It is feasible that some energy drift could have occurred when switching from the low loss to the core loss region when measuring the spectra from the particles, and that could explain the peak positions not matching those of Cr(III)/Cr(IV).

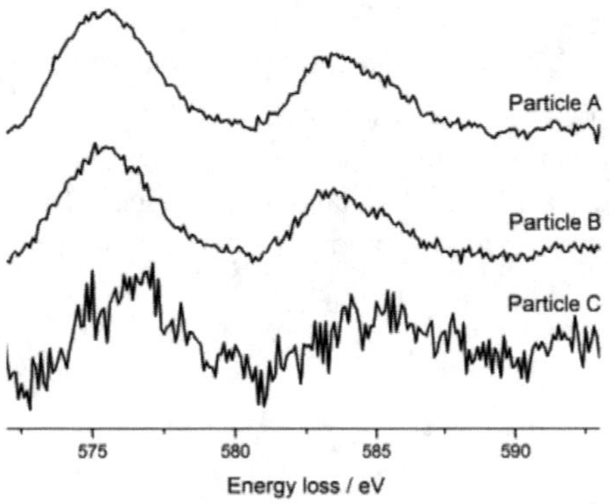

Particle	L_3 peak position / eV
A	575.4
B	575.5
C	576.1

Figure 3 A comparison of the three spectra collected from particles A, B and C (shown in figure 1). The positions of the L_3 peak in each have slightly lower value than those from the standard samples (figure 2), though this may be a result of a different energy calibration method (i.e. using the zero loss peak). Regarding their shape, they most closely resemble those of the Cr(III) materials measured, although particle C may exhibit a shape close to that of Cr(IV).

4. Conclusions

This study has shown that transmission electron microscopy can be used effectively to examine wear debris from metal on metal joint replacements. If the CoCr bulk material forms a passivating chromium oxide layer (as stainless steel does), and this layer is amorphous, then this would explain the observation that the smaller particles were chromium rich and amorphous (and originated from wear at the top surface). Examination of a cross section of the hip replacement will resolve this.

EELS gave even more information about the composition of the particles that were observed. Using reference spectra collected from standards, the particles could be best described as being in the oxidation state Cr(III), with the exception of the smallest particles measured (particle C) which may be either a mixture of Cr(III) and Cr(IV) or wholly Cr(IV). The most reliable indicator as to oxidation state is thought to be the shape of the L_3 peak, as a distinctive difference in peak shape for Cr(II), Cr(III) and Cr(IV) has been shown (figure 2). Errors must be taken into account when performing peak position analysis, unless a thorough and reliable energy calibration method is used.

Although no direct evidence for the mode of failure was found, identification and thorough analysis was carried out on a number of wear particles. It is suggested that the failure mode may well be associated with the fine wear debris. Further TEM and EELS work will try to reinforce this proposal.

References

[1] Sieber H –P, Rieker C B and Kottig. 1998. *Journal of Bone and Joint Surgery (Br)*. **80** B. 46-50.
[2] Willert H -G, Bertram H and Buchhorn G H. 1990. *Clinical Orthopaedics and Related Research*. **258**. 95-107.
[3] Sodiwala V, Bg Rao V and Pillay J. 2006. *Emergency Medicine Journal*. **23**. 639-640.
[4] Doorne P F, Campbell P A, Worrall J, Benya P D, McKellop H A and Amstutz H C 1998. *Journal of Biomedical Materials Research Part A*. **42** (1). 103-111.
[5] Daulton T L, Little B J, Lowe K and Jones-Meehan J. 2002. *Journal of Microbiological Methods*. **50**. 39-54.
[6] Brydson R, Sauer H, Engel W, Thomas J M, Zeitler E, Kosugi N and Kuroda H. 1989. *Journal of Physics: Condensed Matter*. **1**. 797-812.

SEM investigation of heart tissue samples

R Saunders and M Amoroso
Physics Department, University of the West Indies, St. Augustine , Trinidad and Tobago, West Indies.

Abstract. We used the scanning electron microscope to examine the cardiac tissue of a cow (*Bos taurus*), a pig (*Sus scrofa*), and a human (*Homo sapiens*). $1mm^3$ blocks of left ventricular tissue were prepared for SEM scanning by fixing in 96% ethanol followed by critical point drying (cryofixation), then sputter-coating with gold.

The typical ridged structure of the myofibrils was observed for all the species. In addition crystal like structures were found in one of the samples of the heart tissue of the pig. These structures were investigated further using an EDVAC x-ray analysis attachment to the SEM. Elemental x-ray analysis showed highest peaks occurred for gold, followed by carbon, oxygen , magnesium and potassium.
As the samples were coated with gold for conductivity, this highest peak is expected. Much lower peaks at carbon, oxygen, magnesium and potassium suggest that a cystallized salt such as a carbonate was present in the tissue before sacrifice.

1. Introduction

The human heart generates electrical currents produced by the depolarisation-repolarisation processes of its muscle cells. These currents produce an associated time-varying magnetic field that can be measured and recorded non-invasively as the magnetocardiogram (MCG)[1]. Because of developments in the measurement of small magnetic fields the magnetocardiogram can be a very important diagnostic modality in medicine.

Cardiac muscle cells or myocytes show a single nucleus, have a length of approximately 100 μm and are approximately 25μm in diameter. The cardiac cells form a functional synctium or branched network of individual myocytes connected through intercalated discs. The contraction of muscle fibres is normally explained by the sliding filament theory [2]. The close correlation between anatomical structure and physiological function may make it useful to address the Scanning Electron Microscopy (SEM) of the ventricular muscle of the mammalian heart, to better understand how the functions of the structures translate into the cardiac events and possible dysfunctions we may encounter in magnetocardiography. We therefore examined the cardiac muscle fibres of the cow (*Bos Taurus*), pig (*Sus scrofa*) and human (*Homo sapiens*).

2. Method

At least 3 samples of heart tissue were prepared and scanned from each species, using approximately 1 mm^3 blocks of tissue, from left ventricular heart muscle. For the cow and pig tissues, a larger tissue block was excised from the organ on site when the animal was slaughtered and immediately placed into a sterile specimen jar containing 96% ethanol. In the case of the human tissue, a block of tissue was excised from the organ by a pathologist before placing into 96% ethanol. Smaller 1 mm^3 blocks of tissue were cut from these larger blocks with a scalpel. The tissue blocks were then fixed in 96% ethanol followed by critical point drying (cryofixation) using a Tousimis Samdri 795 critical point dryer. They were placed on a

Journal of Physics: Conference Series **241** (2010) 012023 doi:10.1088/1742-6596/241/1/012023

stub and viewed with a light microscope to check the orientation and position of the tissue blocks, sputter-coated with gold using a Denton Vacuum Desk II sputter-coating machine, then examined by electron microscope. Digital photos were taken at increasing magnifications. The samples were stored on stubs in a dark, dry cabinet at 26 ºC.

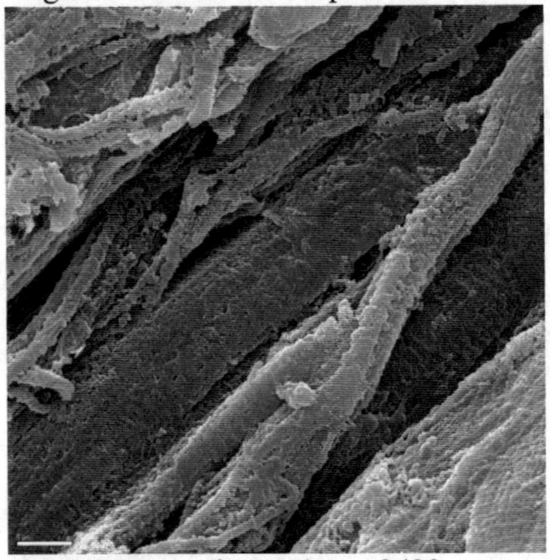

Fig.1: Human heart tissue 2400x Scalebar=5μm

Fig.2: Cow's heart tissue 1200x Scalebar=10μm

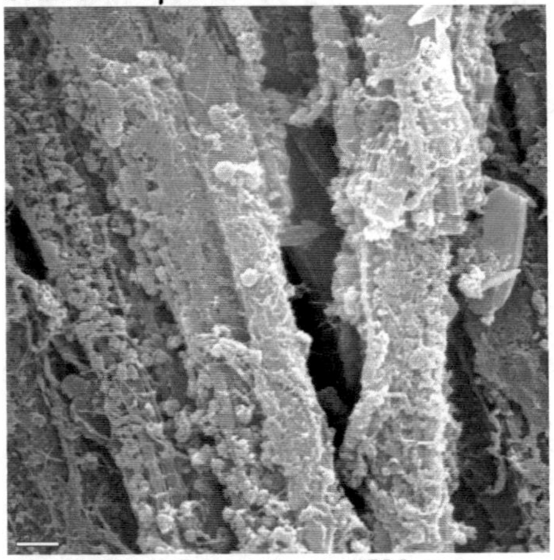

Fig.3: Pig's heart tissue 4580x Scalebar=2μm

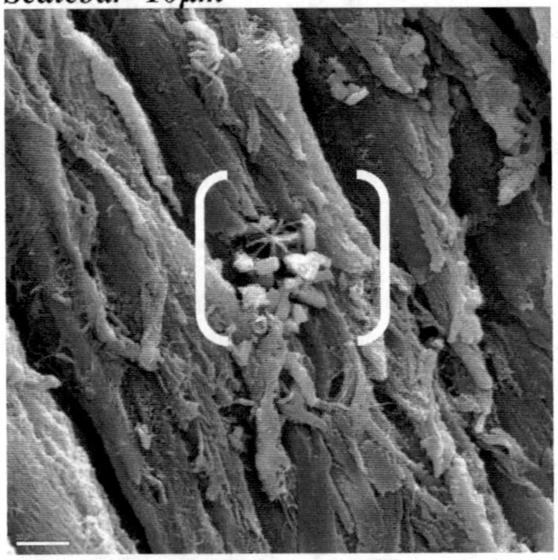

Fig.4: Pig's heart tissue 573x Scalebar=20μm

Fig.5: Pig's heart tissue 2300x. Enlargement of area indicated in Fig. 4: A - elongated triangular prisms; B - irregular prisms and C –granules. Scalebar=5μm

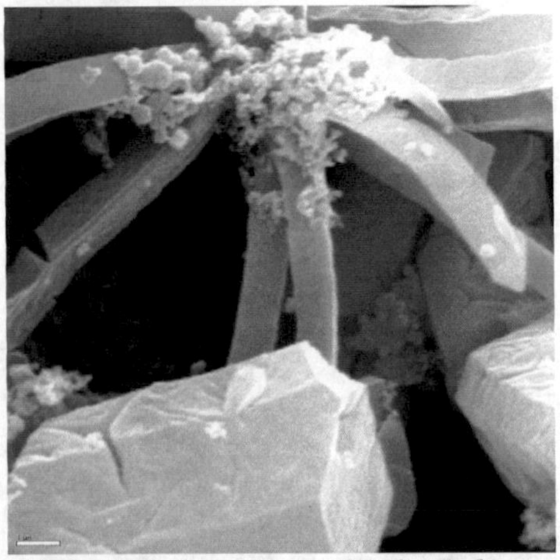

Fig.6: Pig's heart tissue 9150x Enlargement of the area indicated in Fig. 4 showing granules. Scalebar=1μm

3. Results

In each of the SEM images shown in the Fig.s 1 to 6, the scale bar (calibrated in μm) is superimposed on the lower left of the digital image for reference.

In the images of the tissues (Figures. 1,2 and 3), the expected ridged structure of the fibre bundles could clearly be seen; it was possible for structures of 0.5 micrometers to be differentiated visually in the images up to 19200x magnification. At greater magnifications it was difficult to further differentiate features of the specimen surface. In addition to this basic structure, small crystalline structures were found in the heart tissue of the pig (Figures. 4, 5 and 6). These have been labelled as A (elongated triangular prisms), B (irregular prisms) and C (granules) in Figure. 5. These structures were subsequently micro analysed using the EDVAC Genesis V5.2 x-ray analysis probe at two sites and spectral patterns were generated for each session of analysis. The following spectra (Figures. 6 and 7) are representative of those obtained.

In all of the spectral patterns obtained the highest peak always occurred for Au (gold) at 2.00 and 9.6 keV ±0.4 keV. For the oblong solid (as indicated in Figure 5) C and O were the next highest peaks, where C occurred around 0.3keV and O occurred around 0.5keV. Peaks for K and Mg were next highest after that, K occurred at 3.5keV, Mg occurred at 1.2keV. For the rod-like structures (as indicated in Figure 5) C and K were the next highest peaks after Au. The third highest peaks appeared for Na and O, and the fourth highest peak occurred for N.

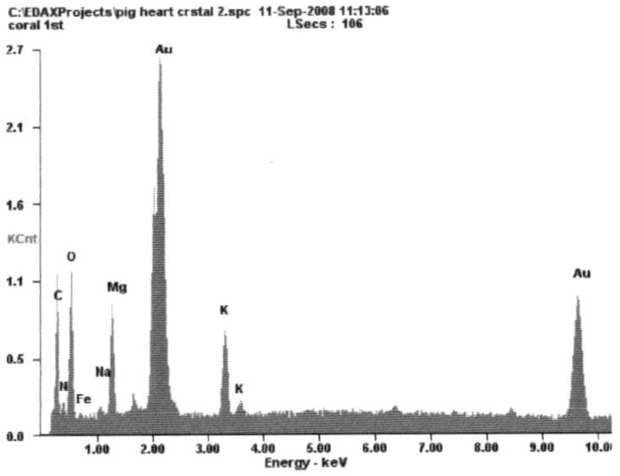

Fig. 6: Spectrum obtained with microanalysis on oblong object

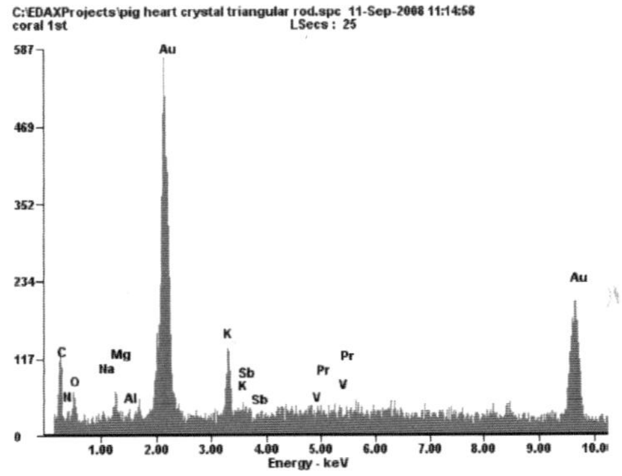

Fig. 7: Spectrum obtained with microanalysis on rod-like object

Table 1. Visual analysis: average dimensions of the solids

Irregular prisms	7 x 1.75 x 2.75 μm ±0.125 μm
Elongated triangular prisms	8.25 μm ±0.125 μm (length), 0.8±0.125 μm (side of face)
Granules	0.66 - 0.45 μm ±0.125 μm diameter

Figure 3b was visually similar in structure and relative dimensions to SEM images depicting collagen-associated calcific deposits in degenerative porcine bioprostheses [2].

4. Discussion

The ridged strands observed in SEM images of cardiac tissue are bundles of myofilaments [3]. The electron microscope examination confirms that the structure of ventricular cardiac tissue of the three large mammals studied is similar in appearance (at magnifications between 70x and 2400x).The presence of crystalline structures in the heart tissue of the pig(Figures 4 and 5) is of interest since it is possible that such structures may be present in the heart tissue of other mammals. This is atypical of healthy mammal heart tissue; during contraction of cardiac myofibrils actin filaments slide inwards between myosin filaments[3] and the presence of crystals in cardiac tissue may impede motion and possibly sever the sliding fibres with resulting pain and loss of function. In the case of this particular sample of pig's heart tissue, the star-shaped structure consists of regularly-shaped solids clustered in what appeared to be a break in a strand of cardiac tissue.

From microanalysis of solids A and B in Figure 5, in all the spectral patterns showed two dominant gold peaks. This was expected as the SEM samples were coated with gold for conductivity. The elemental analysis does not immediately identify the origin of the structure, but is consistent with the presence of crystallized salts. Pathological cardiovascular deposits tend to incorporate sodium, magnesium and carbonate, and are both morphologically and chemically heterogeneous [5].

The difficulty in placing any sort of marker to locate the original sites may have led to analysis of sites that were visually similar to the original sites but may actually have had different chemical compositions; isomorphic crystals are however likely to have the same elemental composition. Identification of the structure by visual comparison was inconclusive due to constraints on the accuracy of the image comparison as well as constraints on the available data used in the comparison. Measuring the approximate dimensions of the structures enabled us to rule out structures having similar shapes but differing in size.

5. Conclusion

The myofibrils in human, porcine and bovine cardiac tissue exhibit a similar ridged structure for cryofixed tissue scanned in the electron microscope. The presence of crystal-like objects found in one particular sample of pig's heart tissue may be due to a pathological state of the animal before sacrifice. As the samples scanned in this study were not classified into groups corresponding to the state of health of the source, it would be interesting to further compare the appearance of tissue from sources with known diseases. The presence of such structures in the heart muscles will affect contraction and hence the associated magnetic field in the magnetocardiogram.

6. Acknowledgements

The authors wish to gratefully acknowledge Mr. David Hinds of the SEM laboratory, EWMSC for his expertise and assistance throughout this project.

7. References

1. G Baule and R McFee, 1963. Detection of the Magnetic Field of the Heart. *American Heart Journal* **66**, 95-6.
2. Y S Lee, 1993. Morphogenesis of Calcification in Porcine Bioprosthesis: Insight from High Resolution Electron Microscopic investigation at Molecular and Atomic Resolution. *Journal of Electron Microscopy* **42,** 156-165.
3. R E Klablunde, 2005. Cardiovascular Physiology Concepts. Lippincott Williams & Wilkins, 42-6.
4. D M K Cooper, 2003. Clinical Xenotransplantation – How close are we? *The Lancet* **362**, 557-59.
5. B B Tomazic , L C Chow, C M Carey, A J Shapiro. 1997. An in Vitro Diffusion Model for the Study of Calcification of Bovine Pericardium Tissue. *Journal of Pharmaceutical Sciences* **86**, No. 12, 1433 – 1438.
6. B B Tomazic, 2001. Physicochemical Principles of Cardiovascular Calcification. *Zeitschrift fur Kardiologie* 90: suppl 3, III68 – III80.

Characterisation of ESEM conditions for specimen hydration control

R Leary and R Brydson

Institute for Materials Research, University of Leeds, Leeds, LS2 9JT, UK

E-mail: pm06rkl@leeds.ac.uk

Abstract. The concept of "calibrated ESEM" – the ability to determine and create the exact conditions within the ESEM required for specimen stability and/or accurate in-situ hydration/dehydration – is an attractive idea. It has the potential to allow true natural state imaging, enhanced analysis and a whole range of new and novel applications. The present work reports on the use of in-situ temperature and humidity sensors to accurately measure and characterise the conditions within an ESEM.

1. Introduction

The environmental scanning electron microscope (ESEM) has enabled major developments in the examination of soft, moist and/or electrically insulating materials, permitting them to be viewed without much of the preparation and pre-treatment associated with conventional SEM (CSEM) [1]. In addition, in-situ experiments may be carried out, such as mechanical deformation, corrosion studies, and the observation of dynamic processes including wetting, swelling, hydration/dehydration and colloidal aggregation.

ESEM is largely hailed as revolutionary because it has the potential to allow samples to be viewed in their natural unaltered state. However, although differing from those associated with conventional SEM, there is still the potential for artefacts or unwanted specimen changes to be created during ESEM analysis. Of particular importance to wet or hydrated specimens is the control of temperature and pressure within the microscope chamber. For stabilisation of such specimens, the appropriate saturated vapour pressure must be maintained which prevents dehydration; or in the case of dynamic hydration experiments, it must be varied in a controlled manner.

2. The need for "calibrated ESEM"

For the full potential of ESEM to be exploited, the state of hydration of a specimen must be carefully controlled so that it remains sufficiently similar to that prior to entry to the ESEM environment. For particularly sensitive substances such as liquids or biological specimens, this will require highly precise and accurate control of the conditions in the microscope chamber [2]. In addition, many authors have reported difficulties with water layers forming over the sample and reducing contrast [3]. With precise control of conditions, the formation of this water layer could be avoided. For dynamic in-situ experiments too, precise and accurate control of the state of hydration may be essential for a comprehensive study to be performed. In both instances it may also be of fundamental importance to know the exact conditions (e.g. temperature, pressure and relative humidity) which the sample is experiencing.

The concept of "calibrated ESEM" is envisaged as the ability to determine and create the exact conditions within the ESEM required for specimen stability and/or accurate in-situ hydration/dehydration. Such control would not only allow more effective imaging of samples, but would vastly improve the quality of dynamic analysis. It could even open up the ESEM to a whole new range of applications. For instance, standard testing that requires specified conditions (e.g. moisture vapour transmission of textiles [4]) could be performed in-situ, enabling simultaneous measurement and analysis of the process on a micro/nanometre scale.

3. Towards calibrated ESEM

The path towards calibrated ESEM is twofold. On the one hand, greater understanding of specimens is required so that the necessary conditions for stability – most notably for hydrated samples, the specific saturated vapour pressure – may be determined. While this is vital information for the microscopist, and such analysis has received greater attention in recent years [5], much of the input for this area must come from the specialist in the particular field with a detailed knowledge of the specimen characteristics.

The second aspect is to obtain a greater knowledge of the ESEM instrument and the environment it creates for the sample. In this aspect, there is still much unknown; relatively, ESEM is still in its infancy when compared to other more established forms of electron microscopy such as conventional SEM and TEM.

The effect of the electron beam on an uncoated native sample in the ESEM and its potential for the introduction of artefacts has been noted and the processes studied in some detail [6]. Consideration of the pumpdown from ambient to observation conditions has also received theoretical and qualitative attention, leading to the almost ubiquitous purge-flood technique [7]. Quantitative monitoring and analysis of the *environment* in the ESEM on the other hand, has received comparatively little coverage in the published literature.

As potentially the most vital aspect of "calibrated ESEM," this may well be the area where significant gains can be made. Aspects of interest include differences in temperature between sample and Peltier stage, electron beam heating effects, temperature, pressure and humidity gradients, and of course knowledge of the exact conditions experienced by the sample itself.

4. Use of in-situ sensors

In this work, dual purpose digital temperature and humidity sensors (Sensirion SHT75, Sensirion, Staefa, Switzerland) were positioned within the chamber of a Philips XL30 ESEM for the purpose of monitoring and recording these parameters during pumpdown and imaging. One sensor was attached directly to the specimen stub, while three further sensors were positioned at progressively greater (variable) distances within the chamber.

Figure 1(a) shows an example of temperature and humidity profiles during a four-cycle purge-flood pumpdown for the sensor attached to the sample stub, as well as a sensor positioned close to the edge of the chamber. Prior to inserting the specimen, the microscope chamber had been flooded to high humidity. The sensors were used to monitor the conditions during various pumpdown regimes, with and without purge-flood cycling, and during subsequent imaging. Qualitative assessment was also made by observing sections of celery, whose open cellular structure made it easy to gauge the success of a pumpdown sequence, and/or to see any dehydration/beam damage during imaging.

With no purge-flood cycling during the pumpdown the sensors showed prolonged periods of low humidity and samples were frequently irreversibly damaged, as illustrated in figure 1(b). The optimum pumpdown sequence was found to consist of eight purge-flood cycles, as recommended by Cameron and Donald [7]. This preserved the cellular network structure of the sample, as shown in figure 1(c).

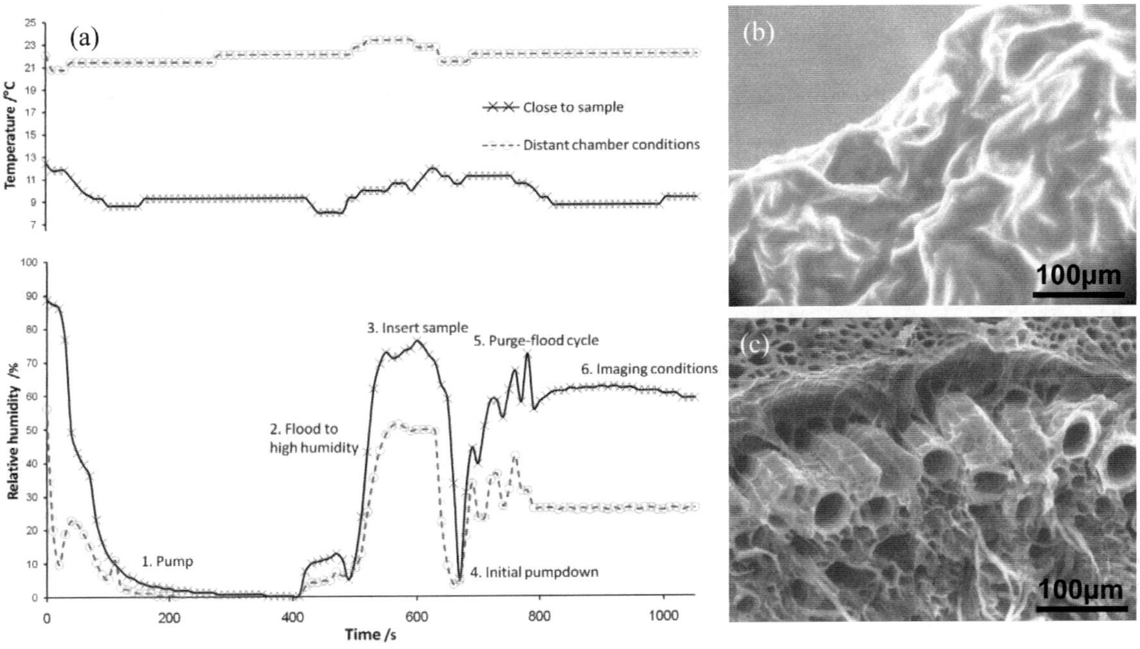

Figure 1. (a) Sensor readings for temperature and relative humidity directly at the sample and further away in the ESEM chamber during a four-cycle purge-flood pumpdown;
(b) celery specimen dehydration resulting from a pumpdown without purge-flood cycling;
(c) successfully preserved celery specimen after an eight-cycle purge-flood pumpdown.

5. Analysis

Although the use of purge-flood cycling during the pumpdown is a well established method for specimen preservation, the use of the sensors has enabled this to be further validated in practice, and has given insight into how it may be more precisely controlled. It is, for instance, clear in figure 1 that there is still a large drop in relative humidity during the initial pumpdown (stage 4). This could be avoided by additive hydration methods such as water drop placement on the Peltier stage adjacent to the sample, and the improvement could be assessed quantitatively using the sensors. While this is again a standard technique, use of the sensors allows its success to be validated and may take on greater importance for more delicate samples than the celery examined here.

Use of the sensors in this work has also shown that the temperature and humidity values may drift and fluctuate on both a short time scale during the course of a single analysis session, and also on a longer timescale over days and weeks. Similar findings have been made by Zimmermann et al. regarding pressure instability [8]. In their work, they ensured calibration of the relative humidity by using an additional in-situ pressure gauge, a thermistor on the Peltier stage and by observing the deliquescence point of salt particles [9]. Similar methods were employed by the authors of this paper but were found to be extremely time consuming and not feasible when analysis of the salt particles is not the main aim of the investigation. It is essential therefore to have in-situ sensors during every use of the ESEM in order to monitor the exact conditions experienced by the sample.

6. Discussion

If ESEM is to develop to become a more precise and sophisticated environment then accurate characterisation of conditions is vital. In addition, this needs to be a routine part of the analysis and the machine itself rather than the addition of make-shift sensors or novel routines. As increased information about the conditions in the ESEM is gained, appropriate changes to operating procedures, imaging parameters and even the design of the ESEM instrument may be implemented. It may well be the case that much of the additional sensing/analytical apparatus currently being investigated (by the authors of this paper and others) may become standard parts of an ESEM instrument.

Furthermore, it may also become possible to develop the machine so that a larger range of conditions can be permitted. Since the time of the work in this paper, this concept has been demonstrated in practice with the development of an improved cryo-ESEM system [10]. However, for many biological specimens, a true natural state would be at body temperature, ~37°C and imaging at this temperature is still to be achieved (although recent work is coming close [11]). With the increasing understanding that may be brought by characterisation of the current ESEM environment, of which the work here is only the beginning, this type of "calibrated ESEM" may not be so far away. The authors of this paper are presently involved in work optimising ESEM conditions for the observation of human dentine, and hope to report new findings elsewhere in the near future.

7. Conclusions

The need for precise knowledge and control of conditions in the environmental scanning electron microscope has been highlighted for both static and dynamic analysis. Greater control of specimen stability and dynamic hydration may be achieved through improved characterisation of the environment within the ESEM. Work to this aim has been initiated through the use of additional in-situ temperature and relative humidity sensors. Further work is in progress.

Acknowledgements

We thank the Nuffield Foundation for financial support and Richard Walshaw for technical assistance during this undergraduate summer project.

References

[1] Muscariello L, Rosso F, Marino G, Giordano A, Barbarisi M, Cafiero G and Barbarisi A 2005 *J Cell Physiol* **205** 328-34

[2] Stokes D J, Rea S M, Best S M and Bonfield W 2003 *Scanning* **25** 181-4
Stokes D J, Thiel B L and Donald A M 1998 *Langmuir* **14** 4402-8

[3] Mestres P, Putz N and Laue M 2003 *Microscopy and Microanalysis* **9** 490-1
Jenkins L M and Donald A M 2000 *Textile Research Journal* **70** 269-76

[4] British Standards Institution 1990. *BS7209:* specification for water vapour permeable apparel fabrics.

[5] Stokes D J 2008 *Principles and Practice of Variable Pressure: Environmental Scanning Electron Microscopy (VP-ESEM)* (John Wiley and Sons Ltd.)
Stokes D J 2003 *Philosophical Transactions of the Royal Society London,* Series A **361** 2771-87

[6] Royall C P, Thiel B L and Donald A M 2001 *Journal of Microscopy-Oxford* **204** 185-95
Kifetew G and Sandberg D 2000 *Wood and Fiber Science* **32** 44-51
Ren Y L, Donald A M and Zhang Z B 2007 *Materials Science and Technology* **23** 105-12
Zankel A, Chernev B, Brandl C, Poelt P, Wilhelm P, Nase M, Langer B, Grellmann W and Baumann H J 2008 *Macromolecular Symposia* **265** 156-65

[7] Cameron R E and Donald A M 1994 *Journal of Microscopy-Oxford* **173** 227-37

[8] Zimmermann F, Ebert M, Worringen A, Schutz L and Weinbruch S 2007 *Atmospheric Environment* **41** 8219-27

[9] Ebert M, Inerle-Hof M and Weinbruch S 2002 *Atmospheric Environment* **36** 5909-16

[10] Waller D, Stokes D J and Donald A M 2008 *Review of Scientific Instruments* **79** 103709-1

[11] Toth M, Uncovsky M, Knowles W R and Baker F S 2007 *Applied Physics Letters* **91** 053122

Reconstruction strategies for structure solution using precession electron diffraction data from hybrid inorganic-organic framework materials

E G Bithell[1,3], C A Merrill[1,2] and P A Midgley[1]

[1]Department of Materials Science and Metallurgy, University of Cambridge, Pembroke Street, Cambridge, CB2 3QZ, UK
[2]Materials Department, University of California, Santa Barbara, CA93016-5050, USA

Email: egb10@cam.ac.uk

Abstract. Precession electron diffraction has received significant recent attention, and has the potential to complement x-ray structure solution methods in situations where the latter are difficult to apply. Certain materials systems may nevertheless present challenges for solution by electron diffraction, and this is particularly true when the unit cell includes light element components. Drawing specifically on our work with hybrid inorganic-organic framework materials, we describe recent approaches to reconstructing their crystal potentials, which have led to significant improvements in the accuracy and quality of the derived potential maps for this type of structure.

1. Introduction

Precession electron diffraction (PED) was proposed by Vincent and Midgley in 1994 [1], and with the development of commercial electron precession systems there has been renewed interest in this technique as a tool for crystal structure determination by electron crystallography. Whilst PED is unlikely to compete with x-ray methods for routine structure determination, it has considerable potential in those cases where the nature of the material is such that x-ray techniques become difficult to apply. This is likely to be the case when samples are available only as very small crystals, when a material can only be synthesised as part of a phase mixture, or when fine scale microstructural features are present (e.g. twinning). In this context we have a particular interest in the application of PED to hybrid inorganic-organic framework materials [2]. Topologically, these compounds are structures in which metal ions act as nodes and linkage is mediated either through inorganic bridges (e.g. metal – oxide – metal) or via organic ligands. Typically, this results in a crystal structure in which a small number of strong scatterers (metal ions) are present on rather widely separated sites, very often at the special positions of the space group. The spaces in between are occupied by groups of much weaker scatterers, and there may be significant voids within the structure.

In order to apply PED successfully to these compounds, both experimental issues and data analysis procedures must be addressed. Rates of electron beam damage are high: complete loss of crystallinity

[3]To whom correspondence should be addressed

Journal of Physics: Conference Series **241** (2010) 012025 doi:10.1088/1742-6596/241/1/012025

within a few seconds is normal and the situation is similar to that encountered in TEM of biological materials. Our goal is therefore to explore the extent to which it might nevertheless be possible to make use of electron crystallographic data, with a particular interest in locating the more weakly scattering elements. Ultimately our aim is to be able to analyse materials where there is little or no prior crystallographic information, and we have therefore sought to develop a procedure which minimises the assumptions which need to made about the structure. For this reason the charge flipping algorithm [3] is particularly attractive, and we have often found it to be more successful with electron diffraction data than established direct methods. It has recently been demonstrated [4] that significant improvements can be made by adding known symmetry information into the charge flipping algorithm, and one might expect this approach to be effective also for the hybrid framework structures.

2. Phasing of structure factors

Figure 1a is a [011] precession electron diffraction pattern from a monoclinic polymorph of copper phosphonoacetate[4] ($P2_1/c$, a=0.8263nm, b=0.7645nm, c=0.8971nm, β =90.52°); figure 1b is a structural model viewed along the same direction and figure 1c is a map of the projected potential for scattering by 300kV electrons. Experimental structure factor moduli were derived from the data in figure 1a by using the peak finding and lattice fitting routines from the Semper image processing package to integrate over a small disc surrounding each peak position. Background levels were subtracted by interpolating between those areas of the pattern which lie between these discs, and the resulting experimental structure factors were corrected for geometric effects using the equations derived by Gjønnes [5]. In order to examine the behaviour of the reconstruction algorithm independently of experimental issues, a set of structure factor moduli for equivalent conditions and exact axial incidence on a 50nm thick crystal was also calculated, using a Bloch wave code written by Zuo [6] and adapted by its author for precession conditions.

In order to assign phases to these structure factors, the charge flipping algorithm requires a minimal set of input parameters. In particular, the procedure for handling the charge flipping parameter (δ) must be specified, and this determines either the level or the proportion of the reconstructed electron

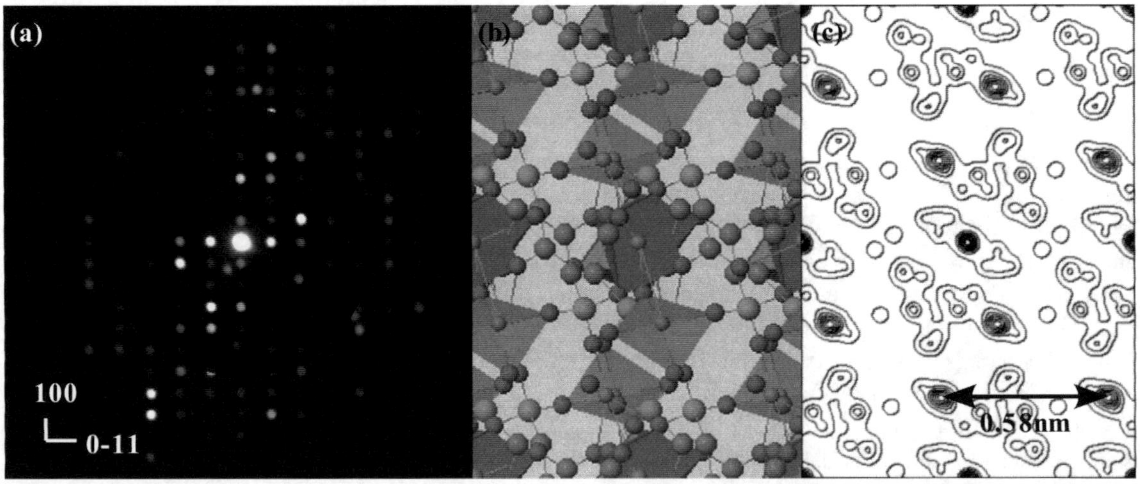

Figure 1 (a) [011] PED pattern obtained at 300kV from copper phosphonoacetate with a precession semiangle of approximately 20mrad; (b) structure model in the same orientation showing the Cu-O groups as polyhedra and the remaining organic elements as individual atoms; (c) projected potential: the strong peaks corresponding to the Cu atoms and the weaker peaks to C, P and O.

[4]Crystallographic data (CCDC 697181) can be obtained from The Cambridge Crystallographic Data Centre via www.ccdc.cam.ac.uk/data_request/cif

density which is flipped at each iteration. Optimal values for δ allow for good exploration of the possible phase space for the structure factors and in this case an effective choice proved to be to flip the 75% of pixels with lowest values at each iteration. No assumptions were made about symmetry, except that for the experimental data an average value of the measured structure factor modulus was used for **+g** and **-g** reflection pairs in order to maintain the required Friedel's law relationship.

Figure 2 is a montage of six different reconstructions obtained by phasing the calculated data. We note that there are significant differences in the details of the reconstructions: in particular, where we expect to see copper atoms in linear groups of three, the absence of enforced symmetry in the phasing tends to strengthen one of the end positions and weaken the other. We would normally expect this issue to be addressed effectively by enforcing appropriate symmetry elements. However we have found that even when symmetry is included, it remains difficult to obtain charge flipping solutions for these structures which both converge and also deliver reproducible results. If we attempt to constrain the reconstruction to include a centre of symmetry, the origin is frequently wrongly determined (as lying between the two stronger Cu positions), and the resultant phasing identifies two rather than three heavy atom positions. A typical result is shown in figure 3a.

The reason for this appears to be instrinsic to the nature of these structures. As noted above, there are normally a small number of metal ions in each unit cell, often located on special positions. The symmetry associated with these strong scatterers is usually significantly reduced by the presence of the light element ligands, and small deviations from high symmetry are common. Even when the presence of particular symmetry elements is beyond doubt, the symmetry-enforcing stage of the algorithm frequently fails to generate reproducible results, usually because there are insufficient high order symmetry elements to unambiguously determine the origin. This is true whether tested on our own coding of the algorithms [4], or on other implementations [7].

3. Strong scatterer registration

It is both difficult to perform a conventional space group determination for these materials (because of their electron beam sensitivity), and easy to overestimate the symmetry if one relies on lattice parameter measurements and perceived pattern symmetry alone. However, given that at least some of the strong scatterers are normally located on special positions, it is usually straightforward to make a visual determination of an appropriate origin for the unit cell from a single phasing. We might therefore expect to be able to locate light elements by repeating the phase assignment many times, registering each converged solution against an external reference containing only the positions of the strongest scatterers, and summing the registered results to reinforce correct phasings.

We have investigated this approach using the Superflip code [7], which by default includes the required registration and averaging operations. In this case we repeated the charge-flipping procedure

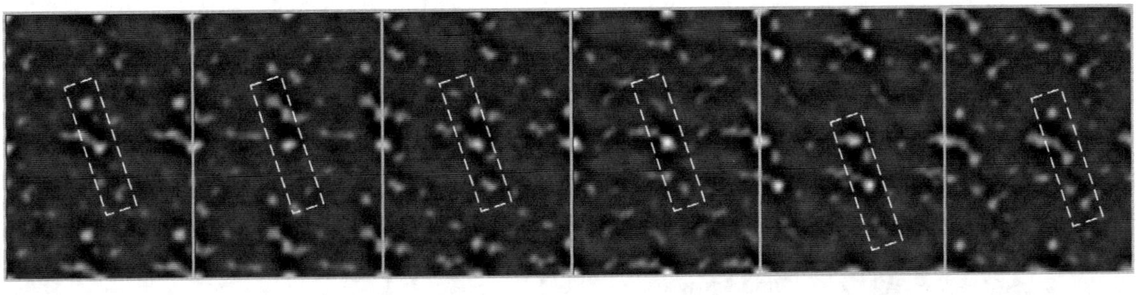

Figure 2 Phased reconstructions of the [011] projected potential for copper phosphonoacetate obtained from different initial conditions (randomised starting phases) for the charge flipping algorithm. The rectangular boxes identify qualitatively similar features in the potential: in the last two views on the right the box is displaced because the algorithm has located a different origin.

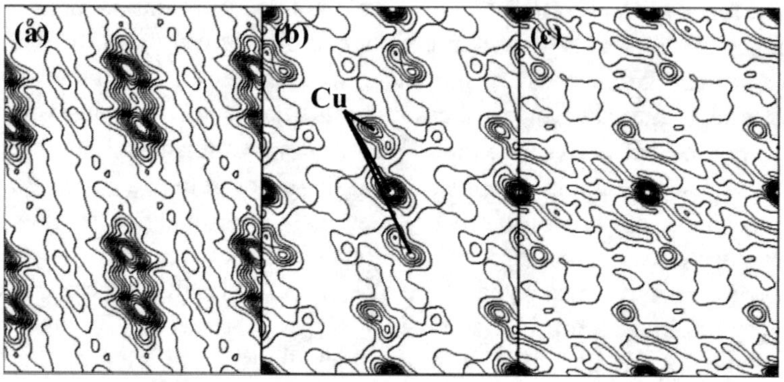

Figure 3 Phased reconstructions using: (a) alignment by symmetry;(b) strong scatterer registration; (c) experimental data with strong scatterer registration.

1000 times, of which 134 converged to a stable solution (higher convergence rates can be achieved but at the expense of detail in the reconstruction). It proved to be sufficient to create a reference file containing only the Cu atom situated at the origin: figure 3b is the result of registering each solution against this reference in order to determine the best-fit origin, and summing the set. All three copper atom positions are now correctly located (compare figures 3b, 1b and 1c), and subsidiary features have been generated around the lighter element sites. It is interesting to note that there is little deviation from the expected centrosymmetry even though this was not enforced by the algorithm. Finally therefore (figure 3c), we can examine the experimental data when phased and reconstructed using the same procedure. Unsurprisingly, the quality of the reconstruction is lower, but the metal atoms have again been located, and as before there are clear features around the lighter element positions.

4. Conclusions

The charge flipping algorithm has been shown to be an effective tool for phasing structure factor moduli determined by precession electron diffraction for hybrid inorganic-organic materials. However we note that the symmetry of the metal sublattice is often higher than that of the structure as a whole, and for this reason the most successful approach has proved to be to register a series of reconstructions against a trial structure containing only the strongest scatterers on the highest symmetry positions, followed by averaging the results to reduce the influence of spurious phasings. This allows the metal atom sites to be located, and there are indications from both calculated and experimental data that it might be possible to identify the approximate locations of lighter element groups.

Acknowledgements

Dr E G Bithell is grateful for support in the form of a Daphne Jackson Fellowship funded by Lucy Cavendish College, Cambridge, the Thriplow Charitable Trust and the Isaac Newton Trust. C A Merrill thanks Unilever plc for financial support, and Prof P A Midgley thanks the EPSRC (grant #EP/E03725). This work also made use of MRL Central Facilities supported by the MRSEC Program of the National Science Foundation under award DMR05-20415.

References

[1] Vincent R and Midgley P A 1994 *Ultramicroscopy* **53** 271-82
[2] Forster P M and Cheetham A K 2003 *Topics in Catalysis* **24** 79-86
[3] Oszlányi G and Süto A 2004 *Acta Cryst. A* **60** 134-41
[4] Eggeman A S, White T A and Midgley P A 2009 *Acta Cryst A* **65** 120-7
[5] Gjønnes K 1997 *Ultramicroscopy* **69** 1-11
[6] Zuo J M 1998 *Annual Reports of the HVEM Laboratory Kyushu University* **22** 3-10
[7] Palatinus L and Chapuis G 2007 *J. Appl. Cryst.* **40** 786-790

Acicular building blocks in the corallites of *Porites lutea*

E Brown[1]**, T Rixen**[2] **and R Kröger**[1]

[1] Department of Physics, University of York, Heslington, YO10 5DD, UK
[2] Centre for Tropical Marine Ecology, Bremen, Germany

Email: rk523@york.ac.uk

Abstract. A detailed study of the crystal building blocks of *Porites Lutea* corallites was performed using transmission electron microscope (TEM) and applying a preparation method that preserved the original microstructure of the skeleton. Bundled acicular aragonite crystals could be identified and were found to be of high structural quality. These tens of μm long needle shaped crystals were in general electron transparent - indicating a thickness well below 500 nm - and covered by an amorphous, possibly organic matrix. The growth of the orthorhombic crystal needles was found to proceed preferentially along <100> directions with the <010> and <001> growth directions being inhibited. The width distribution of these needles shows a multimodal behaviour, which could indicate a multi-stage nucleation process throughout the skeleton formation.

1. Introduction

Living coral polyps build reefs by laying down a calcium carbonate skeleton. Combining bioorganic material and inorganic matter in this way is termed biomineralisation, a process that is common to many marine organisms, particularly molluscs [1]. Biomineralisation is of particular interest to materials scientists because the minerals grow more efficiently (at lower temperatures and pressures) than their geological counterparts [2]; coral skeletons have recently been proposed as a basis for low-cost, self-regenerating coatings [3] and as a matrix for bone grafts [4]. Investigations into biomineralisation in corals have previously focused on the mesostructure of the skeleton [5], but a full understanding of the mechanism behind the nucleation and growth of the mineral phase requires a thorough study of the relevant features of the skeleton microstructure. Key features that have been identified are the centres of calcification (COCs), which are widely considered to be sites of crystal nucleation [6]. It is assumed that acicular (needle-like) crystals are forming the building blocks of the skeleton [7] by radiating out from the COCs and intergrow with those from neighbouring COCs to form the bulk skeleton. However, the detailed microstructure of the crystals is largely unknown. Successive studies have debated their symmetry [8,9], preferred growth orientation [1,10] and the importance of organic material (about 10% wt [1]) in the matrix; the combination of diffraction and high resolution imaging possible with TEM makes it an ideal tool to answer these questions.

TEM analysis of coral skeletons has previously been accomplished with the use of replicas [11] and, more recently, powdered skeleton has been successfully analysed [3]. These techniques are not completely satisfactory: The information provided by replicas is topographical and inverted, whilst laser ablation causes microstructural detail to be lost and introduces artefacts. Also the application of focused ion beam thinning was reported for red corals [12] but the actual building blocks of the coral skeleton could not be retained. Hence, manufacture of a TEM sample where structural features are preserved has rarely been reported. Therefore we developed a thorough preparation method for this purpose and carried out the structural characterization by TEM.

2. Method

Coral specimens of the genus *Porites lutea* were obtained ready-prepared from the Centre for Tropical Marine Ecology (CTME), Bremen. The polyps were soaked in fresh water and removed from the skeleton by brushing. The skeleton was air-dried, with no subsequent chemical treatments being employed. A reliable TEM sample preparation technique allowing the preservation of structural features was developed, based on the established tripod polishing technique. The skeleton was stabilised by soaking it in Gatan G1 epoxy resin and curing it at less than 150 °C. A slice perpendicular to the skeletal growth direction was taken using a diamond-coated saw and single layers of glass and silicon were fixed to either end using M-bond epoxy resin. The sandwich was sliced into samples using a diamond-coated saw. A single sample was temporarily fixed onto a tripod holder using a thin layer of Crystalbond 509 and polished to ~20 μm and in the final stage of thinning a bevel was applied. The sample was then ion milled using a Precision Ion Polishing System (PIPS), to produce an electron transparent area. The sample morphology was studied using a scanning electron microscope (FEI Sirion). Phase contrast and bright field imaging as well as diffraction imaging was performed using a JEOL 2011 TEM at an acceleration voltage of 200kV with an ultimate point resolution of 1.9 Å.

3. Results and discussion

An optical micrograph of coral skeleton sliced perpendicular to the growth direction is shown in figure 1. The marked dark spot is a COC with the crystals of interest in this paper radiating out from it.

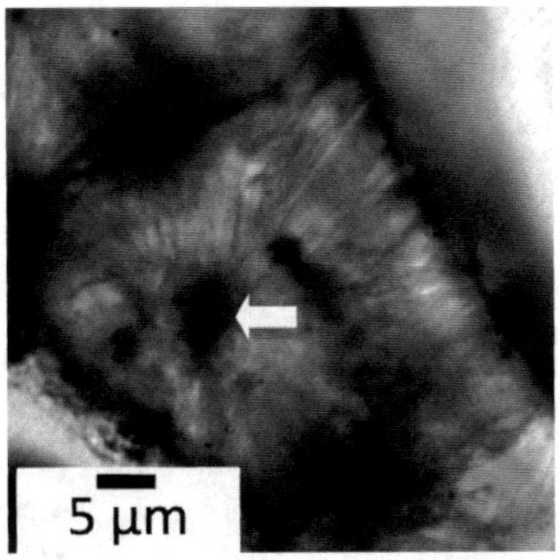

Figure 1. Optical micrograph of a slice of coral skeleton. The dark spot marked with an arrow is a COC, with needle-like crystals arranged in a fan-shaped structure.

An SEM image of the overall structure is displayed in Fig. 2 (a) revealing stacks of aragonite platelets forming the corallite microstructure. Figure 2 (b) shows a bright field image of needle-like crystals which can be attributed to neighbouring COCs. There is clear evidence of intergrowth, which lends weight to competitive growth theories that have previously been discussed [13]. There is evidence of amorphous material surrounding the needles, which may have an organic origin. A study of nacre [14] has indicated the presence of voids of organic material within the aragonite platelet structure. There is no evidence of such voids in the high quality coralline aragonite needles, suggesting that a different growth process for the coral skeleton formation. The sample was tilted into the [010] zone axis and a diffraction pattern of a single needle was obtained (figure 2 (b) inset). Analysis indicates that the needles are monocrystalline and are made of aragonite (a polymorph of calcium carbonate) with orthorhombic symmetry. Some forbidden reflections were observed in the experimental diffraction pattern; this is likely to be a result of multiple scattering as the sample is still relatively thick, but some authors have indicated that forbidden reflections in aragonite may be indicative of a lower triclinic symmetry [9]. From the SEM and the TEM investigations we conclude that the crystals are flat in [010] direction, which indicates that crystal growth in this direction is inhibited compared to the [100] direction.

Figure 2. **(a)** Scanning electron micrograph showing bundled layers of aragonite crystals. **(b)** Bright field transmission electron micrograph of these needle-like crystals. **Inset.** Selected area diffraction pattern from a single needle tilted into the [010] zone axis. The black arrow indicates the [100] growth direction of some of the needles.

Figure 3 shows a magnified image of a single aragonite needle. High resolution TEM (HRTEM) was performed (inset), which show a high crystalline quality. Again, an amorphous phase surrounding the needle is evident.

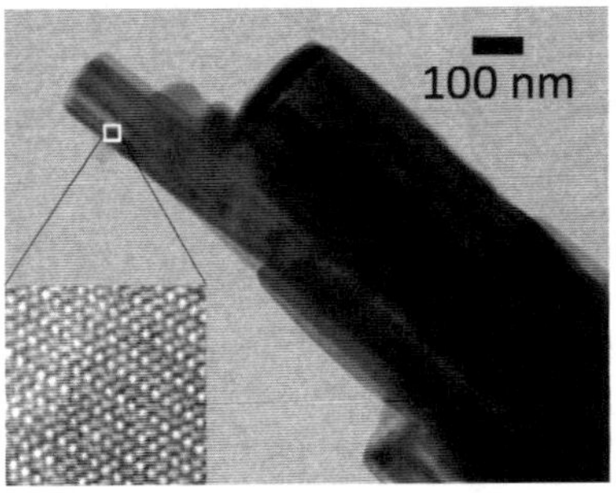

Figure 3. Bright field transmission electron micrograph at higher magnification. **Inset.** HRTEM image taken from an area similar to the one indicated by the box on the main image (not to scale).

An organic amorphous phase has previously been identified as a key component of the biomineralisation proce and is known to be a source of lattice distortions [8]. In corals, a variety of roles have been suggested for organic components at various stages of skeletal growth: as a barrier to prevent damage to the coral polyp [5] or as a mucus containing seed crystals for nucleation [11]. The width-distribution in <001> directions of individual crystalline needles was determined from the TEM images. The resulting distribution is shown in figure 4; it is interesting to note that there is a strong multi-modal distribution. This could be indicative of a multi-stage nucleation process with the wider needles originating at COCs and the narrower needles originating from secondary, unidentified nucleation sites. Alternatively, needles of different widths could originate from separate calcification centres.

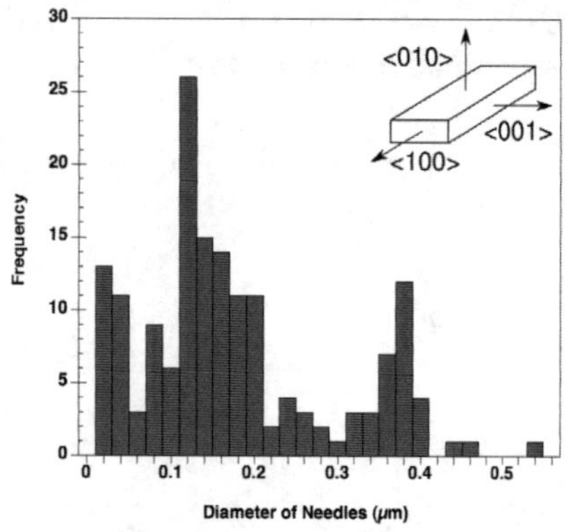

Figure 4. Histogram showing the distribution of aragonite needle diameters, *d*. A multi-modal distribution is evident. **Inset.** Schematic of a simplified aragonite needle showing the main identified crystallographic orientations.

The main crystallographic directions on an aragonite crystal are sketched in figure 4 (inset). A cuboid-shaped crystal has been used for simplicity, but the probability of faceting should not be discounted. The fastest growth is seen along the <100> direction; a result that has also been reported for nacre [15]. However, the strong growth inhibition observed along the <010> direction that causes platelet-shaped crystals to form is not observed in nacre and also voids that can be observed in the nacre structure could not be identified in the corallites crystals.

4. Conclusion

The microstructure of high quality crystalline needles considered to be building blocks of the coral skeleton was successfully analysed with TEM using a sample preparation method based on the tripod polishing technique. The needles were found to be platelets encased in an amorphous phase and were formed from orthorhombic aragonite with a preferred growth direction along <100>. It is suggested that the amorphous phase plays an important role in promoting or inhibiting growth along particular crystallographic directions.

References

[1] R. L. Hayes and N. I. Goreau, in *Proceedings of the 3rd International Coral Reef Symposium*, edited by D. L. Taylor (Rosenstiel School of Marine and Atmospheric Science, Miami, 1977), p. 439.
[2] S. Mann, J. Chem. Soc., Dalton Trans., 3953 (1997).
[3] D. Nave, S. Rosenwaks, R. Vago, and I. Bar, J. App. Phys. **95**, 8309 (2004).
[4] G. Guillemin, J.-L. Patat, J. Fournie, and M. Chetail, J. Biomed. Mat. Res. **21**, 557 (2004).
[5] A. L. Cohen and T. A. McConnaughey, Rev. Mineral. & Geochem. **54**, 151 (2003).
[6] J. Cuif and Y. Dauphin, Biogeosciences **2**, 67 (2005).
[7] M. M. Ogilvie, Phil. Trans. Roy. Soc. London B **187** (1896).
[8] E. N. Caspi, B. Pokroy, P. L. Lee, J. P. Quintana, and E. Zolotoyabko, Acta. Cryst. B **61**, 129 (2005).
[9] D. J. M. Bevan, E. Rossmanith, D. K. Mylrea, S. E. Ness, M. R. Taylor, and C. Cuff, Acta. Cryst. B **58**, 448 (2002).
[10] D. Dahan, R. Vago, and Y. Golan, Mat. Sci. & Eng. **C23**, 473 (2003).
[11] I. S. Johnston, Int. Rev. Cytol. **67**, 171 (1980).
[12] D. Vielzeuf, J. Garrabou, A. Baronnet, O. Grauby and C. Marschal, *Am. Mineralogist* **93**, 1799 (2008).
[13] D. J. Barnes, Science **170**, 1305 (1980).
[14] K. Gries, R. Kröger, C. Kübel, M. Fritz, and A. Rosenauer, Acta Biomaterialia, in press (2009).
[15] K. Gries, R. Kröger, C. Kübel, M. Schowalter, M. Fritz, and A. Rosenauer, Ultramicroscopy **109**, 230 (2009).

Journal of Physics: Conference Series **241** (2010) 012027 doi:10.1088/1742-6596/241/1/012027

Dark-field electron holography for the mapping of strain in nanostructures: correcting artefacts and aberrations

M J Hÿtch[1], F Houdellier, F Hüe[2] and E Snoeck
CEMES-CNRS, 29 rue Jeanne Marvig, 31055 Toulouse, France

E-mail: hytch@cemes.fr

Abstract. We present details of the new electron holographic dark-field technique (HoloDark) for mapping strain in nanostructures. A diffracted beam emanating from an unstrained region of crystal is interfered (with the aid of an electrostatic biprism) with a diffracted beam from the strained region of interest. Geometric phase analysis (GPA) of the holographic fringes determines the relative deformation of the two crystalline lattices. Strain can be measured to high precision, with nanometre spatial resolution and for micron fields of view. Experiments are carried out on the SACTEM-Toulouse, a Tecnai F20 (FEI) equipped with imaging aberration corrector (CEOS), field-emission gun and rotatable biprism (FEI). We operate the microscope in free-lens control with the main objective lens switched off and using the corrector transfer lenses as a Lorentz lens. We will present measurements of strain in test nanostructures and show how artefacts from thickness variations can be removed. Finally, we show our first results using a recently developed aberration-corrected Lorentz mode (CEOS).

1. Introduction

Accurate determination of strain in electronic devices has been the subject of intense work during the last decades. Few techniques are able to provide highly localized and accurate information at the nanoscale. Among these, convergent-beam electron diffraction (CBED) combines the advantages of very small probes and remarkable sensitivity to small variations in the lattice parameter [1]. However, elastic relaxation effects make the analysis extremely difficult, necessitating time-consuming dynamical simulations combined with finite element modelling [2].

Rather than collecting data at isolated points, strain distributions can be mapped in a continuous fashion using high-resolution transmission electron microscopy (HRTEM) [3]. However, HRTEM suffers from limitations due to specimen preparation, field of view and noise. Whilst this technique is highly accurate at the nanometre scale, mapping strain across multilayers and transistor arrays requires fields of view that are not easily accessible to HRTEM, even with a Cs-corrected microscope. Nevertheless, it is possible to map strains in the channel region of transistors with this technique [4].

We present results from the new technique of dark-field electron holography (HoloDark) for measuring strain in nanostructures and electronic devices [5,6]. A hologram is created (figure 1) from the interference between the diffracted beam emanating from an unstrained region of crystal, which serves as the reference, and a beam from the region of interest containing strained crystal using the dark-field holography configuration [7]. A typical example for these two regions would be the

[1] To whom any correspondence should be addressed.
[2] Current address: Dept. Materials Science, University of Cambridge, Pembroke Street, Cambridge, UK

substrate and an active region of a device. The aim is to measure geometric phase differences, from which the deformation can be calculated [3]. Naturally, any other phase contributions should be minimised, notably, dynamic phases due to thickness variations. For this reason, specimens should be prepared with suitably uniform thickness and regions exhibiting bend contours avoided.

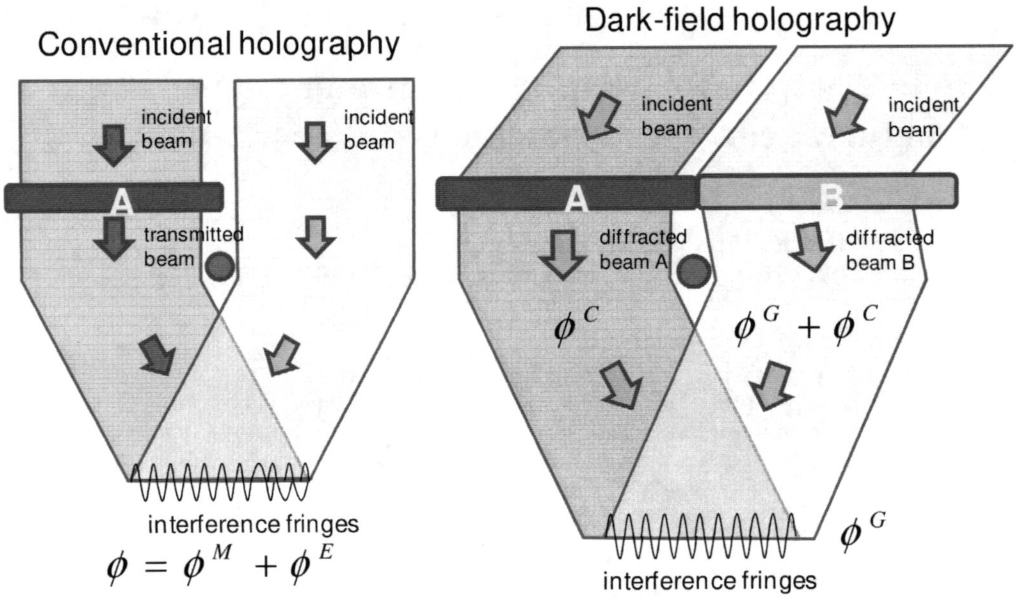

Figure 1. Schematic representation of conventional off-axis holography (left) and dark-field electron holography (right). Transmitted beam in the vacuum, and diffracted beam from unstrained crystal, serve as references for the former and latter respectively.

The technique has a number of advantages over geometric phase analysis (GPA) of high-resolution images for the study of transistors. The specimens do not need to be as thin, being more like those of conventional TEM. Specimens are therefore easier to prepare and the effects of thin-film relaxation reduced. The major advantage, however, is the ability to analyse large regions of crystal at relatively low resolution.

Results will be presented for two different strained-silicon samples. TEM specimens are prepared by focussed ion beam (FIB) to thicknesses of between 100 and 200 nm. Observations are carried out on the SACTEM-Toulouse, a Tecnai (FEI) 200kV TEM equipped with a Cs corrector (CEOS), rotatable biprism and 2k CCD camera (Gatan). The microscope is operated in a pseudo-Lorentz mode using the corrector first transfer lens as a Lorentz lens and with the main objective lens and hexapoles switched off [8]. Results will also be shown with the new aberration-corrected Lorentz mode (CEOS). Typical fringe spacings are 1-2 nm and hologram widths from 300-500 nm allowing lengthwise fields of view of several microns. Strain fields are extracted using a modified version of GPA Phase 2.0 (HREM Research Inc.), a plug-in for DigitalMicrograph (Gatan).

2. Theory

The phase appearing in geometric phase analysis (GPA) is usually presented in the perspective of high-resolution electron microscopy [3]. Here, we need to place it in a wider context: that of the other phases measured in electron microscopy – in particular by electron holography.

The wave function of the fast electron at the exit surface of a non-perfect crystal can be written:

$$\psi(\mathbf{r}) = \sum_g \psi_g(\mathbf{r}) \exp\{2\pi i \mathbf{g}.\mathbf{r}\} \tag{1}$$

where **r** is in the *xy*-plane, conjugate with the image plane, and the phase term due to the forward momentum is implicit. The local Fourier components have an amplitude and phase:

$$\psi_g(\mathbf{r}) = a_g(\mathbf{r})\exp\{i\phi_g(\mathbf{r})\} \tag{2}$$

corresponding to the complex amplitudes of the diffracted beams as a function of position across the exit surface of the crystal [9]. The phases here refer uniquely to the phases of Fourier components in reciprocal space and not those of the wave function in real space. The phase can be written as having four components:

$$\phi_g(\mathbf{r}) = \phi_g^C(\mathbf{r}) + \phi_g^M(\mathbf{r}) + \phi_g^E(\mathbf{r}) + \phi_g^G(\mathbf{r}) \tag{3}$$

crystalline, magnetic, electrostatic and geometric phase. Crystalline phase is due to the (dynamical) diffraction of the fast electron by the crystalline potential, and electrostatic phase includes contributions from the mean inner potential of the material. We choose explicitly the origin to coincide with an axis of symmetry of the crystal. Geometric phase therefore arises from any local displacement, $\mathbf{u}(\mathbf{r})$, of the crystal with respect to this axis:

$$\phi_g^G(\mathbf{r}) = -2\pi\mathbf{g}.\mathbf{u}(\mathbf{r})\,. \tag{4}$$

It is in this context that the setup for dark-field electron holography was imagined [5]. If the diffracted beams can be interfered, the geometric phase component can be measured directly. To measure strain, the requirement is the elimination of the other phase terms, notably crystalline and electric phase. To a first approximation, these two phases do not depend directly on the local strain. Therefore, if the sample is uniformly thick, these terms will constant and will not contribute to the measured strain as this is determined from the gradient of the geometric phase [3]. Nevertheless, where thickness variations are present, the electric phase can be removed, as the example will show.

3. Strained-silicon test sample

Figure 1 shows a test structure for the study of stress liners. It consists of a comb-like periodical array of etched silicon trenches a few hundred of nanometers deep and about 50-100 wide. The Si is covered by a capping layer of a few tens of nanometers thick of Si_3N_4 followed by silicon oxide. The differences in thermal expansion coefficients between the different materials results in a high level of stress in the nitride layer after processing, which will in turn deform the silicon.

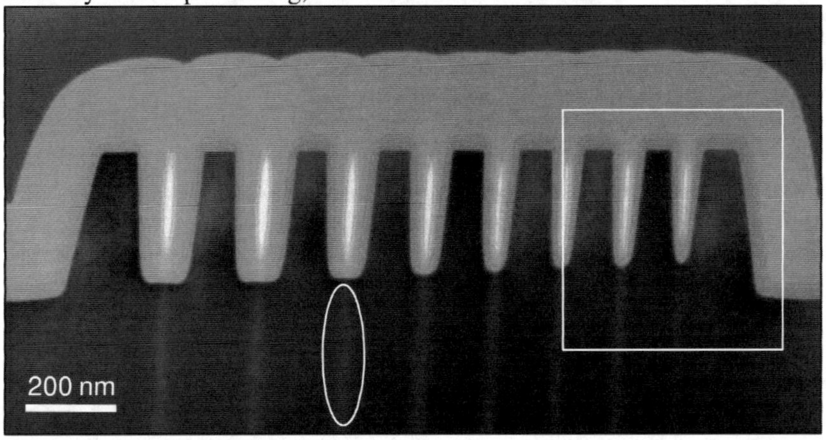

Figure 2. Conventional bright-field image of comb-like Si test structure for Si_3N_4 stress liners. Analysed zone (square) and curtaining effect from FIB specimen preparation (oval) indicated.

Figure 2 shows the analysis of a dark-field hologram obtained using a (111) diffracted beam. A two-beam condition was chosen to maximise the diffracted intensity and to limit multi-beam dynamical scattering effects. The characteristic feature of a dark-field hologram is that there are no interference fringes in the vacuum (see inset figure 2a), excepting the Fresnel fringes from the biprism. The hologram is analysed as usual by masking the side band in Fourier space to produce the phase (figure 2b) and amplitude maps (figure 2c). The deformation of the (111) lattice planes is determined directly by numerical differentiation of the phase (figures 2d and 2e), as in conventional geometric phase analysis.

The major component of the deformation is in the rotation of the (111) lattice planes in the end-most element of the structure. Artefacts from the specimen preparation can also be seen in the substrate that arise from the local variation in specimen thickness from curtaining during the FIB preparation. High frequency artefacts are also present from the Fresnel fringes. We will show how these artefacts can be corrected for later.

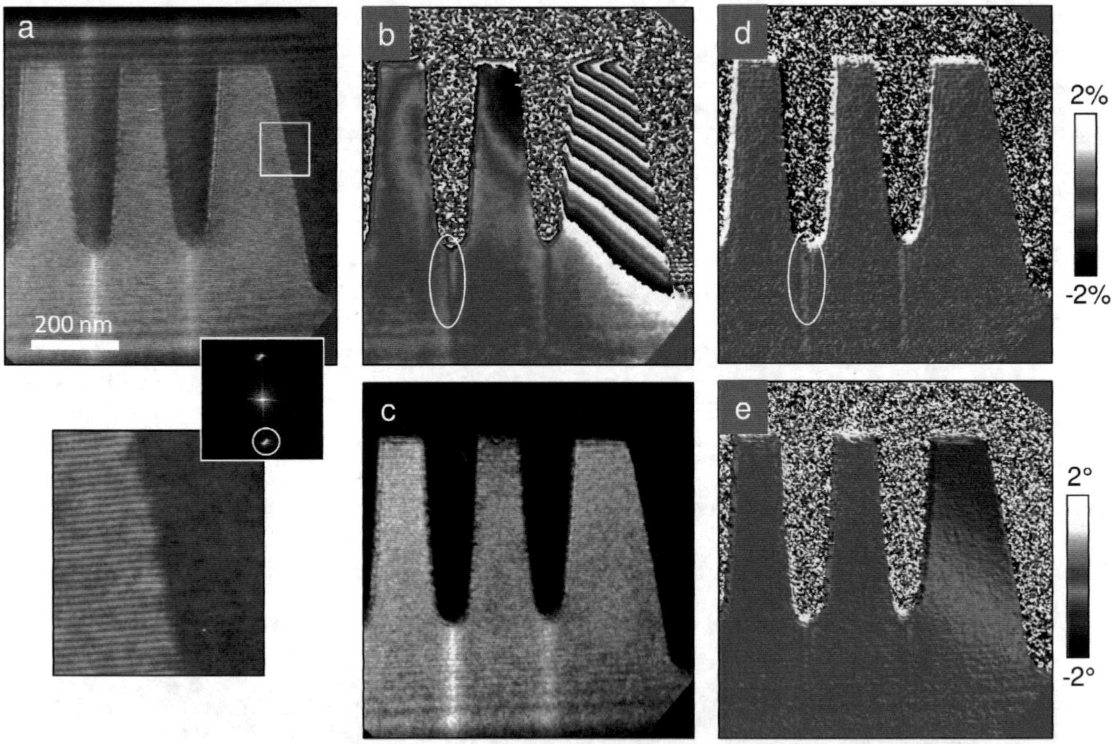

Figure 3. Analysis of dark-field hologram: (a) dark-field hologram from (111) diffracted beam, Fourier transform and zoom inset; (b) phase; (c) amplitude; (d) local variation in planar spacing; (e) local lattice plane rotation.

Figure 4 shows the bright-field and dark-field electron holograms from the same area in a repeat experiment. The bright-field hologram was formed by interfering the transmitted beam from the substrate with the transmitted beam from the strained area, in an exact analogy with the dark-field hologram. This hologram should only contain phase changes due to the crystalline phase and the mean inner potential of the surrounding material. Figure 4e shows the result of the subtraction of the dark-field phase (figure 4c) and the bright-field phase (figure 4d). It can be seen that the artefacts due to the thickness variations have been successfully removed. This is confirmed in the rotation maps (figure 5). For this example we have also removed the Fresnel fringe artefacts by Fourier filtering.

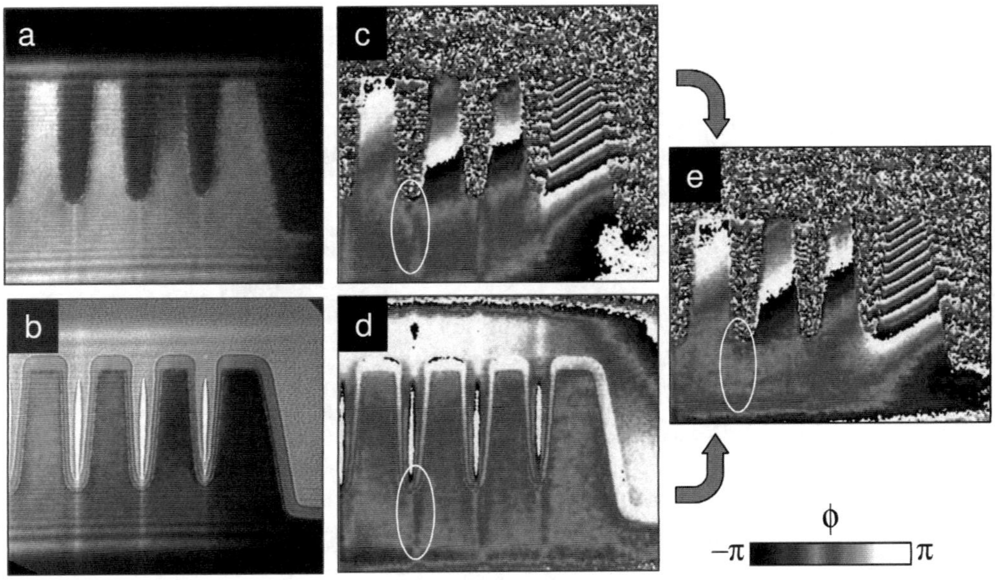

Figure 4. Correction of thickness artefacts: (a) dark-field electron hologram; (b) bright-field hologram; (c) dark-field phase; (d) bright-field phase; (e) subtraction of the dark-field and bright-field phases. Ovals indicate principle artefacts. Fresnel fringes have been removed by Fourier filtering.

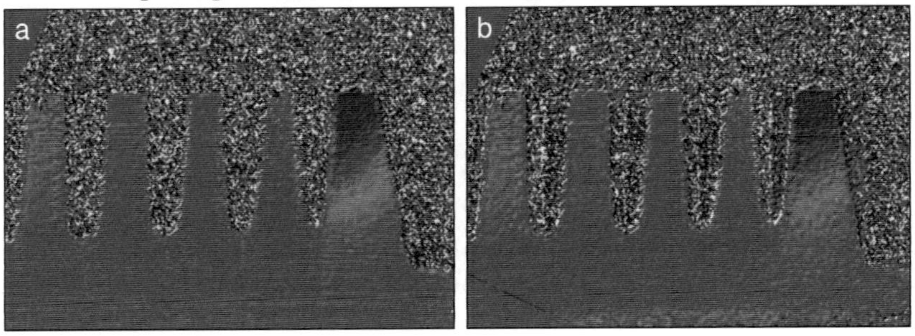

Figure 5. Rotation maps before (a) and after (b) removal of the thickness variation artefacts. The Fresnel fringe artefacts have also been removed from (b) by Fourier filtering.

4. Aberration-corrected Lorentz mode

The first experiment showed results using the pseudo-Lorentz mode with the hexapoles of the corrector switched off. Figure 6a shows the typical problems we have encountered using this mode. The sample is a strain-silicon n-MOSFET transistor, previously studied in detail using the HoloDark technique [10]. The astigmatism and coma are difficult to correct, particularly in the low-light conditions of dark-field imaging. In addition, the minimum steps in defocus are rather large. Nevertheless, excellent information can be obtained from such holograms and we have measured the strain to a precision of 2.10^{-4} at 4 nm spatial resolution [10].

Recently, an aberration-corrected Lorentz mode has been designed and installed on our microscope which utilises the hexapoles to correct the aberrations of the first transfer lens. Results for the same sample are shown in figure 6b. The image is much clearer and well defined. The deformation map has a signal-to-noise 60% higher than the previous example, using identical image processing. Whilst better holograms are possible in the non-corrected mode, they are much more difficult to obtain. The corrected mode in particular promises to increase the spatial resolution of the Lorentz mode, and this we will be investigated on suitable test samples in the future.

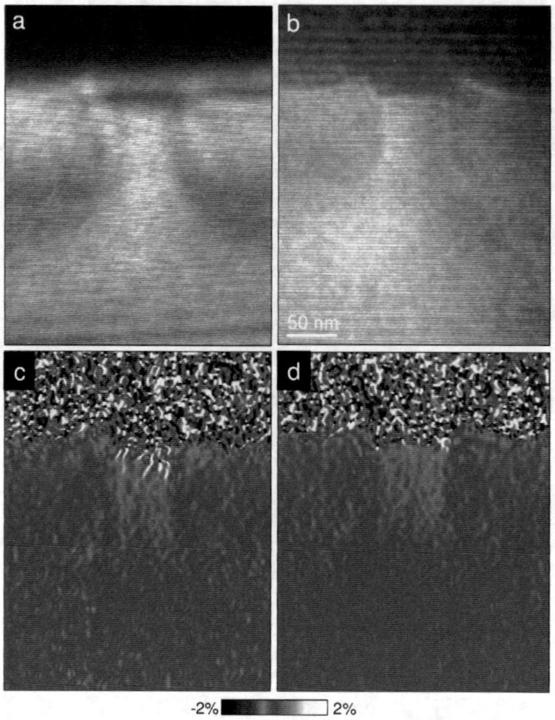

Figure 6. Comparison of pseudo-Lorentz mode (a,c) and aberration-corrected Lorentz mode (b,d). Dark-field electron holograms from the (004) beam (a,b) and corresponding deformation maps (c,d).

5. References

[1] Spence J C H and Zuo J M 1992 *Electron Microdiffraction* (Plenum Press, New York)
[2] Houdellier F, Roucau C, Clément L, Rouvière J-L and Casanove M-J 2006 *Ultramicroscopy* **106** 951
[3] Hÿtch M J, Snoeck E and Kilaas R 1998 *Ultramicroscopy* **74** 131
[4] Hüe F, Hÿtch M J, Bender H, Houdellier F and Claverie A 2008 *Phys. Rev. Lett.* **100** 156602
[5] Hÿtch M J, Houdellier F, Hüe F and Snoeck E 2008 *Nature* **453** 1086
[6] Hÿtch M J, Houdellier F, Hüe F and Snoeck E Patent Pending FR N° 07 06711
[7] Hanszen K J 1986 *J. Phys.* D **19** 373
[8] Snoeck E, Hartel P, Müller H, Haider M and Tiemeijer P C 2006 Proc. IMC16 International Microscopy Congress (IMC, Sapporo) vol. 2 p. 730
[9] Hÿtch M J and Plamann T 2001 *Ultramicroscopy* **87** 199
[10] Hüe F, Hÿtch M J, Houdellier F, Claverie A and Bender H 2009 *Appl. Phys. Lett.* **95** 073103

Acknowledgments
The authors thank the European Union for support under the IP3 project ESTEEM (Enabling Science and Technology through European Electron Microscopy, IP3: 0260019), the French Government (MINEFI) through the NANO2012 initiative (project IMASTRAIN), and the French National Agency for Research (ANR) through its program in Nanosciences and Nanotechnologies (HD STRAIN Project No. ANR-08-NANO-0 32). Laurent Clément and Pierre Morin (STMicro, Crolles) are gratefully acknowledged for supplying the stress liner sample. We are indebted to Peter Hartel, Haiko Müller, Stephan Uhlemann and Max Haider (CEOS) for the design and installation of the aberration-corrected Lorentz mode.

The Effect of Sintering Temperature on Phase, Microstructure and Properties of $Sr_5Nb_4TiO_{17}$

A Manan[1], Y Iqbal[1] and I Qazi[2]

[1]Materials Research Laboratory, Institute of Electronics & Physics, University of Peshawar, Pakistan
[2]Faculty of Engineering Sciences, GIK Institute, Topi, NWFP, Pakistan

Email:dryaseeniqbal@yahoo.co.uk

Abstract. Ceramics with layered perovskite structures are widely used as microwave dielectric components in the telecommunication industry. The potential of $Sr_5Nb_4TiO_{17}$ as a possible Pb- and Bi-free alternative for microwave applications was investigated. $Sr_5Nb_4TiO_{17}$ was prepared by mixed oxide route. Calcined and sintered samples were characterized by X-ray diffraction and scanning electron microscopy. Samples sintered at temperatures ≤1400°C comprised mainly $Sr_5Nb_4TiO_{17}$ along with a small amount of $Sr_2Nb_2O_7$ as a second phase. Single phase dense $Sr_5Nb_4TiO_{17}$ with relatively good electrical properties formed at 1450°C whereas $Sr_6Nb_4Ti_2O_{20}$ formed as second phase at temperatures ≥1500°C. SEM showed an increase in the grain size and variation in grain morphology with increase in the sintering temperature. Further work is underway to optimize the processing and hence the properties of $Sr_5Nb_4TiO_{17}$.

1. Introduction

The fast growth in wireless communication during the last few decades has become possible due to development in ceramics-based dielectric components. The important characteristics required for microwave applications are high dielectric constant ($20<\varepsilon_r<100$) for miniaturization, high quality factor (Q>30,000 at 1GHz) for selectivity and nearly zero temperature coefficient of resonant frequency (τ_f~±2ppm/°C) for thermal stability of the components [1]. A number of compositions with good microwave dielectric properties have been investigated for practical applications; however, there is still a great demand for new materials with improved properties. The structures of some of the ceramics-based dielectric resonators belong to $A_nB_nO_{3n+2}$ type layered perovskite families. For example, $CaLa_4Ti_5O_{17}$ with n = 5 and $CaLa_8Ti_9O_{31}$ with n = 4.5 belonging to $A_nB_nO_{3n+2}$ type layered perovskite family have been reported to have ε_r = 55.2 and 54.9, Q×f = 17359 GHz (at 3.67GHz) and 19345 GHz (at 3.65 GHz), and τ_f = -20 and -6ppm/°C respectively [2]. Substitution of 1 wt% of Ca by Zn in $CaLa_4Ti_5O_{17}$, gave ε_r = 57.6, Q×f = 17100 GHz and τ_f ~ 4.9ppm/°C [3]. Similarly, addition of 0.5 wt% CuO as sintering aid to $Ca_{0.99}Zn_{0.01}La_4TiO_{17}$ resulted in ε_r = 57, Q×f =15000 GHz and τ_f ~ -8.16ppm/°C at a relatively lower temperature [4].

Compounds in the SrO-Nb_2O_5-TiO_2 system also belong to $A_nB_nO_{3n+2}$ (A = Sr, B = Ti, Nb) family with a layered perovskite structure and $Sr_5Nb_4TiO_{17}$ has been reported to have promising values for $T_{C\varepsilon}$ and ε_r, however, with poor quality factor at microwave frequencies [5-6]. $Sr_5Nb_4TiO_{17}$ is anti-ferroelectric at T_c≥860K [7]. At room temperature, the crystal structure of $Sr_5Nb_4TiO_{17}$ is orthorhombic (Pnnm) with lattice parameters a = 5.6614, b = 32.515 and c = 3.9525Å [8]. This compound is iso-structural with $Ca_5Nb_4TiO_{17}$ [9] and adopts $A_nB_nO_{3n+2}$ type perovskite related structure with Sr occupying the A site with coordination of 12 O's, and Nb and Ti occupying the B sites with coordination of 6 O's. This crystal structure can be described as consisting of identical perovskite-like blocks, five corner-sharing BO_6 octahedra thick and separated by an extra layer of oxygen [6]. Here we report preliminary results regarding the effect of sintering temperature on phase, density, microstructure and properties of $Sr_5Nb_4TiO_{17}$ prepared by mixed oxide route.

2. Experimental

$Sr_5Nb_4TiO_{17}$ was prepared by mixing laboratory reagent grade $SrCO_3$, Nb_2O_5 and TiO_2 (Anatase) in stoichiometric ratios. The initial ingredients were mixed and milled in polyethylene mill jars using 0.6 mm diameter Y-toughened zirconia balls as grinding media in 2-propanol for 60h at 200 rpm in a ball mill. The slurry was dried at ~85°C overnight and sieved. Samples were calcined for 2h at 990°C and 1350°C at 10°C/min and re-milled for 30 min to ensure dissociation of agglomerates. 13mm dia. pellets were pressed in a cylindrical steel die at 140 MPa using a manual pellet press. The pellets were sintered at 1400°C, 1450°C, 1500°C and 1550°C for 2h at heating / cooling rate of 10°C/min. XRD of the mixed milled powders and powdered pellets was performed using a JEOL JDX 3532 diffractometer operating at 30 mA and 40 kV, with CuK_α radiation at 2θ= 10 to 70° with step size of 0.03° at 2s/step. The bulk density of sintered pellets was measured using a MD-300S Electronic densitometer. For scanning electron microscopy (SEM), the pellets were cut into 5x3x3 mm^3 pieces using a diamond fine cutting saw and finely polished before thermal etching for 30 min at 10°C/min, at temperatures ~10% less than the relevant sintering temperatures. Gold-coated samples were examined using a JEOL JSM5910 SEM operating at 15-20keV. For electrical measurements, the opposite surfaces of samples were gold-coated to mount electrodes onto them. The dielectric constants and the loss tangent were measured using a 4287A LCR meter at 1 to 100 MHz.

3. Results and Discussion

XRD of the sample calcined at 990°C revealed the formation of $Sr_5Nb_4TiO_{17}$ (JCPDS file # 871170) as the major phase along with a few low intensity XRD peaks due to $Sr_2Nb_2O_7$ (JCPDS file# 70-114). It was difficult to distinguish the peaks due to these phases because of overlapping or close d-values (figure 1a). The decrease in the intensity of distinguishable $Sr_2Nb_2O_7$ peaks upon increase in the calcination temperature to 1350°C (figure 1b) confirmed the formation of second phase in samples calcined at 990°C and 1350°C. These second phase peaks nearly vanished after sintering the sample at 1400°C (figure 1c) and the phase observed after sintering at 1450°C was pure $Sr_5Nb_4TiO_{17}$ (figure 1d). Upon further increase in the sintering temperature to 1500°C and 1550°C, XRD revealed the presence of $Sr_5Nb_4TiO_{17}$ as the major phase; however, a few low intensity peaks due to $Sr_6Nb_4Ti_2O_{20}$ (JCPDS file# 49-544) emerged (figure 1e-f). Levin *et al* [6] also reported the possibility of formation of relevant second phases in the $Sr_n(Nb,Ti)_nO_{3n+2}$ homologous series with n = 5, 6 and 7, at temperatures above 1400°C.

A comparison of the measured apparent density of the pellets sintered at various temperatures indicated that the highest density (4.46 g/cm^3) achieved so far corresponded to 1450°C which is ~86.7% of the theoretical density of $Sr_5Nb_4TiO_{17}$ (table 1). Thus, further work is required to improve the density and hence the electrical properties.

The microstructure of samples calcined at 1350°C, and sintered at 1400, 1450, 1500 and 1550°C are compared in figure 2. The observed grain morphology of samples sintered at 1400°C was irregular in shape with grain size ranging from ~0.2×0.2 μm^2 to ~2.5×5 μm^2 (figure 2a). The large number of irregular shaped grains indicated that they were $Sr_5Nb_4TiO_{17}$ grains, consistent with XRD results. Plate-like single phase $Sr_5Nb_4TiO_{17}$ grains of size ranging from ~1.5×2 μm^2 to ~1.5×36 μm^2 (figure 2b) were observed in the microstructure of samples sintered at 1450°C, consistent with XRD. On further increase in sintering temperature to 1500°C, the microstructure again appeared as consisting of grains of two morphologies with size ranging from 1×1 to 1×12 μm^2 (figure 2c). XRD indicated the presence of $Sr_5Nb_4TiO_{17}$ and $Sr_6Nb_4Ti_2O_{20}$ in this sample. At 1550°C, the size of the grains increased ranging from 2.5×3 μm^2 to 3×18 μm^2 accompanied by a decrease in density.

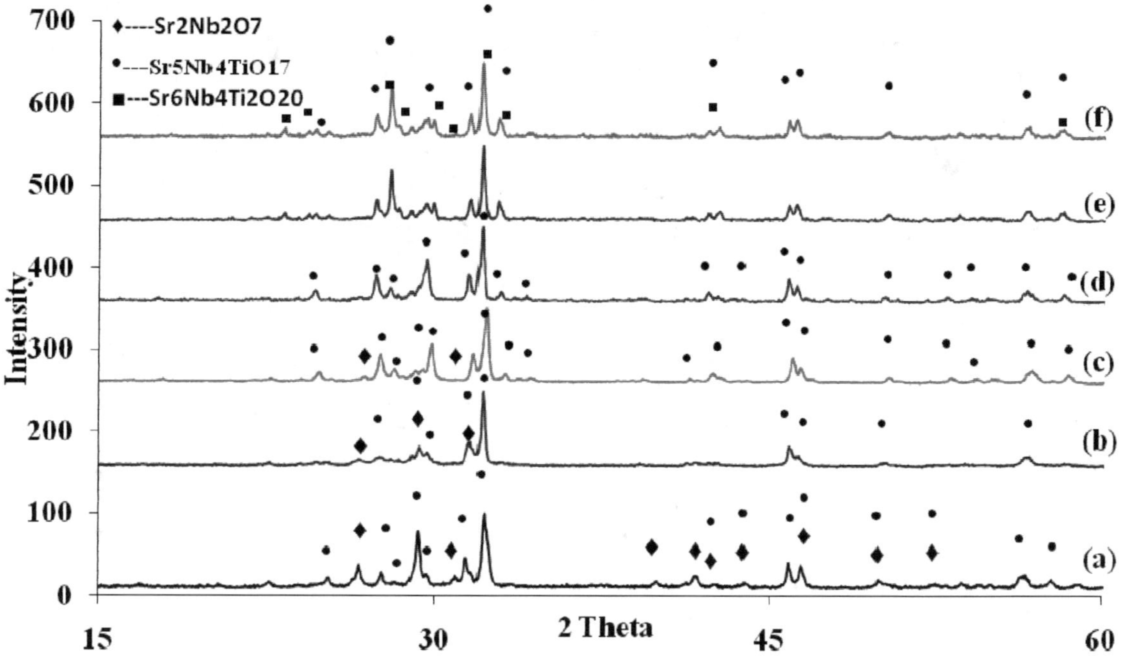

Figure 1. XRD patterns of $Sr_5Nb_4TiO_{17}$, a) calcined at 990°C, b) calcined at 1350°C, and sintered at c) 1400°C, d) 1450°C, e) 1500°C and f) 1550°C, showing the formation of various phases.

Table 1. Dielectric properties of $Sr_5Nb_4TiO_{17}$ at 80 MHz

Temp (°C)	ρ(gcm^{-3})	ε_r at 25°C	Q_o-Value at 80 MHz	tanδ
1400	3.53	6	500	0.002
1450	4.46	13	1290	0.00077
1500	4.43	9	100	0.01

Preliminary results for dissipation factor or its inverse the dielectric quality factor (Q_o=1/tanδ) at 80 MHz for $Sr_5Nb_4TiO_{17}$ sintered at various temperatures are given in table 1. It was observed that the dielectric loss decreased with increase in the sintering temperature. Initially, Q_o increased with increase in frequency from 1 to 15 MHz and then remained almost constant with slight fluctuations at 20 to 100 MHz. The maximum measured Q_o (~1290) for $Sr_5Nb_4TiO_{17}$ corresponded to the dense single phase sample sintered at 1450°C in the frequency range 1 to 100 MHz. This indicated that Q_o increased with increase in density. Although, the density of sample sintered at 1500°C was higher than that sintered at 1400°C, the observed lower value of Q_o for this sample may be due to the formation of second phase ($Sr_6Nb_4Ti_2O_{20}$). As expected, the dielectric constant measured for the sample sintered at 1400°C was lower than the sample sintered at 1450°C and 1500°C, probably due to the low density of the former sample. The dielectric constant for $Sr_5Nb_4TiO_{17}$ in the given frequency range was almost constant and increased with increase in the density. The dielectric constant of $Sr_5Nb_4TiO_{17}$ decreased from 57 reported earlier [5] at a lower frequency (10 kHz) to 13 with increase in frequency to 1 MHz. This suggested that at lower frequencies the electronic, ionic, dipolar and interfacial/surface polarization may have contributed to the dielectric constant, however, at higher frequencies, contributions from extrinsic effects decrease [10].

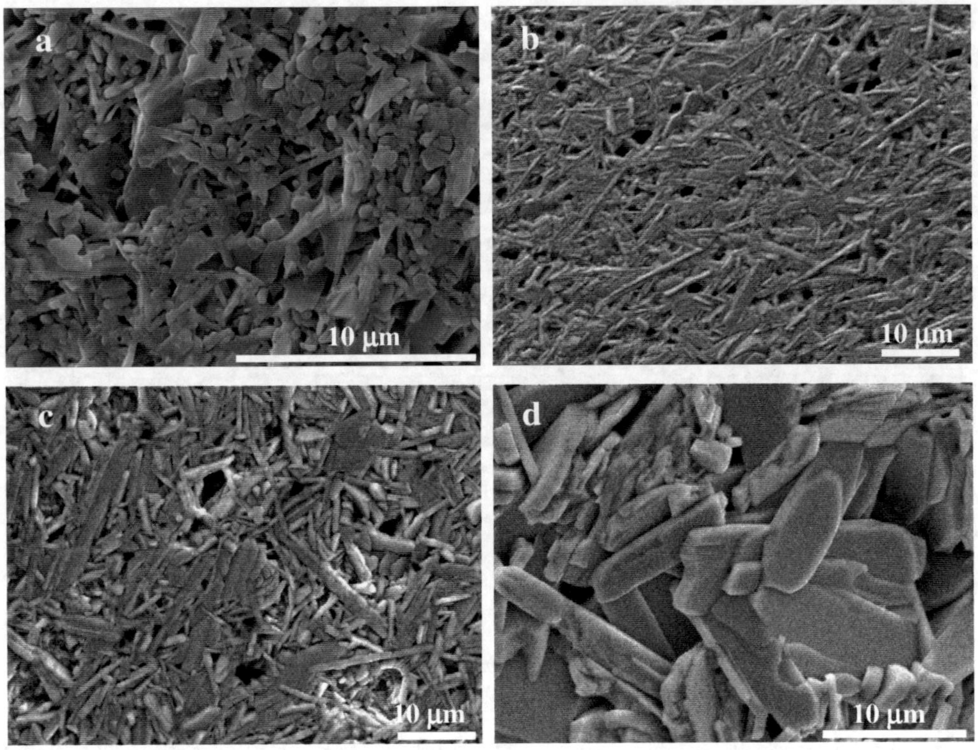

Figure 2. Secondary electron SEM images of samples calcined at 1350°C and sintered at, a) 1400°C, b) 1450°C, c) 1500°C and d) 1550°C, showing variation in grain size and morphology with sintering temperature.

4. Conclusions

Samples calcined at 1350°C and sintered at 1450°C comprised of single $Sr_5Nb_4TiO_{17}$ phase; however, the maximum density achieved was only ~86% of its relevant theoretical density. Thus further studies to improve the density and hence the microwave properties are required. Sintering at temperatures above or below 1450°C resulted in second phase formation, accompanied by a decrease in density. The grain size increased with increase in the sintering temperature which may also be a reason for the observed decrease in density at temperatures above 1450°C.

Acknowledgement

The authors acknowledge the financial support of the Higher Education Commission Pakistan (NRPU Project No. 20-569).

References

[1] Freer R and Azough F 2008 *J. Eur. Ceram. Soc.* **28**, 1433
[2] Jawahar I N, Santha N, Sebastian M T and Mohanan P 2002 *J. Mater. Res.***17**, 3084
[3] Zhao F, Yue Z, Gui Z and Li L 2006 *J. Am. Ceram. Soc.* **89**, 3421
[4] Chen Y C and Tsai J M 2008 *Jpn. J. Appl. Phy.* **47**, 7959
[5] Iqbal Y and Reaney I M 2004 *Ferroelectrics.* **302**, 259
[6] Levin I, Bendersky A, Vanderah T A, Roth R S and Stafsudd O M 1998 *Mat. Res. Bull.* **33**, 501
[7] Isupov V A 1999 *Ferroelectrics.* **220**, 79
[8] Drews A R, Wong-Ng W, Roth R S and Vanderah T A 1996 *Mat. Res. Bull.* **31**, 153
[9] Titov Y A, Belyavina N M, Markiv V Y, Slobodyanik M S and Chumak V V 2005 *J. Alloys Compd.* **387**, 82
[10] Singh A K and Choudh R N P J 2003 *J. Phys. Chem. Solids.* **64**, 1185

ZnO as Sintering Additive in $Sr_2Nb_2O_7$

Y Iqbal, A Manan, M K Safeen, A Khan
Materials Research Laboratory, Institute of Electronics & Physics, University of Peshawar, Pakistan

E-mail: dryaseeniqbal@yahoo.co.uk

Abstract. The effect of 0.3wt% ZnO addition on phase, microstructure and properties of $Sr_2Nb_2O_7$ was investigated. X-ray diffraction (XRD) of samples calcined at 1000°C revealed the formation of the major $Sr_2Nb_2O_7$ phase along with $Sr_5Nb_4O_{15}$ as second phase. After sintering at 1300°C, 1350°C and 1400°C, a two phase ceramic developed comprising $Sr_3Zn_{0.33}Nb_{4.67}O_{15}$ and $Sr_2Nb_2O_7$. The addition of ZnO resulted in the reduction of grain size by ~50% and an optimum sintering temperature reduced by ~50°C that yielded 95% of the theoretical density of $Sr_2Nb_2O_7$. Preliminary results of dielectric properties are presented and compared with previous studies.

1. Introduction

$Sr_2Nb_2O_7$ belongs to the homologous series of $A_nB_nO_{3n+2}$ type compounds with A=Sr, B=Nb and n = 4 [1]. Its Curie temperature is 1342°C, below which it is ferroelectric and possesses interesting piezoelectric and electro-optic properties [2]. $Sr_2Nb_2O_7$ with certain substitutions, e.g. in the form of important solid solutions such as, $Sr_2(Nb, Ta)_2O_7$ and $(Sr, Ba)_2Nb_2O_7$, is used for ferroelectric memory storage devices based on field effect transistors and capacitors [3]. Its crystal structure is orthorhombic with lattice parameters a = 3.97Å, b = 26.86Å and c = 5.72Å, and its space group is $Pna2_1$ [4]. The structure of $Sr_2Nb_2O_7$ comprises slabs, each slab extending parallel to {010} planes and consisting of $4NbO_6$ octahedra with Sr atoms occupying the space between them. These NbO_6 octahedra are known to distort and tilt upon cooling and consequently $Sr_2Nb_2O_7$ undergoes three phase transitions from $Cmcm \rightarrow Cmc2_1 \rightarrow Pbn2_1 \rightarrow Pb11$ at 1342°C, 215°C and -156°C respectively [5].

Most ceramic compounds require high sintering temperatures and therefore, various processing methods are being investigated to reduce the sintering costs and improve the density and hence electrical properties of functional ceramics. These new routes include i) chemical synthesis [6], ii) use of initial materials with fine particle size [7] and iii) addition of low melting fluxes [8]. The high costs and / or laborious processing procedures involved in the first two methods preclude the use of these routes for commercial production of technical ceramics; however, the addition of low melting fluxes in small percentages has been investigated in an attempt to achieve the above targets [9]. For example, La_2O_3 has been used as a sintering aid and highly dense $Sr_2Nb_2O_7$ with 99% of its theoretical density has been fabricated [10]. La doping has been reported to reduce the dielectric loss but resulted in the formation of a second phase, obviously deteriorating the properties. Similarly, the addition of V_2O_5 in low concentrations has been reported to improve the piezoelectric properties of $Sr_2Nb_2O_7$ [11]. ZnO has been found to enhance the density and dielectric properties of other compounds like $0.95MgTiO_3$-

0.05 $CaTiO_3$ and $BaZn_{1/3}Nb_{2/3}O_3$ [12-13] and here we describe our preliminary results regarding the effect of 0.3wt% ZnO on the phase, microstructure and dielectric properties of $Sr_2Nb_2O_7$.

2. Experimental

A 50g batch of $Sr_2Nb_2O_7$ was prepared by mixing reagent grade $SrCO_3$ and Nb_2O_5 and milling in disposable plastic jars using Y-toughened ZrO_2 balls as grinding media in 2-propanole for 24 h. The slurry was dried at ~90°C overnight and calcined at 1000°C at a heating / cooling rate of 10°C/min for 2h. 0.3wt% of ZnO was added to the calcined powders prior to re-milling for 40min. After drying the slurry, the powders were again sieved and pressed into 13 mm diameter pellets at 130 MPa in a tool steel die. The pellets were sintered at 1300°C, 1350°C and 1400°C for 2h at 10°C/min. An electronic densimeter (MD-300S) was used for density measurement. Phase analysis of fired samples was carried out using a JEOL JDX 3500 X-ray diffractometer operating at 30 mA and 40 kV with CuK_α radiation from 2θ =10 to 70° (step size 0.03°) at 2s/step. Microstructure of polished, thermally etched and gold-coated samples was investigated using a JEOL JSM 5910 SEM operating at 15-20 keV. Electrical properties of gold-coated pellets were measured using LCR meter (4287A) at 1 to 100MHz.

3. Results and Discussion

The lattice spacings (d) corresponding to major XRD peaks (labelled as '▲' in figure 1a) from the sample calcined at 1000°C matched JCPDS card# 70-114 for $Sr_2Nb_2O_7$ with maximum intensity peak at 2θ=29.35°. A few low intensity peaks labelled as '■' matching JCPDS card# 480421 for $Sr_5Nb_4O_{15}$ indicated the formation of the second phase as observed previously for samples calcined and sintered at 850°C and 1100°C respectively but for prolonged periods [14]. The d-values corresponding to XRD peaks from samples sintered at 1300°C (figure 1b), 1350°C (figure 1c) and 1400°C (figure 1d) matched JCPDS card# 511868 for $Sr_3Zn_{0.33}Nb_{4.67}O_{15}$ labelled as '•'with maximum intensity peak at 2θ=32.41° and JCPDS card# 70-114 labelled as '▲' for $Sr_2Nb_2O_7$. Note the decrease in the intensity of major $Sr_2Nb_2O_7$ peak at 29.35° accompanied by an increase in the intensity of major $Sr_3Zn_{0.33}Nb_{4.67}O_{15}$ peak at 32.41° with sintering temperature up to 1350°C indicative of second phase formation. At 1400°C, the intensity of major $Sr_2Nb_2O_7$ peak increases again. Although the authors could not find previous studies regarding the addition of ZnO to $Sr_2Nb_2O_7$, ZnO has been reported to react with the base composition such as $LnTiNbO_6$ (Ln = Sm or Dy), forming second phase(s) [15].

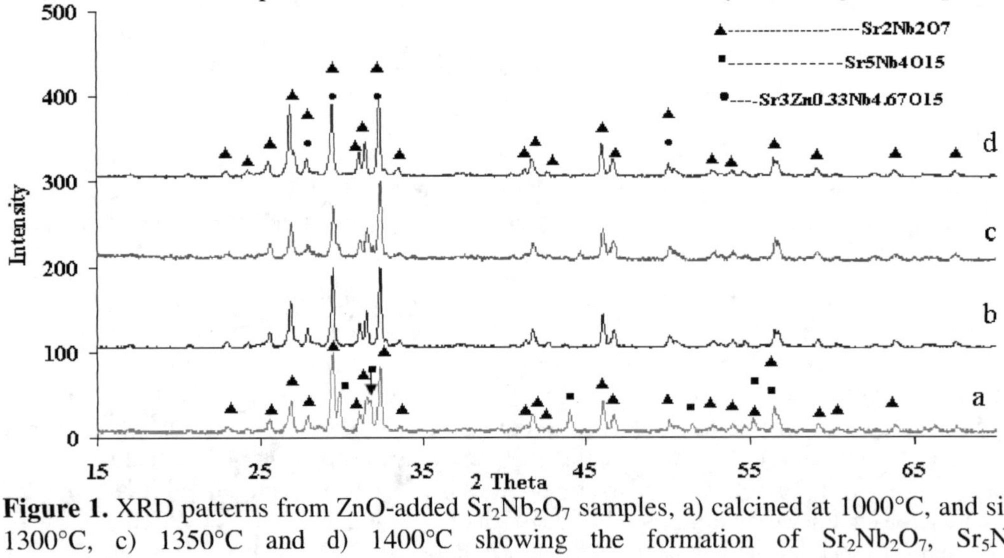

Figure 1. XRD patterns from ZnO-added $Sr_2Nb_2O_7$ samples, a) calcined at 1000°C, and sintered at b) 1300°C, c) 1350°C and d) 1400°C showing the formation of $Sr_2Nb_2O_7$, $Sr_5Nb_4O_{15}$ and $Sr_3Zn_{0.33}Nb_{4.67}O_{15}$ phases.

The highest apparent density measured for ZnO-added $Sr_2Nb_2O_7$ sintered at 1400°C was ~95% of its theoretical density. The achievement of the same density required attrition milling as well as ~50°C higher sintering temperature for $Sr_2Nb_2O_7$ [1].

Secondary electron SEM images of sintered microstructures from pure and 0.3 wt% ZnO-added $Sr_2Nb_2O_7$ samples are compared in figure 2. The observed grain morphology was rod-like and almost similar on the surface and in the bulk for samples without ZnO. The observed grain size varied from ~1×1 to 6×70 μm^2 on the surface and from ~1×1 to 1.5×20 μm^2 in the bulk (figure 2 a-b). The morphology of grains in ZnO-added $Sr_2Nb_2O_7$ sintered at 1300°C appeared as a mixture of irregularly shaped grains and rods with grain size ranging from ~0.5×1 to 1.5×12 μm^2 on the surface and ~0.5×0.5 to 2×10 μm^2 in the bulk (not shown). In the sample sintered at 1400°C, the grain size varied from ~0.8×1.5 to 4×40 μm^2 on the surface and from ~0.5×0.5 to 1.5×5 μm^2 in the bulk (figure 2c-d). This indicated an increase in the grain size with increase in sintering temperature at the surface of the sample only. The average grain size in the bulk did not change much, indicating the influence of ZnO in suppressing the grain growth. A comparison of the microstructures of pure and ZnO-added $Sr_2Nb_2O_7$ demonstrated that the addition of ZnO resulted in a visibly dense microstructure with reduced grain size by at least more than 50%. It is noticeable that the optimum sintering temperature of ZnO-added sample was ~50°C lower than that for pure $Sr_2Nb_2O_7$ which might be a reason for the observed small grain size in addition to the effect of ZnO in suppressing the grain growth.

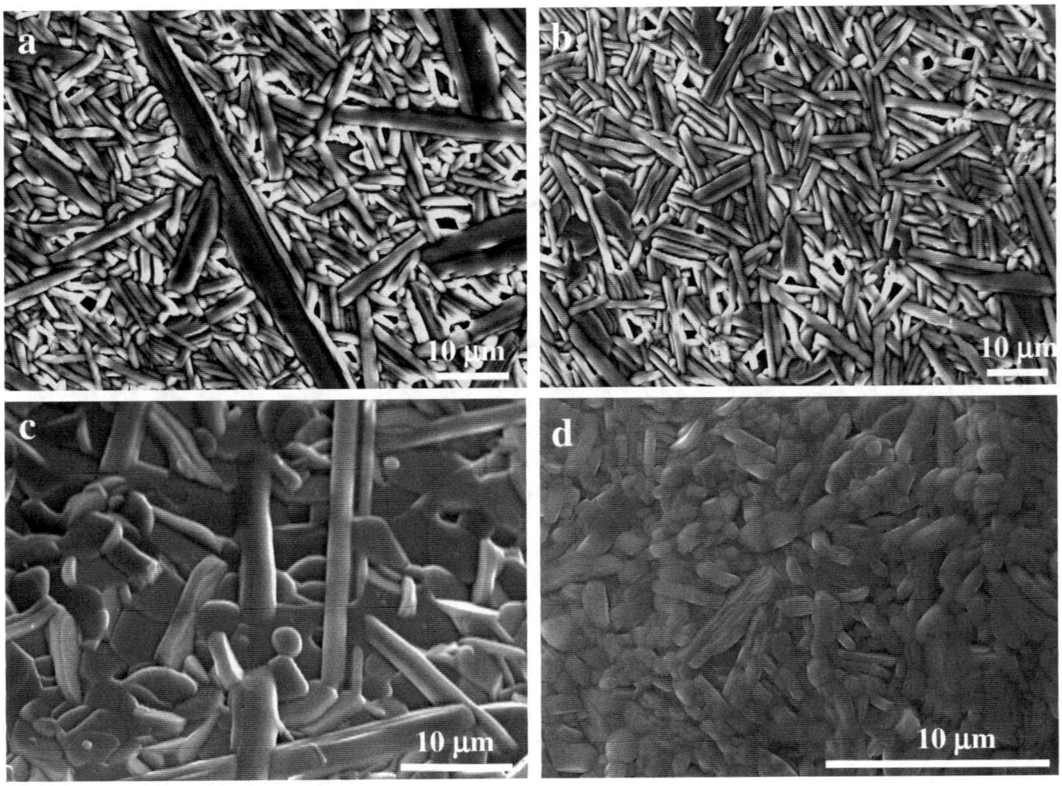

Figure 2. SEM image from $Sr_2Nb_2O_7$ sample sintered at 1450°C showing rod-shaped grains of size ranging, a) from ~1×1 to 6×70 μm^2 on the surface and b) from ~1×1 to 1.5×20 μm^2 in the bulk, and ZnO-added $Sr_2Nb_2O_7$ sintered at 1400°C showing grains of different morphologies of size ranging c) from ~0.8×1.5 to 4×40 μm^2 on the surface and d) from ~0.5×0.5 to 1.5×5 μm^2 in the bulk.

The dissipation factor or its inverse (Q_o=1/tanδ) for ZnO-added $Sr_2Nb_2O_7$ is given in table 1 where δ is the phase angle between voltage and current in the capacitor. The dielectric loss decreased with increase in sintering temperature from 1300°C to 1350°C, obviously due to increased density at high temperature. Upon further increase in sintering temperature to 1400°C, a further decrease in dielectric loss was observed. The maximum value of 'Q_o' was 1687 at 80 MHz. This indicated that Q_o value was higher for the denser sample and was dependent upon the frequency which may be due to a decrease in friction among the dipoles at higher frequencies [16]. Similarly, as expected, the dielectric constant increased with increase in sintering temperature from 1300°C to 1400°C and a consequent increase in density (table 1), however, at a constant temperature, it decreased with increase in frequency. The dielectric constant of samples sintered at 1400°C was lower than that sintered at 1350°C. This unexpected decrease may be due to the increase in ZnO-containing phase at 1400°C than that sintered at 1350°C as evident from XRD (figure 1b-c). The previously reported high dielectric constants of 48 at 10kHz and 42 at 100kHz for $Sr_2Nb_2O_7$ [1, 17] suggests that at lower frequencies the electronic, ionic, dipolar and extrinsic effects contribute to the dielectric constant, however, at relatively higher frequencies, the contribution from interfacial/surface polarization get minimized [18]. The presence of second phase in the samples with 0.3wt% ZnO may also be a reason for the observed decrease in ε_r.

Table 1. Dielectric properties of 0.3wt% ZnO-added $Sr_2Nb_2O_7$ at 80 MHz

Temp (°C)	ρ(gcm^{-3})	ε_r at 25°C	Q_o-Value	tanδ
1300	4.64	13	273	0.0037
1350	4.7	17	714	0.0014
1400	4.98	14	1687	0.0006

4. Conclusions

The effect of addition of 0.3 wt% ZnO on the phase, microstructure and properties of $Sr_2Nb_2O_7$ was investigated. Pure samples calcined at 1000°C were multiphase containing $Sr_2Nb_2O_7$ and $Sr_5Nb_4O_{15}$. The addition of 0.3wt% of ZnO resulted in the formation of highly dense ceramics at lower temperatures than that for pure $Sr_2Nb_2O_7$. The final ceramic was two-phase comprising $Sr_3Zn_{0.33}Nb_{4.67}O_{15}$ and $Sr_2Nb_2O_7$ crystalline grains. The lowest dielectric loss angle corresponded to the sample sintered at 1400°C. Preliminary observations revealed that the dielectric constant decreased with increase in frequency whereas the dielectric loss angle decreased with increase in frequency which is being investigated.

Acknowledgement

The authors acknowledge the financial support (NRPU 20-569) of HEC, Pakistan.

References

[1] Iqbal Y and Reaney I M 2004 *Ferroelectrics*. **302**, 259
[2] Atuchin V V, Grivel J C, Korotkov A S and Zhang Z 2008 *J. Solid State. Chem*. **181**, 1285
[3] Leitner J, Hampl M, Ruzicka K, Straka M, Sedmidubsky D and Svoboda P 2008 *Thermoch. Acta*. **475**, 33
[4] Scheunemann and Muller-Buschbaum 1975 *J. Inorg. and Nucl. Chem*. **37**, 1679
[5] Hushur A, Akishige Y and Kojima S 2004 *Ceram. Intern*. **30**, 2023
[6] Maclaren I and Ponton C B 1998 *J. Mat. Sci*. **33**, 17
[7] Liang M H, Wu S Y, Hu C T and Li I N 2003 *Mat. Chem. Phys*. **79**, 276
[8] Shirey H M 2002 MS Thesis, University of Pittsburg
[9] Cheng C M, Yang C F, Lo S H and Tseng T Y 2000 *J. Eur. Ceram. Soc*. **20**, 1061
[10] Brahmaroutu B, Messing G L and Trolier-Mckinstry S 2000 *J. Mat. Sci*. **35**, 5673
[11] Seraji S, Wu Y, Limmer S, Chou T, Nguyen C and Forbess G Z 2002 *Mat. Sci. Eng*. **B88**, 73
[12] Chaouchi A, Marinel S, d'Astorg S, Aliouat M 2009 *Ceram. Intern*. **35**, 939
[13] Huang C L, Hou J L, Pan C L, Huang C Y, Peng C W, Wei C H and Huang Y H 2008 *J. Alloy. Compd*. **450**, 359
[14] Renoult O, Boilot J P, Chaput F, Papiernik R, Hubert-Pfalzgraf L G and Lejeune M 1992 *J. Am. Ceram. Soc*. **75**, 3337
[15] Solomon S, Joseph J T, Kumar H P and Thomas J K 2006 *Mater. Lett*. **60**, 2814
[16] Rao P P, Nair K R, Koshy P and Vaidyan V K 2007 *Mater. Lett*. **61**, 4188
[17] Levin I, Bendersky A, Vanderah T A, Roth R S and Stafsudd O M, 1998 *Mat. Res. Bull*. **33**, 501
[18] Singh A K and Choudh R N P J 2003 *J. Phys. Chem. Solids*. **64**, 1185

Nanostructured Reactive Metallic Multilayers

M A Mat Yajid[1,2] and G Möbus[1]

[1] Dept of Engineering Materials, University of Sheffield, Sheffield S1 3JD, UK
[2] Dept of Engineering Materials, Faculty of Mechanical Engineering, Universiti Teknologi Malaysia, 81310, Skudai, Johor, Malaysia

E-mail: g.moebus@sheffield.ac.uk

Abstract. Metallic multilayers have been used in cross-sectional TEM specimen geometry to explore nanostructuring routes using ion and electron beams (top-down) as well as self-organised nanopatterning (bottom-up). Intermetallic alloy formation during heat treatment is studied for binary and ternary layer compositions. A choice of pillars (0D), 1D and 2D layer geometries are used to explore dimensionality effects.

1. Introduction

Metallic multilayer coatings have a wide range of applications in engineering as functional materials such as wear resistant coatings [1], magnetoresistive coatings [2] and x-ray mirrors [3]. The most recent applications are in the field of nanoscale layers for ignition devices or oxygen-free energy sources in nanolab experiments [4, 5]. As reported in our earlier work for a binary Cu-Al system [6, 7], we found by both ex-situ and in-situ heating experiments that, starting at ~ 250-300°C a tetragonal phase, θ-$CuAl_2$, is formed, whereas at high heating temperature Cu_9Al_4 is expected. However, it was found that due to the original layer thickness ratio between Cu and Al of 1:3, $CuAl_2$ remains the preferred phase up to the melting point with no occurrence of Cu_9Al_4.

In extension of our earlier work, we now characterize the reaction sequences and products of a Cu-Al-Cu-Ti multilayer system, fabricated by a combination of sputtering (Al) and evaporation (Cu,Ti), with respect to potential ternary alloy formation. Furthermore, we explore how further reduction in dimensionality of the system influences the reaction sequence, e.g. via preferred surface diffusion in hetero-nanowire or nanopillar specimens. Energy dispersive X-ray spectroscopy (EDX), Electron Energy Loss Spectroscopy (EELS), Selected Area Electron Diffraction (SAED) and High-resolution Electron Microscopy (HREM) are the preferred techniques to confirm the reacted phases.

2. Experimental procedures

All the TEM samples were prepared using conventional cross-sectional sample preparation, involving mechanical grinding and polishing followed by precision ion polishing system milling (PIPS-Gatan). The ion milling was conducted at 5 keV energy and between 3-6°C milling angles. Two methods of nanopillar fabrication have been applied. Firstly, we use focused ion beam (FIB, JEOL 6500F dual-beam FEGSEM) and secondly we use e-beam drilling/cutting with a JEOL 3010 TEM at 300kV with LaB_6 source. In-situ heating experiments were performed using a Philips EM430T TEM at 300kV, equipped with a Gatan single tilt heating holder. Heating increments were applied over 45 min. up to ~302°C with current step increments of 25mA. For the ex-situ heating approach, the sandwiched multilayer sample is annealed inside a tube furnace filled with argon gas with a two-step reaction process (to prevent Al melting), firstly at 400°C for two hours, then at 850°C for one hour. All the post-in-situ and ex-situ TEM micrographs were imaged using a JEOL JEM-2010F at 200kV equipped with Oxford Instruments ISIS energy dispersive X-ray (EDX)-mapping control and a Gatan GIF spectrometer for electron energy loss spectroscopy (EELS) chemical analysis.

3. Results and discussions

Figure 1 shows several contrast inverted bright field (BF) TEM images during in-situ heating up to 302°C of the cross-sectional Cu-Al-Cu-Ti layer system on Al_2O_3 substrate. Part of the specimen has been FIB milled into pillar shapes (JEOL/Orsay Fabrika dual beam; Ga beam of 3.5pA at 30kV and aperture #2 for ~3 min.). At room temperature (Fig 1(a)) all four metallic phases can be sorted by their scattering contrast in a grey level sequence of Cu – Ti – Al, according to their atomic number. At around 240°C the Cu-Al-Cu part of the layer system starts to react via interdiffusion, excluding the Ti layer. The transformation process at the nanopillars regions is completed at 285°C, and no further change to this region is observed at any higher temperature. On the other hand, the completion of the intermetallic reaction at the one-dimensional regime is slightly delayed to 302°C, as seen on the very top end of the image. The entire in-situ heating experiment took about ~ 1 hour.

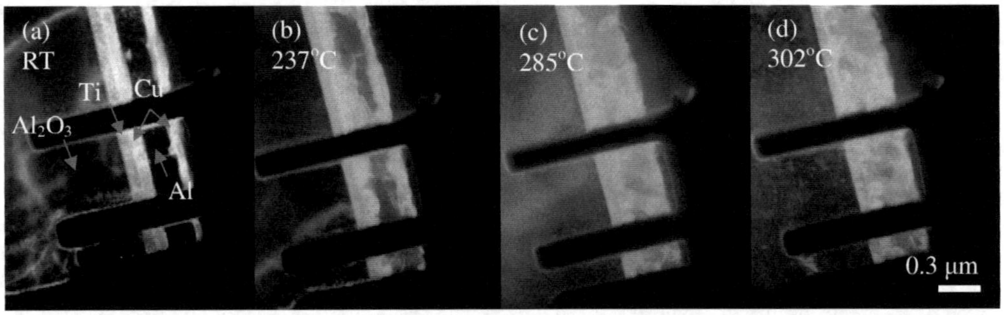

Figure 1. (a-d) TEM micrographs (contrast inverted bright field) of the in-situ heating experiment from room temperature to 302°C for a Cu-Al-Cu-Ti multilayer sample (Philips EM430, 300kV).

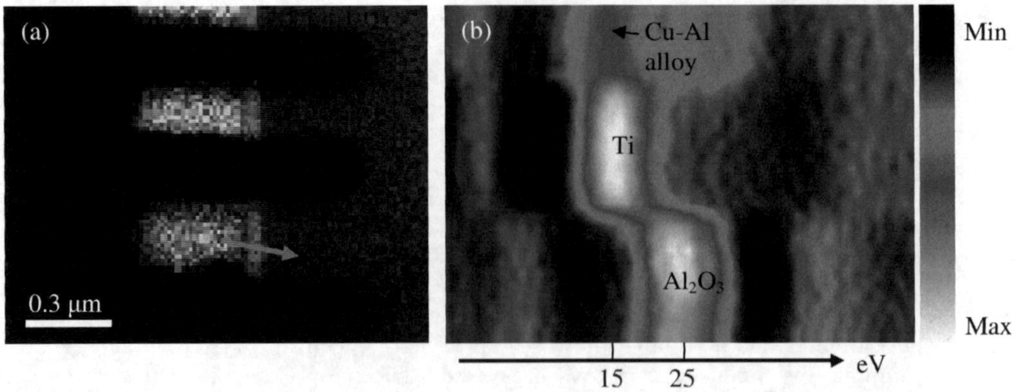

Figure 2. Chemical mapping of the in-situ heat treated Cu-Al-Cu-Ti multilayer. (a) RGB chemical phase map of nanopillars with red = Ti-Kα, green = Cu-Kα and blue = Al-Kα and (b) EELS spectrum

image (linescan along red arrow of (a), total length ~300nm): second derivative plasmon map with corresponding phases: Cu-Al alloy at 15.6 eV, Ti at 17.1 eV and Al_2O_3 at 25 eV.
Figure 2 shows the post-processed results of an in-situ experiment, after return from ~300°C to room temperature and change to an analytical FEGTEM (JEM 2010F), using a different sample than for Fig. 1. Figure 2 (a) illustrates the EDX map of the three layer system (Cu-Al-Cu) which fully transformed into a homogeneous Cu-Al alloy. Along the line in Fig 2(a), an EELS spectrum imaging line scan in the plasmon energy range has been performed. To increase visibility of the three phases (alloy, Ti, substrate), the plasmon spectra have been converted into 2nd derivative maps (using IDL, see figure 2 (b)). The benefit of this technique is to enhance phase-to-phase contrast by means of resonance sharpness in addition to resonance position. The Cu-Al alloy appears to have a plasmon maximum at ~ 15.6 eV, whereas the Ti and Al_2O_3 are at ~ 17.1 and 25 eV respectively.

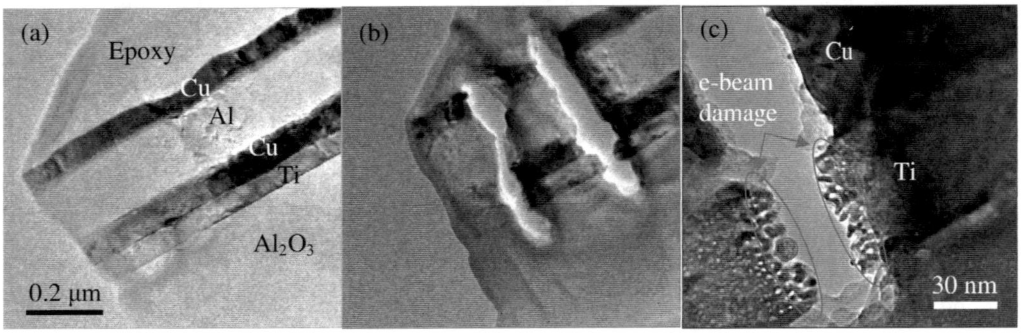

Figure 3. TEM micrograph of Cu-Al-Cu-Ti multilayer sample before (a) and after drilling (b-c).

Figure 3 (a-c) shows an alternative nanofabrication method for metal nano-pillars using e-beam instead of FIB. For the cutting of lines to convert the metallic layers into pillars, the condenser astigmatism is changed to produce an ultra-astigmatic line-focus. The continued irradiation with this line-focus over ~ 2 hours resulted in full cuts through all metal layers with Ti being the last one to separate. The diameter of the final nanopillars is ~ 80-150 nm which depends on the size of condenser aperture (CA) and specimen thickness. Since the size of the cut is given by the total multilayer length of 390 nm, the largest CA with mean convergence angle of ~ 32.0 mrad (~ 120 µm in diameter) became necessary. A beam drift problem caused some curvature in the produced nanopillars (fig 3(b)). However, this process could be minimised by reducing the drilling length for every session.

One motivation for e-beam cutting is the prospect of less damage (e.g. no Ga implantation) than with FIB. However, we also found beam induced transformations along the drilling area, mainly in Ti (Figure 3 (c)). Apart from some carbon contamination effects at the border of the drilling area in Cu we find that the original Ti layer has reacted into ultra-fine round Ti multi-grains with ~ 5 nm in diameter (e.g. red circle). This could possibly be due to irradiation induced "quasi-melting" and re-solidification, a process occurring far below the bulk melting temperature under high e-beam irradiation.

As the upper temperature limit of in-situ heating prevented the Ti layer from participating in any reaction, we extended our study by ex-situ heat treatment of the cross-sectional prepared TEM specimen of the original Cu-Al-Cu-Ti system. Figure 4 shows the Cu-Al-Cu-Ti sample after furnace heat treatment up to 850°C. Figure 4 (a) illustrates via EDX chemical mapping (JEM 2010F) that the original 4 layer system has been transformed into an ordered alternation layer of two different phases, presumably via spinodal decomposition of an intermediate metastable ternary high-temperature phase. The EDX point measurement quantification from both phases indicated a preliminary result for a binary Cu-Al alloy and a ternary Cu-Al-Ti alloy phase. By investigation using HREM plus EDX and diffraction analysis (not included here) we found consistent identification of the binary alloy as Cu_9Al_4

and the ternary alloy as Cu_2AlTi. The HREM image of the interface separating the two phases gives an orientation relationship with parallel zone axes of <1 1 1> and <0 1 1> for Cu_9Al_4 and Cu_2AlTi alloys respectively (figure 4 (b)).

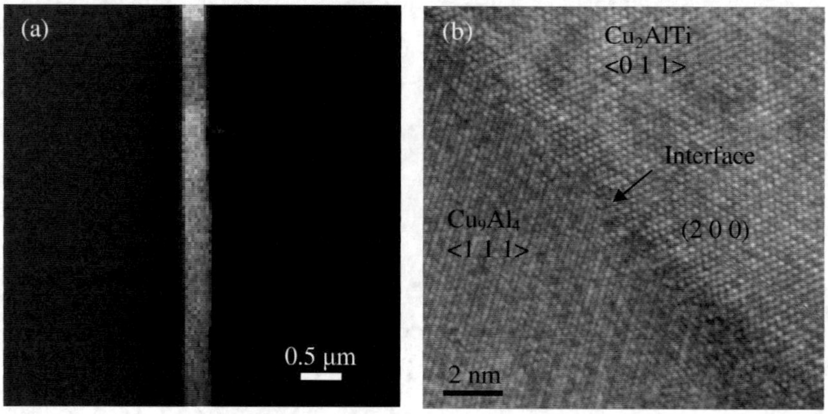

Figure 4. (a) RGB image of the Cu-Al-Ti sample after ex-situ heat treated up to 850°C with red = Ti-Kα, green = Cu-Kα and blue = Al-Kα. (b) HREM micrograph of Cu_9Al_4 – Cu_2AlTi interface with O.R. <1 1 1> (1 0 1) Cu_9Al_4 and <0 1 1> ~ 14° off (2 0 0) Cu_2AlTi.

4. Conclusions

We have demonstrated nanopillar fabrication using both FIB and electron beam cutting starting from a conventional cross-sectional TEM specimen of a multi-layer metal system. It was found that annealing induced transformations of the multilayer and the intermetallics formation can then be studied in parallel within the layered and pillared structures. Formation of functional metallic nanodots would be one application of this top-down nanofabrication route. Alternatively, appropriate choice of the layer-thicknesses inside the multi-layer system to avoid coincidence with a single intermetallic phase could be used for self-organised nanostructuring into alternating nano-phases inside a TEM specimen by means of phase separation following high-temperature annealing. Possible applications for such reactive multilayer films include joining and packaging and also ignition devices.

Acknowledgements

We are thankful for funding contributions from Universiti Teknologi Malaysia/Government of Malaysia (M.A.M.Y.) and EPSRC (G.M.). We also acknowledge Dr T. Wagner from Max-Planck-Institute for Metals Research, Stuttgart, Germany for providing the metallic multilayer samples.

References

[1] PalDey S, Deevi S C 2003 *Materials Science and Engineering A* **342**, 58
[2] Baibich M N, Broto J M, Fert A, Van Dau F N, Petroff F, Etienne P, Creuzet G, Friederich A, Chazelas J 1988 *Physical Review Letters* **61**, 2472
[3] Montcalm C, Sullivan B T, Ranger M, Slaughter J M, Kearney P A, Falco M, Charles M, Chaker M 1994 *Optics Letters* **19**, 1004
[4] Gavens A J, Van Heerden D, Mann A B, Reiss M E, Weihs T P 2000 *Journal of Applied Physics* **87**, 1255
[5] Menon L, Patibandla S, Bhargava Ram K, Shkuratov S I, Aurongzeb D, Holtz M 2004 *Applied Physics Letters* **84**, 4735.
[6] Mat Yajid M A, Möbus G 2009 *Microscopy and Microanalysis* **15**, 54.
[7] Mat Yajid M A, Wagner T, Möbus G 2008 *phys. stat. sol. (RRL) - Rapid Research Letters* **2**, 7.

Journal of Physics: Conference Series **241** (2010) 012031 doi:10.1088/1742-6596/241/1/012031

High-resolution TEM characterization of ZnO core-shell nanowires for dye-sensitized solar cells

G Divitini[1], N O V Plank[2], H J Snaith[3], M E Welland[2], C Ducati[1]

[1] Department of Materials Science, Pembroke Street, Cambridge, CB2 3QZ, UK
[2] Nanoscience Centre, Department of Engineering, University of Cambridge, CB3 0FF, UK
[3] Clarendon Laboratory, Department of Physics, University of Oxford, OX1 3PU, UK

E-mail: gd322@cam.ac.uk

Abstract. Recently ZnO nanowire films have been used in very promising and inexpensive dye-sensitized solar cells (DSSC). It was found that the performance of the devices can be enhanced by functionalising the nanowires with a thin metal oxide coating. This nm-scale shell is believed to tailor the electronic structure of the nanowire, and help the absorption of the dye. Core-shell ZnO nanowire structures are synthesised at low temperature (below 120°C) by consecutive hydrothermal growth steps. Different materials are investigated for the coating, including Mg, Al, Cs and Zr oxides. High resolution TEM is used to characterise the quality of both the nanowire core and the shell, and to monitor the thickness and the degree of crystallisation of the oxide coating. The interface between the nanowire core and the outer shell is investigated in order to understand the adhesion of the coating, and give valuable feedback for the synthesis process. Nanowire films are packaged into dye-sensitised solar cell prototypes; samples coated with ZrO_2 and MgO show the largest enhancement in the photocurrent and open-circuit voltage and look very promising for further improvement.

1. Introduction

The current methods for converting solar power into electricity have many drawbacks, such as high cost and low versatility. Dye sensitized solar cells (DSSCs) can have low fabrication cost and different fabrication techniques, that can lead to lightweight, flexible and transparent solar cells. ZnO is a good candidate material for such cells [1] because of its stability and the possibility to build nanowire (NW) films with high surface area. Still, work has to be done to optimize the overall behaviour, particularly at the interface between ZnO and the charge generator. It was found that the addition of a suitable coating to the NWs improves the charge injection into the ZnO [2]. Moreover, the synthesis techniques need tuning in order to achieve reproducible, tailored films. In this paper we use high resolution electron microscopy to investigate the growth of ZnO NWs, and study the morphology of various external shells in core-shell structures. This allows a better understanding of the results of the characterization of the devices, including information on the interaction among dye, shell and ZnO.

2. Experimental details

Core-shell nanowires were grown by hydrothermal growth [1][3]. A Zn metal layer (100 nm) was deposited by sputtering on clean ITO (Indium Tin Oxide) glass substrates. The NW hydrothermal growth was carried out from a solution obtained mixing 0.025 M zinc nitrate hydrate and 0.025 M hexamethylenetetramine (HMT) in water. The solution was kept at 95°C for 120 minutes, after which the substrates were rinsed with deionised water and dried on a hotplate at 100°C. A second hydrothermal growth procedure was then used to fabricate a core-shell structure on the ZnO nanowires. Oxides of different metals have been used as materials for the outer shell, including Mg, Zr, Al and Cs. It should be noted that, being a completely low temperature deposition, this process can be carried out on a variety of substrates, including flexible polymeric ones. This is a significant advantage with respect to other techniques employing a high temperature (>300°C) annealing step.

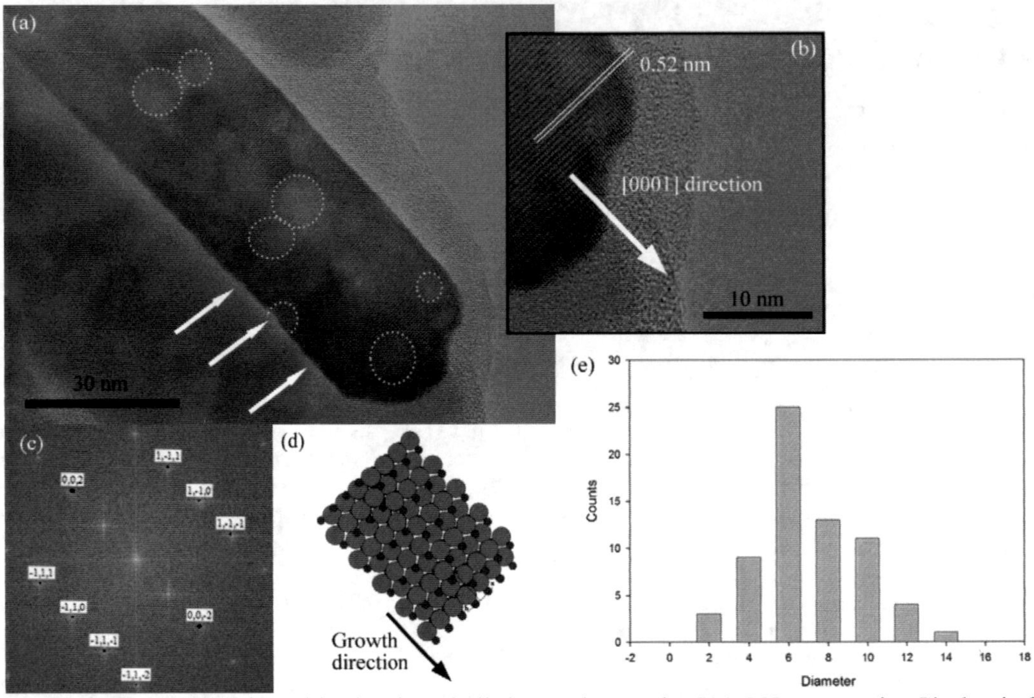

Figure 1. (a) ZnO nanowire with alumina shell, imaged near the [11-20] zone axis. Circles indicate some of the dips on the surface, arrows point out dips on the sidewalls of the NW. (b) Detail of the tip, showing growth along the [0001] direction. (c) FFT of the crystal lattice compared with simulated diffraction pattern. (d) Crystal model. (e) Size distribution of the dips on the surface of NWs.

The morphology of the samples has been verified by imaging with a LEO 1530 VP scanning electron microscope (SEM). The nanowire films were removed from the substrates and dispersed on holey C Cu TEM grids. Transmission electron microscopy (TEM) was used to characterize the core-shell nanowire structures using a high resolution dedicated JEOL 4000EX (400kV acceleration voltage) and a FEI Tecnai F20 microscope (200kV).

The ZnO-MgO and ZnO-ZrO nanowires were then used to build DSSCs (using D101 and D149 dyes) in order to test the performances that could be achieved. ZnO NWs, ZnO-MgO and ZnO-ZrO samples were submerged in a solution containing the dye and rinsed with anhydrous acetonitrile. The hole transporter was 2,2',7,7'-tetrakis(N, N-di-methoxyphenyl-amine)-9,9'-spirobifluorene (Spiro-OMeTAD). 50 nm thick silver counter electrodes were deposited by thermal evaporation.

3. Characterization

The grown NW films have been imaged by SEM: 2 hours hydrothermal growth resulted in well aligned homogeneous NWs approximately 600 nm long and 60 nm wide.

High resolution transmission electron microscopy (HRTEM) was used in order to understand the morphology of the NWs and characterize the different core-shell structures. All samples have shown highly crystalline ZnO NWs, whose structure is completely preserved through the hydrothermal shell fabrication step. Fast Fourier Transform (FFT, figure 1-c) analysis of the high resolution images of the lattice confirmed that the growth occurs in the direction of the c axis of the wurtzite unit cell, the [0001] direction (figure 1-d), as reported in other papers [3][4].

The imaged NWs did not display uniform contrast, but -in bright field TEM mode- presented lighter patches of 2-14 nm in diameter randomly distributed on all surfaces, as seen in figure 1. They could also be seen from the side as dips on the walls of NWs. Such patches are visible on parts of the sample that have just been shortly exposed to the beam, and appear beneath the coating, therefore they can't be ascribed to beam damage. The most reasonable explanation is that these areas are shallow cavities formed during the hydrothermal growth process, caused either by defects in the crystal lattice or by impurities in the growth solution.

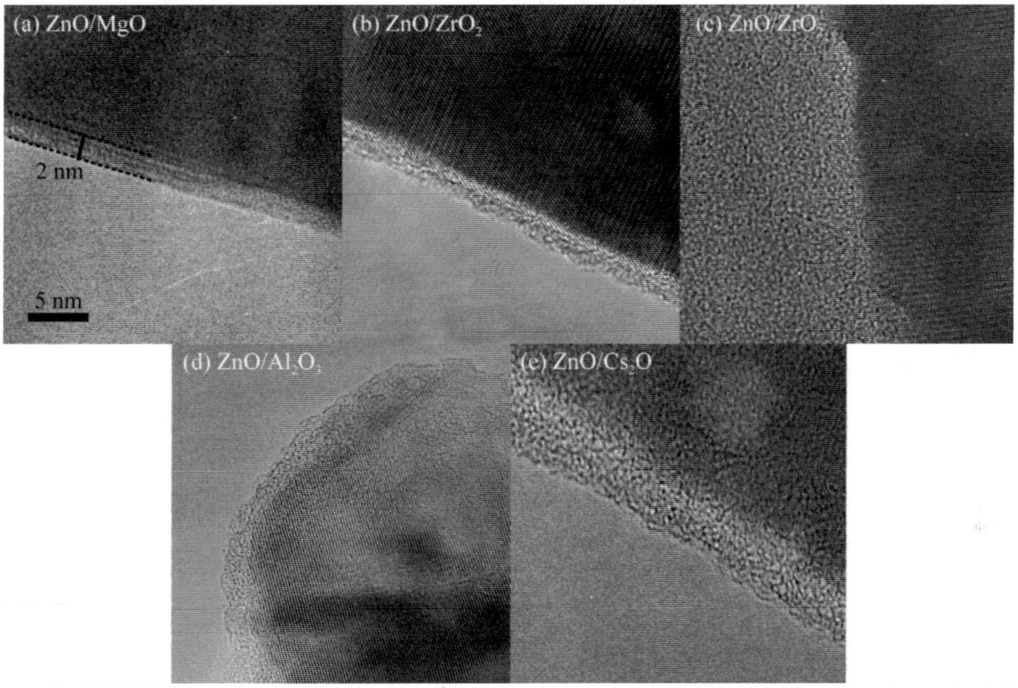

Figure 2. HRTEM images showing the different surface coatings of ZnO nanowires. In (c) the amorphous region to the left is the carbon film supporting the specimen, while the coating is probably less than 1 nm thick and not clearly distinguishable from the NW. (a) MgO, (b) and (c) ZrO_2, (d) Al_2O_3, (e) Cs_2O.

The MgO coating on the nanowires had been previously characterized by the group [1]. Further analysis showed that the coating is highly reproducible, is amorphous and around 2-5 nm thick. Coatings made with other oxides (ZrO_2, Cs_2O, Al_2O_3) resulted in homogeneous coverage of the ZnO surface, proving the flexibility of this growth process. The ZnO-ZrO_2 core-shell NWs showed a regular coating with a thickness ranging from sub-nanometre to 8 nm. The shell is homogeneous for all the length of the NWs. A similar structure has been found for the ZnO-Al_2O_3 samples. The Cs_2O coatings were slightly less homogeneous (up to 5 nm). On all the samples that have been examined by HRTEM the outer shell has been found to be amorphous. On a few NWs, though, the shell was very thin (few atomic layers): in this condition the contrast in the images is also compatible with a crystalline coating matching the underlying ZnO lattice. This has been observed on zirconia (figure 2-c) or magnesia (figure 2-a) coated ZnO NWs. On a larger scale, in some cases the external oxide coating filled the gaps between nanowires. This reduces the surface area that can be put in contact with the electrolyte and limits the overall efficiency of the device. Optimization of the hydrothermal shell growth should reduce this phenomenon and increase performances of obtained DSSCs.

4. Performance and discussion

The functionalized ZnO films have been packaged into DSSCs and tested under an AM 1.5 G solar simulator. A significant performance improvement has been found for the MgO and ZrO_2 devices over the pristine NW ones [5]. The ZrO_2 coating improved the short circuit current from 0.72 to 2.14 mA/cm_2. The best achieved device efficiency η was 0.71% with ZnO-ZrO_2 core-shell NWs [5]. These results are an improvement over the ones obtained by our group [1] with ZnO and ZnO-MgO nanowires prepared with the same technique. The improvement is mainly due to two reasons. On one hand, as shown in figure 3, the oxide layer over the NWs allows to "bandgap engineer" the device, optimizing the charge separation and the injection of electrons into the ZnO core. On the other hand, the ruthenium based dyes that have been employed in the past are less suitable for ZnO devices. Such dyes have the tendency to agglomerate, limiting charge transfer [6]. Small molecule organic based dyes have shown here to be more suitable for ZnO, leading to a boost in overall performance.

5. Conclusions

ZnO NWs have been fabricated with different oxide coating via hydrothermal synthesis. The addition of the outer shell significantly increases device performances. Zirconia coating was the most effective coating, leading to a three-fold improvement in short circuit current and bringing the device efficiency to 0.71%. The whole fabrication process is carried out at temperatures below 100°C and is thus fully compatible with polymeric substrates. Further optimization and improvement in the fabrication and the packaging of the devices can lead to commercially interesting cheap and versatile solar cells.

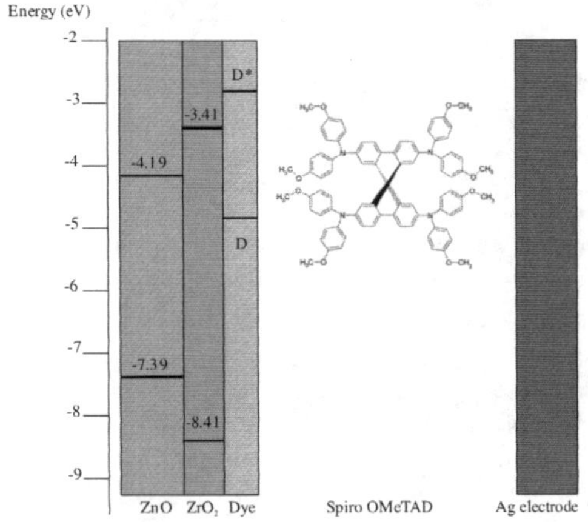

Figure 3. Energy structure of a DSSC. The presence of the ZrO_2 layer helps reducing electron-hole recombination and improving the overall efficiency (from [5]).

Acknowledgements

CD acknowledges the Royal Society for funding. We thank EPSRC and the Isaac Newton Trust for the use of the TECNAI microscope.

References

[1] Plank NOV, Snaith HJ, Ducati C, Bendall JS, Schmidt-Mende L and Welland ME 2008 *Nanotechnology* **19** 465603
[2] Olson DC, Shaheen SE, White MS, Mitchell WJ, van Hest MFAM, Collins RT, Ginley DS 2007 *Adv. Funct. Mater.* **17** 264-269
[3] Vayssieres L 2003 *Advanced Materials* **15** 464-466
[4] Li Q, Kumar V, Li Y, Zhang H, Marks TJ, Chang RPH 2005 *Chem. Mater.* **17** 1001-1006
[5] Plank NOV *et al.*, Improving the efficiency of ZnO nanowire solid state dye sensitized solar cells using organic dyes, *to be submitted to Nanotechnology*
[6] Keis K, Lindgren J, Lindquist S and Hagfeldt A 2000 *Langmuir* **16** 4688-4694

Effect of laser annealing on defected silicon solar cells

AHJ Al- Mousawy

Department of Physics, College of Education (Ibn Al-Haitham),
University of Baghdad, Aadamiyah, Baghdad, Iraq.

E-mail: ajalmousawy@yahoo.com

Abstract. The effect of pulsed Nd-YAG laser annealing (λ = 1.06 μm), on the efficiency of defected polycrystalline silicon solar cells has been studied. Cells performance measurement and microscopic structure examinations of the surface using optical and scanning electron microscopy (SEM) have been done before and after irradiation with different laser fluences (2.3 – 25 J/cm^2) and for successive times until a steady state has been reached. It is found that the annealing causes a change in the topography of the cells, resulting in an increase in grain size at low fluences and beginning of crystal growth at higher fluences up to an optimum value of 7.5 J/cm^2. These processes cause an improvement in the electrical output characteristics of the defect-rich solar cells; hence an increase in the cell efficiency is obtained. The maximum change in efficiency value is found to be about 5 times greater than its initial value. For fluences of more than 16 J/cm^2, SEM micrographs show slightly deeper melt front penetration which affects the junction interface and a reduction in light output characteristics of the cells has been detected.

1. Introduction

The development of laser processing technologies has shown great potential for innovative advances of cell fabrication [1-2]. The junction characteristics of solar cells could also be improved by using a suitable laser annealing power [3]. The lattice strain at the grain boundaries could be minimized, which is a consequence of an increase in electrical conductivity after laser treatment [4].

The goal of this study was to investigate the possibility of recovery of damaged polycrystalline Si solar cells after laser annealing by scanning electron microscopy (SEM). The photovoltaic performance and the efficiency of these cells are considered.

2. Experimental Work

Different types of polycrystalline Si solar cells, enriched in defects (or damaged) during manufacturing, have been selected. The laser irradiation has been performed with Nd-YAG pulsed laser (λ = 1.064 μm) with a Gaussian profile and 10 Hz repetition rate. The fluences used were in the range 2.3 – 25 J/cm^2. The sample was put on a movable stage and irradiated with successive overlapping pulses through the damaged region only. Microscopic examinations before and after laser irradiation were done using a JEOL SEM (JSM-6400).

3. Results and Discussion

In order to define the damaged region, an optical microscope was used to examine the surface of the solar cell before any laser irradiation. Figure 1a shows an optical micrograph of the interface between the undamaged and the damaged (defect-rich) region. Figure 1b represents an SEM micrograph of the defect-rich region irradiated with fluence 4 J/cm^2

revealing the melt region caused by the annealing. Figure 1c is the defect-rich region irradiated with 6 J/cm^2, showing the increase in grain size while micrograph 1d is the defect-rich region irradiated with 9.6 J/cm^2, revealing recrystallization processes occurring due to annealing. The grain sizes, before and after irradiations for other fluencies, calculated from similar micrographs, are tabulated in Table 1.

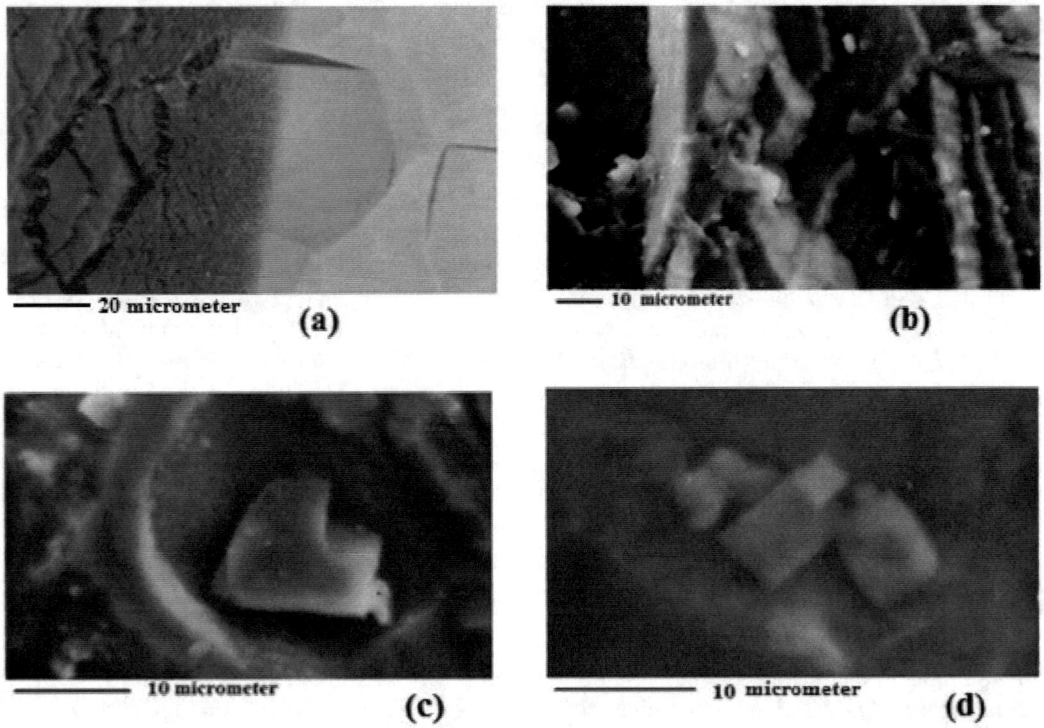

Figure 1. (a) The optical micrograph showing the interface between defect-rich and the undamaged regions in the polycrystalline cell before irradiation. (b) SEM micrograph representing the defect-rich region irradiated with fluence 4 J/cm^2. (c) SEM micrograph for 6 J/cm^2. (d) SEM micrograph for 9.6 J/cm^2.

Table 1. The grain size in μm for different fluences.

Fluence (J/cm$^{2)}$	Average grain size (μm)
before irradiation	10
2.3	13.5
4	25
7.5	35
8.7	26
19.6	20

The change in AM1efficiency $\Delta\eta = \eta_{\text{after}} - \eta_{\text{before}}$ after laser annealing with different fluences is shown in Figure 2, where AM1 (air mass one) is solar radiation of 925 W/m^2 and η is the solar cell efficiency. The measurements in this figure have been repeated many times, applying post treatment annealing (PTA) for different times until a steady state had been reached. It is clear from Figure 2 that, in general, there is an improvement in the solar cell performance, which appears as an enhancement of the efficiency for most of the fluences in the range 2 – 16 J/cm^2. However, there appears a peak in $\Delta\eta$ at the particular fluence of 7.5 J/cm^2. Although $\Delta\eta$ increases with increasing the PTA until a steady state is reached, the position of the peak is not affected by PTA. However, for fluences more than 16 J/cm^2 a slightly deeper melt front affects the cell junction and causes a reduction in the efficiency and solar cell deficiency.

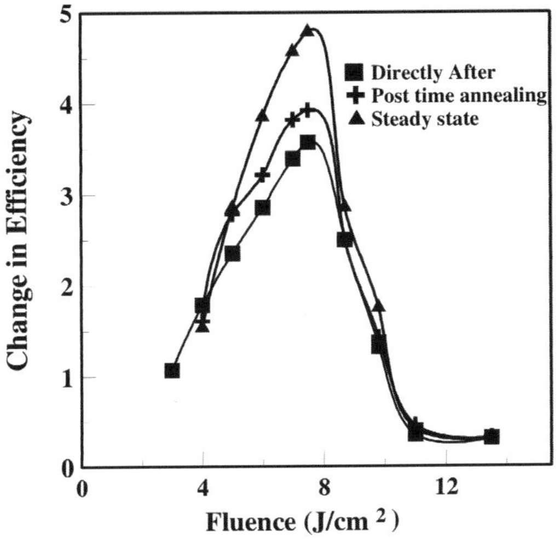

Figure 2. The change in efficiency $\Delta\eta$ due to laser irradiation with different fluences at different times after annealing.

It is well known that grain boundaries play an important role in solar cell performance [4], due to the large densities of defects and segregated impurities at the grain boundaries. The boundaries are regions of strong recombination of electron and holes. This in turn reduces the charge collection efficiency of the solar cell [5, 6]. Therefore, the increase in grain size caused by the laser annealing of the damaged region, apparent from Table 1, could have a special effect on the cell performance as is clear from Figure 2. However, the enhancement of efficiency with PTA may indicate that metastable states are introduced during laser annealing process.

4. Conclusions

Laser annealing is a superior method for removing lattice damage even after manufacturing of the solar cell. Melting of the surface region followed by liquid-phase epitaxial regrowth is the most likely mechanism for the annealing process. Redistribution of the impurities and a reduction in the defected area may also result from laser annealing. Thus the efficiency of a damaged solar cell can be improved by using a proper optimum laser fluence for annealing.

Acknowledgement
The author is indebted to Prof. Abdul Jabbar A. Mukhlus, for providing all the facilities and Dr. Shatha M. Al-Hilly for fruitful discussion.

References

[1] Wood RF and Young RT 1984 *Solar Cells* **12** 245
[2] Bergmann RB 1999 *Appl. Phys.* **A69** 189
[3] Chang HC and Juang MH 1993 *Material Chemistry and Physics* **33** 124
[4] Hovel HJ 1975 *Semiconductor and Semimetals Solar Cells* **11** (Academic Press)
[5] Kuo CC 2009 *Laser Physics* **19** 143
[6] Yuan Z, Lou Q, Zhou J, Dong J, Wei Y, Wang Z, Zhao H and Wu G 2009 *Optics and Laser Technology* **41** *380*

Quantitative high resolution electron microscopy image matching applied to the strontium titanate $\Sigma 3(112)$ grain boundary

K J Dudeck and D J H Cockayne

Department of Materials, University of Oxford, Parks Road, Oxford, OX1 3PH, UK

E-mail: karleen.dudeck@materials.ox.ac.uk

Abstract. Quantitative image matching has been employed in order to compare experimentally observed grain boundary structure to predicted model structures for the strontium titanate ($SrTiO_3$) $\Sigma 3(112)$ grain boundary. Initially, quantitative image matching of image regions containing bulk crystal were used to determine the specimen and microscope parameters, including specimen thickness and tilt, and image defocus. Using the fitted values for thickness and defocus, images have been simulated from two proposed model structures of the $SrTiO_3$ $\Sigma 3(112)$ grain boundary and quantitatively matched to experimental images.

1. Introduction

High resolution electron microscopy is an established tool for characterization of defects in crystalline materials [1]. However, due to the complex electron–specimen interactions, image interpretation is not always intuitive. With the onset of aberration correction, image interpretation has become more straight–forward in many cases; however even with aberration correction, the image pattern observed near defect regions can be complicated and therefore image simulation may be necessary. Further, in order to provide reliable structural information for comparison with model structures, quantitative image comparison is often required. In this paper quantitative image matching has been employed in order to differentiate between two predicted model structures. Using parameters established from bulk crystal images, it has been possible to quantitatively assess the match between simulated images from two model structures of the $SrTiO_3$ $\Sigma 3(112)$ grain boundary in order to determine the structure which agrees best with experiment.

2. Experimental Methods

A $10 \times 10 \times 1$ mm piece of bulk $SrTiO_3$ $\Sigma 3(112)$ bicrystal material, which had been prepared by diffusion bonding, was obtained from a commercial source. The bulk material was heat treated as described in reference [2] and TEM specimens were prepared by using the methods described by Strecker *et al* [3]. Experimental data was collected using a JEOL 2200 MCO microscope operating at 200 kV. The microscope is fitted with CEOS aberration correctors [4] for both the image and probe forming optics. The third-order spherical aberration (C_3) was set to optimal conditions to balance fifth-order spherical aberration, $C_3 = -6\ \mu m$ [5]. In certain circumstances, negative values of C_3 have the additional benefit of improving contrast of light elements such as oxygen in a multicomponent system such as $SrTiO_3$[6]. Twenty member focal series data sets were collected and three images from a focal series were selected for quantitative image matching first of the bulk crystal and then of the grain boundary.

3. Bulk Crystal Fitting

In order to determine specimen and imaging parameters, crystalline sub–regions of the experimental image were compared to a series of images of a bulk crystal model simulated using the multislice method [7] implemented in the QSTEM package [8]. Following the definitions of Hÿtch and Stobbs [9], the mean squared difference between images f and g can be described as the sum of differences due to intensity, contrast and pattern mismatch between the two images. The pattern mismatch between f and g can be described by the cross–mean C_{fg}:

$$C_{fg} = \frac{\overline{fg} - \bar{f}\bar{g}}{\sigma_f \sigma_g} \tag{1}$$

where $\bar{f}$ and σ_f^2 are the mean and standard deviation of pixel intensities in the image (for a complete derivation of equation 1 see reference [9]). In the case when the pattern of the two images are similar, C_{fg} will be close to unity and in this case it is useful to define the angle of the cross–mean, θ_ρ [10]:

$$\theta_\rho = \cos^{-1}(C_{fg}) \tag{2}$$

In this case, the θ_ρ value will approach zero as image similarity increases. For this work, the θ_ρ value was used as a comparison metric because it is directly related to the pattern mismatch between the two images (for alternative metrics see references [11, 12]). The θ_ρ criterion has the additional benefit that it increases linearly in proportion to the difference between two images [10]. The difference could be between the defocus, thickness, crystal tilt, etc. of the two images.

Parallel MATLAB code was written to carry out the matching over a wide range of fitting values using sub–regions of the experimental images from bulk crystal in the grains on either side of the boundary. Following the procedure outlined in references [9, 13], the fitting process was broken down into four steps: defocus–thickness, crystal tilt, beam tilt and astigmatism fitting. The Parallel MATLAB code was run on a 64–node cluster within the Oxford e–Research Centre.

In the case of beam tilt, the best–fit beam tilt had magnitude of zero in all cases. This is likely to be due to the fact that beam tilt could be compensated slightly by the crystal tilt. Figure 1 presents the experimental sub–regions from three different images along with simulated images with the best–fit defocus and thickness values, while table 1 shows the best–fit simulation parameters for each image along with corresponding θ_ρ values. In the case of the defocus–thickness fitting, the code determined the best–fit thickness using all three images simultaneously, which effectively forces the best–fit thickness to be constant for all images. For all three defocus values, the change to the pattern of the simulated image due to incorporating crystal tilt and astigmatism was small and therefore these effects were not included in the simulations shown in figure 1 or for subsequent grain boundary simulations. In addition, because the model structures were produced from density functional theory relaxation [14], it was not possible to incorporate the best–fit crystal tilt values.

4. Grain Boundary Model Comparison

Using the best–fit simulation parameters determined from the bulk crystal fitting, images were simulated from two model structures proposed in the literature [14]. In one configuration the grain boundary was mirror symmetric with no translation, while the other had mirror–glide symmetry (see reference [14] for details). Figure 2 shows the experimental grain boundary images

Table 1. Best–fit thickness, defocus, crystal tilt and astigmatism values, with corresponding θ_ρ values, for three images from the focal series. In all cases the best–fit beam tilt had a magnitude of zero.

Thickness (Å) ±10 Å	Defocus (Å) ±10 Å	Tilt About ⟨100⟩ (mrad) ±2 mrad	Tilt About ⟨110⟩ (mrad) ±2 mrad	Astigmatism Magnitude (Å) ±10 Å	Astigmatism Azimuth (^{o}) ±10°	θ_ρ (rad)
	120	2	5	20	130	0.557
25	190	1	5	40	130	0.377
	330	1	5	20	80	0.627

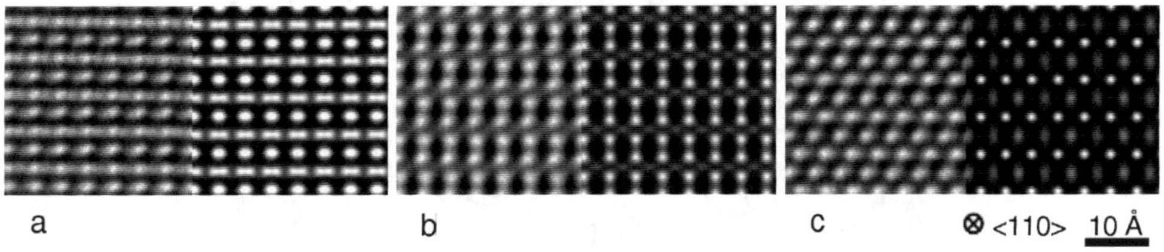

Figure 1. Experimental sub–regions taken from one side of the grain boundary with corresponding best–fit simulations. Simulated images shown in a–c have defocus values of 110, 120 and 330 Å, respectively, and thickness of 25 Å.

along with simulated images from the two model structures, with corresponding θ_ρ values shown below each of the simulated images. From both visual inspection and quantitative comparison of the images, the mirror symmetric model is found to be a relatively better fit to the experimental data. This result was confirmed by comparison with several different segments of the grain boundary (including data sets taken from several different thin regions of the specimen).

5. Conclusions

We have used quantitative image matching including eight fitting parameters to fit imaging parameters for images of bulk crystal near a defect region. Using the fitted parameters, we have simulated images from two proposed model structures of the $SrTiO_3$ $\Sigma3(112)$ grain boundary. Both quantitative and qualitative analysis have shown that the mirror symmetric configuration is the best–fit to the experimentally observed grain boundary structure. Quantitative defocus–thickness matching is also an effective method for confirming the specimen thickness in order to determine if advanced techniques such as focal series reconstruction can be applied [15, 16]. In the case of the thickness value fitted here, the images would be suitable for focal series reconstruction, which could be used for further analysis [2].

Acknowledgments

The authors thank Ute Salzberger at the Max Planck Institute for Metals Research in Stuttgart, Germany for assistance with sample preparation and Christoph Koch for valuable discussions. Nicole Benedek is thanked for providing the DFT model coordinates. The Oxford e–Research Centre at the University of Oxford is acknowledged for assistance with the Parallel MATLAB implementation. Financial support by the European Commission under contract number NMP3-CT-2005-013862 (INCEMS) is acknowledged. KJD would like to thank the Natural Sciences and Engineering Research Council of Canada and University of Oxford Clarendon Fund for financial support.

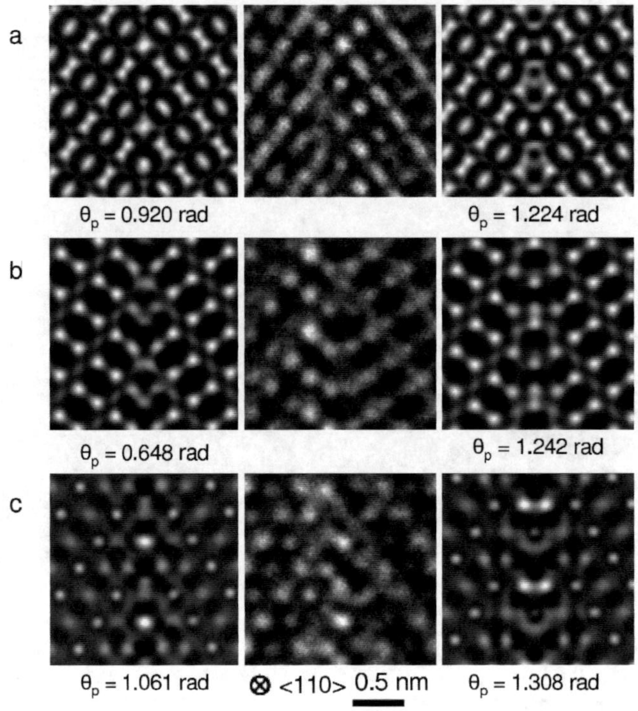

Figure 2. Experimental GB images (centre column), and simulated images from the mirror symmetric model structure (left column) and mirror–glide symmetric model structure (right column). Rows a–c correspond to defocus values 120, 190 and 330 Å, respectively. θ_ρ values corresponding to the model structures are shown below the image for each defocus value. All images are simulated with thickness of 25 Å.

References

[1] Rainforth W M 2004 *Adv. Imaging Electron Phys.* **132** 167–246
[2] Dudeck K J, Benedek N A, Finnis M W and Cockayne D J H 2009 *in preparation*
[3] Strecker A, Mayer J, Baretzky B, Eigenthaler U, Gemming T, Schweinfest R and Rühle M 1999 *J. Electron Microsc.* **48** 235–244
[4] Haider M, Uhlemann S, Schwan E, Rose H, Kabius B and Urban K 1998 *Nature* **392** 768–769
[5] Chang L Y, Kirkland A I and Titchmarsh J M 2006 *Ultramicroscopy* **106** 301–306
[6] Jia C L, Lentzen M and Urban K 2004 *Microsc. Microanal.* **10** 174–184
[7] Cowley J M and Moodie A 1957 *Acta Cryst.* **10** 609–619
[8] Koch C T 2002 *Determination of Core Structure Periodicity and Point Defect Density Along Dislocations* Ph.D. Arizona State University
[9] Hÿtch M J and Stobbs W M 1994 *Microsc. Microanal. Microstruct.* **5** 133–151
[10] Hÿtch M J and Stobbs W M 1994 *Ultramicroscopy* **53** 191–203
[11] Möbus G and Rühle M 1994 *Ultramicroscopy* **56** 54–70
[12] Kauffmann Y, Recnik A and Kaplan W D 2005 *Mater. Charact.* **54** 194–205
[13] Hÿtch M J 1991 *Quantitative High Resolution Electron Microscopy* Ph.D. University of Cambridge
[14] Benedek N A, Chua A L S, Elsässer C, Sutton A P and Finnis M W 2008 *Phys. Rev. B: Condens. Matter* **78** 064110
[15] Saxton W O 1988 *Scan. Microsc.* 213–224 suppl. 2
[16] Van Dyck D, Op De Beeck M and Coene W 1993 *Optik* **93** 103–107

An EELS sub-nanometer investigation of the dielectric gate stack for the realization of InGaAs based MOSFET devices

P Longo[1*], MC Holland[2], GW Paterson[1], AJ Craven[1] and IG Thayne[2]
[1]Department of Physics and Astronomy, University of Glasgow G12 8QQ, UK
[2]Department of Electronics & Electrical Engineering, University of Glasgow, G12 8LT, UK

E-mail: p.longo@physics.gla.ac.uk

Abstract. In this paper, a subnanometer investigation of the Ga_2O_3/GdGaO dielectric gate stack deposited onto InGaAs is presented. Results regarding the influence of the growth conditions on the interface region from a chemical and morphological point of view are presented. The chemical information reported in this paper has been obtained using electron energy loss spectroscopy (EELS) that was carried out in a scanning transmission electron microscope ((S)TEM) showing both spatial and depth resolution.

1. Introduction

Until recently, increased Metal Oxide Semiconductor Field Effect Transistor (MOSFET) performance has been obtained essentially by the reduction of all dimensions of the basic MOSFET structure. As a result, the traditional SiO_2 gate dielectric has become so thin that further reductions would lead to a dramatic increase in the gate leakage current due to tunnelling. This was mitigated to some extent by using Si(ON), but the International Technology Roadmap for Semiconductors (ITRS) has shown the requirement for high-k dielectrics on Si solutions such as HfO_2-based dielectrics for the 45nm node [1]. As the 22nm node is approached, further problems are envisaged and search for new materials is essential. III-V materials present higher mobility than Si and appear to be a promising candidate for n-type devices as well as Ge for p-type devices and the cointegration of both with Si should lead to the realization of the ultimate Complementary Metal Oxide Semiconductor (CMOS). One of the issues that limited, at the early stages, the development of III-V MOSFET devices was the difficulty of creating a suitable insulating layer that keeps the leakage current low and the Fermi level unpinned. However in the case of GaAs, Passlack et al. [2] have shown that a template layer of Ga_2O_3 grown on a As-terminated GaAs surface by molecular beam epitaxy (MBE) unpins the Fermi level. Using Ga_2O_3 alone gives an unacceptably high leakage current [3]. This can be suppressed by incorporating Gd into the dielectric stack after the first ~1nm giving gadolinium gallium oxide GdGaO [4]. Using this approach, world leading MOSFETs have been made on GaAs-based heterostructure incorporating a buried InGaAs channel [5]. This dielectric stack is usually grown by molecular beam epitaxy (MBE) using three different fluxes on the GaAs substrate. The Ga_2O_3 template layer is grown from polycrystalline Ga_2O_3 in a high temperature effusion cell and it sublimes forming a flux of Ga_2O and O_2 [6]. The presence of Ga_2O is believed to unpin the Fermi Level. [6]. The GdGaO is grown using a flux of Gd, Ga_2O and additional molecular oxygen from a gas source. The quality of the resulting gate stack was evaluated using characterization techniques such as photoluminescence (PL), C-V and magneto-transport measurements. The morphology and chemistry were investigated using (scanning) transmission electron microscopy (S)TEM and, in particular, electron energy loss spectroscopy (EELS). The combination of EELS and STEM allows the acquisition of one or more spectra at different positions across a selected region. This technique is called EELS spectrum imaging (SI). Subnanometer resolution can be achieved, although particular care is needed as the Ga_2O_3/GdGaO dielectric gate stack is electron beam sensitive. Nonetheless EELS SI offers improved lateral and similar depth resolution for investigating composition and chemistry when compared to other techniques such as secondary ion mass spectrometry (SIMS), x-ray photoelectron spectroscopy (XPS) and Rutherford backscattering spectroscopy (RBS).

However, to scale down such devices to the size needed beyond the 22nm node, it is essential to increase the mobility and achieve sufficiently low access resistance. To increase the mobility, an increase in the In content seems the most promising approach, although this raises issues with the dielectric. At the moment, it is not clear what type of oxide on top the InGaAs would best suit the realization of MOSFETs with various groups exploring atomic layer deposition of Al_2O_3 [7], HfO_2 [8], ZrO_2 [9], chemical vapour deposited amorphous silicon [10] and MBE grown GdGaO [11]. Based on the success of Ga_2O_3/GdGaO dielectric gate stack on GaAs, it is relevant to explore this oxide on high In concentration III-V surfaces and this is the subject of the paper.

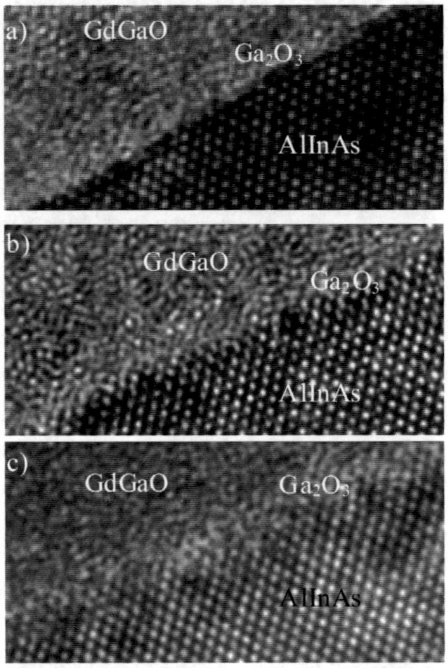

Figure 1 High-resolution TEM images of the interface AlInAs/InGaAs/Ga_2O_3/GdGao in Sample 1 (a), Sample 2 (b) and Sample 3 (c).

2 Experimental methods

Electron energy loss spectroscopy (EELS) carried out in a nanoanalytical electron microscope, has been shown to be an excellent method to characterize Ga_2O_3/GdGaO dielectric gate stacks [12,13]. After developing a way to determine the composition in thick GdGaO layers (>40nm) [12] and thin GdGaO layers [13], a spatial resolution study of the elemental distribution across the InGaAs/Ga_2O_3/GdGaO was started and preliminary results have been published elsewhere [14]. This paper deals with the high-resolution analysis of the elemental distribution across the interface in AlInAs/InGaAs/Ga_2O_3/GdGaO. The heterostructure, grown by MBE, consists of n^+InP substrate followed by ~1.15μm lattice matched $Al_{0.48}In_{0.52}As$ capped with two monolayers of $In_{0.52}Ga_{0.48}As$. The oxide stack is then grown onto the heterostructure and it consists of ~1nm of Ga_2O_3 followed by 25nm of GdGaO. Three samples named as Samples 1-3 have been analysed. They all have the same heterostructure but the oxide growth was carried out at different substrate temperatures. Sample 1 has the lowest temperature while Sample 3 the highest. TEM specimens were prepared by standard cross-sectioning which involves cutting, grinding, dimpling and ion milling using a Gatan PIPS at 4kV for the initial stages of the process and 0.5kV for the final ones. To avoid beam damage, the samples were milled in sample holders cooled to liquid nitrogen temperature. The final polishing process was carried out using a Technorg Linda Gentle Mill at 0.2kV. Conventional high-resolution TEM images were acquired using a FEI Tecnai F20 equipped with a Schottky field emission gun (FEG) and STEM capability. The EELS analysis was carried out in STEM mode using a probe size of 0.5nm, a half convergence angle (α) of 9mrad, a half collection angle (β) of 40mrad, a dispersion of 1eV/channel in a 1340channels CCD and an energy loss range from 400 to 1740eV. This ensures In $M_{4,5}$-edges at 443eV, O K-edge at 532eV, Ga $L_{2,3}$-edges at 1115eV, Gd $M_{4,5}$-edges at 1185eV, As $L_{2,3}$-edges at 1323eV and Al K-edge at 1560eV are all collected in the same spectrum. The contribution to the spectrum from the stray scattering in the Schottky FEG was minimized using a small non-beam defining C1 aperture and the extraction voltage and the gun lens setting were chosen in order to shift the stray to ~800eV where it does not interfere with the Ga $L_{2,3}$-edges and O K-edge at 1115 and 532eV respectively. The spectrum imaging plug-in of Gatan Digital Micrograph was used to record 12x10 spectrum images across the substrate dielectric interface region with a pixel size of 0.55nm and 2.7seconds acquisition time per pixel. A spatial drift reference area was also applied in order to minimize the effects of drift that usually occur in a TEM microscope. The correction was set to be carried out every 3 pixels. The analysis was repeated across three different areas in each sample which showed consistency. The Ga_2O_3/GdGaO dielectric oxide stack is particularly electron beam sensitive therefore the electron dose per pixel needs to be limited. In order to increase the signal to noise ratio, EELS spectra along lines parallel to the interface were summed up as described elsewhere [12].

3. Results and discussion

Figure 1 is a high-resolution TEM image of the interface region in Samples 1-3. The interface in Sample 3 appears much rougher than in Samples 1 and 2 which show, to some extent, similar roughness.

The As $L_{2,3}$-edge intensity was extracted using a power law background. O K and In $M_{4,5}$-edge intensities were extracted using a multiple linear least fit (MLLS) procedure using the shape of the O K-edge and the In $M_{4,5}$-edges respectively extracted from the bulk GdGaO layer and the AlInAs substrate in regions far from the interface. The Al K-edge at 1560eV is quite hard to extract when the As is present. This is mainly due to two reasons, the presence of the As L_1-edge at 1520eV and more importantly the presence of the As $L_{2,3}$-edges at 1323eV that boost the background under the Al K-edge. In addition, the Al K-edge is quite noisy as it belongs to the high energy portion of the EELS spectrum. The signal cannot be improved simply by increasing the acquisition time as the Ga_2O_3/GdGaO is electron beam sensitive.

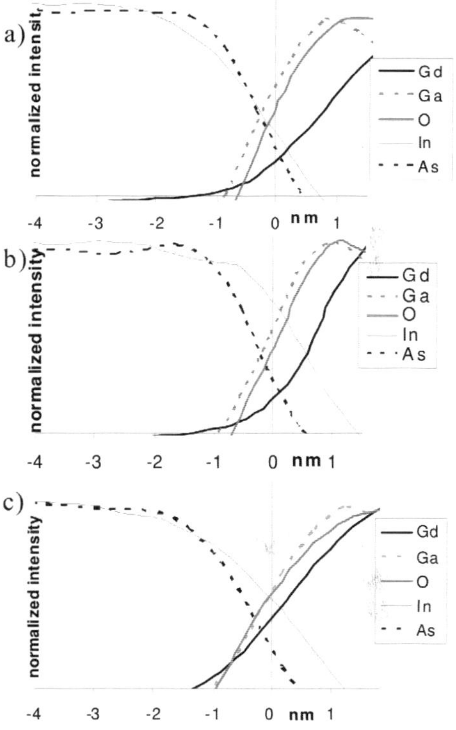

Figure 2 Gd, Ga, O, In and intensity profile from the region across the interface semiconductor/oxide in Sample 1 (a), Sample 2 (b) and Sample 3 (c).

Figure 2 shows the O, Gd, In, Ga and As edges normalised to the same maximum intensity in Samples 1-3 extracted from the AlInAs/InGaAs/Ga_2O_3/GdGaO interface. The 0 on the x-axis is set to be the boundary between the substrate and the oxide stack and on the graph is roughly at half the intensity of the O profile. The distance between the Gd and the O profile shows the presence of the Ga_2O_3 layer which appears to be ~1nm in Samples 1 2 and much narrower in Sample 3. Looking at Figure 2a, the As and In edge intensity profiles in the substrate region are very close to each other. Looking at Figure 2b instead, the In profile appears to be shifted slightly further toward the interface. In Sample 3, all the line profiles appear to be less steep than in Samples 1 and 2 leading to a bigger elemental intermixing across the interface as shown in Figure 2c. It is clear that the substrate temperature employed during the oxide growth, has a major effect and certainly causes the elemental distribution across the interface to vary. In the template layer region where the concentration of As is zero, the concentration of the O can be reasonably assumed as 60%. In this region, as shown by Figure 2, there is $(Ga, Gd, In)_2O_3$. The metal fractions were quantified using Ga_2O_3, $Gd_3Ga_5O_{12}$ and In_2O_3 standards in a similar way as described in [12]. In Samples 1-3 the Ga atomic fractions were respectively 21%, 18%, 7%, the Gd were 5%, 4%, 11% and the In were 14%, 18% and 22%. Analysis of the three areas along the interface suggested uncertainties of ~1-2%. Hence the roughening observed in Figure 1c corresponds to the high degree of intermixing seen in the template layer at the highest substrate temperature. The substrate temperature employed during the oxide growth as shown in Figure 2 influences the In migration toward the Ga_2O_3 template layer as well as the Gd. Such changes on the chemistry of the dielectric-semiconductor interface are growth dependent the understanding of which will play a crucial role as the device size shrinks.

4. Conclusions

Electrical properties are very sensitive to the elemental distributions in the dielectric gate stack and the nature of the interfaces III-V substrate/oxide. The type of analysis presented in this paper is important in optimizing the performance of III-V MOSFET devices, which seems to be related to the quality of the dielectric stack and in particular the oxide-semiconductor interface. The paper also shows that the intensities of characteristic edges such as In, Ga, O and In from the template layer region can be easily converted to atomic fractions using standards of known composition. Variations between the results from other areas along the interface suggest a random error in the atomic fraction of ~1-2% that, given the small pixel size, represents a good result.

References

[1] http://www.electronicsweekly.com/ARTICLES/2007/01/30/40617/Intel+IBM+embrace+high-k+metal+gates+for+45nm.HTM

[2] Passlack M, Yu Z, Droopad R, Bowers B, Overgaard C, Abrokwah J and Kummel AJ, 1999 *J. Vaccum Science & Technology B17*, **1**, 49-52

[3] Passlack M, Medendorp N, Gregory R, and Braddock D, 2003 *Appl. Phys. Lett.*, **83**, 5262,

[4] Paterson GW, Wilson JA, Moran DAJ, Hill R, Long AR and Thayne IG, 2006 *Materials Science Engineering B*, **135**, 277-281

[5] Hill RJW, Moran DAJ, Li X, Zhou H, Macintyre D, Thoms S, Asenov A, Zurcher P, Rajagopalan, Abrokwah J, Droopad R, Passlack M and Thayne IG, 2007 *IEEE Electronic Devices Letters*, **28**, 12,

[6] Hale MJ, Yi S, Sexton JZ, Kummel AJ and Passlack M, 2003 *Journal of Chemical Physics*, **119**, 6719

[7] Xuan Y, Wu YQ, Ye PD, 2008 *IEEE Device Letters*, **29**, 4, 294-296

[8] Zhao H, Shahrjerdi D, Zhu F, Kim H-S, Ok I, Zhang M, Yum JH, Banerjee SK, Lee JC, 2008 *Appl. Phys. Lett.*, **92**, 253506

[9] Goel N, Heh D, Kovenishnikov S, Ok I, Oktyabrsky S, Tokranov V, Kambhampati R, Yakimov M, Sun Y, Pianetta P, gaspe CK, Santos MB, Lee J, Datta S, Majhi P, Tsai W, 2008 *Proceedings of IEDM 2008*, 363-366

[10] Sun Y, Kiewra EW, De Souza JP, Bucchignano JJ, Fogel KE, Sadana DK, Shaidi, 2009 *IEEE Electron Devices Letters*, **30**, 1

[11] Lin TD, Chiu HC, Chang P, Tung LT, Chen CP, Hong M, Kwo J, Tsai W, Chang YC, 2008 *Appl. Phys. Lett.*, **93**, 033516

[12] Longo P, Craven AJ, Scott J, Holland MC and Thayne IG, 2008 *Springer Proceedings in Physics*, **120**, 317-320

[13] Longo P, Craven AJ, Holland MC Moran DAJ and Thayne IG, 2009 *Microelectronic Engineering*, **86**, 214,

[14] Longo P, Paterson GW, Holland MC, Craven AJ and Thayne IG, 2009 *Microelectronic Engineering*, **86**, 1568

Acknowledgements

The authors acknowledge Mr Brian Miller of the University of Glasgow for preparing the TEM cross-sections Engineering Physical Sciences Research Council (EPSRC) and for funding this work through Grant EP/F002610.

Electron Microscopy and Analysis Group Conference 2009 (EMAG 2009) IOP Publishing
Journal of Physics: Conference Series **241** (2010) 012035 doi:10.1088/1742-6596/241/1/012035

Double Aberration-corrected TEM/STEM of tungstated zirconia nanocatalysts for the synthesis of paracetamol

K Yoshida [1,5], NR Shiju[2], DR Brown[2], ED Boyes [1,3,4] and PL Gai [1, 3, 5]
[1] The York JEOL Nanocentre, [3] Department of Physics, [4]Department of Electronics,
[5] Department of Chemistry, University of York, Heslington, York, YO10 5DD.
[2] Department of Chemical and Biological Sciences, University of Huddersfield, Huddersfield, HD1 3DH.

E-mail: pgb500@york.ac.uk

Abstract. We report highly active tungstated zirconia nanocatalysts for the synthesis of paracetamol by Beckmann rearrangement of 4-hydroxyacetophenone oxime. Double aberration-corrected (2AC)-TEM/STEM studies were performed in a JEOL 2200FS FEG TEM/STEM at the 1 Angstrom (1 Å = 0.1 nanometer) level. Observations at close to zero defocus were carried out using the AC-TEM as well as AC-STEM including high angle annular dark field (HAADF) imaging, from the same areas of the catalyst crystallites. The studies from the same areas have revealed the location and the nanostructure of the polytungstate species (clusters) and the nanograins of zirconia. The AC (S)TEM was crucial to observe the nanostructure and location of polytungstate clusters on the zirconia grains. Polytungstate clusters as small as 0.5 nm have been identified using the HAADF-STEM. The nanostructures of the catalyst and the W surface density have been correlated with paracetamol reaction studies. The results demonstrate the nature of active sites and high activity of the tungstated zirconia nanocatalyst, which is an environmentally clean alternative to the current homogeneous process.

1. Introduction

Paracetamol (N-acetyl-p-aminophenol) is one of the most widely used analgesics. The rearrangement of ketoximes to amides or lactams in the presence of acid catalysts is commonly known as Beckmann rearrangement, relevant to the synthesis of paracetamol. The most prominent example for this reaction is the preparation of ε-caprolactam, the monomer used in the production of nylon 6, from cyclohexanone oxime [1-3]. Similar to the case of cyclohexanone oxime rearrangement, the use of solid acid catalysts offers significant advantages over homogeneous acid catalysts. However, much less work has been reported on solid catalysts which is compounded by the difficulty in identifying active surface sites in the reaction. When liquid sulphuric acid is used as a catalyst, including with promoters, N-acetyl-p-aminophenol is recovered from the reaction mixture by neutralization of oleum with aqueous ammonia, resulting in the formation of a large amount of ammonium sulfate. Rearrangement using an ion exchange resin as catalyst and acetic acid as a solvent under an inert atmosphere is reported as well as thionyl chloride in liquid sulphur dioxide [4]. However, sulphuric acid is toxic, corrosive and environmentally unsafe and the use of liquid phase requires special equipment. The use of an insoluble acid catalyst will allow easy separation workup (no neutralization step is required), catalyst recycling and will avoid equipment corrosion and contaminant wastes. Though this has been largely attempted for the Beckmann rearrangement of cyclohexanone oxime to

caprolactam, very few reports are published for the rearrangement of 4-hydroxyacetophenone oxime to paracetamol.
In this study, we have investigated tungstated zirconia catalysts for liquid-phase Beckmann rearrangement of 4-hydroxyacetophenone oxime to N-acetyl-p-aminophenol. Although zirconia-supported oxoanions have been examined as solid acid catalysts recently in industrial isomerization and alkylation processes instead of the highly corrosive and pollutant liquid acid, insights into the nanostructure and optimisation of the catalysts have remained a challenge. Tungstated zirconia possess some advantages such as higher stability under high-temperature treatments and reductive atmospheres, lower deactivation rates, and easier regeneration. Here we report nanostructures of novel tungstated zirconia (WZ) nanocatalysts for the liquid-phase Beckmann rearrangement for clean synthesis of paracetamol /acetaminophen.

2. Experimental Procedure

Tungstated zirconia nanocatalysts were synthesised with different W loadings from 5wt% to 35 wt% (hereafter referred to as 35%WZ). X ray-diffraction (XRD) data of the samples calcined at 973 K have shown that zirconia exists as a tetragonal phase. Double aberration-corrected (2AC)-JEOL 2200FS (S)TEM, configured to have a larger gap (HRP) objective lens polepiece [5] to enable tilting of the sample for alignment into a zone axis orientation as well as heating, was used for nanostructural studies.

3. Results and discussion

Figure 1a shows an AC-TEM image of 15%WZ, illustrating overlapping zirconia nanocrystals with random crystallographic orientations. The crystallite size of the larger nanocrystals in the image was more than 10 nm. The optical diffractogram obtained from the nanocrystals could be indexed based on the tetragonal zirconia structure with a = 0.3607 nm and c = 0.5175 nm. Many of the nanocrystals were observed to be in [001] zone axis orientation as indicated by the selected area FFT/ optical diffractogram in figure 1a, and a few in <110> orientation. Because the nanocrystallinity plays a key role in catalysis, we used dark-field STEM to determine the nanocrystal sizes. Figure 2(a) and Figure 2(b) are STEM images of 15%WZ and 35%WZ. These images revealed crystallites less than 5 nm in diameter constituting the larger zirconia nanocrystals. Average size of the ZrO_2 crystallites was estimated as 2.6nm (with sample variance σ^2 =0.334 nm^2). Figure 1b shows the same crystal as in Figure 1a, slightly defocused, revealing the presence of nanometer scale polytungsate clusters (dark clusters indicated by arrows and identified by tungsten oxide clusters by lattice imaging and nanoanalysis), uniformly distributed throughout the zirconia nanocrystals. Careful examination revealed that the majority clusters were segregated to the zirconia nanocrystal boundaries, as indicated in the schematic in Figure 1(c). Ultra high-resolution TEM (1.2M) clarified the 15%WZ nanocrystals are mosaic crystals with low angle grain boungary. A typical case of α = 182.14 degree are presented in Figure 3. The correlation with catalytic activity has shown that the clusters form active Brønsted sites leading to high activity of the tungstaed zirconia nanocatalyst. Z-contrast analysis [6,7] using aberration corrected- HAADF STEM further illustrates lattice images of tungstated zirconia particles, shown in figure 4(a-d). The areas shown by squares in (a) are enlarged in (b), (c) and (d). Small dots (indicating bright contrast) of less than 1 nm are observed on single crystalline zirconia particle and are identified as tungsten oxide clusters. Their concentration is found to be higher at the surface observed in the profile imaging in (c). In contrast, the 35% WZ catalyst sample showed much larger polytungstate domains and zirconia grain sizes.

4. Concluding remarks

The nanostructures of the tungstated zirconia catalysts for clean synthesis of paracetamol have shown clusters of active tungsten trioxide clusters located at the boundaries of zirconia nano-grains. This heterogeneous process is an environmentally clean alternative to the current homogeneous process.

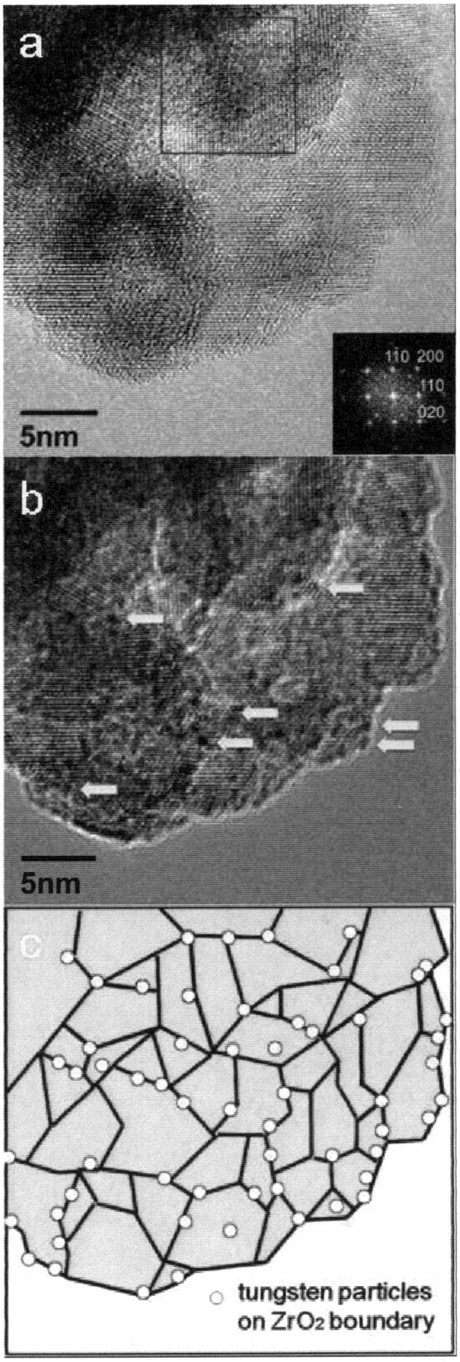

Figure 1. (a) Cs corrected HRTEM and selected area FFT patterns of 15 wt% WZ catalyst. (b) a 30nm under focused transmission electron micrograph of 15 wt% WZ catalyst. (c) Schematic diagram of deposition mechanism.

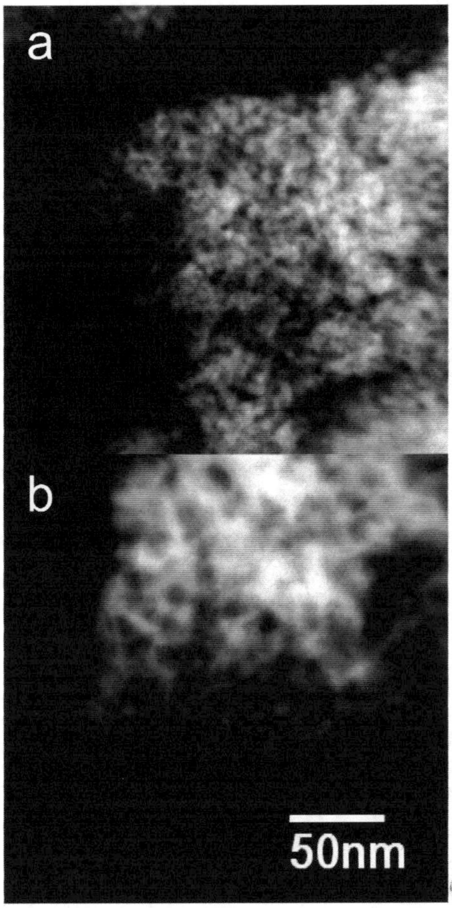

Figure 2. STEM images of WZ catalysts. W concentrations were analysed as (a) 15wt% and (b) 35wt%. Average size of the ZrO_2 crystallites were estimated as (a) 2.6nm and (b) 15.5nm.

Figure 3. Spherical aberration corrected-TEM studies of 15 wt% WZ mosaic crystals

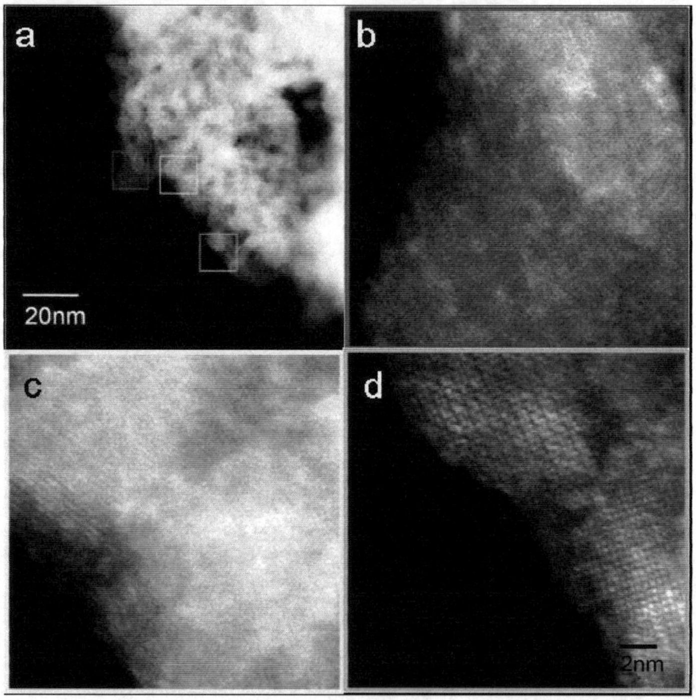

Figure 4. Spherical aberration corrected- HAADF STEM images of 15 wt% WZ catalyst. The areas shown by squares in (a) are enlarged in (b), (c) and (d).

Acknowledgments
We thank the University of York, JEOL and Yorkshire Forward / European Regional Development fund (ERDF) for sponsoring the Nanocentre. KY thanks the JSPS/JFCC for a Visiting Fellowship.

References
[1] Bellussi G and Perego C 2000 *CATTECH* **4** 4
[2] Dahlhoff G, Niederer J P M and Hoelderich W F 2001 *Catal. Rev.* **43** 381 and references therein
[3] Ichihashi H et al.2003 *Catal. Surv. Asia* **7** 261 and references therein
[4] Davenport K G and Hilton C B, US Patent 4524217 (1985) to Celanese Corp
[5] Gai P L and Boyes E D 2009 *Microsc. Res.Tech.* **72** 153
[6] Isaacson M and Crewe A V 1977 *Proc. Natl.* Acad. Sci. **74** 1802
[7] Howie A 1979 *J. Microc.* **117** 11

Importance of characterising the cocatalyst structure in the development of photocatalysts for the splitting of water

N S Hondow,[1] R Brydson,[1] Y H Chou[2] and R E Douthwaite[2]

[1] Institute for Materials Research, University of Leeds, Leeds, LS2 9JT, UK
[2] Department of Chemistry, University of York, York, YO10 5DD, UK

n.hondow@leeds.ac.uk

Abstract. In the development of new photocatalysts the surface of materials with some known catalytic activity is modified by the addition of cocatalysts such as partially oxidised nickel. The exact structure of these cocatalyst nanoparticles potentially has a large effect on the properties of the material as they are expected to act as electron traps and increase the catalyst activity. In the partially oxidised nickel system the cocatalyst structure is expected to consist of a metallic nickel core surrounded by a nickel oxide shell, however there is often little characterisation data to support this. In the present TEM study of the supposed $NiO/Ni/InTaO_4$ system it was found that there was diffusion or migration of the indium possibly from the photocatalyst into the cocatalyst, which is suggested to be a nickel indium oxide surrounding a mixed nickel indium alloy nanoparticle.

1. Introduction

The conversion of solar energy is a desirable option in the development of alternative fuel sources, in particular for the photocatalytic conversion of water into hydrogen and oxygen. An ideal material suitable for this type of reaction has yet to be reported, though several promising developments have been made. An ideal photocatalyst would exhibit certain characteristics, such as long term stability, optimum absorption of the solar spectrum, and the ability to oxidize and reduce water to O_2 and H_2 respectively. Stable oxide semiconductors have shown the best results, with the possibility of manipulating the valence and conduction bands of the materials through alteration of the composition and structure allowing the required redox behaviour. However, the materials previously developed in this field have found limited commercial success for various reasons including a low overall efficiency, as they only utilize the high energy UV periphery of the solar spectrum [1].

Photocatalysts with an appropriate energy band gap split water by absorbing a photon, creating separated electrons and holes in the conduction and valence bands. If the potential difference is greater than that required to split water (1.23 eV) the charges move to the surface and react with water to produce H_2 and O_2. Despite much research on photocatalysis, in particular on materials such as titanates, tantalates and niobates [2], there are at present no materials with a band gap near the theoretical value, and as a consequence methods to try and improve these materials are undertaken. Cocatalysts, such as Pt, RuO_2 and NiO, are intentionally added to the surface of the photocatalyst to act as electron traps and hydrogen evolution sites, so increasing the activity of the catalyst [3].

One examined system has been the addition of partially oxidized nickel to $InTaO_4$ [3]. The nickel source is added to $InTaO_4$, then reduced to form metallic Ni, followed by oxidation to form a nickel oxide shell [3]. It is believed in this case that the Ni-NiO structure is important in the separation of the holes and electrons that are produced after the absorption of a photon in the $InTaO_4$ photocatalyst. The nickel section of the cocatalyst is thought to be useful for the transport of the electrons from the photocatalyst to the NiO surface, where the reduction of water to hydrogen takes place. The presence of the NiO is essential as a metallic nickel surface would actually catalyse the recombination of the produced H_2 and O_2, the backwards reaction of water splitting [3].

Despite the importance implicated in this cocatalyst structure, and the extensive amount of catalyst testing conducted upon this system, there is often little direct evidence of the suggested $NiO/Ni/InTaO_4$ structure. Consequently we have investigated this system with electron microscopy techniques, attempting to determine the cocatalyst structure.

2. Experimental

The Ni-modified $InTaO_4$ photocatalyst was prepared via a literature method [3], specifically a 1 wt% cocatalyst was achieved through the addition of $Ni(NO_3)_2$, and the material calcined in air for 1 hour at 350 °C, reduced in H_2 for 2 hours at 500 °C and oxidized in O_2 for 1 hour at 200 °C.

TEM analysis was conducted on an FEI CM200 FEG TEM running at 197 kV equipped with an Oxford Instruments Energy Dispersive X-ray (EDX) spectrometer and a Gatan Imaging Filter. Samples were prepared by dispersing in methanol, with a drop placed on a holey carbon coated copper grid. SEM samples were coated with a thin layer of platinum and examined in a LEO 1530 Gemini FEG SEM with Oxford instruments EDX spectrometer. Powder X-ray Ray Diffraction (XRD) was performed on a Bruker D8 powder diffractometer. Photocatalyst testing was performed (0.1 g of catalyst, 110 mL H_2O) with a Xe lamp (300 W) and the evolved gases were analyzed by gas chromatography-mass spectrometer for 20 hours.

3. Results and Discussion

The partially oxidized nickel doped $InTaO_4$ system was characterized four times during the synthesis and testing process; the initial unmodified $InTaO_4$, the reduced system $Ni/InTaO_4$, the oxidized system $NiO/Ni/InTaO_4$, and then this latter material after photocatalytic testing. Powder X-ray diffraction patterns of the nickel modified materials only showed reflections attributable to the photocatalyst $InTaO_4$. SEM analysis was performed to measure the size of both the photocatalyst and added cocatalyst, with an average $InTaO_4$ particle diameter measured at 256 nm, and the smaller cocatalyst particles generally being in the range of 15 to 40 nm in diameter.

Initial TEM examination of the nickel added materials show the smaller cocatalyst nanoparticles on the larger photocatalyst particles (figure 1). The material after oxidation appears to have a core-shell structure (figure 1(b)) as opposed to the particles after only reduction (figure 1(a)).

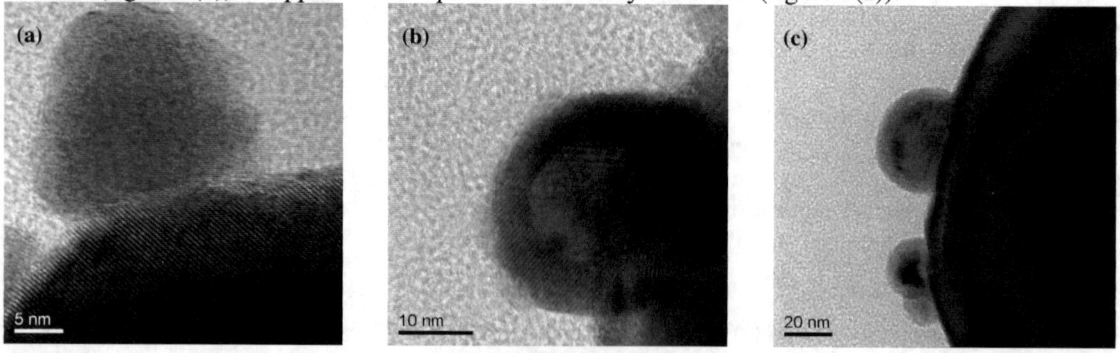

Figure 1. TEM images of 1 wt% Ni added to $InTaO_4$ (a) after reduction (b) after oxidation (c) after photocatalyst testing

The material retains this core-shell structure even after photocatalytic testing, as viewed in figure 1 (c). The cocatalyst particles in the material after testing are not significantly different from those in the oxidized sample in either size or crystallinity, indicating that they are stable under the reaction conditions. This is of particular importance, as it has previously been found that during photocatalytic testing the material can be deactivated due to nickel being eluted and redeposited [4].

The requirement to confirm the presence of particular elements within the materials led to studies by EDX spectroscopy. Initial analysis confirmed the presence of oxygen, nickel, indium and tantalum, but in an attempt to more precisely determine the elemental distribution through the material, dark field STEM and EDX maps were recorded (figure 2). In the EDX maps produced from the material after oxidation, it is possible to see the locations of Ni, In and Ta. The nickel signal (figure 2 (b)) is concentrated in areas corresponding to the small cocatalyst particles, while the tantalum signal (figure 2 (d)) corresponds closely to that of the photocatalyst particles, the same particle shape being seen in the Ta EDX map and the STEM image. However something different is seen in the indium EDX map (figure 2 (c)), here is a similar coverage to that shown in the tantalum EDX map, but there also appears to be several locations where there are concentrated areas, which correspond with the nickel signal and the small cocatalyst particles.

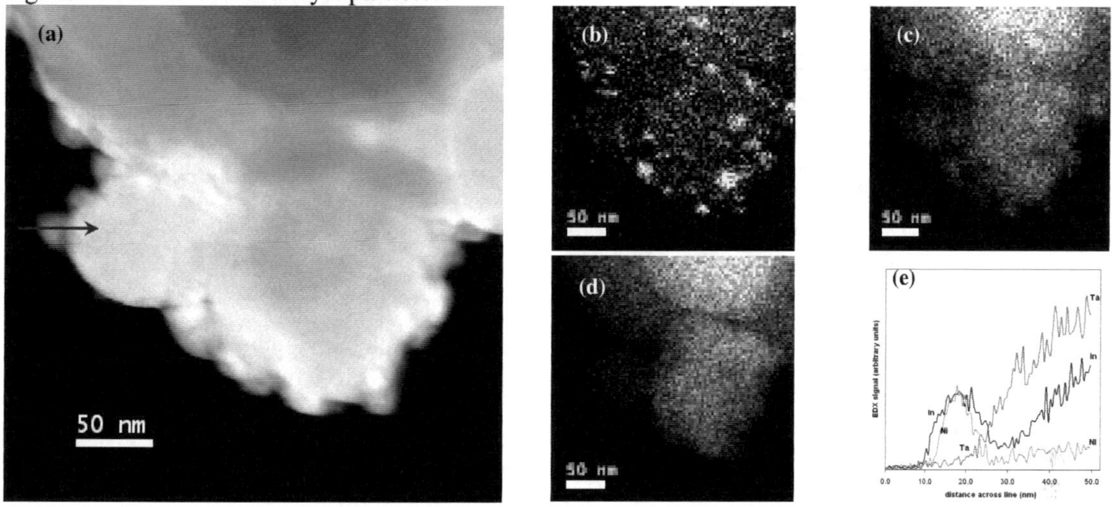

Figure 2. The material after oxidation (a) DF STEM image, with arrow showing location of linescan (b) Ni K EDX map (c) In L EDX map (d) Ta M EDX map (e) linescan of Ni, In and Ta across a nanoparticle into the photocatalyst

This suggestion of indium being present in the cocatalyst particles as well as nickel was confirmed by an EDX linescan across one of the cocatalyst particles, as indicated by the arrow in figure 2 (a). The EDX linescan (figure 2 (e)) shows both nickel and indium, in similar concentrations across the cocatalyst particle. The tantalum signal is only apparent once the electron beam is in the photocatalyst particle, at which point the nickel signal drops off and the indium signal mimics that of the tantalum, but at a lower level. This indicates that the cocatalyst unexpectedly contains both nickel and indium, while the photocatalyst contains only indium and tantalum; this is also found to be the case in the reduced material. There are known nickel indium alloys, such as Ni_3In, Ni_2In, NiIn and Ni_2In_3, which can form at temperatures similar to those used in the preparation of these materials [5], and studies on thin films of nickel and indium have shown diffusion at quite low temperatures [6]. The exact source of this indium is unconfirmed; it may come from unreacted In_2O_3 from the $InTaO_4$ synthesis, however as this mixed Ni-In system is found in all examined cocatalyst particles it is more likely that the source of indium may be the neighbouring photocatalyst particle. This migration or diffusion of atoms from a material to a supported catalyst has been discussed before, in terms of strong metal-support interaction [7], for instance in which titania has been observed to migrate over supported Pt.

TEM images of the oxidized materials (both before and after testing) indicated a core-shell structure, initially believed to be NiO/Ni, and with this new evidence of indium in the cocatalyst, differences between the two layers were investigated. The presence of oxygen was confirmed with an EFTEM map (figure 3 (b)), showing oxygen to be mainly present in the photocatalyst and edge of the cocatalyst. The spinel structure, $NiIn_2O_4$, has previously been reported [8] however its preparation from In_2O_3 and NiO was performed under harsh conditions and the reaction did not go to completion. The lattice fringes that are present in the shell of the oxidized material have been measured (figure 3 (c)) and do not correspond to NiO or to any of the reported nickel-indium alloys. No crystallographic details are available for nickel indium oxides, although data is available for $CuIn_2O_4$ and $Cu_2In_2O_5$ which may have similar structures, and these do have lattice spacings similar to those measured in this system.

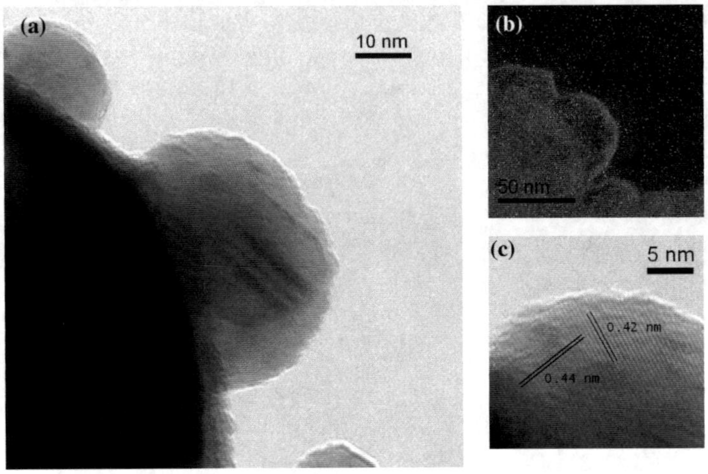

Figure 3. Nominal $NiO/Ni/InTaO_4$ system: (a) TEM image (b) EFTEM oxygen map showing oxide shell on cocatalyst (c) area of TEM image showing lattice fringes in the oxide shell

4. Conclusions

Considerable importance is placed on the cocatalyst structure in photocatalytic systems. In this study we have examined what was believed to be a $NiO/Ni/InTaO_4$ system. Despite the material being synthesised in conditions similar to those previously examined in the literature, it was determined that indium, possibly from the photocatalyst migrated or diffused into the added cocatalyst, forming an inner mixed nickel-indium alloy surrounded by a nickel indium oxide. The exact composition and structure of the cocatalyst has not been determined at this point, but hopefully with further experiments comprising electron energy loss spectroscopy and electron diffraction this should be accomplished. Diffusion of indium from the underlying photocatalyst may have important effects in terms of the magnitude of the band gap and hence photon absorption.

Acknowledgements
Acknowledgments to EPSRC for funding (grant number EP/E018262/1).

References

[1] Kudo A, Kato H and Tsuji I 2004 *Chem. Lett.* **33** 1534-9
[2] Osterloh F E 2008 *Chem. Mater.* **20** 35-54
[3] Zou Z, Ye J, Sayama K and Arakawa H 2001 *Nature* **414** 625-7
[4] Kato H and Kudo A 2003 *Catal. Today* **78** 561-9
[5] Massalski T B 1986 *Binary Alloy Phase Diagrams* (Ohio:American Society for Metals) p 1387.
[6] Platzer R, Wöhrmann U, Ding X L, Fink R, Krausch G, Voigt J, Wesche R and Schatz G 1990 *Hyperfine Interact.* **60** 1003-6
[7] Tauster S J 1987 *Acc. Chem. Res.* **20** 389-94
[8] Tasaki J, Izushi T and Ito T 1985 *Adv. Ceramics* **15** 169-74

Journal of Physics: Conference Series **241** (2010) 012037 doi:10.1088/1742-6596/241/1/012037

A TEM nanoanalytical investigation of Pd/Ge ohmic contacts for the miniaturization and optimization of n-InGaAs MOSFET devices

P Longo[1*], W Jansen[2], C Merckling[3], J Penaud[4], M Caymax[3], IG Thayne[2] and AJ Craven[1]
[1]Department of Physics and Astronomy, University of Glasgow G12 8QQ, UK
[2]Department of Electronics & Electrical Engineering, University of Glasgow, G12 8LT, UK
[3]Imec, Kapeldreef 75, B-3001 Leuven, Belgium
[4]Riber, 31 Rue Casimir Perier, 95873 Bezons Cedex, France

E-mail: p.longo@physics.gla.ac.uk

Abstract. In this paper, a nanoanalytical investigation of electron beam evaporated PdGe ohmic contacts onto an n^+ $In_{0.53}Ga_{0.47}As$ layer using electron energy loss spectroscopy (EELS) is presented. The chemical information reported in this paper has been obtained using EELS spectrum images (SI) that allow not only depth resolution but also spatial resolution which is essential in such inhomogeneous systems.

1. Introduction

As the size of III-V devices decreases, ohmic contacts and their performance become increasingly important. In addition to low resistance, these contacts need to show thermal stability, reproducibility, as well as lateral and depth uniformity. AuGeNi ohmic contacts have been widely used as they show relatively low specific contact resistivity. Using this approach world leading GaAs MOSFET devices have been realised [1]. However as reported in [2], such contacts have the drawback of poor uniformity due to the Au diffusion into the III-V substrate. Obviously this strongly limits them for scaling purposes. In addition, with the quest for co-integration of III-V and silicon MOSFETs for ultimate CMOS applications, silicon compatible III-V devices process modules, which are gold-free, are required. PdGe ohmic contacts have demonstrated a contact resistance as low as 10^{-6} Ωcm^2 which is comparable to that shown by AuGeNi approaches. They also show better uniformity along the interface with the III-V substrate [3,4]. PdGe contacts are formed via a solid state reaction between a layer of Pd on the III-V substrate followed by a thicker one of Ge. During the annealing process, the Ge reacts with all the Pd forming PdGe. However some of it will be also incorporated into the InGaAs, preferentially on Ga sites as there is an excess of Ga vacancies [5]. This will determine the formation of a n-type layer beneath the ohmic contact and this is an important issue when n-type devices are being fabricated. The choice of the thickness of Pd and Ge layers is such that it ensures the Ge penetrates across the entire Pd layer forming PdGe and reacting with the surface of the III-V substrate. The behaviour of this type contact will be affected by the nature of the species present at the interface and the degree of roughness along the interface [6]. The contacts of this study are capped with Ti/Pt that acts as barrier to upwards Ge duffusion during the annealing process [7].

The morphology and the chemistry of such ohmic contacts are investigated using scanning transmission electron microscopy ((S)TEM) and, in particular, EELS for the analytical study.

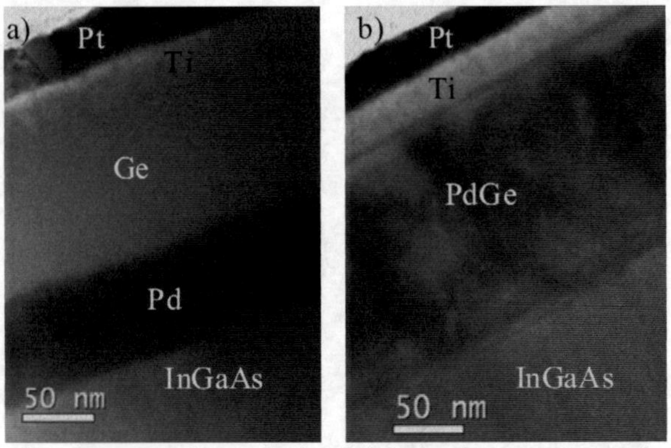

Fig. 1 TEM micrograph of the ohmic contact stack: (a) unannealed sample; (b) annealed

2. Experimental methods

The 70nm Pd, 100nm Ge, 30nm Ti and 40nm Pt ohmic contact stack was deposited by electron beam evaporation onto a 500nm thick $In_{0.53}Ga_{0.47}As$ layer silicon doped to $1x10^{19}$atoms cm^{-3}, grown on a semi-insulating InP substrate. The sample was then cleaved and one half went through a rapid thermal annealing for 10 seconds at 400 °C. TEM specimens of both the annealed and unannealed samples were prepared by standard cross-sectioning that involves cutting, grinding, dimpling and ion milling using a Gatan PIPS at 4kV for the initial stages of the process and 0.5kV for the final ones. The final polishing was carried out using a Technorg Linda Gentle Mill at 0.2kV. Conventional TEM images were taken using an FEI Tecnai F20. The chemical analysis was carried out in a scanning transmission electron microscope (STEM) using a probe size of 0.7nm and probe half convergence angle of 9mrad. EELS data were recorded using a Gatan Enfina spectrometer with a half collection angle of 40mrad and a dispersion of 1eV/channel and an energy loss range from 260eV to 1600eV. This ensures Pd$M_{4,5}$-edges at 335eV, In $M_{4,5}$-edges at 443eV, Ga $L_{2,3}$-edges at 1115eV, Ge $L_{2,3}$-edges at 1217eV and As $L_{2,3}$-edges at 1323eV are collected in the same spectrum. Where possible, the edges where extracted after removal of the background using a power law model fitting in front of the edge. However edges such as Ga $L_{2,3}$ at 1115eV and the Ge $L_{2,3}$ at 1217eV overlap. This made the edge extraction by simple removal of the background impossible. This problem can be overcome by using the multiple linear least square (MLLS) fit method available in the Gatan Digital Micrograph package. The spectrum imaging plug-in of Gatan was employed to collect spectrum images starting in the bulk InGaAs and extending ~25nm into the contact region using a step size of 1.8nm. This provides insight into the chemistry at the interface region. An acquisition time of 8 seconds per pixel was used. A spatial drift correction area was applied in order to minimize the effects of drift that usually occur in a TEM microscope. The drift correction was carried out every 3 pixels. Although this material is electron beam resistant, the analysis was carried out using the sub-pixel scanning which is available in Gatan Digiscan package. This consists of scanning the beam within the pixel area while acquiring the EELS spectrum. Thus the electron dose is spread over the pixel area minimizing the dose and any C contamination. The presence of C whose K-edge is at 284eV affects the background of Pd $M_{4,5}$-edges making its extraction

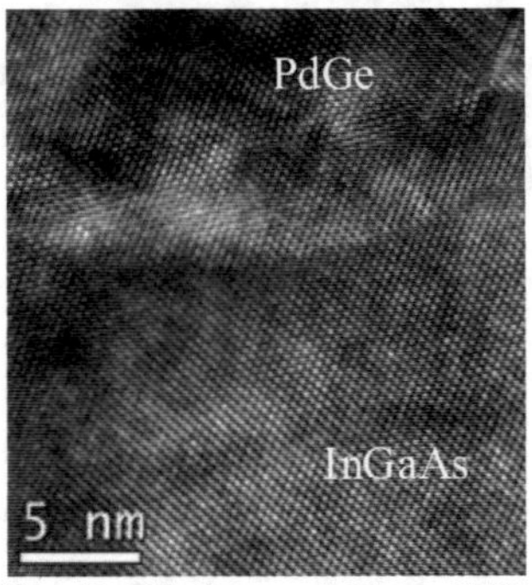

Fig. 2 High-resolution TEM image of the interface region in the annealed sample.

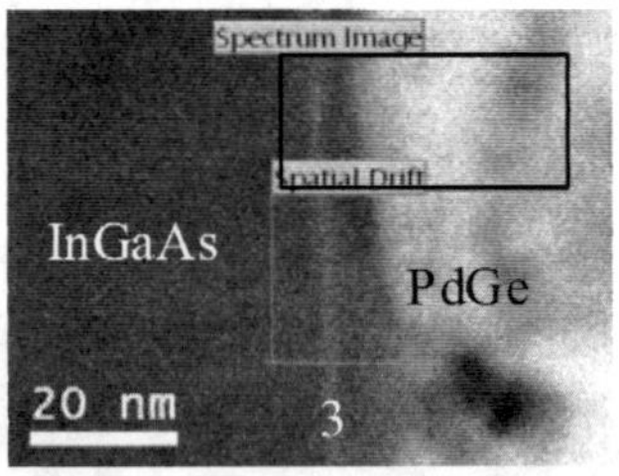

Fig. 3 ADF STEM survey image of the annealed sample. The black box is the area used for the acquisition of the EELS SI.

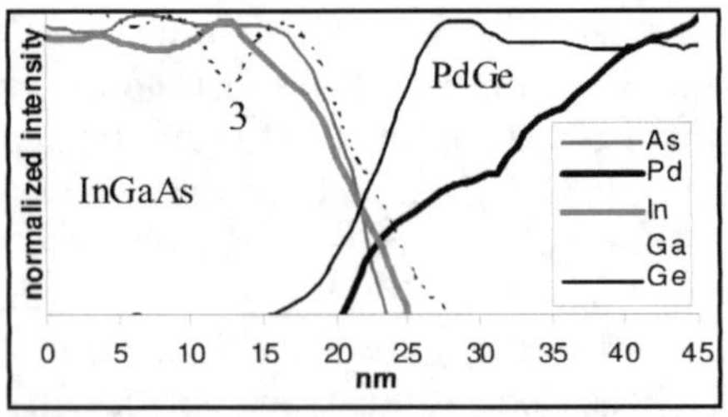

Fig. 4 EELS edge intensity profiles extracted from the region in the black box.

unreliable. Therefore, it is essential to limit the C contamination.

3. Results and discussion

Figs. 1a,b show bright field images of the whole ohmic contact region in the unannealed and annealed samples respectively. The unannealed sample does not show, at a large scale, any sign of major reactions between the contact and the substrate across the whole contact region. Ge and Pd layers appear to be amorphous although some lattice fringes can be observed at positions more than 10nm from the substrate when the image is taken at higher magnification. On the other hand, in the annealed sample, most of the Ge has reacted with all the Pd forming a PdGe granular structure as shown in Fig. 1b. At the Ti side, some amorphous Ge remains and there is a (Ti, Ge) reaction layer at the Ti boundary. On the substrate side the PdGe approaches the substrate. The interface is now fairly rough and it is also quite difficult to distinguish where the original interface was before the annealing. A marker layer would be useful. Fig. 2 is a high-resolution TEM bright field image of the contact/substrate interface taken from the annealed sample. It confirms that most of the interface region next to the substrate has crystallized with the same orientation as the InGaAs substrate. For the annealed sample, the EELS spectrum images were taken across the selected region shown in Fig. 3. The signal to noise ratio was improved by summing the spectra in lines parallel to the interface Fig. 4 shows the In, Ga, Ge, As and Pd edge intensities extracted from that region. The background fitting windows were chosen in regions of the EELS spectrum where perturbations are not significant. However, their use to process the whole spectrum results in negative counts for the Pd, In and As in some regions. Ge has either directly penetrated the Pd layer or propagated along grain boundaries and diffused into the InGaAs more strongly than the Pd. Ga, In and As have diffused into the contact region with Ga propagating further. In the InGaAs region, a 2nm layer rich in In and deficient in Ga has been detected. This corresponds to the bright line marked 3 in Fig. 3 and the peak in In and dip in Ga marked 3 in Fig. 4. To further explore the mechanism of formation of the In rich layer, two more wafers were prepared. In the first one, referred to as Ge only, 100nm of Ge followed by 30nm of Ti and 40nm of Pt were deposited onto the same InGaAs substrate. In the second one, referred to as Pd only, 70nm layer of Pd followed by Ti and Pt were deposited. Both wafers were cleaved and one half went through the same annealing process described above. Fig. 5 is a high-resolution TEM micrograph of the interface region between the InGaAs and the Ge layer in the Ge only sample after annealing. The interface appears to be very smooth. It seems that the annealing process has not caused major variations in the morphology of the interface nor most likely, of the chemistry. Figs. 6a,b show the In, Ga, Ge and As edge intensities extracted from the interface region in the Ge only unannealed and annealed samples respectively. In the unannealed sample the In, As and Ga lines drop all together and symmetrically with respect to the Ge line. This suggests little or no

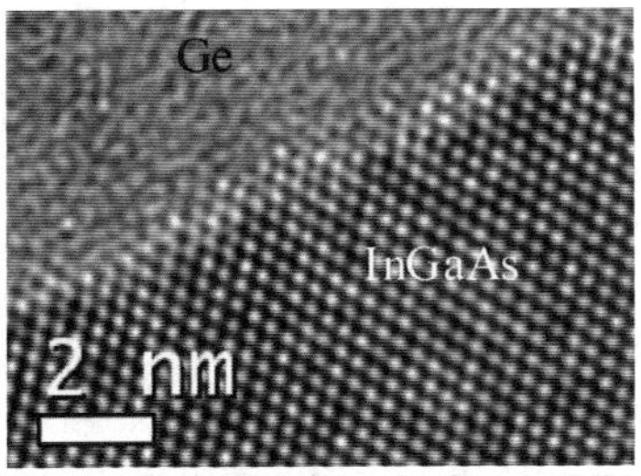

Fig. 5 High-resolution TEM image of the interface region in the Ge only annealed sample.

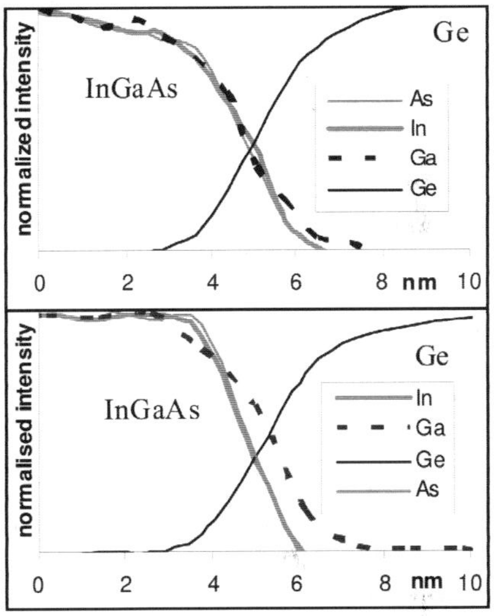

Fig. 6 EELS edge intensity profiles extracted from the interface region in the Ge only sample: (a) unannealed; (b) annealed.

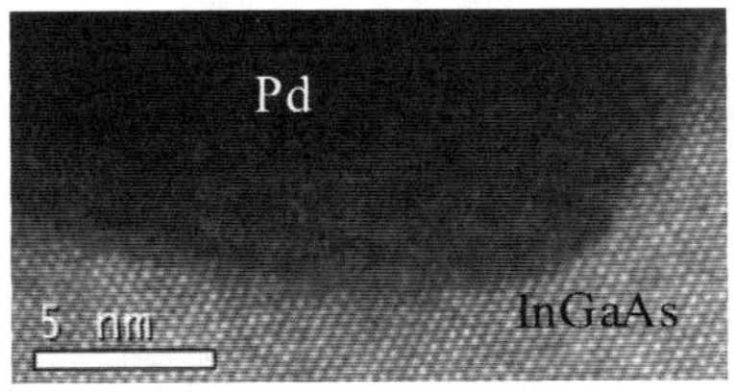

Fig. 7 High-resolution TEM image of the interface with the substrate in the annealed Pd only sample.

interpenetration with beam spreading or roughness causing the width of the transition. However, in the annealed sample the Ga appears to diffuse further to the right making the Ge layer p-doped and this is not good if n-devices are being fabricated. Here it is not clear if the Ge has really diffused into the InGaAs to replace the Ga. Fig. 7 shows a high-resolution TEM micrograph of the interface between the InGaAs and the Pd layer in the Pd only sample after annealing. The interface appears to be fairly sharp but reasonably rough with the presence of granular structures most likely containing Pd that extends into the InGaAs. Here the EELS line spectrum images were taken across the selected region shown in Fig. 8, using 2.5nm step size. Fig. 9 shows the In, Ga, As and Pd lines extracted from the selected region in Fig. 8. As and Ga have penetrated into the Pd layer more strongly than the In and diffused across the whole Pd layer. The region marked A is the InGaAs substrate. In region B, there is a high concentration of Pd with significant Ga and some In, both decreasing with distance from the InGaAs, while the As increases toward region C. Region C is mainly As with some Ga but no Pd or In and corresponds to the darker layer marked C in Fig 8. Region D is Pd with only Ga and As and extends to the Ti layer. Clearly the interaction of Pd with InGaAs is quite complex. When there is a Ge layer on top of the Pd, there will be a competition between the interaction of the Pd with the Ge and that with the InGaAs. Thus there may be a link between the potential to form the In containing layer B in Figs 8 and 9 with the In rich layer marked 3 in Figs. 3 and 4.

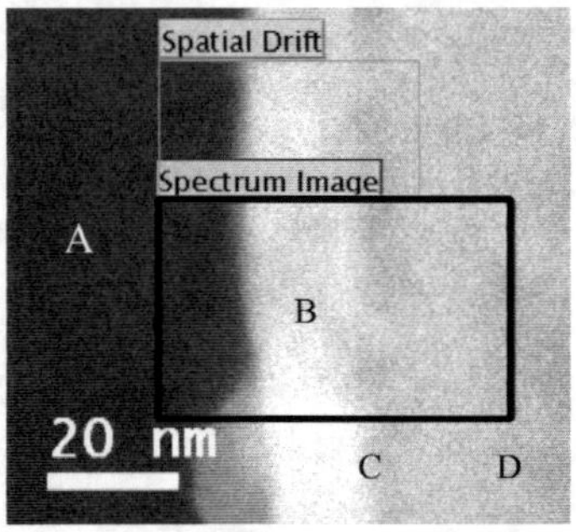

Fig. 8 ADF STEM survey image used for the acquisition of the EELS SI in the annealed Pd only sample.

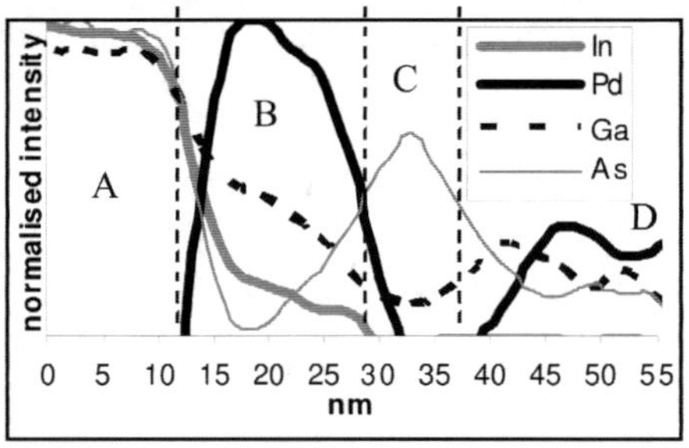

Fig. 9 Edge intensity profiles extracted from the selected region in Fig. 8.

4. Conclusions

The results presented in the paper show that it is possible to analyse with spatial and depth resolution inhomogeneous systems such as PdGe ohmic contacts at a nanometer-level using EELS SI. This allows to understand in details the reactions occurring during the contact stack formation and in particular to address the role of both the Ge and the Pd.

References

[1] Hill RJW, Moran DAJ, Li X, Zhou H, Macintyre D, Thoms S, Asenov A, Zurcher P, Rajagopalan, Abrokwah J, Droopad R, Passlack M and Thayne IG, 2007 *IEEE Electron Devices Letters*, **28**, 12
[2] Tsuchimoto J, Shikata S and Hayashi H, 1991 *J Appl. Phys.* **69**, 6556
[3] Wang LC, 1995 *J Appl. Phys.* **77**, 2007
[4] Kim T, Lee L, Mum JK and Kim H, 1997 *Appl. Phys. Lett.* **71**, 2656
[5] Wang LC, Zhang B, F Fang, Marshall ED, Lau SS, sands T and Kuech TF, 1988 *J Mater. Res.* **3**, 922
[6] Marshall ED, 1987 *J Appl. Phys.* **62**, 942
[7] Wang YG, Wang D and Ivey DG, 1998 *J. Appl. Phys.,* **84**, 1310

Acknowledgements

The authors acknowledge Mr Brian Miller of the University of Glasgow for preparing the TEM cross-sections, Engineering Physical Sciences Research Council (EPSRC) via the Grant EP/F002610 and the EU FP7 Dual Logic Project for funding this work.

Journal of Physics: Conference Series **241** (2010) 012038 doi:10.1088/1742-6596/241/1/012038

Study of incommensurate phases in Lanthanum-doped zirconium-rich Lead Zirconate Titanate ceramics.

R Villaurrutia[1], I MacLaren[1] and A Peláiz-Barranco[2]

[1]Department of Physics and Astronomy, University of Glasgow, Glasgow G12 8QQ, UK.
[2]Facultad de Fisica/IMRE, San Lazaro y L, Univ. de la Habana, C. Habana 10400 Cuba.

E-mail: r.villaurrutia@physics.gla.ac.uk

Abstract. The microstructure and nanostructure of zirconium-rich $Pb(Zr,Ti)O_3$ ceramics doped with small amounts of La which are right at transition between the ferroelectric (FE) and antiferroelectric (AFE) orderings has been examined using Transmission Electron Microscopy (TEM) imaging and diffraction. In this region, the La doping frustrates the formation of simple FE or AFE phases and promotes long period ordered phases (2-3 nm) with ordering along <110> with unit cells incommensurate with the primitive cubic unit cell. We show that the domain structure in these materials is closely related to that previously observed in AFE $PbZrO_3$. Moreover, precision measurements of crystallographic tilts at domain boundaries using Kikuchi diffraction methods also confirms the close relationship to $PbZrO_3$. The domains also contain a nanostructure perpendicular to the long-period ordering direction, but the reasons for the appearance of this nanostructure remain unclear.

1. Introduction

Lead Zirconate Titanate (PZT) ceramics are exciting functional materials with very important technological applications on account of their ferroelectric and piezoelectric properties. In recent years there has been particular interest in the Zr-rich end of the $Pb(Zr,Ti)O_3$ (PZT) phase diagram where ferroelectric (FE) and antiferroelectric (AFE) ordering schemes compete. It has been reported that when doping with Lanthanum, these ceramics undergo a phase transition from a rhombohedral FE to an orthorhombic AFE phase. Dai *et al.* [1] in their classic study showed that this transition can be reached in PZT with a Zr:Ti ratio of 90/10 and a few at % of La. Structurally, this results in the appearance of some unusual satellite reflections parallel to 110 directions in selected area diffraction patterns (SADP) and were interpreted as evidence of a long period ordered incommensurate phase. In the past few years, several articles have been published concerning electron microscopy investigations of such incommensurate antiferroelectric phases in La or Sn doped PZTs [2-4], but less attention has been given to either the domain structure of such phases or to their atomic structure, although the domain structure of commensurate $PbZrO_3$ has been investigated previously [5]. In the present work, we explicitly study the domain structure of incommensurate AFE phases found in La-doped PZT with ratios of 2/90/10, 3/90/10 and 4/90/10, investigate its relationship to the commensurate $PbZrO_3$ phase.

2. Experimental Methods

Samples with compositions given by $(Pb_{1-x}La_x)(Zr_{0.9}Ti_{0.1})_{1-x/4}O_3$ where x = 2, 3 and 4% were made by the well-known mixed oxide approach and powders were calcined at 800°C, cold-pressed, and sintered at 1250°C for 1h in a covered platinum crucible. They were prepared for TEM analysis using standard methods of slicing, grinding, disc cutting, dimpling, ion beam thinning and carbon coating to minimise charging. Transmission electron microscopy was performed using either a FEI Tecnai T20 TEM operated at 200 kV or a FEI Tecnai TF20 TEM operated at 200 kV.

3. Results

3.1 Domain structure and nanostructure in the incommensurate phase.

Figure 1 shows dark field images of one grain recorded under two different diffraction conditions. Figure 1a is a strong beam dark field image and conventional domains, similar to those seen in many ferroelectric compositions, are clearly visible; whilst these domains are often wedge shaped and can show significant variations in boundary plane, they tend to lie close to {101} planes (as would be expected for both rhombohedral and orthorhombic structures). Figure 1b is recorded in a very weak diffraction condition, and shows a streaked nanostructure within the domains: each domain has its own unique streaking direction which was always perpendicular to the appearance of satellite reflections in the diffraction pattern along a given 110 direction. These satellite spots correspond to those previously seen by Dai *et al.* [1] and must be related to the long-period incommensurate AFE structure and are clearly inconsistent with a rhombohedral PZT structure. Observation of this nanostructure is thus convincing evidence that an area is *not* rhombohedral and ferroelectric.

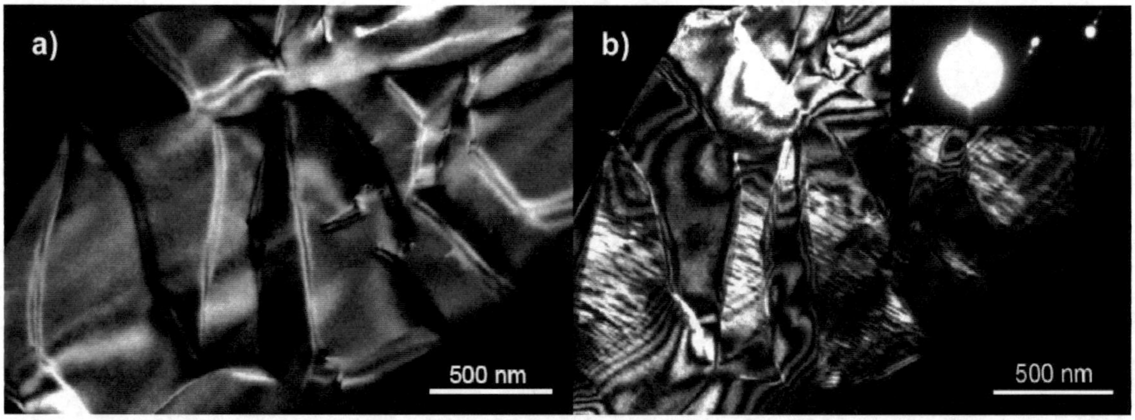

Figure 1. Dark field TEM images of the domain structure in a PLZT 2:90:10 sample; a) Strong beam diffraction condition showing domain contrast only; b) weaker diffraction condition showing nanostructure in one set of domains.

3.2 Misorientation analysis of the microdomains

Figure 2 shows images of the domain and nanostructure in a 4:90:10 sample together with two processed Kikuchi patterns taken at points 3a and 4, respectively. The misorientations between points 1 and 3, 2 and 4, 4 and 5 and 5 and 6 give a mean value of 0.68 ± 0.06 °. This unit cell rotation at the boundary can be used to measure local c/a ratios [6], which matches well to expectations for the orthorhombic AFE phase (giving a local c/a ratio of 0.988, compared to an expected value of 0.991 based on published lattice parameters for pure $PbZrO_3$ [7]). Determining misorientation axes for such small tilts using conventional orientation determination from Kikuchi patterns is known to be very sensitive to small errors to the point of giving completely meaningless results [8] and thus we do not report the misorientation axes. We have, however in a separate selected area diffraction experiment confirmed that the rotation axis is <010> as would be expected. Thus, we can confirm that the phase really is closely to the one observed in $PbZrO_3$.

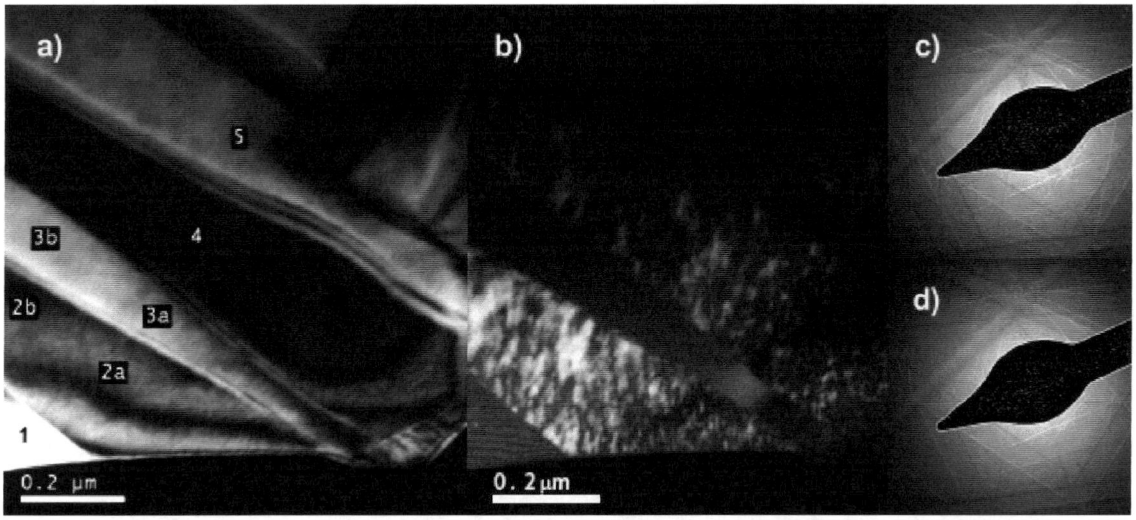

Figure 2. Structures in a PLZT 4:90:10 sample: a) Strong beam dark field TEM image of the domain structure; b) weaker beam TEM image of the nanostructure in domains 2 and 4; c) and d) processed Kikuchi patterns from points 3a and 4.

3.3 Satellite reflections in diffraction patterns and incommensurate lattice parameters

Figure 3 shows a [001] diffraction pattern taken from a single domain where weak satellite spots around the principal reflections are observed. These satellite reflections indicate a long period ordering of the crystal structure. The periodicity of such satellite spots was about 22.6 Å, about 7.7 ± 0.2 longer than the (110) plane spacing for this structure of ~ 2.93 Å. So typically, these materials have very large unit cells, with periodicities that often incommensurate with the spacings of the primitive perovskite cell, and usually with a periodicity of 7-8 (110) spacings. These long periodicities are thought to arise due to competition between antiferroelectric and ferroelectric ordering schemes in these intermediate compositions.

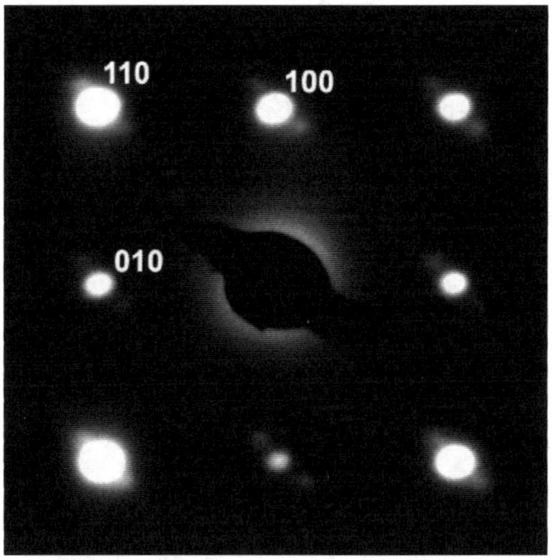

Figure 3. [001] Selected area diffraction pattern from the nanostructured phase from a single domain of the 4/90/10 composition showing the satellite spots parallel to 110.

3.4. 60° Domain boundaries

The domain boundaries considered above must be 60° domain boundaries where the polarisation orientation is rotated by an angle of 60° at the boundary [5], in addition to a near-90° rotation of the *c*-axis which causes the unit cell misorientation discussed in section *2.2*. Figure 4 shows one such domain boundary imaged close to a <111> direction as well as a weak beam dark field image where the streaked nanostructure is strongly evident. It is clear that this streaked nanostructure rotates by an angle of 120° at the boundary. Figure 4c shows a selected area diffraction pattern from domain 1 and Figure 4d from domain 2; Figure 4e shows a diffraction pattern of an area comprising both domains. It is clear that the long period incommensurate ordering in the diffraction patterns rotates by 120° across the boundary, as one would expect for a 60° domain boundary [5].

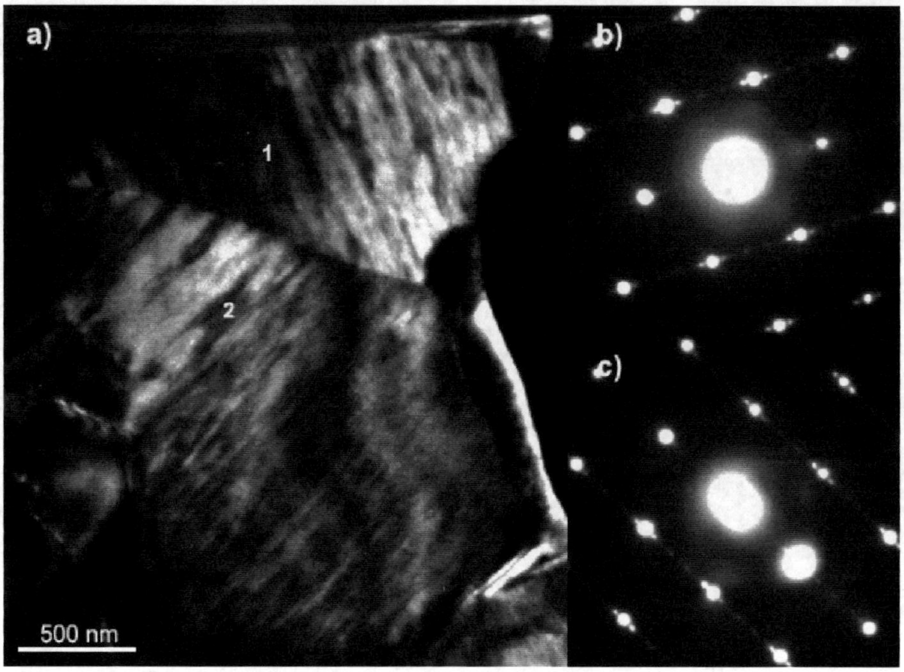

Figure 4. a) Dark field TEM image of two domains divided by a 60° boundary, b) <111> SADP from domain 1, c) <111> SADP from domain 2.

4. Conclusions

We have shown conclusively that an incommensurate antiferroelectric phase in La-doped PZT with a Zr/Ti ratio of 90/10 and La content in the range 2-4 cat.% is closely related to the commensurate antiferroelectric phase of $PbZrO_3$ in a number of ways:

- Microdomains are formed close to (101) planes which are correspond to the 60° boundaries observed previously in $PbZrO_3$
- The slight tetragonality of the primitive perovskite cell causes a small unit cell rotation at these boundaries of the order of 0.5° about <010> to ensure a good fit, and this tetragonality is similar to that in $PbZrO_3$
- Long period orderings are found along [110] directions of the order of 7-8 (110) spacings, and usually incommensurate with the primitive unit cell, but clearly related to the ordering along [110] of 4 (110) spacings found in $PbZrO_3$.
- There is always a nanostructure found within the domains perpendicular to this [110] long period ordering direction. The nature of this nanostructure and the reasons for it giving contrast in dark field images remain unclear, but are under investigation.

References

[1] Dai X H, Xu Z, Li J F and Viehland D 1996 *J. Mater. Res.* **11** 626.
[2] Knudsen J, Woodward D I and Reaney I M 2003 *J. Mater. Res.* **18** 263.
[3] Cai Y, Phillipp F, Zimmermann A, Zhou L, Aldinger F and Rühle M, 2003 *Acta Mater.* **51** 6429.
[4] He H and Tan X 2005 *Phys. Rev. B* **72** 024102.
[5] Tanaka M, Saito R, and Tsuzuki K 1982 *Jpn. J. Appl. Phys. Part 1* **21** 291.
[6] Farooq M U, Villaurrutia R, MacLaren I, Kungl H, Hoffmann M J, Fundenberger J J and Bouzy 2008 *J. Microsc.* **230** 455.
[7] Corker D L, Glazer A M, Dec J, Roleder K and Whatmore R W 1997 *Acta Cryst* B**53** 135.
[8] Bate P S, Knutsen R D, Brough I and Humphreys F J 2005 *J. Microsc. (Oxford).* **220** 36.

TEM Investigation of MgO Thin Films for Magnetic Tunnel Junction Application

V Harnchana[1], A T Hindmarch[2], A P Brown[1], R M Brydson[1] and C H Marrows[2]
[1]Institute for Materials Research, University of Leeds, LS2 9JT, UK
[2]School of Physics and Astronomy, University of Leeds, LS2 9JT, UK

E-mail: sms5vh@leeds.ac.uk

Abstract. TEM analysis of sputter-deposited MgO has been performed using a plan view and cross-section approach. The influence of the sputtering parameters including target to substrate distance (TSD), sputtering power and Ar pressure on the structure and composition of deposited films has been investigated by analysing in plan view. Optimum deposition conditions have been identified and MgO layers deposited under these are partly crystalline within an amorphous matrix. Cross-sectional TEM imaging of MgO deposited under optimum conditions onto CoFeB in a basic MTJ structure (CoFeB/MgO/CoFeB trilayer), shows that in the initial deposition, at the bottom interface of the trilayer, amorphous MgO forms and this then becomes partly crystalline (but with no preferred texture) with continued deposition, towards the top interface of the layer. TEM- energy dispersive X-ray (EDX) and electron energy loss spectroscopy (EELS) composition analyses suggest that the inhomogeneous oxide film structure could be caused by the deposition of disordered, non-stoichiometric MgO.

1. Introduction

CoFeB/MgO based magnetic tunnel junctions have attracted increasing attention due to the demonstration of large tunnelling magnetoresistances (TMR) which are promising for non-volatile and high density data storage device applications. The key point is the development of a CoFe(100)[110]/MgO(100)[100]/CoFe(100)[110] orientation in the as-annealed structures because crystalline MgO lattice matching of CoFeB at the interfaces acts as a spin filter allowing only particular electron waves to coherently tunnel across the junction. TMRs of over a thousand percent have been theoretically predicted for such coherent lattice matching structures [1-3]. Nevertheless, TMRs of only hundreds of percent have been reported so far in devices prepared by sputter-deposition, with the shortfall being attributed to the fabrication of non ideal structures.

In order to improve fabrication, the crystallization and growth mechanisms of the active trilayer and the relationship to the sputtering parameters should be investigated. Here we present a fundamental study of thin films sputter-deposited on carbon support films commonly used for TEM (using the combination of dc-rf magnetron sputtering). Plan view TEM imaging, selected area electron diffraction (SAED) and energy dispersive X-ray (EDX) spectroscopy, and electron energy loss spectroscopy (EELS) were employed to study the resulting nanostructure of the films both as-deposited and post-annealing. The MgO sputtering parameters such as target substrate distance (TSD), sputtering power and Ar gas pressure have been optimized to produce the best crystallinity and microtexture of the deposited films. A cross-section of an active trilayer prepared under these optimum deposition conditions has also been analysed by TEM.

2. Experiment

The samples were produced by using dc-rf magnetron sputtering with a base pressure of 10^{-7} mTorr. MgO thin films were sputter-deposited using a sintered MgO target with 99.99 % purity, 50 mm diameter, supervac grade obtained from the Testbourne Company. The effect of sputtering parameters including target to substrate distance (TSD), Ar gas pressure and sputtering power were examined. The TSD was originally at 15 cm, and was then adjusted to 11 cm. Ar gas pressures were 1, 5 and 10 mTorr. Sputtering gun powers were 75, 125 and 175 watts. The plan view samples were prepared by directly depositing on continuous carbon film TEM support grids (Agar Scientific Ltd.). A multilayer sheet structure of a simple MTJ was also produced for cross-sectional TEM analysis. The stack sequence of this MTJ is Ta[7]/CoFeB[4]/MgO[12]/CoFeB[8]/Ta[5], the numbers in parenthesis are estimated thickness in nanometers. The cross-sections were prepared by conventional TEM sample preparation using ion milling. Nanostructural analysis was performed using a Philips CM200 field emission gun TEM fitted with an Oxford Instruments ultrathin window ISIS energy dispersive X-ray (EDX) analyser and a Gatan Imaging Filter (GIF 200). Composition was monitored in the TEM with spot-EDX (the Mg/C and Mg/O ratios were determined from the ISIS processing software using virtual standards for the C, O and Mg $K\alpha$ X-ray peaks) and EELS. Bulk energy loss spectra were taken in diffraction mode (image coupled) with an approximately 0.18 μm diameter selected area aperture (SAD) inserted, leading to a collection semi-angle of 6 mrad, and a convergence semi-angle of approximately 1 mrad. EEL spectra were acquired with an energy dispersion of 0.3 eV/channel and an energy resolution of 0.8eV.

3. Result

3.1. Target to substrate distance (TSD) effect

TEM images of the MgO films deposited on carbon film supports at the two TSDs are shown in figure 1. Both films were sputtered at 175 watts and 1 mTorr Ar pressure and both films have MgO crystallites approximately 8-10 nm in size embedded in an amorphous matrix. Denser clusters of agglomerated crystallites can be observed at the 11 cm TSD. The EDX Mg/C ratio increases at the shorter TSD suggesting a higher deposition rate at 11 cm TSD. Both films were deposited for the same length of time and yet despite the increased deposition rate (and therefore thickness of the film) at 11 cm TSD, a continuously crystalline MgO layer has not been achieved.

The differences in film crystallinity could be because at the shorter TSD the deposited species have, high surface mobility and agglomerate, as seen in figure 1 e) and f). When a substrate is further away from a sputtering target the sputtered atoms experience a higher number of collisions before approaching the substrate and therefore on average have lower kinetic energy on deposition. The EELS oxygen *K*-edges in the two samples are also compared in figure 2, the edge of the 15 cm film exhibits less fine structure than that of the 11 cm film and it has been reported that this loss of the EELS extended fine structure is indicative of disorder or fewer surrounding oxygen atoms [4]. A shoulder just above the edge onset is also present in the 15 cm film (marked by the arrow) but not the 11 cm one. This feature is believed to be due to gap states that reflect band tailing in the conduction band and has been attributed to vacancies or other structural defects in a MgO/Fe structure [4-5]. The oxygen *K*-edge of the 11 cm film matches very closely that of bulk MgO. Therefore we suggest that there is a significant concentration of defects in the film deposited at 15 cm TSD.

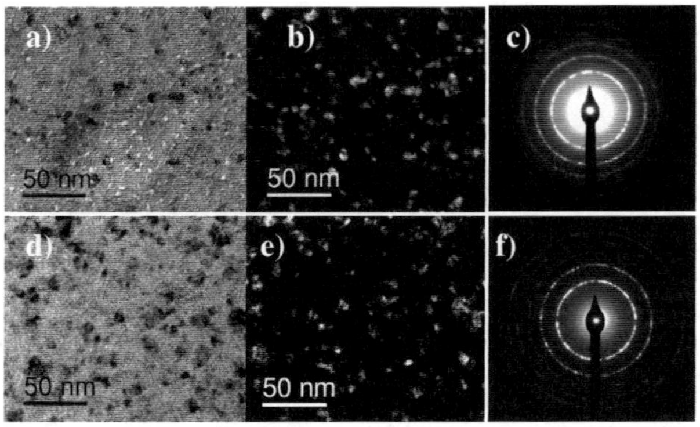

Figure 1. a), b), c) are the bright field, conical dark field images, and selected area diffraction pattern of the film deposited at 15 cm TSD respectively, d), e), f) are those of the 11 cm TSD. The number of crystallites in the film deposited at 11 cm is larger than that deposited at 15 cm.

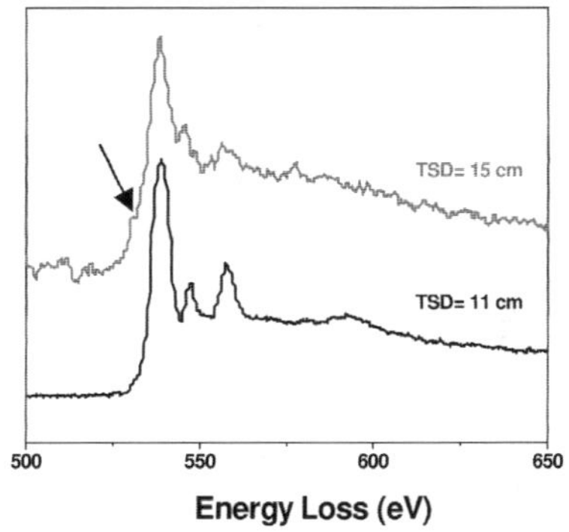

Figure 2. Background stripped Oxygen *K*-energy loss edges of the MgO films deposited at 15 and 11cm TSDs. The edge at 15 cm TSD has a poorly defined fine structure and an arrow shows a shoulder at the pre-edge. The edge at 11 cm TSD is consistent with bulk MgO [6].

3.2. Sputtering power and Ar pressure effect

The EDX Mg/C ratio increases with increasing sputtering power and decreases with increasing Ar pressure (figure 3 and 4 respectively). Assuming the carbon films to be of approximately constant thickness suggests that the deposition rate is directly proportional to the sputtering power but inversely proportional to the Ar pressure. At a TSD of 11 cm, 1 mTorr Ar pressure and 75 watts a very thin amorphous MgO film was deposited (not shown here). Very few numbers of crystallites form when sputtering at 125 watts, and much larger numbers are observed at 175 watts (as shown in figure 1b). Varying the Ar gas pressure at 175 watts from 1 to 5 mTorr does not significantly change the film structure apart from producing slightly less crystallite agglomeration at 5 mtorr. Only isolated crystallites are observed at 15 mTorr Ar Pressure. EDX Mg/O values for all of the films are non-stoichiometric, but approach stoichiometry in the films grown at high power and low Ar pressure.

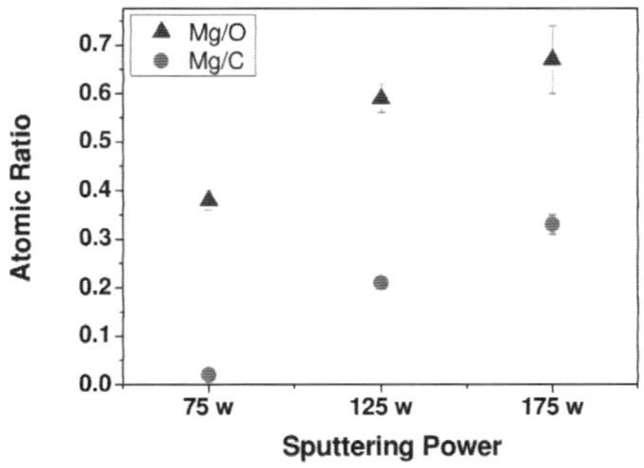

Figure 3. Variation in EDX Mg/O and Mg/C ratios of the MgO grown at a TSD of 11 cm and 1 mTorr Ar pressure with various sputtering powers; 75, 125, 175 watts. The ratios rise consistently with increasing power.

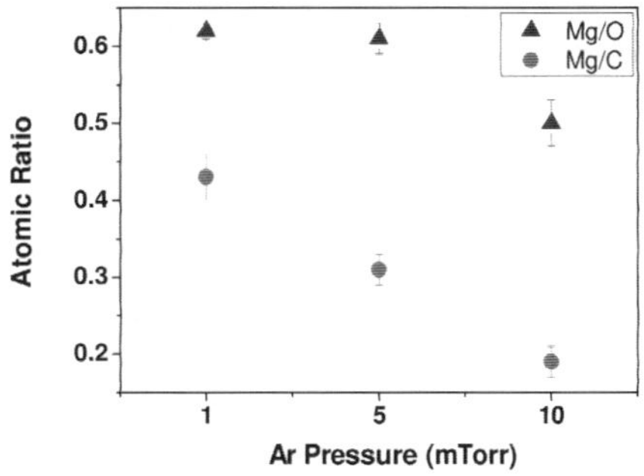

Figure 4. Variation in EDX Mg/O and Mg/C ratios of the MgO grown at a TSD of 11 cm and 175 watts sputtering power with various Ar pressures; 1, 5, 10 mTorr. The ratios fall with increasing Ar pressure

We believe that the relative amount of disordered/amorphous material in the deposited MgO films is consistent with the change in Mg/O ratio, suggesting that the amorphous phase is non-stoichiometric.

3.3. Cross-sectional TEM

Cross-sectional TEM images of the MTJ active trilayer fabricated using the optimum sputtering conditions determined above show an MgO layer, approximately 10 nm thick (figures 5). The MgO barrier was made deliberately thicker than a working MTJ device to be comparable to the MgO films produced for plan view analysis. This MgO barrier is amorphous in the first few nanometres above the bottom CoFeB electrode (initial stage of deposition): no lattice fringes were visible despite tilting $\pm 10°$ away from the interface plane. Crystalline MgO grains are then evident further up the layer increasing in number towards the top (final stage of MgO deposition). The oxygen intensity profile determined using EELS (white line in figure 5b) shows an oxygen rich content in the first few nanometres of MgO deposition. A textured, crystalline MgO layer should develop immediately upon deposition on the smooth surface of the amorphous electrode. Clearly this is not happening and may be a result of disorder due to an initially oxygen rich deposition or surface roughening of the bottom electrode by O ions released by the sputtering of the MgO [4]. Finally, introducing a thin Mg layer prior to the MgO deposition may compensate for the initial oxygen-rich deposition and prevent the oxidative roughening of the bottom electrode [4].

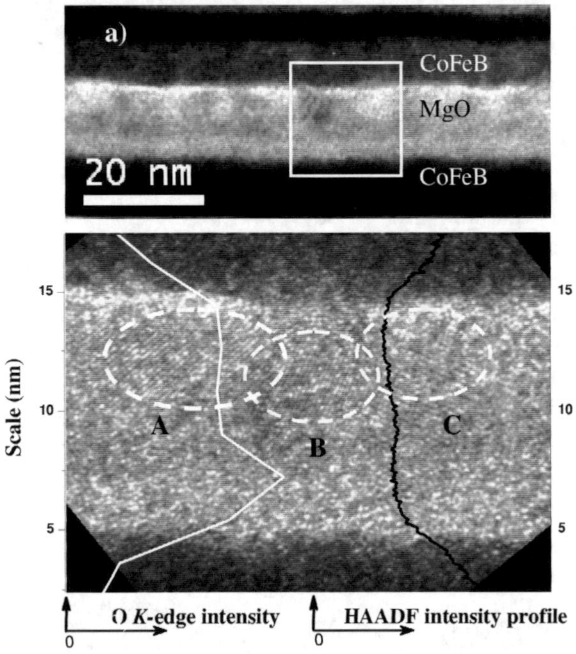

Figure 5. TEM cross-section of the MTJ prepared by the optimum sputtering conditions. a) low magnification image showing a rougher MgO/CoFeB top interface than the bottom one. b) a magnified inset of a), MgO crystallites are seen towards the top of the junction whilst the first few nanometers at the bottom interface are amorphous or disordered, the white line on the left shows the oxygen *K*-edge intensity which indicates an initial O-rich deposition, producing disordered MgO. The black line is the intensity profile of the associated scanning(S)TEM-high angle annular dark field (HAADF) image that indicates approximately uniform TEM-specimen thickness within the barrier. Measured lattice spacings in regions A, B and C suggest, A =MgO (200) B= MgO (200) or CoFe(110), and C= MgO(111), $d_{MgO(200)}$ = 2.1 Å, $d_{CoFeB(110)}$ = 2.0 Å and $d_{MgO(111)}$ = 2.4 Å.

4. Conclusion

This work has demonstrated the sensitivity of sputtering parameters on the structure and composition of sputter-deposited MgO barriers, destined for magnetic tunnel junction application. All of the deposited MgO layers consist of MgO crystallites in an amorphous matrix. Films with a greater fraction of amorphous material exhibit poorer MgO stoichiometry. Shortening the target to substrate distance, increasing the sputtering power and lowering the Ar gas pressure produce films that have more crystallinity, less defects, improved texture and increased MgO stoichiometry.

References

[1] Butler W H, Zhang X G, Schulthess T C, and MacLaren J M, 2001 *Phys. Rev. B* **63** 054416.
[2] Mathon J and Umersky A, 2001 *Phys. Rev. B* **63** 220403R.
[3] Zhang X G, and Butler W H, 2004 *Phys. Rev. B* **70** 172407.
[4] Cha J J, Read J C, Buhrman R A, and Muller D A, 2007 *App. Phys. Lett* **91** 062516.
[5] Mather P G, Read J C, and Buhrman, 2006 Phys. Rev. B 73, 205412.
[6] Rez P, Weng X, and Hong M, 1991 *Microsc. Microanal. Microstruct.* **2** 143.

Electron Microscopy and Analysis Group Conference 2009 (EMAG 2009) IOP Publishing
Journal of Physics: Conference Series **241** (2010) 012040 doi:10.1088/1742-6596/241/1/012040

Microstructural investigation of strontium titanate films grown by interval pulsed laser deposition

YY Tse[1], SRC McMitchell[2], TJ Jackson[2], IP Jones[1] and A Genc[3]
[1] School of Metallurgy and Materials, University of Birmingham, Birmingham, B15 2TT, UK
[2] School of Electronic, Electrical and Computer Engineering, University of Birmingham, Birmingham, B15 2TT, UK.
[3] Department of Materials Science and Engineering, Ohio State University, 2041 College Rd, Columbus, OH 43210

E-mail: y.y.tse@bham.ac.uk

Abstract. Strontium titanate ($SrTiO_3$) films were deposited on (001) magnesium oxide (MgO) substrates by pulsed laser interval deposition (PLID). The growth modes of the films were monitored by in-situ reflection high-energy electron diffraction (RHEED). The microstructures were investigated by transmission electron microscopy (TEM). It has been shown that PLID can enhance the two dimensional (2D) growth of this system, which has the large lattice misfit of 7.9%. The addition of a mono TiO_2 buffer layer to the MgO substrate surface further improved the 2D growth mode. TEM results revealed that the films grown in 2D mode with large lattice mismatch were dominated by antiphase boundaries (APBs) and holes. All the holes were connected by antiphase boundaries. The holes were rectangular in shape, with facets along the [100] and [010] directions, and they ran though the whole film thickness. Threading and misfit dislocations have also been identified as growth defects in the films grown with and without a TiO_2 buffer layer. However, the density of defects varies with growth mode.

1. Introduction

$SrTiO_3$ (STO) is an alkaline earth titanate which is widely used in microelectronics. The unique dielectric properties of STO thin films at low temperatures are important in microwave applications [1]. Conventional pulsed laser deposition (PLD) is popular for growing complex oxide films because it can maintain good stoichiometric transfer of materials from target to substrate [2]. However, the high density of defects formed during deposition of STO thin films leads to a higher dielectric loss in thin films than in bulk material - in the order of about 100 [3]. Therefore, it is of technical importance to grow the film with fewer defects. Forced layer by layer growth will give an atomically flat interface and reduce the density of the growth defects of the films. The novel pulsed laser interval deposition (PLID) technique was introduced to achieve "true" layer by layer growth in homo- and closely matched hetero-epitaxy [4, 5]. In PLD, laser pulses at a fixed rate were used during deposition, but in PLID, rapid bursts of laser pulses were used, each burst being sufficient to deposit a monolayer of material. The growth mode was intensively investigated by in-situ reflection high-energy electron diffraction (RHEED) and layer by layer growth was confirmed via the temporal evolution of the RHEED intensity [4, 5]. The purpose of this work was to study the microstructure features of heteroepitaxial films with large misfit (~ 7.9%) by PLID.

2. Experiment

STO thin films with nominal thickness of 40 nm, with and without a TiO_2 buffer layer, were deposited on (001) MgO by PLID equipped with in-situ RHEED. Deposition parameters and detailed studies of the RHEED are described elsewhere [**6**]. Plan view and cross section TEM samples were prepared by tripod polishing to electron transparency. TEM and high resolution TEM were carried out using a Tecnai F20 microscope operated at 200kV. High resolution scanning TEM (HRSTEM) work was carried out using a probe corrected FEI Titan operated at 300kV with a probe size of about 0.1 nm. The high angle annular dark field (HAADF) collection angle is ranged from 46 – 325 mrad. DigitalMicrograph was used to analyse the HRSTEM images.

3. Results and Discussion

The evolution of the RHEED intensity during growth was used to confirm the layer by layer growth mode of the films [6]. The purpose of growing the film with a TiO_2 buffer layer is to wet the MgO surface, ensuring that the initial layer deposited is TiO_2 instead of SrO, which has a higher interfacial energy [**7**]. The addition of TiO_2 buffer layer will further promote the layer by layer growth mode. A film without a TiO_2 layer was grown to compare the growth modes and subsequent microstructures. Figure 1 shows a pair of cross-sectional two beam bright field images taken with (a) g = -200 and (b) g = 002 of an STO film grown without a TiO_2 buffer layer. The film was of uniform thickness with a few threading dislocations (marked by the black arrows in figure 1a) when compared to films grown by standard PLD [8]. APBs were observed and indicated by the white arrows in figure 1b.

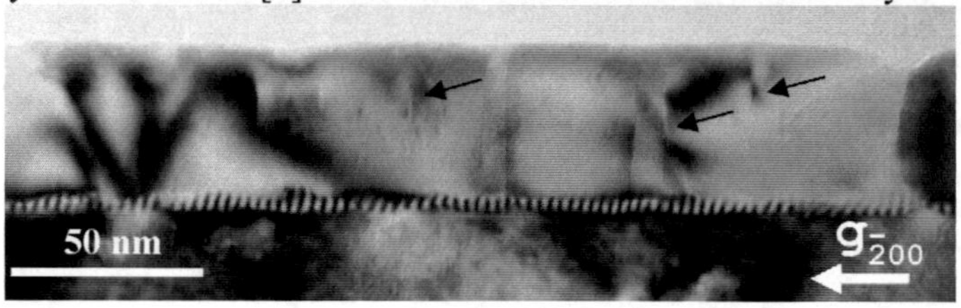

(a)

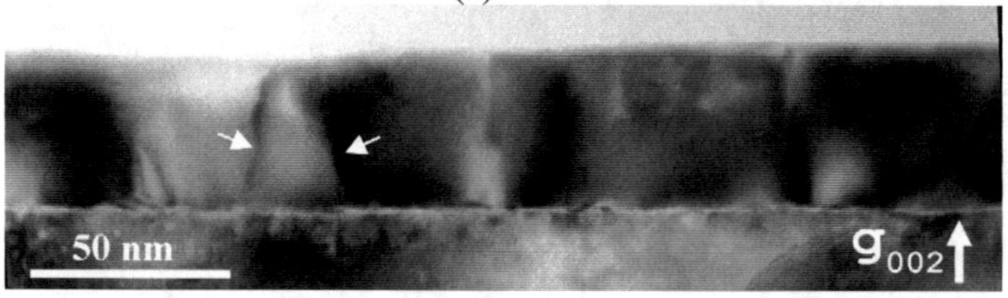

(b)

Figure 1 Cross-sectional two beam bright field images of film without TiO_2 buffer layer taken with (a) g = -200 and (b) g = 002. The APBs are marked by white arrows and threading dislocations are marked by black arrows.

Figure 2a shows a HRSTEM HAADF image taken with B = [001] showing the interface and the inset is the FFT filtered image of the region marked by square. Figure 2b shows the intensity profile of the Sr, Ti and Mg atoms from the film region to the substrate. The HAADF image of the interface was not clear due to sample drift, specimen thickness and the different scattering power of the STO and MgO. With the aid of the intensity profile, it was possible to determine the initial atomic layer of the film in specific regions. The intensity drop from the Sr to a Ti atom was estimated using 10 different areas from three different HRSTEM images, and is about 25 ± 5%. As shown in figure 2b, the diagonal spacing between the Sr and Ti atoms is 0.26 nm. Assuming the relative intensity drop near the interface was similar to those in the film region, it was expected to see the Ti peak at the position marked by the dotted line with intensity 64 ± 4 a.u. (the intensity of the corresponding Sr atom from the intensity profile is 85 a.u.). However, there is no peak in that particular position, which indicates that the first atomic layer above the MgO substrate was Sr. Albeit TiO_2 is energetically favourable to form during deposition, there exists some regions covered by SrO as the initial deposited atomic layer. This result was consistent with the AFM observation on a four monolayer STO film on MgO, which showed the presence of SrO inclusions [**8**]. Figure 3 shows the intensity profile of Sr, Ti and Mg obtained from the line drawn across the interface indicated in the inset HAADF image. Most of the interfacial areas were covered by TiO_2 as the initial atomic layer since TiO2 is energetically favourable to form on MgO substrate. The variation of Mg-Mg distance close to the interface might be due to the large misfit strain and the atomic steps on the surface of the MgO. From the RHEED result [6], the growth mode of the film without a TiO_2 buffer layer was not entirely two dimensional: the growth mode changed to three dimensional in the later stages of deposition. The existence of SrO inclusions may cause the nucleation of three dimensional growth in the later part of the deposition.

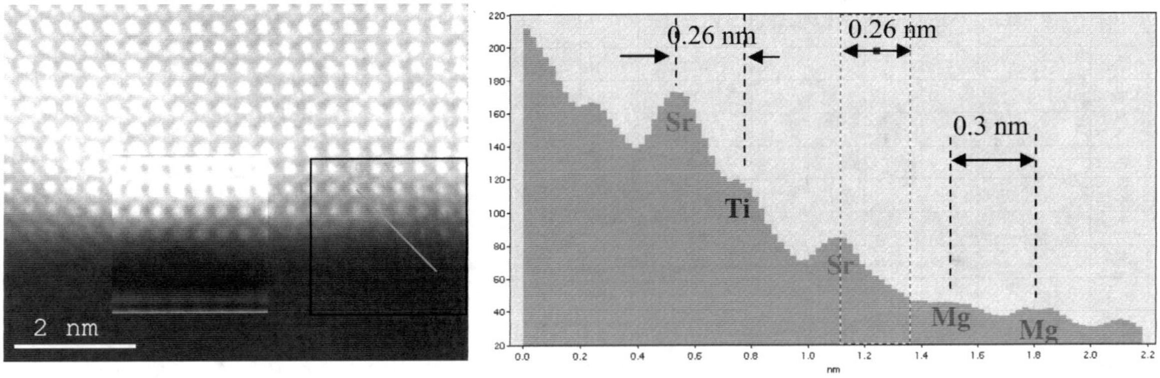

(a) (b)

Figure 2 (a) Cross-sectional HRSTEM HAADF image taken with B = [001] showing the SrO as the termination layer of a film grown without a TiO2 buffer layer. The inset is the FFT filtered image of the area marked by the square. (b) The intensity line profile from the film to the substrate and the line position is illustrated in figure 2a.

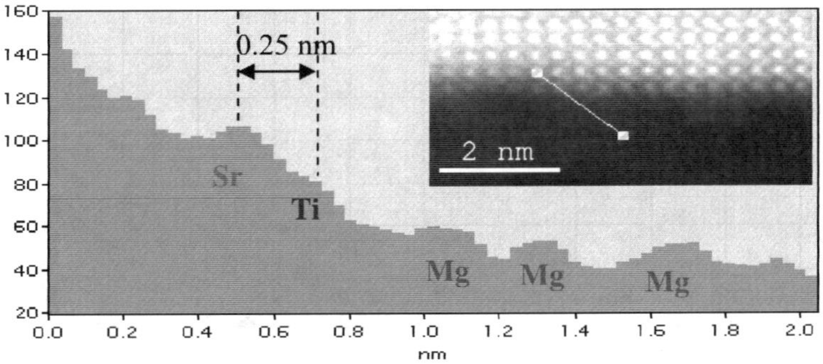

Figure 3 The intensity line profile from the film to the substrate. The line position is shown in the inset HRSTEM HADDF image of the interface taken with B = [001].

Therefore, to promote the 2D growth mode further, an artificial TiO_2 buffer layer was deposited prior to the STO film to wet the MgO surface. Figure 4a shows a cross sectional bright field image taken with g = 200. More APBs were observed in the STO film with a buffer layer and the film has even less threading dislocations than the film without a buffer layer. Figure 4b shows the HRTEM of the interface of the STO film with a TiO_2 buffer layer. The interface is sharp since there is no second phase or amorphous layer but the roughness of 0.5 to 2 nm was observed (figure 4b). The steps on the MgO surface are highlighted by the dotted line in figure 4b. It is interesting to note that the atomic column contrast was good on both STO and MgO sides but became unclear at the interface. This may be due to either the roughness of the MgO substrate in the direction parallel to the electron beam or the fact that the interface was strained. For similar reasons, there is no clear enough HRSTEM HAADF image to carry out the intensity profile analysis for the initial atomic layer of this film with a TiO_2 buffer layer. More work will be continued to verify that issue. As indicated by the arrows in figure 4a, there is a gap running from the substrate to the top of the film was observed. To confirm the existence of the holes, a plan view TEM sample was prepared. Figure 5 shows the holes were of rectangular/square shape and the edges of the hole are on {100} planes. The holes were connected by the APBs. A tilting series suggests that most of the APBs can be characterised with displacement vector R = 1/2[010] or 1/2[100]. When ordered planar faults were transposed from one set to another set of {100} plane, the interface would form a zig-zag boundary as shown in the inset to figure 5. RHEED results suggested the film was forced in a layer by layer growth mode through its entire thickness by PLID despite the large mismatch (~7.9%). The microstructure revealed by TEM showing a high density of APBs, holes and low density of threading dislocations suggests that a high level of strain was developed during deposition. Therefore, fracture caused by strain may occur in <010> directions during cooling the film from 850°C to room temperature after deposition. High resolution HAADF STEM taken in plan view shows the fracture might occur at the junction of two or three 1/2[010] APBs where the APBs energy is the highest.

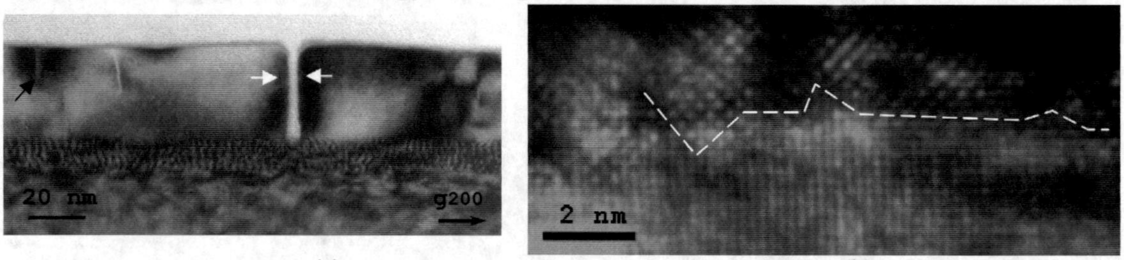

(a) (b)

Figure 4 (a) TEM bright field image of STO film with TiO_2 buffer layer taken at g = 200. The hole is marked by white arrows and the threading dislocation is marked by black arrow. (b) HRTEM of STO and MgO interface. The steps on the MgO substrate are highlighted by the dotted line.

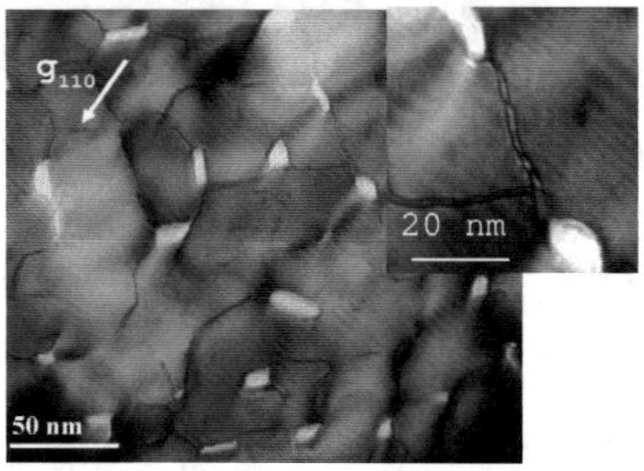

Figure 5 Two beam bright field plan view image of STO film with TiO_2 buffer layer. Image taken with g = 110 and showing holes are connected by APBs. Inset is a zoom in of the zig-zag APBs.

4. Conclusions

There exist small regions of SrO as the initial atomic layer in the STO film without a TiO_2 buffer which may nucleate three dimensional growth. Instead of forming a large number of threading dislocations, the film grown in layer by layer mode has a tremendous amount of 1/2[010] or 1/2[100] APBs and holes.

Acknowledgement

The authors thank Professor Hamish Fraser (Ohio State University) for invitation to use Titan for the HRSTEM work.

References

[1]Vendik OG, Hollmann EK, Kozyrev AB and Prudan AM 1999 *J. of Superconductivity* **12**, 325.
[2]Venkatesan T and Sharma RP 1996 *Mater. Sci. Eng.*, **B41**, 30.
[3]Ryen L, Wang X, Petrov P, Carlsson E, Helmersson U and Olsson E 1999 *J. Appl. Phys.* **85**, 3976.
[4]Koster G, Rijinders GJHM, Blank DHA and Rogalla H 1999 *Appl. Phys. Letts.*, **74**, 3729.
[5]Blank DHA, Koster G, Rijnders AJHM, Van Setten E, Slycke P, Rogalla H 2000 *J. of Crys. Growth* **211**, 98.
[6]McMitchell SRC 2008 *PhD Thesis* The University of Birmingham.
[7]McKee RA, Walker FJ, Specht ED, Jellison Jr GE, and Boatner LA 1994 *Phys. Rev. Lett.*, **72**, 2741.
[8]Tse YY, McMitchell SRC, Jackson TJ, Liu YH and Jones IP 2008 *Ferroelectrics*, **368**, 49.

HRTEM and EELS of nanoantenna structures fabricated using focused ion beam techniques

Ai Leen Koh[1,*], Ondřej Tomanec[2], Michal Urbánek [2], Tomáš Šikola[2], Stefan A. Maier[3] and David W. McComb[1]
[1] Department of Materials, Imperial College London, London SW7 2AZ, UK
[2] Institute of Physical Engineering, Brno University of Technology, Czech Republic
[3] Department of Physics, Imperial College London, London SW7 2AZ, UK

E-mail: a.koh@imperial.ac.uk

Abstract. This paper describes the use of focused ion beam (FIB) techniques to fabricate Au nanoantenna structures for plasmonics applications. High-resolution transmission electron microscopy (HRTEM), diffraction and electron energy-loss spectroscopy (EELS) techniques are applied to investigate the structures of the initially patterned film, fabricated nanoantenna structures, and to map localized surface plasmon resonance (LSPR) modes in the nanostructures.

1. Introduction

Metal nanostructures can act as light concentrators in the visible and near-infrared regime due to the excitation of localised surface plasmons (LSPs), which are resonant waves of the free electrons in the metal coupled to the electromagnetic field [1,2]. Such nanostructures show considerable potential for applications ranging from optical sensing to enhancing the absorption in thin-film solar cells [3-7].

The nanoantenna structures described in this paper were produced using the "top-down" technique of FIB-lithography. They were characterised using a (scanning) transmission electron microscopy ((S)TEM) owing to the latter's ability to resolve objects to the sub-nanometre level. Because LSPs can also be generated by high energy electrons, a STEM equipped with an EEL spectrometer and monochromator is also a powerful method to map out the underlying LSPR modes in these nanostructures with high energy (~ 0.2 eV) resolution.

2. Method

Au nanoantennas were fabricated using a dual beam FIB/scanning electron microscopy (SEM) system (Tescan Lyra). The process begins by the patterning of Au film of thickness 20 nm on a silicon-supported Si_3N_4 membrane of dimensions 0.5mm × 0.5mm and thickness 50 nm. A 100μm × 100μm region was then identified on the membrane, within which arrays of nanoantennas were created by milling away parts of the Au film using FIB. Analyses of the nanoantennas were carried out using an

* To whom any correspondence should be addressed.

FEI-Titan (S)TEM operated at 300kV. TEM imaging and diffraction techniques were applied to determine the nanostructure morphology and film quality after ion-milling. EELS spectrum imaging (SI) [8] was performed in monochromated mode (with energy resolution 0.2eV) and focusing on the low energy-loss regime to map out the LSPR modes of the nanostructures.

3. Results and Discussion

3.1. Nanoantenna structure fabrication

Figures 1(a) is a 30kV secondary electron (SE) image showing the 3mm-diameter Si frame and the 0.5mm × 0.5mm Si_3N_4 membrane after the deposition of Au film with thickness 20nm. The optical image in the reflectance mode (Figure 1b) shows that the membrane was distorted after deposition, possibly due to the stresses induced during the process. TEM bright field (BF) images of the Au film (Figure 1c) reveal that it is non-continuous and consists of random lines of voids. Selected area diffraction (SAD) analyses (Figure 1c inset) show that the film is polycrystalline in nature and of fcc structure. After deposition, a 100µm × 100µm region was identified for nanoantenna structure fabrication (Figure 1d). Arrays comprising three-by-three cells (Figure 1e) were created by milling the Au- Si_3N_4 substrate with 30keV Ga+ ions. The cell sizes in the arrays are of different sizes ranging from 500nm to 1µm. The nanoantenna structures are residual Au film in the middle of the cells that had been retained after the ion milling process. They are supported by the silicon nitride membrane which is still retained after the ion milling process. It is estimated that 10nm of Si_3N_4 film was removed from the ion milling step.

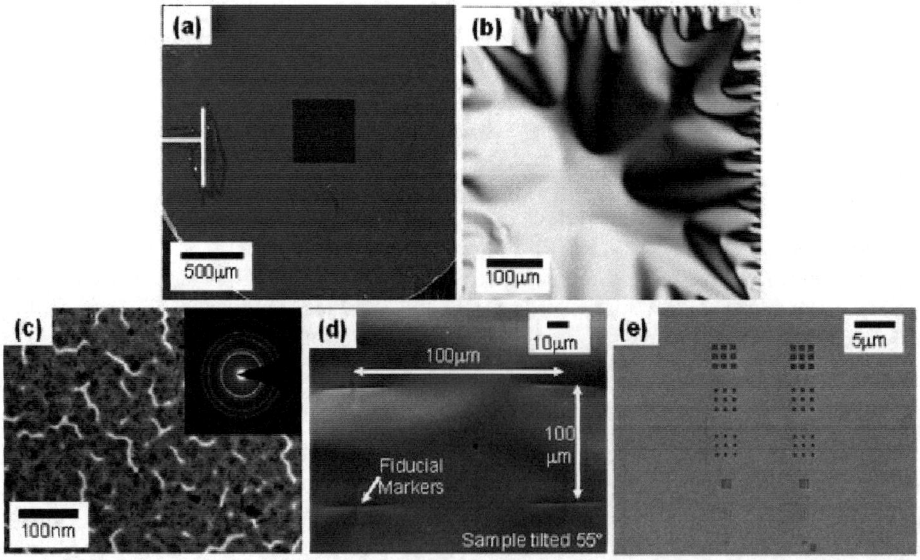

Figure 1. (a) SE image at 30kV showing 3mm-diameter Si frame with a 0.5mm×0.5mm SiN window on which Au had been patterned. (b) Optical reflectance image showing distortions in the membrane after Au patterning. (c) TEM BF image of a region of patterned Au. The SAD pattern (inset) shows that the film is polycrystalline. (d) SE image at 30kV, showing a 100µm×100µm region of the Si_3N_4 membrane in which fiducial markers had been made by ion milling. The sample has been tilted by 55°. (e) SE image at 30kV showing arrays, each comprising nine cells, which had been ion milled.

3.2. TEM Analyses of Nanoantenna structures

TEM analyses were performed to determine the structure and morphology of the nanoantennas. Figure 2(a) shows a low-magnification, BF image of an array consisting of nine cells. Figure 2b(i) to (ix) are higher-magnification TEM BF images of the nanoantenna structures found in the cells. At the present, the discontinuity of the patterned Au film (as seen in Figure 1(c)) may have an influence on final nanostructure morphology, and improvements in the film patterning process are being developed. Diffraction contrast in these images suggests that polycrystallinity of the Au film was largely preserved after ion milling.

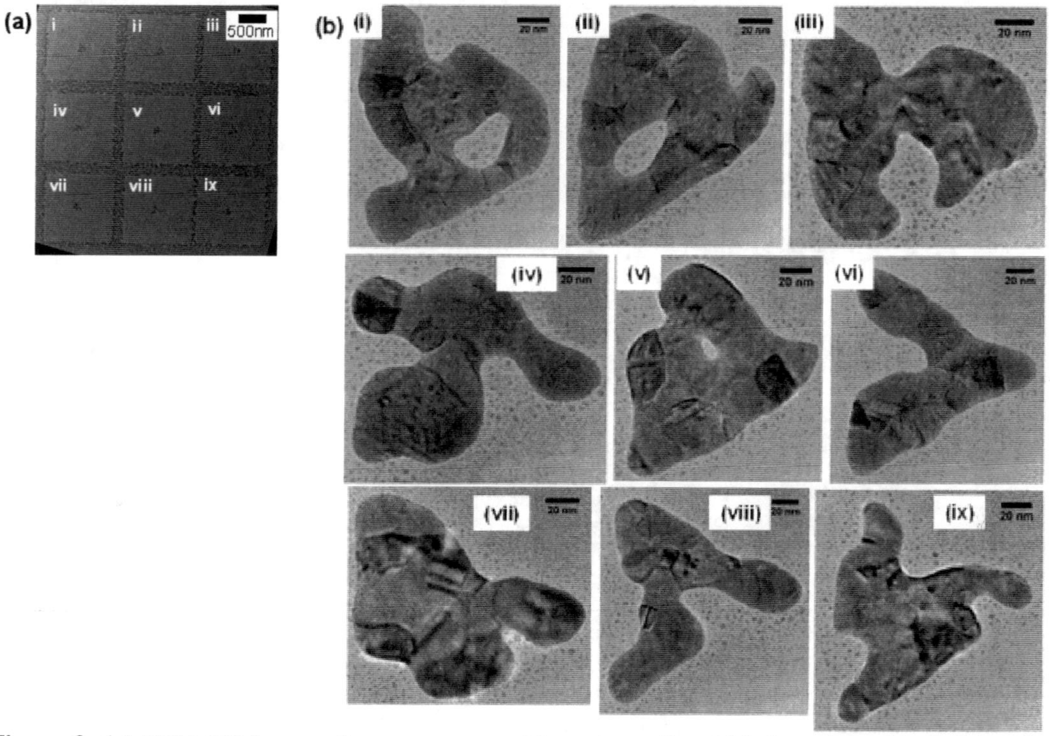

Figure 2. (a) TEM BF image of an array comprising nine cells which have been labelled i to ix. Figures b(i) to b(ix) are corresponding TEM BF images of the nanoantenna structures found in the array of nine cells shown in a(i) to a(ix).

3.3. EELS analyses of structures

Monochromated STEM-EELS SI was applied to map up the underlying LSPR modes in the nanoantennas. Figure 3(a) shows a high angle annular dark field (HAADF)-STEM image of a structure on which EELS SI was performed. The SI comprises 18 pixels by 27 pixels, with pixel size of 1.4nm × 1.4nm. The acquisition time per pixel was 300ms. The ZLP was removed by applying the Reflected Tail Model using Digital Micrograph. Figure 3(d) shows the extracted EELS spectra (after zero loss peak removal) from (1) Si_3N_4 membrane, (2) edge and (3) core of the nanostructure. The extracted spectra show different LSPR modes at each of these areas. However, signal-to-noise ratios were found to be low, possibly due to specimen damage during its initial fabrication process. Below 1.2eV, the falling tail of the ZLP still remained in the extracted EEL spectra, as this routine is not sufficient to completely remove the ZLPs. Investigations to accurately fit and remove the ZLPs are currently underway.

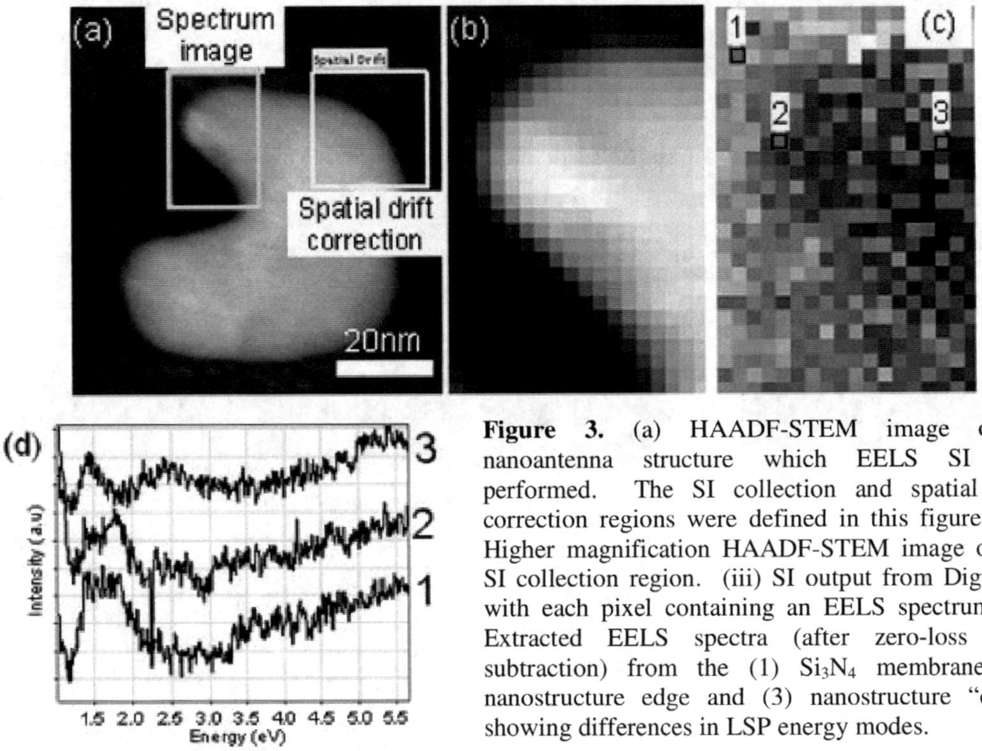

Figure 3. (a) HAADF-STEM image of a nanoantenna structure which EELS SI was performed. The SI collection and spatial drift correction regions were defined in this figure. (b) Higher magnification HAADF-STEM image of the SI collection region. (iii) SI output from Digiscan, with each pixel containing an EELS spectrum. (d) Extracted EELS spectra (after zero-loss peak subtraction) from the (1) Si_3N_4 membrane, (2) nanostructure edge and (3) nanostructure "core", showing differences in LSP energy modes.

4. Conclusion

In summary, we have shown that FIB techniques can be used to fabricate nanoantennas for plasmonics applications. TEM analyses show that the patterned Au film is not continuous and can be amorphised during ion milling. We have since taken steps to improve the quality of the deposited film (via the introduction of a seed layer), and will report our findings in a separate publication. While different LSPR modes can be distinguished from various parts of a nanostructure using EELS SI, their signal-to-noise ratios and tunability can be improved upon refinements in the fabrication process.

Acknowledgements

This work is supported by the Engineering and Physical Sciences Research Council (EPSRC) and the Royal Society International Collaboration Grant.

References

[1] Ritchie R 1957 *Phys. Rev.* **106** 874
[2] Stern E. and Ferrell R 1960 *Phys. Rev.* **120** 130
[3] Kneipp H, Kneipp K, Itzkan, Dasari R R and Feld M S 2002 *J. Phys.: Condens. Matter* **14** R597
[4] Nelayah J, Kociak M, Stephan O, Garcia de Abajo F J, Tencel M, Henrard L, Taverna D, Pastoriza-Santos I, Liz-Marzan L M and Colliex C 2007 *Nature Phys.* **3** 348
[5] Hirsch L R, Stafford R J, Bankson J A, Sershen S R, Rivera B, Price R E, Hazle J D, Halas N J and West J L 2003 *Proc. Natl Acad. Sci. USA* **100** 13549
[6] Barnes W L, Dereux A and Ebbesen T W 2003 *Nature* **424** 824
[7] Maier SA 2006 *IEEE Journal of Selected Topics in Quantum Electronics* **12**(6) 1671
[8] Egerton R F Electron *1996 Energy Loss Spectroscopy in the Electron Microscope* (New York: Plenum Press) p 32

Journal of Physics: Conference Series **241** (2010) 012042 doi:10.1088/1742-6596/241/1/012042

Sol-gel synthesis and TEM-EDX characterisation of hydroxyapatite nanoscale powders modified by Mg, Sr or Ti

S A Qaisar, M Bilton[1], R Wallace, R Brydson, A P Brown, M Ward, S J Milne.
Institute for Materials Research, SPEME, University of Leeds, Leeds LS2 9JT, UK
E-mail: pre5mb@leeds.ac.uk

Abstract. Chemically modified hydroxyapatite nanoparticles have been prepared by a sol-gel route. Analysis by TEM-EDX confirmed that the dopants, Mg, Sr or Ti had been incorporated into the product. However each sample displayed variability in the level of dopant incorporation in different particles within the same sample. Particle sizes were ≤ 100 nm for samples modified by Mg or Ti. Doping with Sr produced an elongated particle morphology, with dimensions ~ 200nm x 50 nm.

1. Introduction

Nanosized particles of synthetic hydroxyapatite (HAp), $Ca_5(PO_4)_3OH$, are of interest for a number of applications, including tissue scaffolds, drug delivery and the occlusion of dentinal tubules for the treatment of sensitive teeth [1,2].

Natural HAp, present as the main mineral component of bone and teeth, is a complex solid solution composition, in which Ca, P and OH ions are substituted by small amounts of a variety of different cations and anions. In order to enhance mineralisation reactions involving synthetic HAp it is desirable to devise preparation routes which enable ions present in natural HAp to be incorporated into the product. This would result in a material more closely matching the composition of natural HAp, leading to improved bioactivity.

Magnesium and strontium are two of the important divalent substituents readily taken up by bone. Strontium has been shown to increase bone formation and reduce resorption [3]. Doping synthetic HAp with Sr has been shown to change the phase content, increase the amount of tricalcium phosphate relative to HAp, and to promote the formation of an amorphous phase. Magnesium also affects phase content and osteointergration [4,5].

The substitution of Ti into HAp has been less widely studied. Titanium doped HAp biomaterials are of interest because of the possibility of an additional antibacterial function. A hydrothermal synthesis route has been used in the past to attempt Ti substitution on Ca lattice sites for photocatalytic applications [6]. There were uncertainties as to the amount of Ti incorporated into the needle-like HAp crystallites. Co-precipitation methods have also been used, and the product showed photocatalytic and bactericidal response [7].

The present paper reports results of TEM-EDX analysis of powders of HAp produced by a sol-gel route in which up to 5 % of the calcium starting reagent was replaced by magnesium, strontium or titanium reagents. The approach indicates that significant variations can arise in the amount of dopant incorporated into different particles within a sample.

2. Experiment

Sols were prepared from calcium nitrate and triethyl phosphite, according to procedures reported in reference 8. For Sr and Mg modifications, 5 mol % of the calcium reagent was substituted by Mg or Sr nitrate. In the case of Ti modified samples, 3 mol % of the calcium was substituted by titanium di-isopropoxide bis pentanedionanate. The ratio of Ca:Ti substitution was 2:1 in order to maintain charge balance. After ageing the sols, followed by evaporation of solvent at 60 °C, the samples were dried to form a gel. This was then heat-treated at 650 °C for 2 h.

Powder characterisation was carried out using a FEI CM200 field emission gun TEM operating at 197 kV coupled with an Oxford Instruments ultra thin window ISIS energy dispersive X-ray (EDX) system and a Gatan imaging filter (GIF 200). For TEM sample preparation, powders were ultrasonically dispersed in methanol prior to being drop cast onto a holey carbon support film. Spot-EDX in the TEM allows for the detailed monitoring of powder composition as determined from the ISIS processing software using virtual standards for the Ca and P $K\alpha$ X-ray peaks. EDX data were acquired with a probe current of ~ 1 nA over areas about the size of each measured particle such that the total electron fluence was ~ 10^6 electrons nm^{-2} (following the method and results described in Eddisford et al, 2007 [9] and Bilton et al. [10], these proceedings).

3. Results

3.1 Un-doped HAp

Particles were ≤ 100 nm in size (Figure 1). X-ray powder diffraction patterns showed only peaks due to hydroxyapatite. The individual nanoparticles were grouped into micron scale agglomerates.

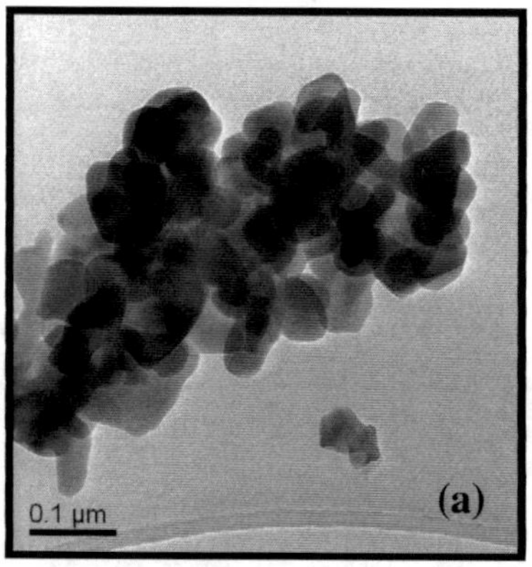

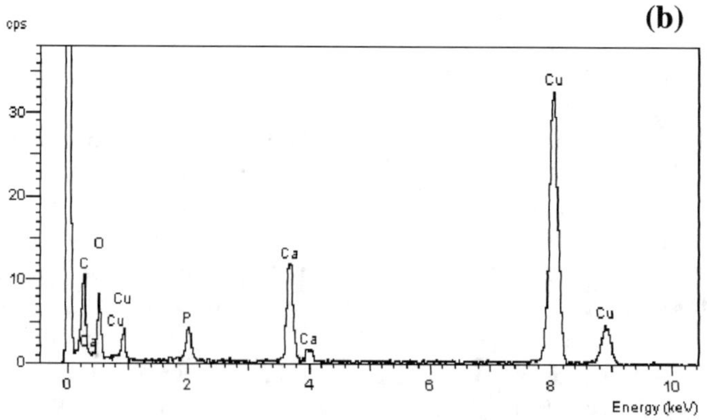

Figure 1: (a) Bright field TEM micrograph of unmodified hydroxyapatite powder; (b) Corresponding EDX spectrum

3.2. Magnesium doping

The TEM micrograph of the Mg- modified sample shown in Figure 2a shows particle sizes up to 200-300 nm. Elemental analysis by TEM-EDX, (Fig 2b), confirmed the presence of Mg, along with Ca and P in the particles examined, but the ratio of Mg/Ca was higher in some of the smaller < 50 nm particles. There were also some particles in which Mg was not detected.

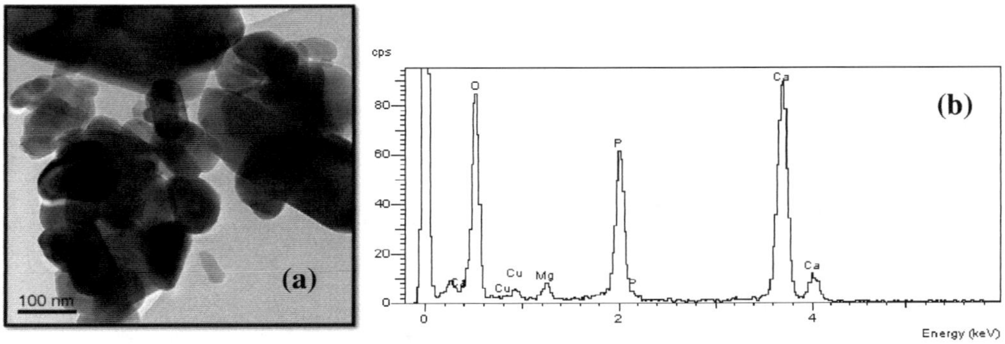

Figure 2: (a) TEM micrograph of Mg- modified hydroxyapatite powder : (b) corresponding TEM-EDX spectrum

3.3 Strontium doping

Figure 3a shows that Sr modification produced particles that were more elongated than un-doped or Mg doped HAp compositions. Dimensions of most particles were around 200 nm x 50 nm. Ten particles were examined by EDX; eight showed Ca, Sr and P, with Sr/Sr+Ca atomic ratios of around 4 (compared to a nominal ratio of 5 in the as-prepared starting sol), Fig 3b. One particle showed a very strong Sr signal with little evidence of Ca or P; another showed only Ca and P with a peak ratio of 1.7, consistent with hydroxyapatite (Ca/P 1.67). There was evidence from X-ray powder diffraction of tricalcium phosphate and calcium oxide secondary phases.

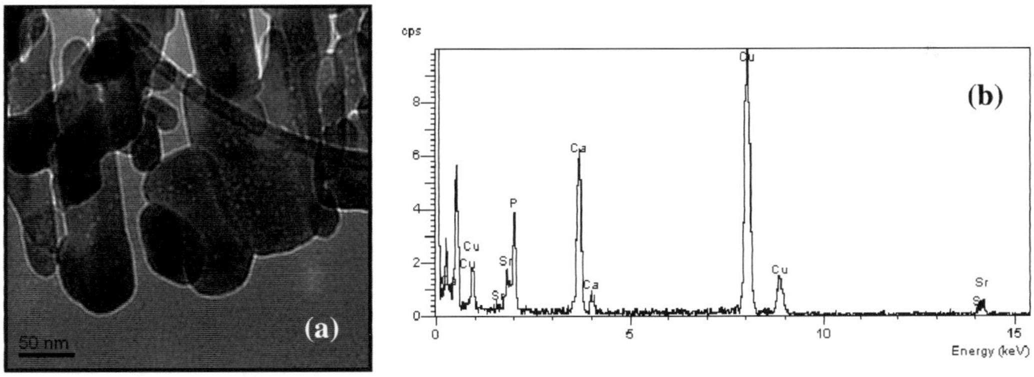

Figure 3: (a) TEM micrograph of Sr-modified hydroxyapatite powder; (b) corresponding TEM-EDX spectrum

3.4 Ti doping

The Ti modified sample (Fig 4) displayed a greater variability in composition than for the Mg or Sr modified samples. Typical micrographs and EDX spectra are shown in Fig 4. TEM-EDX indicated: i) a Ti modified HAp composition; ii) a HAp phase with no Ti evident by EDX; iii) a Ti-rich phase also containing calcium; iv) a Ca–rich phase with little EDX evidence of other elements. Crystallites were mostly 50-100 nm in size, except for the Ca-rich phase, where the crystallites were much smaller, ~ 10 nm.

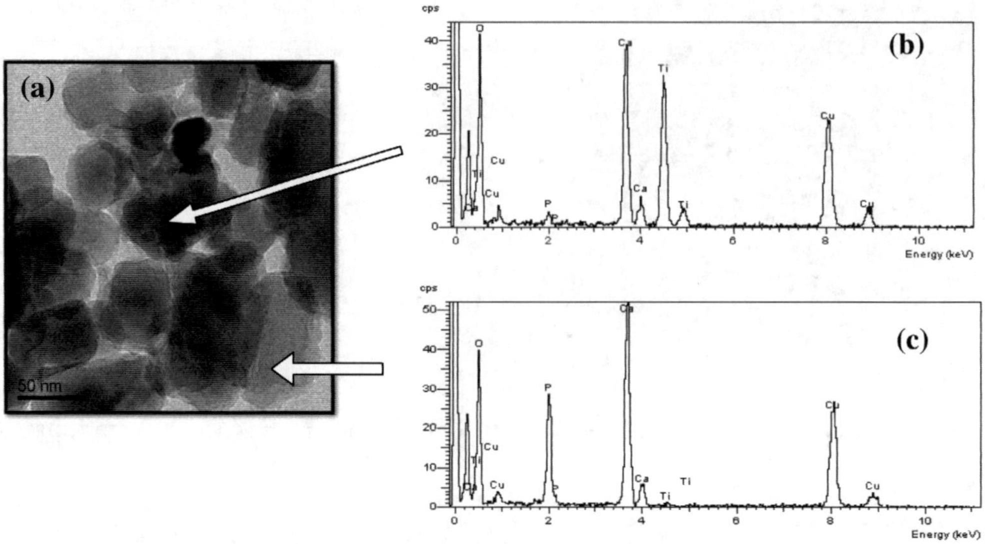

Figure 4: (a) TEM micrograph of Ti modified hydroxyapatite powder; (b) and (c) examples of TEM –EDX spectra illustrating variability in elemental content between different particles

4. Conclusions

TEM-EDX show that that each doped sample is heterogeneous in terms of inter-particle composition. Each of the doped sol-gel compositions produced nanoparticles in which the dopant species was incorporated, but some particles indicated no evidence of dopant substitution. Titanium modification produced the greatest chemical heterogeneity. Sr doping led to elongated particles, around 200 x 50 nm, whereas other compositions produced equiaxed particle morphologies, this elongation associated with Sr doping is thought to be due to the increase in ionic radius from Ca to Sr. Refinements to process conditions, in particular the efficiency and duration of reagent mixing will be examined in future. This may allow for more intimate mixing. However the volatility of the P reagent triethyl phosphite at room temperature is a major concern for this specific sol-gel system in terms of achieving batch to batch reproducibility. The results highlight the importance of TEM-EDX analysis for the characterisation of doped synthetic hydroxyapatite powders.

5. References

1. Earl J S, Wood D J, Milne S J., Amer. Cer. Soc. Bulletin. 85,22-5 (2006).
2. Earl J S, Wood D J, Milne S J., J Nanoscience and Technology (in press))
3. Renaudin G, Laquerriere P, Filinchuk Y, Jallot E, Nedelec J M., J Mat Chem. 18, 3593-3600 (2008)
4. Qi G, Zhang S, Khor K A, Liu C, Zeng X, Weng W, Qian M., Thin Solid Films 516, 5176-5180 (2008)
5. Qi G, Zhang S, Khor K A, Liu C, Zeng X, Weng W, Qian M, Thin Solid Films 516 , 5172-5175 (2008)
6. Anmin H, Ming L, Chengkang C, Dali M., J Mol Cat A: Chem 267, 79-85 (2007)
7. Wakamura M, Hashimito K, Watanabe T., Langmuir 19, 3428 -3431 (2003)
8. Liu D M, Troczynski T, and Tseng W J., Biomaterials, 23 1227 (2002)
9. Eddisford, P., A. Brown, and R. Brydson, *Electron Microscopy and Analysis Group Conference 2007 (EMAG 2007)*. 2007, Journal of Physics: Conference Series. IOP Publishing.
10. Bilton, M., A.P. Brown and S.J. Milne. *Sol-gel synthesis and characterisation of nano-scale hydroxyapatite.* in this proceedings*Electron Microscopy and Analysis Group Conference 2009 (EMAG 2009)*. 2009, these proceedings

HAADF-STEM imaging of CuAg core-shell nanoparticles

C. Langlois[1*], Z.W. Wang[2], D. Pearmain[2], C. Ricolleau[1] and Z.Y. Li[2*]

[1]Materials and Quantum Phenomena Laboratory MPQ-CNRS, 10 rue Alice Domont et Léonie Duquet, 75013 Paris (FRANCE)

[2]Nanoscale Physics Research Laboratory, School of Physics and Astronomy, University of Birmingham, Birmingham B15 2TT (UK)

[*] Corresponding authors. cyril.langlois@univ-paris-diderot.fr, z.li@bham.ac.uk

Abstract. Scanning transmission electron microscopy (STEM) is used in this study to determine the location of silver in bimetallic CuAg nanoparticles. The Z-contrast effect in STEM is particularly useful in our case because of the large difference in Z number between the two materials. However, the channeling effect that occurs when electrons are transmitted through a crystalline nanoparticle near a zone axis orientation prevents a direct interpretation of the STEM image contrast in terms of silver location. Two simple procedures are presented in this paper to estimate the contribution due to channeling, by varying the detection angle and by tilting the specimen a few degrees from its original orientation.

1. Introduction

Nanoparticles comprising two different metallic elements offer an additional degree of freedom for altering their physical properties compared to monometallic nanoparticles. This could potentially enable a wide range of new materials and properties. However, as physical properties of nanoparticles depend strongly on their atomic arrangements and structures and because in most cases nanoparticles on a specimen can exhibit size/structure distribution, it is crucial to obtain the localized elemental and structural information of a given bimetallic nanoparticle. For example, for the CuAg core-shell nanoparticles that we have prepared, the location of silver within the particles (core-shell or side-segregated) is particularly important because the optical properties of such particles are linked to the plasmon resonance frequency of silver, which often depends on its environment.

Scanning transmission electron microscopy (STEM) has been demonstrated possessing capability of exploring the three-dimensional shape and structure of nanoparticles at atomic-scale [1]. It also has potential to correlate structures with electronic and chemical properties of nanomaterials [2,3]. We focus in this paper on the use of STEM with high-angle annular dark field (HAADF) detector for a direct determination of chemical composition within bimetallic CuAg nanoparticles through an incoherent Z-contrast imaging interpretation. Here, we present two simple procedures developed to rule out the possible artifacts from the true chemical information by controlling the so-called channeling effect that occurs when electrons cross a crystal near zone axis orientation. The effects of the detection angle and small sample tilting on the contrast of HAADF-STEM images are considered.

2. Experimental details

Bimetallic CuAg nanoparticles are prepared under ultra-high vacuum conditions by pulsed laser deposition at MPQ-CNRS laboratory. Copper nanoparticles are first synthesized by condensing Cu atoms onto a thin amorphous carbon film, supported by an electron microscopy grid. Silver is then ablated and deposited on the pre-existing Cu nanoparticles. As neither Cu nor Ag wet easily on the amorphous carbon film, nanoparticles are formed instead of a continuous film. For this study, two samples were prepared. Sample A has been obtained with a deposition of Cu and Ag, with nominal thicknesses of 2 and 1 nm respectively. Sample B is a reference sample with elemental Cu nanoparticles only, obtained simultaneously during the same experiment. For both samples, the substrate temperature was held at 350°C during the deposition. Finally, a 2 nm layer of alumina was deposited to cap the nanoparticles in order to prevent their contamination.

A 200kV FEI Tecnai F20 scanning transmission electron microscope with a field-emission electron source at the University of Birmingham was employed for ex-situ examination. The microscope is fitted with a Fischione model 3000 HAADF detector. The camera length used for this study ranges from 520 to 52 mm, corresponding to the inner collection angle from 12 to 120 mrad.

3. Results and discussion

3.1. Channeling effect

STEM channeling takes place in thin crystals oriented in a zone axis parallel to the electron beam. This effect corresponds to the tendency for the electron beam to stay close to the atomic columns when it crosses the sample [4]. This maximizes the scattered intensity of the crystal and would make the crystal appear brighter than a corresponding crystal of same thickness and composition but in an off-axis orientation. Without taking any particular experimental care, the channeling effect would prohibit a direct interpretation of the images in HAADF-STEM mode in terms of chemical localisation by Z-contrast. This is particularly true for the crystalline CuAg nanoparticles prepared for this study. To illustrate this phenomenon, we carried out HAADF-STEM imaging on both CuAg and Cu nanoparticle samples.

Figure 1 shows a comparison between HAADF-STEM images for the two specimens, acquired using the same experimental conditions with a large camera length of 520 mm. This camera length corresponds to an inner angle of 12 mrad and an outer angle of 60 mrad on the HAADF detector. For the CuAg sample (Figure 1a), strong intensity variations between different particles can be noticed, and also variations of the intensity within a single nanoparticle. The latter could be attributed to the segregation of Cu and Ag inside the nanoparticle, because of the large difference in atomic number between Cu (Z=26) and Ag (Z=47). However, similar contrast has been identified on images of sample B that has only elemental Cu nanoparticles on. This comparison demonstrates that these intensity variations are not solely due to the Z-contrast. Indeed, a large proportion of the copper nanoparticles of sample B are not monocrystalline [5], and often one of the crystallites composing a copper nanoparticle can be in Bragg orientation relative to the incident electron beam. Channeling effect is then the most probable reason for explaining such a sharp variation of contrast when no chemical contrast presents. The another possible mechanism of intensity variation is strain contrast [6], which is important in this lattice mismatched CuAg bimetallic nanoparticles but does not applys to Cu nanoparticles on the amorphous substrate.

In our present study, we need to avoid channeling effect because here the aim is to get insight on the position of silver atoms over all the area of interest. Because each nanoparticle possesses its own crystallographic orientation, the channeling effect occurs in a different orientation for each of the nanoparticles and it would not be straightforward to deduce useful information about the chemical composition of the bimetallic nanoparticles through simple intensity variation analysis. Care needs to be taken to circumvent the channeling effect.

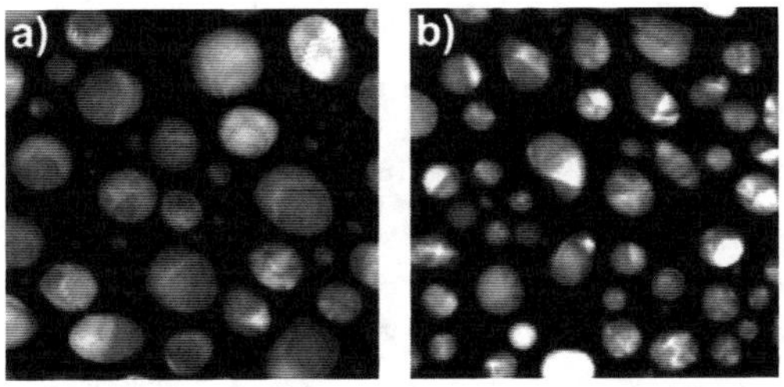

Figure 1. HAADF-STEM images acquired with a camera length of 520 mm (inner collection angle of 12 mrad). a) Sample A with Cu and Ag deposited. b) Sample B with only Cu. Field of view for both: 101 nm.

3.2. Scattering angle variation

Figure 2 shows four HAADF-STEM images of the same area of sample A, CuAg nanoparticles, acquired with different collection angles from ADF to HAADF conditions. Whereas a majority of the nanoparticles have kept their internal intensity contrast variation unchanged when the camera length is varied, it is interesting to note that a few particles undergo a contrast inversion (particles arrowed on the figures). This means that, for small collection angles (large camera length), the Bragg-oriented part of the particle gives an intensity by channeling that is larger than the intensity due to the atomic number contrast. This contrast inversion shows that the collection angle used for Figure 2d is large enough to cancel a large part of the diffracted intensity and to restore a contrast linked to the local chemical composition within the nanoparticle.

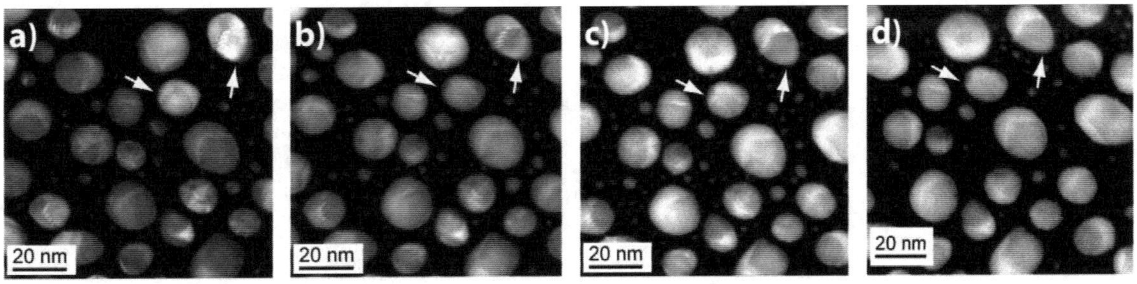

Figure 2. The effect of the detection angle on HAADF image contrast (a) CL = 520mm. (b) CL = 285mm. (c) CL = 100mm. (d) C L= 52mm.

Applying this procedure, it is then possible to discriminate between particles in Bragg orientation giving rise to high intensity through channeling for small collection angles, and off-axis nanoparticles for which the Z-contrast theory can be safely applied for localizing silver because of its big difference in atomic number compared to copper. Clearly, the signal-to-noise ratio of images acquired with such large collector angles is weak. Therefore, optimum conditions need to be carefully assessed and chosen. For bimetallic particles with two metals close in their atomic number, the decreased signal to noise ratio could mask the intensity difference arising from chemical composition.

3.3. Specimen Tilting

The channeling effect is very sensitive to the lattice orientation [7]. It is found that one or two degrees tilt of the specimen is enough to decrease the intensity of on-axis nanoparticles and recover a contrast linked to the local chemistry. Figure 3 displays an example of the specimen tilt effects on CuAg nanoparticles, where intensity variation in the arrowed particle is obvious.

This procedure has some advantages compared to the use of a small camera length. By keeping a reasonable scattering angle, the intensity detected does not decrease quickly as the specimen is tilted, thus the signal-to-noise ratio is largely preserved. Small nanoparticles can then be investigated without loosing too much resolution. However, the nanoparticles prepared through the approach we take are oriented randomly along the electron beam, therefore, each particle needs to be examined individually.

Again, among all the particles within the area under consideration, only a few of them may show a contrast that can unambiguously be assigned to silver. It is worth of pointing out that the specimen tilting, whatever the amplitude of the tilt is, can easily result in a loss of the region of interest at such high magnification. The specimen should be at an exact eucentric position, and the backlash of the goniometer should be cancelled by pre-tilting the specimen in the same direction of tilt.

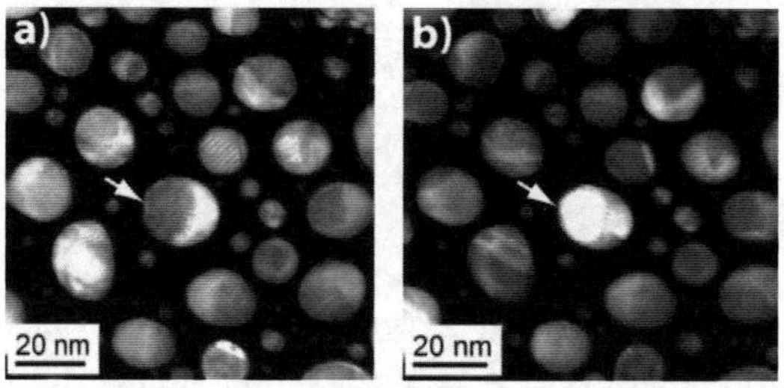

Figure 3. The effect of a small sample tilt on the STEM image contrast (CL = 520mm). (a) 0 degree. (b) 2 degrees.

4. Conclusion

In this paper we detail the use of two different procedures to estimate the channeling effect in STEM observations of CuAg nanoparticles. The first one is to increase the collection angle, in order to decrease any contribution from Bragg diffraction. The second procedure, mostly used for bulk specimen, consists in tilting the specimen out of a zone axis, and relies on the fact that the intensity due to channeling varies rapidly with the tilt angle. Both procedures are applied to the CuAg nanoparticles. It appears that increasing the collection angle by decreasing the camera length can restore a contrast linked to the position of silver atoms but decreases the signal-to-noise ratio of the images. On the contrary, acquiring two (or more) images with a reasonably large detector angle at two slightly different specimen orientations allows one to interpret (with precautions) the contrast in terms of chemical composition qualitatively for each particle. Extra-care is necessary when tilting the sample to keep the area of interest in the field of view. A compromise between the two procedures would be to acquire two images at different tilts with the smallest acceptable detector angle.

Acknowledgment

CL and ZYL are grateful to the Egide (France) and British Council (UK) for supporting this study through the Alliance project. Prof. Jun Yuan from University of York (UK) is gratefully acknowledged for helpful discussions about the channeling effects in HAADF-STEM imaging.

References

[1] Li ZY, Young NP, Di Vece M, Palomba S, Palmer RE, Bleloch AL, Curley BC, Johnston RL, Jiang J and Yuan J 2008 *Nature* **451** 46-49
[2] Li ZY, Yuan J, Chen Y, Palmer RE and Wilcoxon J 2005 *Appl. Phys, Lett.* **87** 243103
[3] Jiang J, Yuan J and Bleloch A 2007 *Appl. Phys. Letts.* **91** 113107-3
[4] Reimer L 1985 *Scanning Electron Microscopy* ed. by P.W. Hawkes (Springer-Verlag, Berlin) pp 247-249
[5] Langlois C, Alloyeau D, Le Bouar Y, Loiseau A, Oikawa T, Mottet C and Ricolleau C 2007 *Faraday Discussion* **138**, 375-391
[6] Liu C-P, Twesten RD and Gibson JM 2001 *Ultramicroscopy* **87** 79-88
[7] Yu Z, Muller DA and Silcox J 2008 *Ultramicroscopy* **108** 494-501

TEM analysis of Si-passivated Ge-on-Si MOSFET structures for high performance PMOS device technology

DJ Norris[1], IM Ross[1], AG Cullis[1], T Walther[1], M Myronov[2], A Dobbie[2], T Whall[2], EHC Parker[2], DR Leadley[2], B De Jaeger[3], W Lee[3] and M Meuris[3].

[1] Dept. of Electronic and Electrical Engineering, University of Sheffield, Sir Frederick Mappin Building, Mappin Street, Sheffield, S1 3JD, UK

[2] Dept. of Physics, University of Warwick, Coventry, CV4 7AL, UK

[3] IMEC, Kapeldreef 75, B-3001 Leuven, Belgium

d.j.norris@sheffield.ac.uk

Abstract. In this paper, we present a transmission electron microscopy analysis of novel Ge-on-Si MOSFETs using a JEOL 2010F and an aberration-corrected JEOL 2200FSC. A key feature of these devices is the incorporation of a very thin (~1nm) Si passivation layer on top of the Ge virtual substrate, which is partially oxidised to form SiO_2 (~0.5nm) prior to depositing HfO_2 dielectric. We will show that the thin SiO_2 layer is not purely amorphous but has some degree of crystal ordering due to being bonded to crystalline materials. Moreover, we will examine the presence of small monolayer variation in Si/SiO_2 interface roughness.

1. Introduction

In recent years, there has been an increased drive to incorporate strained Ge into conventional Si-based complementary metal-oxide-semiconducting (CMOS) device structures. The effects of such pseudomorphic straining are to enhance carrier (hole) mobility in these types of devices [1].

One particular set of device structures that have been investigated are Ge-on-Si based MOS field-effect-transistors (MOSFETs). Here, a relaxed layer of pure Ge, or $Si_{1-x}Ge_x$ with high Ge content x, onto a (001) Si wafer to produce a Ge-on-Si virtual substrate (VS) and MOSFET devices can then be processed onto these wafers [2]. One of the issues associated with fabricating devices onto these wafers is the need to form, on the wafer surface, a stable dielectric prior to fabricating the gate electrode. One of the drawbacks of Ge-based devices, however, relates to the poorer thermal oxide quality compared with that of Si, which has been found to give rise to a higher interface trap density [3]. Methods have been developed, therefore, to reduce this through Ge passivation [4]. One possible route for such passivation involves depositing an ultra-thin epitaxial Si layer on the surface of the Ge substrate which is then partially oxidised, prior to the growth of a high-k gate dielectric [4]. Issues to be considered, however, relate to the uniformity of the Si/SiO_2 interface, as roughness can lead to carrier scattering, and also whether Ge segregation during growth of the epi-layer will affect passivation and consequently the interface trap density [5].

In this paper, we present a TEM/STEM analysis of MOSFETs on a Ge-on-Si VS. Such devices incorporate a narrow partially oxidised Si passivation layer onto which MOSFET devices are processed. As device dimensions shrink to the sub-nm level, it is necessary to perform analysis at even higher resolution using aberration corrected microscopy techniques.

2. Experimental Details

Ge p-MOSFETs were fabricated using a silicon-compatible process flow on (001) Si wafers with a thick (~2µm) relaxed Ge layer grown on the surface. After the p-well and channel implants, the Ge-on-Si wafer was passivated by growing a thin epi-Si layer (~1nm) which was then partially oxidised at room temperature in a water bath with ozone; see [4]. To avoid further oxidation, a thin layer of HfO_2 (~4nm) was then deposited by atomic layer deposition onto the SiO_2 before growing the TaN (~10nm) and next the TiN (~70nm) metal layers. The remaining processes, i.e. the dry metal gate etch, As halo's, BF_2 extension implants, spacer definition, B implant, NiGe formation in source and drain regions and metal contact fabrication are all detailed elsewhere [6].

The processed wafers include a narrow region of device lines of MOSFETs of various gate lengths which could be processed into TEM specimens. [110] cross-sectional specimens for TEM, were prepared in the usual manner with 3mm disks thinned mechanically to ~30-40µm followed by Ar^+ ion thinning to electron transparency. Finished specimens were then analysed using a JEOL 2010F field-emission gun instrument, operating at 197kV, and equipped with a Gatan imaging filter and STEM analysis using bright-field (BF) and annular dark-field (ADF) detectors. Specimens were also analysed using a double aberration corrected JEOL 2200FSC microscope operating at 200kV.

3. Results and Discussion

The low magnification image in Figure 1(a) provides an overview of the MOSFET device structures. The thickness of the Ge VS is 1.63µm and the surface is uniformly flat. These particular devices have a 325nm gate length and are spaced ~160nm apart. A higher magnification BF STEM image is shown in Figure 1(b), and shows the various components which make up the device structure. The dark layer at the base of the gate electrode is the heavy TaN layer (~14nm) under which are the HfO_2 (3.6nm) and SiO_2 (0.5nm) gate dielectrics. On top of this is a TiN (~70nm) metal electrode above which is a SiO_2 'hard mask'. In the source and in the drain regions, the NiGe (~12nm) is clearly visible.

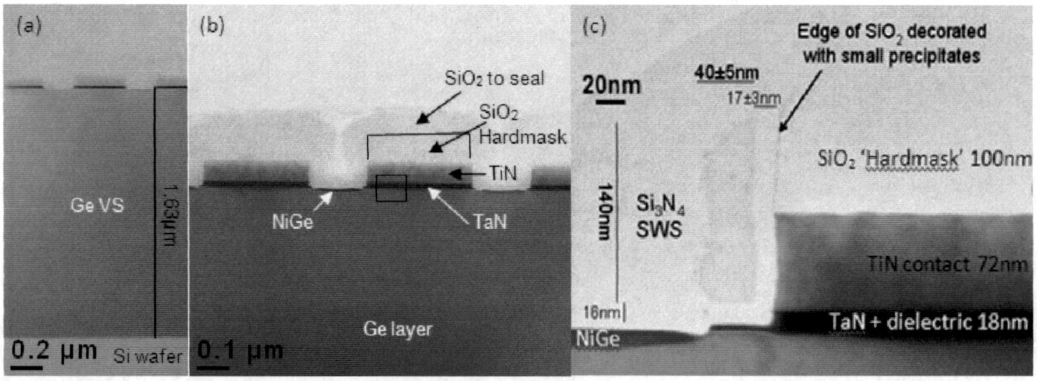

Figure 1. (a) Low magnification BF STEM image of device structures on Ge VS; (b) higher magnification BF STEM image with architecture of device defined; (c) even higher magnification image of side-wall spacer and NiGe of larger gate length device.

An even higher magnification BF STEM image of the edge of a large device is provided in Figure 1(c). Here, the Si_3N_4 + SiO_2 side-wall-spacer (SWS) is visible with dimensions of ~55-60nm wide and ~150nm high. The important region in terms of device performance is the passivation and dielectric region beneath the gate electrode (region boxed in Figure 1(b)) which we will now focus upon.

The dielectric region, of a shorter gate-length device, was imaged using planar illumination under HREM conditions, using a JEOL 2010F microscope, and is shown in Figure 2(b). One feature of the present device structures is the incorporation of a Si passivation layer grown on top of the Ge VS. This is partially oxidised to form a very thin SiO_2 layer, which (from Figure 2(b)) was measured to be 0.5±0.2nm, in line with previous estimates [2]. A closer examination of this image shows that the SiO_2 layer is bounded by the crystalline Ge/Si VS and the crystalline HfO_2 layer which seem to have a common set of lattice fringes across the SiO_2. This would suggest that the SiO_2 layer is not purely amorphous but instead has some degree of crystalline order due to the fact that it is so thin and is bonded on each side to crystalline materials. Indeed, from Figure 2(b) there appear to be lateral regions within the SiO_2 layer which exhibit some degree of crystallinity such as that marked (x). Fourier transforms of the crystalline HfO_2 and Ge VS were obtained to determine lattice spacings and orientations, and these are shown in Figures 2 (c) and (d), respectively. From these, {111} lattice spacings of 0.305nm (HfO_2) and 0.336nm (Ge) were measured. Also, there seems to be an orientational relationship between the HfO_2 and the Ge VS where $d_{111}(HfO_2)$ is near parallel to d_{111}(Ge).

It is difficult to confirm the presence of the residual unoxidised Si from the phase contrast image in Figure 2(b). An attempt was made, therefore, to determine the presence of a residual Si layer by ADF (Z-contrast) imaging, and such an image is shown in Figure 2(a). The width of the dark region indicated corresponds to the sum of the thicknesses of the residual Si and SiO_2 layers (1.2±0.2nm). Hence, a comparison of these images shows that the thickness of the residual Si passivation layer is 0.7±0.3nm, again in line with previous estimates [2]. However, the point resolution of the JEOL 2010F is 0.19nm, which is insufficient to resolve individual monolayers of spacing d_{004}=0.14nm in Ge.

We have then performed HREM experiments using a JEOL 2200FSC aberration corrected instrument, to look in more detail at the morphology of the $Si/SiO_2/HfO_2$ regions. A high resolution image obtained from a similar region to that shown in Figure 2(b) is shown in Figure 3(a). For this, the spherical aberration (C_s) was set to –3µm and images were obtained in overfocus so that atomic columns in thin regions appear with positive contrast (white on black), as opposed to classical Scherzer conditions [8]. The beam tilt coma was corrected to ~20nm amplitude and all other aberrations were near their optimum for this instrument to give a uniform phase plate up to an angle of ~30mrads, which corresponds to an information transfer out to $(0.08nm)^{-1}$ in reciprocal space. From the image in Figure 3(a), the Ge dumb-bells are clearly resolved at 0.14nm spacing and a Fourier transform of the image (Figure 3(b)) shows the presence of the {551}Ge reciprocal lattice position (circled), confirming information transfer out to 0.08nm.

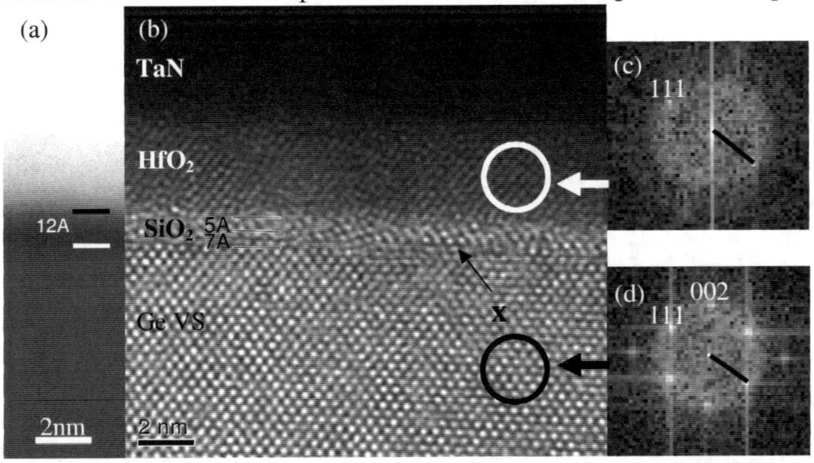

Figure 2. ADF image (a) and lattice image (b) of dielectric region under the gate terminal. (c) and (d) are power spectra of the HfO_2 and Ge VS, respectively.

An examination of the Si/SiO_2 boundary shows that there is monolayer scale roughness. Bi-layer step edges appear at positions (A)-(D) indicated, demonstrating a non-uniform variation in the Si passivation layer thickness. The origin of such roughening is still unclear; however, it may be due to the fact that the Ge VS surface exhibited small scale roughness or surface undulations prior to

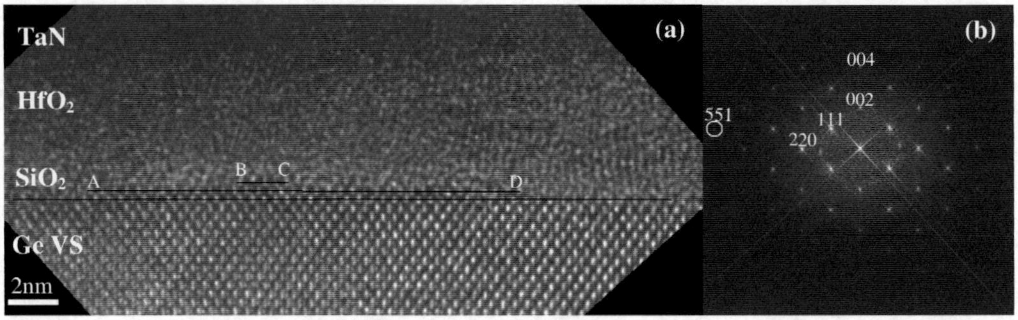

Figure 3 (a) HREM image of $Si/SiO_2/HfO_2$ region with dark lines inserted to show the monolayer-scale roughness at the Si/SiO_2 boundary; (b) power spectrum of the image with $\{551\}_{Ge}$ (circled) showing information transfer down to 0.08nm in the horizontal direction. In particular, along the vertical, monolayers of d_{004}=0.14nm are clearly resolved. A, B, C, D indicate bilayer steps.

deposition of the Si passivation layer. Oxidation of this layer may therefore lead to a Si/SiO_2 interface which displayed some degree of roughness. This, of course, has implications on device performance where enhanced roughness of this boundary may contribute to greater carrier scattering at high source/drain bias.

One further concern is whether there is any Ge present within the residual Si and SiO_2 layers. It is known that during growth, Ge may segregate from the surface of the Ge substrate through the epitaxially grown Si passivation layer [7] and may be incorporated into the processed SiO_2 layer, or accumulate at the lower Si/SiO_2 boundary by 'snowploughing'. However, verifying this is difficult since the layers concerned are extremely narrow. Further work is in progress to determine the Ge distribution more precisely, using the JEOL 2200FSC instrument in ADF-STEM mode.

4. Conclusions

Ge-on-Si MOSFET structures have been analysed using TEM-based techniques using two different electron microscopes; a JEOL 2010F and a double aberration corrected JEOL 2200FSC. The latter instrument confirmed an information transfer of 0.08nm in one direction and resolved monolayer dumb-bells in Ge of 0.14nm spacing. A feature of the present device structures was the use of Si passivation which required the deposition of a very thin epi-layer of Si (1nm) on the Ge-on-Si wafer, which was then partially oxidised. High resolution TEM images showed that the thin SiO_2 layer is not likely to be purely amorphous but instead has some degree of crystalline ordering due to it being only a few monolayers wide and bonded on each side to crystalline materials. Moreover, the Si passivation layer exhibits monolayer scale roughness which could impact on device performance through enhanced interface roughness scattering. Our results confirmed Si passivation with the residual Si passivation layer being 0.7±0.3nm and the SiO_2 layer thickness 0.5±0.2nm.

Acknowledgements

The authors would like to thank the Engineering and Physical Sciences Research Council for their financial support of this work under grant number EP/F033893/1 "Renaissance Germanium".

References

[1] On Chui C et al. 2002 *IEEE Electron Dev. Lett.* **23**, 473
[2] Kaczer B et al. 2007 *Microelectronic Engineering* **84**, 2067
[3] Houssa M et al. 2007 *Microelectronic Engineering* **84**, 2267
[4] De Jaeger B et al. 2005 *Microelectronic Engineering* **80**, 26
[5] Leys F E et al. 2006 *Materials Science in Semiconductor Processing* **9**, 679
[6] Millard J et al. 2008 *IEDM Tech. Dig.*, 873
[7] Norris D J et al. 1999 *J. App. Phys.* **86**, 7183
[8] Lentzen M 2004 *Ultramicroscopy* **99**, 211

Effect of phosphate on the morphology of hydrothermally synthesised β-FeOOH and α-Fe_2O_3 nanoparticles

Trevor Almeida, Mike Fay[1], Yanqiu Zhu[*] and Paul D Brown

Division of Materials, Mechanics and Structures, Faculty of Engineering, &
Department of Mechanical, Materials and Manufacturing Engineering; &
[1]Nottingham Nanotechnology and Nanoscience Centre,
University of Nottingham, University Park, Nottingham, NG7 2RD, UK.

Abstract. β-FeOOH and α-Fe_2O_3 nanoparticles have been grown from an $FeCl_3$ solution precursor in the presence of an ammonium dihydrogen phosphate surfactant through the process of hydrothermal synthesis. β-FeOOH production varied from small, rounded nanoparticles (~ 45 nm) to larger nanorods (~ 120 nm in length) depending on the stage of synthesis at which the surfactant was added. Larger α-Fe_2O_3 nanorods (~ 70 nm wide and ~ 450 nm in length) were synthesised from a β-FeOOH suspension at 200°C upon phosphate addition prior to initial β-FeOOH precipitation. The anisotropic growth of α-Fe_2O_3 was associated with the preferential adsorption of PO_4^{3-} anions onto α-Fe_2O_3 crystal planes parallel to the c-axis. Phosphate addition to the suspension subsequent to the initial β-FeOOH precipitation inhibited the phase transformation to α-Fe_2O_3 at 200°C and assisted with increasing the β-FeOOH particle size.

1. Introduction

One-dimensional (1D) nanostructures have attracted considerable attention due to their novel magnetic properties which are greatly dependent on nanorod size and shape [1]. Ferromagnetic hematite (α-Fe_2O_3), the most thermodynamically stable phase of iron oxide, is of particular interest as a multifunctional nanostructured material with potential applications in catalysis, gas sensing, magnetic recording, water treatment, lithium-ion batteries and pigments; owing to its non-toxicity, low processing cost and high resistance to corrosion [2]. The approach of hydrothermal synthesis (HS), in particular, offers effective control over the size and shape of α-Fe_2O_3 nanoparticles at relatively low reaction temperatures and short reaction times, providing well-crystallized reaction products with high homogeneity and definite composition. Dilute iron (III) chloride ($FeCl_3$) solution is a simple precursor for the formation of α-Fe_2O_3 nanoparticles, whereby an intermediate phase of β-FeOOH (akaganeite) is produced prior to α-Fe_2O_3 precipitation [3]. Chloride, sulphate, phosphate and hydroxide ions have been shown to play a decisive role in the anisotropic growth of hematite [4]. Phosphate anions have been especially effective for the growth of 1D α-Fe_2O_3 nanorods, despite utilising relatively low concentrations [5]. However, not much has been reported on the effect of phosphate on the initial development of β-FeOOH, in the context of the anisotropic growth of α-Fe_2O_3 nanostructures. Here, the growth of β-FeOOH and α-Fe_2O_3 nanostructures is described, produced through the HS of an $FeCl_3$ precursor solution, with the phosphate surfactant added at different stages of the process to appraise its effect on the morphology and growth of the reaction products.

[*] Author to whom correspondence should be addressed: yanqiu.zhu@nottingham.ac.uk

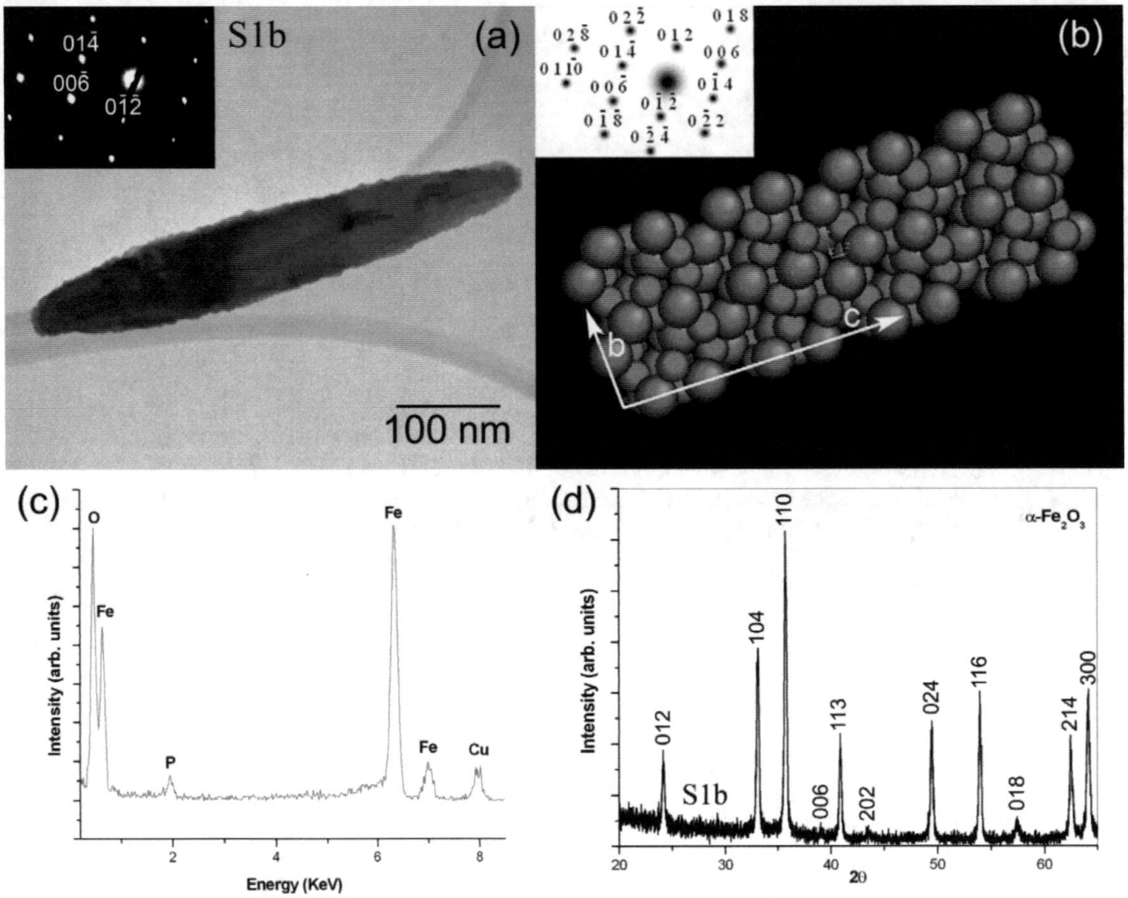

Figure 1. (a) BF TEM image (with SAED pattern inset) of an individual α-Fe_2O_3 nanorod (sample S1b). (b) Carine model of the α-Fe_2O_3 crystal structure (with simulated diffraction pattern inset). (c) Associated EDX spectrum acquired from an individual α-Fe_2O_3 nanorod (sample S1b), cleaned with acetone. (d) Associated XRD pattern from sample S1b, confirming the presence of the α-Fe_2O_3 phase.

2. Experimental

β-FeOOH and α-Fe_2O_3 nanoparticles were prepared through the hydrothermal reaction of 0.2 ml 45% pure $FeCl_3$ aqueous solution, further diluted in 40 ml of distilled water. The reactant solution was mixed with and without 3 mg of ammonium dihydrogen-phosphate (99.999% $NH_4H_2PO_4$) surfactant (samples S1a and S2a, respectively), mechanically stirred in a 125 ml Teflon-lined steel autoclave, and then sealed and inserted into a temperature controlled furnace at a reaction temperature of 110˚C for 2 hours. The autoclave, once removed from the furnace, was allowed to cool down to room temperature naturally. The experiment was repeated and after cooling, sample S1a was reinserted into the furnace again at 200˚C for 2 hours (sample S1b), whilst 3 mg of $NH_4H_2PO_4$ was added to sample S2a before resealing and reinsertion into the furnace at 200˚C for 2 hours (sample S2b). The synthesised reaction products were deposited straight from solution onto single crystal silicon substrates for bulk structural characterisation using a Siemens D500 X-ray diffractometer. For the purpose of TEM investigation, the HS product suspensions were deposited straight onto lacey carbon / copper mesh support grids for conventional diffraction contrast bright field (BF) imaging, performed using a Jeol 2000fx operated at 200 kV. Energy dispersive X-ray (EDX) analysis provided information on the elemental constituents, whilst selected area electron diffraction (SAED) allowed for phase identification and an appraisal of the relationship between nanorod morphology and crystallographic orientation, which was also modelled using Carine software [6].

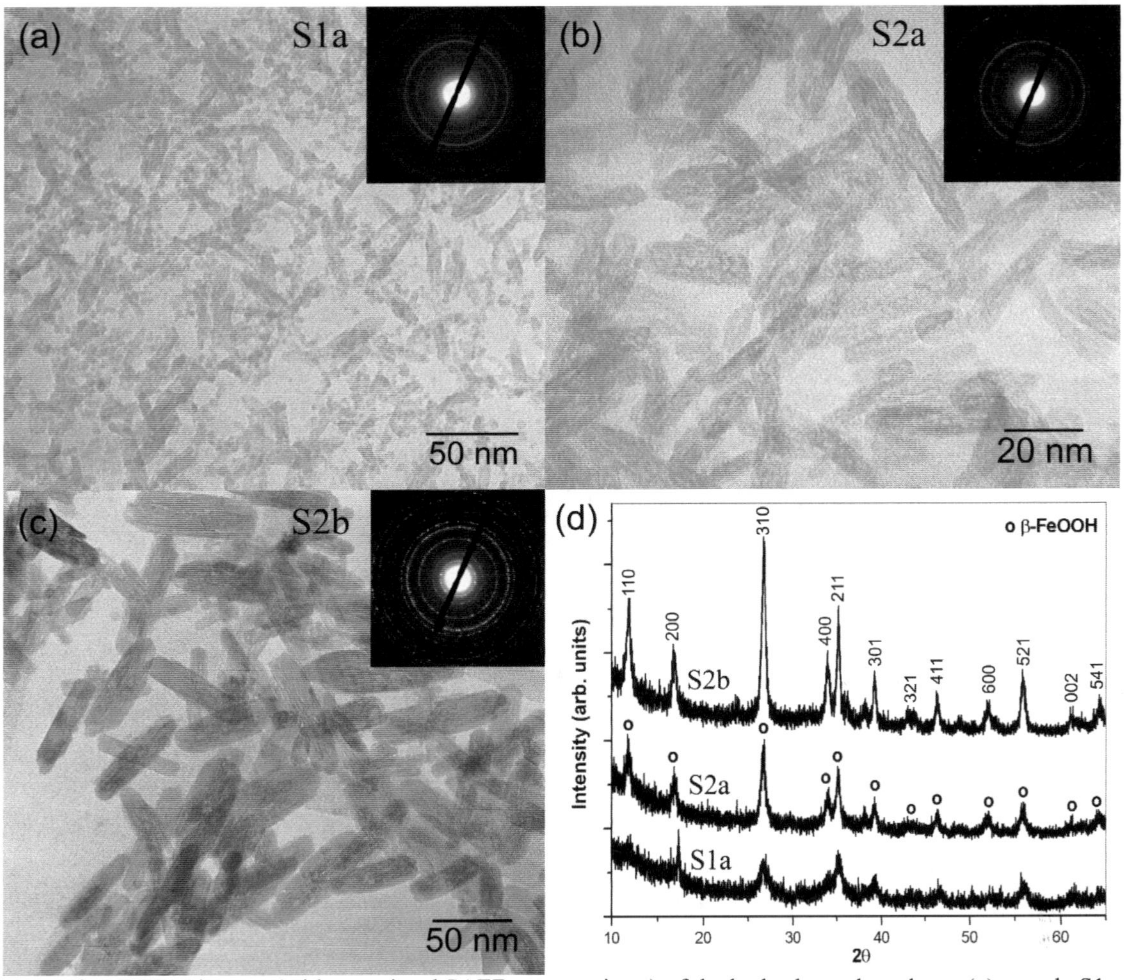

Figure 2. BF TEM images (with associated SAED patterns inset) of the hydrothermal products: (a) sample S1a; (b) sample S2a; and (c) sample S2b. (d) Associated XRD patterns.

3. Results

Figure 1 presents a BF TEM image (with associated SAED pattern inset), a crystal structure model, an EDX spectrum and an associated XRD pattern from sample S1b. The XRD pattern (Figure 1d) confirmed that the sample was comprised solely of the α-Fe_2O_3 phase. TEM investigation confirmed the development of lenticular α-Fe_2O_3 nanorods, ~ 75 nm wide and ~ 450 nm in length, in this instance (Figure 1a). In all cases, the major axis of these well-defined, single crystalline nanorods was found to be perpendicular to the {006} set of crystal planes (Figure 1a). A computer model of the α-Fe_2O_3 crystal structure (Figure 1b) confirmed that the nanorod major axis corresponded to the c-axis of the α-Fe_2O_3 unit cell. Further, EDX investigation indicated the presence of a small amount of phosphorus within the bulk of such nanorods (*e.g.* Figure 1c), considered to be residual from the phosphate surfactant.

BF TEM images of samples S1a, S2a and S2b are presented in Figures 2a, 2b and 2c, respectively. The associated XRD patterns of Figure 2d confirmed all these samples to comprise of β-FeOOH. Similarly, EDX of sample S1a again revealed a presence of phosphorus. Application of the Scherrer equation suggested an increase in relative average particle size of samples S1a, S2a and S2b, respectively. However, the Scherrer equation does not take into consideration particle morphology.

TEM investigation of sample S1a confirmed this suspension to be a mixture of small and round, as well as larger rod-like, β-FeOOH nanoparticles, up to ~ 15 nm wide and ~ 45 nm in length (Figure 2a). Conversely, sample S2a (Figure 2b) exhibited only the β-FeOOH nanorods (~ 20 nm wide and ~ 60 nm in length). Sample S2b displayed larger, well-crystallised β-FeOOH nanorods, up to ~ 40 nm wide and ~ 120 nm in length (Figure 2c). It is worthy to note the initial precursor of 0.2 ml 45% pure $FeCl_3$ aqueous solution further diluted in 40 ml distilled water with and without 3 mg of $NH_4H_2PO_4$ heated straight up to 200°C, under the same HS conditions, produced lenticular and agglomerated rhombohedral α-Fe_2O_3 nanostructures, respectively [7].

4. Discussion

This combined TEM and XRD investigation confirmed that samples grown with and without phosphate at 110°C (samples S1a and S2a), and without phosphate at 110°C then with phosphate at 200°C (sample S2b) all comprised the β-FeOOH phase, whilst only sample S1b grown in the presence of phosphate throughout, at 110°C and then 200°C, comprised the α-Fe_2O_3 phase. Hence, it is evident that the stage at which the $NH_4H_2PO_4$ surfactant was added to the initial suspension had a dramatic effect on the crystalline phase of the reaction products synthesised at 200°C, *i.e.* the presence of the phosphate was required at the outset for the efficient formation and development of α-Fe_2O_3 nanorods via the intermediate β-FeOOH phase.

Sample S1a synthesised at 110°C in the presence of the surfactant comprised a mixture of small, rounded nanoparticles and nanorods. Sample S2a, synthesised at 110°C in the absence of a surfactant, mainly consisted of slightly larger nanorods, whilst TEM investigation of sample S2b, synthesised at 200°C subsequent to the addition of $NH_4H_2PO_4$ to sample S2a, revealed larger still, well-crystallised β-FeOOH nanorods. It is apparent that the initial addition of surfactant to sample S1a acted to inhibit the growth of the developing β-FeOOH nanoparticles, as compared with sample S2a. It is noted that the precursor solution exhibited an initial pH of ~ 2.4, being much lower than the zero point charge of β-FeOOH (pH 8), and hence it is possible that the inhibition of particle growth arises from strong electrostatic attraction and the adsorption of high levels of PO_4^{3-} anions.

Indeed, strong surface PO_4^{3-} anion adsorption onto the β-FeOOH nanorods of sample S2a, after interruption of the HS process, is considered to stabilise these established β-FeOOH nanorods at the elevated reaction temperature of 200°C. This delay to the timing of surfactant addition provides an element of control over the HS process, preventing the phase transformation of β-FeOOH in sample S2b to α-Fe_2O_3, whilst also assisting with increasing the size of the β-FeOOH nanorods.

TEM and SAED investigation and simulation of an individual α-Fe_2O_3 nanorod (sample S1b) showed the major axis to be perpendicular to the {006} set of planes, *i.e.* coincident with the crystallographic c-axis. It is suggested that the surfactant is preferentially adsorbed onto the faces parallel to the c-axis, thereby promoting anisotropic growth.

In summary, it has been shown that it is possible to control that size and phase of either β-FeOOH or α-Fe_2O_3 nanorods, using HS, through appropriate choice of temperature and the timing of phosphate surfactant addition.

References

[1] Bate G, 1991 *J. Magn. Magn. Mater.* **100**, 413
[2] Cornell R M and Schwertman U 1996 *The Iron Oxides* (Weinham: VCH)
[3] Matijevic E, Scheiner P 1977 *J. Colloid Interface Sci.* **63,** 509-524
[4] Sugimoto T, Muramatsu A *et al* 1993 *J. Colloid Interface Sci.* **158,** 420-428
[5] Sugimoto T, Muramatsu A 1996 *J. Colloid Interface Sci.* **184,** 626-638
[6] CaRine Crystallographie, CaRine Version 3.0 copyright 1989-1996
[7] Almeida *et al*, manuscript in preparation

Acknowledgements

The authors would like to thank the EPSRC for financial support.

Journal of Physics: Conference Series **241** (2010) 012046 doi:10.1088/1742-6596/241/1/012046

EELS and ELNES studies of nano-scale nitride multilayers deposited by unbalanced magnetron sputtering

C.R. Seabourne[1], I.M.Ross[2], W.M.Rainforth[2], A.J. Scott[1], B.G. Mendis[3] and P.Eh.Hovespian[4]

[1]Institute for Materials Research, University of Leeds, Leeds, LS2 9JT, United Kingdom

[2]Department of Engineering Materials, University of Sheffield, Mappin Street, Sheffield S1 3JD, United Kingdom

[3]SuperSTEM, Daresbury Laboratory, Daresbury WA4 4AD, United Kingdom

[4]Materials and Engineering Research Institute, Sheffield Hallam University, Sheffield, S1 1WB, United Kingdom

i.ross@shef.ac.uk

Abstract. High spatial resolution chemical analysis and imaging of a CrAlYN/CrN multi-layer coating was performed using sub-nanometer probe electron energy loss spectroscopy in a spherical aberration corrected scanning transmission electron microscope. Analysis indicated the CrN layers to be near stoichiometric with a relative Cr/N ratio of 1.05±0.1 while the CrAlYN layers gave a Cr/N ratio of 0.59±0.02. Analysis of the energy loss near edge structure at the nitrogen K-edge was performed. The experimentally determined fine edge structure in electron energy loss spectra were compared with theoretically determined spectra, calculated using electron density functional theory. For the CrN layers the best match between the direct experimental analysis and the simulated edges corresponds to stoichometric CrN consistent with the quantitative analysis while the best match for the CrAlYN layers was to $(Cr_{0.5}Al_{0.5})N$.

1. Introduction

Nano-scale multilayer coatings, deposited by unbalanced magnetron sputtering, have attracted significant attention in recent years due to their ability to extend component lifetime and durability. In particular, coatings based on CrAlYN/CrN with a multilayer periodicity of the order of a few nanometres have been shown to offer excellent oxidation resistance and tribological performance [1]. The mechanical properties and oxidation resistance of these coatings depends on the nanometer scale microstructure. Hence, in order to understand the origin of the properties of the coating, a full understanding of the nano-chemistry is important. Since the CrAlYN and CrN components are isostructural and mutually miscible, the interface between the layers is expected to be coherent. However, commercial coating systems used to deposit such coatings do not employ shutters and therefore intermixing between layers is to be expected. This will have an effect on the sharpness of the interface between layers.

For many metal-nitride materials the fine structure at the nitrogen K-edge is known to be sensitive to the local atomic environment and consequently the N k-edge can be used as a fingerprint for identification [2]. Hence, in this study we present a characterisation through the application of electron energy-loss spectroscopy (EELS) and an investigation of the energy-loss near edge structure (ELNES) at the nitrogen K edge for the individual CrAlYN and CrN layers.

2. Experimental methods

The nano-scale multilayer coating, investigated in this study was deposited at the Nanotechnology Centre for PVD Research at Sheffield Hallam University UK using unbalanced magnetron sputtering (UBM) in an industrial size Hauzer HTC 1000-4 PVD coater [1]. Following a metal ion etch of the substrate surface to improve adhesion and coating quality, a base layer of CrAlN ~0.3μm thick was applied by UBM followed by the multilayer coating, comprising alternating layers of CrAlYN and CrN to provide a total coating thickness of approximately 5μm.

Cross sectional TEM specimens were prepared in the conventional manner by mechanical polishing followed by Ar^+ ion milling to electron transparency. Analysis of the multilayer structure was performed in a Vacuum Generators dedicated STEM equipped with a Nion spherical aberration corrector and a Gatan Enfina EELS spectrometer (UK Super-STEM). In this instrument the EELS data was acquired at 100kV with a beam convergence half-angle of 24 mrad and a spectrometer collection angle of 19 mrad. High angle annular dark-field (HAADF) images were recorded with a detector acceptance angle of 70-210 mrad. The probe correction algorithm was set up to tune aberrations up to 3rd order to typically achieve a probe with FWHM of 0.13nm, although analysis of HAADF FFT has shown information can routinely be achieved out to sub 1A. Individual layers within the multilayer were examined by positioning a ~ 0.1nm probe over the centre of the layers and the energy loss near edge structure (ELNES) at the N K and Cr $L_{2,3}$ edges recorded with a spectrometer dispersion of 0.1eV/pixel.

Computational simulations of the ELNES at the nitrogen K-edge utilised the Wien2k all-electron density functional theory (DFT) code, which employs the linear augmented plane wave (LAPW) method for electronic structure calculations [3]. In this method, the theoretical cell is separated into non-overlapping spheres and interstitial space which is modeled by a plane wave basis-set. As is usual for DFT codes, the Kohn-Sham formulism is used, in which an approximation must be chosen for the exchange-correlation potential. In all calculations reported the generalized gradient approximation (GGA) was used for this purpose [4]. Wien2k predicts EELS results directly via an expression for the double differential scattering cross section, the methodology employed having been reported in detail [5]. The structures of the end members of the cubic Fm3-m series of CrN and AlN were chosen with lattice parameters 0.4140nm and 0.4050nm respectively. Vegard's rule (using a linear relationship based on cell volumes) was then used to find plausible lattice parameters for the intermediate compositions ($Al_{0.25}Cr_{0.75}N$, $Al_{0.375}Cr_{0.625}N$ $Al_{0.5}Cr_{0.5}N$, $Al_{0.375}Cr_{0.625}N$, $Al_{0.75}Cr_{0.25}N$), nitrogen being the common element in all cases. Using the calculated lattice parameters, unit cells were constructed, with calculations being in the ground state. A theoretical core-hole was therefore not included indeed it has been shown that for systems related to CrN the core-hole has only a small effect on the predicted ELNES result [6]. Wien2k was then used to find predicted nitrogen K edges. 1000 k-points were used in total in the reducible Brillouin zone, with RK_{MAX} of 7, l_{max} of 12, and energy and charge convergence criteria of ±0.0001Ry. When there was more than one nitrogen atom present in the cell, an 'average' K edge was determined. In the figures presenting the theoretical results, two types of broadening were applied. Firstly, a nominal energy-independent Lorentzian broadening (of magnitude 1eV), and secondly an energy-dependent final state lifetime broadening as reported by Moreau *et al* [7].

3. Results and Discussion

Fig. 1a shows a conventional bright-field TEM image of the CrAlYN/CrN multi-layer revealing a columnar microstructure with a strong [220] preferred orientation. The bi-layer thickness of the multi-layer measured directly from the TEM images and was found to be ~3.8±0.2nm in good agreement with previous values derived from XRD analysis. The C_s corrected HAADF STEM image in Fig. 1b shows the individual layers in more detail, the brighter contrast layers corresponding to the CrN layers. Lattice fringes, corresponding to (111) planes (~0.24nm) can be observed in Fig 1b highlighting the structural coherency between layers. It is also clear from the HAADF image that the interfaces between the layers are not abrupt consistent with the shutter free deposition process [1].

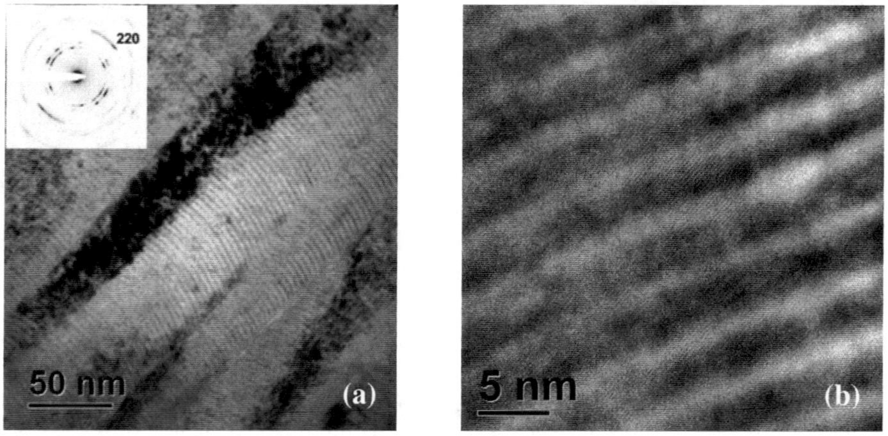

Fig. 1. (a) Bright-field TEM image of the multilayer region of the CrAl(Y)N/CrN coating (Philips EM420), (b) HAADF-STEM image of the individual multilayers.

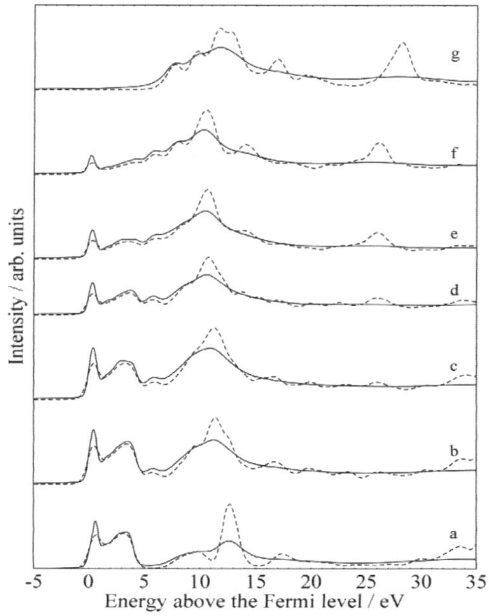

Fig. 2. Calculated ELNES at the N K-edge for a) CrN, b) $Al_{0.25}Cr_{0.75}N$, c) $Al_{0.375}Cr_{0.625}N$, d) $Al_{0.5}Cr_{0.5}N$, e) $Al_{0.375}Cr_{0.625}N$, f) $Al_{0.75}Cr_{0.25}N$ and g) AlN (where the solid line is with energy-dependent final state lifetime broadening, and the dashed line is a simple linear broadening).

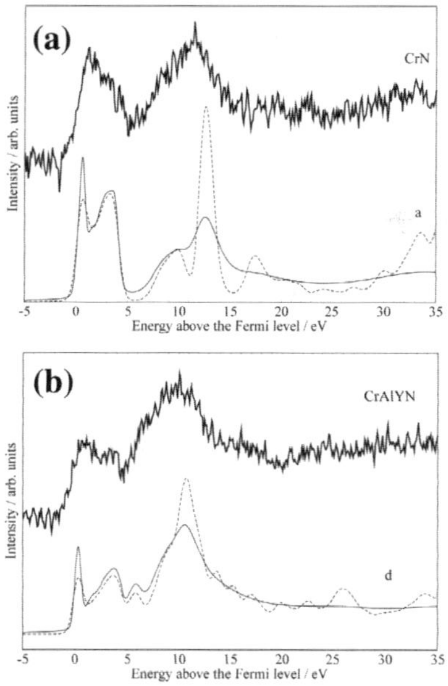

Fig. 3. Experimental N K-edge ELNES and corresponding best fit calculated model structure for (a) nominal CrN layer, (b) nominal CrAlYN layer.

Previous secondary neutral mass spectroscopy studies for the bulk CrAlYN/CrN multilayer coating revealed a relative composition of 29.5±1.5at% Al, 30.0±2at% Cr, 39.0±3at% N and 0.5±0.1at% Y [8]. Measurement of the relative N and Cr intensities of EEL spectra derived from the individual layers revealed a Cr/N ratio of almost unity in the nominal CrN layers (1.05±0.01) that is a slight excess of Cr while in the CrAlYN layers the N concentration is significantly higher with a Cr/N ratio of 0.59±0.02.

The results of the calculated ELNES at the nitrogen K-edge are shown in Fig. 2 and reveal some interesting trends. Moving from systems that mostly contain Cr to those mostly containing Al, the intensity of the peaks just above the Fermi level diminishes. These peaks are likely to be as a consequence of chromium *d* / nitrogen *p* orbital interactions which will generate additional states above the Fermi level not observed when aluminium begins to dominate. The experimental ELNES for the CrN, CrAlYN layers were compared with all the various predictions. Fig 3 shows the corresponding best fit of the experimental CrN and CrAlYN ELNES (shown in bold) to the calculated edges. Considering the nominal CrN layer (Fig. 3a), the shape and peak positions are comparable to those reported by Mitterbauer *et al* [2] for bulk CrN and the best match with the current Wien2K calculations was also that of stoichiometric CrN. This is also in good agreement with the relative Cr/N ratio of 1.05±0.1 obtained from the quantitative EELS data. For CrAlYN (Fig. 3b) the best match to the simulated edges was $(Cr_{0.5}Al_{0.5})N$, that is, where there is twice as much nitrogen as chromium. This is consistent with the relative Cr/N ratio of 0.59±0.02 revealed in the quantitative EELS analysis. The results are also consistent with Ding and Zeng [9] who reported that for CrAlN deposited by unbalanced magnetron sputtering the nitrogen content of the coatings was around 1:1 with respect to the metal content, and that incorporation of aluminium proved beneficial towards the formation of a near stoichiometric composition within the coating.

5. Conclusions

The compositional distribution within a CrAlYN/CrN multilayer with a bi-layer thickness of ~3.8±0.2nm has been investigated by electron energy loss spectroscopy in a spherical aberration corrected STEM. Experimental ELNES at the N K-edge was obtained from each of the multilayer components. Matching theoretical and experimental near edge fine edge structure in electron energy loss spectra, combined with quantitative chemical analysis, indicated that the layers comprise CrN in a near stoichiometric proportion (Cr/N ratio of 1.05±0.01), and CrAlYN, with a composition close to $Cr_{0.5}Al_{0.5}N$ (Cr/N ratio of 0.59±0.02). The results show that overall the qualitative agreement between the experimental and theoretical ELNES results is good.

References

[1] Hovsepian P Eh, Reinhard C, Ehrasarian A P 2006 *Surf. Coat. Technol.* **201** 4105
[2] Mitterbauer C, Hébert C, Kothleitner G, Hofer F, Schattschneider P and Zandbergen H W 2004 *Solid State Comm.* **130** 209
[3] Schwarz K, Blaha P, and Madsen G K H, 2002 *Comp. Phys. Commun.* **147** 71
[4] Perdew J P,Burke K, and Ernzerhof M 1996 *Phys. Rev. Lett.* **77** 3865
[5] C Hébert, J Luitz, and P Schattschneider 2003 *Micron* **34** 219
[6] Scott A J, Brydson R, MacKenzie M, Craven A 2001 *J, Phys. Rev. B.* **67** 245105
[7] Moreau P, Boucher F, Goglio G, Foy D, Mauchamp V and Ouvrard G 2006 Phys. Rev. B. **73** 195111
[8] Reinhard C, Ehiasarian A P and Hovsepian P Eh 2007 *Plasma. Process. Polym.* **4** S910
[9] Ding X Z and Zeng X T , 2005 *Surf. Coat. Technol.* **200** 1372

Acknowledgements

The Authors gratefully acknowledge Dr Andrew Bleloch and Dr Mhairi Gass (UK SuperSTEM), Prof. Rik Brydson (University of Leeds) for helpful advice regarding EELS modeling and the European Commission (INNOVATIAL Project: NMP3-CT-2005-515844).

Journal of Physics: Conference Series **241** (2010) 012047 doi:10.1088/1742-6596/241/1/012047

Quantitative high-resolution microscopy of size-controlled metallic nanodots on titania photocatalysts

K. Yoshida[1,5], S. Tanemura[1], T. Torimoto[2], N. Tanaka[2], E.D. Boyes[3,4] and P.L. Gai[3,5]
[1] Nanostructures Research Laboratory, Japan Fine Ceramics Center, 2-4-1 Mutsuno, Atsuta-ku, Nagoya, 456-8587, Japan
[2] Department of Crystalline Materials Science and Ecotopia Science Institute, Nagoya University, Furo-cho,Chikusa-ku, Nagoya, 464-8603, Japan
[3] Departments of Physics and [4]Electronics, The York JEOL Nanocentre, The University of York, Heslington, York TO10 5BR,UK
[5] Departments of Chemistry and [3]Physics, The York JEOL Nanocentre, The University of York, Heslington, York TO10 5BR,UK

E-mail: pgb500@york.ac.uk

Abstract. We applied the cryo-HAADF-STEM tomography (cryogenic Tecnai G2 Plara) for analysing three-dimensional nano-structures of platinum (Pt) and gold (Au) nanodots synthesised on titanium dioxide. The structural parameters such as particle size, density and volume were successfully evaluated and compared between macroscopic properties (methylene blue dissolution and CO oxidation). The atomic structures of metallic clusters on the TiO_2 surface were also carefully investigated by using double aberration corrected TEM/STEM (double AC JEOL JEM-2200FS) in order to study the quantum size effect of these noble metals.

1. Introduction

Oxides supporting precious metals act as oxidation catalysts. Platinum supported on titania (Pt/TiO_2) is an especially high-active oxidation photocatalyst. Photocatalytic processes such as aqueous photo-degradation, antibacterial activity and environmental cleanup are already applied to a lot of products familiar with the public. Since the early 19th century, many researchers have tried to develop methods to achieve high photocatalytic activity by controlling nano-structures of platinum-supported titania, e.g., producing fine particles or porous configurations with large surface areas and the size of Pt clusters on the TiO_2 surface as effective reaction sites [1]. However, analysing nm-sized Pt dots on TiO_2 particles of several hundred nm's diameter has been difficult by spectroscopic methods including X-ray diffraction.

Transmission electron microscopy (TEM) is one of the most effective methods to analyse such supported nanodot structures, though information obtained from images is limited to two-dimensions (2D). In 1992, "electron tomography" was proposed in order to obtain three-dimensional (3D) structural information from 2D TEM images [2]. In electron tomography, 3D structures of samples are reconstructed by many TEM images taken from various directions, in the same manner as the computed tomography (CT) by X-rays. In recent years, electron tomography using high angle annular dark field scanning TEM (HAADF-STEM) images is shown to be a more powerful method to observe 3D structures of crystalline materials than using TEM images [3-6].

In the present study, cryogenic high-angle annular dark field scanning transmission electron microscopy (Cryo-HAADF STEM) tomography has been used to obtain 3D structural information of metallic nanodots on TiO_2 particle surfaces.

2. Experimental Procedure

The platinum (Pt) and gold (Au) nanodots were prepared on titanium dioxide by the photo-electrocrystallization in the two-phase liquid solution as water and hydrophobic ionic liquid. Titanium oxide powder (Ishihara Sanngyou Co.), hydrophobic ionic liquid N,N,N-Trimethyl-N-propylammomium bis(trifluoromethanesulfonyl)imide (TMPA TFSI, Kanto Reagent) and tetrachloroauric acid were aged after adding water. Subsequently, the mixture was irradiated with UV light (10W × 2, λ = 310 ~ 400nm), while stirring it at 300K for 5 minutes. The mixture was then filtered and the recovered solid was thoroughly washed with ethanol and acetone, and dried in an oven with UV irradiation for 12 hours.

The STEM images were recorded with tilt angles ranging from -76 to +74 degrees with 1 degree increments in cryogenic HAADF-STEM (FEI Tecnai G2 Plara). A HAADF detector of 100 mrad as the inner angle was used for obtaining Z-contrast images. By using a Howie detector, which excludes Bragg scattering, one can use actively incoherent thermal diffuse scattering (TDS) electrons and obtain Z-contrast STEM tilt series [7]. The IMOD suite of programs [8,9] was used for 3D re-construction of the HAADF-STEM tilt series. After adjusting experimental data, including image sizes, orientation and image drift, 3D reconstruction was performed. By calculating three-dimensional coordinates and intensities of the images, the pore size, volume fraction of pores and specific surface area of the porous structures could be analysed in real space using IMOD and in-house programs. In addition, the atomic structure of Pt clusters on the TiO_2 surface and the interface structure were carefully investigated by using a double aberration corrected TEM/STEM (double AC JEOL JEM-2200FS). The 2AC-JEOL 2200FS (S)TEM, configured to have a larger gap (HRP) objective lens polepiece [10] to enable tilting of the sample for alignment into a zone axis orientation, was used for nanostructural studies.

3. Results and discussion

Figure 1 shows a tilt series of the Pt/TiO_2. Figure 1(a)-(g) are HAADF-STEM images corresponding to the tilted sample with tilt angles of -70, -35, 0, 35 and 75 degrees from the position of zero degree. Figure 2(a)-(g) are HAADF-STEM images of Au/TiO_2 with tilt angles of -60, -30, 0, 30 and 70 degrees from the position of zero degree.

Figure 3 shows reconstructed 3D structures of a Pt/TiO_2 and Au/TiO_2 prepared in water solution. We can clearly observe fine Pt nanodots about 1-3 nm diameter located on the TiO_2 surface without any visible artefacts, which can occur from a missing wedge or digital quantum noise. After 3D reconstruction based on the back-projection method, we have quantitatively analysed Pt nanodot structures, including size, volume and number of nanodots for each unit area on the titania surface by using the IMOD and in-house measurement programs. The 3D morphologies of nanodots are shown in Table 1. Average particle size of Pt nanodots was found to be less than 3nm. The Pt nanodots are small compared to the TiO_2 support of 150nm in diameter. It is usually not possible to see nanodots by HRTEM because of strong Fesnel's interference from the thick supporting material.

Figure 4 is a spherical aberration corrected (Cs-corrected) TEM image of the Pt/TiO_2 sample. Pt lattice spacing of (111) and (200) planes were observed at white arrowed areas. Crystal size observed was in good agreement with that quantified by HAADF-STEM tomography. The high-resolution observation have elucidated the single crystallite of Pt nanoparticle (atomic number; Z = 78), which exhibit stronger contrast than the anatase substrate (atomic number, gray atom approximation; Z = 12.7).

4. Conclusion

In the present study, electron tomography was applied to the 3D analysis of metallic nanodots on titania photocatalysts. The shapes and 3D distribution of nanodots (Pt particles of 2.89 nm and Au particles of 18.4 nm in diameter) on the modified TiO_2 surface were successfully reconstructed. The volume fraction and the density of the nanodots were also accurately measured in real space. The results and the data of Methylene Blue degradation indicate that smaller dots, produced via a solution containing ionic liquids have considerably greater number of effective reaction sites than nanodots produced by the conventional electrocrystallization method. Our results have shown that 3D analysis of nanostrucutures using electron tomography is an effective method of studying the influence of the synthesis process on the photocatalytic activity of metal nanodot/TiO_2 photocatalysts.

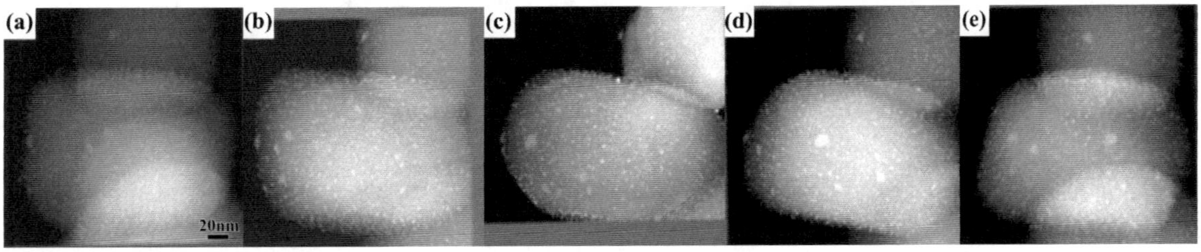

Figure 1. Tilt series of the Pt/TiO_2.

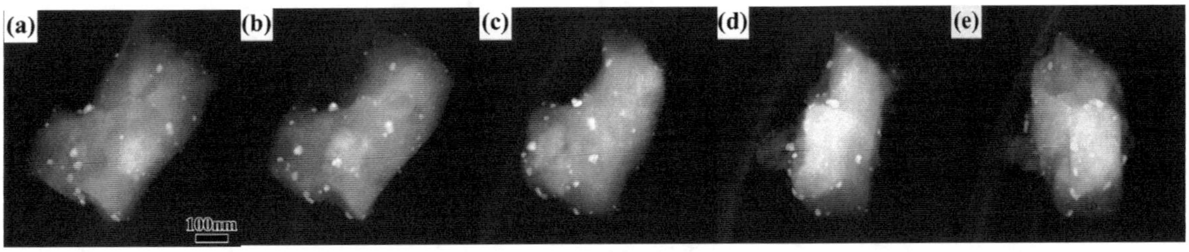

Figure 2. Tilt series of the Au/TiO_2.

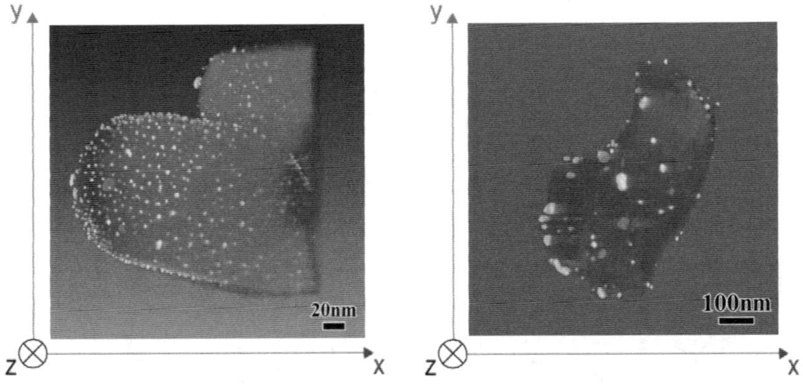

Figure 3. Reconstructed 3D structure of the Pt/TiO_2(left) and Au/TiO_2 (right). The black, red and blue arrows show the incident axis of the electron beam (Z axis), the rotation axis of tomography (Y axis) and another X axis, respectively.

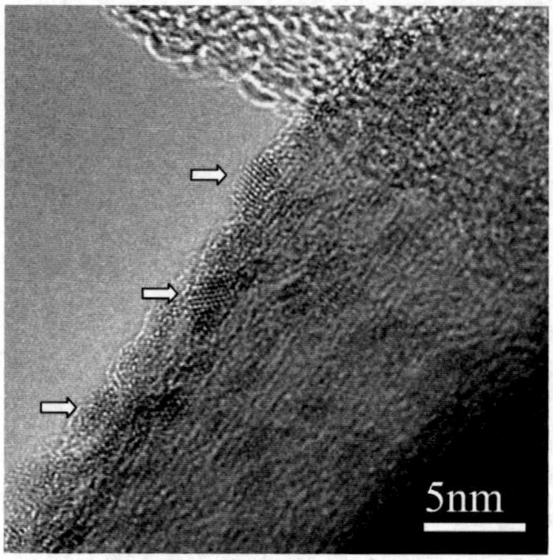

Figure 4. TEM image of the Pt/TiO_2 using double Cs corrected thermal field-emission HRTEM (JEOL: JEM-2200FS).

Table 1. Structural parameters of metal nanodots synthesized in water and water/ IL.

Solvent	Supported Metal	Particle size	Density	Volume
Water	Au	18.4nm	$1.80 \times 10^{16}/m^2$	$9.64 \times 10^{-3} g/m^2$
	Pt	2.89nm	$2.75 \times 10^{16}/m^2$	$17.7 \times 10^{-3} g/m^2$
Water + IL	Au	3.70nm	$3.10 \times 10^{16}/m^2$	$3.52 \times 10^{-3} g/m^2$
	Pt	2.20nm	$4.50 \times 10^{16}/m^2$	$2.68 \times 10^{-3} g/m^2$

5. Acknowledgments

We thank the University of York, JEOL and Yorkshire Forward / European Regional Development fund (ERDF) for sponsoring the Nanocentre. KY thanks the JSPS/JFCC for a Visiting Fellowship.

6. References

[1] M. Haruta 1997 *Catalysis Today* **36** 15.
[2] J. Frank, Electron Tomography :Three Dimensional Imaging with the Transmission Electron Microscope, Plenum, New York, 1992.
[3] P. A. Midglay et al. 2003 *Ultramicroscopy* **96** 413.
[4] H. Friedrich, M. R. McCartney and P. R. Buseck 2005 *Ultramicroscopy* **106** 18.
[5] L. Gonzalez-Rovira, J. M. Sanchez-Amaya, M. Lopez-HaroM, E. del Rio, A. B. Hungr, P. A. Midgley, J. J. Calvino, S. N. Bernal and F. J. Botana 2009 *Nano Lett.* **9** 1395.
[6] X. Xu, Z. Saghi, R. Gay and G. Mobus 2007 *Nanotech.* **18** 225501
[7] A. Howie 1979 *J. Microsc.* **117** 11.
[8] J. R. Kremer, D.N. Mastronarde and J.R. McIntosh 1996 *J. Struct. Biol.* **116** 71.
[9] D. N. Mastronarde 1997 *J. Struct. Biol.* **120** 343.
[10] P.L. Gai and E.D. Boyes 2009 *Microsc. Research. Tech.* **72** 153.

Electron microscopy of AlGaN-based multilayers for UV laser devices

L. Lari, H. Amari, T. Walther, J. Bai, T. Wang and A. G. Cullis

Dept. Electronic & Electrical Engineering, University of Sheffield, Sheffield, UK

E-mail: l.lari@sheffield.ac.uk

Abstract. UV-Laser technology is one of the most challenging and important areas in nitride semiconductor research. A new growth technology, using a thin GaN interlayer approach, has recently been developed for UV-LEDs in our department. Improved optical quality of so grown LEDs has been demonstrated. A similar growth approach has now been used to grow an UV-laser structure that has shown stimulated emission (lasing) at ~340 nm via optical pumping. We report here a study of the microstructure, the epitaxial quality and the elemental distributions of such an aluminium gallium nitride-based heterostructure using a combination of various electron microscopy techniques.

1. Introduction

Ultraviolet (UV)-Laser technology is a very important area in nitride semiconductor research. GaN/AlGaN-based heterostructures are studied for various optoelectronic applications from biological agent detection to data storage and laser printing. The information density for these latter applications is defined by the laser spot area. This is limited by diffraction to the square of the emission wavelength [1]; hence the efforts towards producing shorter wavelength laser diodes (LDs) going to the deep-UV. Recently a laser device emitting at 342 nm has been demonstrated [2]. Conditions to achieve good emission properties are a good control of the elemental distribution during growth and a reduced density of crystal defects. We report here the structural and elemental analysis performed on a GaN-AlGaN multilayer, following a preliminary metal-organic chemical vapour deposition (MOCVD) growth project aiming to produce an UV-laser device emitting below 340 nm.

2. Sample description

Vertical low pressure MOCVD was used to grow the sample on α-alumina (sapphire). The initial thermal cleaning of the substrate under a molecular flux of hydrogen was followed by the deposition of a 0.5 μm thick AlN buffer layer at high temperature under group III-rich condition. A thin GaN interlayer (~20nm) was grown to reduce the dislocation density, followed by an $Al_{0.75}Ga_{0.25}N$/GaN superlattice used to prevent cracking. An undoped $Al_{0.29}Ga_{0.71}N$ layer (~3.5μm) was then deposited followed by a silicon doped n-AlGaN cladding layer (~1μm), keeping the aluminium concentration fixed. Afterwards an n-$Al_{0.15}Ga_{0.85}N$ guiding layer (100nm) was grown. The active region consisted of a multiple quantum well (MQW) structure with 3 $Al_{0.05}Ga_{0.95}N$ thin layers separated by $Al_{0.15}Ga_{0.85}N$ layers. On top of these, a p-type $Al_{0.35}Ga_{0.65}N$ (15nm) blocking layer (to decrease electron flow), and then a p-type $Al_{0.15}Ga_{0.85}N$ guiding layer (100 nm) and a p-type $Al_{0.29}Ga_{0.71}N$ cladding layer (500 nm) were deposited. Finally, a 40 nm p-type GaN layer was grown to yield a good ohmic contact.

3. Methods

A TEM cross-section sample was prepared gluing face to face two pieces of material. Grinding with SiC paper and final polishing with diamond powder was applied to the sample before Ar^+ ion milling to electron transparency. The analysis was performed on a JEOL 2010F TEM/STEM microscope operated at 197 keV and equipped with an Oxford Instrument Si:Li Pentafet EDX detector with ultra thin window and Link ISIS 300 acquisition software, annular dark field (ADF) (55 mrad inner angle) and bright field (BF) detectors coupled with a Gatan Digiscan system. A focused ion beam (FIB) AlN sample of (100±20) nm thickness was used for EDX and EELS calibration.

4. Results and Discussion

4.1. ADF imaging

The ADF image in Fig. 1a) shows the full device structure investigated in this work. The arrow points along the growth direction. The first dark band from the bottom represents the 0.5 μm AlN buffer layer. This appears darker because of its lower mean atomic number with respect to Al_2O_3 (substrate) and AlGaN (large middle layer). The sample appears to be crack-free, giving credit to the strain balance function of the $Al_{0.75}Ga_{0.25}N$/GaN superlattice. The various layers within the sample are planar and uniform except for the GaN interlayer that grows in form of islands as observed in [3]. The $Al_{0.75}Ga_{0.25}N$/GaN superlattice, as shown in Fig. 1b), covers the GaN islands. The contrast of the large cladding layer is quite uniform, except for some lines parallel to the growth direction ((0001) of wurtzite structure) that are due to misfit dislocations (MDs). Conversely, as shown in Fig. 1c), the guiding layer preceding the MQW was found to have a fluctuating Al concentration, resulting in a series of brighter and darker bands perpendicular to the growth direction that are visible from Z-contrast in ADF imaging. This fluctuation is also present in the p-type $Al_{0.15}Ga_{0.85}N$ guiding layer deposited above the blocking layer. This variation is strongly correlated to the local Al concentration as discussed in section 4.3.

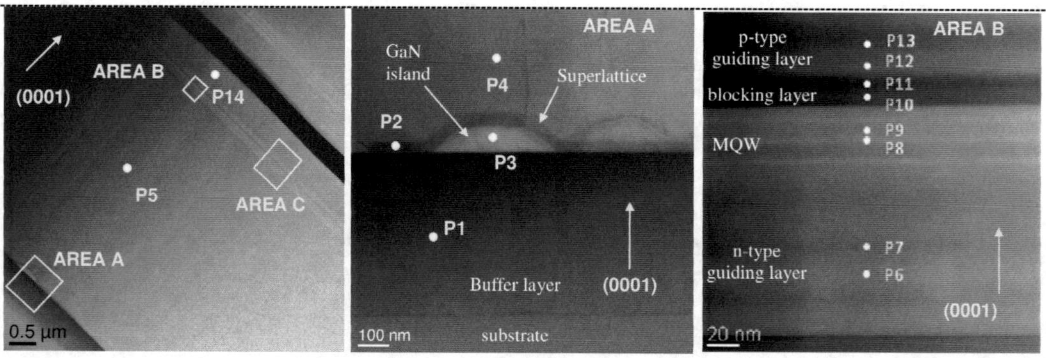

a) b) c)

Figure 1. (a) ADF image including the device structure in overview. (b) ADF image of the Area A including the AlN/sapphire interface and the superlattice area. c) ADF image of the Area B with the MQW area. Inset are the positions of the EDX point analyses reported in Table 1.

4.2. Thickness measurement by EELS

The sample thickness t was estimated using Electron Energy Loss Spectroscopy (EELS). Low loss energy spectra were acquired over the energy window from -56eV to +456eV close to the position P5 in Fig. 1a) using a convergence angle of α=(10.0±0.5)mrad and a collection angle of β=(2.1±0.2)mrad. The corresponding thickness was obtained via the Kramers-Kronig (K-K) sum rule [4] using the EELS routines of GATAN Digital Micrograph. This method was tested with an AlN FIB sample

(100±20)nm thick which gave a thickness of (150±5)nm and an elastic mean free path of λ=106nm. The results for the laser structure are reported in the Table 1 together with the position where the spectra were acquired. Starting from a thickness of 403nm near the AlN substrate, the thickness decreases towards the active layer to about 300nm but then increases at the blocking layer to 560nm. This thickness variation could have been caused by a preferential thinning rate during the ion milling process associated with an incidence angle of ~6 degrees with respect to the sample plane.

4.3. EDX analysis

A preliminary analysis of the AlN FIB reference sample was undertaken to test the ISIS 300 X-ray quantification routine [5] on a spectrum acquired under the same acquisition conditions as for the LD sample. By varying the nominal input thickness it was found that the expected ratio of 1:1 for Al:N was obtained only at t=0 while for the true sample thickness t=150nm the apparent N concentration reported was 55at%. This indicates both, a relative error in the k-factor above 10% and a faulty calculation of the absorption. Lacking a reference sample containing Al, Ga and N at known ratios to be used for a calibration of the k-factors and the absorption coefficients, we have used for our analysis two approaches. A first quantification was obtained by using the ISIS software under thin film approximation (input t=0nm) and a second quantification by using an internal calibration of the k-factor assuming the nominal composition in the middle of the cladding layer (point P5 in Fig. 1a). This position is far enough from the substrate to limit the Al contribution from it. The k-factor was obtained using the Cliff-Lorimer (C-L) equation $C_{Al}/C_{Ga} = k_{Al/Ga} \cdot I_{Al}/I_{Ga}$, where C is the elemental concentration in atomic percent (at%), and I the area under the full-width-at-half-maximum of the peak of interest (background subtracted with the software ISIS). Only Al-K and Ga-L peaks were considered in order to have similar absorption for the changes in thickness across the sample. For the spectrum in position P5 we found a value of $k_{Al/Ga}$=1.56±0.04.

Table 1. Result of EDX point analyses of Al fraction in $Al_xGa_{1-x}N$ and EELS thickness measurement with K-K sum rule. Positions P1-P14 refers to Fig. 1.

Position	Nominal Al fraction x $Al_xGa_{1-x}N$	Al fraction x (ISIS software for t=0 nm)	Al fraction x (ref. C-L for t=272nm)	t (nm) (EELS K-K)
P14	0.29	0.319±0.007	0.312±0.004	556
P13 Dark	0.15	0.183±0.005	0.179±0.008	
P12 Bright	0.15	0.156±0.004	0.152±0.008	
P11 Dark	0.35	0.435±0.005	0.427±0.042	560
P10 Bright	0.35	0.415±0.005	0.407±0.025	
P9 Dark	0.15	0.174±0.005	0.170±0.008	
P8 Bright	0.05	0.147±0.004	0.143±0.007	
P7	0.15	0.181±0.005	0.177±0.008	
P6	0.15	0.169±0.004	0.165±0.008	303
P5	0.29	0.297±0.004	Ref.	272
P4	0.29	0.303±0.005	0.296±0.004	403
P3	0.00	0.118±0.004	0.115±0.007	
P2	0.75/0.0	0.650±0.005	0.64±0.11	
P1	1.00	0.998±0.005	1.0±0.6	

A series of EDX point analyses was performed along the whole length of the sample to asses the Al concentration in each layer at the positions indicated in Fig. 1 (see Table 1). The AlN buffer (P1) contains no Ga, as expected. The superlattice (P2) gives a value of x=0.65, a value intermediate between its $Al_{0.75}Ga_{0.25}N$ and GaN components. The interlayer islands in Fig. 1b) consist mainly of GaN; the 10at% Al fraction (P3) is probably a contribution from the surrounding superlattice partly covering them along the beam direction. The Al content is close to the nominal values in the various layers. Larger values than nominal are found for the Al-rich blocking layer and the quantum wells. The beam broadening, estimated using the ISIS software for a sample thickness of 500nm, is ~90 nm. This is much larger than the QW thickness (~4nm) and indicates there is a significant contribution from the neighbouring layers. The concentration fluctuations visible in the active area of the LD shown in Fig. 1c) are evident from the EDX data (P6-P7, P10-P11, P12-P13) where a darker area in ADF corresponds to a higher Al fraction in the material. More accurate quantification could be obtained from a thinner sample in order to reduce the beam broadening; also more thickness uniformity would improve the method of internal calibration.

4.4. Two-beam analysis

Dark film images obtained in two beam conditions with g=(11-22) are shown in Fig. 2a,b). Along this orientation all dislocation components are visible. A high density of MDs is generated, as expected, at the interface between the sapphire substrate and the AlN buffer. The GaN interlayer layer was found to filter the number of dislocations propagating to the undoped $Al_{0.29}Ga_{0.71}N$ layer, as in [3]. Some of the MDs form a dense network, as indicated by arrows in Fig. 2a), others propagate towards the active region of the device passing through the MQW, blocking layer and reaching the top GaN layer (see Fig. 2b). In addition, dislocations are also formed directly within the guiding layer and MQW active region, as shown in the ADF image of Fig. 2c). The overall good material quality of this heterostructure was confirmed by the stimulated emission (not reported here) at ~340 nm wavelength, obtained via optical pumping based on a control sample.

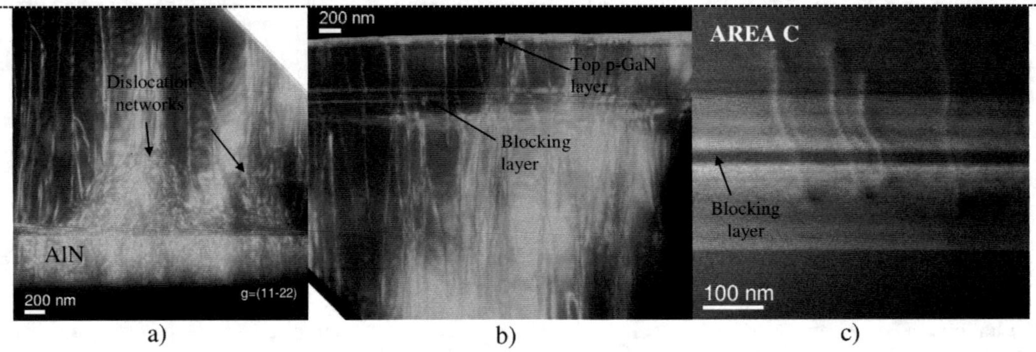

Figure 2. TEM DF images obtained selecting two beam conditions with exciting vector g=(11-22), near the buffer layer (a) and the LD active region (b). c) ADF image showing dislocations generated in the MQW, from the area C in Fig. 1a)

5. Conclusions

A GaN-AlGaN laser diode structure showing stimulated emission at ~340 nm via optical pumping on a control sample was investigated using electron microscopy imaging and analytical techniques. Elemental concentrations in the multilayers were found in agreement with the nominal device structure within the limit of the techniques used for the analysis. An additional unwanted concentration modulation was revealed by Z-contrast imaging and EDX analysis. Misfit dislocations, reduced in density by the GaN interlayer, were found to propagate through the cladding layer up to the top layer together with dislocations generated in the active region.

Acknowledgements

The authors would like to thank Dr. I. M. Ross for fruitful discussions. Financial support by EPSRC grants EP/F02374X/1 and EP/F03363X/1 is gratefully acknowledged.

References

[1] J-Y Duboz and J-C Garcia in *Low-Dimensional Nitride Semiconductors*, edited by B. Gil, Oxford University press, (2002, Oxford) chapter 1
[2] H. Yoshida *et al.* Nature Photonics **2**, (2008) 551
[3] J. Bai *et al.* Applied Physics Letters **91**, (2007) 131903
[4] R. F. Egerton: *Electron Energy-Loss Spectroscopy in the Electron Microscope*, Plenum Press (1996, New York) p. 307
[5] EDX LINK ISIS 300 software Rev. 3.2 manufactured by Oxford Instruments pcl

A HAADF Investigation of AlAs GaAs interfaces using SuperSTEM

M Finnie[1,3] A J Craven[1,2] and P D Robb[1,2]
[1]Department of Physics & Astronomy, University of Glasgow, G12 8QQ, UK
[2]SuperSTEM Laboratory, Daresbury Laboratory, WA4 4AD, UK

E-mail: m.finnie@physics.gla.ac.uk

Abstract. Z-contrast imaging using a high-angle annular dark field detector can be used to characterise III-V heterostructures. GaAs/AlAs heterostructures were grown using MBE and prepared for TEM using a cross-sectional method. SuperSTEM 1 was used to investigate both the GaAs-on-AlAs and the AlAs-on-GaAs interfaces as a function of specimen thickness. The analysis of the images showed that the apparent interface widths varied with thickness in an unexpected manner. The GaAs-on-AlAs widths remained constant with thickness while the AlAs-on-GaAs interface widths increased. The actual interfacial width can be a result of either surface stepping during MBE growth or inter-diffusion of the column III atoms. To assist the interpretation of this result a series of interfacial models were created and explored using a frozen phonon multislice simulation. The models consisted of terraced, vicinal and diffused interfaces. The results show that while the width of the diffused interface reduces with specimen thickness, the widths of the stepped interfaces vary in a more complex manner. This variability is due in part to the different channeling depths of the Ga and Al columns, but the measured width is also affected by the depth of the material before the first step and whether GaAs or AlAs is on top. These effects can cause the apparent interfacial width to vary markedly from the actual width. Even measuring the width itself is non-trivial.

1. Introduction

The ability to control the growth of layers during molecular beam epitaxy (MBE) is vital for the development of sophisticated semiconductor devices. MBE techniques can currently produce very thin and abrupt layers with good control over parameters such as thickness, doping and composition [1]. With the advent of aberration correction it is now possible to study these interfaces at the atomic level using HAADF STEM imaging. HAADF imaging has been used at SuperSTEM to investigate the widths of [110] orientated GaAs/AlAs interfaces [2]. This paper reports multislice calculations used to aid interpretation of the experimental data.

2. Experimental Results

The experimental interfaces investigated were grown on a [001] orientated GaAs substrate by MBE. The samples consisted of a superlattice with repeats of 9 mono-layers of GaAs and AlAs. The samples were prepared using a standard cross-section method orientated in the [110] direction. The preparation method consisted of mechanical dimpling, ion milling and a final low kV polish. The experimental STEM images[4] were recorded using SuperSTEM 1with a convergence angle of 24mrad at 100kV giving a probe FWHM of ~0.1nm. The HAADF collection angles were 70-210mrad.

In the case of [110] orientated GaAs and AlAs the atomic columns align to form a series of dumbbells, see figure (1a), each dumbbell containing one As column and a type-3 column. For GaAs this type-3 column

[3] To whom any correspondence should be addressed.
[4] SuperSTEM images courtesy of P D Robb.

would consist solely of Ga atoms while for AlAs it would consist of only Al. If the column is at the interface and has stepping or diffusion the column would contain a combination of both these atom types.

The HAADF detector collects the incoherent high angle scattered electrons. This scattering is Rutherford like and in the simplest approximation the scattering depends on the square of the atomic number, Z [3]. The composition of a particular dumbbell can therefore be estimated from the column ratio. This is defined as the ratio of scattering from the type-3 column to that from the As column after background subtraction. Thus, in principle, HAADF images can be used to give an indication of the interfacial atomic structure. However the scattering of the STEM probe (and corresponding HAADF signal) depends greatly on the arrangement of atoms within the dumbbells. The electron density on an atomic column varies with depth, therefore the resulting HAADF signal will be sensitive to the depths of any compositional change. Because of this, it is necessary to compare experimental images to image simulations.

2.1. Column Ratio Profiling

The development of a consistent method for extracting column ratio data from HAADF images was developed [4]. This method was an attempt to standardise and automate the procedure of background removal, column ratio calculation and the subsequent interface width measurements. Figure (1a) shows the original HAADF image. Figure (1b) is a plot of the variation of the HAADF signal across a GaAs/AlAs interface from an averaged line profile from the region indicated in figure (1a). The solid trace is the HAADF signal, the dotted trace is the background and the dashed trace corresponds to the background subtracted dumbbell signal. The positions of the on column and background locations used in subsequent calculations are shown in figure (1c).

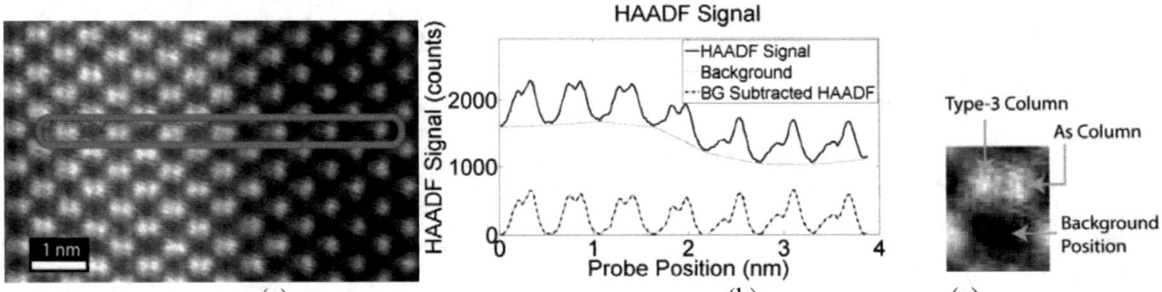

(a) (b) (c)

Figure 1. a) A HAADF image of GaAs/AlAs interface; b) line profile (solid), background (dotted) and background subtracted dumbbells (dashed).; c) definition of positions used in calculations.

A background subtracted profile can be calculated for each row in the image, the dumbbell offset of each alternate row produces a set of interleaving line traces. Each alternate line trace is averaged together to form a complete column ratio profile. One such profile can be seen in figure (2a).

2.2. Width Measurement

The interface width is measured by least squares fitting an error function to the column ratio data, figure (2a). The width of the interface can then be defined as the distance between the 95% and 5% points on this curve. These values are bounded by the horizontal broken lines in figure (2a). This method can result in a width measurements of fractional values of a dumbbell, which can be interpreted as the averaged width in the [110] projection.

Figure (2b) is a plot of the experimental interfacial widths as a function of specimen thickness measured using this method [2]. 'GaAs on AlAs' corresponds to the interface where the GaAs layer was grown on an AlAs surface during MBE. Likewise 'AlAs on GaAs' corresponds to the interface where AlAs was grown on a GaAs surface. Figure (2b) indicates a clear asymmetry between these types of interfaces. The GaAs on AlAs interface widths are scattered around a constant width and appear to be independent of specimen thickness. However, the AlAs on GaAs interface widths increase with increasing specimen thickness.

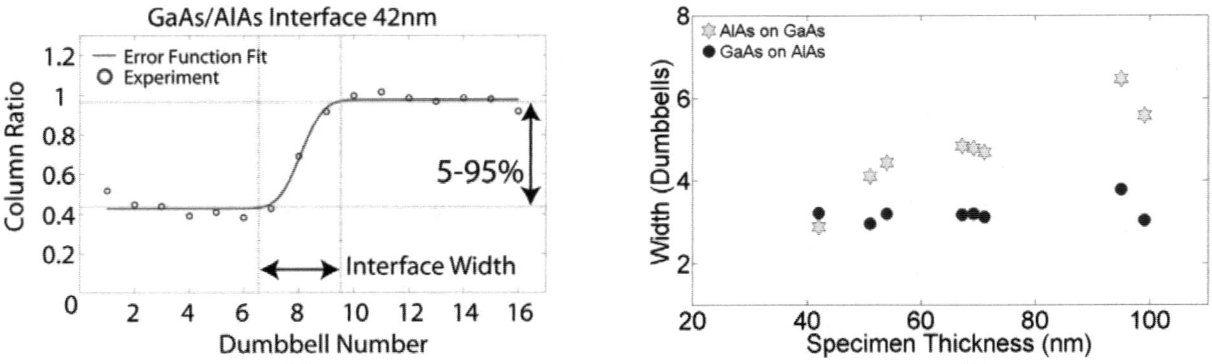

(a) (b)

Figure 2. a) Averaged column ratio profile with a fitted error function. b) plot of 5-95% widths as a function of thickness.

To develop a greater understanding of these results a series of interfacial models were developed that could describe this behavior. A simple diffusion model was formed where the interface width was due to intermixing of the type-3 atoms. A repeating terraced structure was also considered where the interface structure resembles a saw tooth with a characteristic step length. Another hypothesis was that the initial substrate was vicinal before commencement of MBE growth (the substrate polishing tolerance being ±0.5°). When the TEM sample is prepared and orientated this would cause the interfaces to also be vicinal and therefore the STEM probe positioned at the interface would project through one material before the other. These proposed interfaces were explored using a multislice code starting with an idealised perfect interface that can also be used as a control and a comparison for more complex structures.

3. Model Configuration

The interface model was constructed using a supercell with 14 x 12 x 300 unit cells (5.60 x 6.78 x 119.93 nm). The SuperSTEM probe was modeled including aberrations up to fifth order and a convergence angle of 24mrad giving a FWHM in agreement with the experimental probe. The simulations were calculated to thicknesses of 120nm and the effects of thermal diffuse scattering were included using the frozen phonon method, incoherently summing the results from 60 atomic configurations.

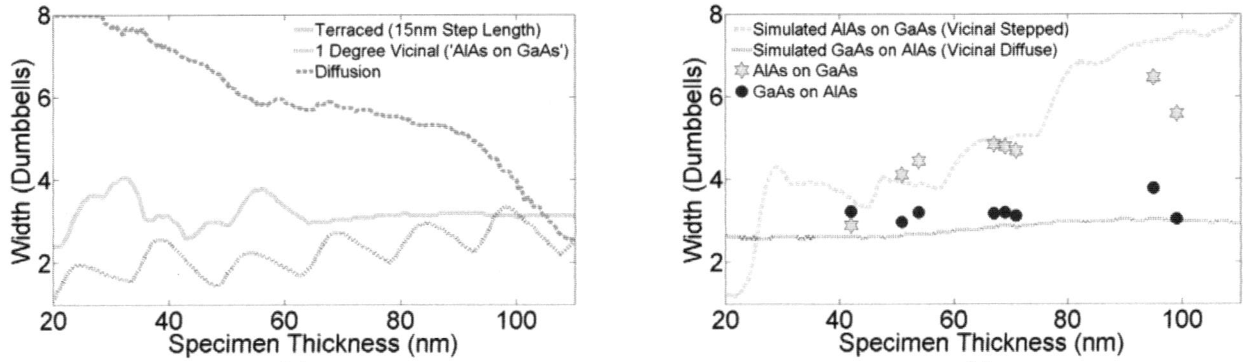

(a) (b)

Figure 3. a) Plot of the interface width versus thickness for three models; b) Comparison of the experimental data to the best fit model.

4. Model Results and Discussion

Figure (3a) shows the results for three models of the 'AlAs on GaAs' interface. The diffusion model generally showed a narrowing as a function of thickness. The degree of narrowing was dependent on the rate of concentration change across the interface, with a gradual change in concentration giving the highest degree of narrowing. For the terraced interface the measured interfacial width was very sensitive to changes in specimen thickness below 60nm. For thicknesses above 60nm the sensitivity to further changes in column composition was greatly reduced. Neither model reproduced the experimental data. Thus it became apparent that a degree of vicinality was required to describe the experimental data.

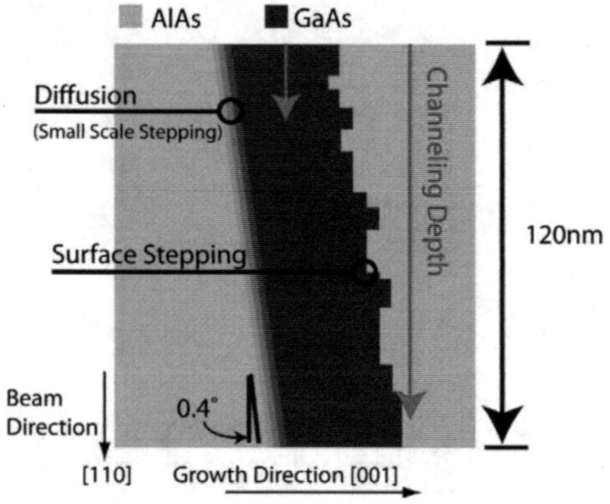

Figure 4. A model of the interfaces which gives good agreement with the experiment. It involves a 0.4° vicinal interface giving GaAs at the entrance for the GaAs on AlAs layer and AlAs at the entrance for the AlAs on GaAs interface. In addition diffusion and stepping are required respectively.

The width of a vicinal interface inclined at ϑ to the normal increases linearly with specimen thickness. However the apparent width resulting from the above procedures is normally less than this and typically has a component that oscillates with thickness, see figure (3a). It also depends whether the probe is incident on GaAs or AlAs with the former case having no overall increase with thickness and the latter case increasing with thickness. This is the result of the channelling depth on the Ga column being ~20nm whereas that on the Al is >120nm. The results are also very sensitive to the exact thickness of the layers involved.

While this is encouraging, it is not, in itself, enough to get good agreement. Figure (4) is the model which has given the best agreement to date. Here the value of ϑ is 0.4 °.

However, at the GaAs on AlAs interface on the left, it is necessary to have some inter-diffusion (or terracing with a short step length) over a width of 2 dumbbells. At the AlAs on GaAs interface, it is necessary to have terracing with a random step length in the range 5–10nm [5]. The predicted profiles for this model are shown in figure (3b) along with the experimental data. The agreement is good but other models may prove to give equally good agreement. Thus the interpretation may not be unique. On the other hand, it is consistent with earlier work in which the GaAs on AlAs interface was described having microscopic roughness, while the AlAs on GaAs interface was described as atomically abrupt but stepped [5].

In conclusion, HAADF images of interfaces in which a large range of *Z* is present need careful interpretation backed by appropriate modelling.

Acknowledgments
The authors would like to thank Martin Holland for growing the MBE materials, Brian Miller for specimen preparation and the superSTEM team. This work was funded by the Engineering and Physical Sciences Research Council research grant (GR/S41036) and DTA Studentship (M.F.).

References

[1] Cheng K Y 1997 *Proceedings of the IEEE*, **85**, 11.
[2] Robb P D 2007 *PhD Thesis* University of Glasgow.
[3] Pennycook S J and Jesson D E 1991 *Ultramicroscopy* **37**, 14.
[4] Robb P D and Craven A J 2008 Ultramicroscopy, **109**, 1.
[5] Ikarashi N, Ishida K 1996 *Journal of Materials Science: Materials in Electronics* **7**, 285-295.

Journal of Physics: Conference Series **241** (2010) 012050 doi:10.1088/1742-6596/241/1/012050

Imaging the structure of activated carbon using aberration corrected TEM

Peter J.F. Harris[1], Zheng Liu[2], Kazu Suenaga[2]

[1] Centre for Advanced Microscopy, J.J. Thomson Physical Laboratory, University of Reading, Whiteknights, Reading RG6 6AF, UK.

[2] Nanotube Research Center, National Institute of Advanced Industrial Science and Technology (AIST), Tsukuba 305-8565, Japan.

Corresponding author: p.j.f.harris@reading.ac.uk

Abstract. The precise atomic structure of activated carbon is unknown, despite its commercial importance. Here we show that the structure of a commercial activated carbon can be imaged directly using aberration corrected transmission electron microscopy. Images are presented both of the as-produced carbon and of the carbon following heat- treatment at 2000°C. In the 2000°C carbon clear evidence is found for the presence of pentagonal rings, suggesting that the carbons have a fullerene-related structure.

1. Introduction

Activated carbon is used on a huge scale in gas and water purification, metal extraction, medicine and many other applications, but its atomic structure is unknown. Rosalind Franklin showed in the 1950s that carbons produced by the pyrolysis of organic materials fall into two classes, which she called graphitizing and non-graphitizing [1]. The kind of carbon from which activated carbon is derived is non-graphitizing, meaning that it cannot be transformed into crystalline graphite even at temperatures as high as 3000°C. Franklin put forward a model of non-graphitizing carbon based on small graphitic crystallites joined together by cross-links, but did not explain the nature of these links. A more recent suggestion [2 - 5] is that non-graphitizing carbon has a structure related to that of the fullerenes, which consists of curved fragments containing pentagons and other non-hexagonal rings in addition to hexagons, as shown in Fig. 1. Such a structure would explain the microporosity of the carbon, and many of its other properties. However, obtaining direct experimental support for this hypothesis is difficult. Both X-ray and neutron diffraction have been extensively applied to non-graphitizing carbons, and in some studies the data has been interpreted in terms of a structure containing non-hexagonal rings [6]. But definitive proof that the atoms are bonded in pentagonal or hexagonal rings cannot be obtained using diffraction methods. Until relatively recently, imaging the atomic structure of carbons using transmission electron microscopy would also have been impossible, as the resolution was greater than the C-C bond length (0.142 nm). However, the development of aberration correctors has improved the resolution of TEM to the point where the direct imaging of carbon networks becomes possible. Here we apply this technique to a commercial activated carbon, namely Norit GSX. In addition to the as-received carbon, samples heated in Ar to 2000°C are imaged.

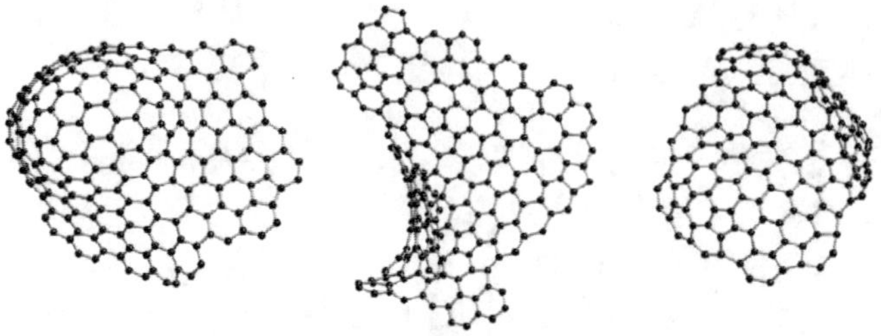

Figure 1: Illustration of curved carbon fragments containing pentagonal and heptagonal rings, in addition to hexagons.

2. Experimental Methods

Specimens were prepared for TEM by dispersing the carbon in iso-propyl alcohol, mixing ultrasonically and depositing onto holey carbon support films. Images showing the overall morphology of the carbons were recorded using the Reading University JEOL 2010 microscope, operated at 200kV. Atomic resolution imaging was carried out using the Tsukuba JEOL 2010F instrument, operated at 120kV. This has a post-specimen aberration corrector (CEOS), giving a point resolution of better than 0.14 nm at 120 kV. The Cs was set to a value in the range from 0.5 to 5 µm. Images were digitally recorded using a Gatan 894 CCD camera under a slightly under-focus condition (Δf = -1 to -6 nm). Image simulations were carried out using a standard multi-slice software package (Mactempas), with structural models prepared using the molecular modelling software DTMM, and appropriate values of Cs and defocus.

3. Results

Micrographs of typical areas of the as-received and 2000°C carbons are shown in Figs. 2 (a) and (b) respectively. The as-received carbon has a disordered and porous microstructure consisting mainly of tightly curled single carbon layers. In the 2000°C carbon, the structure is still disordered, but with larger pores, bounded by more perfect carbon layers.

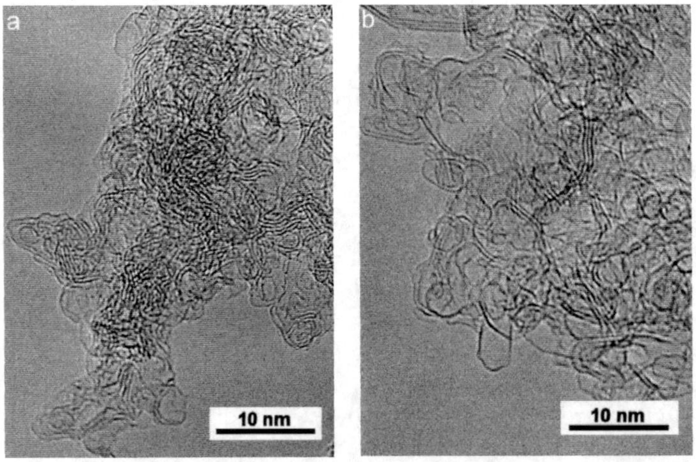

Figure 2: TEM micrographs showing general appearance of (a) fresh activated carbon, (b) activated carbon following heat treatment at 2000°C.

A typical atomic resolution image of the fresh carbon is shown in Fig. 3 (a). Here the bright spots represent the centres of the individual rings of carbon atoms. In most cases these bright spots form hexagonal arrangements, but pentagonal arrangements are also seen. In Fig. 3 (b) the bright spots have been highlighted, with the position of a possible pentagonal ring shown by a black dot. The images we have obtained of the fresh carbon are not yet of sufficient quality to justify comparisons with simulations based on models of the structure. Nevertheless, our results demonstrate that atomic level imaging of this kind of highly disordered carbon is possible.

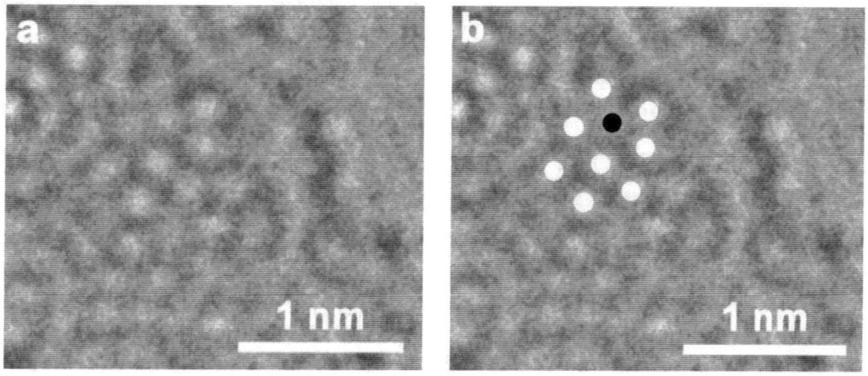

Figure 3: (a) Aberration-corrected HRTEM micrograph of fresh activated carbon, (b) same image, with white dots indicating hexagonal rings and black dot indicating possible pentagon.

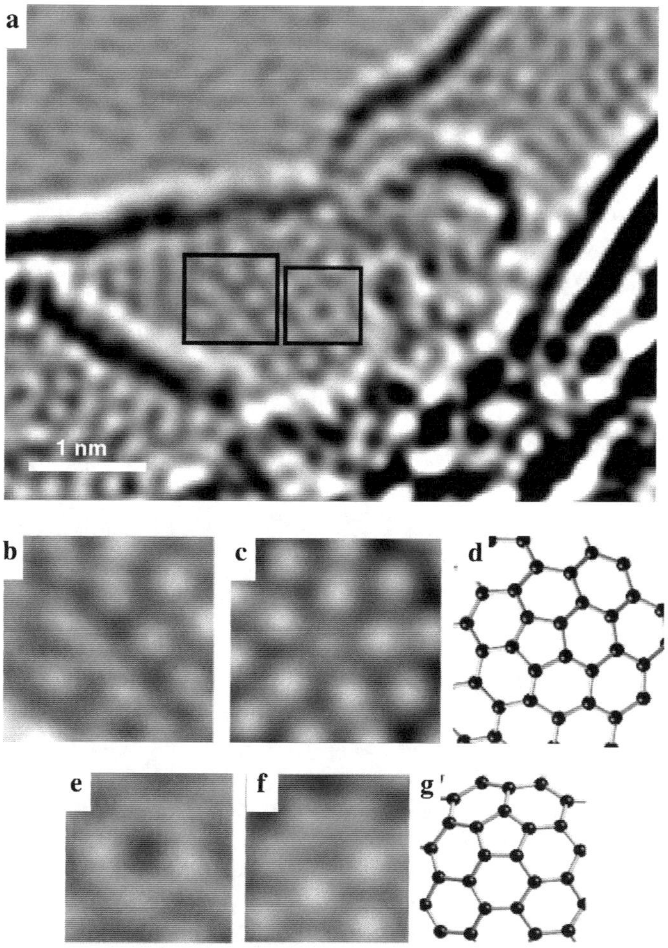

Figure 4: (a) Aberration-corrected micrograph of small area of the 2000°C-heated carbon. (b) Enlarged region showing pentagonal arrangement of spots. (c) Simulated image of structure shown in (d). (e) Second region showing pentagonal arrangement. (f) Simulated image of structure shown in (g).

Recording atomic resolution images of the 2000°C-heated carbon was considerably easier than for the as-prepared carbon, due to the greater degree of crystallinity. The carbon sheets in the heat-treated carbon were frequently just a single layer thick, and in such cases images could be obtained which showed the true atomic structure. In some areas there was clear evidence for the presence of five-membered rings. Figure 4 (a) shows a small region in which at least two such rings are apparently present. The area enlarged in Fig. 4 (b) shows an arrangement of 5 bright spots surrounding a central spot. A good match was obtained with the simulated image in Fig. 4 (c), which was obtained from the structure in Fig. 4 (d). Here, the pentagon is oriented approximately parallel to the plane of the image. A second area which contains a pentagonal structure is shown in Fig. 4 (e). In this case the central pentagonal ring is not visible, and we believe this is because the ring is tilted away from the plane of the image. Support for this comes from the reasonable match which can be seen between the image and the simulated image in Fig. 4 (f), obtained from the structure in Fig. 4 (g). We recognise that the agreement between the images and simulations shown here is not exact, but this is to be expected since we are not dealing with perfect crystalline structures, but with disordered materials. Taking this into account, we believe that the agreement is sufficiently good to provide convincing evidence for the presence of pentagonal carbon rings.

4. Discussion

We have shown here that aberration-corrected TEM can be used to image the atomic structure of a conventional carbon material, specifically the commercial activated carbon Norit GSX. Images of the fresh carbon contained evidence of hexagonal rings, and possibly non-hexagonal ones, but were difficult to interpret due to the highly disordered structure. The 2000°C-heated carbon contained larger networks which were much more readily imaged, and showed clear evidence of pentagonal rings. Our results therefore support the idea, put forward more than ten years ago [2, 3] that this type of carbon has a fullerene-related structure. The idea that microporous carbons have a fullerene-like structure has important implications for the modelling of adsorption on such carbons. Traditionally, such modelling exercises have utilised structural models derived from graphite, in which all the atoms are in hexagonal rings. The carbon pores are then assumed to have a slit-like shape, confined by parallel planes. If the fullerene-like models are correct, these ideas may have to be modified. Indeed, theoretical studies have already been carried out which show that a model structure containing fullerene-related elements provides a better basis for understanding adsorption on activated carbon than the traditional models [e.g. 7, 8]. A fuller description of the present work has been published [9].

Acknowledgement

This work is partially supported by CREST and Grant-in-Aid from MEXT (19054017).

References

[1] Franklin R E 1951 *Proc. Roy. Soc.* A **209** 196
[2] Harris P J F and Tsang S C 1997 *Phil. Mag.* A **76** 667
[3] Harris P J F 1997 *Internat. Mater. Rev.* **42** 206
[4] Harris P J F, Burian A and Duber S 2000 *Phil Mag Lett* **80** 381
[5] Harris P J F 2005 *Crit. Rev. Solid State Mat. Sci.* **30** 235
[6] Hawelek, L, Koloczek J, Brodka A, Dore J C, Honkimaeki V and Burian A 2007 *Phil. Mag.* **87** 4973
[7] Terzyk A P, Furmaniak S, Gauden P A, Harris P J F, Włoch J and Kowalczyk P 2007 *J. Phys: Condens. Mat* **19** 406208
[8] Terzyk A P, Furmaniak S, Gauden P A, Harris P J F and Włoch J 2008 *J. Phys: Condens. Mat* **20** 385212
[9] Harris P J F, Liu Z and Suenaga K 2008 *J. Phys: Condens. Mat* **20** 362201

Journal of Physics: Conference Series **241** (2010) 012051 doi:10.1088/1742-6596/241/1/012051

Low-electron fluence STEM-EELS analysis of ferrihydrite nano-particles approaching the Ångstrom scale

G Vaughan[1]*, R Brydson[1], A Brown[1] and K Sader[2]
1 Institute for Materials Research, University of Leeds (UK)
2 SuperSTEM, Daresbury Laboratories, Warrington (UK)

E-mail: pmgmv@leeds.ac.uk

Abstract. We present here the preliminary results from a modified low electron-fluence (< 10^7 electron nm^{-2}) mode of electron energy-loss spectroscopy (EELS) combined with scanning transmission electron microscopy (*viz.* Smart-acquisition) as applied to ferrihydrite nanoparticles. Ferrihydrites of two different crystallinities (2-line and 6-line ferrihydrite) have been synthesised and characterised by bulk techniques and then comparing Fe-EELS from the *surface* and *bulk* of individual particles (crystallites of 2-6 nm diameter). Thus far we have been able to isolate the first 0.1-0.5 nm of *surface* EEL-signal specific to both oxygen and iron and detect no differences between the surface and bulk within a particle. The Fe-EEL signal of the 6-line ferrihydrite is consistent with previously published work however we do observe a broadening in the Fe-L_3 peak of the less crystalline 2-line ferrihydrite and tentatively attribute this to distortion of the iron co-ordination in the more disordered structure.

1. Introduction

The aims of this study are, first, to establish a set of experimental parameters for the characterisation of the surface and bulk iron rich mineral cores of the ferritin protein complex. Second but by no means a less important goal, is it to thoroughly characterise ferrihydrite (FHY) nanoparticles which are thought to constitute the principle phase within ferritin mineral cores (FMCs) [1]. The main analytical technique employed in this study is scanning transmission electron microscopy (STEM) combined with electron energy-loss spectroscopy (EELS) as this allows us to probe structure and chemistry ultimately down to the Å scale. A modified mode of spectrum acquisition that continuously averages similar specimen regions as identified by prior high angle annular dark field imaging is used here. This technique limits the effects electron beam-damage and facilitates the comparison of average *surface* to *bulk* EELS analyses. The technique, termed smart-acquisition, is described in greater detail elsewhere in these precedings [Sader et al.].

Of the approximately 14 known iron oxyhydroxides, FHY is considered to be the most poorly crystalline (crystallite sizes of 2-6 nm). Its nominal bulk formula is typically quoted as $Fe_5HO_8{\cdot}4H_2O$ following Towe & Bradley 1973 [2], however careful TEM analysis of single particles has shown the water contribution to this formula to be variable depending on both the synthesis route followed and subsequent handling of the sample prior to examination [3]. Analytical probes of atomic structure, such as X-ray diffraction (XRD), show FHY to be generally structurally disordered and to exhibit a

* To whom any correspondence should be addressed.

range of crystallinities. There are generally accepted to be two forms of FHY which in order of increasing crystallinity are 2-(X-ray) Line FHY and 6-Line FHY (2LFHY and 6LFHY), based on their XRD profiles. It is the general consensus that 6LFHY is simply a more crystalline version of 2LFHY [4]. Within this structure iron is nominally ferric and octahedrally coordinated with neighbouring oxygen atoms. It has been suggested that perhaps the surfaces of individual crystallites are structurally different to the bulk [5].

One important factor to consider when attempting (S)TEM characterisation of FHY is that it is sensitive to the high energy beam-electrons. The damage mechanism is not well understood however it is apparent that both iron's oxidation number and coordination number are reduced [6]. A full assessment of 2LFHY's damage susceptibility to 100 keV electrons has not yet been carried out however, using 200 keV electrons preliminary results (unpublished) have shown it to be no more prone to damage than 6LFHY.

The $Fe\text{-}L_{2,3}$ electron energy-loss ionization edge is dominated by the dipole-allowed transitions (i.e. where $\Delta l = \pm 1$) of *2p* initial-state electrons to *3d* final-states. It is characterised by a peak-doublet at with an edge-onset of ca. 708-710 eV and peak-to-peak separation of 12.4 eV. Superimposed on this basic motif is a fine-structure (ELNES) which reflects the valence of iron as well as the coordination number, geometry and nature (e.g. chemical-species) of binding ligands. The intensity ratio of the two principle $Fe\text{-}L_{2,3}$ peaks has been show to be a sensitive probe of $Fe(III)/\Sigma Fe$ in mixed valence Fe-(oxy)hydroxides and Fe-silicates [7]. The nature of this ELNES is also affected by any distortions of coordination-sphere geometry since there will then exist multiple Fe-O bond lengths surrounding a given site.

2. Experimental methods

All STEM-EELS work has been carried out using a VG HB501 STEM retro-fitted with a Nion spherical aberration corrector and a Gatan Enfina electron energy loss spectrometer at the Daresbury Laboratories, UK (SuperSTEM1). Beam convergence angles of 24 mrads and probe diameters of 0.13nm were used. When performing EELS the spectrometer collection angle was 19 mrads. Beam currents are estimated to be approximately 100 pA. Using this set-up we have been able to acquire spatially resolved core-loss spectra at electron-fluences of ca. 10^7 electrons.nm^{-2}.

Briefly the Smart-Acquisition EELS technique used here proceeds in the following manner; first regions of interest are identified in high-angle annular dark field STEM images and a binary *mask* image created; masked regions of interest, which can have arbitrary shapes and do not necessarily need to be connected, are then converted into a series of beam scan-coordinates; the beam is then scanned over these areas whilst collecting a single EEL spectrum. This modified technique spreads the electron-fluence over an extended area of sample thereby increasing the signal-to-noise ratio in the resulting spectrum for a given electron-fluence level when compared to traditional STEM-EELS.

The FHY samples used in this study were prepared following standard methods [8]. The success of the FHY synthesis has been confirmed by comparison of high-resolution TEM (HRTEM), X-ray powder diffraction (XRD), energy-dispersive X-ray spectroscopy (EDS) and Fourier transform infrared spectroscopy (FTIR) results with those in the literature [8, 9].

3. Data analysis

We use here the linear least-squares fitting (LLS) of four well characterised reference minerals to our experimental Fe-$L_{2,3}$ edges to estimate the fractional compositions of octahedrally and tetrahedrally coordinated ferric and ferrous iron. Goodness of fit is determined by both visual inspection and by the χ^2 value (a measure of difference between experimental spectra and calculated fits) generated by the LLS fitting routine. This method and the reference standards used are described in greater detail in

[10]. In addition to LLS we obtain estimates of *Fe(III)/ΣFe* by analysis of the integrated white-line intensity ratio I(L_3):I(L_2) as per van Aken et al [7].

The final data set used for surface to bulk analysis consists of the following: For analysis of 2LFHY, 7 *surface-bulk* mask pairs were selected for the measurement of the *Fe* $L_{2,3}$-ELNES and 9 *surface-bulk* pairs were selected for analysis of the O:Fe atomic ratio; For 6LFHY these corresponding numbers were 5 and 8 respectively. In all cases *surface-bulk* pairs have been chosen to allow for comparison on a particle-to-particle basis. Only *Fe-L* spectra having a SNR better than 5 were selected for ELNES analysis. A software routine generates an EELS *scan-mask* of each particle's surface from an initial HAADF image and the quality of alignment has been assessed by overlaying the *mask* with the HAADF image from which it was derived. By analysing the pixel intensity distribution about the masked regions we have selected spectra which best isolate the *surface* signal. Masks sampling > 0.5nm into the particle's bulk were rejected. The same was true for those masks having a significant offset with respect the particle's *surface*. Therefore we are confident that the surface spectra are taken from the outer 0.5 nm of the FHY particles and respective bulk spectra from further within.

4. Results

There little evidence of a difference between surface and bulk specific EEL signals for both 2LFHY and 6LFHY. Quantification of EELS gives iron to oxygen ratio of Fe:O ≈ 0.65 ± 0.1 for both 2LFHY and 6LFHY. This result is consistent, within experimental errors, with a dehydrated form of the generally accepted formula for FHY: $Fe_5HO_8{\cdot}4H_2O$ and previous EELS work on FHY [3].

Table 1. Average EELS-derived estimates and standard-deviations (STDEV) of coordination number and valence of iron in 2LFHY and 6LFHY. The STDEV has been quoted in *lieu* of measurement errors to illustrate the variability in the results.

	Fe ELNES fitting for both surface and bulk				Van Aken	EELS Quant. - Oxygen %	
	($^{[6]}Fe^{3+}$) Haematite	($^{[4]}Fe^{3+}$) Fe-Orthoclase	($^{[6]}Fe^{2+}$) Hedenbergite	($^{[4]}Fe^{2+}$) Hercynite	Fe^{3+}/ΣFe	*Surface*	*Bulk*
2LFHY	36 ± 14%	18 ± 12%	47 ± 17%	-	0.8 ± 0.10	65 ± 8%	63 ± 2%
6LFHY	93 ± 2%	1 ± 2%	7 ± 2%	-	0.93 ± 0.05	59 ± 6%	67 ± 3%

Fe-ELNES analysis reveals that for 6LFHY both the surface and bulk Fe is predominantly (i.e. 93% ± 10%) ferric in octahedral coordination with neighbouring oxygen atoms. The remaining Fe is also predominantly in O_h symmetry but now in a reduced ferrous state. This observation is backed up by the results obtained using the Van Aken method.

Fe-ELNES analysis of the 2LFHY samples yields variable results. In general the ferric content is lower than that of 6LFHY for both surface and bulk measurements. This is confirmed by analysis using the Van Aken method. As for 6LFHY ligand coordination is predominantly O_h however there a slight increase in both the T_d and O_h ferrous component.

5. Discussion

The lack of any clear evidence for differences between *surface* and *bulk* measurements may be interpreted in one of two ways. Firstly there may in fact be no (detectable) difference between these two cation locations or secondly there may be a deficiency in the current sampling capabilities (i.e. we are only isolating the first 0.5nm) of *surface*-masks. In order to resolve this question we need to collect more data for which only the first few atomic layers (i.e. ca. 0.1 nm) are sampled in *surface* measurements and repeat the analysis.

The success of the LLS fitting, as assessed by the values of χ^2, was found to be different for 6LFHY and 2LFHY. In the case of 6LFHY both surface and bulk spectra can be reasonably reproduced from a linear combination of the reference standards. However, in the case of 2LFHY fitting results have proved to be more variable and fits less satisfactory. In the case of 6LFHY, LLS fitting gave one distinct best-fit for a given experimental spectrum. For 2LFHY there were several similar LLS fit solutions with similar χ^2 values. In each case the solution with the lowest χ^2 value was chosen. This difficulty in fitting may be due to an observed increase in width of the L_3 peak for 2LFHY. This increase, which is of variable magnitude, forces the fitting algorithm to provide spurious fits to the data. Despite being a *best-fit* in the least squares sense (i.e. a global minimum in χ^2) it is clear from visual inspection that the LLS fitting can reproduce ELNES features not present in the experimental spectra (at least at the current SNR levels).

The origin of this increased L_3 peak width is unclear and may be due to one of several reasons. First the damage susceptibility of 2LFHY has not yet been carefully assessed for 100 keV electrons and may be greater than for 6LFHY (which shows little or no propensity to damage at the fluences used here) [6]. If this susceptibility is significant our measured spectra may include some type of broadening due to alteration of the 2LFHY structure. Despite not being able to rule this possibility out we believe this not to be the case as we have shown (unpublished) 2LFHY to be no more damaging than 6LFHY for 200 keV electrons.

Second and probably more likely, the measured spectra may affected by genuine bond distortions in the octahedral symmetry about the ferric iron centre and/or an admixture of lower coordination Fe-O environments in the structure. Consequently, the variability in the L_3-width may reflect the inherent structural variability of 2LFHY; something which is hinted at by XRD albeit on a more bulk level. This result also suggests that the fitting of relatively well crystalline reference standards to systems in which there is a significant degree of distortion may be invalid. A series of experiments to investigate this effect on the L_3-edge could be envisaged using Fe-(oxy) hydroxides of more established valence and coordination (such as goethite or lepidocrocite) with a range of fine-particle sizes and so varying degrees of structural distortion.

6. Conclusion

We have successfully applied a novel low-electron fluence STEM-EELS acquisition technique to the spatially resolved (≤0.5nm) structural and chemical characterisation of FHY nanoparticles. We do not observe a difference between the *surface* and *bulk* specific signals and results are consistent with previous work on FHY. This method and the subsequent data analysis used herein have been shown to be suitable for characterisation of morphologically irregular nanoparticles such a FHY and FMCs.

References

1. Massover, W.H., 1993. Micron, **24**(4): p. 389-437.
2. Towe, K.M. and W.F. Bradley, 1967. Journal of Colloid and Interface Science, **24**(3): p. 384-92.
3. Mavrocordatos, D. and D. Fortin, 2002. American Mineralogist, **87**(7): p. 940-946.
4. R. M. Cornell, U.S., ed. 2003: Wiley-VCH.
5. Manceau, A. and W.P. Gates, 1997. Clays Clay Miner., **45**(3): p. 448-460.
6. Pan, Y., et al., 2006. Micron, **37**(5): p. 403-411.
7. van Aken, P.A. and B. Liebscher, 2002. Physics and Chemistry of Minerals, **29**(3): p. 188-200.
8. U.Schwertmann, R.M.C., ed. 1991: VCH Verlagsgesellschaft mbH.
9. Janney, D.E., J.M. Cowley, and P.R. Buseck, 2000. Clays and Clay Minerals, **48**(1): p. 111-119.
10. Pan, Y.-H.,2006, Institute for Materials Research, University of Leeds (UK)Thesis

Journal of Physics: Conference Series **241** (2010) 012052 doi:10.1088/1742-6596/241/1/012052

Sol-gel synthesis and characterisation of nano-scale hydroxyapatite

M.Bilton[1], A.P. Brown, S. J. Milne
Institute for Materials Research, SPEME, University of Leeds, Leeds LS2 9JT, UK
E-mail: pre5mb@leeds.ac.uk

Abstract. Hydroxyapatite (HAp) forms the main mineral component of bone and teeth. This naturally occurring HAp is in the form of nano-metre sized crystallites of $Ca_{10}(PO_4)_6(OH)_2$ that contain a number of cation and anion impurities, for example CO_3^{2-}, F^-, Na^+, Mg^{2+} and Sr^{2+}. Synthetic nano-sized HAp particles exhibit favourable biocompatibility and bioactivity and in order to better match the composition to natural HAp there is great interest in producing a range of chemically modified powders. In this study, two HAp powders have been synthesised via a water-based low-temperature sol-gel method and a third, commercial powder from Sigma-Aldrich have been analysed. Subsequent powder calcination has been carried out within the temperature range of 500-700 °C and the products characterised by bulk chemical analysis, X-ray diffraction and electron microscopy. Energy dispersive X-ray spectroscopy (EDX) in the TEM has been used to assess the composition of individual HAp particles. In order to do this accurately it is first necessary to account for the sensitivity of the HAp structure and composition to irradiation by the high energy electron beam of the TEM. This was done by monitoring the estimated Ca/P ratio derived from TEM-EDX of stoichiometric HAp under increasing levels of electron fluence. A fluence threshold (at a given beam energy) was established below which the measured Ca/P ratio can be considered to be stable. Subsequent elemental analysis at or below this threshold has enabled the variation in composition between particles both within and between synthesis batches to be accurately assessed. Compositional variability between particles is also evident, even in the commercial powder, but is far greater in the powders prepared by the sol-gel method.

1. Introduction

Hydroxyapatite (HAp) is a calcium phosphate ceramic. Naturally occurring HAp is the main mineral component of bone and teeth, in the form of nano-metre sized crystallites of $Ca_{10}(PO_4)_6(OH)_2$ containing a number of cation and anion impurities, for example CO_3^{2-}, F^-, Na^+, Mg^{2+} and Sr^{2+}. Chemically pure, stoichiometric HAp has a composition with a Ca/P ratio of 2.151 by atomic weight, a Ca/P atomic ratio of 1.67, and within a pH range of 4.2-8.0 has a greater stability in aqueous media compared to alternative calcium phosphate ceramics [1]. HAp exhibits excellent biocompatibility and as such is utilised in many biomedical applications related to bone substitution, including prosthesis coatings, dentistry, and as matrices for controlled drug release. HAp is also used in many non-biomedical applications such as packing media for column chromatography, gas sensors, catalysis and as host materials for lasers [2, 3].

Many applications of synthetic HAp require careful control of the particle size, shape and phase. A commonly used approach to assess HAp stoichiometry is to identify the crystal structure by X-ray diffraction (XRD) and the average composition by X-ray Fluorescence (XRF). One can also use transmission electron microscopy (TEM) combined with elemental analysis such as energy dispersive X-ray (EDX) spectrometry to characterise the structure and composition of powders. The advantage of the microscopic technique is that individual particles can be characterised, enabling compositional variability, that may not be detected by the relatively bulk techniques of XRD and XRF, to be identified. However experiments investigating the impact of a progressively increasing electron fluence (the number of electrons applied per unit area) on the Ca/P atomic ratio in HAp irradiated in the TEM have shown that a fluence above 10^7 electrons nm^{-2} at 200 keV induces a radiolytic damage process involving phosphorus and oxygen loss, amorphisation and eventual re-crystallisation to calcium oxide (CaO) [4, 5]. In this study we will establish an electron fluence threshold below which the composition of HAp can be analysed in the TEM by EDX without significant elemental alteration. We will then assess the compositional variability between particles in two synthetic HAps and compare this to characterisation with XRD.

2. Experiment

Two powders, calcined at 500 °C and 700 °C respectively, have been synthesised via a sol-gel method outlined by Liu [2]. The method involves the dilutions of triethyl phosphite in anhydrous ethanol and a small amount of distilled water for hydrolysis. Once hydrolysis is complete (signified by a transformation of an opaque solution to transparent), a stoichiometric amount of calcium nitrate, dissolved in anhydrous ethanol, is added drop-wise with continuous stirring. Further stirring and subsequent oven drying results in a white gel which is ground with a mortar to a fine powder and heated in a furnace (for 2 hours) at various temperatures. For comparative purposes a third reagent grade HAp powder has been purchased from Sigma-Aldrich.

Samples were examined in a FEI CM200 field emission gun TEM operating at 197 kV and fitted with an Oxford Instruments ultra thin window ISIS energy dispersive X-ray (EDX) system and a Gatan imaging filter (GIF 200). For TEM sample preparation, all HAp powders were ultrasonically dispersed in methanol and drop cast onto a holey carbon support film (Agar Scientific Ltd). Powder composition was monitored in the TEM with spot-EDX and specifically the Ca/P ratio (as determined from the ISIS processing software using virtual standards for the Ca and P $K\alpha$ X-ray peaks), with a specimen tilt angle of 15° and take off angle of 20°. The beam current at a fixed condition (spot size 5-8 and largest (200 μm) condenser aperture) was estimated by using an FEI calibration curve based on the measured brightness (i.e. exposure time) on the (large) fluorescent TEM viewing screen: current (in nA) = 4.875/(Exposure time for an emulsion setting of 2.0). Bulk compositional and phase analysis has also been performed using powder X-ray diffraction (XRD) with a Philips/PanAlytical X'pert Diffractometer.

To determine a threshold-fluence, whereby accurate and representative data can be obtained without damage to the powders, experiments have been carried out using the Sigma-Aldrich HAp powder. First, the irradiated area has been changed to provide approximate electron fluences across the range ~10^5-10^9 e nm^{-2}. At each fluence, spot-EDX measurements of 30 second total duration were taken over a particle cluster to obtain the Ca/P ratio. Secondly, 3 EDX measurements were taken across a cluster of particles, at a fixed irradiation area, to measure the effects of cumulative fluence over time. 30 second EDX measurements were taken after 10, 100 and 1000 seconds respectively to provide data across the fluence range 10^6-10^8 e nm^{-2}. Upon establishing a threshold fluence, compositional data for the three powders has been sought. EDX data from 50 individual particles of each powder were measured to obtain Ca/P ratios, at a total fluence of ~10^6 e mn^{-2} per particle.

3. Results & Discussion

In general all three samples consisted of particles 50-100 nm in length and 10-50 nm in diameter. Powder XRD patterns of all three samples can be indexed to a HAp phase. There are little or no impurities detectable in the Sigma Aldrich powder plots but there is a significant presence of CaO in the 700 °C and to a lesser extent the 500 °C sol-gel powders (figure 2 *a*, *c* and *e*).

3.1. Initial experiments: electron fluence induced damage

Exposure of aggregates of the Sigma-Aldrich HAp to the electron beam results in a dramatic increase in the estimated Ca/P ratio as the electron fluence exceeds ~10^7 e nm^{-2} (figure 1*a*); consistent with reference [4]. Time-resolved variation in the fluence confirms that the Ca/P ratio is not completely stable below 10^7 e nm^{-2}, the ratio changes from 1.42 to 1.53 between 10^6 e nm^{-2} and 10^7 e nm^{-2} (figure 1*b*). The signal to noise ratio (i.e. the intensity ratio of the P *Kα* peak/ peak noise) in these spectra changes from 8 to 10 respectively. To obtain reliable compositional statistics without significant alteration of the HAp subsequent measurements were made at ~10^6 e nm^{-2}.

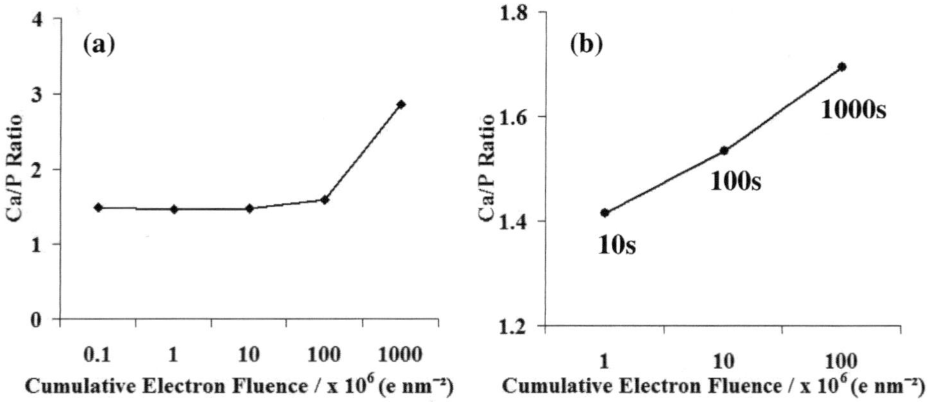

Figure 1: (a) EDX Ca/P ratio of the Sigma-Aldrich HAp as a function of electron fluence (b) EDX Ca/P ratio of the Sigma-Aldrich HAp as a function of cumulative fluence over time, measured at 10, 100 and 1000 seconds respectively

3.2. Secondary experiments: compositional data across 50 individual particles

Using TEM-EDX and operating within the fluence threshold of ~10^6 e nm^{-2}, compositional data across the three HAp powders (Sigma Aldrich and sol-gel powders calcined at 500 and 700 °C) has been measured over 50 individual particles per sample, (figures 2*b*, *d* and *f*). All three powders display some variation in particle composition, with the Sigma-Aldrich powder having the least variability (all particles exhibiting Ca/P ratios < 2) and the 700 °C sol-gel HAp the most (with up to 40% of the particles exhibiting Ca/P ratios of > 2). This variability is reflected in the reported standard deviations of the Ca/P ratios, (figure 2). The mean Ca/P ratios suggest the production of a non-stoichiometric HAp phase in all cases. The Sigma-Aldrich powder displays a non-stoichiometric composition (below the desired 1.67) however this is in agreement with data from previous literature [6]. The mean Ca/P ratios for the powders calcined at 500 and 700 °C are above 1.67 and consequently suggest the presence of a Ca-rich phase; this is consistent with XRD findings however no discrete CaO particles were detected in the sample of 50. Bulk XRF is being undertaken to provide an independent measure of the average compositions. Although CaO was detected by XRD, no discrete CaO particles were found in the sol-gel powders by TEM-EDX; significant numbers of Ca-rich particles (Ca/P ratios of > 2) were present in the sol-gel powders but to be truly representative, more than 50 particles per powder should be measured.

XRD indicates that near stoichiometric HAp is the only crystalline phase in the sample, other than small amount of CaO, and the particles measured were generally single crystal. This may be explicable by the presence of an amorphous Ca-rich phase intimately mixed with near stoichiometric HAp particles. This amorphous phase may crystallize to CaO at increasing calcination temperature. More detailed TEM analysis will be undertaken to confirm this hypothesis alongside attempting removal of any amorphous material with a selective etch.

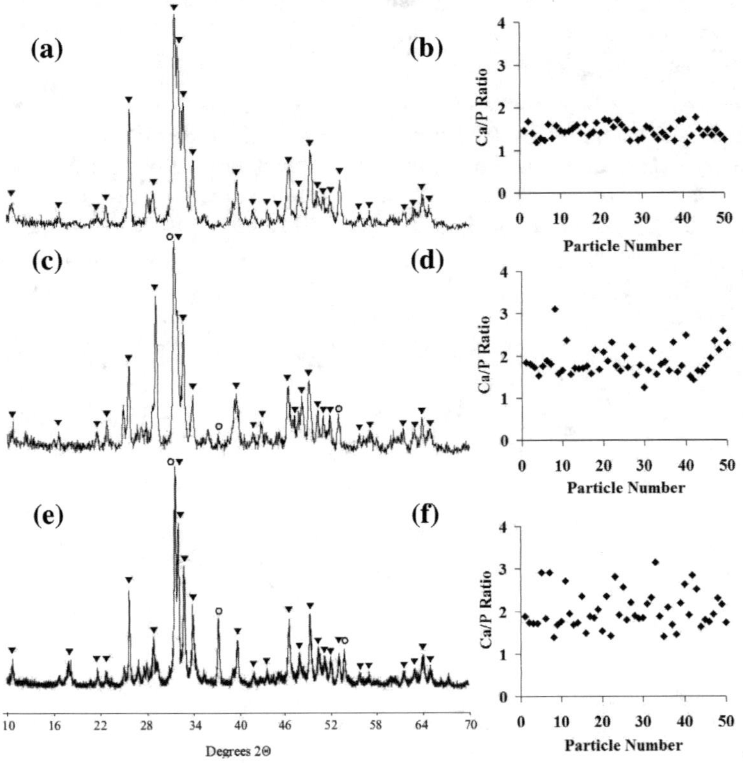

Sigma-Aldrich HAp	
Avg Ca/P Ratio	Standard Deviation
1.46	0.16

500 °C HAp	
Avg Ca/P Ratio	Standard Deviation
1.87	0.34

700 °C HAp	
Avg Ca/P Ratio	Standard Deviation
2.02	0.45

Figure 2: Powder XRD and corresponding TEM-EDX of 50 individual particles, with noted Ca/P ratio averages and standard deviations, taken at the threshold fluence for (a) & (b): Sigma-Aldrich HAp, (c) & (d): 500 °C HAp and (e) & (f): 700 °C HAp. '▼' denotes HAp peaks indexed to JCPDS file number 01-074-0566 [7], 'O' denotes peaks indexed to cubic CaO; JCPDS file number 01-077-2010.

4. Conclusion

Radiolytic damage of hydroxyapatite induced by an electron beam has been investigated resulting in the confirmation of a critical threshold fluence [4]. Operating within this limit, TEM-EDX compositional data from individual particles has been recorded over three HAp powders, one purchased from Sigma Aldrich and two powders produced by a sol-gel method, calcined at 500 and 700 °C. The results highlight the processing variability in the sol-gel preparation. Refinement of the synthesis is underway to produce more stoichiometric HAp that will be characterised by the same robust method. This synthesis and characterisation route will then be utilised to develop chemically modified HAp for biomaterial application.

Acknowledgments

This work has been supported by a DTA scholarship courtesy of the EPSRC.

5. References:

1. Best, S.M., et al., *Bioceramics: Past, present and for the future.* Journal of the European Ceramic Society, 2008. **28**(7): p. 1319-1327.
2. Liu, D.M., T. Troczynski, and W.J. Tseng, *Water-based sol-gel synthesis of hydroxyapatite: process development.* Biomaterials, 2001. **22**(13): p. 1721-1730.
3. Ergun, C., *Effect of Ti ion substitution on the structure of hydroxylapatite.* Journal of the European Ceramic Society, 2008. **28**(11): p. 2137-2149.
4. Eddisford, P., A. Brown, and R. Brydson, *Identifying and quantifying the mechanism of electron beam induced damage and recovery in hydroxyapatite*, in *Electron Microscopy and Analysis Group Conference 2007 (EMAG 2007).* 2007, Journal of Physics: Conference Series. IOP Publishing.
5. Bres, E.F., et al., *Hrem Study of Irradiation Damage in Human Dental Enamel Crystals.* Ultramicroscopy, 1991. **35**(3-4): p. 305-322.
6. Earl, J., *Hydroxyapatite Nanoparticles and Dentine Tubule Infiltration*, in *Institute of Materials*. 2007, University of Leeds. p. 199.
7. Sudarsanan K and R.A. Young, *Significant Precision in Crystal Structural Details: Holly Springs Hydroxyapatite.* Acta Cryst, 1969. **B25**(1534).

Electron Microscopy and Analysis Group Conference 2009 (EMAG 2009) IOP Publishing
Journal of Physics: Conference Series **241** (2010) 012053 doi:10.1088/1742-6596/241/1/012053

Atomic structure-colour relationship in natural diamonds

I S Godfrey[1,2], U Bangert[2]

[1]SuperSTEM Laboratory, STFC Daresbury Laboratory, Warrington, WA4 4AD, UK
[2]School of Materials, The University of Manchester, Manchester M1 7HS, UK

Abstract. Colour is a physical attribute that can be very difficult to characterise in diamond and consequently it receives regular attention from scientists working in the gem industry. In this work we compare natural brown (the most common colour) and colourless type IIa diamonds containing only trace quantities (< 1 at. ppm) of nitrogen. Numerous attempts have been made to trace the origin of brown tints in natural diamond, with the most likely culprits, i.e. dislocations and nitrogen impurities, ruled out through the application of various analytical techniques. Consequently more emphasis has recently been placed on the study of smaller defects in the diamond structure and their influence on colour. The focus of this research work is the analysis of vacancy defects having a size of the order of 1nm using aberration corrected scanning transmission electron microscopy (AC-STEM). The small electron probe size and depth of focus afforded by this technique allows such defect structures together with their position to be resolved far more readily than with conventional HR-TEM. Small-scale contrast variations are apparent in the lattice images of brown and not of colourless diamonds. These features have been compared to simulated phase contrast images of vacancy clusters in diamond. In addition, both experimental and simulated defocus series indicate that such features are not restricted to the surface of the specimen.

1. Introduction

Some chemically pure natural diamonds have a brown colouration that can only be attributed to inherent crystal lattice defects. One hypothesis is that the brown colour is caused by the presence of single vacant atoms sites and vacancy clusters (small groups of missing atoms). These are believed to form as a result of the movement of dislocations through the diamond lattice, which effectively removes the covalent bonds in localised regions to form sp^2 type bonds. Dislocation movement occurs during the plastic deformation of the diamond whilst exposed to extreme pressure and temperature beneath the earth's surface. An extensive survey of defocused bright field (BF) HR-TEM images has revealed the presence of contrasting light and dark features in a number of natural type IIa brown coloured diamonds [1]. In addition, positron annihilation measurements have proven that vacancy clusters exist only in natural brown diamond and are absent from the colourless type [2] [3]. A recent review of research into the cause of brown colour also supports the vacancy cluster hypothesis [4].

Electron energy loss spectroscopy (EELS) measurements have shown small areas (~1nm) where there is increased π-bonding intensity in natural brown diamond; π-bonding indicates a graphitic type structure and could indicate the presence of vacancy clusters [5].

In this work, using AC-STEM, we reinforce the vacancy cluster theory by assessing accurately the size and depth of the small contrasting features and in the process give far more direct evidence that they are groups of vacancies. Accompanying image simulations indicate that vacancy clusters are visible and show the contrast variation that can be expected.

Experimental

2.1 High Resolution STEM Imaging

Thin electron transparent specimens of untreated brown and colourless natural type IIa diamond were assessed using the high resolution bright field STEM imaging mode on the aberration corrected SuperSTEM at Daresbury Laboratory. The probe size in SuperSTEM is around 1 Å and it can therefore achieve atomic resolution. It is also ideal for observing small scale defects due to its short depth of focus (about 8nm), resulting from the large convergence angle (27mrad). Features of interest such as dislocations and other discontinuities in the crystalline structure can be observed more readily in bright field images due to phase contrast. Phase contrast imaging exploits the change in phase that electrons experience when they pass through and interact with the specimen. The SuperSTEM also provides a simultaneous high angle annular dark field (HAADF) image which highlights localised changes in specimen thickness and also heavier impurities, if they are present. A 40s plasma clean was carried out on all samples prior to loading them in the microscope.

A series of BF and HAADF images were acquired at various positions across the specimens in an attempt to build up a representative picture of their defect structures. In addition, a number of defocus series were acquired in order to observe the way in which the contrast of the features changes with varying amounts of defocus.

2.2 Image Simulation

Image simulation can be a useful aid to understanding the detailed structure that is observed in experimental images. The image simulations employed here use the multi-slice method [6] which transmits and propagates the electron wave function through successive slices of the crystal structure, integrates the electron intensity over the detector area and outputs the proportion of the incident intensity detected as a function of probe position. For STEM image simulation, the multi-slice calculation is performed for each position of the incident electron probe. Experimental parameters were selected for the 100kV VG HB501 SuperSTEM with Nion aberration corrector. BF image simulations of a 54 vacancy cluster in 30nm of perfect diamond crystal lattice were produced over the defocus range, -20nm to +20nm, in steps of 5nm. The angular range of the BF detector was 0-6 mrad.

2. Results and Discussion

3.1 STEM Images

The high magnification BF images of brown and colourless diamond shown in figure 1 reveal some differences between the crystal structures of the two types. The images of the brown specimen show localised regions of different contrast that are absent from those for the colourless specimen. The differences become more apparent at lower magnification as demonstrated by the images shown in figure two, where the magnification has been decreased by a factor of 2. The darker patches that are observed for the brown diamond in figure 2A are of the order of 1nm in size, which correlates with the positron measurements and also agrees with the HR-TEM observations that reveal similar features in a number of natural brown diamonds. Although there is evidence of micro-charging and central beam damage in figure 2B, the contrasting dark patches are not in evidence for the colourless diamond: the micro-charging phenomenon is most problematic for natural colourless diamond specimens in AC-STEM and could provide a clue to the difference in the electronic states found in brown and colourless diamonds, namely a lack of sp^2 bonds, which would be present on the surfaces of vacancy clusters.

By comparison, the HAADF images taken at both magnifications are not intuitive, since channelling of the electron probe along the [110] zone axis will have a marked effect on the intensity detected at the exit surface and the contrast from a defect will depend on its position within the thickness of the specimen.

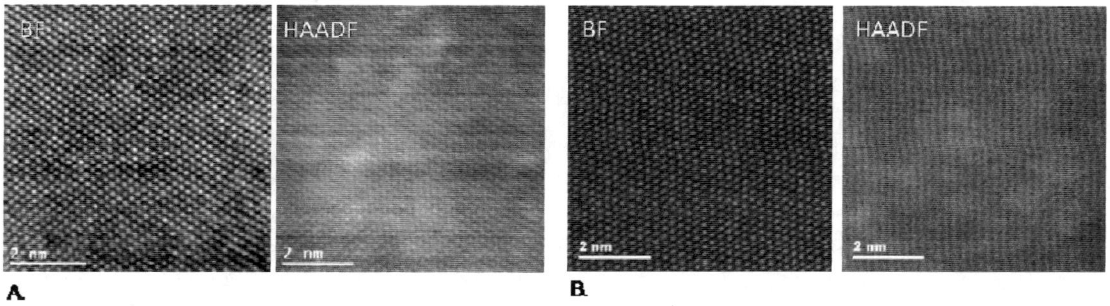

Figure 1. High magnification BF and HAADF STEM images of (A) brown diamond (B) colourless diamond

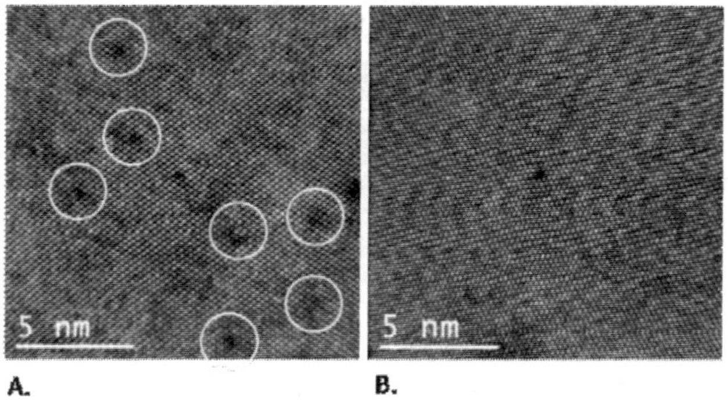

Figure 2. Lower magnification bright field STEM images of (A) brown diamond and (B) colourless diamond. The dark patches of contrast are circled.

The interpretation of these contrast variations is complicated by the possibility that they could originate from surface contaminants. A useful way of determining whether a feature is restricted to the surface or extends further into the specimen is to perform a through-focal series. This involves acquiring a number of images at various defocus values, as displayed in figure 3. The short depth of focus of the STEM probe allows the position and extent of a feature in the specimen to be investigated. The experimental series shown in figure 3 was acquired over a defocus range of -12nm to +16nm, starting at underfocus and moving to overfocus, with focus being at the beam entrance surface. The thickness of the specimen in this region was measured using electron energy loss spectroscopy (EELS) and found to be approximately 80nm. Prominent light and dark features are observed to gain and lose contrast as the defocus becomes more positive. This effect is not unexpected as the contrast displayed by a vacancy defect will vary depending on its position relative to the electron probe and therefore will not necessarily be visible when the surface of the specimen is in focus. As an example, the circled feature is absent at underfocus, becomes more visible at focus and then gradually darker with increasing overfocus. The fact that the feature is invisible at around -12nm defocus means that it is in-focus at a depth probed 12 nm below the beam entrance surface and hence is positioned at this depth; this effect is comparable with that observed in the image simulations shown in fig. 4. The feature may therefore be disregarded as a surface asperity.

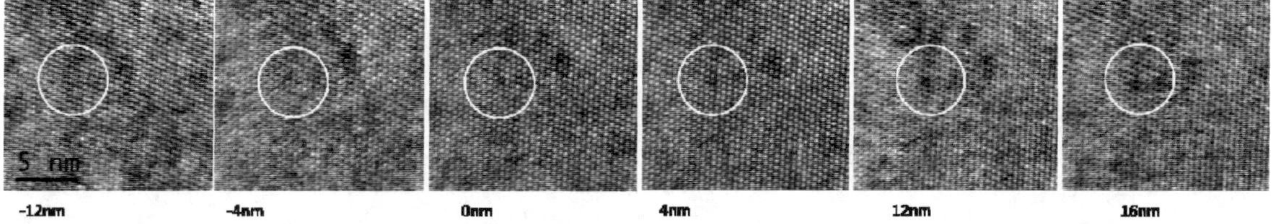

Figure 3. Bright field through-focal series of natural brown diamond demonstrating how the contrast of the circled feature becomes darker on moving from underfocus to overfocus.

3.2 Image Simulations

A comparison can be made between the small scale features highlighted in the STEM images and the contrasting central feature that appears in the simulated through-focal series shown in figure 4. The 54 vacancy cluster shown in the simulation lies at a depth of 7nm below the beam entrance surface. On moving from underfocus to overfocus the contrast of the simulated cluster is observed to change from light to dark. In addition, very little contrast is seen when the electron beam is focused on the defect rather than on the specimen surface, i.e. when the defocus is set to -5nm and -10nm. The change in appearance of the simulated cluster with defocus correlates with that observed in the experimental images (figure 3). Although the feature in figure 3 does not display a complete reversal of contrast it does appear less prominent at underfocus and this agrees with the simulation. Furthermore, if the disappearance of the feature in both the experimental and simulated images can be correlated, then the approximate position of the defect within the volume of the specimen can be determined. For example, one could ascertain from the simulated images that defects lying between 5nm and 10nm below the surface of the specimen would be largely invisible to the electron probe.

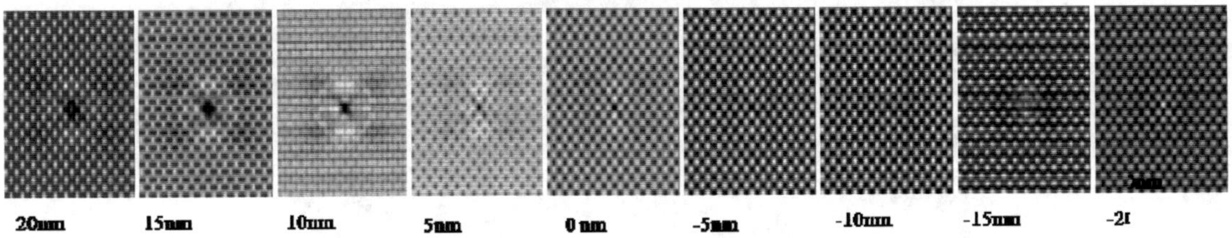

Figure 4. Simulated bright field focus series of a 54 vacancy cluster in 30nm of pure diamond lattice.

3. Conclusion

There is evidence to suggest that the patches of contrast observed in BF STEM images of brown diamond are vacancy clusters. When electrons from the incident probe interact with the vacancies it is expected that they will change phase and thereby impart a degree of contrast to a corresponding localised region of the BF image. The scale of the observed features is in keeping with the findings of positron annihilation experiments. Through-focal series show that the contrasting features are not confined to the surface of the specimen. The small depth of focus afforded by the SuperSTEM allows defects to be imaged at various positions through the thickness of the specimen; this is an advantage over the conventional TEM. Image simulations predict that vacancy clusters should be visible over a range of defocus in bright field STEM images.

4. Acknowledgements

The authors are grateful to the DTC and EPSRC for funding the experimental work and to the staff at the SuperSTEM Laboratory for their assistance with STEM and EELS training.

5. References

[1] Barnes R, 2007 PhD Thesis Manchester
[2] Tuomisto F, Mäki J-M, Kelly C J, Fisher D and Martineau P M, Origin of the brown colouration in high purity natural diamond, Submitted to *Physical Review Letters* on 27.11.2008
[3] Avalos V and Dannefaer S, 2003 Vacancy-type defects in brown diamonds investigated by positron annihilation, *Physica B* **340-342**, 76-79
[4] Fisher D, 2009 Brown diamonds and high pressure high temperature treatment, *Lithos*, doi: 10.1016/j.lithos.2009.03.005
[5] Barnes R and Bangert U, 2006 *Journal of Physics: Conference Series* **26** 157-160
[6] Kirkland E J, Advanced Computing in Electron Microscopy, Plenum, New York, 1998.

Study of the effect of annealing of In(Ga)As quantum dots

Yang Qiu, ZY Zhang, RA Hogg, AG Cullis and T Walther

Department of Electronic & Electrical Engineering, University of Sheffield, Mappin Street, Sheffield S1 3JD, UK

Y.Qiu@shef.ac.uk

Abstract. The size, structure and chemistry of In(Ga)As quantum dots as-grown by molecular beam epitaxy and after annealing at temperatures of 700 and 750°C have been studied by scanning transmission electron microscopy. The influence of the annealing temperatures on the degree of interdiffusion is investigated in detail by studying the size change of the quantum dots, and the values of diffusivities and activation energies for interdiffusion are derived.

1. Introduction

InGaAs based quantum dot (QD) devices have been developed for infrared device applications and demonstrated high performance. Post-growth annealing has been used to tune the wavelength of emission of quantum well (QW) lasers and QW photodetectors [1, 2, 3]. In recent years, post-growth annealing processes have also been used in order to improve QD devices. Hence, it is important to study the effects of annealing on the QDs.

2. Experimental

Eight layers of InAs/InGaAs QDs were grown in a molecular beam epitaxy (MBE) system on Si-doped GaAs (100) substrates. Nominally, 0.8nm InAs was deposited to form QDs, followed by 5nm InGaAs deposition to reduce the strain. Then, 33nm undoped GaAs barrier layers were deposited. Post-growth annealing processes (PAP) were applied to the sample for 5mins at 700°C or 750°C, respectively. To study the microstructure of these samples, specimens for cross-sectional transmission electron microscopy (TEM) were prepared by standard methods, glueing together stacks of material, then cutting, grinding, polishing and finally Ar^+ ion milling them until perforation. We applied annular dark-field (ADF) scanning transmission electron microscopy (STEM) using a JEOL 2010F field-emission transmission electron microscope equipped with a scan unit, providing an electron beam ~0.25nm in size at ~10mrad semi-angle of beam convergence and 0.5mm spherical aberration constant.

3. Results

ADF and bright field (BF) STEM imaging have been used to study the density and the size of the QDs. Some images are shown below, with the growth direction pointing from bottom left to top right. Figure 1 shows an overview of the sample (a) and ADF-STEM images at higher magnification with weak lattice fringe contrast of the QDs of the as-grown sample (b), the sample annealed at 700°C (c) and 750°C (d).

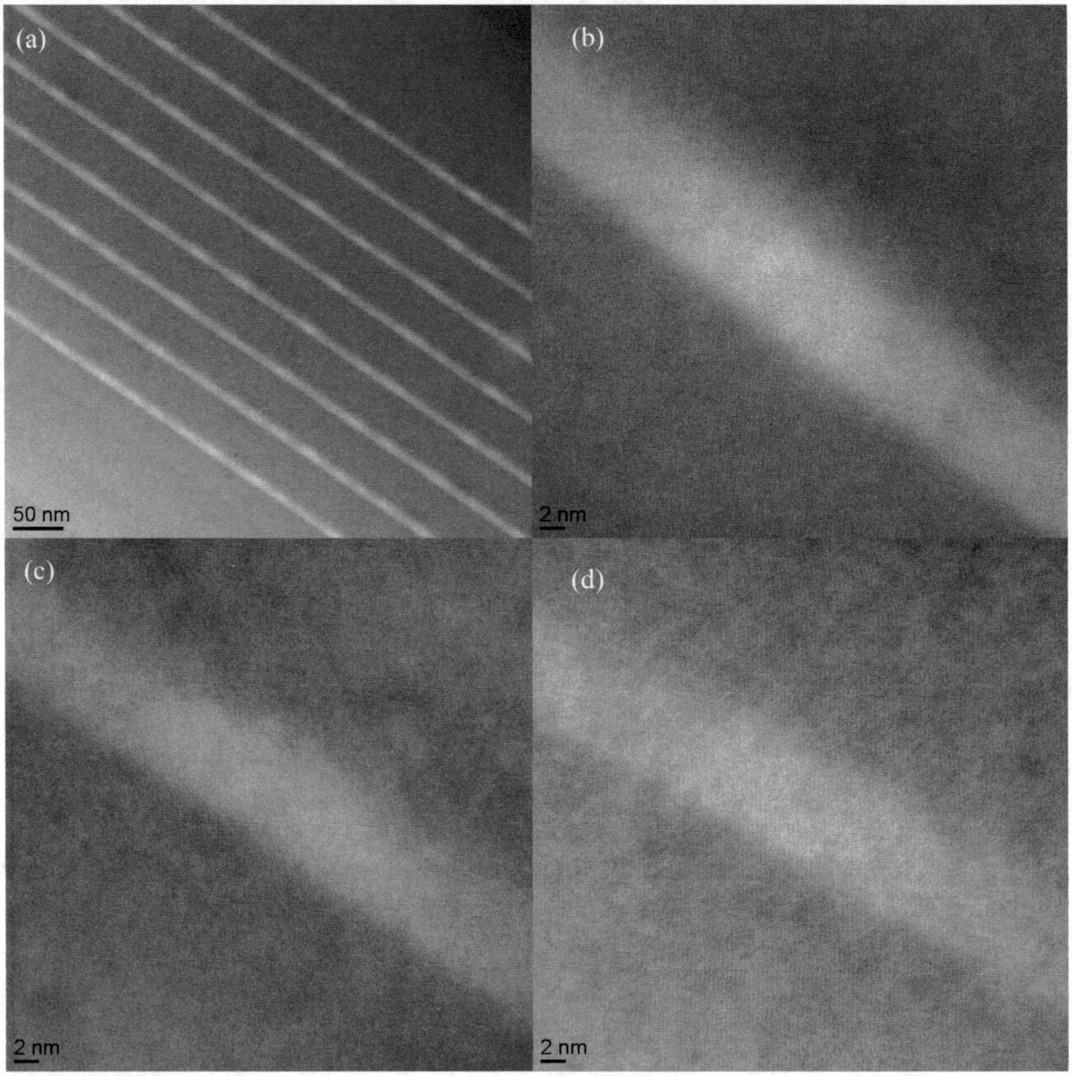

Figure 1. ADF overview of the as-grown sample at 120kX (a); ADF at 1500kX magnification of as-grown sample (b), the sample annealed at 700ºC (c) and annealed at 750ºC (d)

4. Discussion

4.1. Size of Quantum Dots

Around 50 QDs have been investigated for each sample using ADF imaging at 1500kX magnification to determine their heights and lengths. The intensity in ADF imaging is approximately proportional to the square of the average atomic number ("Z-contrast") if the inner collection angle is very large (high-angle ADF) and the specimen uniformly flat over the field of view. Hence, maps of the chemistry can be obtained for systems such as InGaAs where only two types of atomic species can interdiffuse (here: In and Ga on the group-III sub-lattice). Neglecting contrast reduction due to finite collection angles [4] we can approximate a chemical map by calculating the square root of the ADF image intensity. Neglecting the subtraction of the background the detector produces when the electron beam is blanked, typically ~25% for the settings we used, will decrease the contrast ratio of InGaAs to GaAs by a few % but not noticeably change the form of intensity profiles evaluated. The heights and the lengths of the QDs have been measured as full widths at half maximum (FWHM) values of intensity line profiles across these maps. The ratios of length to height (aspect ratios) have also been calculated in order to understand the annealing effect.

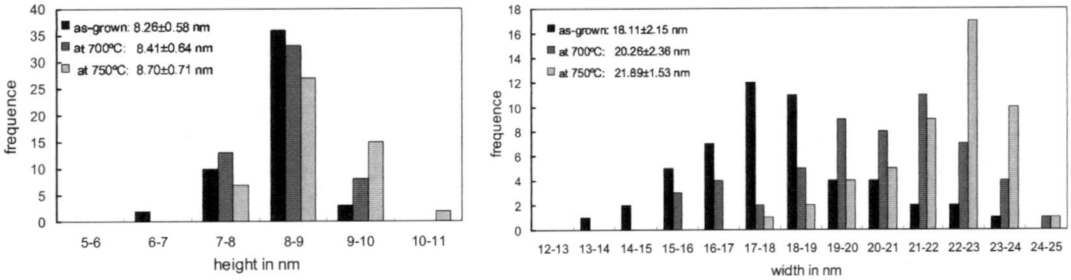

Figure 2. QD heights (left) and lengths (right) before and after anneal at different temperatures

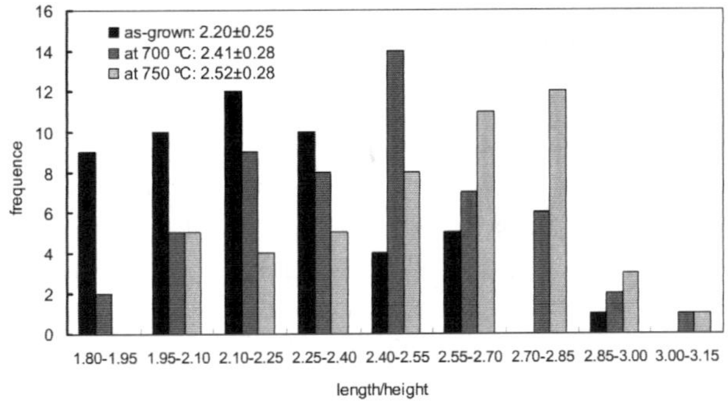

Figure 3. QD aspect ratios before and after anneal at different temperatures

The histograms demonstrate that interdiffusion mainly occurs within the (001) growth plane, rather than along the [001] growth direction. Using Gaussian approximations of the profiles we can estimate the diffusivity, *D*, for different anneal temperatures as well as the activation energy, *E*, for interdiffusion along and perpendicular to the growth direction [5].
The relationship between full width at half maximum (FWHM) and standard deviation σ is:

$$\mathrm{FWHM}=2\left(2\ln 2\right)^{1/2}\sigma \approx 2.355\,\sigma \tag{1}$$

The diffusion lengths l_i at temperatures T_i (T_1=973K, T_2=1023K) can then be calculated using σ_i :

$$l_i = [2(\sigma_i^2 - \sigma_0^2)]^{1/2} \tag{2}$$

Perpendicular to the growth direction we measure σ_0=7.79nm, σ_1=8.60nm and σ_2=9.30nm for as-grown, annealed at 700ºC and 750ºC, respectively. The diffusion length, l_i, during anneal for a time t is related to the diffusivity, D_i, by:

$$l_i = (2D_i t)^{1/2} \tag{3}$$

From equations (2) and (3), we can evaluate the diffusivity, D_i, at temperature T_i by:

$$D_i = \frac{\sigma_i^2 - \sigma_0^2}{t} \tag{4}$$

Thus, in the (001) plane, D_1=5.0×10^{-20} m^2s^{-1} for annealing at 700ºC and D_2=9.1×10^{-20} m^2s^{-1} for 750ºC. Furthermore, the diffusivity, D, can also be expressed in terms of the diffusion constant, D_0,by:

$$D_i = D_0 \exp\left(-\frac{E}{kT_i}\right) \tag{5}$$

where E is the activation energy, k the Boltzmann constant and T the anneal temperature. From this,

$$\frac{D_1}{D_2} = \exp\left(\frac{E(T_1 - T_2)}{kT_1T_2}\right) \tag{6}$$

Thus, the activation energy is:

$$E = \frac{kT_1T_2 \ln \frac{D_1}{D_2}}{T_1 - T_2} \tag{7}$$

Within the (001) plane the activation energy for lateral interdiffusion of QDs is calculated as E_{lat}=1.0eV. Along the growth direction, the In(Ga)As-on-GaAs interface is rather abrupt and the chemical profile may be approximated by just one side of a Gaussian curve, extending into the InGaAs overlayer so that we can evaluate σ by using the half width at half maximum (HWHM) instead:

$$HWHM = (2\ln 2)^{1/2}\sigma \tag{8}$$

From this we obtain σ_0=7.02nm, σ_1=7.14nm and σ_2=7.39nm for as-grown, anneal at 700°C and 750°C, respectively. Equation (4) gives $D_1=0.61\times10^{-20}\,m^2s^{-1}$ for annealing at 700°C and $D_2=1.8\times10^{-20}\,m^2s^{-1}$ for 750°C. Using equation (7), the activation energy for vertical interdiffusion can be calculated as E_{vert}=1.9eV, which is about twice as large compared to E_{lat} and only marginally lower than an earlier measurement of activation energy for vertical interdiffusion in low pressure chemical vapour deposited SiGe/Si layers [6].

4.2. Line density: spacings between nearest quantum dots

More than 300 distances between neighbouring quantum dots have been investigated by ADF and BF STEM overview images taken at lower magnifications of 50kX, 60kX, 120kX, 150kX and 200kX. Figure 4 compares the histograms for the as-grown and annealed samples. At 750°C some of the quantum dots exhibit so strong interdiffusion with the surrounding InGaAs that they disappear from the statistics, causing an additional peak at about twice to three times the original period.

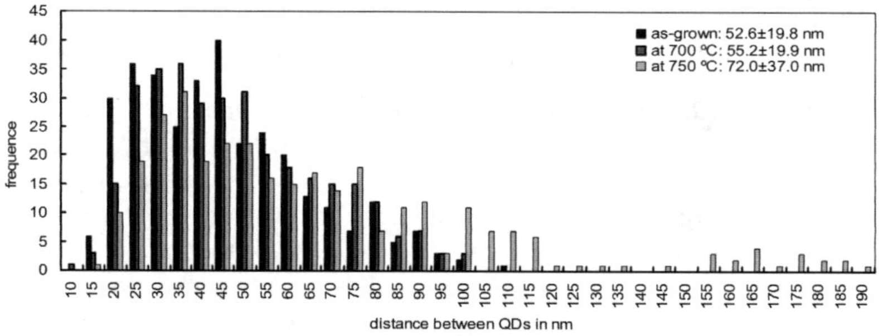

Figure 4. Spacing between neighbouring quantum dots

5. Conclusion

GaAs/InAs/InGaAs QDs of as-grown and annealed, nominally undoped materials have been studied. The post-growth annealing process causes interdiffusion of In and Ga atoms. The interdiffusion mainly happened within the (001) plane rather than along the [001] growth direction. Diffusivities and activation energies have been calculated along both directions. Results show the diffusivity in the (001) plane, perpendicular to the growth direction, is 5 to 8 times higher than along [001] and the activation energy is halved for lateral interdiffusion. The spacings between neighbouring QDs show that due to interdiffusion some QDs disappear.

References

[1] Fu L, Tan HH, McKerracher I, Wong-Leung J and Jagadish C 2006 *Appl. Phys.* **99** 114517
[2] Tan HH and Jagadish C 1997 *Appl. Phys. Lett.* **71** 2680
[3] Johnston MB, Gai M, Li N, Chen ZH, Lin XQ, Li N, Ln W, Shen SC, Fu L, Tan HH and Jagadish C 1999 *Appl. Phys. Lett.* **75** 923
[4] Walther T 2006 *J. Microsc.* **221** 137
[5] Walther T 1996 *PhD thesis*, University of Cambridge, UK, p. 144
[6] Walther T, Humphreys CJ and Robbins DJ 1997 *Defect and Diffusion Forum* **143-147** 1135

Journal of Physics: Conference Series **241** (2010) 012055 doi:10.1088/1742-6596/241/1/012055

Angstrom analysis with dynamic in-situ aberration corrected electron microscopy

P.L. Gai [1,2,3] **and E.D. Boyes**[1,3,4]
The University of York, [1]The York JEOL Nanocentre, Departments of [2]Chemistry, [3]Physics and [4]Electronics, Heslington, York, YO10 5BR, UK

E-mail: pgb500@york.ac.uk

Abstract. Following the pioneering development of atomic resolution in-situ environmental TEM (ETEM) for direct probing of gas-solid reactions, recent developments are presented of dynamic real time in-situ studies at the Angstrom level in an aberration corrected electron microscope. The in-situ data from Pt-Pd nanoparticles on carbon with the corresponding FFT/optical diffractogram (OD) illustrate an achieved resolution of < 0.11 nm at 500 °C and higher, in a double aberration corrected JEOL 2200 FS TEM/STEM employing a wider gap objective pole piece and gas tolerant TMP column pumping system. Direct observations of dynamic biofuel catalysts under controlled calcinations conditions and quantified with catalytic reactivity and physico-chemical studies show the benefits in-situ aberration correction in unveiling the evolution of surface active sites necessary for the development efficient heterogeneous catalysts. The new results open up opportunities for dynamic studies of materials in an aberration corrected environment and direct future development activities.

1. Introduction

Many technological processes are carried out in dynamic reaction environments. Therefore, for a full understanding, dynamic in-situ data related to the real world are needed under controlled conditions of vacuum/gas/vapour/liquid environment and temperature. The direct observation of the sequence of nanostructural evolution under dynamic reaction conditions is a powerful structural tool in the chemical and materials sciences. Dynamic in-situ experiments access metastable states and the mechanisms of reactions, phase transformations and nanostructure development, as well as contributing key information to synthesis, processing and stability studies. They have proved particularly informative in heterogeneous catalysis and solid state defect phenomena in controlled atmospheres [1-4], semiconductor studies in high vacuum [5] and metallurgy [6]. To be useful they need to be conducted under controlled conditions of temperature, and where appropriate environment, and of course always with a minimal, mechanistically non-invasive influence of the electron beam. This means the microscope must be able to accept a hot stage, if heating is the driving force for change being investigated, and with all other conditions held constant. In-situ E(S)TEM under controlled reaction conditions provides dynamic information on processes and performance critical defect structures which cannot be obtained so directly and readily by other techniques.

Recently, atomic resolution-ETEM has been pioneered by Gai and Boyes for probing of gas-solid reactions directly at the atomic level under controlled atmosphere conditions [3,4]. Temperature, time and pressure resolved studies with high precision are possible. Briefly, the whole EM column in a high resolution 300 kV CM30T high resolution (S)TEM has been extensively modified with the introduction of a fully integrated and permanently mounted ECELL system for atomic resolution studies of dynamic gas-solid reactions and for in-situ nanosynthesis [4,7]. Highlights of this development include a novel ETEM design with objective lens pole-pieces incorporating radial holes for the first stage of differential pumping; in addition, the controlled environment ECELL volume is the regular sample chamber of the ETEM and is thus integral to the ETEM [4]. 'Blank' calibration experiments (without the beam and the beam switched on only to record the final reaction end point and compared to in-situ data) are used to ensure that there are no deleterious effects of the electron beam [1]. The novel design of the in-situ atomic resolution-ETEM [4] has been adopted by TEM manufacturers FEI for commercial production and later versions of this in-situ ETEM instrument (including the CM 200-300 series, Tecnai and Titan (including the aberration corrected versions) have been installed in other laboratories around the world (ref: American Chemical Society's Chemical and Engineering News and Materials Research Society (MRS) Bulletins) [8-10].

Aberration correction is particularly beneficial in dynamic in-situ experiments because there is rarely, if ever, the opportunity to record from a scene which may be continuously changing, a systematic through focal series of images for subsequent data reconstruction. In these applications it is necessary to extract the maximum possible information from each single image in a continuously changing sequence. It is also desirable to limit the electron dose exposure to ensure minimally invasive conditions and to avoid secondary effects such as contamination. Based on these considerations we proposed aberration corrected dynamic in-situ electron microscopy [11,12]. Both the in-situ development and its applications to nanoparticles and biofuels are described below.

Nanoparticles of Pt, Pd and Au, only a few nanometres (nm) in size, are important in technological chemical processes and sintering impacts their performance. Bimetallic systems such as Pt-Pd offer better performance compared to monometallic systems in technological applications such as energy conversion, catalysis, nanoelectronics, chemical and biological sensing and photonics [13]. Biofuels are promising renewable energy sources and vegetable oils have been proposed as feed stocks [14]. Biodiesel is one such clean and renewable biofuel. Solid catalysts (e.g. alkaline earth oxides) offer advantages, especially pertaining to biodiesel separation and the opportunity for continuous process operation. Although solid catalysts have great promise in plant oil triglyceride transesterification to biodiesel, the identification of active sites and operating surface nanostructures during processing is required for the development of energy efficient heterogeneous catalysts for biofuel synthesis [14].

2. Experimental Procedures

2.1.1. We sputtered Pt-Pd nanoparticles (at various ratios including at 80-20 wt%) on carbon, which are of interest in catalysis and fuel cell technology, to an average thickness of 2.5nm with particle sizes of ≤ 1 to 2 nms on carbon, following methods similar to those reported [15]. We prepared high quality nanocrystalline-MgO (nano-MgO) practical powder catalysts using magnesium hydroxide methoxide precursor ($Mg(OH)(OCH_3)$) at 300°C [14].

2.1.2. To observe changes in the nanostructure as a function of operating temperature under controlled in-situ conditions in an aberration corrected environment, we employed the following procedures.
The need to accommodate special specimen holders in an aberration corrected machine has been one of the more important criteria driving the specification of the JEOL 2200FS double aberration corrected (2AC) FEG TEM/STEM operating at 200 kV in the new Nanocentre at the University of York. Since aligning the sample into a zone axis orientation is a prerequisite for atomic resolution electron microscopy of crystalline materials, an increased specimen tilt range was also desired. Both these conditions benefit from the larger gap (HRP/Midi) objective lens polepiece [11,12], shown in the

table below. The Cs (C3) aberration correctors [16] on both the STEM probe forming and TEM image sides of the instrument were used to provide the desired expanded specimen geometry with minimal effect on the 1Å imaging performance of the system. The effect of the increased polepiece gap on Cs in this range is much more evident (x2) than on Cc (x1.2); and of course Cs is now corrected (Cs~0).

Polepiece Type	Gap Range (mm)	Uncorrected Cs (mm)	Uncorrected Cc (mm)	Cc with Cs AC (mm)	Std Hot Stage fits ?
UHR	2.2 – 2.5	0.5 – 0.6	1.0 – 1.2	1.4	No
Midi	4.3 – 5.4	1.0 – 1.2	1.2 – 1.4	1.6	Yes

The advantages of the configuration we have adopted, include promoting a contrast transfer function (CTF) extending to higher spatial frequencies and resolution in the data; allowing image recording at close to zero defocus, including to strengthen and simplify interpretation of information at internal interfaces and external nanoparticle surfaces; analysing small (<2nm) nanoparticles and clusters on supports, using high resolution TEM as well as HR STEM; facilitating HAADF STEM and extending HAADF STEM resolution to 1Å (0.1nm) and below [17]. As well as benefiting from improved resolution at 1Å and below, it becomes important to be able to set the conditions to avoid the previously intrusive contrast transfer function (CTF) and defocus sensitive oscillations in image contrast in the spatial resolution range from 1Å to 3Å. This is where atomic neighbourhoods in crystals lie in low index projections and it is especially important in studying surfaces of ultra-small nanoparticles, atomic scale defects and internal interfaces. It is a key topic of interest in nanoparticle studies, e.g. in considering the possible origin of sintering mechanisms important in heterogeneous catalyst design including control of processing, activation, operation and deactivation mechanisms. (Calibration experiments were performed according to procedures described in section 1).

3. Results and Discussion

We show in both theory (www. maxsidorov.com), with contrasting CTFs shown in figure 1, and in practice (figure 3), it is possible with Cs aberration correction to combine demonstrated spatial resolution around 1Å with the HRP lens polepiece with the large ~ 4.3 mm gap [11,12] needed to accommodate a standard hot stage (Gatan model 628); and with it in operation (using Gatan Digital Micrograph). This is an example of using aberration correction to combine the limited added space required for in-situ experiments with a high level of imaging performance with which such facilities were previously incompatible; and thereby to extend considerably application specific and relevant TEM and STEM experimental capabilities.

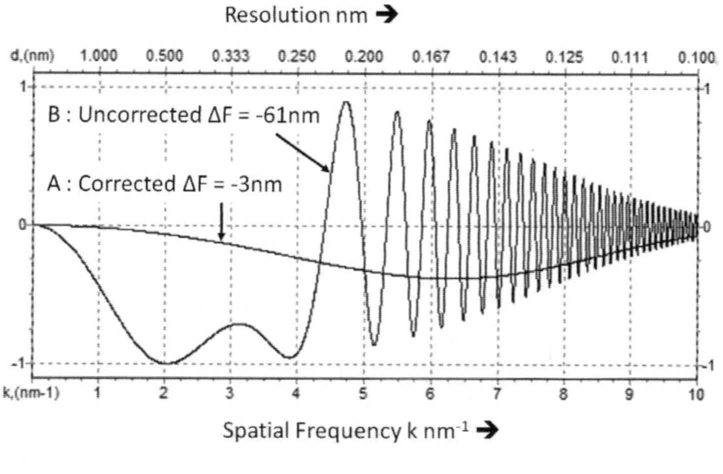

Figure 1: Calculated contrast transfer functions (CTFs) for uncorrected (B) and aberration corrected (A) imaging conditions of the HRP version of the double aberration corrected JEOL 2200FS FEG TEM/STEM STEM at the University of York (UK).

Journal of Physics: Conference Series **241** (2010) 012055 doi:10.1088/1742-6596/241/1/012055

The system is in practice stability limited (CTF envelope terms) and some of the practical steps necessary to deliver this powerful combination of capabilities will be covered elsewhere with further examples of the new tool in action. These considerations quickly set a limit to how far the lens gap can be stretched without beginning to compromise performance too seriously; taking into consideration realistically attainable stabilities in internal electronics and mechanics, and in key external environmental factors.

Our instrument is further configured with 3 turbomolecular pumps (TMPs) as the main column vacuum system to be tolerant to out-gassing hot samples, and as a key foundation for controlled gas environment experiments [11, 12]. The 'stretched' instrument configuration is considered to be a pre-requisite for further in-situ developments adding additional facilities and capabilities [4,3,9] to the exciting and still relatively new generation of aberration corrected TEM and STEM machines. In our case both capabilities are combined on the remotely controlled JEOL 2200 FS platform (figure 2).

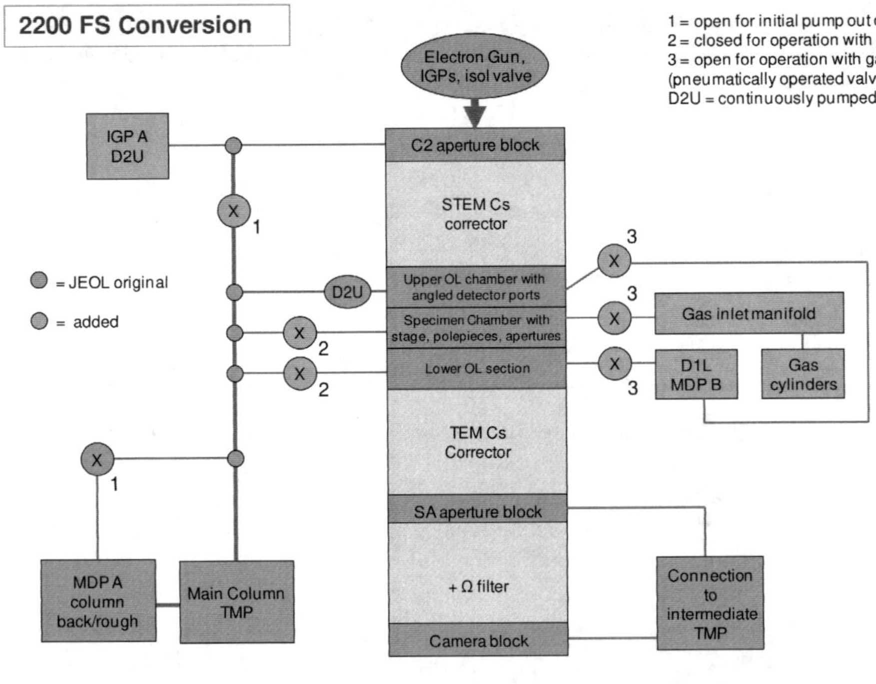

Figure 2. A schematic of the 2200FS conversion at the York Nanocentre. The instrument is configured with turbo-molecular pumps as shown.

TEM correction montage procedures (Zemlin tableaux) and image resolution to 1Å have been maintained with the hot stage, including with power connected and the samples held at elevated temperatures; so far to ~ 900°C. Figure 3(a) shows an in-situ AC-TEM image of Pt-Pd nanoparticles on carbon support at room temperature (RT) and (b) shows the corresponding FFT/optical diffractogram (OD). The particles are in the size range of ≤ 1nm to 2 nm, with the carbon support contributing strongly to the diffraction. Figure 3(c) and (d) show in-situ hot stage images of a selected Pt-Pd nanoparticle on carbon at 500 °C, and the corresponding FFT/optical diffractogram with < 0.11 nm resolution, respectively. In-situ AC HAADF-STEM studies showed this is possible (with some difficulty).

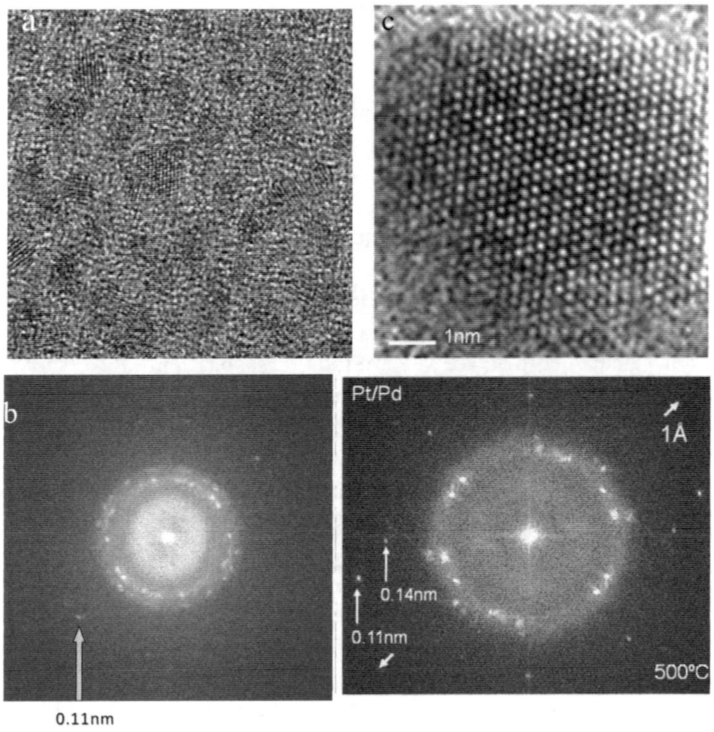

Figure 3. Left: (a) Pt-Pd nanoparticles on carbon at room temperature (RT) ; (b) the corresponding FFT/optical diffractogram (OD). The particles are in the size range ≤ 1nm to 2 nm with the carbon support contributing strongly to the diffraction; right: (c) In-situ hot stage image of a selected Pt-Pd nanoparticle on carbon at 500 °C ; (d) the corresponding FFT/OD in with < 0.11 nm resolution.

In-situ studies of nanocatalysts for biofuels:

MgO is a commercially important alkaline earth oxide, and a strong candidate for use in solid base catalysed triglyceride transesterification. It adopts a cubic structure with a = 0.4212 nm. Our as-prepared nano-MgO showed randomly orientated, overlapping single crystallites of only 3-4 nm mean

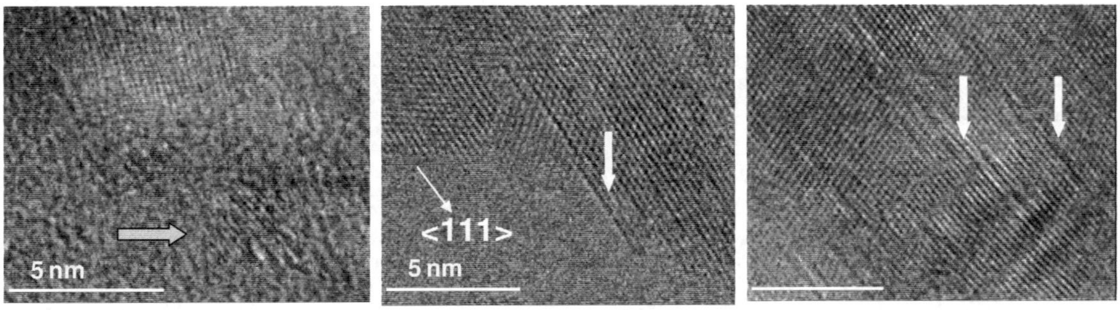

Figure 4 : Real time dynamic in-situ AC TEM studies: (a) single and overlapping nanocrystals of only about 3 to 5 nms (arrowed) are observed at room temperature (RT); (b) nucleation of defects at 500 °C in calcinations in surface profile imaging. The defects are at or near the surface (e.g. indicated by arrows). They are viewed in (110) projection and are along <111> directions ; (c) profile image (at the same magnification as in (b)) at 700 °C shows the defects have increased in concentrations and have grown along their length extending into the bulk (some are indicated by arrows).

width, embedded within an amorphous medium, with crystallites primarily in (110) and (001) type orientations. We have carried out systematic direct observations of dynamic nano-MgO catalysts in these orientations under controlled calcination conditions using in-situ AC-TEM at the 0.1 nm level and quantified with parallel reactivity (catalytic turn over frequency) and chemical studies [14]. Figure 4 shows real time in-situ, dynamic AC-TEM studies of the calcination of (110) nano-MgO: (a) at room temperature (RT); (b) extended defects are observed at the catalyst surface at 500 °C ; (c) the growth of the defects is observed at 700 °C. We have used high resolution and diffraction contrast imaging under dynamic conditions to determine the displacement vector of the defects. The analyses have indicated that they are partial screw dislocations formed by crystal glide shear mechanism along <111> and the shear or the displacement vector **R** of the type, ± a/2 <111> lies in the plane of shear (figure 5). The correlation between the atomic scale glide shear defects with the catalytic activity implicate coplanar anion vacancies in active sites in the transesterification of triglycerides to biodiesel.

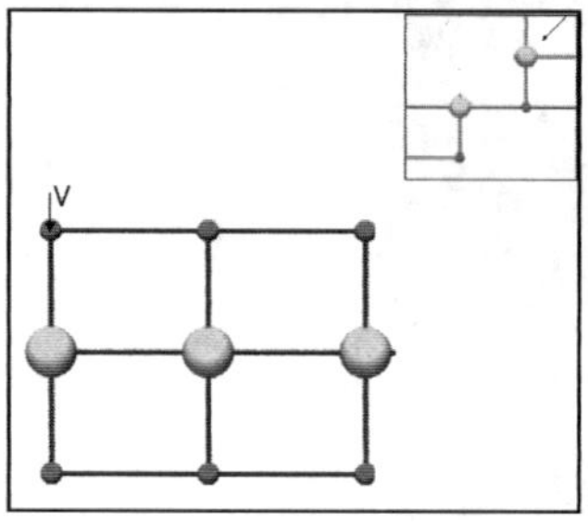

Figure 5. Structure modification to account for the defects along <111> in in-situ AC-TEM implicates coplanar anion vacancies (indicated at V) between Mg atoms. Inset : A possible glide shear model along <111> (arrowed) for the defect mechanism is shown in part in the inset, consistent with the experimental observations: projection of one layer in an idealised magnesium oxide onto (110) indicating the defect boundary

4. Conclusions

Dynamic in-situ studies of Pt-Pd nanoparticles and MgO nanocatalysts for biofuels within an aberration corrected-TEM modified with a wide gap objective polepiece have been conducted at high temperatures while maintaining 0.1 nm resolution. Preliminary experiments suggest that dynamic in-situ AC experiments are more readily performed with TEM than STEM.

Acknowledgements

We thank the University of York, JEOL and Yorkshire Forward / European Regional Development fund (ERDF) for sponsoring the Nanocentre and I. Wright for assistance with sample preparation.

References:

[1] Gai P L 1992 *Catal. Rev.Sci. Eng.* **34** 1
[2] Gai P L 1981 *Phil.Mag.* **43** 841 and Gai P L et al 1990 *Nature* **348** 430
[3] Gai P L et al 1995 *Science* **267** 661
[4] Boyes E D and Gai P L 1997 *Ultramicroscopy* **67** 219
[5] Sinclair R et al 1981 *Nature* **290** 386
[6] Butler E and Hale K F 1981 *Dynamic Experiments* (North Holland) 1
[7] Gai P L 1997 *Acta Cryst.* **B53** 346 and Gai P L 1998 *Adv.Mat.* **10**, 1259
[8] Haggin J 1995 *Chem. Eng. News* (ACS) **73** 39
[9] Gai P L, Boyes E D, Helveg S, Hansen P L, Giorgio S and C Henry 2007 *MRS Bull.* **32** 1044
[10] Gai P L, Sharma R and Ross F 2008 *MRS Bull.* **33** 107
[11] Gai P L and Boyes E D 2008 *14th Euro Micr Congress Proc. Instrumentation and Methods* **1** 479
[12] Gai P L and Boyes E D 2009 *Micr.Res.Tech.* **72** 153
[13] Yen C et al 2007 *Energy Fuels* **21** 2268
[14] Gai P L, Montero J, Lee A F, Wilson K and Boyes E D 2009 *Catal.Lett.* (in press)
[15] Gai P L and Smith B C 1990 *Ultramicroscopy* **34** 17
[16] Haider M et al 1998 *Nature* **392** 768
[17] Batson P, Delby N and Krivanek O 2002 *Nature* **418** 617

Cyclic deformation and nano-contact adhesion of MEMS nano-bridges by in-situ TEM nanomechanical testing

A J Lockwood[1], M S Bobji[2], R J T Bunyan[3] and B J Inkson[1]

[1] Dept. Engineering Materials, University of Sheffield, Sheffield, S1 3JD, UK
[2] Dept. of Mechanical Engineering, Indian Institute of Science, Bangalore, India
[3] MEMS Division, QinetiQ, Malvern, WR14 3LG, UK

Email: *a.lockwood@sheffield.ac.uk*

Abstract. MEMS nano-bridges fabricated by FIB have been deformed in-situ in the TEM. The polysilicon bridges show high levels of flexibility but also, at increased indentation depths, residual plastic deformation after fully unloading the bridges. Here, a significant number of cycles were applied to the centre of a bridge by a W-probe. This resulted in the formation of an adhesive contact with the W-probe. On unloading the nano-bridge regained its original shape and then deformed upwards, adhered to the W-probe. A significant high tensile force of $-17\mu N$ was required to sever the nano-contact. Analysis of the W-probe and polysilicon nano-bridge indicate that carbon migration along the W-probe and local contact heating due to the associated fatigue cycles were responsible for the adhesive bond, with initial carbon contamination layer on the W-probe of 2nm, thickening during the loading cycles to 25nm.

1. Introduction

The mechanical response of nanoscale structures and nano-contacts, including those used in microelectromechanical systems (MEMS), are not simple to characterise individually because of the difficulty in locally selecting and applying a mechanical load to the area or object of interest. Here, a specifically designed nanomanipulation and nanoindentation/nanomechanical test holder has been built for in-situ transmission electron microscope (TEM) studies, which allow a sharp probe to be brought into close proximity and to contact a structure. The holder uses an ultra miniaturised slip-stick coarse positioner and fine piezoelectric tube to manipulate a probe inside the TEM [1]. Fine position (within $<2\mu m$) and the final programmable cyclic or single loading indentation process is achieved by using a fine piezoelectric tube with quartered external electrodes [2].

There are many mechanisms documented regarding MEMS failure [3-7]. One in particular is stiction, which is caused by static pull down of components [3, 4]. This is generally seen in devices with low stiffness and very high surface to volume ratios, such as cantilevers or RF switches. Stiction failure is not generally seen in MEMS nano-contacts or devices designed with high stiffness that are able to self release from the attractive forces of their contact pads.

Here we use a W-probe to form a metallic-silicon nano-contact with a polysilicon MEMS nano-bridge typically seen in some MEMS applications [8]. Nano-scale movement of the W-probe is used to cyclically deform the nano-bridge over >3000 cycles.

2. Sample preparation

Polysilicon MEMS cantilevers of micron dimensions were fabricated in a standard 300nm thick process by QinetiQ. For in-situ TEM to be successfully performed, these cantilevers were post patterned and machined by 5keV focused ion beam (FIB) to generate a set of uniform suspended bridge type structures with a minimum feature size ~100nm (Figure 1(a)). The polysilicon nano-bridge investigated here had dimensions of x=3040nm and z=115nm, supported by a leg at either end with dimensions of x=180nm and z=1220nm. The as deposited thickness (y-direction) was 300nm and formed a columnar grain structure with an average grain size of 50–100nm.

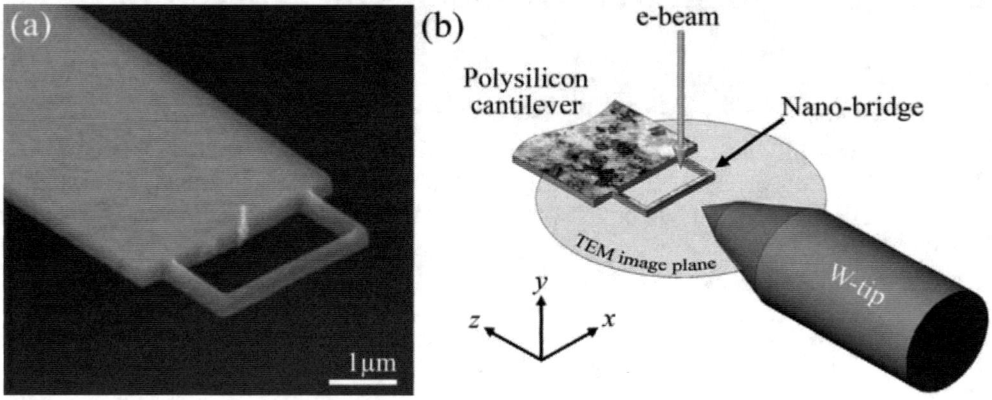

Figure 1. (a) An SEM image of a nano-bridge after machining by FIB and (b) a schematic illustration showing an in-situ TEM nanoscale mechanical test of a nano-bridge.

3. In-situ experimental procedure

The coarse positioner of the in-situ nanomanipulator was used to bring a tungsten tip with an end radius of ~25nm into close proximity (<1µm) of a polysilicon bridge structure. Fine position and precise alignment was performed by altering the applied voltage across the opposing electrodes of the fine piezo tube. A programmed indentation was performed along the z-axis by applying a uniform voltage across all quadrants of the piezoelectric tube with some small degree of x, y-scan offset to ensure the tip trajectory remained perpendicular to the bridge cross-beam (Figure 1(b)).

3.1. Single loading of a nano-bridge

Two displacement controlled indentations of +250nm at a rate of ~5nms^{-1} were applied to a polysilicon bridge (Figure 2(a)). The first indentation cycle (shown in light grey) resulted in the bridge deforming by a maximum of +205nm with a relatively elastic response, regaining its original shape and position after unloading. The second indentation cycle (shown in dark grey) showed increased deformation resulting in the bridge displacing by +215nm. The unload curve did not match the loading curve as closely as the first indentation, and the bridge developed a residual plastic deformation of ~+25nm.

3.2. Cyclic loading of a nano-bridge

A third indentation was made on the nano-bridge (Figure 2(b)), and reached a maximum load of ~40µN (Figures 2(b) and 3(a)) before it showed some evidence of plasticity in the load-displacement curve, resulting in a gradual drop in load to ~24µN and final indentation displacement of +235nm (Figure 3(b)). At this point, a large number of sinusoidal cycles were applied to the structure (applied position: +250nm↔+150nm, 600× at ~1Hz and 2750× at ~2Hz). During the cyclic loading which took over 25 minutes to perform, some piezo drift resulted in an unloading of the bridge from +235nm to +176nm (Figure 3(c)), equivalent to <3nm/min. After this, the bridge was fully unloaded at a rate of ~20nms^{-1}. During unloading of the nano-bridge, the load suddenly dropped by 7µN over 4nm (Figures

2(b) and 3(d)) and changed from compressive to tensile at 103nm displacement from the original beam position. At the point of zero applied load, we can estimate the residual deformation at the centre of the cross-beam to be ~+100nm.

The load applied during the tensile unloading portion of the curve was calculated to reach a maximum tensile value of –17μN where it was seen that the beam had been pulled past its original position by –100nm (Figure 3(e)) and distance –200nm from the zero load position. As the contact fractured, the bridge instantly sprung back to a position of residual plastic deformation of +50nm (Figure 3(f)). This is less than the estimated residual deformation of +100nm before the tensile loading to the beam, and indicates that the applied tensile loading causes microstructural changes in the bridge (Figure 3(f)). No further time dependant recovery was evident after several minutes of zero applied load.

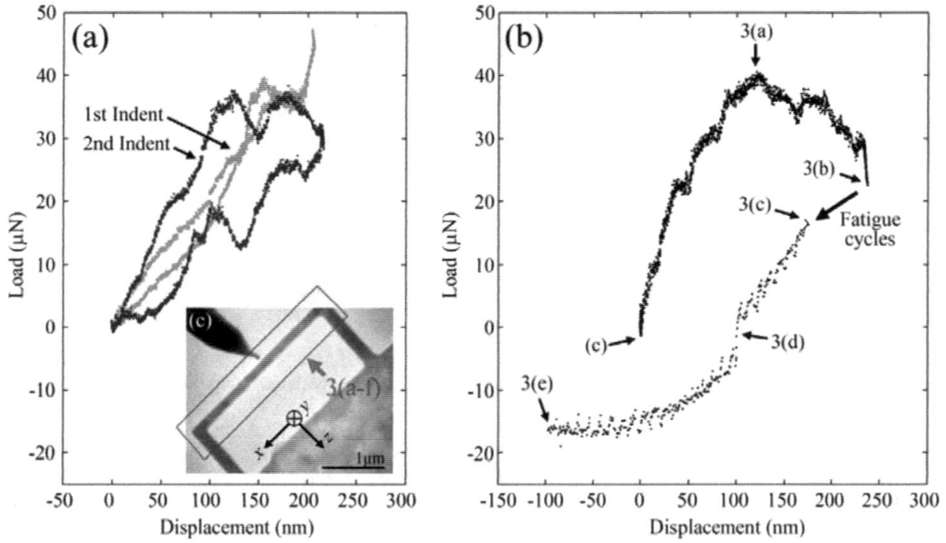

Figure 2. (a) Load-displacement curves for 2 single consecutive indentations of a polysilicon nano-bridge, (1st indent: light grey, 2nd indent: dark grey), (b) load and unload curves for the bridge with applied fatigue cycles at peak displacement (labels signify where Figure 3 images are taken) and insert (c), the original nano-bridge profile before deformation.

As previously seen [2, 9], the residual plastic deformation present after fully unloading the bridge was evident as a gentle residual curvature along the entire length of the cross-beam, suggesting that the deformation was more actively widespread than just the local contact region (Figure 3(f)). Maximum lateral deformation (x-axis) of the two supporting legs was also measured to be ±56nm, but no residual deformation was seen as they regained their original morphology on unloading.

4. Formation of a probe – nano-bridge bond

The tensile section of unloading was a result of the tip becoming strongly adhered to the beam. This adhesion has only been seen with *cyclic* loading, and has not been observed in any single loading test. The level of adhesion excludes mechanisms such as van der Waals or electrostatic attraction being proposed as the dominant adhesive force and here we can say that the two objects are mechanically bonded to one another. Throughout the entire sequence the tip did not show any significant penetration into the cross-beam (<4nm), and there was no evidence of delamination of the cross-beam. TEM imaging does however show a significant increase in thickness of the carbon shell around the tip. Initially this carbon rich layer is measured to be <3nm in thickness and fairly uniform throughout, but during cyclic loading, increases up to +25nm.

After separation, no additional carbon was observed on the surface of the cross-beam at the contact point and all the migrated carbon remained attached to the tip. It should also be noted that the W core of the tip did not deform at any point throughout the applied cyclic loading and the only morphology difference was the increased thickness carbon shell.

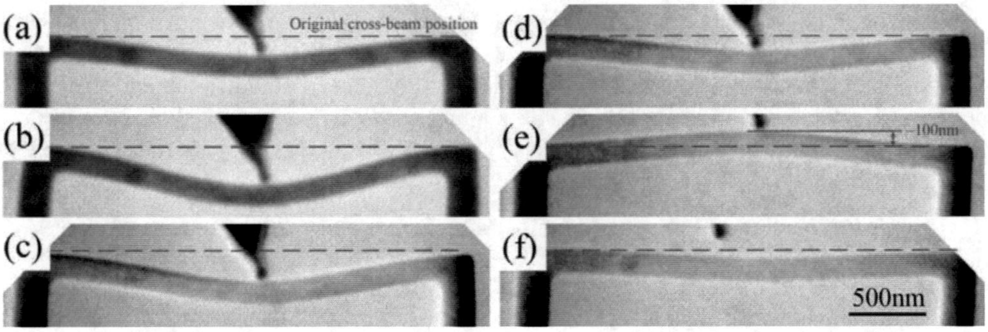

Figure 3. A sequence of images extracted from a video recording showing the deformation stages during the third indentation and fatigue cycles. (a) The bridge at peak load, (b) after some plastic deformation and load reduction prior to the applied fatigue cycles, (c) the nano-bridge after the applied fatigue cycles, (d) during unloading at the point of zero applied load, (e) maximum tensile load and (f) residual deformation after full separation.

5. Conclusions

A W-probe was brought into contact with a polysilicon MEMS nano-bridge to form a metallic-silicon nano-contact. The probe was used to deform a flexible polysilicon MEMS nano-bridge in both single indentation and programmed cyclic loading modes. During single indentation and unloading, the beam initially behaves elastically with only some minor residual plastic deformation <+25nm. However, during applied cyclic loading over more than 25 minutes, significant carbon migration around the W-probe increased the nano-contact area by a factor of more than 30×, resulting in the tip becoming strongly adhered to the nano-bridge and allowing the probe to pull the cross-beam with a tensile force of –17μN during unloading past the original bridge position by –100nm, and –200nm past the point of estimated residual plastic deformation.

Here we have demonstrated that carbon contamination, which is present on all surfaces exposed to air, can have a significant influence on the performance of MEMS/NEMS structures and their contacts. Applied cyclic loading and the associated heating can lead to significant contact adhesion between MEMS contacts of only a few nanometres in radius.

6. References

[1] Bobji M S, Ramanujan C S, Pethica J B and Inkson B J 2006 *Meas. Sci. Technol.* **17** 1324-1329
[2] Lockwood A J, Wedekind J, Bunyan R J T and Inkson B J 2009 *Nanotechnology (submitted).*
[3] Tas N, Sonnenberg T, Jansen H, Legtenberg R and Elwenspoek M 1996 *J. Micromech. Microeng.* **6** 385–397
[4] Lumbantobing A and Komvopoulos K 2005 *J. MEMS.* **14**(4) 651-663
[5] Alsem D H, Stach E A, Muhlstein C L and Ritchie R O 2005 *Appl. Phys. Lett.* **86** 041914-3
[6] Alsem D H, Stach E A, Dugger M T, Enachescu M and Ritchie R O 2007 *Thin Solid Films.* **515**(6) 3259-3266
[7] Muhlstein C L, Stach E A and Ritchie R O 2002 *Acta Materialia.* **50** 3579-3595
[8] Van Kessel P F, Hornbeck L J, Meier R E and Douglass M R 1998 *Proc. IEEE.* **86**(8) 1687-1704
[9] Lockwood A J, Bunyan R J T and Inkson B J 2008 *Proc. EMC2008, Aachen.* **1** 495

Electron Microscopy and Analysis Group Conference 2009 (EMAG 2009) IOP Publishing
Journal of Physics: Conference Series **241** (2010) 012057 doi:10.1088/1742-6596/241/1/012057

A novel tripod-driven platform for in-situ positioning of samples and electrical probes in a TEM

BD Medford[1], BL Rogers[1], D Laird[1], N Berdunov[1], AJ Lockwood[2], T Gnanavel[2], W Guan[2], J Wang[2], G Möbus[2], BJ Inkson[2] and PH Beton[1*]
[1] School of Physics & Astronomy, University of Nottingham, Nottingham, NG7 2RD, UK
[2]Department of Engineering Materials, Sheffield University, Sheffield S1 3JD, UK

E-mail: peter.beton@nottingham.ac.uk

Abstract. We present a design for a novel coarse positioning system based on a tilting platform which is positioned using linear slip/stick motors. The design differs from common arrangements of stacked *x, y,* and *z* motors, and also ball mounted slip/stick motors, by allowing easy access along the central axis of the microscope holder. The drive motors are highly compact and co-linear and may be easily incorporated in an off-axis configuration, leaving a central cylindrical region with an approximate diameter of 3mm which is available to accommodate screened electrical wiring and optical fibres. We show that the tripod can be used to manoeuvre two metallic tips towards each other in-situ in a TEM in nanometre-scale lateral steps.

1. Introduction

The use of in-situ motors to induce displacements of samples within the column of a transmission electron microscope (TEM) has attracted great interest over recent years [1-6]. A common application of these motors is their use to investigate the interaction of samples which are brought into close proximity, or even mechanical contact with each other. This permits investigations of mechanical deformation [1,2] and also, by incorporating appropriate electrical contacts, local conduction properties [5]. In addition the use of contacts has led to the development of in-situ scanning tunnelling microscopy [4-6]. The combination of these two forms of microscopy, STM-TEM, opens up many new areas for possible investigation and highlights of recent research in this area include the observation of the modification of nanotubes when a current is passed through them [5].

There are several designs of coarse motors which have been demonstrated to be compatible with operation in a TEM. The most common approach is to simply stack three motors with independent motion along mutually orthogonal (*x*-, *y*- and *z*-) axes. The slip/stick mechanism, in which a (temporally) short impulse initiates movement in a clamped rod or plate, is used to actuate motion in each individual motor. An alternative, and highly novel approach reported by Svensson et.al. [4], makes use of a spring loaded mechanism to clamp an electrical or mechanical probe to a metal coated sphere. In this case also the slip/stick mechanism [7] is used to initiate motion. The motors described

* To whom any correspondence should be addressed.

above are designed to control the three dimensional positioning of samples, while more recent developments have led to the addition of rotational stages which allow samples to be rotated around the axis of the TEM sample holder [8].

In this paper we describe a new approach to the design of motors which are compatible with operation in a TEM. A particular aim is to develop a design which allows easy access between the sample and connections to external circuitry. This access is required to route wires or optical fibres, depending on the nature of the TEM application, and in our design we avoid placing the component parts of the motor on the axis of the TEM holder leaving this region available for the required connections. This is achieved by using three parallel motors which control, by tilting, a spring loaded platform on which there is the volume available to place miniaturised components such as heaters, mechanical actuators, or, as we demonstrate, the tip of a scanning tunnelling microscope.

2. Construction of TEM insert

The main consideration for the design process was to construct a movable platform that is compatible with the JEOL 2000 and 3000 series microscopes. The supporting framework of the insert was constructed from stainless steel (grade 304), while all other materials, such as phosphor bronze and macor, were selected for their properties of high ridigity, density and, in a bid to reduce thermal drift, low coefficients of thermal expansion.

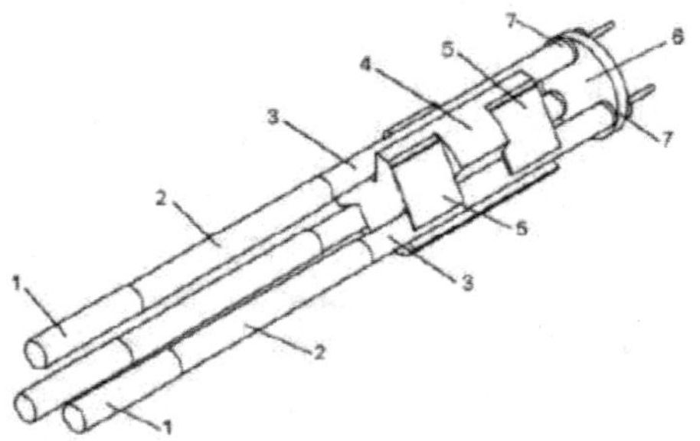

Figure 1. Schematic of motor assembly. Numbered components are discussed in the text.

The coarse approach mechanism was designed to fit within a 13mm diameter cylindrical volume which is bored from a stainless steel rod with an outer profile which matches a standard TEM sample holder. The design, shown in figure 1, is based upon a hammer-action piezoelectric drive [9] consisting of a mass (1), a piezoelectric transducer (2), and a quartz rod (3). Slip-stick motion is initiated by the application of a triangular voltage pulse with a typical magnitude of 50-100V and duration 50ms giving rise to displacements ~10nm per pulse. Three such motors are supported, parallel to each other, within grooves machined on the circumference of a phosphor bronze central support (4). The quartz rods are held in place with the use of clamping plates (5), which fix the rods in position and prevent any unwanted movement. At the end of the three quartz rods a plate is attached via flexible linkages (7), so that the plate may pivot on the ends of each of the sliding rods. Thus, by simultaneously applying identical voltages to all of the piezoelectric drives, the plate may be moved towards or away from the central support. However, by driving only selected piezoelectric drives, the plate can be tilted to a desired angle. Tilting angles of ± 5-10° are readily attainable, equivalent to displacements in x and y of $\pm$ 1mm for an object with a length ~1cm which is attached to the plate. Using this novel mechanism the three parallel linear motors are used to control the positioning of a probe or sample in three dimensions.

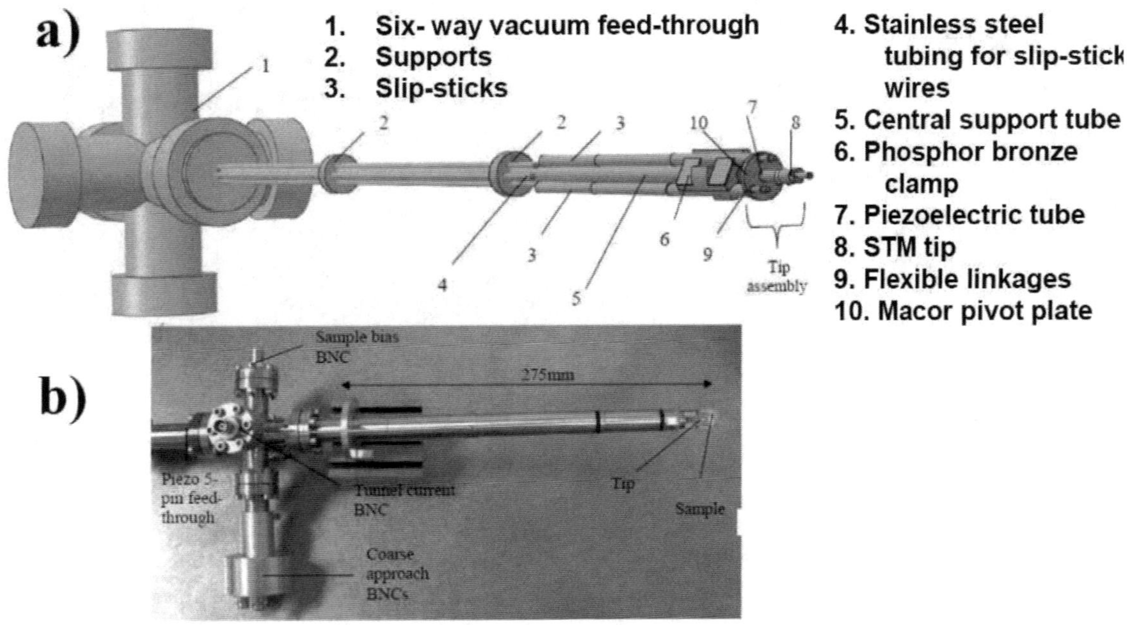

Figure 2. Diagram showing integration of motors into the sample holder: (a) schematic of construction; (b) assembled sample holder – see text for detailed explanation of layout.

The attachment of the motors to the TEM sample holder is shown schematically in figure 2(a) and includes a tip assembly on the platform which is appropriate for scanning probe microscopy. A six-way stainless steel cross (1), which is supplied commercially and incorporates industry-standard CF flanges, is attached to a central stainless steel supporting rod (5). We have configured the current design in a form which accommodates the screened wiring required for an STM and this runs parallel to the central rod within several earthed stainless steel tubes. These tubes provide electrical shielding for the wiring which reduces the capacitive coupling with the connections to the (external) current pre-amplifier. The central rod and the adjacent tubes run through ceramic supports (2) which provide additional mechanical stability. A phosphor-bronze clamp for the slip-stick (6) is attached to the end of the stainless steel rod. As described below this clamp provides grooves for three linear slip-stick motors (3) which provide an adjustable tripod support for a MACOR plate (10). A piezoelectric tube, intended as an STM scanner (7), and to which an STM tip (8) may be attached, is mounted on the plate.

The assembly in figure 2(a) may be inserted into the bored TEM sample holder and in figure 2(b) we show a photograph of the completed assembly. The tubulation in the six-way cross has an internal diameter of 16 mm, while the flange diameter is 34 mm. The remaining ports of the six-way cross are used to mount the electrical feedthroughs as identified in figure 2(b). The wires that make electrical contact with the slip-stick motors are kapton coated copper wires (0.24*mm OD*) and are contained within stainless steel tubes (1*mm ID*) (4). The 3*mm OD* stainless steel central tube (5) is available to route wiring etc to the tilting platform in a compact and well-screened arrangement. The arrangement in figure 2 is configured for use as an STM for which the wiring runs through several smaller stainless tubes to minimise electrical coupling of signals. Discussion of the performance of the STM will be presented elsewhere.

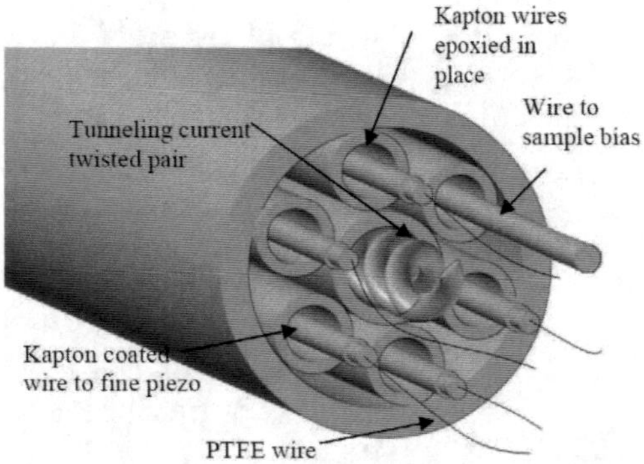

Figure 3. shows an arrangement of shielded wiring which may run up the central support rod.

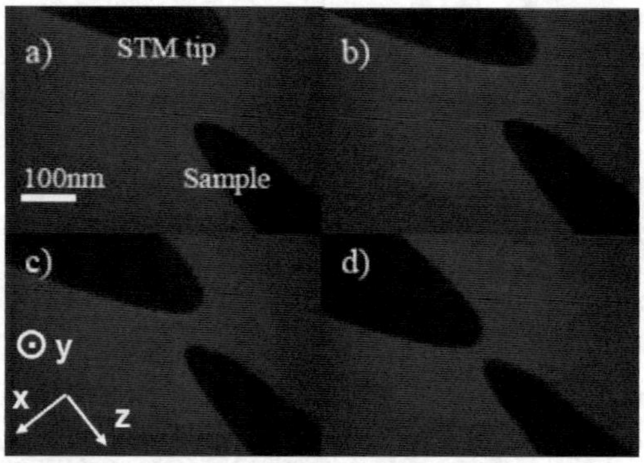

Figure 4. (a)–(d) Frames taken from a TEM video showing the alignment of the STM tip to the sample tip using the novel coarse approach mechanism.

3. Verification of motor actuation

To demonstrate the positional control of our newly developed coarse approach mechanism, a tungsten tip was guided to a small region using this mechanism. Two tungsten (W) tips were electrochemically etched and one was mounted on the sample stage (and remained stationary) while the other was mounted on the tilting platform. The sample holder was loaded into the TEM (having demonstrated vacuum compatibility to $\sim10^{-9}$Torr), and using the TEM for guidance, the STM tip was positioned as close as possible to the sample tip - using only the coarse approach mechanism. Alignment along the *x*- and *z*-axes (respectively orthogonal and parallel to the axis of the sample holder) was achieved by simply guiding the tip using TEM images. However, *y*-axis (parallel to electron beam) alignment was more difficult, as this direction is parallel to the TEM electron beam axis, and was achieved by monitoring the relative focus of the two tips in the TEM images.

Figure 4 illustrates that the STM tip can be accurately positioned to within 20 *nm* of a region of interest, in 3 dimensions, using this coarse approach mechanism. This is easily within the range of a fine positioning system which can be mounted on the platform and used for more precise positioning.

4. Summary

A novel coarse positioning system has been developed which is based on a tilting platform and incorporates screened axial electrical connections. We demonstrate that this mechanical design is compatible with operation within a TEM column and can also be used for high precision spatial positioning (~20*nm*).

Acknowledgments

This project was funded by EPSRC under grant number GR/S85702/01.

References

[1] Wall MA and Dahmen U 1998 *Mic. Res. Tech.* 42 248
[2] Bobji MS, Ramanujan CS, Pethica JB and Inkson BJ 2006, *Meas. Sci. Technol.* **17** 1324
[3] Lockwood AJ and Inkson BJ 2009, J.Phys. D **42** 035410
[4] Svensson K, Jompol Y, Olin H and Olsson E 2003, *Rev. Sci. Instr.* **74** 4945
[5] Svensson K, Olin H and Olsson E 2004, *Phys. Rev. Lett.* **93** 145901
[6] Naitoh Y, Takayanagi K and Tomitori M 1996, *Surf. Sci.* **358** 208
[7] Pohl DW 1987, *Rev. Sci. Instr.* **58** 54
[8] Xu XJ et.al. 2008 *Micros. Res. Tech.* 71 773
[9] Woodburn CN et.al. 1993, *Meas. Sci. Technol.* **4** 535

In-situ heating studies of gold nanoparticles in an aberration corrected transmission electron microscope

M J Walsh[1,4], K Yoshida[2], P L Gai[1,2] and E D Boyes[1,3]
[1]Department of Physics, University of York, Nanocentre, York, YO10 5DD, UK
[2]Department of Chemistry, University of York, Nanocentre, York, YO10 5DD, UK
[3]Department of Electronics, University of York, Nanocentre, York, YO10 5DD, UK
[4]Author to whom any correspondence should be addressed.

E-mail: mw527@york.ac.uk

Abstract. Gold nanoparticles have a high catalytic activity for CO oxidation at low temperatures providing they remain less than 5nm in diameter. Their structure and stability and the growth processes that occur during heating have been investigated using Angstrom resolution *in-situ* double aberration corrected transmission electron microscopy with a JEOL JEM-2200FS. Colloidal Au nanoparticles suspended in water, with mean diameters of 2nm and 5.6nm, have been deposited onto 3.5nm thin carbon supported on holey carbon grids and onto Si_3N_4 membranes. Dynamic *in-situ* high resolution AC-TEM images show competitive sintering processes on the different supports. Whilst the 5.6nm particles were observed to be very stable on the carbon, the 2nm particles showed sintering predominantly through particle migration and coalescence, with particle migration occurring as early as ~200°C, peaking at ~500°C. In contrast Au nanoparticles on Si_3N_4 membranes were observed to coalesce at ~180°C, before Ostwald Ripening became the dominant growth process at higher temperatures. It is believed that atoms and small clusters migrate away from their original particle before becoming trapped on the Si_3N_4 substrate. These trapped clusters then build up to form the small particles observed, before having sufficient energy to continue to migrate and join another larger particle at ~500°C.

1. Introduction

Despite the known catalytic properties of platinum and other noble metals, gold was traditionally believed to be inert until the pioneering work of Haruta *et al* [1] and Hutchings *et al* [2] in the 1980's. Haruta *et al* went on to show that when reduced to less than 5nm in diameter, gold is in fact a more active catalyst than other noble metals for many applications, and that these Au nanoparticles exhibit extremely high activity for CO oxidation at low temperatures [3]. The exceptional catalytic activity of such gold nanoparticles has direct applications in automotive and industrial emission control [4, 5] whilst also being of great interest for the removal of CO from hydrogen streams in PEM fuel cells [6] and within the fuel cell itself as an electrocatalyst [7].

Further studies have investigated the properties of Au nanoparticles key for high catalytic activity, with differing conclusions. Perimeter sites [8], charge transfer [9] and low-coordination atoms [10-12] have all been attributed to enhance activity. However, it is commonly agreed that for Au nano-catalysts, particle size plays a crucial role [13-17], with an optimum diameter occurring at 3.2nm for CO oxidation [17]. Also the effect of the support has been widely investigated, with much of the focus on metal oxide supports such as TiO_2 and Fe_2O_3. Grunwaldt *et al* found significantly different CO oxidation activities for Au nanoparticles with the same mean diameter on different supports [18], and later related this to a change in support interactions leading to a difference in the number of low-coordinated gold sites [19]. Also it is clear that the substrates defects or vacancies play a significant role in the ripening process of catalytic nanoparticles, especially under atmospheric conditions [17].

Journal of Physics: Conference Series **241** (2010) 012058 doi:10.1088/1742-6596/241/1/012058

One of the major problems with the application of nano-catalysts is their inherent instability and tendency to sinter and deactivate [20, 21]. Currently, the stability of promising new catalysts needs lengthy testing under reaction conditions before being further developed and implemented in industry. In order to speed up this process and improve our ability to develop highly active but stable catalysts, a greater understanding of the sintering processes is needed. It is generally accepted that there are two main mechanisms that give rise to cluster and particle growth. These are migration and subsequent collision and coalescence of particles, and Ostwald Ripening, in which one particle will grow at the expense of another through inter-cluster transport [21-23].

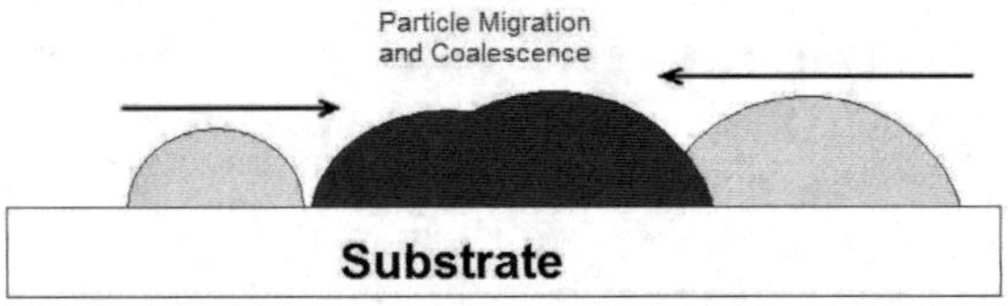

Figure 1. Particles migrate, collide and coalesce in order to minimise surface energy.

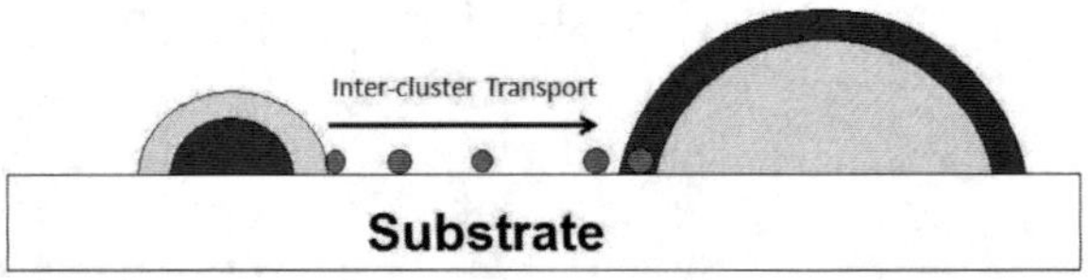

Figure 2. Ostwald Ripening in vacuum conditions. One particle shrinks whilst the other grows.

The earliest models of such sintering mechanisms were based on the Gibbs-Thompson relation, in which particle energy is dependent on particle size [21], and this relationship is still used today in current atomistic models [17, 24]. However, studies have shown that the stability of metal atoms decreases more rapidly than predicted when the particle diameter approaches the few nanometres range [25]. With a rises in the proportion of surface atoms, the melting temperature falls and the mobility of the surface atoms increases. Surface atoms also start to behave more independently, as the average number of bonds between atoms decreases and the band structure is weakened [26].

2. Experimental Procedure

The angstrom resolution double aberration corrected JEOL JEM-2200FS with wider gap (HRP) objective lens polepiece in the Nanocentre at the University of York is uniquely equipped to carry out *in-situ* studies of the structure and stability of small nanoparticles [27]. The nature of *in-situ* work means that it is rarely possible to obtain a complete through focal series of images. By including in the CTF of each image frame a complete range of spatial frequencies, and by minimising delocalisation effects by operating at close to zero defocus, AC-(S)TEM allows us to accurately observe inter-atomic spacings, crystal structures, interfaces and external surfaces from a single image, whilst the wider gap pole piece gives hot stage capabilities and an increased specimen tilt range [27-29]. Also, AC-TEM under low electron dose imaging and 'blank' calibration experiments (without the beam and with the beam switched on only to record the final reaction end point and compared to *in-situ* data) are used to ensure no effects of the electron beam [30].

The use of 3.5nm thin carbon as a well defined substrate on holey carbon grids gives reduced background noise and improved resolution of Au nanoparticles; especially for those less than 2nm in diameter. Colloidal Au nanoparticles suspended in water, with mean diameters of 2nm and 5.6nm, have been deposited onto the thin carbon and onto Si_3N_4 membranes. Initial AC-HRTEM studies into the structure of the particles were carried out, before dynamic *in-situ* AC-TEM was used to find the developing particle size distributions and particle number density as a function of temperature. Finally *in-situ* AC-HR(S)TEM was used to study the structure of the Au particles throughout the sintering process.

3. Results and Discussion

It was found that the 2nm Au colloid suspension contained particles in a size range of 1-8nm, see figure 3, whereas the larger 5.6nm Au colloids showed a much narrower size distribution. Particles less than 2nm in diameter were observed to be predominantly single crystals, whereas as above this multiply twinned particles (MTP) were much more common. Figure 3 shows a classical MTP on the right, a decahedron made up of five tetrahedral planes, bound by (111) twin planes [31] and a small, single crystal particle on the left. Figure 4 is an aberration corrected ultra high resolution TEM (AC-UHRTEM) image showing clear inter-atomic spacings, crystal structure and external surface detail.

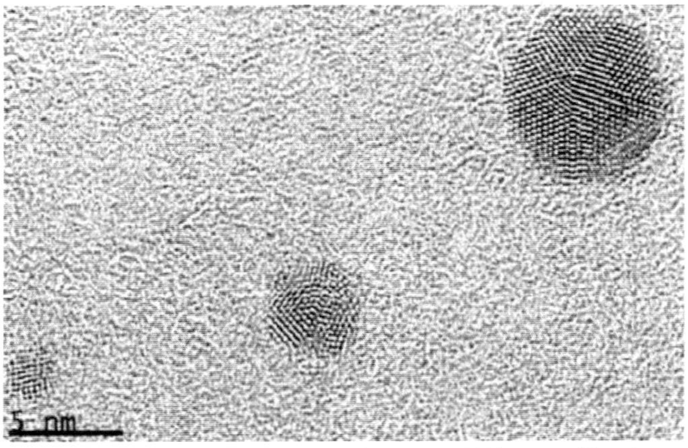

Figure 3. AC-HRTEM image taken at x500k magnification, showing a range of particle sizes and structures observed.

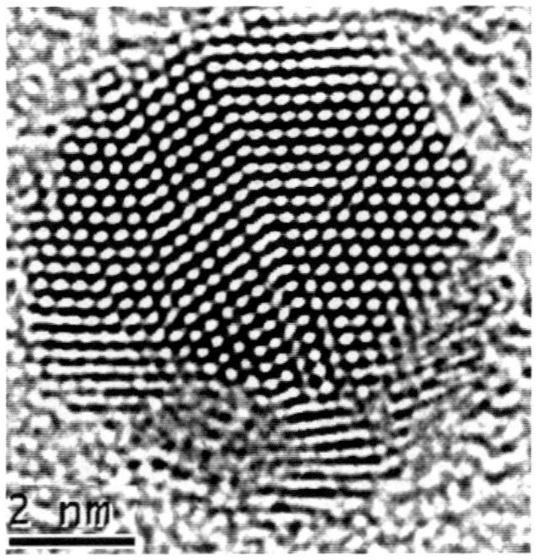

Figure 4. AC-UHRTEM image taken at x1.2M magnification

Samples prepared using 2nm Au colloids supported on thin carbon showed clear sintering, predominantly through particle migration and coalescence, with an increase in mean particle diameter from 3.37nm to 5.22nm and a 53% reduction in the number of particles present in the area observed in figure 5. However the larger 5.6nm Au colloids deposited on to the same carbon support proved very stable, with little to no sintering observed throughout heating. In contrast, the same colloidal suspension deposited on to Si_3N_4 membranes showed firstly particle migration and coalescence at ~180°C, followed by sintering predominantly via inter-cluster transport. It is clear that between 200-500°C there is a significant increase in the number of small particles and clusters approximately 1nm in diameter (circled in red in figure 6). Atoms or small clusters may leave a particle in order to reduce the original particles surface energy; however before joining another larger particle the clusters appear to become trapped on the Si_3N_4 surface, thus leading to the increase observed in figure 6.

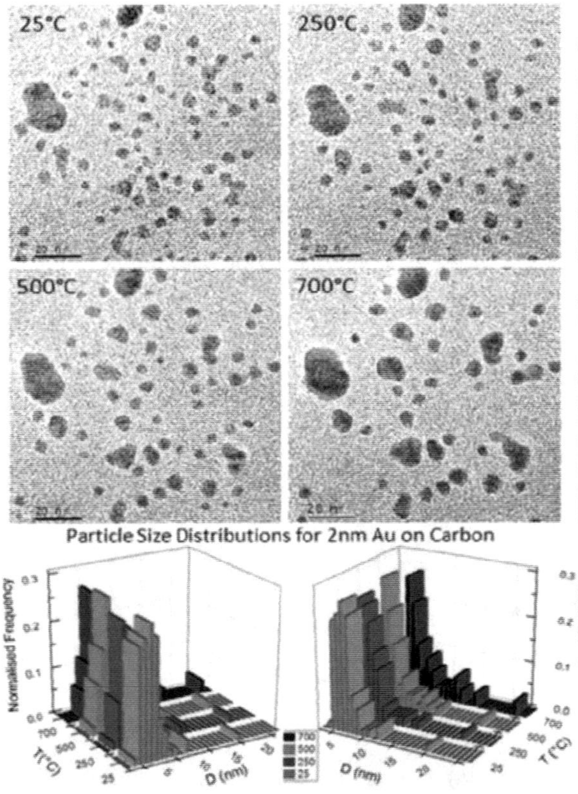

Figure 5. *In-situ* images and particle size distributions for 2nm Au supported on thin carbon

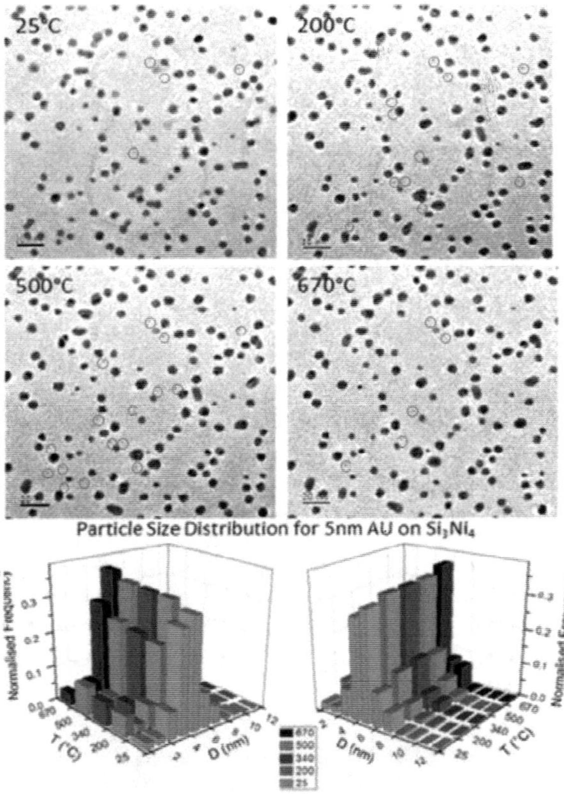

Figure 6. *In-situ* images and particle size distributions for 5nm Au supported on Si_3N_4

Above 500°C, the trapped clusters and small particles have sufficient energy to continue to migrate across the surface and rejoin a larger particle, (as seen in the marked reduction of 1nm particles present) thus completing the process of Ostwald Ripening. From 200°C onwards both the total number of particles present and the mean particle diameter is constant, within error, whilst the standard deviation of the particle size increases from 1.15nm to 1.36nm, before reducing back to 1.09 nm at 670°C, consistent with the above conclusions.

Studies of the structure of the particles after heating showed many particles had undergone surface reconstruction and internal recrystallisation in order to minimise surface area and surface energy. Many larger particles that would have been typically multiply twinned were observed to have become single crystals, with an average increase in crystal size, whilst agglomerates of particles that had coalesced were observed to minimise their surface area by reconstructing towards more spherical shapes, as can be seen in both figures 5 and 6.

Acknowledgements
We would like to thank the University of York, regional development agency Yorkshire Forward, the European Union through the European Regional Development Fund and JEOL (UK) Ltd, for the facilities at the Nanocentre at the University of York.

References

[1] **Haruta M, Koybayasi T, Sano H and Yamada M** 1987 *Chemistry Letters*, 405- 08.
[2] **Hutchings G J** *J. Catal* 1985, **96**, 292-95.
[3] **Haruta M, Tsubota S, Koybayashi T, Kageyama H, Genet M J and Delmon B** 1993 *J. Catal*. **44**.
[4] **Dwyer F G.** 1972 *Catal. Rev.* **261,** 6.
[5] **Haber J and Bielanski A** 1991 *Oxygen in Catalysis.* New York : Dekker, p. 221.
[6] **Goodman D W and Choudhary T V** 2002 *Topics in Catalysis*, **21**, 25-34.
[7] **Corti C W, Holiday R J and Thompson D T** 2005 *J. App. Catal.* A, **291**, 253-61.
[8] **Haruta M** 2004 *J. New. Mater. Electrochem. Syst.*, **163**, 7.
[9] **Hammer B and Molina L M** 2003 *Phys. Rev. Lett.*, **90**, 206102,
[10] **Remediakis I N, Lopez N and Norskov J K** 2005 *Angew. Chem.* **44**, 1824
[11] **Lopez N, Norskov J K, Janssens T V W, Carlsson A, Puig-Molina A, Claussen B S and Grunwaldt J D** 2004 *J. Catal*. **86,** 225.
[12] **Rashkeev S N, Lupini A R, Overbury S H, Pennycook S J and Pantelides S T** 2007 *Phys. Rev.* B, **76**. 035438,
[13] **Valden M, Lai X and Goodman D W** 1998 *Science*, **281**, 1647
[14] **Bond G C and Thompson D T** 1999 *Catal. Rev. Sci. Eng.* **41**, 319
[15] **Choudhary T V, Sivadinarayana C, Chusuei C, Datye A K, Fackler Jr J P. and Goodman D W** 2002 *J. Catal*. **207**, 247
[16] **Valden M, Pak S, Lai X and Goodman D W** 1998 *Catal. Lett.* **7**, 56.
[17] **Goodman D W and Lai X L** 2000 *J. Mol. Catal.* A. **162**, 33-50.
[18] **Grunwaldt J D, Kiener C, Wogerbauer C and Baiker A** 1999 *J. Catal*. **223**, 181.
[19] **Grunwaldt J D, Maciejewski M, Becker O S, Fabrizioli P and BaikerA** 1999 *J. Catal.* **458,** 186.
[20] **Campbell C T** 1997 *Surf. Sci. Rep*. **1**, 227.
[21] **Gjostein N A and Wynblatt P** 1976 *Acta Metall.* **24.** 1165
[22] **Dadyburjor D B and Ruckenstein E** 1983 *Rev. Chem. Eng.* **251**, 1.
[23] **Bartholomew C H, Fuentes G A and Salinas-Rodriguez E** 1997 *Catalyst Deactivation.* Amsterdam : Elseiver Science, 573-92.
[24] **M J Jak *et al*** 2001 *Surface. Sci.* **28,** 474.
[25] **Campbell C T, Parker S C and Starr D E** 2002 *Science*, 811-14.
[26] **Bond G C.** 2002 *Catalysis Today*, **72**, 5-9.
[27] **Boyes E D and Gai P L** 2009 *Microsc. Res. Tech.*
[28] **Boyes E D and Gai P L.** 2005 *Microsc. Microan.* **11**, 1526-27.
[29] **Gai P L.** [ed.] Hutchinson J L and Kirkland A. *Roy. Soc. Chem.* Ch. 7, 268-90.
[30] **Gai P L.** 2006 *Proc. Int. Microsc. Cong.* **1**, IMC16, 53-54.
[31] **Marks L D and Smith D J.** 1981 *J Cryst Growth,* **54**, 425-30.

In-situ TEM deformation of aluminium nanopillars

R J Milne, A J Lockwood, B J Inkson

University of Sheffield, Dept. of Engineering Materials, Sheffield, S1 3JD, UK

E-mail: r.milne@sheffield.ac.uk

Abstract. We demonstrate here that compressive and cyclic loads can be applied to individual nanostructures enabling the observation of nanoscale deformation and fatigue. We demonstrate the capability of the new NanoLAB cyclic nanotesting instrument with nanoscale mechanical testing of individual Al nanopillars with diameters of less than 350nm. The thin Al nanopillars are seen to deform via a buckling mode where single crystal Al is turned into polycrystalline Al.

1. Introduction

Following the recent trend of shrinking dimension sizes for mechanical devices such as MEMS/NEMS (micro/nano-electromechanical systems), it has been clearly shown that mechanical properties of nanostructures and nanostructured materials vary considerably from their bulk counterparts. These size-scale effects have shown that as the dimensions of an object decrease to the micron-scale and below, they have increased flow stresses and increased strength [1-7]. It is thought that both the severely limited volume and the increased surface area to volume ratio have a significant effect on the dislocation generation and propagation mechanisms which can operate.

One of the major problems in testing individual nano-objects' mechanical properties is that nano-resolution positioning is essential, linked with a method of observing and measuring their response in-situ during a test. Novel scanning probe instruments have been developed which allow a range of mechanical testing methods to be implemented in-situ in both SEM and TEM [8, 9]. Our NanoLAB instrument is based on a generic piezoelectric positioner which can be adapted with the addition of various nanotools [10]. In this work, a series of nanoscale Al pillars have been repeatedly indented in real-time in-situ in a TEM, with the aim to increase our knowledge of deformation mechanisms in aluminium single crystal nanopillars. The repeated indentation of these pillars allows a large amount of deformation to be applied to the nanopillars whilst allowing the microstructure to rearrange itself during and between indentations in order to relieve stress.

2. Experimental

Large crystals of pure aluminium (99.999%) were grown by cold rolling and a heat treatment. These samples were thinned down and electrochemically etched, in a solution of Perchloric acid and Ethanol in a 1:4 ratio, to give a finely polished edge to the samples. On this edge, figure 1(a), pillars with diameters of less than 350nm were machined using a Focused Ion Beam (FIB) microscope. The heights of the pillars were between 1 and 2μm, giving aspect ratios between 1:3 and 1:6, figure 1(b). A conductive flat punch diamond indenter was used to uniaxially compress the pillars in a 300kV JEOL 3010 microscope. A NanoLAB piezo-driven indentation drive was used to apply a local load which has sub-nm positioning movement. Software developed for use with the indenter allows adjustment of the direction of travel to give the correct axial alignment. A calibrated force sensor [10] was used during the indentations and a tracking programme was used to determine the displacements from the recorded video.

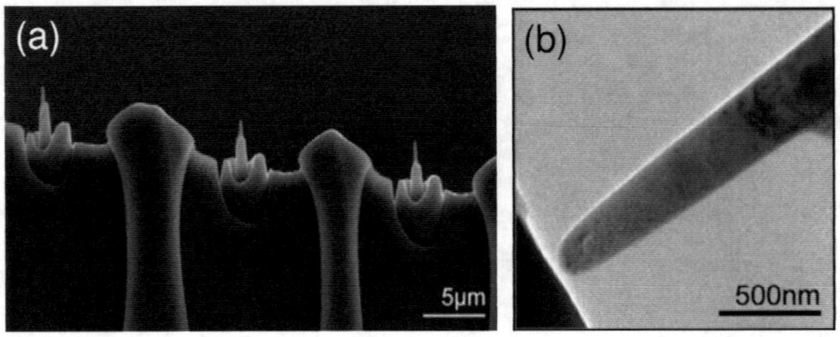

Figure 1. (a) SEM image of pillars on etched edge before testing. (b) TEM image of an Al nanopillar prior to testing.

3. Mechanical Compression of Al Nanopillars

A 350nm diameter aluminium pillar was axially compressed by displacement controlled indentation of 580nm at a speed of approximately $15nms^{-1}$ (figure 2(a)). The pillar was unloaded, and then two subsequent compression cycles of applied displacement 470nm and 570nm were carried out (figures 2(b, c)). The aluminium nanopillar is observed in-situ to undergo severe mechanical deformation. Figure 2(d) shows the load-displacement curves for the three cycles of indentation. The third cycle shows three significant jumps in load (I, II, III), due to large changes in the microstructure of the nanopillar under load. This can be seen in figure 3(a) to (f), which was recorded in-situ during the compression test.

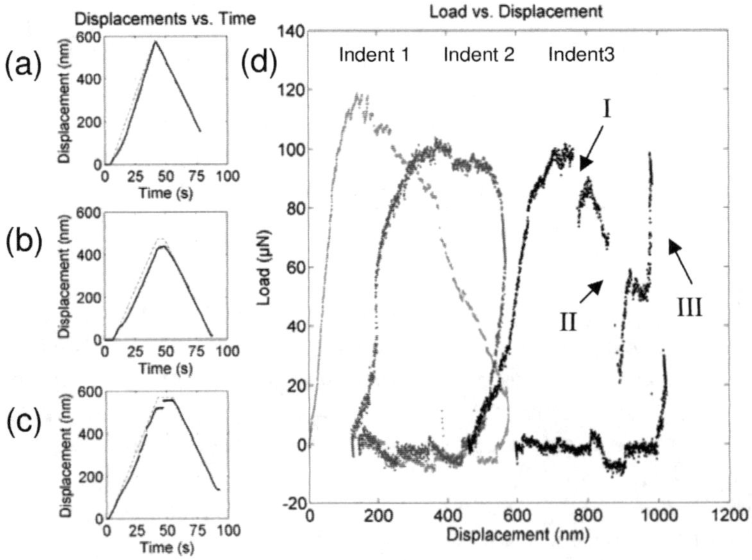

Figure 2. (a), (b) and (c) show the displacement vs. time curves for the three separate indents, 1, 2, and 3 respectively. The dotted line indicates the applied displacement. (d) Overlaid load vs. displacement curves for the three indents. Jumps in load I, II and III are discussed later.

During indent 1, whilst approaching a load of 120μN, the pillar started to buckle away from the axis due to its relatively long length. The nanopillar shown here had an initial aspect ratio of 1:4, and was only constrained at the base. The tip of the pillar then underwent a small lateral translation across the diamond indenter which resulted in the introduction of a slight bend in the tip of the pillar. Subsequent indentations caused the pillar to deform further from the original axis as the applied load was increased. The lateral deformation resulted in new grain formation and straightening of the pillar.

During the indentations, loads of ~100μN were reached before activating significant plasticity, in the form of dislocation flow, which was sufficient to reduce the applied load (figure 2(d)). In the load-displacement curve for indent 3, the three significant jumps (reduction) in load (I-III, figure 2(d)) can be associated with significant rapid microstructural events. These appear to be the collective rearrangement of dislocations to form low and high angle grain boundaries to reduce stress (see §4, 5). After the first two of these events (jumps I and II), there is a subsequent small initial increase in the load, before dislocation activity re-establishes and the load again begins to decrease. At a total displacement of 960nm the load, and therefore the compressive stress in the pillar again begins to build up. A significant lateral displacement of material, corresponding to Jump III, occurs to relieve this stress.

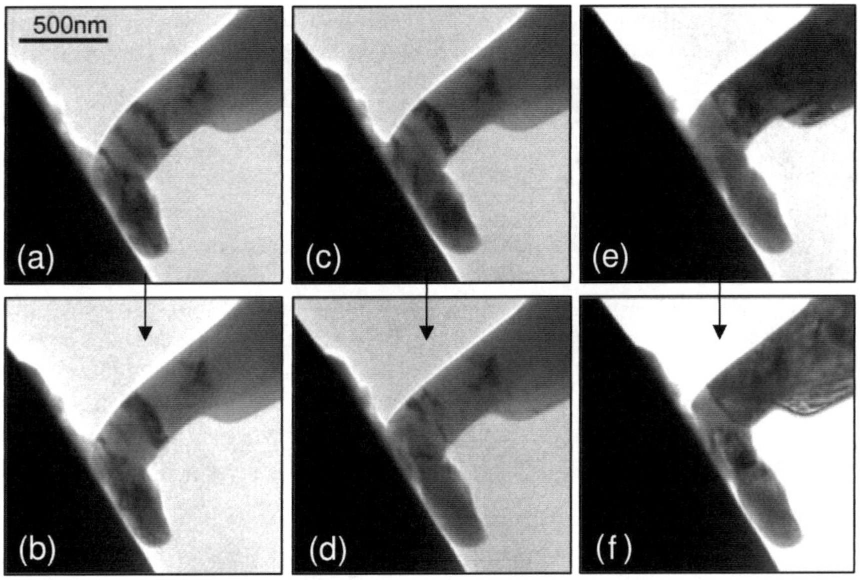

Figure 3. TEM images extracted from a recorded video taken from before and after the displacement jumps seen in the third indentation cycle. Jump I: images (a)→(b), jump II: images (c)→(d), and jump III: images (e)→(f). The time between the first and last image in each pair is 0.12 seconds.

4. Nanopillar Deformation

Under the uniaxial compression a bowing of the pillar was first seen with Bragg contours moving down the pillar. A zone of high compressive stress formed at the tip of the pillar which was in contact with the diamond indenter. To release compressive stress, the pillar deformed by bending the tip over and straightening the remaining, *shortened*, pillar (figure 4). On unload, after indent 2, and with the loss of contact with the indenter, some time-dependant recovery was seen with the pillar straightening itself by approx. 75nm. During the subsequent indentation there was an increase in the size of the sideways displaced material with increasing indentation depth. Within the displaced material, TEM contrast was seen to change in discrete areas, due to locally different Bragg scattering and thickness. This is also representative of the low angle lattice tilt expected during low angle sub-grain formation. As the applied force reaches a maximum the severe lattice distortion leads to dislocation rearrangement and the formation of new grains (figure 4), generating a polycrystalline pillar tip. Figure 4(b) shows grains protruding from the surface of the deformed pillar after testing.

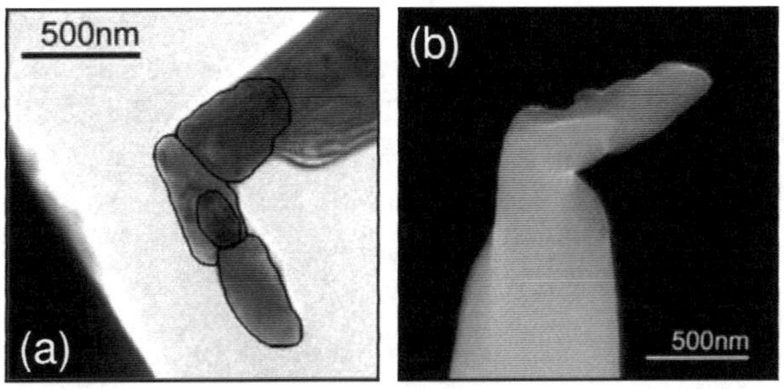

Figure 4. (a) TEM image of grains formed after the sequence of indentations. (b) SEM image, taken at 45°, of protruding grains and deformation along the top of the pillar after indent 3.

5. Deformation Mechanisms

To release stress in the bent pillar, dislocation flow occurs within the zone of highest compressive stress at the tip. Dislocations are also seen to propagate in the less stressed main shaft of the pillar. Rapid jumps in displacement are observed as the new grains are formed within the deformed volume. Post mortem evaluation in the SEM shows a three dimensional arrangement of these grains as a protrusion away from the original axis of the pillar, figure 4(b). The grain formation enables straightening of the remainder of the main shaft of the pillar which now contains less material. The constraint at the base of the pillar leads to dislocation pile-up at the pillar base and the creation of a high-angle grain boundary which stops further dislocation flow into the substrate, figure 3(e-f). In other pillar investigations dislocations are seen to flow up the shaft towards the deformed area, as well as from the deformed area towards the base, to reduce stress.

The deformation in these 350nm diameter pillars differ from deformation mechanisms seen in wider, stiffer, pillars, such as slip bands and shearing [1-3]. Plastic flow due to localised slip has been seen in single crystal aluminium alloy pillars (140-180nm diameters) resulting in 'mushrooming' [7]. This would have the effect of spreading the load over a greater area, thus reducing the applied stress, as seen in this experiment; however the formation or growth of grains was not observed. The localised slip seen to cause the 'mushrooming' was explained due to the taper of the pillar.

The growth of existing grains under deformation has been seen in aluminium thin films [11] and grain rotation was seen in nanoscale grained aluminium [12], however the formation of new grains through deformation has not been previously reported by in-situ experiments. Thin film single crystals have previously been observed to form new grains after ex-situ nanoindentation [13]. The grains formed here during the in-situ compression of the nanopillars also appear to have undergone grain rotation as the deformation progressed.

6. Conclusions

During in-situ real-time uniaxial compression tests of nanopillars undertaken in the TEM microscope, it is seen that rapid microstructural changes in the Al nanopillars can be directly related to features in the force-displacement curves. The mechanisms of deformation were analysed and Al nanopillars are seen to reduce stress by grain formation and re-arrangement in a three dimensional manner. As the diameter and length of the Al pillars are clearly important factors in the deformation mechanisms it would be prudent for future work to investigate deformation mechanisms as a function of geometry.

References

[1] Uchic MD, Dimiduk DM 2005 *Mater. Sci. Eng. A* **400–401** 268-278.
[2] Greer JR, Oliver WC, Nix WD 2005 *Acta Materialia* **53** 1821-1830.
[3] Dimiduk DM, Uchic MD, Parthasarathy TA, 2005 *Acta Materialia* **53** 4065-4077.
[4] Haque MA, Saif MTA 2002 *Experimental Mechanics* **42** (1) 123-128.
[5] Frick CP, Clark BG, Orso S, Schneider AS, Arzt E 2008 *Mater. Sci. Eng. A* **489** 319-329.
[6] Wu B, Heidelberg A, Boland JJ 2005 *Nature Materials (Letters)* **4** 525-529.
[7] Ye J, Mishra RK, Minor AM 2008 *Scripta Materialia* **59** 951-954.
[8] Boji MS, Ramanujan CS, Pethica JB, Inkson BJ 2006 *Meas. Sci. Technol.* **17** 1324-1329.
[9] Lockwood AJ, Inkson BJ 2009 *J. Phys. D: Appl. Phys* **42** 035410.
[10] Bobji MS, Pethica JB, Inkson BJ 2005 *J. Mater. Res.* **20** (10) 2726-2732.
[11] Minor AM, Asif SAS, Shan Z, Stach EA, Cyrankowski E, Wyrobek TJ, Warren OL 2006 *Nature Materials(Letters)* **5** 697-702.
[12] Jin M, Minor AM, Stach EA, Morris Jr. JW 2004 *Acta Materialia* **52** 5381-5387.
[13] Inkson BJ, Steer T, Möbus G, Wagner T 2000 *J. Microscopy* **201** (2) 256-269.
[14] This work was funded by the EPSRC Basic Technology Project; Grant GR/S85689.

In-situ TEM study of dislocation patterning during deformation in single crystal aluminum

P Landau[1], R Z Shneck[1], G Makov[2], A Venkert[2]

[1] Department of Materials Engineering, Ben-Gurion University, P.O. Box 653, Beer-Sheva, 84105, Israel
[2] Department of Physics, NRCN, P.O Box 9001, Beer- Sheva, 84190, Israel

E-mail: venkert@bgu.ac.il

Abstract. The evolution of dislocation patterns in single crystal aluminum was examined using transmission electron microscopy (TEM). In-situ tensile tests of single crystals were carried out in a manner that activated double slip. Cross slip of dislocations, which is prominent in all stages of work hardening, plays an important role in dislocation motion and microstructural evolution. In spite of the limitations of in-situ straining to represent bulk phenomena, due to surface effects and the thickness of the samples, it is shown that experiments on prestrained samples can represent the early stages of deformation. Transition between stage I and stage II of work hardening and evolution during stage III were observed.

1. Introduction

Patterning of dislocations into cell structures is common to many metals during plastic deformation during stage III of work hardening. The extensive research devoted to this phenomenon has revealed many of its characteristics [1-5].

The first stage of work hardening is characterized by dislocation glide, referred to as easy glide, where only one slip system is active. The transition between stage I and stage II of work hardening is believed to evolve due to cross slip of dislocations as a result of different barriers for dislocation glide, eventually causing the formation of dislocation tangles [1-3]. The tangles form dislocation cells by dislocation's rearrangement into self-screening structures, in order to minimize the stored energy per unit length of the dislocation lines. Stage III is known as dynamic recovery by the increase of dislocations within cell boundaries, increasing misorientation between adjacent cells and a decrease in cell size [1, 6]. In all stages of work hardening cross slip plays an important role in the dislocation's rearrangement into different patterns [5]. The ability of the dislocations to cross slip depends on the stacking fault energy and the activation energy for cross slip determines the ease at which typical dislocation patterns will form [7-8].

In this research in-situ transmission electron microscopy experiments were performed in order to study mutual interactions between dislocations and the collective behavior of dislocations to form typical patterns during deformation. Moreover, the validity of in-situ TEM experiments relative to bulk experiments was discussed.

2. Experimental

High purity (99.999%) aluminum single crystals [011] discs were cut into dog-bone shaped samples. The samples were prestrained in an Instron type machine to different strains (10-15%). The samples were prestrained in an orientation that assured the activation of two slip systems in <221> directions [4]. From these samples 5.50x1.68mm^2 plates were cut and polished to 100μm. Final thinning was achieved by jet electro-polishing in a solution of 70% methanol and 30% HNO_3 at a temperature of –15°C.

In- situ tensile tests were performed in a JEOL- 1200EX microscope equipped with a *double tilt* stage (x=±30°, y=±10°). A detailed description of the experimental setup is given in [9].

3. Results

The initial microstructural evolution following low levels of strain, shown in figure 1, consisted of a stage I to stage II transition microstructure. The dislocations observed were mostly easy glide dislocations. The dislocations were jogged and amongst them some dislocation loops were noticeable. Figure 1(a) shows the typical microstructure of a sample prestrained to 10% in the $[\bar{1}\bar{2}2]$ direction. This direction was chosen in order to activate only two slip systems. The dislocations were characterized using invisibility criterion in two beam conditions as $\frac{1}{2}[1\bar{1}0]$ and $\frac{1}{2}[101]$. The tensile axis (TA) is marked on the micrograph.

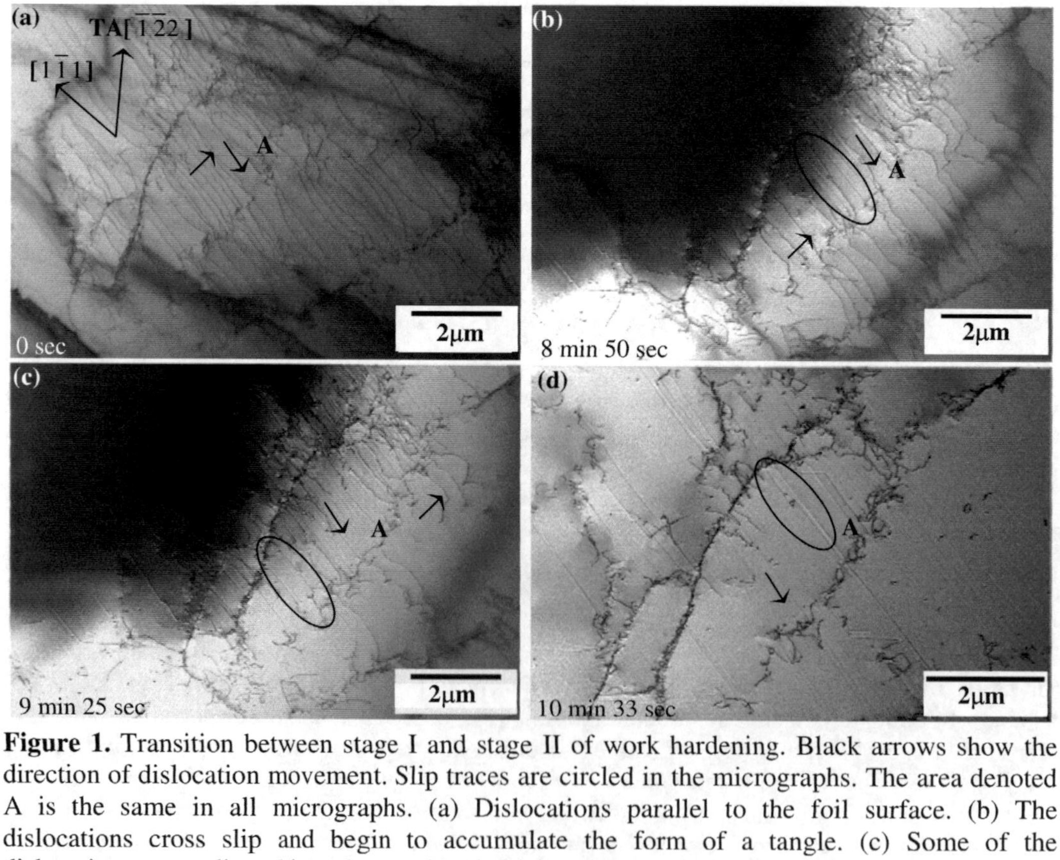

Figure 1. Transition between stage I and stage II of work hardening. Black arrows show the direction of dislocation movement. Slip traces are circled in the micrographs. The area denoted A is the same in all micrographs. (a) Dislocations parallel to the foil surface. (b) The dislocations cross slip and begin to accumulate the form of a tangle. (c) Some of the dislocations cross slipped into the tangle. (d) Dislocation tangle formed at area denoted A.

When force was applied, the dislocations bow out in $[\bar{2}\bar{1}1]$ trace direction and cross-slip towards the area denoted as **A** in Figure 1 (same area in all frames) to form a tangled dislocation boundary. Pre-existing dislocation loops and dislocations act as pinning points and barriers for dislocation movement. The direction of the movement and the cross slip can be observed via the slip traces (circled on the micrographs).

Figure 2 shows the evolution of a dislocation boundary by the cross slip of dislocations and the accumulation of dislocations within a cell boundary at stage III of work hardening in single crystals. The sample was prestrained to 13% in the $[1\bar{2}2]$ direction (TA) and the dislocations were characterized as $\frac{1}{2}[110]$ and $\frac{1}{2}[10\bar{1}]$. As force is applied, dislocations cross slip towards the boundary denoted as **A** on figure 2(a), from both sides of the boundary. The activation of two slip systems is noticeable in the two sets of slip traces inside the cells. A change in contrast between the upper cell and the lower cell in figure 2 is observed with the increase of strain, comparing figures 2(a) and 2(c), as a result of the misorientation between the two volume elements. These processes continue until the volume of the cell is depleted of dislocations.

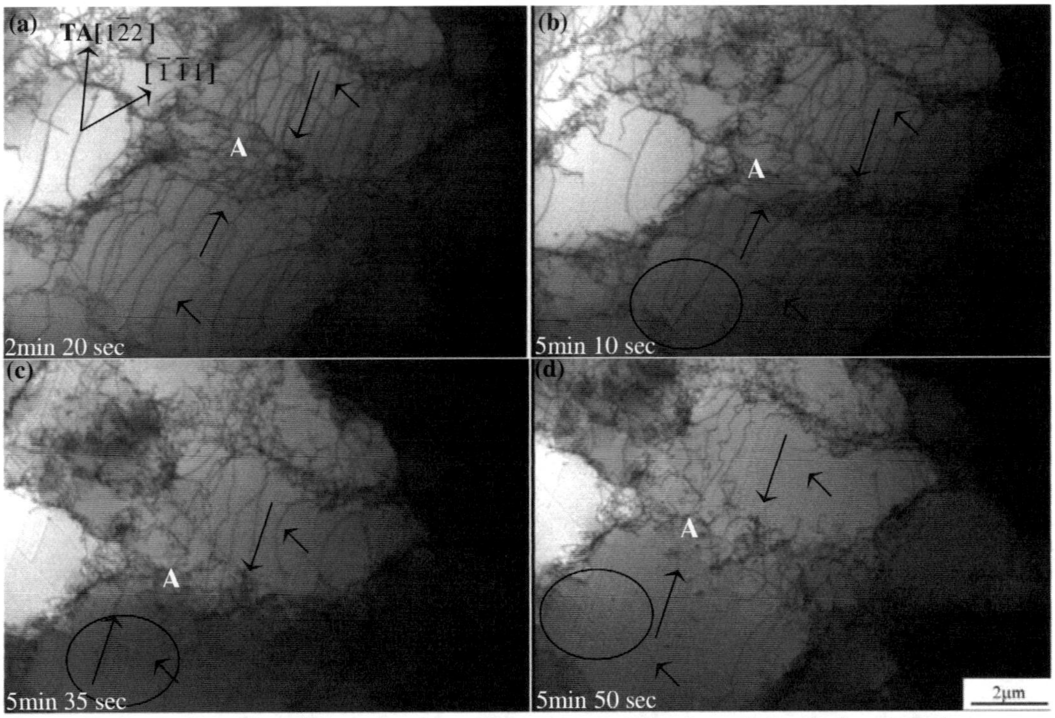

Figure 2. Evolution of a dislocation boundary at stage III of work hardening. Black arrows show the direction of dislocation movement. Area denoted A is the same area in all micrographs. (a)- (d) frames showing the depletion of the cell's volume from dislocations into the cell's boundary by cross slip of dislocations. Slip traces are circled in the micrographs.

In both cases, it is seen that cross slip of dislocations is necessary for the formation of typical dislocation patterns. The processes observed in the in-situ experiments conform to the common belief that cross-slip of dislocations is necessary for the formation of typical dislocation patterns.

4. Discussion

In-situ tensile experiments were performed on [011] single crystal aluminum in <221> orientation in order to study dislocation interactions and the formation and evolution of typical dislocation patterns. Furthermore, the possibility of imaging bulk deformation processes in thin TEM samples was examined. The samples were prestrained prior to in-situ experiments in order to form an initial microstructure that minimizes surface effects. In-situ tensile experiments show the transition between stage I and stage II of work hardening by the cross-slip of dislocations to form dislocation tangles and cross slip of dislocations in stage III to form dislocation cells. This evolution was extensively modeled and experimentally observed, mostly in post- mortem observations of deformed samples [1-8].

In order to get a comprehensive understanding of dislocation interactions, dynamic behavior during deformation processes and correct sequence of events during microstructural evolution, post-mortem observations are not merely enough, and in-situ observations are essential to get a full picture. However, there are key limitations to in-situ TEM experiments that make the dynamic processes observed questionable in whether they reflect bulk behavior [10-12]. These limitations emanate from the limited thickness of the sample and surface effects. In spite of the above mentioned limitations, we have shown that the microstructural evolution of thin *prestrained* TEM samples is similar to that of bulk samples. The transition between stage I and stage II of work hardening occurs via the cross slip of gliding dislocations. Furthermore, the evolution of the cellular structure is by the depletion of the cell's interior.

5. Conclusion

In-situ tensile experiments can contribute to the understanding dislocation interactions and microstructural evolution sequence during deformation. We have shown that dislocation cross slip is essential for the transition between stage I and stage II of work hardening, in the formation of tangled dislocation boundaries and the depletion of cell's interior during stage III. The processes observed in in-situ experiments conform to the common knowledge and modeling concerning dislocation pattern formation and evolution in bulk samples.

6. References

[1] Kuhlmann-Wilsdorf D 1989 *Mat. Sci. Eng.* A **113** 1
[2] Kuhlmann-Wilsdorf D 1987 *Mat. Sci. Eng.* **86** 53
[3] Sevillano J Aernoudt E 1987 *Mat. Sci. Eng.* **86** 35
[4] Kawasaki Y 1994 *Strength of Materials* (ed. Oikawa et al. The Japan Institute of Metals) 187
[5] Jackson P J 1985 *Prog. Mater. Sci.* **29** 139
[6] Swann P R 1963 *Electron Microscopy and Strength of Crystals* (ed. G Thomas and J Washburn, Interscience, NY,) 131
[7] Landau P, Shneck R Z, Makov G, Venkert A 2007 *J. Mat. Sci.* **42** 9775
[8] Landau P, Shneck R Z, Makov G, Venkert A 2009 *Mat. Sci. Eng.* in press
[9] Dlabacek Z, Gemperle A, Gemperlova J 2004 *Proceeding of the 13th European Microscopy Congress* (ed. D. Schryvers, J.-P. Timmermans) **1** 417
[10] Fujita H 1967 *J. Phys. Soc. Japan* **23** 1349
[11] Tabata T, Yamanaka S, Fujita H 1978 *Acta Metall.* **26** 405
[12] Hirsch P B, Howie A, Nicholson R B, Pashley D W , Whelan M J 1965 *Electron Microscopy of Thin Crystals* (Butterworth & co., London, England) 78

Acknowledgments

The authors would like to thank Dr. Viera Gartnerova from the Institute of Physics of the ASCR, Prague, Czech Republic, for her help in conducting the TEM in- situ experiments.

Elemental mapping in scanning transmission electron microscopy

L J Allen[1], A J D'Alfonso[1], S D Findlay[2], J M LeBeau[3], N R Lugg[1] and S Stemmer[3]

[1] School of Physics, University of Melbourne, Victoria 3010, Australia
[2] Institute of Engineering Innovation, School of Engineering, University of Tokyo, Tokyo 113-8656, Japan
[3] Materials Department, University of California, Santa Barbara, CA 93106-5050, U.S.A.

E-mail: lja@unimelb.edu.au

Abstract. We discuss atomic resolution chemical mapping in scanning transmission electron microscopy (STEM) based on core-loss electron energy loss spectroscopy (EELS) and also on energy dispersive X-ray (EDX) imaging. Chemical mapping using EELS can yield counter-intuitive results which, however, can be understood using first principles calculations. Experimental chemical maps based on EDX bear out the thesis that such maps are always likely to be directly interpretable. This can be explained in terms of the local nature of the effective optical potential for ionization under those imaging conditions. This is followed by an excursion into the complementary technique of elemental mapping using energy-filtered transmission electron microscopy (EFTEM) in a conventional transmission electron microscope. We will then consider the widely used technique of Z-contrast or high-angle annular dark field (HAADF) imaging, which is based on phonon excitation, where it has recently been shown that intensity variations can be placed on an absolute scale by normalizing the measured intensities to the incident beam. Results, showing excellent agreement between theory and experiment to within a few percent, are shown for Z-contrast imaging from a sample of $PbWO_4$.

1. Introduction

Using scanning transmission electron microscopy (STEM) 2D atomic resolution chemical mapping in a specimen can be obtained using electron energy loss spectroscopy (EELS) [1-3] and, more recently, energy-dispersive X-ray (EDX) analysis [4]. We will show that this is also possible in energy-filtered transmission electron microscopy (EFTEM), though more care needs to be taken to ensure that atomic resolution information is present and simulations may be needed to decode the atomic resolution information in the data [5]. Z-contrast imaging also allows for some degree of quantitative elemental mapping through the strong atomic number dependence of the signal. A complete understanding of such images requires detailed quantitative comparisons between simulations and experiments, which has only recently become possible [6,7].

2. Chemical mapping using EELS

Since atomic resolution EELS was first demonstrated in the early 1990's using STEM [8-10], there has been rapid progress. Aberration correction in the electron optics has allowed probe sizes of less than 0.1 nm and this, combined with increases in EELS detector efficiency, has allowed chemical mapping of crystals in two dimensions [1-3]. Such maps are not always directly interpretable [1,11]. The nonlocality of the interaction is expected to reduce with increasing detector collection angle (see, for instance, Ref. [3], which uses a detector angle as large as 60 mrad on a Nion UltraSTEM system). However, elastic scattering effects may still hamper simple image interpretation. This can be seen in figure 1. The O signal in <111> $Bi_{0.5}Sr_{0.5}MnO_3$ shown in (a) is consistent with the placement of those atoms, albeit rather delocalized. However the Mn signal in (b) is a minimum on the Mn columns. Simulation shows that this unexpected result arises because the probe placed between the columns interacts significantly with the delocalized transition potential and this yields more signal than that when the probe is placed on column, where thermal diffuse scattering depletes the probe. Using a system with an even larger detector angle, Wang et al. [11] obtained a contrast reversal when images were constructed from differing regions of the energy loss spectra, as shown in figure 2. This case is particularly interesting: the elastic scattering is common to both images, so the differences are purely due to the transition potentials at the lower energy loss window being more delocalized, reaching further into the open channels between the Si dumbbells.

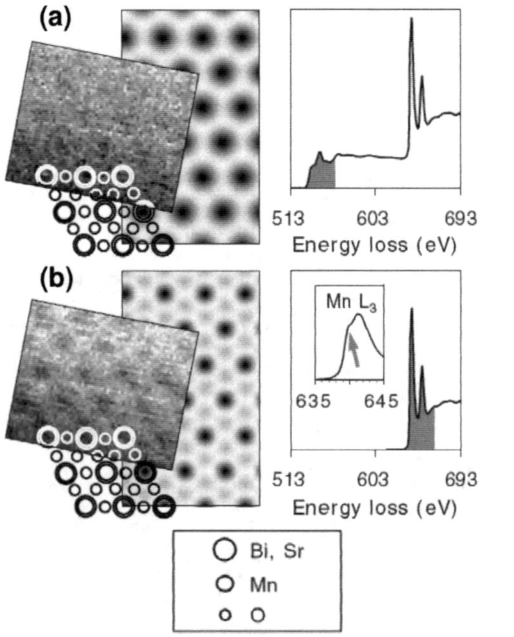

Figure 1. Experiments and simulations of $Bi_{0.5}Sr_{0.5}MnO_3$ oriented along the <111> zone axis. (a) O K edge and (b) Mn $L_{2,3}$ edge STEM EELS images. The EELS maps were generated by integrating over a 30 eV window above the respective ionization thresholds (shown in grey shading on the spectra to the right, which are summations of all the spectra in the adjacent EELS maps). The simulations assume a 40 nm thick sample.

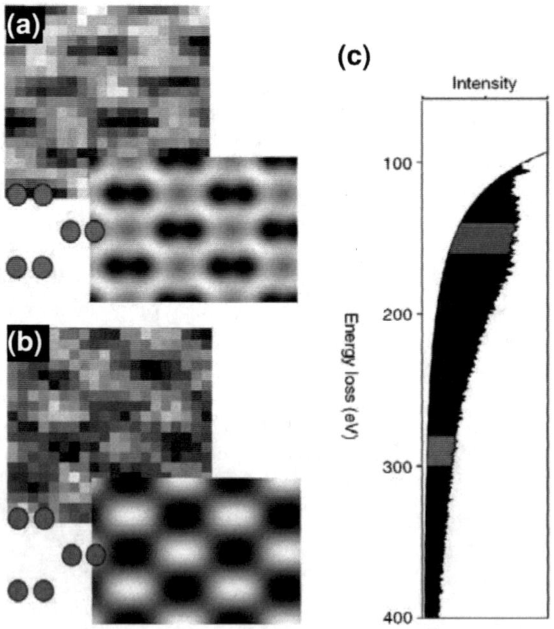

Figure 2. Comparison between experimental and simulated images of <011> Si. (a) 143–163 eV energy filtered EELS, and (b) 280–300 eV energy filtered EELS images. (c) A typical acquired Si EELS spectrum showing the background fit curve subtraction. All images were acquired simultaneously.

3. Chemical mapping using EDX

One advantage that STEM EDX has over STEM EELS is that the inelastic interaction is effectively local, like that in high angle annular dark field (HAADF). This is because all ionization dynamics are sampled: EDX is effectively equivalent to STEM EELS with a detector that covers the whole solid angle. The effective large detector averages over the phase information and leads to incoherent imaging [12,13] and the resulting image can usually be directly interpreted. The effective local scattering potentials for Ti K-shell x-ray emission and the sum of the Sr K- and L-shell potentials for x-ray emission along the <110> direction in <001> $SrTiO_3$ are compared to that for HAADF in figure 3 and exhibit a similar localization. In figure 4 we show line scans for a test sample of <001> $SrTiO_3$ based on STEM EDX and using an FEI Titan 80-300 operating at 300 keV [4]. The sample thickness was estimated to be 100 nm (a specimen of this thickness was needed to ensure sufficient x-ray counting statistics). As the probe was not aberration corrected, the probe forming aperture semi-angle was limited to 9.6 mrad and the probe was under-focused by 56 nm into the specimen. The EDX detector subtended a collection angle of 0.13 sr. Theoretical simulations are in good agreement with the experimental data [4].

Following the formulation outlined in Ref. [14], simulations confirming the data were performed using a hybrid of the frozen phonon (FPh) algorithm to account for thermal scattering [15] and the mixed dynamical form factor (MDFF) [16] to model ionization. The assumption is made that the total cross section for all x-ray emission events corresponding to filling a hole in the K- or L-shell is proportional to the total cross section for K- or L-shell ionization respectively.

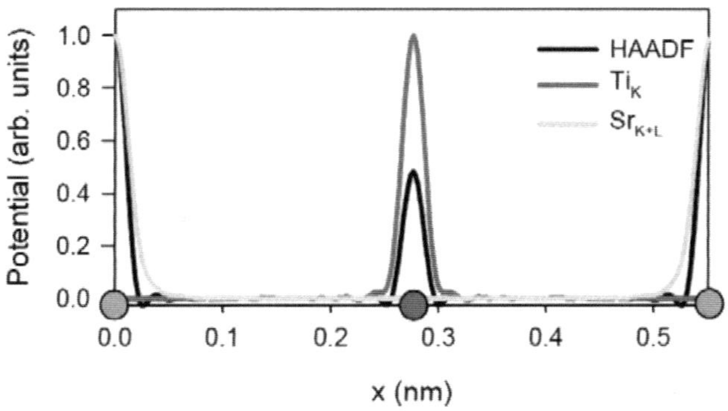

Figure 3. The effective scattering potential for HAADF for a detector spanning 60-160 mrad (black), Ti K- shell x-ray emission (red) and the sum of the Sr K- and L-shell potentials for x-ray emission (blue) for EDX. The scattering potential is for the <110> direction for a single unit cell of $SrTiO_3$. The Sr and Ti sites are indicated by the blue and red circles respectively. Each line plot has its maximum value normalized to one.

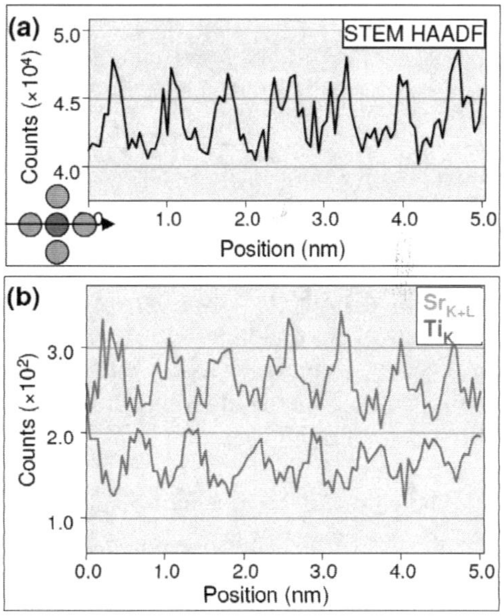

Figure 4. (a) Reference HAADF line scan image of the $SrTiO_3$ sample along <110>. (b) EDX line scans for Ti (red, below) and Sr (blue, above). The maxima in the EDX line scans correlate with the expected atomic positions.

4. Atomic resolution EFTEM

EFTEM based on inner-shell ionization can contain atomic resolution information [5]. EFTEM has the advantage that the spatial information needed for elemental mapping is acquired in parallel, allowing for precision measurements of column locations. Experimental data taken on a 6 nm thick specimen of LaB_6 viewed down the [001] zone axis and monitoring the lanthanum $N_{4,5}$ edge are shown compared to a first principles calculation in figure 5. Calculations show that the transition potentials in the vicinity of the $N_{4,5}$ edge of La (threshold energy 99 eV) are such that an EFTEM image of a suitably thick specimen of LaB_6 contains information on the location of the La columns. The simulation confirms that the brightest points in the image correspond to the positions of the La atoms. Previous work makes such a direct interpretation somewhat surprising [17,18], though the specimen is quite thin. The localization of the transition potentials is such that elastic contrast is only weakly preserved in the EFTEM image and the degree of localization is sufficient that the preservation of elastic contrast idea is not very helpful in understanding the images.

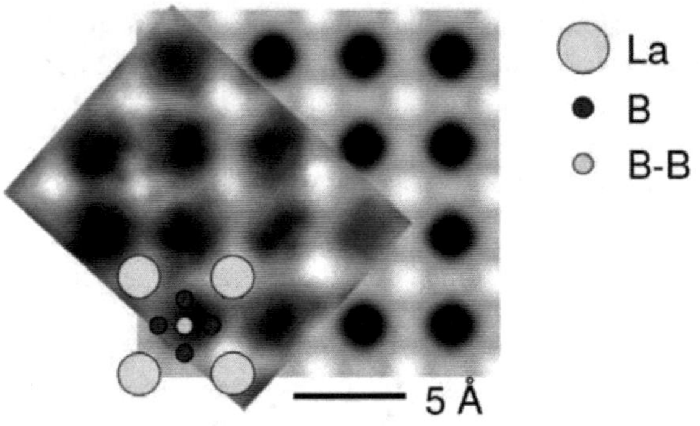

Figure 5. Experimental EFTEM image and theoretical simulation (behind) for La in LaB_6 for the experimental conditions described in the text. The projected structure is indicated, and we distinguish between columns containing one (B) and two (B-B) B atoms per unit cell.

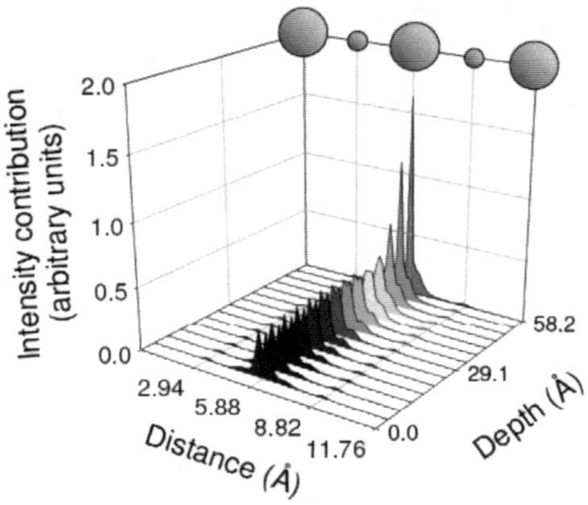

Figure 6. Line scan along [110] of the contribution to the total exit surface intensity of each La atom (fourteen in total) from a single column. The depths of the atoms in the specimen are indicated relative to the entrance surface (at 0 Å).

The calculations in figure 6 support the thesis. The signal detected at the exit surface (after further elastic scattering) from each atom is localized around the column in question. The signal from La is providing direct information as to the location of the La columns and is therefore chemical information and not phase contrast.

5. Quantitative HAADF or Z-contrast imaging

Atomic resolution HAADF or Z-contrast imaging in STEM is now finding application in the exploration of interfaces and defects at the atomic level. To first order, image contrast is sensitive to the atomic number, Z, of the various atomic species in the specimen. More subtly, the Debye-Waller factors of the atoms in the specimen and the detailed evolution of the elastic probe wave function also influence the HAADF images. LeBeau and Stemmer [19] made a significant advance in developing a method that allows HAADF images to be placed on an absolute scale, thus unlocking the quantitative potential of this form of imaging, especially when combined with first principles image simulations.

The method was first demonstrated on sample of $SrTiO_3$ [6]. Excellent agreement between experiment and calculations was obtained for a range of specimen thicknesses provided that the simulations account for thermal diffuse scattering (TDS) to all orders i.e., all multiple thermal scattering event sequences, which the frozen phonon model effectively does.

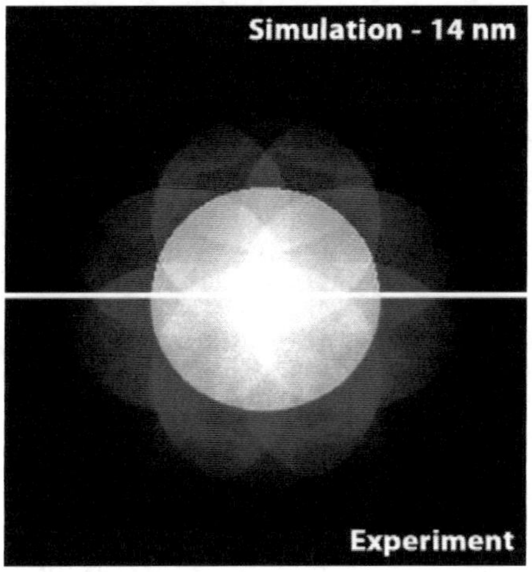

Figure 7. Simulated (top half) and experimental (bottom half) PACBED pattern along [100] for a 14 nm thick region of $PbWO_4$.

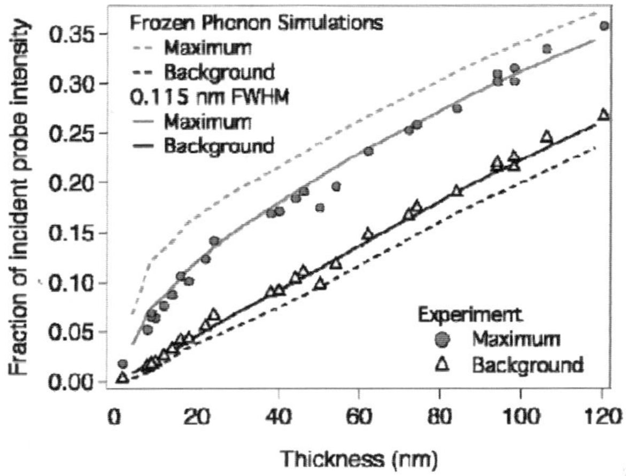

Figure 8. Comparison of experimental (symbols) and simulated (line) image signals as a function of thickness. The maximum signal and minimum ("background") signal are shown. The dashed and solid lines represent simulations without and with the effects of spatial incoherence taken into account by convolution with a 0.115 nm FHWM Gaussian.

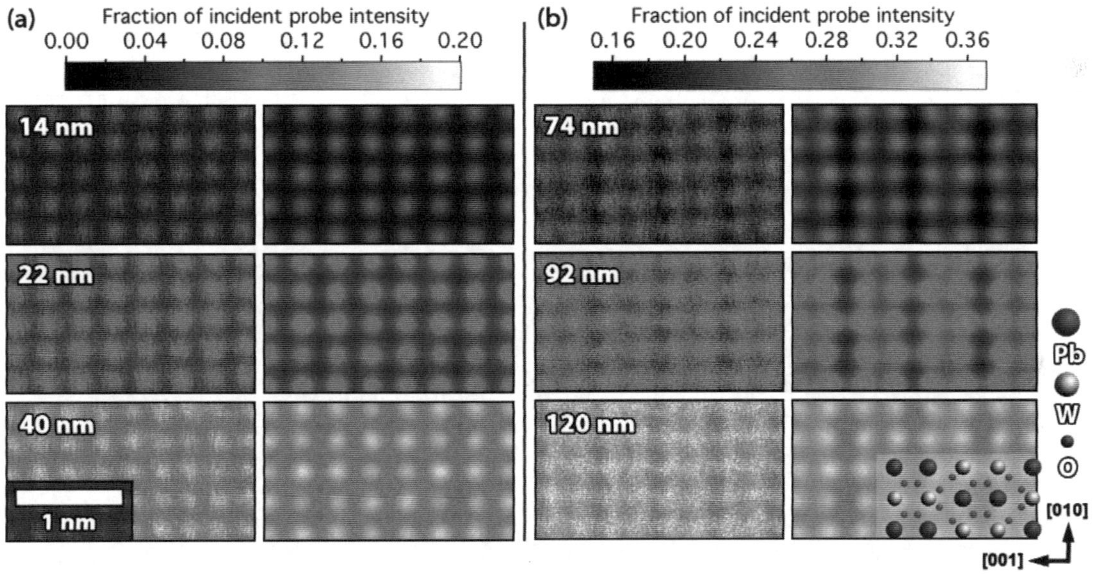

Figure 9. Experimental (left panels) and simulated (right panels) images for different thicknesses (see labels). The simulated images account for finite source size (using a 0.115 nm FWHM Gaussian distribution). Note that different intensity scales are used in (a) and (b) for clarity. The uncertainties in the experimental thicknesses are ± 2-4 nm and simulated thicknesses are within ± 2 nm of the values given.

A further study was recently carried out to see if this agreement between experiment and simulation would be maintained for crystals containing heavier elements than those in $SrTiO_3$ [7]. A single crystal sample of lead tungstate ($PbWO_4$), which contains two atoms with large charge ($Z = 82$ for Pb, $Z = 74$ for W) was used. A key step in the experiment was the accurate determination of thickness using position averaged convergent beam patterns, as illustrated in figure 7. Figure 8 shows a comparison of the experimental and simulated image signals. To model the influence of spatial incoherence, the simulated images were convolved with a Gaussian with a 0.115 nm FWHM to obtain excellent agreement between experiments and simulations. Figure 9 shows the experimental and simulated images for different thicknesses, demonstrating excellent quantitative agreement, and opening the way for quantitative Z-contrast imaging.

Acknowledgments

L.J.A. acknowledges support from the Australian Research Council. S.D.F. is supported as a Japan Society for the Promotion of Science (JSPS) fellow. The research at UCSB was supported by the U.S. National Science Foundation (grant number DMR-0804631). J.M.L. also thanks the U.S. Department of Education for a fellowship under the GAANN program (grant number P200A07044). The work made use of the UCSB MRL Central facilities supported by the MRSEC Program of the National Science Foundation under award No. DMR 0520415.

References

[1] Bosman M, Keast V J, Garcia-Munoz J L, D'Alfonso A J, Findlay S D and Allen L J 2007 *Phys. Rev. Lett.* **99** 086102
[2] Kimoto K, Asaka T, Nagai T, Saito M, Matsui Y and Ishizuka K 2007 *Nature* **450** 702
[3] Muller D A, Fitting Kourkoutis L, Murfitt M, Song J, Hwang H, Silcox J, Dellby N and Krivanek O 2008 *Science* **319** 3
[4] D'Alfonso A J, Freitag B, Klenov D and Allen L J 2009 *Phys. Rev.* B submitted for publication
[5] Lugg N R, Freitag B, Findlay S D and Allen L J 2009 *Ultramicroscopy* submitted for publication
[6] LeBeau J M, Findlay S D, Allen L J and Stemmer S 2008 *Phys. Rev. Lett.* **100** 206101
[7] LeBeau J M, Findlay S D, Wang X, Jacobson A J, Allen L J and Stemmer S 2009 *Phys. Rev.* B **79** 214110
[8] Browning N D, Chisholm M F and Pennycook S J 1993 *Nature* **366** 143; Corrigendum 2006 **444** 235
[9] Muller D A, Tzou Y, Raj R and Silcox J 1993 *Nature* **366** 725
[10] Batson P E 1993 *Nature* **366** 727
[11] Wang P, D'Alfonso A J, Findlay S D, Allen L J and Bleloch A L 2008 *Phys. Rev. Lett.* **101** 236102
[12] Dwyer C 2005 *Ultramicroscopy* **104** 141
[13] Findlay S D, Oxley M P and Allen L J 2008 *Microsc. Microanal.* **14** 48
[14] Findlay S D, Oxley M P, Pennycook S J and Allen L J 2005 *Ultramicroscopy* **104** 126
[15] Kirkland E J 1998 *Advanced Computing in Electron Microscopy* (New York and London: Plenum Press)
[16] Allen L J, Findlay S D, Oxley M P and Rossouw C J 2003 *Ultramicroscopy* **96** 47
[17] Stallknecht P and Kohl H 1996 *Ultramicroscopy* **66** 261
[18] Verbeeck J, Schattschneider P and Rosenauer A 2009 *Ultramicroscopy* **109** 350
[19] LeBeau J M and Stemmer S 2008 *Ultramicroscopy* **108** 1653

Journal of Physics: Conference Series **241** (2010) 012062 doi:10.1088/1742-6596/241/1/012062

Electronic structure of oxide fuels from experiment and first principles calculations

J A Aguiar[1,2 *], N Grönbech-Jensen[2], A Perlov[3], V Milman[3], S P Gao[4], C J Pickard[5], N D Browning[1,2]

[1] Lawrence Livermore National Laboratory, 3000 East Avenue, Livermore, California, 94520, USA
[2] University of California Davis, 1 Shields Avenue, Davis, California, 95616, USA
[3]Accelrys Inc, 334 Science Park, Cambridge CB4 0WN, UK
[4] Department of Materials Science, Fudan University, Shanghai 200433, People's Republic of China
[5] University College London, Gower St, London, WC1E 6BT, UK

E-mail: aguiar9@llnl.gov

Abstract. Energy loss spectra from a variety of cubic oxides are compared with *ab-initio* calculations based on the density functional plane wave method (CASTEP). In order to obtain agreement between experimental and theoretical spectra, unique material specific considerations were taken into account. The spectra were calculated using various approximations to describe core-hole effects and electron correlation. The calculations are based on both the generalized gradient approach and the local spin density approximation when dealing with the correlation in to show qualitative agreement with the sensitive oxygen *K*-edge spectra in ceria, zirconia, and urania. Comparison of experimental and theoretical results let us characterize the main electronic interactions responsible for both the electronic structure and the resulting EELS spectra of the compounds in question.

Introduction

In the past few years, there have been growing efforts to apply modern first principles materials modelling tools as a framework for accelerated materials design and development. One area where such modelling efforts can make unique contributions is in the understanding of the evolution of the mechanical properties of advanced nuclear fuels (key to many future reactor designs). Establishing a damage evolution model for oxide fuels currently suffers from inaccurate models where there is lack of data due to a lack of experimental studies. As damage mechanisms need to be extrapolated for the expected 50 year lifetimes of reactors, the absence of models that accurately address the energetic contributions of all the induced defects limits the applicability of the theoretical models. Direct validation of the theoretical techniques, in their application to oxides, requires direct experimental measurements of structure, composition and electronic interactions at the nanometer scale – in

* To whom any correspondence should be addressed.

particular to observe the atomic and electronic structure of relevant point defects and their extended complexes. Model validation can be achieved by acquiring images and an electron energy loss spectrum (EELS) in a scanning transmission electron microscope (STEM). As we are interested in oxide fuels, the oxygen *K*-edge is chosen to compare experiment and theory directly, and here we attempt a first validation of the theory by studying three fluorite structured oxides - ceria, zirconia, and urania. Within the density functional theoretical framework we employ a variety of calculations, but for the purposes of this presentation both the generalized gradient (GGA) and local spin density approximation (LSDA+U) with unique materials considerations using the CASTEP code is considered. The materials considerations include correlated f-states, magnetic ordering, and in view of the direct comparison to STEM/EELS the core-hole effect.

Methods

1.1. Experimental Analysis

Standard ceria, zirconia, and urania powder samples were dispersed on 3 mm standard holey carbon TEM grids. The experimental spectra were obtained from the double corrected FEI Titan at Lawrence Livermore National Laboratory (LLNL) operating in STEM mode at 300 keV. EELS spectra were obtained with a Gatan Tridiem electron energy loss image filter. The energy resolution is 0.7 eV based on the full width half maximum of the zero-loss peak. The acquisition time for each spectrum is 5 seconds with 0.1 eV/pixel dispersion.

1.2. Theoretical Treatment

All calculations presented below are carried out using the CASTEP code [1,2]. CASTEP uses a plane wave basis set and pseudopotentials within the density functional theory formalism. Perdew, Burke, and Erzenhof (PBE) version of the generalized gradient approximation (GGA) was used to describe ceria and zirconia. To describe localized *f*-orbitals of U, in urania we employed the LSDA+U approach [3] with an effective Hubbard U value of 6 eV to achieve an antiferromagnetic state. Ultrasoft pseudopotentials were generated using the "on the fly" formalism in CASTEP, and one core electron was removed from a relevant core level when performing core–hole calculations [4]. All core–hole calculations were carried out using supercells sufficiently large to eliminate unphysical interactions between periodic images. Previous experience from all-electron calculations gives 8–10 Å as a recommended distance between periodic images. The supercells used in the current study corresponded to the distances between atoms with core holes of 10.822, 10.14, and 10.94 Å for ceria, zirconia, and urania respectively. Crystal symmetry is reduced when a core-hole is introduced in a supercell. All calculations were carried out by using reduced symmetry rather than by ignoring point group symmetry altogether; even a reduced symmetry allows calculation speed to be increased by considering only *k*-points in the irreducible part of the Brillouin zone. All calculations furthermore show a converged antiferromagnetic ground state.

Results

Normalized measured electron energy loss oxygen *K*-edge spectra in ceria, zirconia, and uranium dioxide are compared below with the theoretically predicted spectra from the CASTEP code. All spectra were aligned based on the position of the first experimental peak.

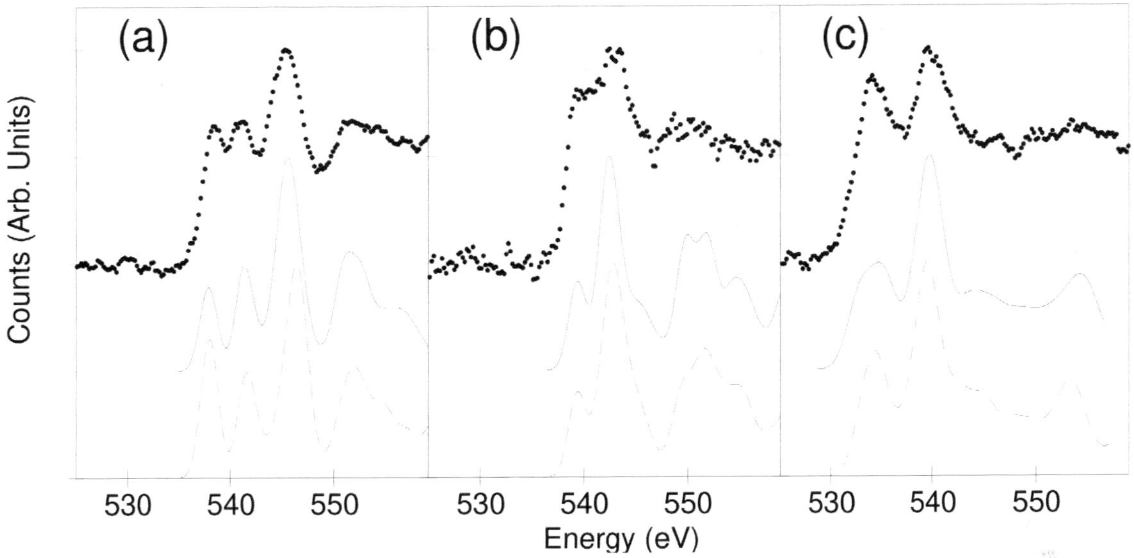

Figure 1. Oxygen *K*-edge spectra for ceria (a), zirconia (b) and urania (c). Experiment (dots) is compared to CASTEP calculations with (solid line) and without the core-hole effect (dashed line).

Discussion

Building in complexity we have examined the electronic structure of non-correlated and correlated fluorites. When treated correctly across these fluorites structures we predict insulating ground state as confirmed by the analysis of their density of states. EELS *K*-edge spectra that incorporate information about oxygen partial density of *p*-states and an energy dependent matrix element were calculated for three transition metal oxides. The resultant spectra were broadened with a Gaussian (width of 0.8 eV for ZrO_2 and CeO_2, and 1 eV for UO_2) to simulate experimental resolution. Across the fluorite oxides, we compare calculated and experimental spectra to quantify the effect of taking core hole into account.

The calculated spectra of ceria, zirconia, and urania show broad agreement with the experiment with or without the core-hole effect. Taking into account the core hole shows a marginally better agreement in some cases, but it appears to be material specific. The core hole effect is suggested in the literature to increase the qualitative agreement with experiment by taking into account the fundamental electronic excitation from 1*s* to a *p*-like state [5]. The theoretical spectrum for ceria overestimates the first peak intensity unless core hole effect is taken into account (Figure 1a). In the zirconia case, the near edge region of the spectrum remains largely unchanged by the incorporation of the core hole effect, but the post peak around 550 eV shows better agreement with experiment when the core hole effect is taken into account (Figure 1b). In the correlated urania case, the core hole effect broadens the first peak and post edge peak to show increased agreement with experiment (Figure 1c). However, in all cases, even calculations without the core hole effect produce correct peak positions and qualitatively correct spectra compared with the experimental EELS results.

It is important in electronic structure calculations for correlated materials to use a Hubbard U value that corrects the DFT results sufficiently. In the case of urania the requirement is to predict an insulating ground state, so the correction must open the band gap and tune the overlap between the *p* and *f*-partial density of states above the Fermi level. Without a Hubbard U correction for urania, first-

Journal of Physics: Conference Series **241** (2010) 012062 doi:10.1088/1742-6596/241/1/012062

principles calculations predict a metallic ground state [3]. The ongoing issue involves the value of U-J that one must use to predict agreement across a variety of material properties such as band gap and spectra. The U-J value of 6 eV used in this study produces good agreement with the experimental EELS spectrum as demonstrated in Figure 1c. Dudarev *et al.* used a slightly smaller value of U-J = 4 eV [6]; it appears that either value predicts correct spectroscopic character associated with the oxygen *K*-edge for urania.

Across the fluorite structures, predicting the electronic structure based on the band structure method in CASTEP shows broad agreement with experiment in terms of peak positions and relative intensities. Incorporation of the core-hole effect has only a marginal effect on the spectroscopic character for the fluorites considered in this study.

Conclusion

The oxygen *K*-edge EELS spectra for the oxides agree qualitatively with the theoretical electronic structure calculations based on the CASTEP code. Taking into account the core-hole effect shows even better qualitative agreement with experimental EELS spectra. Further theoretical and experimental work will address the influence of structural defects on EELS spectra in these materials.

Acknowledgments

The authors would like to credit US Department of Energy Nuclear Energy Research Initiative for Consortia contract number DR-FG07-071D14893.This work performed under the auspices of the U.S. Department of Energy by Lawrence Livermore National Laboratory under Contract DE-AC52-07NA27344. The authors would also like to credit the following collaborators Mark Asta, Warren Pickett, John Bradley, Zurong Dai, Sean McNary.

References

[1] Clark S J, Segall M D, Pickard C J, Hasnip P J, Probert M I J, Refson K and Payne M C 2005 *Z. Kristall.* **220** 567
[2] Accelrys 2008 Materials Studio 4.4
[3] Anisimov V I, Zaanen J, and Andersen O L 1991 *Phys. Rev.* **B44** 943
[4] Gao S-P, Pickard C J, Perlov A and Milman V 2009 *J. Phys.: Condens. Matter* **21** 104203
[5] Mo S-D, Ching W Y V 2000 *Phys. Rev.* **B62** 7901
[6] Dudarev S L, Castell M R, Botton G. A, Savrasov S Y, Muggelberg C, Briggs G A D, Sutton A P and Goddard D T 2000 *Micron* **31** 363

Computational EELS modelling of magnesium oxide systems

C R Seabourne[1*], A J Scott[1], R Brydson[1], V Harncharna[1], S G Wang[2], R C C Ward[2], C Wang[3] and A Kohn[3]
[1]Institute for Materials Research, University of Leeds, Leeds LS2 9JT UK
[2]Department of Physics, University of Oxford, Oxford OX1 3PU UK
[3]Department of Materials, University of Oxford, Oxford OX1 3PH UK

E-mail: chmcrs@leeds.ac.uk

Abstract. An area of contention in ELNES simulations using density functional theory (DFT) has been whether geometry optimisation is entirely necessary to obtain useful results. This issue is addressed for bulk magnesium oxide by comparison of results with and without energy minimisation of structures. An MgO (001) surface was investigated. Experiments were carried out to determine how the predicted oxygen K edges varied in different layers extending from the surface into bulk. Similarly the comparison of the MgO oxygen K edge in bulk and at the interface of Fe(001)/MgO(001) is of interest for spintronic devices. For this system, the oxygen K edge shows pre-edge intensity as compared to bulk MgO, which we attribute to additional states available due to transition metal *d* / oxygen *p* orbital interactions at the interface.

1. Introduction

This work considers the energy-loss near edge fine structure (ELNES) for oxygen K edges in various MgO systems. All calculations presented have been carried out using the *ab-initio* density functional theory (DFT) code CASTEP [1].

2. Methodology

In an EELS ionisation event, an electron is promoted from a core-state to a higher lying conduction band. This leaves a 'core-hole' in its place. CASTEP uses a plane wave basis set, and pseudopotentials to approximate exchange and correlation effects. The projector augmented wave (PAW) approach can be used to model core-holes using CASTEP [2-4]. In these experiments, it was considered important to converge the key user-controlled parameters in a DFT code; namely the kinetic energy cut-off and choice of k-point mesh density [5]. A single cell of MgO was constructed in the cubic rock-salt structure ($a=b=c$ = 4.21Å, $\alpha=\beta=\gamma$ = 90°), and for one of the oxygen atoms a single electron core-hole was put in place. The oxygen K edge was then converged against the key DFT code parameters. For the kinetic energy cut-off, the value was changed sequentially in 100eV intervals from 300eV to 600eV. Upon each increase, for each data point obtained (on a consistent set of energy axis values) the percentage variation in the edge intensity was found, and those percentage variations were averaged into a single value. The parameter was considered to be converged when the average percentage variation was 1.25% or less. In the EELS section of the calculation, the k-point mesh was increased from 12x12x12 to 48x48x48 in 'steps' of 12x12x12, with the same convergence criterion applied.

* To whom any correspondence should be addressed.

Journal of Physics: Conference Series **241** (2010) 012063 doi:10.1088/1742-6596/241/1/012063

Table 1. Chosen calculation parameters for a single MgO cell with a core-hole in place.

Parameter	Chosen value
Kinetic energy cut-off	600eV
EELS calculation k-point mesh	24 x 24 x 24 (actual spacing – 0.0099Å^{-1}), modified to a spacing of 0.0197Å^{-1} for surface and interface studies.
Core-hole	Single electron core-hole in place

For the subsequent large interface calculations, it was observed that this set of k-points was not computationally practical, so was revised as is explained in Table 1. For bulk MgO the observable differences (when the spectra were plotted) between results correlating to the 'ideal' and 'revised' set of parameters were negligible.

3. Results

The role of geometry optimisation was considered. The dimensions of a single cell of MgO were systematically varied (with a core-hole in place), and system energies found. The unit cell dimensions that resulted in the lowest system energy were then considered to be 'geometry optimised'. The resultant unit cell dimensions obtained were $a=b=c$ = 4.47Å, $\alpha=\beta=\gamma$= 90°. In these calculations the electron taken from the core was placed into a conduction state so one might not necessarily expect a contraction of the unit cell. A predicted oxygen K edge was found for this system, and was observed to be at least qualitatively similar to the result for the experimental lattice parameters, though did not display the doublet feature discussed subsequently. Therefore, it was considered acceptable to use experimental lattice parameters in the remaining work. In Figure 1 therefore, a non-geometry optimised single cell result (with a single electron core-hole in place) is compared to experiment [6]:

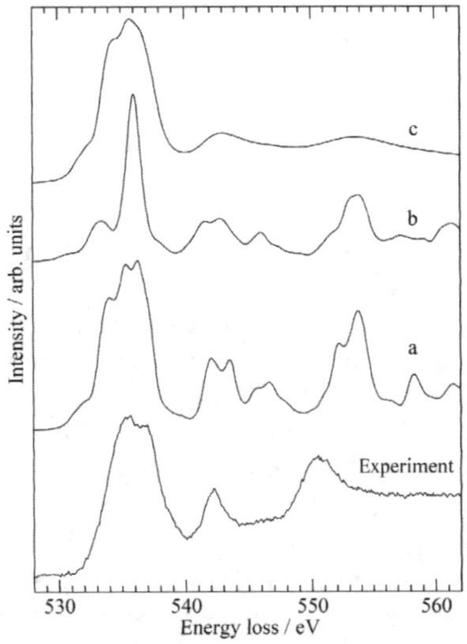

Figure 1. Comparison of theoretical and experimental results for bulk MgO [6]. a) Theoretical result shown with the addition of a 0.8eV Gaussian broadening, but no lifetime-broadening scheme, with a core-hole in place, b) shows a comparison with the ground state result. c) The addition of an energy-dependent final state lifetime broadening scheme to result a), with a core-hole in place. [7]. The experimental result was obtained using a VG HB501 STEM retro-fitted with a Nion spherical aberration corrector and a Gatan Enfina electron energy loss spectrometer at the Daresbury Laboratories, UK (SuperSTEM1). The FWHM of the zero-loss peak was 0.7eV.The theoretical spectra have been aligned to the first peak of the experimental result.

The qualitative agreement with experiment is reasonable, especially for the 0.8eV Gaussian-broadened result. The 'doublet' nature of the first peak in the spectrum is reproduced for example. Furthermore the theoretical result shows qualitative agreement with Lindner et al's seminal paper on MgO ELNES [8]. As can further be observed by comparing results a) and b), the ground state result is qualitatively similar to the core-hole result, displaying similar energy separations in terms of the main sets of peaks – though the doublet feature as seen in experiment is not observed. As mentioned previously, this was also not observed with the geometry optimised result. However, the agreement was considered sufficiently close to use ground state calculations for the surface and interface systems.

For an (001) surface of MgO (with a mixed oxygen and magnesium terminated surface, vacuum thickness 22.7Å), predicted oxygen K edges were found at different positions; at the surface (B), atom (C) (a distance of 4.21Å from a surface) and atom (D) (a distance of 6.32Å from a surface). These calculations were carried out in the ground state. Figure 2 shows the results. As can be observed, by a distance of at most 6.32Å into the surface the result matches that of 'true' bulk (on this occasion calculated in the ground state). These results led to an investigation of Fe (001)/MgO(001) interfaces.

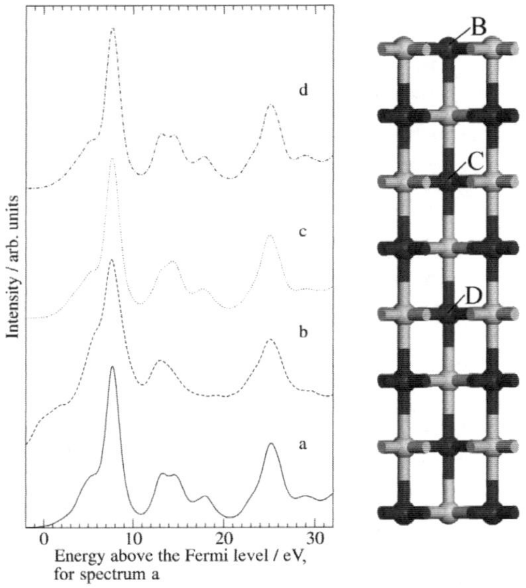

Figure 2. a) Standard bulk MgO predicted oxygen K edge, using the revised parameters (see Table 1) in the ground state. b, c and d are atomic positions as labelled. All results are subject to energy-independent broadenings of 0.2eV Lorentzian and 0.4eV Gaussian, aligned by the first peak of spectrum a. Spectra c and d begin to rapidly approach the bulk result. Spectrum b (the surface oxygen result) shows a pre-edge shoulder.

Magnetic tunnel junctions (MTJs) based on Fe/MgO interfaces can act as magnetic sensors or as part of non-volatile memory elements for example [9]. There has been some suggestion in the literature that these interfaces may not be entirely sharp, and a FeO monolayer might form [10,11]. Therefore, one might postulate two possible structures - depicted in Figure 3; a clean interface and an 'oxidised' one. Various ground state calculations were performed. Firstly for both structures, deep in the MgO 'slab' oxygen K edge predictions were obtained, and it was verified that these matched the simulation for the true bulk result (spectrum a in Figure 2). Indeed, for both the clean and oxidised interfaces (spectra b and c) the MgO 'slab' results were a good match for the true bulk result – suggesting a valid construction of the theoretical cells. Therefore, at interface positions 1, 2 and 3 predicted oxygen K edges were found, as are also shown in Figure 3. For results 1-3, there is additional 'pre-edge' intensity as compared to bulk. This can be argued to be as a consequence of oxygen $2p$ / Fe $3d$ orbital interactions. These interactions will create additional lower-lying unoccupied energy states, hence the 'pre-edge' intensity. For the 'oxidised' interface structure (spectra 2 and 3), particularly interesting results are observed. As compared to spectrum 1 (for the clean interface), spectra 2 is more 'bulk-like' in appearance, but at position/spectra 3, the pre-edge intensity is very high as compared to all the other results.

Experimental results have been added to Figure 3, which have been subject to multi-variate statistical analysis [10-13]. These results are perpendicular line scans of an interface, with 0.3eV energy resolutions obtained at SuperSTEM [14]. The first experimental result is a line scan from the MgO 'slab' of the MTJ. As can be observed, this result matches theory reasonably well. The second experimental result is from a line scan at the interface. In this instance the best match (based on comparison of peak ratios for the pre-edge peak and the first peak in the 'bulk' MgO spectrum) is with the predicted result at position '1' of the clean interface. Further work will be carried out on rationalising experimental results for other interfaces, via comparison with theoretical models.

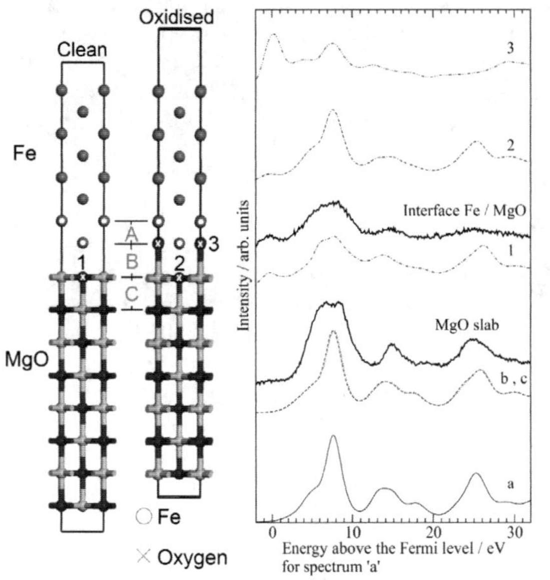

Figure 3. a) Standard bulk MgO predicted oxygen K edge, using the revised parameters (see Table 1) in the ground state. b) The 'bulk-slab' result for the clean interface c) is the 'bulk-slab' result for the oxidised interface. Spectra 1-3 correspond to the labelled oxygen positions at the Fe/MgO interfaces as shown. All results are shown with a 1.0eV Gaussian broadening. Theoretical results are aligned by the first peak of spectrum a. Experimental results have been aligned relative to the closest matching theoretical results. The experimental results have a 0.3eV energy resolution and have been subject to multi-variate statistical analysis [12,13]. Distances; A = 0.149nm, B = 0.235nm, C = 0.224nm [10,11]. For each of the unit cells, a=b = 0.292nm, c= 3.209nm.

4. Conclusions

Good agreement with experiment was observed for the oxygen K edge in bulk MgO, without geometry optimisation, thus saving significant computational effort. These studies were then extended to an MgO (001) surface. Two possible Fe/MgO interfaces were investigated, predicted results being compared to experiment with reasonable matches obtained – though further work is required.

References

[1] Clark S J, Segall M D, Pickard C J, Hasnip P J, Probert M I J, Refson K and Payne M C 2005 Zeitschrift für Kristallographie 220 567-570.
[2] Blöchl P E 1994 Phys. Rev. B 50 17953-17979.
[3] Pickard C J, *Ab Initio* Electron Energy Loss Spectroscopy Ph.D. thesis (1997).
[4] Gao S P, Pickard C J, Payne M C, Zhu J and Yuan J 2008 Phys. Rev. B 77 115122.
[5] Seabourne C R, Scott A J, Brydson R and Nicholls R J 2009 Ultramicroscopy, in press.
[6] Thanks to G. Vaughan, University of Leeds.
[7] Moreau P, Boucher F, Goglio G, Foy D, Mauchamp V and Ouvrard G 2006 Phys. Rev. B 73 195111.
[8] Lindner Th, Sauer H, Engel W, Kambe K 1986 Phys. Rev. B 33 22-24.
[9] Yuasa S, Nagahama T, Fukushima A, Suzuki Y and Ando K 2004 Nature Mater. 3 868-871.
[10] Wang C, Wang S, Kohn A, Ward R C C and Petford-Long A K 2007 IEEE Transactions on Magnetics 43 2779-2781.
[11] Wang S G, Ward R C C, Du G X, Han X F, Wang C and Kohn A 2008 Phys. Rev. B 78 180411.
[12] Trebbia P and Mory C 1990 Ultramicroscopy 34 179-203.
[13] Lozano-Perez S 2007 Microsc. Microanal. 13 (Suppl S02) 1258-1259.
[14] Thanks are given for the assistance of B. Mendes.

Solving for the phase of STEM probes in real space.

K M Atkinson [1], F Sweeney and J M Rodenburg

University of Sheffield, Department of Electronic and Electrical Engineering, Mappin Street, Sheffield, S1 3JD, UK

[1] E-mail: elp06kma@sheffield.ac.uk

Abstract. The combination of a Gerchberg-Saxton phase retrieval algorithm with a genetic algorithm is used to retrieve the phase of a TEM or STEM probe and pinpoint its defocus. Results from modelled data and initial experimental work are included.

1. Introduction

The intensity distribution of a probe in TEM or STEM can easily be recorded but the corresponding phase is unknown. Knowledge of this information is essential in using phase retrieval methods such as ptychography. Although in STEM it is possible to calculate aberration constants from the Ronchigram, errors in these values can change when moving from a defocused probe to a focused probe. We have attempted to by-pass these problems by retrieving the phase from a series of several intensity measurements at varying levels of defocus.

2. Phase Retrieval

As the intensity images of the probe are recorded without any known background support this rules out using many phase retrieval methods, however we can use Gerchberg-Saxton method [1]. This technique uses two intensity images recorded at different distances (defoci) from an object. An estimate of the phase is then combined with the known amplitude of the first image to produce a complex wave. This wave is propagated to the second image plane. The resulting amplitude is then compared with the actual amplitude recorded in the second plane and the difference used to form an error metric. The estimated phase is retained whilst the estimated amplitude is discarded and replaced by the correct amplitude for the second plane. The wave is then propagated back to the first plane and the process repeated. By continuing to iterate between the two images the correct phase is reached.

Here we have extended this technique to include more than two images, propagating from one to the next until the final image in the sequence is reached at which point we propagate the wave directly back to the first image. Tests on simulated probe images showed that the use to four of five images rather than just two significantly increased the speed at which the algorithm converged.

3. Genetic Algorithm

The Gerchberg-Saxton algorithm relies on the propagation distance between images been known accurately. If incorrect propagation distances are used the phase estimate may still improve over successive iterations however the error metric will never reach zero. When recording images the microscope can provide an estimated defocus however this is unlikely to be accurate enough. We must therefore search through a range of potential defocus values close to this estimate for each probe image so there will be many possible combinations of values for a sequence of several images.

We have chosen to employ a genetic algorithm to search through this range of combinations. The algorithm performs the Gerchberg-Saxton phase retrieval on a selection of combinations for a set number of iterations and compares the resulting error metrics to rank the selected combinations from best to worst. The next generation of combinations is then created by the techniques of crossover and mutation [2]. It should contain many combinations similar to the best combinations from the previous generation in order to examine these sections of the search space more thoroughly, however it should also contain some significantly different combinations to avoid being caught in local minima. Over successive generations this algorithm should identify the most suitable combination of defocus values to use in phase retrieval.

4. Results

Initially the combined Gerchberg-Saxton genetic algorithm (GS-GA) was tested with modelled data to establish if it was working correctly and how robust it was under different circumstances. Afterwards some initial experimental data was evaluated using the algorithm.

4.1. Testing with modelled data

The model consisted of four probe images with differing defocus values, -600nm, -330nm, +30nm (focus) and +345nm respectively (where a positive value indicates overfocus and a negative value indicates underfocus). These probe images are all generated from an aperture with a semi-angle of 5mrads. The algorithm is supplied with amplitude images of each of these probes, an initial guess of -600nm, -300nm, 0nm and +300nm for the defocus values with a search range of +/-50nm for each value and was run over 200 generations.

Table 1 shows the results of these tests. When the modelled probes have no aberrations the algorithm has incorrectly fixed on an underfocused value rather than an overfocused one for probe 3. This is not surprising as without the presence of any aberrations the probe intensity will be identical at the same distance either side of focus. If the search range for the defocus value of probe 3 is reduced or spherical aberration is introduced then this problem no longer occurs. The presence of spherical aberration or any other aberration who's effects differ either side of focus will eliminate this problem in experimental data.

Table 1 – Results from GS-GA using modelled data, including the effects of spherical aberration and noise.

	Defocus (nm)			
Probe characteristics	**Probe 1**	**Probe 2**	**Probe 3**	**Probe 4**
True defocus values	-600.00	-330.00	30.00	345.00
Unaberrated probes	-617.33	-347.33	-47.30	327.70
Probe 3 restricted to overfocus	-606.15	-336.15	23.83	338.82
Spherical aberration, Cs=0.5mm	-607.45	-337.46	22.61	337.62
Extra 50 generations with full size probe images	-595.52	-325.53	-25.54	349.47
Cs=0.5mm and Poisson noise	-594.66	-331.24	25.58	345.96
Cs=0.5mm and salt and pepper noise	-598.36	-290.93	-14.60	307.75

The modelled probes were originally generated as images of 1024x1024 pixels. To speed up the running time of the algorithm these were reduced to 128x128 pixels. Using the original larger images for an extra 50 generations at the end of the algorithm did give a slight improvement in accuracy although the results showed only the first 5 of these generations were necessary.

The effect of noise was also simulated. Surprisingly the results generated in the presence of Poisson noise are actually the most accurate results in Table 1 however the results from salt and pepper noise simulations show that large quantities of noise can reduce the accuracy of the algorithm.

Interestingly the result for the unaberrated probes in Table 1 seems to accurately predict the difference in defocus between consecutive probes (except for the erroneous result for probe 3) but the defocus values are incorrect by -17.3nm. This indicates that the algorithm may be most effective in calculating the difference in defocus between two consecutive probes. With this in mind the algorithm was run again with the same test data but was altered to output the difference in defocus between consecutive probes rather than specific defocus values. The algorithm was given an initial guess that all probes were separated by 300nm +/- 100nm. The results of this are shown in Table 2.

Again we can see that without any aberrations probe 3 produces erroneous results however with some spherical aberration introduced the results appear to be more accurate than those in Table 1. Unfortunately this approach does seem to be less resilient to noise.

Table 2 – Results from GS-GA searching only for difference in defocus between consecutive probes.

	Defocus change (nm)		
Probe characteristics	Probe 1-2	Probe 2-3	Probe 3-4
True defocus changes	270.00	360.00	315.00
Unaberrated probes	270.03	299.91	375.00
Probe 3 search range 350+/-50nm	269.97	300.03	375.00
Probe 3 search range 375+/-25nm	269.97	360.02	315.02
Spherical aberration, Cs=0.5mm	269.97	359.99	315.02
Cs=0.5mm and Poisson noise set1	263.23	282.26	390.29
Cs=0.5mm and Poisson noise set2	282.08	365.62	217.21
Cs=0.5mm and salt and pepper noise	306.50	268.11	317.22

4.2. Experimental data

Data was obtained on a JEOL 2010F microscope at the University of Sheffield. Sequences of probe images were obtained, beginning with a highly underfocused probe and incrementing the defocus in steps of 32nm or 64nm, recording the image at each point until a highly overfocused probe was reached. Selections of four probes were taken from these sequences to form data sets. An estimate for the defocus of a probe in the sequence was obtained from the relative change in defocus between that image and the image obtained closest to focus. These estimates were supplied to the algorithm to form an initial guess around which to search.

Table 3 shows the results from a typical pair of overlapping data sets. Probe 1 and 2 in data set A are the same as probes 3 and 4 in data set B. By including a pair of probes in two different data sets we were able to test whether the results were consistent with one another. Each data set was put through two versions of the algorithm, one to find specific defocus values for each probe and the other to look at only the difference in defocus between probes. We can see from Table 3 that the results from both versions of the algorithm are self consistent for each data set. We found this to be true of most but not all data sets tested so far. Unfortunately the results from the two data sets are not consistent with each other despite containing two of the same probes.

Journal of Physics: Conference Series **241** (2010) 012064 doi:10.1088/1742-6596/241/1/012064

The convergence of the algorithm was monitored over each generation. For modelled data the error metric decreased rapidly over the first 50 generations then further small decreases were observed up to the 200th generation when the algorithm terminated. For the experimental data however a decrease was only observed over the first 20 generations then the error varied up and down a little each generation with the overall trend being a plateau. This indicates that the Gerchberg-Saxton part of algorithm is struggling to estimate the phase. The most likely cause of this and the inconsistent results shown in Table 3 is that when the defocus is changed by altering the current through the objective lens this changes the convergence angle significantly enough to disrupt the algorithm [3]. Subsequent testing with modelled data showed the algorithm convergence would be significantly impaired by only a 1% change in convergence angle over a defocus range of 950nm. Further work is required to determine the scale of this effect experimentally.

Table 3 – Typical experimental results obtained from GS-GA.

	Defocus (nm)	Probe 1	Probe 2	Probe 3	Probe 4
Data set A	Estimated defocus	2080.00	2240.00	2400.00	2560.00
	Result	2001.14	2210.92	2386.23	2569.09
Data set B	Estimated defocus	1760.00	1920.00	2080.00	2240.00
	Result	1690.83	1887.37	2055.90	2236.61

	Defocus change (nm)	Probes 1-2	Probes 2-3	Probes 3-4
Data Set A	Estimated difference	160.00	160.00	160.00
	Result	206.61	176.45	180.38
	Difference from result above	209.78	175.31	182.86
Data Set B	Estimated difference	160.00	160.00	160.00
	Result	197.68	168.95	177.42
	Difference from result above	196.54	168.53	180.71

Figure 1 shows an example of the final phase retrieved for probe1 in data set B. It also shows the final amplitude calculated after the last propagation in the algorithm and how this compares to the recorded amplitude for this probe.

Figure 1 – Real amplitude of probe 1, data set B (left), retrieved amplitude (centre) and phase (right) produced by the GS-GA algorithm.

5. Further work

We hope to improve this algorithm by being able to incorporate a varying convergence angle, perhaps as a free variable for each probe which can be outputted by the algorithm together with the defocus values.

References

[1] Gerchberg R W and Saxton W O 1972, *Optik*, **35** 237-246

[2] Chakrabarti S et al 2008, *Data Mining: Know It All*, Morgan Kaufmann, 221-292

[3] Atkinson K M, Sweeney F and Rodenburg J M 2008, *J.Phys: Conf. Ser.*, **126** 012092

Journal of Physics: Conference Series **241** (2010) 012065 doi:10.1088/1742-6596/241/1/012065

Noise limit on practical electron ptychography

T B Edo[1], F Sweeney, C Lui and J M Rodenburg
University of Sheffield, Department of Electronic and Electrical Engineering, Mappin Street, Sheffield S1 3JD, UK

[1]Email: elp08tbe@sheffield.ac.uk

Abstract. Ptychography is a wavelength-limited phase retrieval method that promises to provide sub-0.5Å resolution in the electron microscope, even if the lens employed (which is required only to form an illumination spot) can itself only achieve rather modest resolution. The fidelity of a practical specimen reconstruction using ptychography hinges on various experimental factors, the major one of which is the source brightness. Achieving a high degree of coherence of an electron beam requires source demagnification which reduces the electron counts reaching the detector for a given exposure time. Specimen drift, damage and contamination also limits the practical exposure time. In this paper we investigate the effect of the counting statistics required for a good quality ptychographic reconstruction

1. Introduction

Ptychography can reconstruct the transmission function of a complex object (both phase and amplitude) from the diffraction patterns collected in the Fraunhofer diffraction pattern or the back focal plane of a microscope [1]. A requirement of the technique is that the system used to acquire the diffraction data must be set up to be in the most coherent configuration. To achieve this configuration in a field emission gun electron microscope, we change the potentials presented to the first focusing anode and make the aperture size of the probe-forming lens as small as possible. In doing so, we reduce the total effective number of electrons impinging on the sample. The resulting diffraction patterns are therefore noisier; however the visibility of interference effects in the diffraction pattern is increased.

To improve the signal-to-noise ratio we can increase our exposure time; however, the longer it takes to complete the experiment the more problems we have with sample drift, which introduces an uncertainty in our probe positions required for our reconstructions. An increase in the number counts per unit area on the specimen for a given illumination size will mean a reduction in the spatial coherence. A deeper insight into the counting statistics is necessary to make informed decisions about the compromises that are inevitable (i.e. counting statistics, electron probe dwell time and area of illumination on the sample). In this paper we shall investigate through numerical simulation the relationship between counting statistics and the quality of theoretical ptychographical reconstructions. In particular we assume that the total number of counts throughout the entire experiment for a given field of view is conserved; A field of view (spanned by a grid of 5x5 probe positions using a 90 nm probe defocus, and by a 16x16 grid using a 45 nm defocus) was recovered using the same electron

[1] To whom any correspondence should be addressed.

counts for both calculations). A common area of illumination is present for two adjacent probe positions (a fundamental requirement for the expression of ptychographic phase in the diffraction patterns [2]), this region is referred to as the 'overlap region' and in the case of our simulations this is equivalent to 87% of a probe diameter.

2. Effect of count statistics on reconstruction

It is not obvious whether the total count in the ptychographical data set determines the achievable resolution or the counts in a single diffraction pattern. To investigate this, we employ the setup shown in figure 1.

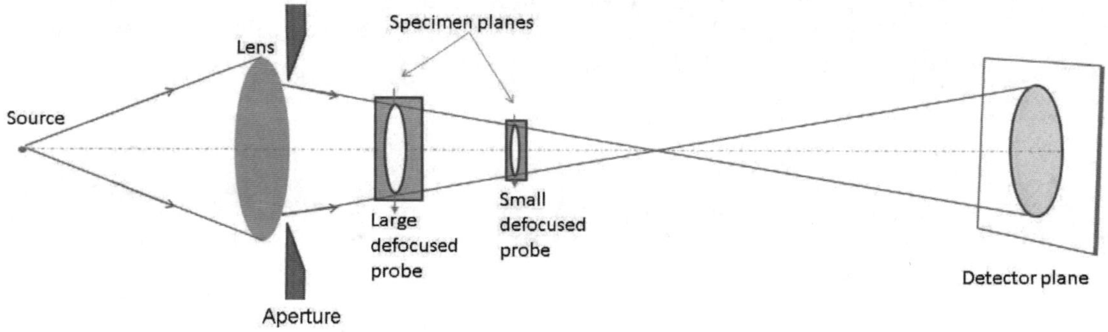

Figure 1. Experimental setup. The recovered object resides at either of the defoci planes for different experiments.

In order to recover a given field of view, a largely defocused probe with a long exposure time or a smaller probe with short exposure time can be used. We put the total counts in the entire diffraction data sets to the same value; hence the data set for a small probe would have fewer counts per diffraction pattern but contain more diffraction patterns as indicated in figure 2.

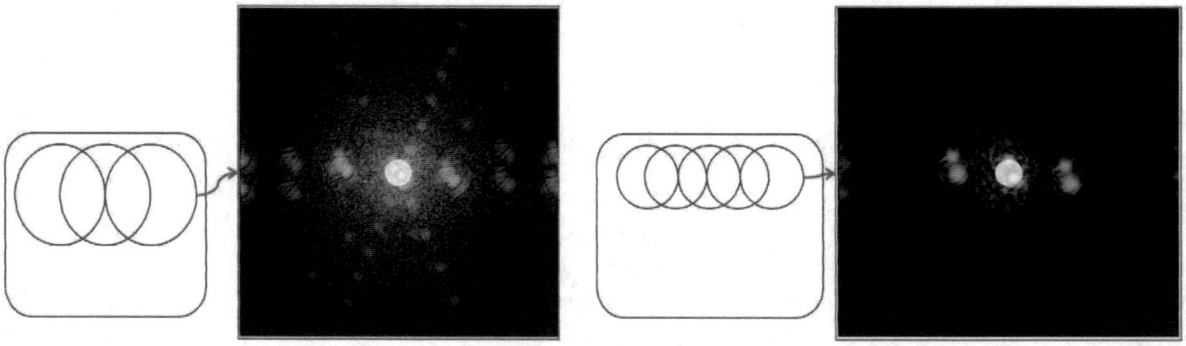

Figure 2. Pytchographic data set from probes of different defoci values

Employing a smaller probe with short exposure time per diffraction pattern reduces the effect of sample drift per diffraction pattern. The effective specimen drift should be mostly manifested in probe positions deviations from expected values which can in principle be accounted for during reconstruction [3]. This is not the case for larger probes with long exposure time as the effects of drift blurs out the spatial frequency information in each diffraction pattern thus reducing the maximum achievable resolution. Figure 3 and figure 4 show the probes used in the ptychographic reconstruction.

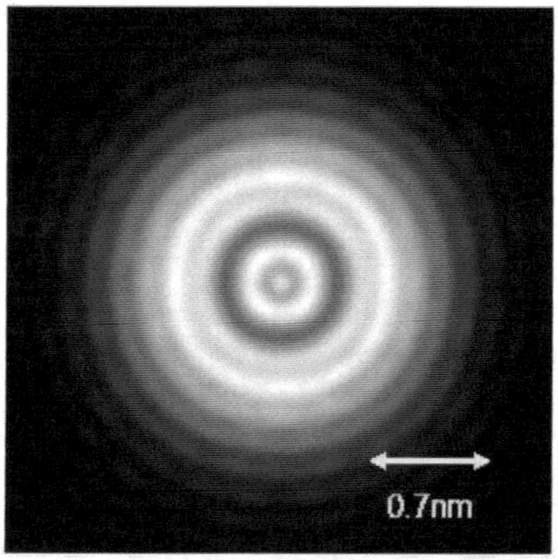

Figure 3. Probe profile (defocus = 90 nm) 10 mrad aperture, 200 kV acceleration voltage and 0.01 nm pixel size

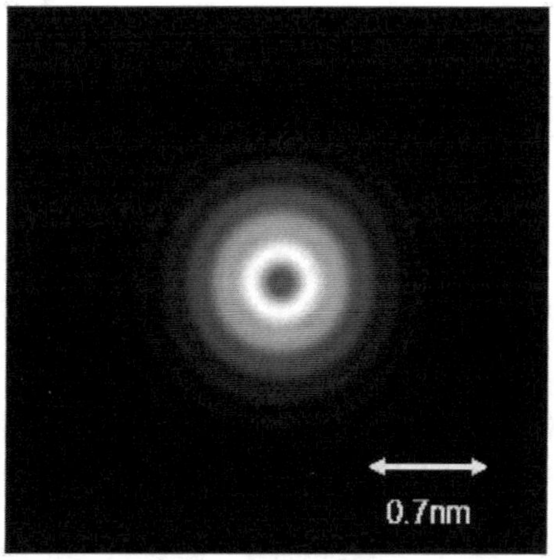

Figure 4. Probe profile (defocus = 45 nm) 10 mrad aperture, 200 kV acceleration voltage and 0.01 nm pixel size

To account for the effects of an amorphous support, multislice calculations with a model of amorphous carbon was used [4]. Test features, analogous to small gold particles, but with varying periodicities, were placed upon this amorphous background. The smallest periodicities represent atomic planes separated by 0.06 nm. Currently, the resolution of ptychography is limited by drift, probe position uncertainties and noise. All calculations were undertaken with a 200 kV accelerating voltage. The maximum scattering angle captured by the detector (which had 512x512 pixels) was 64 mrad.

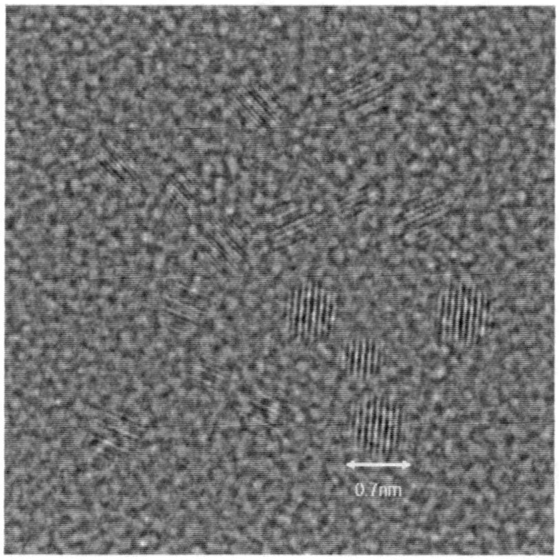

Figure 5. Magnitude of complex exit wave of specimen on a 4 nm amorphous carbon support, together with synthetic test particles.

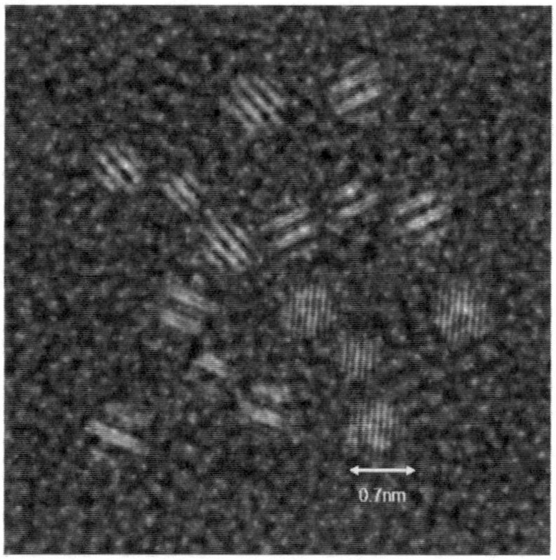

Figure 6. Phase of complex exit wave of specimen on a 4 nm amorphous carbon support

By modelling Poisson statistics for the recorded diffraction patterns, it was discovered that a count of about 10 electrons per detector pixel resulted in a reliable ptychographic reconstruction. Therefore achieving a resolution of say 0.1nm would require at least 10 counts per pixel in the corresponding spatial frequencies. Put simply, a strongly scattering specimen is easier to recover than a weak scattering specimen. The average dark field counts of the diffracted beams using a 90 nm probe data set was set to about 10 electrons per detector pixel. This corresponds to a count of 1 electron per detector pixel in the 45 nm probe data set using the same sampling condition.

The total counts/power required to reproduce the aforementioned count statistic is $\sim 10^7$ electrons. Experimentally, a highly coherent configuration with a beam current of 10pA produces a count of $\sim 10^8$ electrons for an exposure time of 1 minute (as measured on a JOEL 2010F). Figure 5 and figure 6 show the amplitude and phase of the complex exit-wave from the test specimen as it would appear under a plane wave illumination.

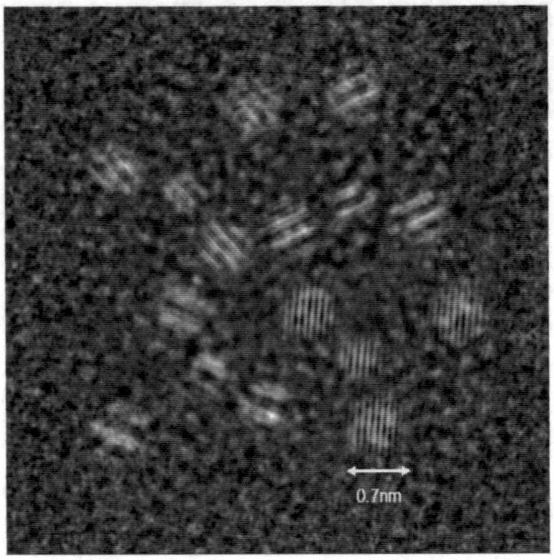

Figure 7. Recovered object phase (90 nm defocus)

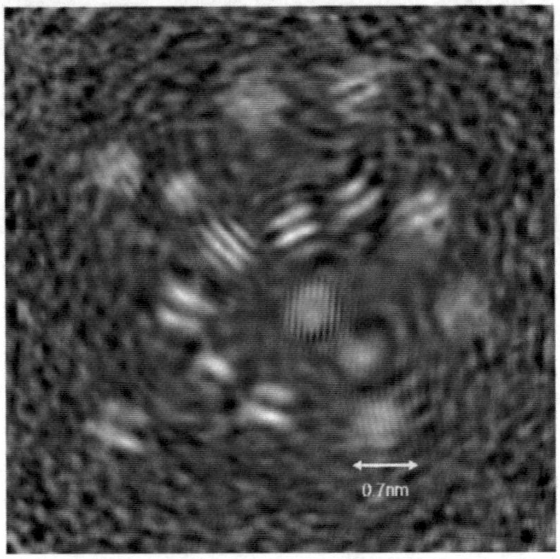

Figure 8. Recovered phase (45 nm defocus)

Figure 7 and figure 8 show the recovered phases from the two data sets. These results suggest that the counting statistic required for a good ptychographic reconstruction is easily achievable in a field emission gun electron microscope.

3. Conclusions

In this paper we showed that it was better to use a large defocused probe to generate a ptychographic diffraction data set as the counting statistic required for a reliable reconstruction is attainable in most modern electron microscopes.

The theoretical simulations presented here show that indeed it is possible to achieve a reasonable reconstruction when the number of counts in the dark-field is low (10 counts). This gives us confidence that we shall be able to carry out experiments with shorter exposure times, and thereby reduce the detrimental effects of drift and contamination encountered when carrying out ptychographical-type experiments in the transmission electron microscope.

4. References

[1] J M Rodenburg, A C Hurst and A G Cullis, *Ultramicroscopy*, Volume 107, Issues 2-3, (2007), 227-231
[2] J M Rodeburg Ptychography and related diffractive imaging methods, in: P Hawkes (Ed.), Advances in Imaging and Electron Physics, vol. 150, 2008
[3] J. M. Rodenburg, H. M. L. Faulkner, *Ultramicroscopy*, Volume 103, Issue 2, (2005), 153-164
[4] G Mountjoy *Journal of Non-Crystalline Solid* 293-295 (2001) 458-463

Local structures of two metallic glasses with good plasticity

Guoqiang Li, Konstantin B. Borisenko, David J. H. Cockayne
Department of Materials, University of Oxford, Parks Road, Oxford OX1 3PH, United Kingdom

E-mail: guoqiang.li@materials.ox.ac.uk

Abstract. The local structures of two metallic glasses with good plasticity, i.e., $Al_{89}La_6Ni_5$ and $Cu_{49}Hf_{42}Al_9$, were studied by reduced density functions (RDFs) from electron diffraction in a transmission electron microscope. Atomic models were refined against the RDF data using the reverse Monte Carlo (RMC) method and density functional theory (DFT) calculations. This approach allows the contributions of different bond types to the total $G(r)$ to be examined. For $Al_{89}La_6Ni_5$, RDFs from different local areas showed three distinct amorphous structures while RDFs from different local areas in $Cu_{49}Hf_{42}Al_9$ were identical, but with a broad spread of Hf-Hf bond lengths. The coexistence of three distinct amorphous states in $Al_{89}La_6Ni_5$ and the wide variation of Hf-Hf first neighbour distances in $Cu_{49}Hf_{42}Al_9$ could be associated with the high plasticity for both glasses.

1. Introduction

Unlike metals, metallic glasses lack structural imperfections such as dislocations and grain boundaries. As a consequence, metallic glasses are resistant to corrosion and wear, and have good mechanical properties including extraordinarily high strength, hardness and elasticity [1,2]. However, most metallic glasses have poor ductility, which limits industrial and structural applications. Nevertheless, some metallic glass systems such as Cu-Hf-Al and Al-La-Ni, do exhibit good plasticity [3,4,5]. In metallic glasses where there is no long-range order, local atomic arrangements play an important role in determining properties. Detailed investigation of local atomic structure of these glasses is therefore essential to provide a basis for understanding their properties and building models for structure-properties correlation, which is the aim of this work.

In this study, reduced density function (RDF) analysis based on electron diffraction data has been used to determine local structures of chosen small volumes in $Al_{89}La_6Ni_5$ and $Cu_{49}Hf_{42}Al_9$ metallic glasses, and atomic models have been refined against the RDF data using the RMC method and DFT calculations.

2. Experimental

Master alloys of $Al_{89}La_6Ni_5$ and $Cu_{49}Hf_{42}Al_9$ (in at. %) were kindly prepared by the Institute of Metal Research, Chinese Academy of Science by arc melting the pure elements together on a water cooled copper plate under a Ti-gettered, ultra-high purity argon atmosphere. The master alloys were re-melted several times to ensure homogeneity. $Al_{89}La_6Ni_5$ melt spun metallic glass ribbons, about 30 μm in thickness and 5 mm in width, were produced using a Buhler single-roller melt spinner [4], and $Cu_{49}Hf_{42}Al_9$ metallic glass ingots of 3 mm in diameter were formed using a copper mould [5].

Transmission electron microscope (TEM) samples of both metallic glasses were prepared by electropolishing. The TEM samples were then cleaned in a plasma cleaner to remove any surface contamination, followed by low angle ion milling at 300 V (Fischione 1010 Low Angle Ion Milling & Polishing System) to remove any possible surface oxide layer before observation in a JEOL 3000F field emission gun TEM operating at 300 kV. Selected area diffraction (SAD) was used to collect the electron diffraction patterns from a large number of different areas in both samples. The selected area was a disc of 100 nm diameter in each case. The diffraction patterns were recorded on a 1024×1024 pixel CCD camera and corrected for dark current and pixel to pixel gain variations. All the diffraction patterns were obtained using constant lens settings for the TEM, which enabled the calibration against an Au polycrystalline sample to be maintained through data collection to a stability of better than 1%.

3. Results and discussion

High resolution electron microscopy (HREM) was carried out on both samples. As a demonstration, figure 1a shows a typical HREM micrograph from $Al_{89}La_6Ni_5$ metallic glass, and figure 1b is the SAD pattern collected from this local area. No detectable crystalline clusters were observed in either metallic glass, and the diffraction pattern was a set of broad diffraction rings as usually seen for amorphous materials.

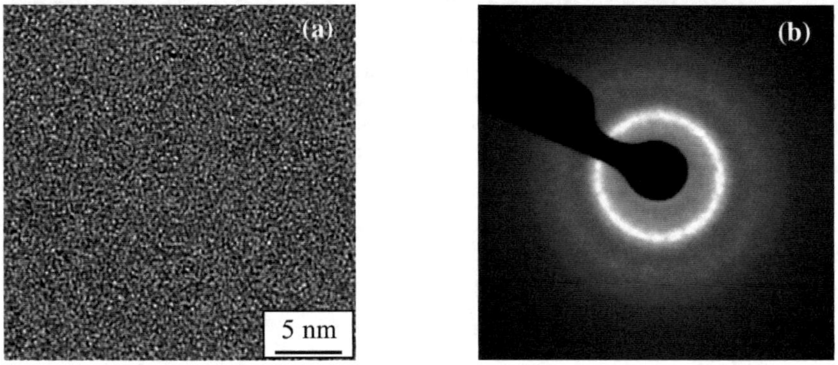

Figure 1. (a) HREM image from $Al_{89}La_6Ni_5$ metallic glass, and (b) the corresponding SAD pattern collected from this local area.

The electron diffraction patterns taken from the both metallic glasses were converted to RDFs by techniques developed by Cockayne et al. [6,7]. We note that scattering intensity profiles, $I(q)$, were recorded well above the background intensity level to scattering vectors as high as q=220 nm^{-1}. The maximum scattering vector in the present study was much higher than the typical 120-150 nm^{-1} used in X-ray and neutron diffraction experiments. This means that in the present work we can obtain a high spatial resolution. The fit of the experimental $I(q)$ to the theoretical intensity from independently scattering atoms of the same material, was excellent even at high q, and $I(q)$ oscillated around this theoretical curve at low q, giving confidence that the sample was thin enough for the single scattering approximation to be satisfied [7], which has also been confirmed by electron energy loss spectroscopy (EELS) measurement of the sample thickess of about 30 nm.

RDFs from various local areas on both samples were obtained in this way. For $Al_{89}La_6Ni_5$, RDFs from different areas showed three distinct amorphous structures with the first/second peaks located at 2.76/3.32 Å, 2.90/3.46 Å, and 2.96/3.55 Å, as shown in figure 2a. We denote these three as types A, B and C. On the other hand, RDFs from different local areas in $Cu_{49}Hf_{42}Al_9$ were identical, as shown in figure 2b, with the first two peaks at 2.68 Å and 3.23 Å, respectively.

To understand the origin of peaks in these RDFs, a Reverse Monte Carlo (RMC) simulation program similar to that described by McGreevy *et al.* [8] was used to refine the atomic structure, by comparing the experimental reduced scattering intensity with that computed for the atomic models of both metallic glasses. The energies of each RMC-refined model were then optimised using density functional theory (DFT), leading to energetically stable atomic structures. Figure 3a shows the good agreement between experimental type A RDF of $Al_{89}La_6Ni_5$ metallic glass and the corresponding $G(r)$ computed from the energy-optimised atomic model.

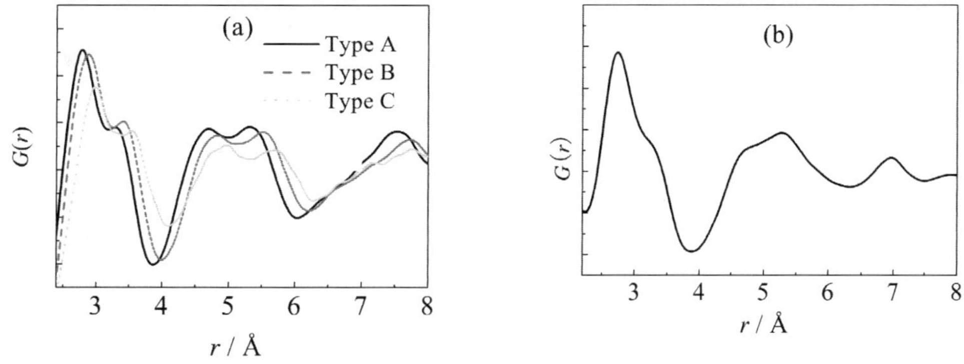

Figure 2. RDFs from different areas in the studied metallic glasses. $Al_{89}La_6Ni_5$ metallic glass shows three distinct RDFs depending on location (a), while RDF for $Cu_{49}Hf_{42}Al_9$ is identical everywhere (b).

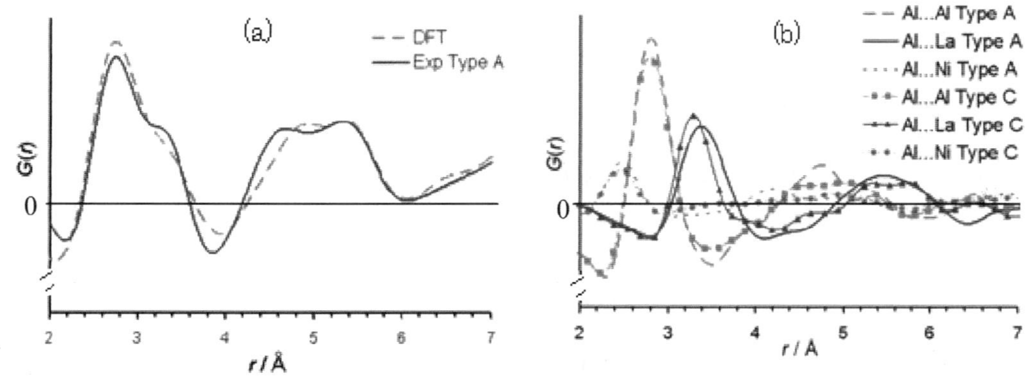

Figure 3. (a) Comparison between the RDF computed from the atomic model and the experimental one for type A, and (b) partial RDFs for types A and C of $Al_{89}La_6Ni_5$ metallic glass.

Figure 3b shows the partial RDFs resulting from this DFT energy minimised optimisation for types A and C of the $Al_{89}La_6Ni_5$ metallic glass. The DFT optimisation has been carried out to each experimental RDF, separately. Therefore, types A, B, and C RDFs have their corresponding partial RDFs. Differences between the model $G(r)$ and the experimental $G(r)$, and between different types A, B and C, can be considered in terms of bond types as shown in Figure 3b. For example, the pronounced shoulder at 3.2 Å in the experimental $G(r)$ suggests an increased number of Al-La bonds compared with the model. In the case of types A and C, Figure 3b suggests a shortening of the Al-La bond length between A and C. Type B is of intemediate structures between types A and C. The coexistence of three distinct amorphous structures in $Al_{89}La_6Ni_5$ glass points to a possibility of structural transformation between these three structures under external stress, which might be a reason for the plasticity of the glass.

A similar analysis of G(r) for $Cu_{49}Hf_{42}Al_9$ metallic glass (Figure 2b) and the partial RDFs (not shown) suggests that the first nearest neighbour distance at 2.68 Å is from Hf-Cu atomic pairs while the shoulder at about 3.2 Å comes from Hf-Hf pairs with bond length 3.23 Å The second nearest neighbours from Hf-Cu and Hf-Hf atomic pairs also contribute the most to the second main peak of the G(*r*) between 4 and 6 A. The Cu-Cu atomic pairs with the nearest neighbour distance at 2.39 A, on the contrary, do not show a strong contribution, even though the concentration of Cu in this ternary metallic glass is high. This suggests that Cu atoms are mainly coordinated with Hf atoms forming predominantly Cu-Hf bonds. Comparison of the RMC+DFT refined structure with a model prepared by random close packing of hard-sphere atoms suggested that the refined structure has a much wider distribution of Hf-Hf first neighbour distances than the structure obtained by random close packing of hard-sphere atoms. This wide variation of Hf-Hf first neighbour distances in $Cu_{49}Hf_{42}Al_9$ could be related to the observed high plasticity for $Cu_{49}Hf_{42}Al_9$ metallic glass.

4. Summary

The local structures of $Al_{89}La_6Ni_5$ and $Cu_{49}Hf_{42}Al_9$ metallic glasses were studied by RDF analysis based on electron diffraction data. No crystalline clusters could be seen in high resolution images of either metallic glass. For $Al_{89}La_6Ni_5$, RDFs from different local areas showed three distinct amorphous structures with the first/second peaks located at 2.76/3.32 Å, 2.90/3.46 Å, and 2.96/3.55 Å. RMC combined with DFT refinements using these RDFs allowed a detailed study of the structures. On the other hand, RDFs from different local areas in $Cu_{49}Hf_{42}Al_9$ were identical, with the first nearest neighbour distances for Cu-Cu, Hf-Cu, and Hf-Hf atomic pairs at 2.39 Å, 2.68 Å, and 3.23 Å, respectively. Comparison of the RMC+DFT refined structure with a model prepared by random close packing of hard-sphere atoms suggested a wide variation of Hf-Hf first neighbour distances in the material. Coexistence of three distinct amorphous states in $Al_{89}La_6Ni_5$ and the wide variation of Hf-Hf first neighbour distances in $Cu_{49}Hf_{42}Al_9$ could be related to the high plasticity of these glasses.

References

[1] He Y, Poon S J and Shiflet G J 1988 *Science* **241** 1640
[2] Inoue A and Takeuchi A 2002 *Mater Trans JIM* **43** 1892
[3] Saksl K, Jovari P, Franz H, Zeng Q S, Liu J F and Jiang J Z 2006 *J Phys: Condens Matter* **18** 7579
[4] Guo H, Yan P F, Wang Y B, Tan J, Zhang Z F, Sui M L and Ma E 2007 *Nat. Mater.* **6** 735
[5] Jia P, Guo H, Li Y, Xu J and Ma E 2006 *Scripta Mater.* **54** 2165
[6] Cockayne D J H and McKenzie D R 1988 *Acta Cryst A* **44** 870
[7] Cockayne D J H 2007 *Annu Rev Mater Res* **37** 159
[8] McGreevy R L 2001 *J Phys: Cond Matter* **13** R877

Towards quantitative analysis of core-shell catalyst nano-particles by aberration corrected high angle annular dark field STEM and EDX

H E[1], P D Nellist[1], S Lozano-Perez[1] and D Ozkaya[2]

[1] Department of Materials, Oxford University, Parks Road, Oxford OX1 3PH, UK
[2] Johnson Matthey Technology Centre, Blounts Court Road, Sonning Common, Reading RG4 9NH, UK

Email: haibo.e@materials.ox.ac.uk

Abstract. Core-shell structured heterogeneous catalyst nano-particles offer the promise of more efficient precious metal usage and also novel functionalities but are as yet poorly characterised due to large compositional variations over short ranges. High angle annular dark field detector in a scanning transmission electron microscope is frequently used to image at high resolution because of its Z-contrast and incoherent imaging process, but generally little attention is paid to quantification. Energy dispersive X-ray analysis provides information on thickness and chemical composition and, used in conjunction with HAADF-STEM, aids interpretation of imaged nano-particles. We present important calibrations and initial data for truly quantitative high resolution analysis.

1. Introduction

Core-shell structured nano-particles offer the promise of more efficient precious metal usage and also novel functionalities for heterogeneous catalyst applications. Characterisation of individual nano-particles is vital for understanding their catalytic properties. However, mapping such short range compositional variations in nano-particles is challenging.

Using a high angle annular dark field (HAADF) detector in the scanning transmission microscope (STEM), it is possible to directly image the atomic configuration of small nano-particles. This technique has often been chosen for heavy elements on light supports for the strong correlation between atomic number and imaged intensity, known as Z contrast. The major advantage of HAADF STEM over conventional high resolution transmission electron microscopes (HRTEM) is its incoherence; there are no contrast reversals often seen in HRTEM due to the phase contrast nature of the imaging [1].

However, though it offers many advantages, meaningful quantitative compositional information is less than straight-forward to extract from HAADF images because intensities are dependent on both composition and thickness. Some information on the composition and thickness of the sample is necessary before interpretation of an image can be made. Energy dispersive X-ray (EDX) analysis can map composition over a broad range of atomic numbers but is often not favoured as a high resolution analysis technique due to its poor spatial resolution and long acquisition times. Recent developments in aberration correction microscopy allowing for smaller, higher intensity probes have increased to atomic resolution both imaging and chemical mapping [2].

2. Towards Quantification in HAADF Images

Though much work has been done comparing simulations to experimental data, rigorous quantification is still not standard. LeBeau and Stemmer noted that it is essential to properly calibrate the ADF detector for quantitative analysis of HAADF images [3]. Here we determine the response efficiency of our ADF detector and the behaviour in gain and offset of its related analogue to digital convertors with incident beam currents to allow for a complete interpretation of images.

2.1. Detector Mapping

We mapped the scintillator efficiency of the upper ADF detector of the Oxford-JEOL 2200MCO double aberration corrected S/TEM. The probe was scanned across the detector (see Figure 1). It can be seen that there are variations across the detector, especially at low angles and including the shadow of the electron pipe around the central hole. Thus all HAADF simulations should include a detector efficiency weighting factor to allow for accurate comparison between simulated images and experimental data.

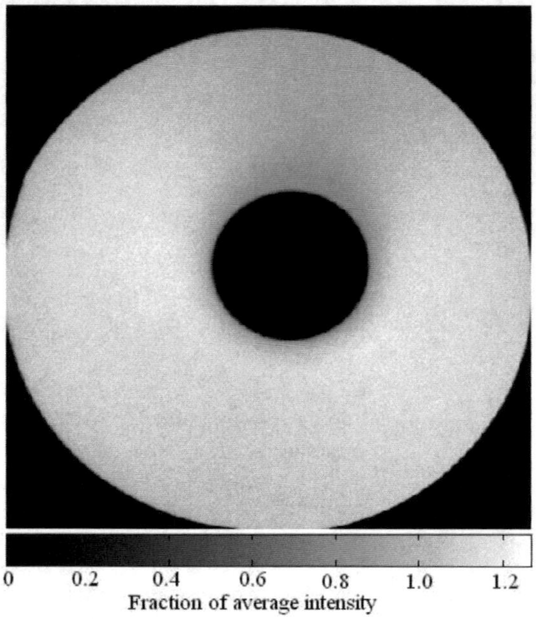

Figure 1: The detector scintillator efficiency as a function of probe position across the detector surface, scaled to mean value across detector. Variations in efficiency are seen across the detector surface, in particular at low angles.

2.2. Gain and Offset

The gain and offset (contrast and brightness) of the amplifiers which convert the signal on the detector to digital values were measured. By adjusting the settings with a beam of constant current, we found that the analogue to digital convertor behaviour varied linearly with offset but the gain behaved non-linearly over a short range of contrast values (See Figure 2).

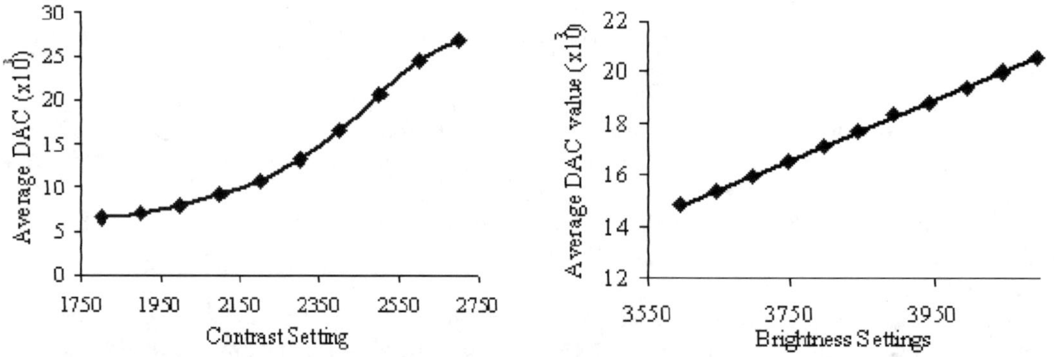

Figure 2 a) Non-linear response of the analogue to digital convertor with contrast (gain) **b)** Linear behaviour of the analogue to digital convertor with brightness (offset).

2.3. Variation of Detector Response to Incident Probe Current

The incident probe current for three spot sizes were measured using a Faraday Cup. Apertures of known size were used to systematically decrease the incident probe current, while brightness and contrast settings were optimised for the highest beam current. The response of the detector to variations in incident probe current is linear (See Figure 3). The intensity of HAADF images can be directly compared with simulation.

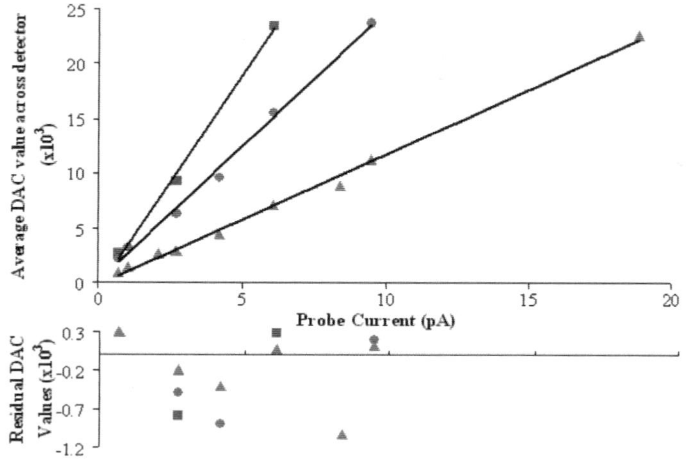

Figure 3 Brightness and contrast settings were optimised for the highest beam current and then kept constant as spot size and apertures were changed to vary the incident current. Three maximum beam currents were used (from spot size 8, 9 and 10 to give 9, 6 and 3 points respectively) to verify detector response across three brightness and contrast settings. The residual is from a linear fit. Variation of detector response to incident probe current is linear.

3. HAADF Imaging

Core-shell nano-particles of palladium core and platinum shell on porous carbon support were deposited onto a TEM grid with lacy carbon film support. The sample was then baked on a bulb for 30 minutes in vacuum at 115°C to remove possibilities of contamination when exposed to the electron beam. High resolution HAADF-STEM images were acquired using the Oxford JEOL 2200MCO. A map of the detector was taken so that the image of the nano-particle could be normalised to a known current.

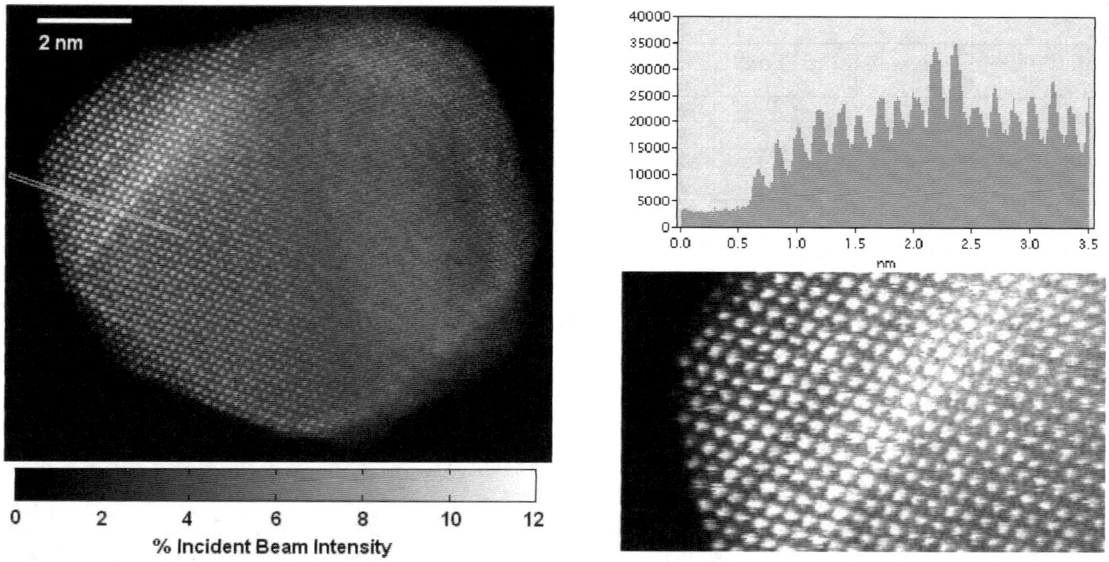

Figure 4 Twinned platinum and palladium nano-particle imaged using the Oxford-JEOL 2200MCO **a)** ADF image normalised to percentage of incident beam intensity. The particle is viewed down <110> zone axis. A twin boundary can be clearly seen down the centre of the particle and atomically sharp terminated edges on the left side. **b)** Line profile of intensity with integration width 3. Sharp increases in intensity indicate presence of heavier platinum as a dramatic increase in local thickness is unlikely. **c)** Detail of left corner of nano-particle.

A local approximation multi-slice simulation method is in development for the specific case of nano-particles, based on the Melbourne STEM program developed by L J Allen, S D Findlay and M P Oxley [4].

4. EDX

Composition and thickness information can be achieved with EDX line scans and mapping (See Figure 5), showing clearly the core-shell structure of the nano-particles. Though EDX by itself does not offer atomic resolution mapping, in combination with high resolution HAADF imaging, exact structures could be determined.

EDX spectra simulation techniques for compositional analysis often do not use accurate self-absorption corrections. We aim to compare the traditional Cliff-Lorimer approach with a composition-independent and mass-thickness-independent zeta-factor method [5].

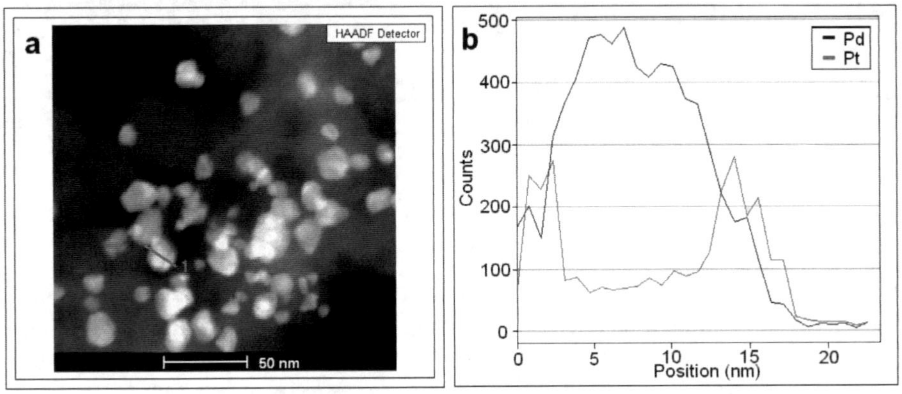

Figure 5 a) HAADF image of a cluster of nano-particles. **b)** EDX line profile taken across a nano-particle (see HAADF image in a), showing clear core-shell structure. (Acquired on a Tecnai F20)

5. Conclusions and Future Work

Calibrations of the Oxford-JEOL 2200MCO are completed so that true quantitative information could be extracted from HAADF images. We show that the detector responds linearly with incident probe current but that care is needed with the contrast settings as the analogue to digital conversion is non-linear with respect to changing gain. Nano-particles were imaged and intensities normalised to incident probe current so that they can be compared to simulation without free scaling parameters.

Simulation techniques for both HAADF images and EDX spectra are under development. Further work is planned towards achieving truly quantitative measurements of shell thickness, proportional coverage and information about facial orientations with a combination of these techniques.

6. Acknowledgements

This work has been funded by the EPSRC and Johnson Matthey as part of a CASE-Award studentship. The authors would like to thank Johnson Matthey Technology Centre for provision of experimental facilities and thank the staff for their help.

References

[1] Pennycook SJ, Jesson DE, McGibbon AJ and Nellist PD 1996 *J. Electron Microscopy* **45** 36-43
[2] Bleloch A and Lupini A 2004 *Materials Today* **7** 42-48
[3] LeBeau JM and Stemmer S 2008 *Ultramicroscopy* **108** 1653-1658
[4] Allen LJ, Findlay SD, Oxley MP and Rossouw CJ, 2003 *Ultramicroscopy* **96** 47-63
[5] Watanabe M and Williams D 2006 *Journal of Microscopy* **221** 89-109

Measuring the contrast in annular dark field STEM images as a function of camera length

T Walther, Y Qiu and AG Cullis

Department of Electronic & Electrical Engineering, University of Sheffield, Sir Frederick Mappin Building, Mappin Street, Sheffield S1 3JD, UK

t.walther@sheffield.ac.uk

Abstract. Image contrast in annular dark-field (ADF) imaging in scanning transmission electron microscopy (STEM) is studied as function of the inner collection angle of the annular detector. The smaller the camera length is, the larger is the effective inner detection angle. As a result, the intensity recorded drops dramatically but the contrast increases with detection angle. For very large angles (high-angle ADF), Rutherford scattering by the atomic nuclei is the dominant scattering mechanism. The intensity is then simply proportional to the product of thickness (=number of scattering atoms) and the square of the atomic number (="Z-contrast"), however, sufficiently high angles are experimentally difficult to obtain. By extrapolating the dependence of the image contrast as a function of camera length (=inverse of the width of the detector hole in the diffraction plane) to zero camera length (=infinite scattering angle) we can quantify the chemical composition in systems like InGaAs/GaAs thin film heterostructures where the concentration of only one element (here: indium) needs to be considered.

1. Introduction

The image contrast in annular dark-field (ADF) imaging in scanning transmission electron microscopy (STEM) [1] depends on the specimen thickness [2, 6], its orientation [3], the chemical composition [4], surface oxidation [5] and also on the inner detection angles of the annular detector [2, 6]. Due to the complexity of contrast formation, energy filtering was applied in the past to investigate inelastic effects, however, such experiments are complicated in that they either need an energy-selecting slit before the ADF detector [7] or an angular selecting aperture before the spectrometer, which can be a centre-stop aperture in front of the energy filter [8] or an off-axis positioned filter entrance aperture [5]. The smaller the camera length is, the larger is the effective inner detection angle and the lower the intensity becomes. For very large angles, i.e. >>100mrad for fast electrons, Rutherford scattering by the atomic nuclei is the dominant scattering mechanism and the intensity is then, over a wide range, proportional to the product of number of atoms scattering (i.e. thickness, t) times the square of the atomic number, Z (="Z-contrast") [9]. However, according to Rutherford's formula the intensity in high-angle annular dark-field imaging (HAADF) falls off with the inverse of the 4th power of the scattering angle [10], so that in effect most of the signal on an annular detector is provided by the innermost ring section near to the detector hole. In high-resolution (HR-)STEM small electron probes are employed and in order to see lattice fringes clearly a rather high signal-to-noise ratio is necessary. Hence, HR-STEM uses lower detection angles, typically ~30mrad. This means the effect of Rutherford scattering is superimposed by the effects of screened low-angle or Bragg scattering, which lowers the effective atomic number dependence of the intensity recorded to $I=ctZ^{\varepsilon}$ with some constant c and some exponent in the range $\varepsilon=1.6,\ldots, 2.0$ [11].

While it has been shown that the lattice fringe contrast in high-resolution ADF STEM can be matched to that of simulations for bulk material of known thickness [12], we proposed a new scheme for evaluating quantitatively image intensities from an interface between two materials in ADF STEM at medium resolution, by measuring the dependence of the image contrast as a function of camera length (=inverse of the width of the detector hole in the diffraction plane) and extrapolating these measurements to zero camera length (=infinite scattering angle, β) [13]. This is somewhat similar to using an ADF detector with multiple concentric ring sections [14] which would offer the advantage of simultaneous read-outs for different values of β. It was shown that for SiGe/Si thin films the chemistry can be reliably determined from the image intensity ratio relative to silicon as a reference, assuming ε=2 for $\beta \to \infty$. Applications to InGaAs/GaAs thin film heterostructures are discussed in the following.

2. Simulations

Figure 1 shows that the exponent ε assumed in the relationship between image intensity and atomic number, $I \propto tZ^{\varepsilon}$, depends on the experimental set-up and generally increases with collection angle, up to 2. Figure 2 shows simple kinematic calculations for the intensity ratio of $In_xGa_{1-x}As$ to GaAs for ε=1.6, 1.7, 1.8, 1.9 and 2.0, assuming the above proportionality. For pure InAs the intensity should be 1.5-1.7 times higher than for GaAs, depending on the detector geometry.

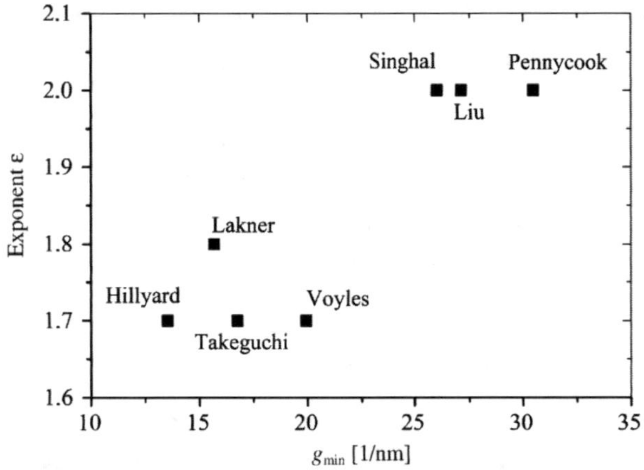

Figure 1. Comparison of exponent ε in the relationship between ADF image intensity, I, and atomic number, Z, in the equation $I=ctZ^{\varepsilon}$, as function of scattering vector, $g=\lambda\beta$, where λ is the wavelength and β the inner detection angle used in the ADF-STEM experiments. Names refer to the first authors of the corresponding publications [15-21].

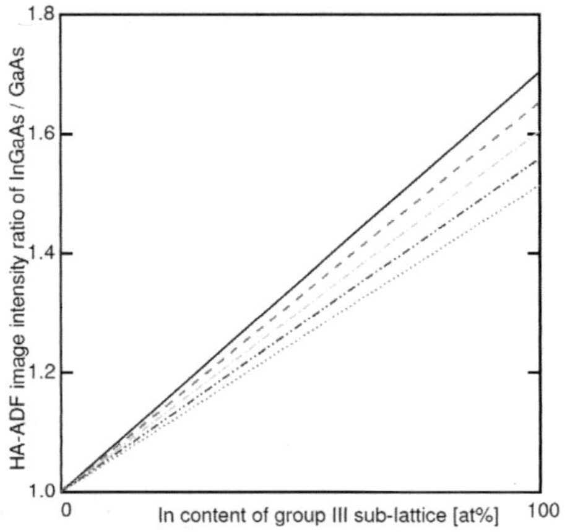

Figure 2. Plot of the ratio of intensities of $In_xGa_{1-x}As$ to that of GaAs as a function of indium content x, assuming different exponents e in the equation $I=ctZ^{\varepsilon}$, from ε=2 (black, solid) to 1.9 (red, dashed), 1.8 (green, dot-dashed), 1.7 (blue, double dot-dashed) and 1.6 (cyan, dotted)

3. Measurements of differently annealed InGaAs/GaAs heterostructures

We have studied with our approach InGaAs/GaAs heterostructures with quantum dots embedded in quantum wells that have been annealed at temperatures of 700 and 750 °C. Cross-sectional samples were prepared by glueing, cutting, grinding, polishing and argon ion milling until perforation under standard conditions. The ADF-STEM experiments were all conducted in a JEOL 2010F field-emission (scanning) transmission electron microscope using the Gatan Digiscan software to acquire images. The intermediate lens current was changed manually to vary the camera length of the detector. In this way, multiple image stripe sections of the layer structures as well as a dark current signal, I_0 (intensity with the electron beam switched off), were acquired under otherwise identical settings. Signal non-linearity was avoided by setting the detector gain in the initial image section with the highest image intensity (longest camera length) to just below saturation and then subtracting the dark current value I_0 from all image intensities recorded. As this microscope does not have an in-column energy filter no energy filtering as in [13] could be used here.

Figure 3. ADF STEM image at 50kX nominal magnification of the InGaAs/GaAs structures annealed at 700 °C. The [001] growth direction points toward the top right corner.

Figure 4. ADF STEM image at 150kX nominal magnification of the same sample. The camera length decreases from 223mm (top) to 147mm (second last strip) during the acquisition.

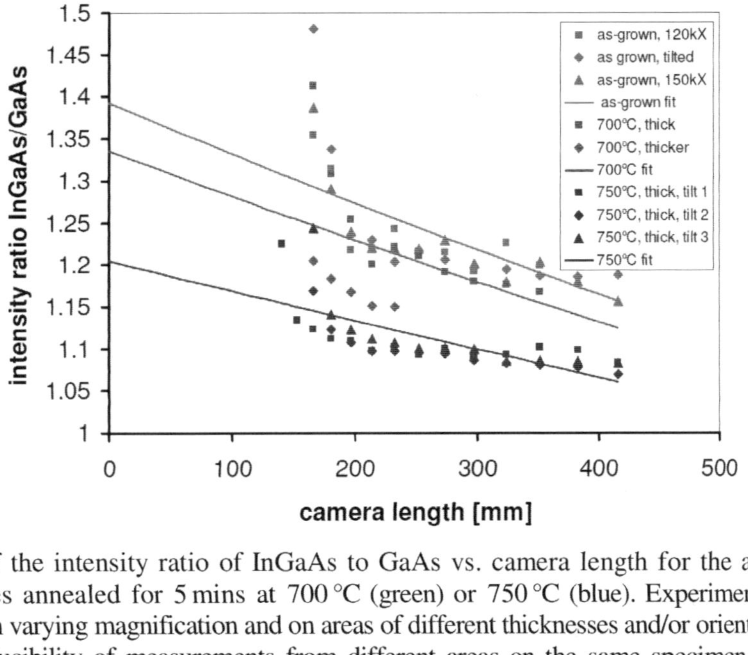

Figure 5. Plot of the intensity ratio of InGaAs to GaAs vs. camera length for the as-grown sample (red) and samples annealed for 5 mins at 700 °C (green) or 750 °C (blue). Experiments were repeated several times, with varying magnification and on areas of different thicknesses and/or orientations. Error bars stating the reproducibility of measurements from different areas on the same specimen are estimated as ±0.03 for 200-420mm camera lengths and ±0.10 for 140-200mm camera length.

Measured and plotted in Figure 5 is the intensity ratio of $In_xGa_{1-x}As$ to GaAs after black leveal correction, which is related to the In content x by $R=1+x(49^{\varepsilon}-31^{\varepsilon})/(31^{\varepsilon}+33^{\varepsilon}) \rightarrow 1+0.7024x$ for $\varepsilon \rightarrow 2$. Table 1 shows that the fits are not as good as expected, with regressional coefficients in the range of 0.7-0.8, instead of above 0.9 as for SiGe [13]. Results from linear and logarithmic fits are similar. For shorter camera lengths the intensity ratios measured increased unsystematically, exhibiting a large scatter of typically ±0.10 and in two cases even producing values >1.7 in thin regions, which could indicate preferential etching. There is also a notable curvature around 250mm camera length for some series, pointing to non-linearities (hysteresis) in the electromagnetic field of the intermediate lens whose current determines the camera length. Hence, the reliability of the fitting procedure only gave the indium content of $In_xGa_{1-x}As$ to within $\Delta x=\pm(0.03\text{-}0.06)$, with a typical relative error of ~10%. This is too high to allow us to calculate the activation energy for interdiffusion along the growth direction: the results of (3.4±1.6)eV obtained from a simple model by estimating the thermal broadening of Gaussian profiles [22] agrees with direct width measurements [23] but is not very meaningful due to its large error. We expect this can be improved one the mechanism of the anomalously high contrast for the shortest camera lengths used is better understood.

Table 1. Result of regressional analysis of the intensity ratios plotted in Figure 5

condition	as-grown	5mins @ 700°C	5mins @ 750°C
intercept *b* from lin. reg. fit	1.389±0.049	1.332±0.042	1.204±0.024
lin. regression coefficient, *R*	0.719	0.737	0.752
intercept *b* from log. reg. fit	1.393±0.038	1.336±0.034	1.205±0.020
log. regression coefficient, *R*	0.731	0.742	0.761
resulting In peak concentration *x*	0.557±0.062	0.476±0.054	0.291±0.031

4. Summary

We have shown that an extrapolation of the plot of ADF intensity ratios vs. camera length to zero yields the intensity ratio expected for infinite scattering angle, which for (pseudo)binary compounds can be converted into chemical composition using Rutherford's formula. This yields consistent data for InGaAs/GaAs only as long as very thin specimen regions are avoided where surface oxidation, preferential surface etching during ion milling and strain relaxation effects can become important.

References

[1] Howie A 1979 *J. Microsc.* **117**, 11
[2] Pennycook SJ et al. 1992 *Scanning Microsc. Suppl.* **6**, 233
[3] Cowley JM and Huang Y 1992 *Ultramicroscopy* **40**, 171
[4] Pennycook SJ and Boatner LA 1988 *Nature* **336**, 565
[5] Walther T and Humphreys CJ 1997 *Inst. Phys. Conf. Ser.* **153**, 303
[6] Treacy MMJ and Gibson JM 1993 *Ultramicroscopy* **52**, 31
[7] Haider M 1989 *Ultramicroscopy* **28**, 240
[8] Bleloch AL, Castell MR, Howie A and Walsh CA 1994 *Ultramicroscopy* **54**, 107
[9] Treacy MMJ, Howie A and Wilson CJ 1978 *Philos. Mag. A* **38**, 569
[10] Rutherford E 1911 *Philos. Mag.* **21**, 669
[11] Hartel P, Rose H and Dinges C 1996 *Ultramicroscopy* **63**, 93
[12] LeBeau JM et al. 2009 *Phys. Rev. B* **79**, 214110
[13] Walther T 2006 *J. Microsc.* **221**, 137
[14] Boulin C and Epstein A 1992 *IEEE Trans. Nucl. Sci.* **39**, 226
[15] Hillyard S and Silcox J 1995 *Ultramicroscopy* **58**, 6
[16] Lakner H, Mendorf C, Bollig B, Prost W and Tegude FJ 1997 *Mat. Sci. Eng. B* **44**, 52
[17] Liu CP, Twesten RD and Gibson JM 2001 *Ultramicroscopy* **87**, 79
[18] Pennycook SJ, Rafferty B and Nellist PD 2000 *Microsc. Microanal.* **6**, 343
[19] Signhal A, Yang JC and Gibson JM 1997 *Ultramicroscopy* **67**, 191
[20] Takeguchi M, McCartney MR and Smith DJ 2004 *Appl. Phys. Lett.* **84**, 2103
[21] Voyles PM, Grazul JL and Muller DA 2003 *Ultramicroscopy* **96**, 251
[22] Walther T, Humphreys CJ and Robbins DJ 1997 *Defect and Diffusion Forum* **143-147**, 1135
[23] Qiu Y at al., these proceedings

Occurrence of the vortex state for magnetic phase plates

M K Hari, M Beleggia[1,2], R M D Brydson

Institute for Materials Research, University of Leeds, Leeds LS2 9JT, UK

E-mail: mb@cen.dtu.dk

Abstract. Tunable magnetic phase plates may be realized as nanorings magnetized in the vortex state, where tunability would be granted by the temperature dependence of the saturation magnetization. Here, we study the statistical occurrence of the magnetic vortex state in circular and pentagonal rings within their magnetic phase diagram. We outline the useful operational range of parameters that may be utilized in practice.

1. Introduction

In addition to a large number of promising applications in the field of magnetic memories [1] and spintronics [2], nanorings may be utilized as magnetic phase plates in electron microscopy to achieve Zernike-type phase contrast [3,4]. In fact, thanks to the Aharonov-Bohm effect experienced by the electrons traveling through a nanoring in the vortex state, a phase shift is induced between the transmitted and scattered beams. The main benefit is visible in-focus contrast associated with weak (or strong) phase objects such as biological samples [5]. Furthermore, it was hinted in [6] that employing Zernike mode may also result in a substantial decrease of the electron dose necessary to image single atoms or molecules, so that imaging radiation sensitive materials becomes more practical.

Aiming at the development of magnetic phase plates we focus here on the stability region of the vortex state in rings within a suitable parameter space. This is a necessary preliminary step to ascertain the range of dimensions and shapes that are compatible with an operational phase plate. Furthermore, we analyze with micromagnetic simulations the statistics of obtaining a vortex from a random initial configuration, and how the statistics varies with shape and size. This is useful to select the appropriate parameters that results in more frequent vortex states and, hence, reliable operability.

2. The phase shift of a magnetic phase plate

The vortex state is the sought magnetization state in the ring as it can induce the correct phase shift between the transmitted and scattered beam. The phase shift associated with a ring in the vortex state can be calculated from the standard expression [7] linking it to the magnetic vector potential

$$\varphi(r) = -\frac{\pi}{\phi_0}\int_{-\infty}^{+\infty} A_z(r,z)\mathrm{d}z = N\pi\begin{cases}1 & r<\overline{R}\\ 0 & r>\overline{R}\end{cases} \tag{1}$$

where $\phi_0 = 2.07\times10^{-15}$ T m^2 is the flux quantum, and we adopted a simplified description of a ring of thickness t and radii R_2 (outer) and R_1 (inner) as a circular flux line of radius $\overline{R}$ carrying N flux quanta

---

[1] To whom any correspondence should be addressed.

[2] *Now at*: Center for Electron Nanoscopy, Technical University of Denmark, DK-2800 Kgs. Lyngby, Denmark.

(see Fig. 1). The phase shift described by equation (1) is what is required to achieve Zernike-type phase contrast: a constant phase equal to zero outside the ring (leaving the scattered electrons unperturbed), and another constant phase equal to $\phi_Z=N\pi$ (N times the phase shift of a flux quantum) inside the ring (inducing a shift to the transmitted electrons). If the vorticity of the ring changes, the phase shift within the inner radius of the ring changes sign, producing "negative" phase contrast.

3. Tunability with temperature and geometrical constraints

As discussed in [8], in electron microscopy the quarter-wavelength phase shift typical of light optics may not always be the optimal choice, as it is only appropriate for weak phase objects. Furthermore, even under controlled growth, we can never be sure *a priori* that the thickness of magnetic material is exactly what is needed to realize the Zernike shift we seek. Hence, tunability of the phase plate is a necessity. While electrostatic phase plates [9] are tunable by controlling some applied voltage, in magnetic phase plates one possibility is to operate the device at different temperatures. In fact, as illustrated above, the Zernike phase shift induced on the electrons is proportional to N, the number of flux quanta enclosed in the ring. Since N is the ratio between the magnetic flux and the flux quantum, and the magnetic flux is the product of magnetization M (measured in Tesla) and ring cross section S, once we fix the ring dimensions we may achieve tunability from the proportionality between ϕ_Z and $M(T)$. Choosing then an appropriate magnetic material with a suitable Curie-Weiss behavior, we could define a temperature interval of operation, say T_1 and T_2, so that the magnetic flux $M(T_1)S\sim M(T_2)S$ increases by a factor of 2. This would provide a tunable magnetic phase plate capable of producing Zernike shifts between $\pi/2$ (higher temperature) and π (lower temperature).

Given the proportionality between magnetic flux and Zernike shift, in term of ring shape we may choose any combination of thickness and width of the ring that satisfies

$$\frac{\phi_0}{4R_2^2 M(T)} \le \tau(1-\sigma) \le \frac{\phi_0}{2R_2^2 M(T)} \tag{2}$$

where $\tau=t/2R_2$ and $\sigma=R_1/R_2$ are the dimensionless shape parameters of the ring. To give an example, suppose we choose to fabricate the phase plate with some yet unidentified Iron compound that has a Curie point around 600 K. Its saturation magnetization at $T=0$ might be, hypothetically, 2 T. If we choose T_1 at room temperature we may expect from a rough mean-field estimate a magnetization around $M(T_1)\sim1.7$ T. Choosing T_2 just below the Curie point, e.g. $T_2/T_c=0.9$, we expect $M(T_2)\sim 0.85$ T. We could then adjust the Zernike shift by $\pi/2$ as we vary the temperature in a 240 K wide range, which does not seem to require extensive technical efforts to be implemented in a microscope.

With such material and temperature range chosen, we now illustrate what the geometrical constrain would dictate. Suppose we fabricate rings with $R_2=150$ nm, which is a reasonable compromise between current patterning capabilities and electron-optical requirements of having the smallest possible hole. Equation (2) now reads $0.0177\le\tau(1-\sigma)\le0.0354$, so that we are free to choose thickness t and width (R_2-R_1) accordingly. For instance, if we set $(R_2-R_1)=50$ nm, which is well within patterning capabilities, we have $\sigma=2/3$ and a thickness constrained to $16\le t\le32$ (in nm).

4. Vortex state statistical occurrence

As a preliminary exploration of this parameter space, we aimed at developing some feeling on the likelihood of finding a stable vortex state using micromagnetic simulations [11]. We found a variety of competing states, the occurrence of which depends strongly on the input parameters of the simulations. Figure 2 shows four observed states: onion, horseshoe vortex and non-vortex. In this case, variety is a problem, as ideally we would like a set of shape parameters, compatible with the geometrical constraints, that results in 100% vortex occurrence. A possibility worth considering is to employ non-circular rings, such as pentagons. The presence of corners, smooth or sharp as they may be depending on patterning resolution during fabrication, may have an effect in the occurrence and stability of vortices. To explore this perspective, we have compared the statistics of vortex occurrence in circular and pentagonal rings starting with the same initial condition of random magnetization. The results are shown in Fig. 2, where a clear statistical improvement is observable for pentagonal rings: almost half (48%) of the simulations terminated in a vortex state (versus 30% for circular rings), with the remaining three states divided unevenly. In comparison, the statistic for circular rings is almost uniformly distributed over the four states.

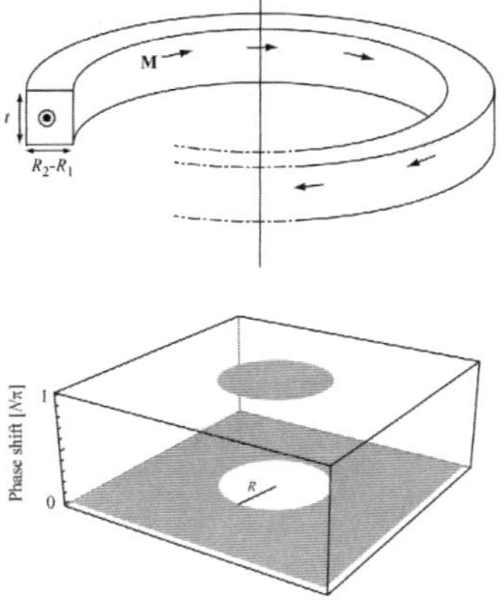

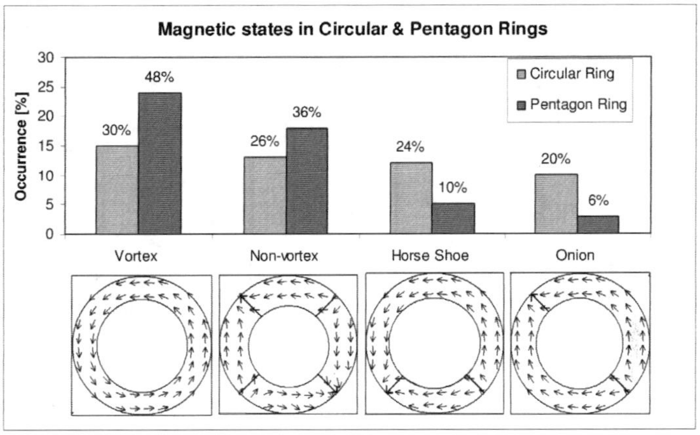

Figure 1. A magnetized ring acting as a phase plate.

Figure 2. Statistics of vortex occurrence from random initial magnetization in circular and pentagonal rings.

There are two additional advantages in pentagonal rings: i) an enlargement of the region in parameter space where a vortex state is expected, which will be discussed in the next section, and ii) the possibility of restoring vortex states by application of a suitable external field, something that is prevented by the symmetry of a circular ring. The presence of corners, acting as pinning sites for domain wall motion, enables us to apply a suitable set of in-plane external fields and turn any non-vortex into vortex states. This analysis will be the subject of future efforts.

5. The magnetic phase diagram

A theoretical phase boundary separating the expected occurrence of in-plane and vortex states in circular rings as we vary the shape parameters (σ, τ) was derived in [10]. However, the simplified considerations therein turned out to be a rough underestimate of a realistic curve, such as one obtained by micromagnetic simulations. The likely reason is that the metastable onion state is located around a deep local minimum of the energy landscape, and it may not be possible for a system with a given magnetic history and at a certain finite temperature to overcome the barriers separating the metastable (onion) and ground (vortex) states.

To establish the operational region of magnetic phase plates, we have calculated a phase diagram through micromagnetic simulations [11] on rings with different shape parameters. Simulations are carried out with the following choice of fixed parameters: Iron (A=21 pJ/m, M_S=2.15 T), damping parameter α=0.1, crystal anisotropy was set to 0 to mimic a polycrystalline ring, outer radius of the ring R_2=150 nm, uniform initial magnetization. From the simulations, we derive simulated phase boundaries that are expected to be more realistic than the idealized theoretical curves presented in Ref. [10]. Simulations were carried out varying the two aspect ratios τ and σ. For every value of σ (which requires a different mask in the LLG simulation software), we increased the thickness progressively until a vortex state was obtained. The aspect ratio τ of the first simulation producing a vortex was assigned to the phase boundary. Further increase of τ always resulted in a vortex. Results are shown in Fig. 3, where we observe that, for pentagonal rings (circles), a substantial fraction of the operational domain (dark shade of grey) identified by the geometrical constraint, equation (2), is compatible with the expectation of a vortex. The curve corresponding to circular rings (squares) is substantially higher in the phase diagram, so that a smaller portion of the operational domain (C+P) is available.

With pentagonal rings, we observe that the phase boundary lies around smaller values of τ. For example, when τ=0.04, we expect a vortex in pentagonal rings, whereas in circular rings we do not. We also observe that, in both circular and pentagonal rings, as σ approaches 1 (thin shell), vortex occurrence seems to lose its dependence on σ. However, for pentagonal rings, most of the phase boundary is actually almost flat (little σ-dependence). We observe that after a critical τ value, any change in σ no longer influences or determines the vortex formation. This may grant us a bit of extra freedom in designing and fabricating a phase plate, as the role of the inner diameter appears not critical. Overall, we note a sizeable enlargement of the constraint-compatible (σ,τ)-region where vortex occurrence is expected, thus favouring the choice of pentagonal over circular rings.

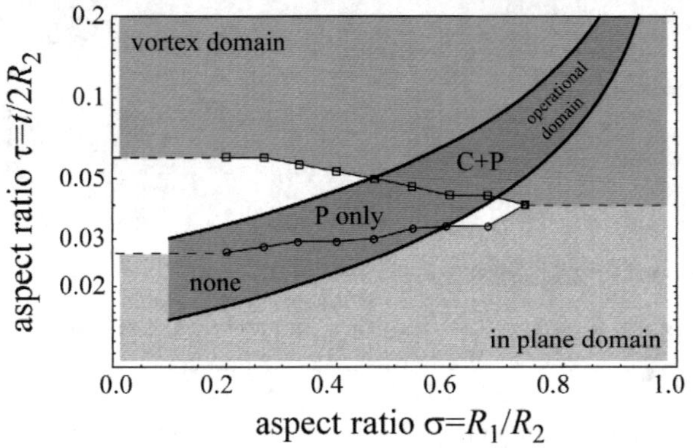

Figure 2. Numerical phase boundary separating vortex and in-plane/onion domains for circular (C) and pentagonal (P) rings. The operational region resulting from geometrical constraints is bounded between the two curves calculated from equation (2) for the hypothetical material as described in the text.

6. Conclusions

In this paper we have analyzed rings that may be potentially employed in the development of tunable magnetic phase plates. LLG simulations and analysis reveal that i) a useful operation domain exists in the (σ,τ) plane compatible with geometrical constraints; ii) the operational domain appears larger when pentagonal rings are examined, suggesting a higher degree of freedom in the choice of shape parameters; iii) the statistical occurrence of the vortex state is more favourable with pentagonal rings; iv) the asymmetry of pentagonal rings results in a sensitivity to in-plane applied field (absent in circular rings) that open a pathway for restoring a vortex from a non-vortex state. From the analysis and simulations carried out on both the circular and noncircular rings, pentagon shape rings emerge as the most favored for the phase plate development. A strategy for vortex state restoration that relies solely on in-plane applied fields is the subject of our current efforts, and will be reported elsewhere.

References

[1] Zhu J, Zheng Y and Prinz G A 2000 *J. Appl. Phys.* **87** 6668
[2] Kikkawa J M and Awschalom D D 1999, *Nature* **397** 139
[3] Hosokawa F, Danev R, Arai,Y and Nagayama K 2005 *J. Electron Microsc.* **54** 317
[4] Lentzen M 2004 *Ultramicroscopy* **99** 211
[5] Gamm B, Schultheiß K, Gerthsen D and Schröder R R 2007 *Ultramicroscopy* **108** 878
[6] Malac M, Beleggia M, Taniguchi Y, Egerton R F and Zhu Y 2008 *Ultramicroscopy* **109** 14
[7] Volkl E, Allard L and Joy D C (editors) 1999 *Introduction to Electron Holography* (New York: Kluwer Academic) pp 211-217
[8] Beleggia M 2008 *Ultramicroscopy* **108** 953
[9] Majorovits E, Barton B, Schultheiß K, Pérez-Willard F, Gerthsen D and Schröder R 2007 *Ultramicroscopy* **107** 213
[10] Beleggia M, Lau J W, Schofield M A, Zhu Y, Tandon S and De Graef M 2006 *J. Magn. Magn. Mater.* **301** 131
[11] Scheinfein M R and Price E A 2003 *LLG User Manual v2.50* (llgmicro@mindspring.com)

Journal of Physics: Conference Series **241** (2010) 012070 doi:10.1088/1742-6596/241/1/012070

Probing the atomic structure of amorphous Ta_2O_5 mirror coatings for advanced gravitational wave detectors using transmission electron microscopy

Riccardo Bassiri[1], Konstantin B. Borisenko[2], David J. H. Cockayne[2], James Hough[1], Ian MacLaren[1], Sheila Rowan[1]

[1] SUPA, Department of Physics and Astronomy, University of Glasgow, Glasgow G12 8QQ, United Kingdom

[2] Deapartment of Materials, University of Oxford, Oxford OX1 3PH, United Kingdom

E-mail: r.bassiri@physics.gla.ac.uk

Abstract. Advanced generations of ground-based gravitational wave detectors will use ultra-low-loss amorphous dielectric multilayer mirror coatings in order to minimise thermal noise, a limiting factor in detector sensitivity. Transmission electron microscopy is a promising way to probe the atomic structure of these coatings in an effort to better understand the causes of the observed mechanical loss (internal friction) and hence thermal noise.

1. Introduction

Long-baseline interferometers such as those of the LIGO [1], Virgo [2] and GEO [3] projects are being used in searches for the effects of gravitational radiation emitted from astrophysical sources on the relative displacements of test masses, which are coated to produce highly reflective mirrors. These gravitational wave detectors are the most sensitive displacement instruments ever to be developed and are able to detect strains in mirror displacements at the level of $10^{-19}\mathrm{m}/\sqrt{\mathrm{Hz}}$, with planned detector upgrades to form 'Advanced' detectors aiming at another factor of 10 improvement in sensitivity [4, 5, 6].

Noise from various sources is a major challenge when attempting to reach such sensitivities. At the highest sensitivity range (from a few 10s Hz to several 100 Hz) thermal noise from the test mass mirrors and their suspensions is a significant limitation. The test mass mirrors are made from ultra pure fused silica (SiO_2) with multi-layer amorphous dielectric mirror coatings used to provide high reflectivity. Significant levels of thermal noise arise from the mechanical losses of the mirror coatings [7]. The coatings currently used are formed from multiple layers, each of $\lambda/4$ thickness layers of Ta_2O_5 and SiO_2 [7].

The fundamental processes which cause mechanical loss in these coatings are not well understood. The aim of this research is therefore to understand the processes causing the mechanical losses on an atomic scale. Here we use transmission electron microscopy and associated analysis techniques to study, for the first time, the local structures and bonding in argon ion-beam-sputtered Ta_2O_5 coatings produced by CSIRO [8] which have been heat treated to a variety of temperatures. This is an important preliminary step towards identifying a quantitative relation between atomic structure and macroscopic coating mechanical loss.

2. Experimental methods

The transmission electron microscope (TEM) gives the ability to probe and characterise the atomic structure of the amorphous mirror coatings using reduced density functions (RDFs). Figure 1 (left) shows an image of a typical multilayer mirror coating, the darker sections are the Ta_2O_5 layers. For the purposes of this research, the samples used for analysis have only a single layer of Ta_2O_5 deposited on a silica substrate. Figure 1 also shows a typical amorphous diffraction pattern one would get when looking at only the Ta_2O_5 layer. It is these diffraction patterns that act as the raw data for the RDFs.

Figure 1. Bright field image of a multilayer coating of dark Ta_2O_5 and light SiO_2 layers (left), CBED amorphous diffraction pattern of Ta_2O_5 coating (right).

The RDF is a statistical representation of where atoms sit with regards to a central atom [9] and can be effectively described as the Fourier transform of the intensity profile one would get from an amorphous diffraction pattern,

$$G(r) = 4\int_0^\infty \varphi(q)\sin(qr)dq, \tag{1}$$

where $q = 4\pi sin\theta/\lambda$, r is the distance from a central atom and $\varphi(q)$ is the reduced intensity function derived from the intensity profile [9]. A detailed guide to the numerical calculation approach for the RDF can be found in the review article by Cockayne [9].

The peaks of this distribution give information about the nearest atomic neighbours such as the peak intensity, which describes the probability of finding a particular atom at a certain distance. Also, thermal vibrations within the atomic structure contribute to the peak widths and may be useful for understanding the measured mechanical loss of the coating which is directly related to the thermal noise.

To gain the maximum amount of data from the RDFs, reverse Monte Carlo modelling can be used to model the amorphous structure. Modelling can also be used to find the interatomic bond angles and co-ordination numbers.

The model is produced by firstly randomly packing a defined number of atoms in a defined area and given the atomic boundary conditions of a typical crystalline phase of the material. It is then compared to the experimental RDF via reverse Monte Carlo simulations where the atoms in the model are re-arranged randomly until they sit in a configuration that agrees with the RDF data, and hence the amorphous phase. Energy optimisation of the model was then carried out using DFT modelling to make sure the atoms are sitting in physically reasonable positions.

Journal of Physics: Conference Series **241** (2010) 012070 doi:10.1088/1742-6596/241/1/012070

3. Results

The RDF and modelling procedure was carried out on three Ta_2O_5 argon-ion beam sputtered coatings that were heat treated at 300, 400 and 600°C. These coated samples were single layers of Ta_2O_5 on a SiO_2 substrate. The process of heat treatment is to leave the sample in an oven under normal atmospheric conditions at a defined temperature for 24hrs.

The mechanical losses of these samples were measured at cryogenic temperatures primarily as a means of using temperature as a probe to study the behaviour of the internal friction of the material, and also because new generations of detectors may work at cryogenic temperatures to further reduce noise sources. The samples show that higher heat treatment temperatures correspond to an increase in the magnitude of loss peaks observed at cryogenic temperatures [10].

The aim of this experiment was to see if there was any difference in the amorphous structures measured using RDFs.

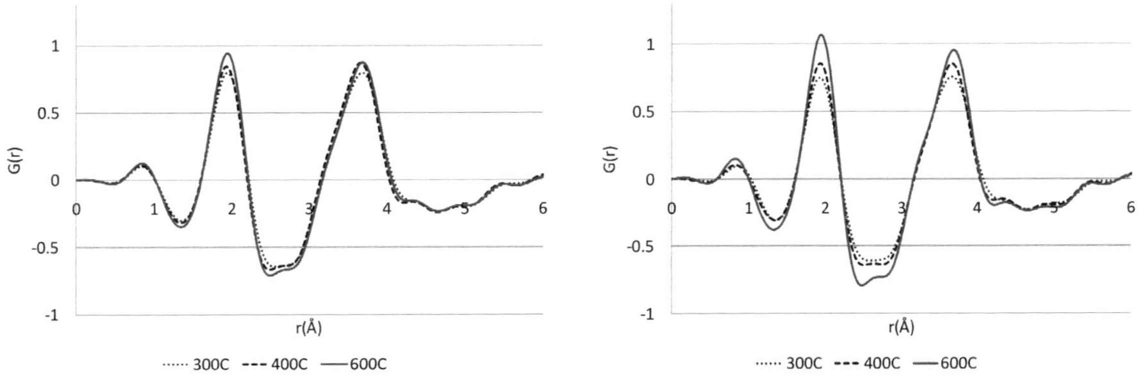

Figure 2. Average RDFs of coatings heat treated at 300, 400 and 600°C (left), Local structure variation showing a greater difference in RDFs at 300, 400 and 600°C (right).

Figure 2 (left) shows three reduced density functions from a tantala layer of mirror coatings heat treated at 300, 400 and 600°C. The averaged RDFs, where RDFs from various areas in the same sample are averaged, show a negligible change as the heat treatment temperature rises. However preliminary investigations into local structural changes, as opposed to averaged areas, suggest that there may be a slight change in the local ordering as the heat treatment temperature rises as can be seen in the right of figure 2. To explain further, some local areas seem to show better defined RDF peaks in samples that have been heat treated at high temperatures suggesting that, at least locally, the heat treatment may be leading to an increase in local ordering. It is worth pointing out that all the coatings looked at had no signs of crystallisation, therefore any differences measured are changes in short range order as these samples all remain amorphous. Further investigation is required to understand the nature of this change and to relate it to the mechanical loss measurements.

Figure 3 shows the amorphous model generated from the reverse Monte Carlo process and energy optimisation based on the 400°C RDF data. Preliminary results from this model show an average Ta to O bond length of 2.10Å with the presence of short Ta to Ta distances of 3.28Å on average.

What is planned next is to relate this model using the reverse Monte Carlo procedure to the local RDF variation from the 300 and 600°C samples and compare the data, looking for any differences that can be related back to changes in the mechanical loss.

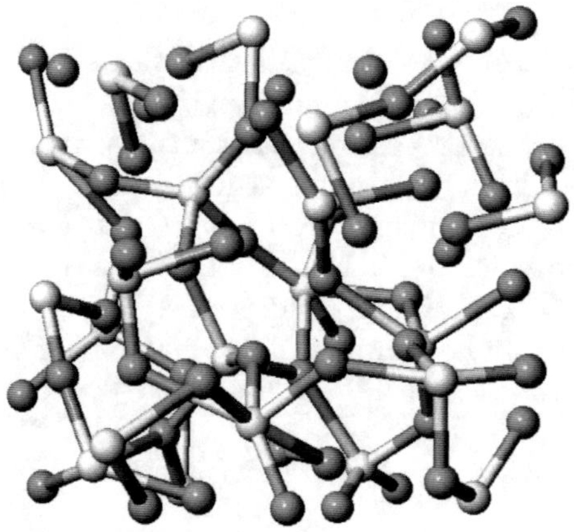

Figure 3. Atomic model of amorphous Ta_2O_5 with dark coloured oxygen and light coloured tantalum atoms, based on the 400°C RDF.

4. Conclusions

We have made major advances in our ability to characterise the atomic structure of amorphous mirror coatings and are working to use this to understand measured mechanical losses. The results from the averaged RDFs of three Ta_2O_5 coatings heat treated at 300, 400 and 600°C show a negligible change in amorphous structure. However, there may be some local changes from area to area in the samples showing an increase in the Ta to O coordination as the heat treatment temperature increases. Reverse Monte Carlo modelling combined with energy optimization of the amorphous structure provides a promising way to probe the maximum amount of information possible from these samples. The main emphasis of future research will be to understand how these results relate to the mechanical loss of these coatings.

5. Acknowledgements

We would like to thank our colleagues in the IGR, GEO, and LIGO Scientific collaborations for useful discussions. We would also like to thank EPSRC, SFTC, SFC, The Leverhulme Trust and the University of Glasgow in the UK and the NSF in the US for financial support.

References

[1] Sigg D and the LIGO Scientific Collaboration 2008 Status of the LIGO detectors *Class. Quantum Grav.* **25** 114041
[2] Acernese F et al 2008 The VIRGO status *Class. Quantum Grav.* **25** 184001
[3] Grote H and the LIGO Scientific Collaboration 2008 The status of GEO 600 *Class. Quantum Grav.* **25** 114043
[4] Smith J R 2009 The path to the enhanced and advanced LIGO gravitational wave detectors *Class. Quantum Grav.* **26** 114013
[5] Acernese F et al 2006 Virgo upgrade investigations *J. Phys.: Conf. Ser.* **32** 2239
[6] Willke B et al 2006 The GEO-HF project *Class. Quantum Grav.* **23** 20714
[7] Harry G M et al 2002 Thermal noise in interferometric gravitational wave detectors due to dielectric optical coatings *Class. Quantum Grav.* **19** 897 - 917
[8] CSIRO, Bradfield Road, West Lindfield NSW 2070, PO Box 218, Australia, www.csiro.au
[9] Cockayne D J H 2007 The Study of Nanovolumes of Amorphous Materials Using Electron Scattering *Annu.Rev.Mater.Res* **37** 159-87
[10] Martin I 2009 *Studies of materials for use in future interferometric gravitational wave detectors* (University of Glasgow) Ph.D. thesis

Prospects of aberration correction for lattice-resolved electron tomography

Z Saghi[1], I M Ross and G Möbus
Dept Engineering Materials, University of Sheffield, Sheffield S1 3JD, UK.

E-mail: z.saghi@shef.ac.uk

Abstract. Electron tomography is a valuable 3D characterization technique, so far limited to a nanometer scale resolution. This paper intends to explore the extension of the technique to the atomic scale. Computer simulations of three HREM technologies are performed on a CeO_2 crystal and the effects of the voltage, focus spread and aberration correction are assessed. Experimental challenges related to the technique are discussed and illustrated on HREM images of a CeO_2 nanoparticle acquired with the JEM 2200FS aberration-corrected TEM. Finally, geometric tomography is proposed as a complementary shape reconstruction technique when only few zone-axis tilt images are used for the lattice-resolved reconstruction.

1. Introduction

Electron tomography (ET) is now routinely employed for the three-dimensional (3D) characterization of nanostructures with a resolution of few nanometers [1]. In order to extend ET into a lattice-resolved technique, the signal recorded in high resolution transmission electron microscope (HREM) images should directly reflect the projected atomic potential of the specimen. This requirement depends on both the specimen nature and the microscope optics. In the present work, we explore the capabilities of three HREM technologies to reconstruct a simulated CeO_2 cone-shaped supercell, and point out some challenges related to the experimental application of the technique, using the aberration-corrected microscope JEM-2200FS-AC.

2. Tomographic reconstructions on simulated tilt series

A cone-shaped CeO_2 model with a maximum thickness of 2.5 nm was generated with Interactive Data Language (IDL, ITT, USA) and tilted from -90° to +90° with 5° increment. The rotation axis was chosen parallel to the symmetry axis of the cone. At each tilt angle, HREM images at Scherzer defocus were calculated by the multi-slice method using the EMS software package [2]. The three microscopes selected were: (i) JEM-3010 (LaB_6, 300kV, Cs=1.1mm), (ii) JEM-ARM1250 (LaB_6, 1250kV, Cs=1.6mm) and JEM-2200FS-AC (FEG, 200kV, Cs=50μm). The obtained tilt series were exported to IDL for the filtered backprojection and the 3D visualization of the reconstruction. The quality of the reconstructions was assessed by comparing the number and position of the retrieved atoms with the ideal CeO_2 model. Figure 1(a,e,i) show one member of the HREM tilt series simulated with the JEM-ARM1250, JEM-2200FS-AC and JEM-3010, respectively. The voxel projection views of the three reconstructed volumes are presented in Figure 1(b,f,j) with the Ce atoms appearing in yellow/red and the oxygen atoms in green for the JEM-ARM1250 and JEM-2200FS-AC

[1] To whom any correspondence should be addressed.

reconstructions (online only), while the two last columns are the horizontal Ce plane and the vertical oxygen plane at highest thickness through the volume. The high voltage of JEM-ARM1250 and the aberration correction of the JEM-2200FS-AC seem to be both important for a reliable reconstruction with the retrieval of both Ce and oxygen atoms at their right positions. However, when analysing the slices through the volume, we notice that the shape of oxygen atoms and the intensity of the Ce atoms are more uniform in the JEM-ARM1250 reconstruction than the JEM-2200FS-AC, suggesting that the voltage has a more significant effect than the aberration correction. The JEM-3010 reconstruction retrieves all Ce atoms, although blurred and distorted, and only the oxygen atoms at the outermost ring of the cone are restored.

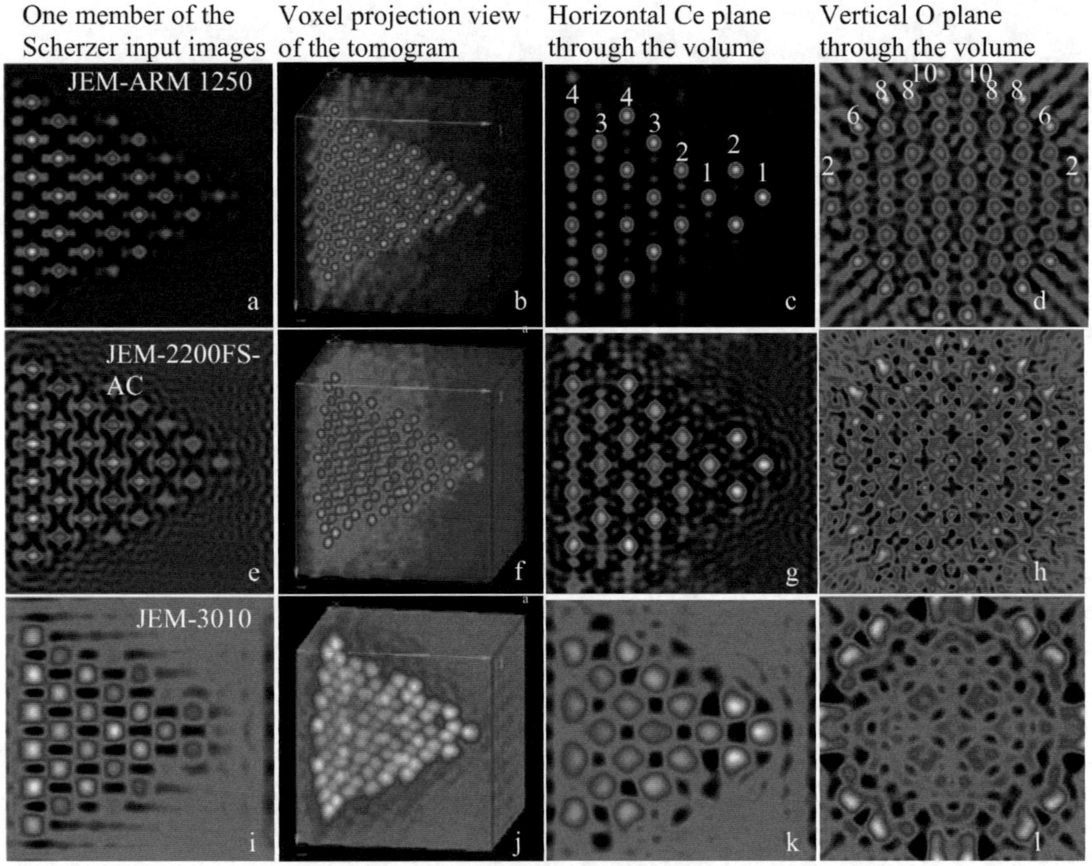

Figure 1 Inverted Scherzer images of a CeO2 supercell, simulated for the JEM-ARM 1250 (a), JEM-2200FS-AC (e) and JEM-3010 (i). Voxel projection (b,f,j) and slices through the Ce plane (c,g,k) and oxygen plane (d,h,l) of the tomographic reconstructions from the Scherzer tilt series. The numbers is (c) and (d) correspond to the number of atoms in each column of the initial model.

3. Experimental challenges related to atomic electron tomography

To achieve a successful experimental implementation of the technique, several difficulties have to be overcome. In this section, we point out some of these challenges on HREM images of CeO_2 nanoparticles acquired with the Sheffield aberration-corrected JEM-2200FS-AC, a combined TEM/STEM microscope operating at 200kV and equipped with a TEM corrector below the objective lens and a STEM corrector above, both built by CEOS GmbH.

For a meaningful reconstruction, many projection views have to be oriented in a zone axis direction, which implies that the particle has to be preoriented in a specific direction with respect to the rotation axis. As an example, Figure 2(a) shows a CeO_2 nanoparticle with its [110] spectrum (b), which could be suitable for atomic tomography. Moreover, it is important to know whether the contrast on the HREM images reflects the number of projected atoms, so as to select which of these two tomography modes is most appropriate: computed (quantitative) tomography or geometric (qualitative) tomography [6]. A line-profile (Figure 2(c)) through the HREM image in Figure 2(a) shows that for columns of Ce atoms and with an aberration corrected TEM, the projection linearity is fulfilled for a thickness up to 5nm. Beyond this threshold, both contrast attenuation and inversion are observed and would lead to various artefacts in the quantitative reconstruction [3,6]. Nevertheless, a qualitative retrieval of the atomic distribution can be expected even from non-linear HREM projections as long as the projected columns appear *all* as bright spots (or dark) and there are no half-spacing lattice fringes. Defocus and Cs are the two parameters to optimize for maximum projection linearity and minimum contrast inversions. The HREM images of a CeO_2 nanoparticle in Figure 3 (a) and (c) confirm that phase contrast at opposite defocus values become exactly inverse at Cs=0, a feature which should be avoided in a tomography experiment but could be used to remove non-linear HREM contrast by subtracting two images of opposite defocus values [4]. Even within one image, contrast inversion due to the high thickness at the center of the particle can be observed: white dots (top-left and bottom-right corner) and black dots (centre) interchange. For an amplitude contrast setting at Cs=0 and z=0nm, this inversion is avoided, however at the cost of lower contrast/higher noise. For tomography of non-round specimens, an additional problem occurs as focus has to be referred to the rotation axis in the centre of particle rather than to the exit plane. Finally, Figure 3 (d-e) compares the two "super-Scherzer-focus" settings for Cs=-50μm and z=+10nm with Cs=+50μm and z=-15nm [7], the latter being preferable for tomography for SNR reasons, as selected for Figure 1(e).

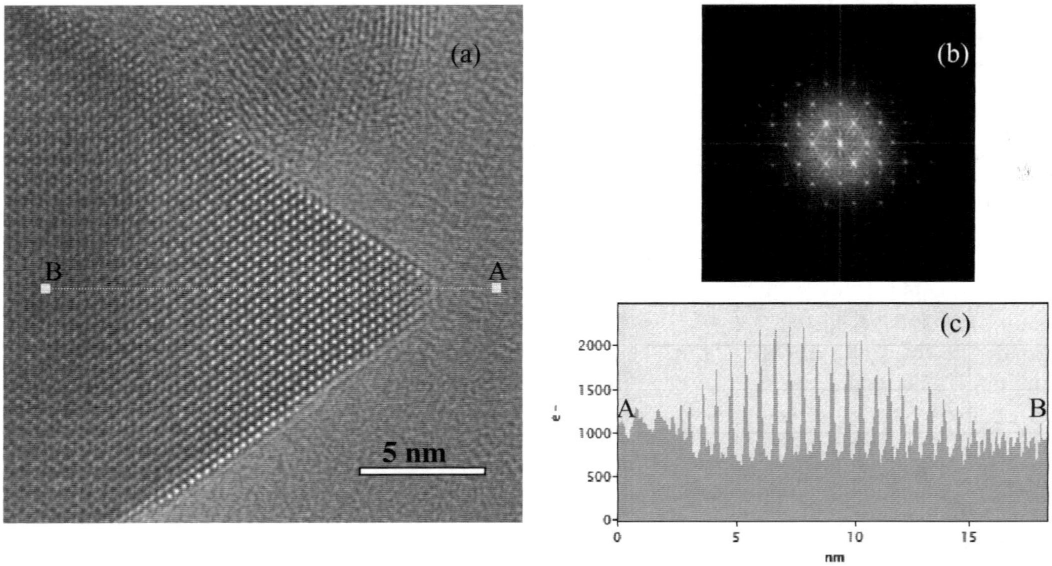

Figure 2 (a) HREM image of a CeO_2 nanoparticle, acquired with 2200FS-AC (Cs=-253nm, Z=16nm). (b) Fourier spectrum of (a). (c) Line-profile along the blue line in (a).

Taking into consideration all these experimental challenges, it is very likely that from the acquired tilt series, only few HREM images can be used for the reconstruction. It is possible to achieve a lattice-resolved tomogram from few zone-axis projections, as reported in [5, 6], but in order to reconstruct the correct 3D shape of the particle, a geometric reconstruction has to complement the atomic one. This consists in extracting the binary support of the HREM images and backprojecting the obtained shadow images. The intersected volume is then the convex envelope of the particle, which is a good approximation of the 3D shape if the particle if convex. By combining the two tomograms of different resolution, both the atomic arrangement and the facet configuration of the crystal can be retrieved [6]. Another advantage of the shadow images is that they can also be used for the precise alignment of the HREM images, since cross-correlation performed on HREM patterns might be ambiguous.

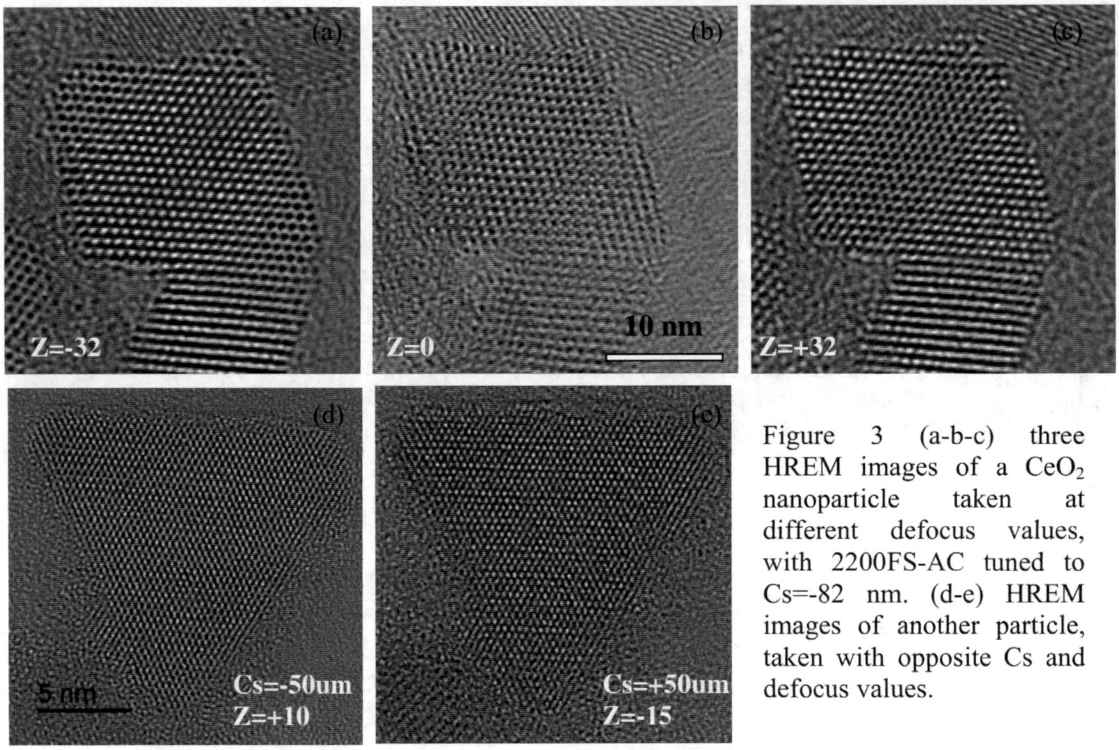

Figure 3 (a-b-c) three HREM images of a CeO_2 nanoparticle taken at different defocus values, with 2200FS-AC tuned to Cs=-82 nm. (d-e) HREM images of another particle, taken with opposite Cs and defocus values.

4. Discussion and Conclusion

In summary, lattice-resolved electron tomography is a promising technique for the 3D structural analysis of crystals. The method has been demonstrated on a simulated CeO_2 crystal, and the correspondence of high-voltage TEM and 200kV Cs-corrected TEM has been shown. First experimental Cs-corrected images of CeO_2 pyramidal corners indicate that the concept is feasible and surface reconstruction or structure deviations at the very tip are rather small. Focus/Cs-settings for each zone-axis image have to be known and meaningful to maximise “structure image” conditions, but not necessarily identical throughout the tilt series.

Acknowledgements
This work was supported by grant GR/S85689/01 from EPSRC, UK

References

[1] Möbus G and Inkson B J 2007 *Materials Today* **10** p18.
[2] Stadelmann P A 1987 *Ultramicroscopy* **21**(2) p131.
[3] Saghi Z, Xu X and Möbus G 2008 *J Microsc.* **232** p186.
[4] Tanaka N 2008 *Sci. Technol. Adv. Mater.* **9** 014111.
[5] Jinschek J R, Batenburg K J, Calderon H A, Kilaas R, Radmilovic V and Kisielowski C 2008 *Ultramicroscopy* **108** p589.
[6] Saghi Z, Xu X and Möbus G 2009 *J Appl. Phys.* in press.
[7] Jia C-L, Lentzen M and Urban K 2004 *Microsc. Microanal.* **10** 174.

In-situ Monitoring of Gas-Assisted Focused Ion Beam and Focused Electron Beam Induced Processing

I. Utke, M. Gabureac, V. Friedli, L. Bernau, J. Michler

Empa, Swiss Federal Institute for Materials Testing and Research, Feuerwerkerstrasse 39, CH-3602 Thun, Switzerland

ivo.utke@empa.ch

Abstract. The state of the art of several in-situ monitoring techniques for gas assisted focused ion beam and focused electron beam induced processing (FIBIP and FEBIP) is presented. The monitoring techniques discussed comprise (a) the stage current and secondary electron signal for topographic information, (b) mass sensors to determine process yields, deposit density, molecule coverage and adsorption, (c) electrical resistance measurements, (d) reflectometry, (e) annular dark field sensing, and (f) SEM integrated mechanical measurements.

1. Introduction

Focused electron beam (FEB) and focused ion beam (FIB) induced processing are well established techniques for local deposition and etching that rely on decomposition of surface adsorbed precursor molecules by irradiation, see fig. 1.

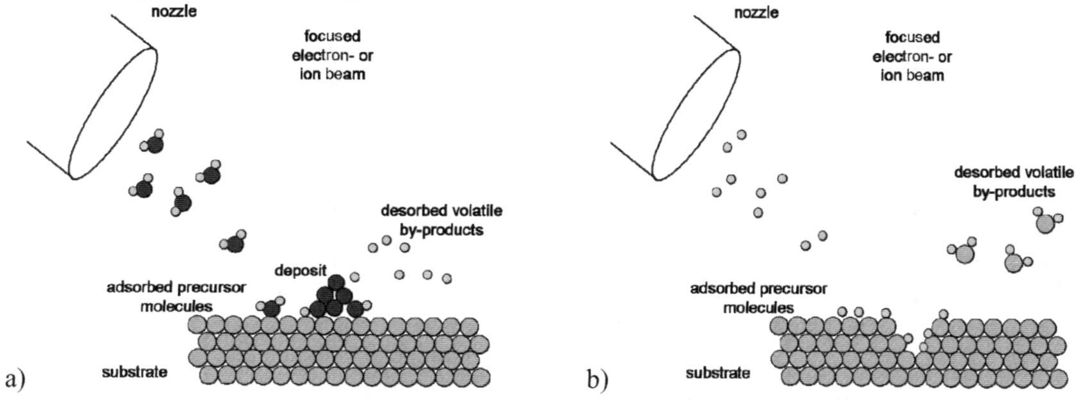

Fig. 1: Principle of gas assisted FEB or FIB processing (for sake of clarity the FIB sputtering mechanism - proceeding without gas assistance – was omitted). Volatile molecules are introduced into the microscope chamber and reversibly physisorb on the substrate surface. Dissociation of molecules is taking place by charged particle (electrons or ions) irradiation a) Deposition: the dissociation reaction results in non-volatile fragments forming the functional deposit, b) Etching: the dissociation products form volatile compounds with the substrate and remove material from the bulk.

These high-resolution nanostructuring techniques have various applications in nanoscience and industry comprising attach-and-release procedures in nanomanipulation, fabrication of sensors (magnetic, optical and thermal) for scanning probe microscopy, fabrication of electrical contacts and devices, and photomask repair of the next generation below the 64 nm node. For a detailed listing of applications as well as the underlying molecule dissociation mechanisms the reader is referred to a recent review [1]. Control of the process with respect to deposit thickness or etch depth is fairly well established, however, a complete physical and chemical understanding of the process is hampered by the lack of suitable means to monitor and to access the numerous interrelated and time-varying process parameters (deposition and etch rate, yield, molecule flux and adsorption/desorption).

This article will discuss in-situ monitoring of gas-assisted focused electron beam and focused ion beam deposition and etching using various signals and related setups. It will not cover in-situ observation of the process using dual or cross beam machines with both a focused electron and ion beam. Generally, such an approach will produce unwanted co-deposits from the observing beam.

2. Monitoring of stage current and secondary electron signal

These signals are collected through the substrate stage and the secondary electron (SE) detector, see fig. 2a, and are related through the current balance: $I_{stage} = I_{beam} - I_{SE} - I_{BSE}$. Both the number of secondary electrons (SE) and backscattered electrons (BSE) change with the sample's topography and material. Consequently, "end-point detection" during FIB or FEB etching can be carried out by monitoring these signals: once a buried layer in the structure is reached by etching the signals will change due to the differing SE and BSE emission of materials and the process can be stopped. This technique is limited to vias holes having a high aspect ratio that still permits a significant fraction of the secondary electrons to leave the surface.

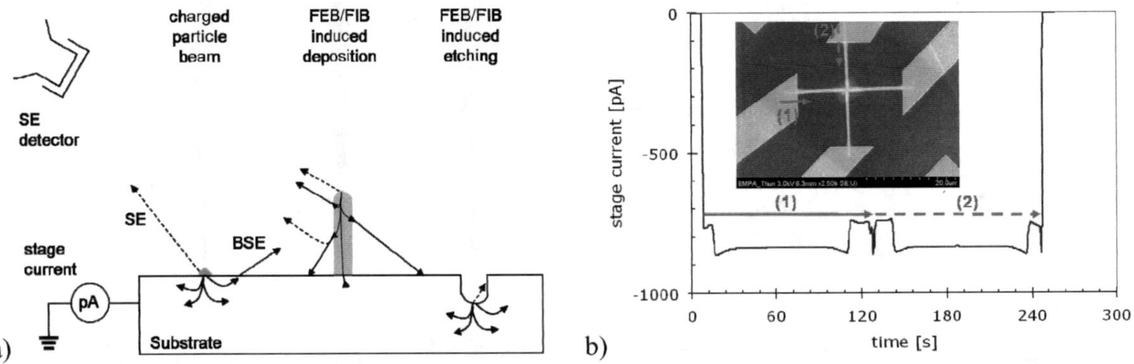

Fig. 2: a) Principle of time resolved stage current monitoring. As the deposit grows, more SEs and BSEs leave the deposit through its side, augmenting the recorded SE signal and diminishing the measured stage current. The inverse holds true for particle beam induced etching. b) Example: stage current during FEB deposition of a cross-shaped deposit ($Co_2(CO)_8$, 25keV, 1000pA). Inset: SEM side view of the entire structure. The arrows indicate the writing direction.

Fig. 2b shows monitoring of the stage current during deposition of cross shaped deposit. The deposition starts on a lithographically pre-defined gold electrode, then the FEB is scanned to connect to the opposing gold electrode. As a result the "horizontal" line deposit is obtained. The deposition process is repeated for the "vertical" line and finally a Hall cross is obtained. Monitoring the stage current during the deposition probes the location of the electrodes (low stage current -750 pA), the substrate SiO_2 (100 nm)/Si (high stage current -850 pA) and the first line deposit (small peak at 190 s). Together with the microscope observation and electrical measurements such a stage current plot is a finger print for a specific deposit geometry and deposit material. If similarly programmed deposition routines have inconsistent stage current profiles, this may indicate irregularities in the beam conditions (focus, beam current oscillations, drifts) or molecule supply (precursor exhaust or bursts). These signals show their full control virtue during FEB or FIB deposition of more complicated 3D structures, like pillars [2], periodic wires [3], and nanoscale artwork [4]. Of note, the stage current control was used to fabricate hole diameters in membranes *smaller* than the focused electron beam [5].

3. Mass sensing

When attempting to monitor the mass during FEB or FIB deposition / etching one should keep in mind that the nanodeposits feature very small masses, typically in the picogram range or below. This range can be addressed by cantilever based sensors [6]. Special care must be taken to stabilize the temperature of the setup in the milliKelvin range. The principle is given in fig. 3a and consists of a cantilever with piezoresistive readout being actuated by a piezoelectric actuator. The frequency response to mass changes must be calibrated due to the unknown cantilever force constant. This can be best achieved by monitoring the mass change while FIB milling a box of easily observable dimensions into the cantilever material. Once this calibration step is performed the mass measurements inside the microscope chamber are quantitative. From a typical dot series deposition shown in fig. 3b it can be seen that the deposition process is highly reproducible and that with increasing pillar height, the deposited mass assumes a linear dependence with time, i.e. the contributing electron fluxes (see bottom inset of fig. 3) become constant.

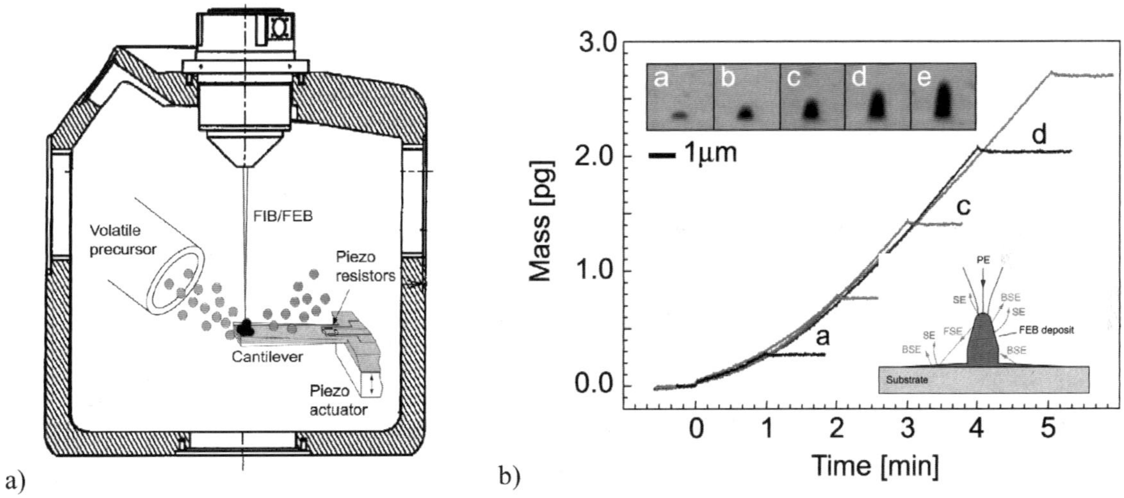

Fig. 3: a) Schematic diagram of the SEM integrated cantilever mass sensor for FIB/FEB induced process monitoring with local precursor supply from a microtube gas injection system. The mass added to or removed from the cantilever is detected as a negative or positive resonance frequency shift, respectively. b) FEB deposits (5kV, 300pA, dot exposure, $Cu(hfa)_2$). Top inset: SEM tilt images of pillars deposited for 1-5 minutes. Bottom inset: contributions to mass deposition.

Furthermore, from molecule adsorption / desorption measurements (without irradiation) the monolayer coverage and the residence time of molecules on the cantilever surface can be determined. Also, deposition or etch yields, in units of deposited or etched atoms per incident electron or ion, are precisely accessible according to eqn.1:

$$Y = \frac{N_A \cdot m / M}{I_P \cdot t / e_0} \qquad (1)$$

where $I_P{\cdot}t$ is the incident dose during exposure time t and beam current I_p, m is the measured deposited or etched mass, and M is the molar mass of the etched or deposited material. N_A is the Avogadro constant and e_0 the elementary charge. The molar mass of the etched material is generally known and etch yields can be derived in-situ. In contrast, for deposits the composition is generally unknown and the molar mass must be determined from composition measurements. Eqn. 1 represents a means to precisely determine yields since it does not require any assumption about the deposit density. The correct deposit density ρ_{dep} of the deposit can be obtained by measuring the deposit volume V_{dep} and $\rho_{dep} = m_{dep}/V_{dep}$. For the $Me_3PtCpMe$ molecule (Me – methyl) was found [6]: 7 % monolayer coverage, a residence time of 30 µs, a deposit density of 11.2 g/cm^3 (irradiation with 30 keV Ga^+-ions) and 4.5 g/cm^3 (irradiation with 5 keV electrons). Only for precursors containing no hydrogen the density can be estimated directly from the deposit composition [7].

4. In-situ electrical resistance measurements

Time-resolved resistance measurements during gas assisted FEB deposition can be performed using pre-deposited electrodes on an insulating substrate. They will be "connected" by scanning the electron or ion beam back and forth between them which will deposit a wire, the resistance of which depends on the deposited material (electron impact molecule dissociation) and the deposited thickness. In-situ resistance measurements during FEB/FIB deposition were reported for the molecules $W(CO)_6$ [8], $Me_3PtCpMe$ [9, 10] and acrylic acid [1]. In-situ measurements of FIB deposited wires from $Me_3PtCpMe$ show the formation of a Ga-phase [11] due to Joule heating, while FEB deposited freestanding wires show electromigration of platinum [12]. An example of a two point electrical resistance real-time measurement while FEB depositing with the molecule $Co_2(CO)_8$ is shown in fig. 4.

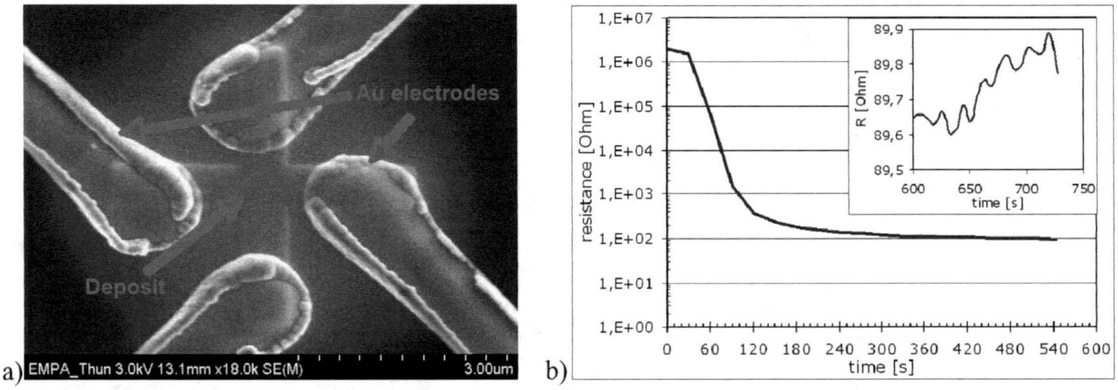

Fig. 4: a) FEBID of $Co_2(CO)_8$ is performed on pre-structured Au electrodes. Electrodes active for in-situ current measurement are marked with red arrows. b) In-situ measurement of resistance evolution during deposition. Inset: evolution of deposit resistance during venting with N_2.

The in-situ measurements permit to study the "electrical thickness" of the deposit. For constant length, width, and homogeneous material the resistance is inversely proportional to the deposited wire thickness. Hence, for a constant deposition rate an inverse relationship between the resistance and the deposition time is obtained. Deviations from this trend point to changes affecting the charge carrier generation and kinetics, for example, by the formation of intermediate species. Also post-irradiation and air exposure phenomena can be studied and related to eventual oxidation processes, see fig. 4b. Two point measurements can be uniquely attributed to the deposit only if the contact resistance between electrodes and deposit is low compared to the deposit resistance. Alternatively, four point probe measurements would cancel contact resistance effects.

5. Reflectometry

Real-time *in situ* reflectometry during FEB deposition requires that a laser beam is allowed to enter into the microscope chamber. The monochromatic laser is then focused on the growing transparent or semi-transparent deposit inside the SEM chamber and the reflected optical signal is collected by a photodiode. The principle is such that the intensity of the reflected laser beam depends on the optical thickness of the deposit and its optical absorption. A periodic signal with deposition time (deposit thickness) is obtained due to constructive and destructive interference. This allows calculating the real and imaginary parts of the diffractive index at the laser wavelength from the period and the amplitude decay after having performed a thickness calibration.

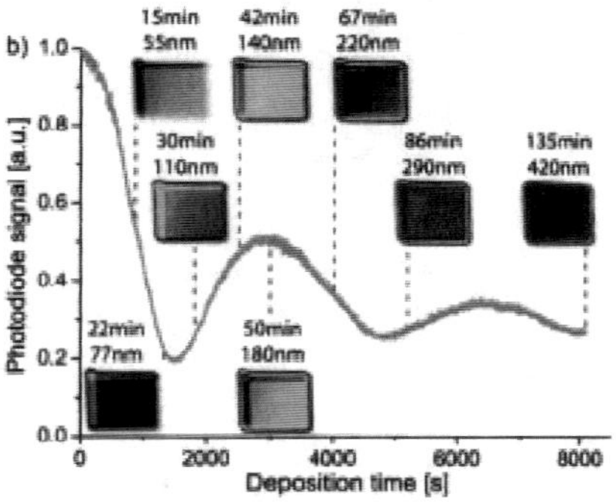

Fig. 5: Reflectometry curve as function of time recorded during focused electron beam induced deposition with $Si(OCH_3)_4$. Insets show the deposit colour as seen in an optical-microscope image and the corresponding deposition time and thickness. Taken from [13].

Comparing the optical thickness to the thickness measured by profilometry at various exposure times reveals if the deposited material is homogeneous or if changes in composition, i.e. the refractive index, occurred during growth. Measurements at 514 nm laser wavelength gave the following refractive indices: 1.51+i0.055 (HCOOH), 1.56+i0.14 ($Si(OCH_3)_4$), and 2.19+i0.013 ($Ti(NO_3)_4$) [13]. The value of the imaginary part is a measure for absorption and indicates that considerable amounts of carbon were co-deposited during SiO_2 deposition from the molecule $Si(OCH_3)_4$. The smallest absorption is evidently obtained from the carbon-free molecule ($Ti(NO_3)_4$). In order to perform such optical measurements the size of the deposit must be at least somewhat larger than the focused laser beam.

6. In-situ annular dark field signal

Recently, in-situ monitoring of the annular dark field (ADF) signal during electron beam induced deposition on electron transparent membranes was used to control the deposit's thickness homogeneity [14]. This signal has a strong atomic number contrast and is assumed to depend linearly on the deposited mass. Using the signal as a feed back in the deposition process proves useful for avoiding enhanced deposition on (pre-deposited) structures having a tilt to the incident electron beam (FEBID proximity effect). With regard to the small noise on the ADF signal it was suggested that this method is capable of resolving the deposition of one single molecule. This would be the highest resolution of all the in-situ techniques employed today.

7. SEM integrated mechanical measurements

These measurements involve integration of nanomanipulation setups, comprising x, y, z stages and corresponding electronics, into the microscope chamber. A detailed overview on such cantilever based force sensors and piezo-driven vibration stages and their applications in scanning electron / ion microscopes can be found in [15]. The determination of Young's modulus follows straightforwardly from the force balance of a reference cantilever which is gently pushed against the nanopillar under consideration, see fig. 6a. When Young's modulus of the nanopillar is known, its density can be determined from vibration experiments as shown in fig. 6b. Mechanical properties of pillar deposits from phenanthrene, $W(CO)_6$, and $Cu(hfa)_2$ were measured from different groups and summarized in [16]. The typical range of Young's modulus is in the order of magnitude of 10 GPa for FEB deposition and up to a few 100 GPa for FIB deposition [17]. Dependencies on electron or ion dose and energy were found.

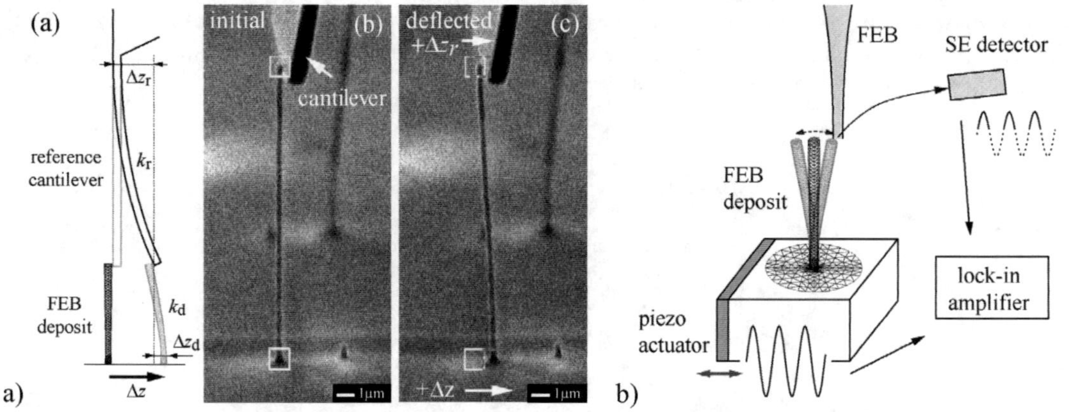

Fig. 6: a) Principle of deflection experiment. Cantilever and nanopillar are deflected according to their force constants *k* when moved against each other. The SEM images are taken from a video sequence and show tracking of the cantilever and substrate displacements. b) Principle of vibration experiment. When the resonance frequency of the pillar is approached the secondary electron detector measures the peak curve since the pillar vibration intersects the focused electron beam. Taken from [16].

Using a tensile stage setup, tensile strengths in the order of 1 GPa were found for FEB deposits from $Co_2(CO)_8$ and $W(CO)_6$ [18].

References

[1] Utke I, Hoffmann P and Melngailis J 2008 *J. Vac. Sci. Technol. B* 26 1197-276
[2] Bret T, Utke I, Bachmann A and Hoffmann P 2003 *Appl. Phys. Lett.* 83 4005-7
[3] Bret T, Utke I, Gaillard C and Hoffmann P 2004 *J. Vac. Sci. Technol. B* 22 2504-10
[4] Bret T, Utke I, Hoffmann P, Abourida M and Doppelt P 2005 *Microelec. Engin.* 83 1482-6
[5] Miyazoe H, Utke I, Michler J and Terashima K 2008 *Appl. Phys. Lett.* 92 043124
[6] Friedli V, Santschi C, Michler J, Hoffmann P and Utke I 2007 *Appl. Phys. Lett.* 90 053106
[7] Utke I, Friedli V, Michler J, Bret T, Multone X and Hoffmann P 2006 *Appl. Phys. Lett.* 88 031906
[8] Porrati F, Sachser R and Huth M 2009 *Nanotechnology* 20 195301
[9] De Teresa J M, Cordoba R, Fernandez-Pacheco A, Montero O, Strichovanec P and Ibarra M R 2009 *Journal of Nanomaterials* 2009 936863
[10] Botman A, Hesselberth M and Mulders J J L 2008 *J. Vac. Sci. Technol. B* 26 2464-7
[11] Reguer A, Bedu F, Tonneau D, Dallaporta H, Prestigiacomo M, Houel A and Sudraud P 2008 *J. Vac. Sci. Technol. B* 26 175-80
[12] Gazzadi G C and Frabboni S 2009 *Appl. Phys. Lett.* 94 173112
[13] Perentes A, Bret T, Utke I, Hoffmann P and Vaupel M 2005 *J. Vac. Sci. Technol. B* 24 587-91
[14] van Dorp W F, Hagen C W, Crozier P A and Kruit P 2007 *J. Vac. Sci. Technol. B* 25 2210-4
[15] Friedli V, Hoffmann S, Michler J and Utke I 2008 *Applied Scanning Probe Methods,* vol 8-10, ed B Bushan, *et al.* (Berlin: Springer) pp 247-87
[16] Friedli V, Utke I, Molhave K and Michler J 2009 *Nanotechnology* 20
[17] Fujita J, Ishida M, Ichihashi T, Ochiai Y, Kaito T and Matsui S 2003 *Nucl. Instr. & Meth. Phys. Res. B* 206 472-7
[18] Utke I, Friedli V, Fahlbusch S, Hoffmann S, Hoffmann P and Michler J 2006 *Adv. Engin. Mat.* 8 155-7

Development of a novel SEM microgripper

K J Briston[1], A G Cullis[2] and B J Inkson[1]

[1] Department of Engineering Materials, University of Sheffield, Mappin Street, Sheffield S1 3JD, UK

[2] Department of Electronic and Electrical Engineering, University of Sheffield, Mappin Street, Sheffield S1 3JD, UK

E-mail: beverley.inkson@sheffield.ac.uk

Abstract. Although there are a number of different microgrippers that have been produced for use in SEM, most of these are made using expensive and time consuming microfabrication techniques such as lithography. In this work a novel type of microgripper has been developed that avoids lithography. Instead, the end-effecters of the gripper are made from a single piece of tungsten wire which is electrochemically thinned and then cut in two with a focused ion beam. Alignment of the two end-effecters is not an issue since they both come from the same wire, so the whole gripper is straightforward and quick to produce and it is possible to shape the end-effecters as desired. Independent electrical connections to each of the end-effecters allow them to be grounded to prevent charging in the SEM beam and give the option of applying a voltage across an object that is picked up. Furthermore, unlike lithographically produced grippers, if the end-effecters become too dirty or damaged it is possible to mount a new piece of tungsten wire and form new end-effecters without having to throw away the whole system. Actuation of the gripper is accomplished using a trimorph bender actuator giving very good resolution (typically ~170nm/V) and a maximum possible span range of over 20μm. The gripper has been successfully demonstrated in picking up and placing copper microspheres in-situ in the SEM.

1. Introduction

With the ongoing interest in the properties of micro- and nano-objects, and the manipulation of these objects into working devices, in-situ scanning electron microscope (SEM) nanomanipulation tools have become increasingly important. Of the nanomanipulation tools in use, the most common is the nanomanipulator fitted with a sharp electrochemically etched tip (used, for example, in [1]). This tool is fine in most cases but can struggle when the micro- or nano-object being manipulated is heavy or has large van der Waals forces between itself and the substrate (e.g. carbon nanotubes lying on a substrate). In situations such as this, electron beam deposition of carbon or another material over the contact between the object and tip can improve the strength of the bond allowing manipulation [1], but this can also contaminate the object.

To allow micro and nano-objects to be manipulated more readily without electron beam deposition, even when large van der Waals forces to the substrate are involved, micro/nanogrippers can be used. Grippers allow larger, opposing forces to be applied to micro and nano-objects making pick-up easier. A variety of different grippers have been manufactured (though not all for use in the SEM) out of materials such as gold [2], stainless steel [3], Ni [4,5], NiTi [3], SU-8 polymer [6,7] and even carbon nanotubes [8] but the majority have been made out of predominately silicon and silicon compounds using standard silicon microfabrication techniques such as lithography (e.g. [9-12]) Even some of those that are not made of silicon make use of silicon microfabrication techniques or similar procedures (e.g. [2,6,7]). While lithography techniques have the advantage that they form part of a batch process, the equipment, especially for x-ray and electron beam lithography used for the smallest features, is expensive [5] and not available to everyone. Moreover, the process can be complex [5] and time consuming, and if the gripper end-effecters become dirty or damaged in any way, the complete gripper system must be replaced.

Opening and closing of the manufactured grippers in the literature is generally achieved using electrothermal (e.g. [2,6,12]) electrostatic (e.g. [8,11]) or piezoelectric (e.g. [4,5,9,10]) actuation, though actuation has also been achieved by other means such as voice coil motors [3]. Electrothermal and electrostatic actuators are often used due to the ease with which they can be integrated within a batch process. However, the thermal and potential gradients they create during operation can be problematic for the object being picked up. Even if these gradients are not a problem, electrothermal and electrostatic actuators have a limited movement range and blocking force compared to piezoelectric actuators [3].

In this paper a novel microgripper is presented for use with a SEM nanomanipulator which is made from tungsten wire using electrochemical and focused ion beam (FIB) techniques. Using this method avoids lithography and makes the manufacturing process simple. The end-effecters are also easy to replace if they become dirty or damaged. Actuation of the presented gripper is accomplished using a trimorph piezo bender actuator (Argillon 427.0086.14A). Using a bender actuator simplifies construction, gives a large range of movement with high resolution and avoids unwanted thermal and electrical gradients.

2. Manufacturing the microgripper

The design of the microgripper in this work is relatively flexible and can be modified depending on the specific application. One thing that can be modified is the length of the trimorph bender actuator. Making the actuator shorter decreases the distance of movement of the gripper for a given voltage applied (increases the resolution) but reduces the maximum opening range as a result. The gripper manufactured in this work is constructed with the bender actuator ceramic cut to a length of 12.5mm. The width and thickness of the bender actuator are 1.5mm and 0.8mm respectively.

The gripper is constructed so that the bender actuator supports one of the tungsten end-effecters while a 12mm×1.5mm×1mm fixed stainless steel arm mounted parallel to the bender actuator supports the other end-effecter. Connections to the bender actuator terminals are made with PTFE coated 0.125mm silver wires.

The end-effecters for the gripper are made from 0.2mm diameter tungsten wire that is cut to a length of approximately 8mm. The key to the design of the gripper is that both end-effecters are made from this same piece of wire, which is bent in the middle and, after securing both ends of the wire on the bender actuator and stainless steel support arm, cut with a FIB. Doing this means that, when the end-effecters are formed by cutting the wire, they automatically line up with each other avoiding a painstaking alignment procedure (see figure 1).

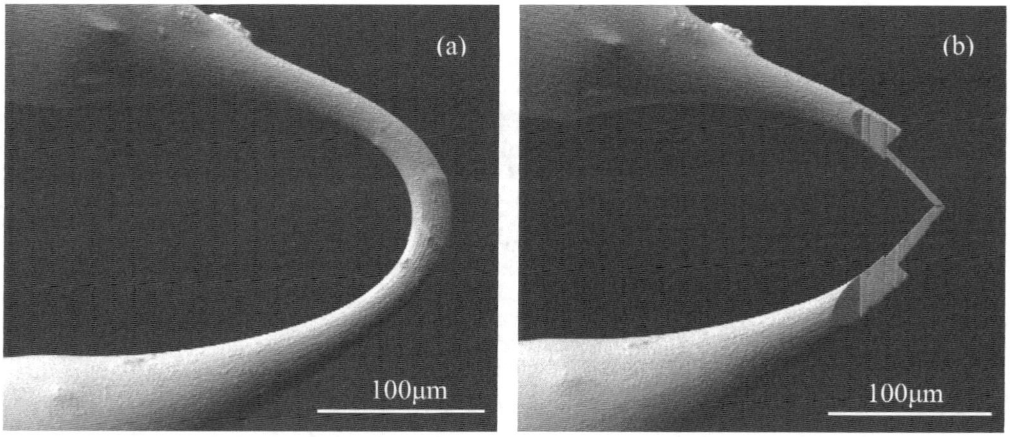

Figure 1. SEM images of the tungsten end-effecters for the microgripper (a) before the initial FIB cut and (b) after full FIB shaping.

In order to reduce milling times in the FIB, the tungsten wire is first electrochemically etched at its central point. The etching is done with 2M NaOH solution using the lamellae method [13]. With this method, a 1cm diameter gold ring is dipped in the etching solution so that an etchant film covers the hole. The tungsten wire is then passed through the hole and positioned so the central point of the wire lines up with the film.

The lamellae etch is performed until the middle section of wire has been thinned to a diameter of <40μm. The wire is then bent at this middle section to give an angle of ~30°. It is glued onto to the gripper body with one side stuck to the bender actuator and one side stuck to the stainless steel arm.

Once the wire is secured in place, electrical connections are made to each side of the wire. These connections allow the 2 arms of the gripper to be grounded, preventing them charging in the SEM beam, and can also be used to apply a voltage across an object picked up between the jaws.

After the etching is completed and the wire is secured, the end of the bent section is cut open in a FIB (Quanta 3D 200) to create the 2 arms which can then be shaped as desired (see figure 1). Shaping is usually performed with the ion beam both perpendicular and parallel to the plane of the bend to produce a microscale gripping surface.

3. Performance Evaluation

Although precise gripper performance is highly dependent on parameters chosen during construction (such as length of bender actuator and width of gap milled in the FIB) a typical gripper has been tested in-situ in the SEM (JEOL 6500F). This was found to have a rate of opening of approximately 170nm/V, becoming fully closed at 35V from an initial gap of 6μm at 0V. The gripper has also been mounted in a Kleindiek MM3A nanomanipulator and used to pick up and place Cu microspheres. Spheres between 1μm and 5μm have all been successfully picked up and placed and figure 2 shows the process for a 2.5μm microsphere.

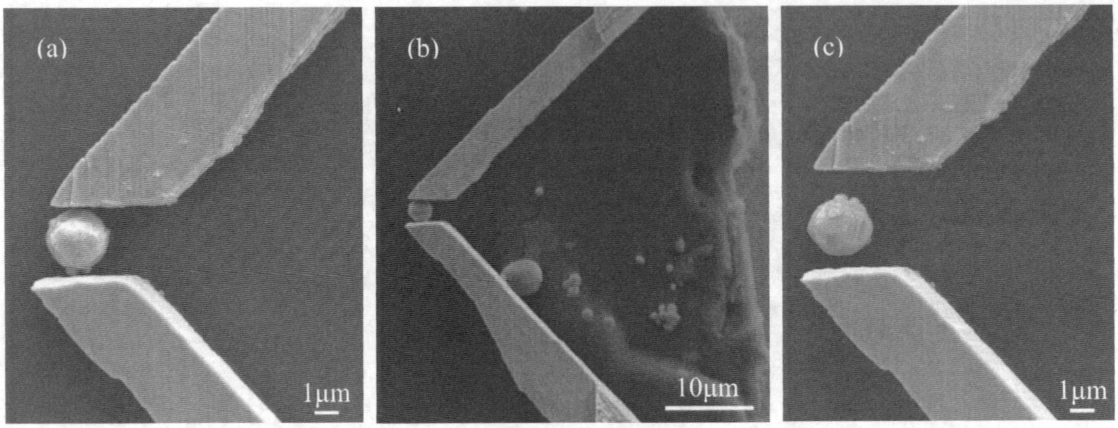

Figure 2. SEM images of the gripper being used to (a) pick up, (b) transport and (c) put down a 2.5μm Cu microsphere.

4. Conclusion

In this work a novel SEM microgripper has been constructed making use of electrochemical etching and FIB techniques to construct the end-effecters from tungsten wire. Since the end-effecters are both made from the same wire which is secured in place and then cut in two using a FIB, alignment of the end-effecters is not an issue making production of the gripper relatively simple. Actuation of the gripper is achieved with a piezo bender actuator which gives good range and resolution, and simplifies construction. The gripper has been successfully demonstrated picking up Cu microspheres.

Acknowledgements

The authors thank the EPSRC for a PhD studentship and funding GR/S85689/01.

References

[1] Peng Y, Luxmoore I, Forster M D, Cullis A G and Inkson B J 2008 *Journal of Physics: Conference Series* **126** 012031
[2] Mølhave K and Hansen O 2005 *J. Micromech. Microeng.* **15** 1265-70
[3] Kim D-H, Lee M G, Kim B and Sun Y 2005 *Smart Mater. Struct.* **14** 1265-72
[4] Carrozzay M C, Menciassi A, Tiezzi G and Dario P 1998 *J. Micromech. Microeng.* **8** 141-3
[5] Fahlbusch St, Mazerolle S, Breguet J-M, Steinecker A, Agnusd J, Pérez R and Michler J 2005 *Journal of Materials Processing Technology* **167** 371-82
[6] Nguyen N-T, Ho S-SA and Low C L-N 2004 *J. Micromech. Microeng.* **14** 969-74
[7] Choi H-S, Lee D-C, Kim S-S and Han C-S 2005 *J. Micromech. Microeng.* **15** 1327-33
[8] Nakayama Y 2002 *Ultramicroscopy* **91** 49-56
[9] Jericho S K, Jericho M H, Hubbard T and Kujath M 2004 *Review of Scientific Instruments* **75** 1280-2
[10] Blideran M M, Bertsche G, Henschel W and Kern D P 2006 *Microelectronic Engineering* **83** 1382-5
[11] Bøggild P, Hansen T M, Tanasa C and Grey F 2001 *Nanotechnology* **12** 331-5
[12] Lu M S-C, Wu Z-H, Huang C-E, Hung S-J, Chen M-H and King Y-C 2007 *J. Micromech. Microeng.* **17** 482-8
[13] Cavallini M and Biscarini F 2000 *Rev. Sci. Instrum.* **71** 4457-60

Journal of Physics: Conference Series **241** (2010) 012074 doi:10.1088/1742-6596/241/1/012074

Energy Filtered Scanning Electron Microscopy: applications to characterisation of semiconductors

C Rodenburg[1], M A E Jepson[1], B J Inkson[1], E G T Bosch[2], C J Humphreys[3]

[1]Department of Engineering Materials, The University of Sheffield, Mappin Street, Sheffield, S1 3JD, United Kingdom
[2]FEI Electron Optics, Achtseweg Noord 5, Acht, 5600 KA Eindhoven, The Netherlands
[3]Department of Materials Science and Metallurgy, University of Cambridge, Pembroke Street, Cambridge, CB2 3QZ, United Kingdom

C.Rodenburg@shef.ac.uk

Abstract. It has been recognised previously that limiting the maximum energy of secondary electrons used for image formation in a scanning electron microscope (SEM) generates contrast not present in standard SEM images. Here we give examples how energy filtered SEM improve the usefulness of SEM in the characterisation of semiconductors.

1. The concept of energy filtering in Scanning Electron Microscopy

In Transmission Electron Microscopy (TEM) energy filtered imaging is now a widely used technique as it is recognised that energy filtered images contain additional information to that of an unfiltered image. It is therefore surprising that energy filtering has not gained much attention in scanning electron microscopy (SEM) applications, where additional information can also be gained. This has been demonstrated using Auger analysis where contrast sufficient to distinguish differently doped regions in a semiconductor was generated purely by energy filtering [1]. In a different application it was demonstrated that contrast between copper and brass can be obtained by secondary electron (SE) energy filtering (EF), whereas there is no contrast in the unfiltered image [2]. Presumably the lack of availability of technical solutions for user controlled EF in main stream SEM instruments is responsible for the lack of a more widespread use of EF SEM in materials characterisation. Here we describe a low pass energy filter integrated in the lens of a field emission SEMs (FEI Sirion / Philips XL30) and present examples for useful applications in particular in the field of semiconductor characterisation.

2. Description and characterisation of low pass energy filter in FEI Sirion / Philips XL30 SEM

The work described here was carried out in an FEI Sirion / Philips XL30 field emission SEM, equipped with a through-the-lens-detection (TTL) option. The TTL system consists of an extractor tube that can be positively biased (Tube bias, *T*). This is used to direct SEs escaping from the specimen towards deflection electrodes, consisting of three layers of electrodes of opposite potential. Each pair generates an electrostatic deflection field perpendicular to the optic axis. The function of the lowest layer is to deflect SEs towards the scintillation detector (Fig.1a). The strength of the deflection field can be adjusted by changing the deflector voltage, *D*. Fig.1b is an acceptance diagram calculated for *T*=250V, *D*=60V at a working distance, *WD*=3mm (standard imaging). In the acceptance diagrams black represent electrons that are detected and white represents no detection. Fig 1b demonstrates that SEs of all energies are detected, although for high energy SEs, detection is only possible for those emitted at angles larger than 40°. Fig 1c shows that reducing the deflector voltage to *D* =6V prevents the detection of SEs with energies larger than 2.5 eV. Adjusting to *D*=20V translates to a reduction of the maximum energy of SE to be detected to 10eV as can be seen from Figure 1d. Hence the TTL does work as a low pass energy filter.

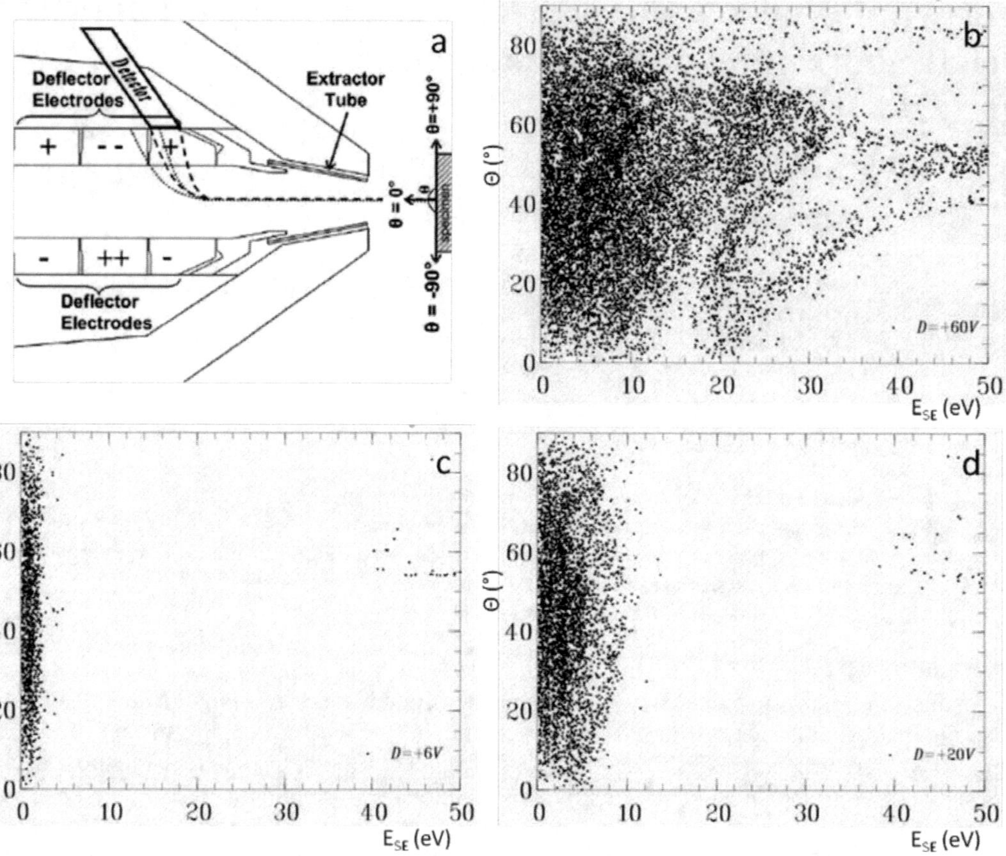

Figure 1 a) Schematic cross-section of TTL detector (after Kazemian et al [3]), b) calculated acceptance diagram for SEs generated by 1keV incident beam, at *WD*=3mm, *T*=250V *D*=60V, c) *D*=6V d) *D*=20V.

3. Applications of EF in SEM

3.1 Voltage contrast:

Voltage contrast is used in the failure analysis of semiconductor devices. It is based on the principle that areas that have no conductive connection will charge and as result have a different potential. Here we demonstrate that EF can considerably enhance the contrast due to potential differences. This is evident by comparing the images in Fig. 2a and b. Both images were obtained from a silicon surface covered by native oxide using a 1keV primary beam, with identical imaging parameters such as working distance and beam current. A bias voltage in the form of a square wave was applied to the specimen during the scan. This is reflected in brighter and darker stripes in the image. The contrast between these stripes represents the contrast caused by a potential difference of 1 V. Fig 2a is an image collected with *D*=20 V and Fig.2b is an image with *D*=6 V. The contrast between the regions biased with -0.5 V and +0.5V is improved by a factor of three with *D*=6V compared to *D*=20V.

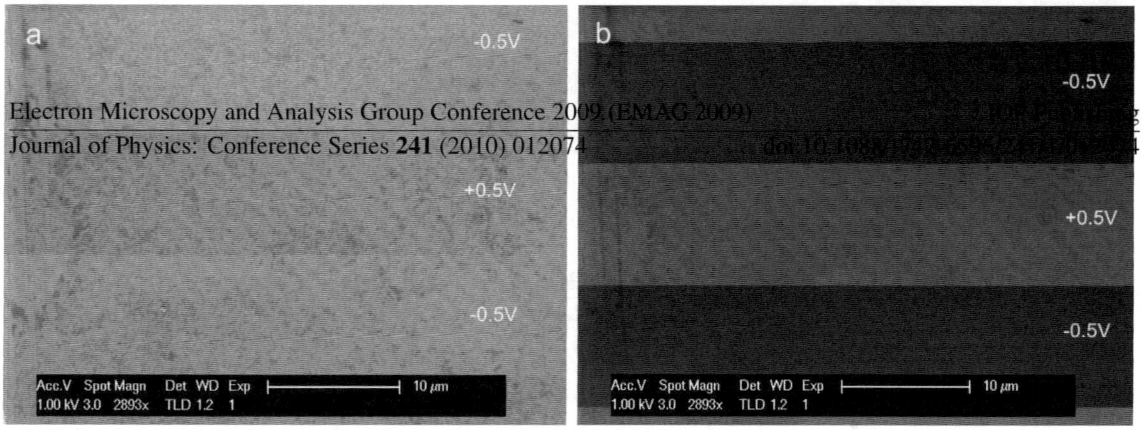

Figure 2 SEM images of a silicon specimen biased with square wave with amplitude alternating between -0.5V to +0.5 V. (a) D=20 V (b) D=6 V.

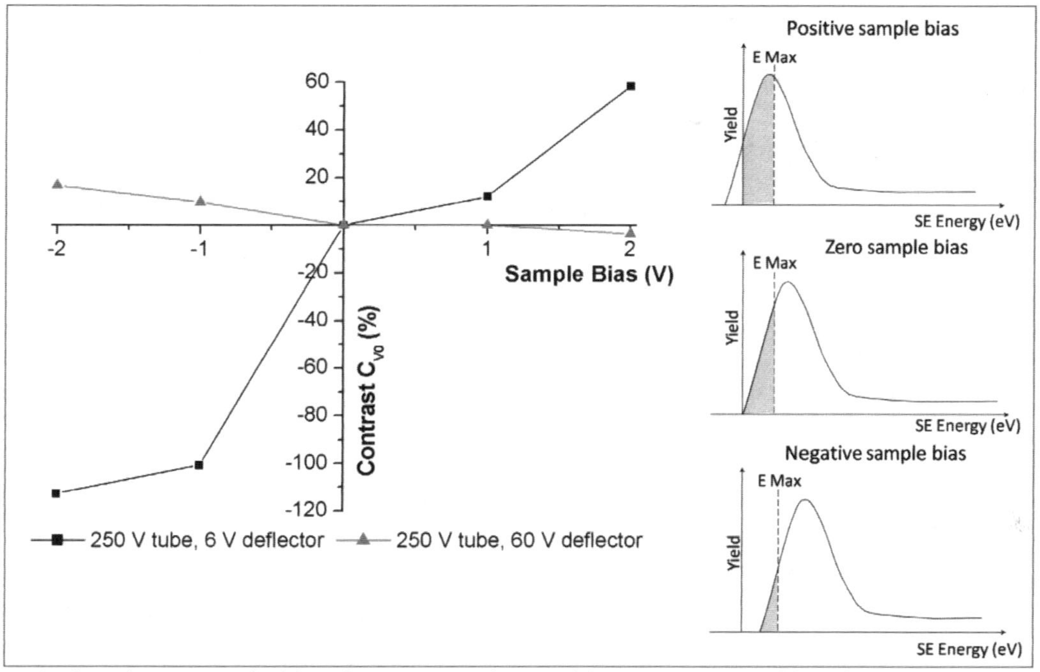

Figure 3 Plot of contrast of a biased copper specimen with respect to an unbiased specimen C_{V0}, for EF and unfiltered case and schematic representation of corresponding SE spectra.
C_{V0}=((I_V-I_0)/0.5(I_V+I_0))x100 where I_V and I_0 represent the intensity in the biased and unbiased specimen images respectively

To understand the origin of the substantial increase in sensitivity to potential differences in EF images, we first consider the effect of biasing a specimen on the SE energy spectrum. If a positive specimen bias is applied, the whole spectrum is shifted to lower energies (see insert in Fig 3). The application of a negative bias results in a shift of the entire SE spectrum to higher energies. The exact effect of this shift on the contrast as a result of specimen bias compared to the unbiased specimen, C_{V0}, depends therefore on the sign of the bias and its magnitude in relation to the shape of the spectrum, in particular the position of the peak. For most materials, the peak of the SE spectrum is located at several eV, with a 50% width of several eV (e.g. Silicon peak 4eV, width 6.5eV) [5]). An applied bias of +0.5 V will increase the number of SEs detected in the energy interval, 0-2.5eV (this interval corresponds to D=6V and is represented as E Max in Fig 3) resulting in an intensity increase in the EF image, and hence in a positive C_{V0} (Fig 2b). In analogy, a negative bias will lead to a negative C_{V0} in the EF image. This is confirmed by the experimental results shown in Fig 3.

Fig 3 also contains experimental results for the unfiltered case, where one would expect a small negative C_{V0} for positive specimen bias, as SEs with very low energies are unable to leave the

specimen. For an ideal detector (detecting electrons of all energies and emission angles) we would expect $C_{V0} = 0$ for a negative specimen bias. However, we observe a small positive C_{V0} as the negative bias affects the angular distribution of the emitted SE and the detector is non-ideal.
Fig 3 clearly demonstrates the superior sensitivity to potential contrast in an EF image, as long as the resulting shifts move the SE peak region in or out of the detected energy interval.

3.2 Dislocations in GaN
It is well known that Gallium nitride (GaN) has great potential for semiconductor devices especially when it comes to high power, low-noise and high-temperature operations. However, the electrical properties of GaN are strongly influenced by the presence of dislocations, and therefore the characterisation of potential changes associated with dislocations is of interest. Analysis is often carried out in TEM (Holography) which limits the analysis to small areas. As demonstrated above, small potential differences can be easily detected in energy filtered SEM and we show here that potential differences associated with dislocations can help to identify dislocations in the SEM. Fig 4 shows unfiltered (Fig. 4a) and filtered (Fig. 4b) images of the surface of p-type Gallium nitride. In the unfiltered image a number of dark spots (some marked by arrows) are visible. In the energy filtered image the contrast of the marked regions is reversed. The contrast reversal indicates that the marked regions have a different potential to the surrounding areas. As dislocations in GaN are known to be associated with potential differences, energy filtered SEM images could be used for easy identification of dislocations across large areas of wafer.

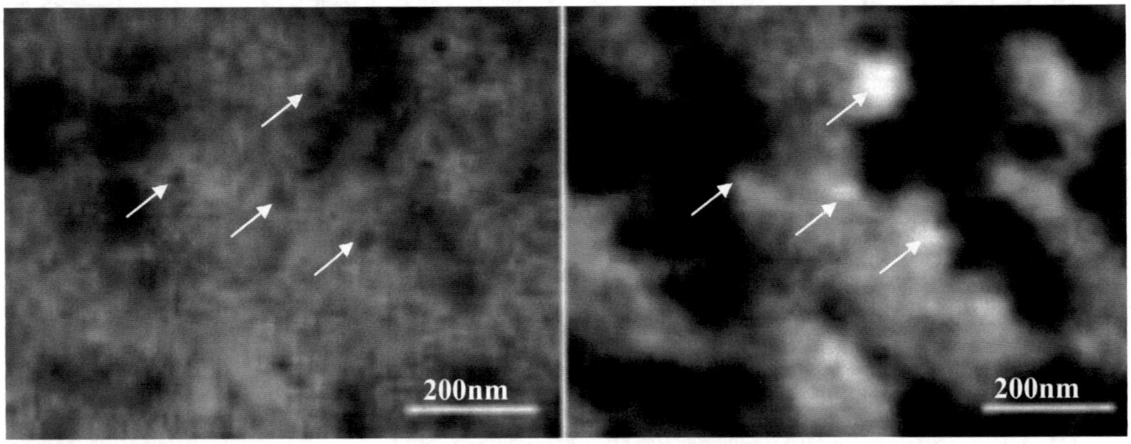

Figure 4 SEM of p-type GaN surface dislocations marked by arrows (a) *D*=60 V, (b) EF: *D*= 4 V. Note that only the contrast related to dislocations is reversed while the overall contrast is not reversed.

4. Conclusion
We have demonstrated that EF SEM can be used to detect potential differences with very high sensitivity. This can be used to detect dislocations in GaN.

Acknowledgments
C. Rodenburg thanks the Royal Society for their support in the form of a Dorothy Hodgkin Fellowship. Thanks also to the EPSRC for supporting this project (EP/E030602/1).

References
[1] Pantel R, Davitaya F A, Ged P and Bois D 1982 Scanning Electron Microscopy **1** (2), 549
[2] Khursheed A and Karuppiah N 2001 *Review of Scientific Instruments* **72** (3), 1708
[3] Kazemian P, Mentink S A M, Rodenburg C and Humphreys C J 2006 *APL* **100** 054901
[4] Cherms D, Jiao C G, Mokhtari H, Cai J and Ponce F A 2002 *phys. stat. sol. (b)* **234** (3), 924
[5] Joy D C, Prasad M S and Meyer H M 2004 *Journal of Microscopy*, **215** (1) 77

Nanoscale sculpting of ferromagnetic structures by electron beam ablation

T Gnanavel, Z Saghi, M A Mat Yajid, Y Peng, B J Inkson, M R J Gibbs and G Möbus
Department of Engineering Materials, University of Sheffield, Mappin Street, Sheffield, S1 3JD, UK

E-mail: t.gnanavel@sheffield.ac.uk

Abstract. We report an electron beam (e-beam) fabrication method to produce sub-5 nm structures, e.g. nanohole-arrays, nanojunctions and nanotips inside a TEM. The method is demonstrated using ferromagnetic nickel cross-sectional TEM specimens as well as electrochemically etched nickel tips. Different e-beam shapes and electron guns are compared, including point versus line focus and field-emission versus LaB_6 guns. As an extension of this 2-D patterning of nanostructures, a 3-D nanofabrication technique has been introduced using a high tilt tomographic holder. An electrochemically etched nickel tip of initially 18 nm in radius is sculpted down to sub-5 nm diameter. In some cases, mostly in 3-D nanofabrication, a protrusion during hole drilling was observed and confirmed by EELS as pure nickel.

1. Introduction, motivation and experimental details

Standard electron beam lithography and focused ion beam fabrication methods are widely used in semiconductor industries to produce nanostructures with a smallest dimension of ~20 nm. On the other hand, scanning probe techniques are used to realize structures on planar surfaces with atomic level precision. Bridging the gap in length-scale of 1-20 nm between these two groups of techniques is still challenging and searching for new fabrication methods is vital. TEM/STEM-based electron beam nanofabrication, known since the early 1980s, has recently been rediscovered as promising approach for nanostructuring of semiconductors, insulators and most recently metal nanostructures [1-6]. We explore advancements of this e-beam structuring technique in various aspects, concentrating on e-beam shape, influence of electron gun type, and sculpting from various viewing directions.

Two test samples have been prepared: A nickel thin film was sputtered on Si substrate and subsequently turned into an electron transparent ribbon prepared by the standard cross-sectional TEM sample preparation method [7]. The other sample is an electrochemically etched nanotip fabricated from nickel microwires using a modified SPM tip making method as detailed elsewhere [8]. The FEGTEM used is a JEM 2010F at 200kV, which provides the finest beam focusing, while the LaB_6 technology is represented by a JEM 3010 at 300kV, which provides the highest beam currents.

Journal of Physics: Conference Series **241** (2010) 012075 doi:10.1088/1742-6596/241/1/012075

2. Results and discussions

2.1. FEGTEM nanofabrication with various beam shapes

Tuning the beam shape via condenser astigmatism and defocus into, e.g. point, line, or slightly underfocused probes with a sharp central maximum (Figure 1) was found to be an effective method to produce differently shaped nanostructures in the FEGTEM. Figure 1a shows two parallel arrays of holes sculpted on a TEM cross-sectional nickel specimen using point focused probe in order to explore the effect of increasing probe diameters. Figure 1b shows experiments to explore astigmatism: a slightly underfocused probe with a sharp central maximum with different diameters is seen to produce triangular shaped holes due to residual three-fold condenser astigmatism. Next to it, a parallel line focused probe is applied, which generates a line perforating the nickel film in a fine sequence of holes, presumably revealing thickness fluctuations. Finally figure 1c shows an array of 3x4 holes of ~4 nm in diameter and separated by ~12 nm, sculpted using a point focused probe.

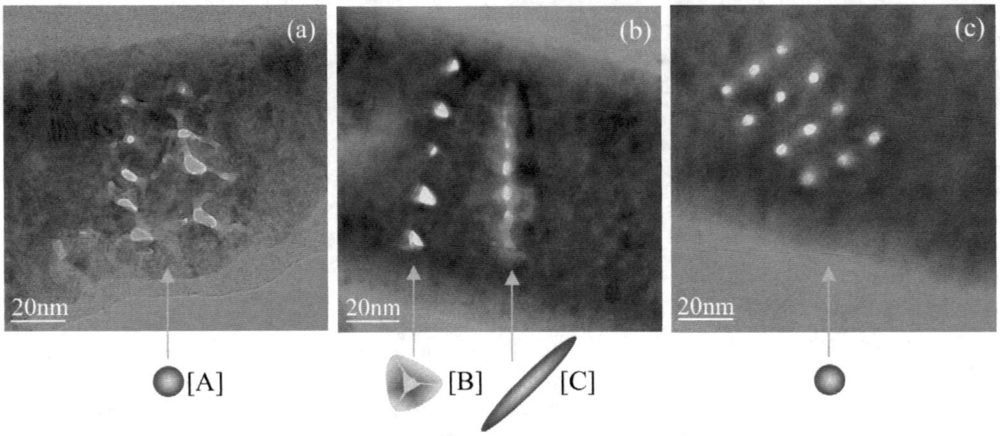

Figure 1: (a-c) Nanopattern sculpting on a nickel cross-sectional TEM specimen using various probe shapes (JEOL JEM 2010F at 200kV). [A] Point focused, [B] Slightly underfocused, [C] Line focused.

2.2. LaB_6 TEM nanofabrication of nanoconstrictions

LaB_6 gun e-beam sculpting was implemented mainly for nanojunction fabrication. Figure 2a shows a hole drilled with 10 nm radius indicating the focusing capability of a thermionic gun using the same type of cross-sectional nickel TEM specimen as for figure 1. Iterating the hole drilling by moving the beam from the centre of the Ni film towards both the laterally connected Si substrate and the specimen hole sides leads to the sculpting of a 2 nm wide nanojunction as shown in figure 2b. The junction thereby not only gets laterally constricted but also thinned, down to possible fracture, as the rather broad Gaussian beam profile of the probe is leading to delocalised ablation. The e-beam irradiation by the highly intense LaB_6 focused probe is also found to change the microstructure of the thin film (Figure 2c). Nano-sized grains of around ~7 nm in diameter were observed along the edge of the hole, indicating beam-induced irradiative quasi-melting, likely far below the nickel melting temperature.

The fabrication of free-standing nanostructures in this way is a method in early stages of development. To evaluate the potential, we explore the comparison to the more established method of patterning of metal thin films in plan-view geometry, deposited on TEM support films, using commercial Raith 150 electron beam lithography system. Some preliminary results are shown in figure 3 for comparison. Particularly attractive seems a combination of both techniques using holey support films which would combine the benefits of ultrafine nanostructuring with the ease of contacting the nanoconstrictions to external circuits.

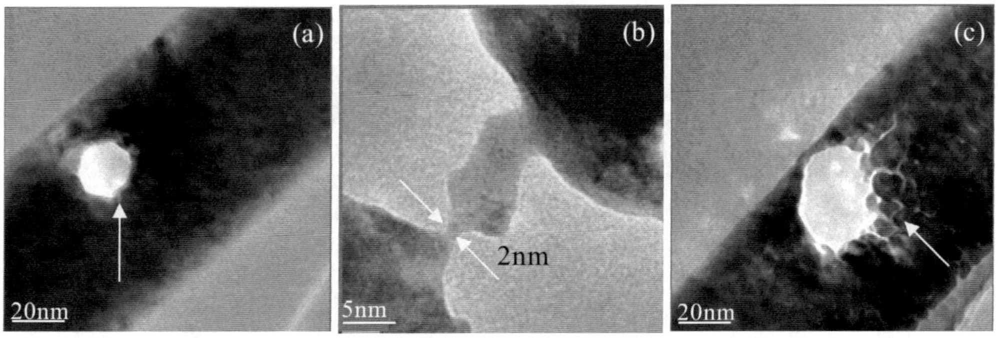

Figure 2: Nanohole, nanojunction sculpting and structural change studies (JEOL JEM 3010 at 300 kV).

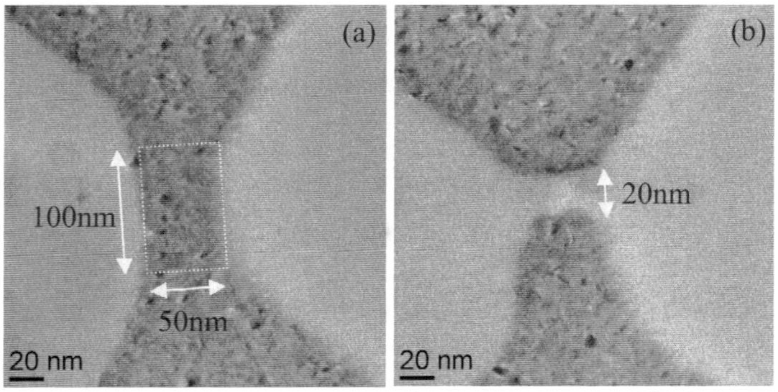

Figure 3: Nickel nanostructures with (a) a 100x50x10 nm^3 constriction and (b) with a gap of ~20 nm, fabricated by conventional e-beam lithography on a Si_3N_4 TEM support grid.

2.3. *Tomographic nanofabrication*

Some nanostructures are intrinsically three-dimensional in design and cannot be fabricated with the planar methods of 2.1/2.2. However, usage of a high tilt tomographic holder [9] provides a solution as demonstrated for sharpening of a nickel nanotip, which after etching would still be too large in tip radius for many applications, such as nanobridge or breakjunction formation experiments.

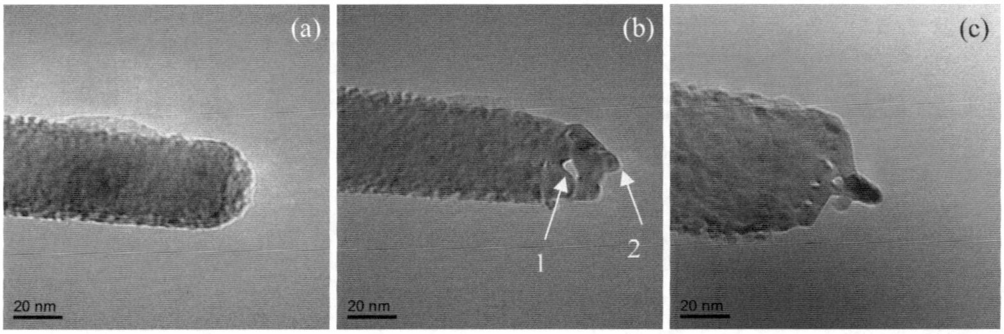

Figure 4: (a) Initial nanotip of 18 nm radius. (b,c) Nanotip sharpened in two mutually perpendicular directions.

The tip (Figure 4a) with initially 18 nm radius was mounted along the rotation axis of the goniometer, perpendicular to the e-beam. First a hole was drilled (arrow 1, Figure 4b) using a point focused probe

for testing purpose. A cut with a line focused probe at the very end of the tip yielded an apparent ~3 nm radius tip (arrow 2, Figure 4b), however in this viewing direction only. The tip was then rotated by 90° for further cuttings from both sides (Figure 4c) to achieve a final radius of the tip in all viewing direction of ~3 nm. The mounting of the tip on a tomography holder for the 3-D nanofabrication allows the combination with tomographic reconstruction via a second tilt series.
In some cases, during the hole drilling on nickel nanotips, protrusions were observed, confined one-sided to the e-beam exit point only. To decide between the two possibilities mentioned in the literature, (i) materials redeposition versus (ii) carbon contamination, a test tip was perforated by several point focused e-beams (Yellow circles, Figure 5a). After rotation by 45° the protrusions at the exits of the holes (Arrow, Figure 5b) became visible and accessible for EELS. They were confirmed as pure Ni, via Ni M-edge (and L-edge, not shown) and the absence of any C-K-edge (Figure 5c).

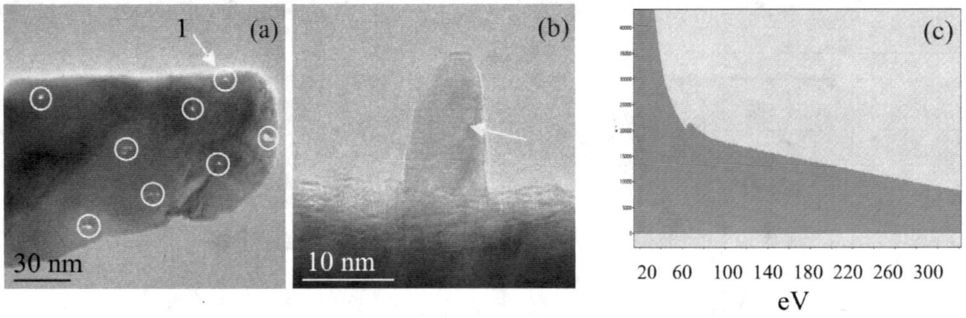

Figure 5: (a) Nanotip with several drilled holes. (b) After rotation of 45° showing the extrusion of hole marked by arrow 1 in figure 5a. (c) Ni-M edge around 70 eV compared with region around 284 eV confirming absence of carbon.

3. Conclusions

2-D and 3-D e-beam nanofabrication were successfully tested and found to be very promising techniques to realize the 1-20 nm structures that are not easily attainable by other fabrication methods known so far. Both type of guns have been proven to provide advantages, the FEGTEM by producing finer beams of different beam shapes and the LaB_6 guns with higher total beam current.

Acknowledgements
This work was supported by an RCUK Basic Technology Grant (GR/S85689/01).

References

[1] Storm A J, Chen J H, Ling X S, Zandbergen H W and Dekker C 2003 *Nature Materials* **2** 537-540
[2] Bysakh S, Shimojo M, Mitsuishi K, Furuya K 2004 *J. Vac. Sci. Technol. B* **22** 2620-27
[3] Zandbergen H W, van Duuren R J H A, Alkemade P F A, Lientschnig G, Vasquez O, Dekker C, and Tichelaar F D 2005 *Nano Letters* **5** 549-553
[4] Fischbein M D and Drndic M 2007 *Nano Letters* **7** 1329-37
[5] T Gnanavel, Z Saghi, Y Peng and G Möbus 2008 *Mcroscopy & Microanalysis* Proc M&M 2008 14(S2) 460
[6] Z Saghi, T Gnanavel, Y Peng, B J Inkson, M R Gibbs and G Möbus 2008 *Appl. Phys. Lett.* **93** 153102
[7] Klarr H-Y and Hsu F-Y 1996 *Material Characterization* **36** 365-369
[8] Albonetti C, Bergenti I, Cavallini M, Dediu V, Massi M, Moulin J F and Biscarini F 2005 *J. Vac. Sci. Technol. B* **23** No. 6 2564–66
[9] Xu X, Saghi Z, Yang G, Peng Y, Inkson B J, Gay R and Möbus G 2007 *MRS Symposia Proceedings* No. 928E *Materials Research Society* Pittsburgh p. 0982-KK02-04

Dopant Contrast in the Helium Ion Microscope: Contrast Mechanism

C Rodenburg[1], M A E Jepson[1], B J Inkson[1] and X Liu[2]

[1]Department of Engineering Materials, The University of Sheffield, Mappin Street, Sheffield, S1 3JD, United Kingdom
[2]Carl Zeiss NTS GmbH, Carl-Zeiss-Strasse 56 , 73447 Oberkochen, Germany.

C.Rodenburg@shef.ac.uk

Abstract. With the continued miniaturisation of semiconductor devices, there is an increasing need for nanoscale characterisation. Dopant mapping in a Low Voltage Scanning Electron Microscope (LV-SEM) was identified as a potential technique to fulfil this need, provided that a small enough probe size (~0.1 nm) could be achieved. It has been shown that He ion beams in Helium Ion Microscopes (HeIM) can be focussed to probe sizes as small as 0.24 nm. As the image in both LV-SEM and HeIM is formed by secondary electrons it is not surprising that HeIM exhibits dopant contrast as recently demonstrated. In this paper we describe similarities and differences between HeIM dopant contrast and SEM dopant contrast and explore implications for the contrast mechanism.

1. Introduction

The shrinking dimensions of electronic devices pose a challenge to the measurement of dopant concentrations with sufficiently high accuracy and spatial resolution for the analysis of two- and three-dimensional structures. Images formed by secondary electrons (SEs) generated in LV-SEM were proposed as a potential solution to this challenge, provided that a sufficiently small probe size could be achieved [1]. The smallest usable charged particle probe to generate secondary electron (SE) images of bulk specimens to date was not obtained in a Scanning Electron Microscope (SEM) but in a Helium Ion Microscope (HeIM) operated at 20 kV [2]. A result of several SEM dopant contrast studies was that increased primary beam energies lead to a reduction in SEM dopant contrast [3, 4, 5]. Therefore it was questionable if any dopant contrast could be observed in the HeIM. Initial results obtained in the first generation HeIM showed that doped areas could be identified but a direct correlation between dopant concentration and SE intensity, similar to that observed in SEM, was not found [6]. A further study employing a second generation HeIM with increased detection efficiency revealed that contrast similar to that in the SEM can be obtained [7]. This promising result opens up the question of why dopant contrast can be observed for high primary beam energies in the HeIM. Therefore, in this paper we investigate differences in dopant contrast behaviour in the HeIM and SEM in order to gain a better understanding of the contrast mechanism involved.

2. Investigation of SEM and HeIM dopant contrast characteristics

2.1. The influence of primary beam energy

Figure 1 a represents the contrast obtained on a cross section of a p-n junction, obtained by cleaving in air (for estimated oxide thickness see [8]), when imaged in the SEM using an Everhart-Thornley detector and an accelerating voltage, U, where $U = 2$ kV. The bright layer reflects the extent of a p-doped area (boron doped to 5×10^{18} atoms cm^{-3}) on an n-type substrate (arsenic doped to 1×10^{19} atoms cm^{-3}). Figure 1 b was also obtained in the SEM but with $U = 25$ kV. The contrast (defined as the intensity difference between the p and n doped regions divided by the average intensity of these regions) is reduced compared to that in figure 1 a. A further reduction in contrast is visible in the SEM image collected with U=30 kV figure 1 c.

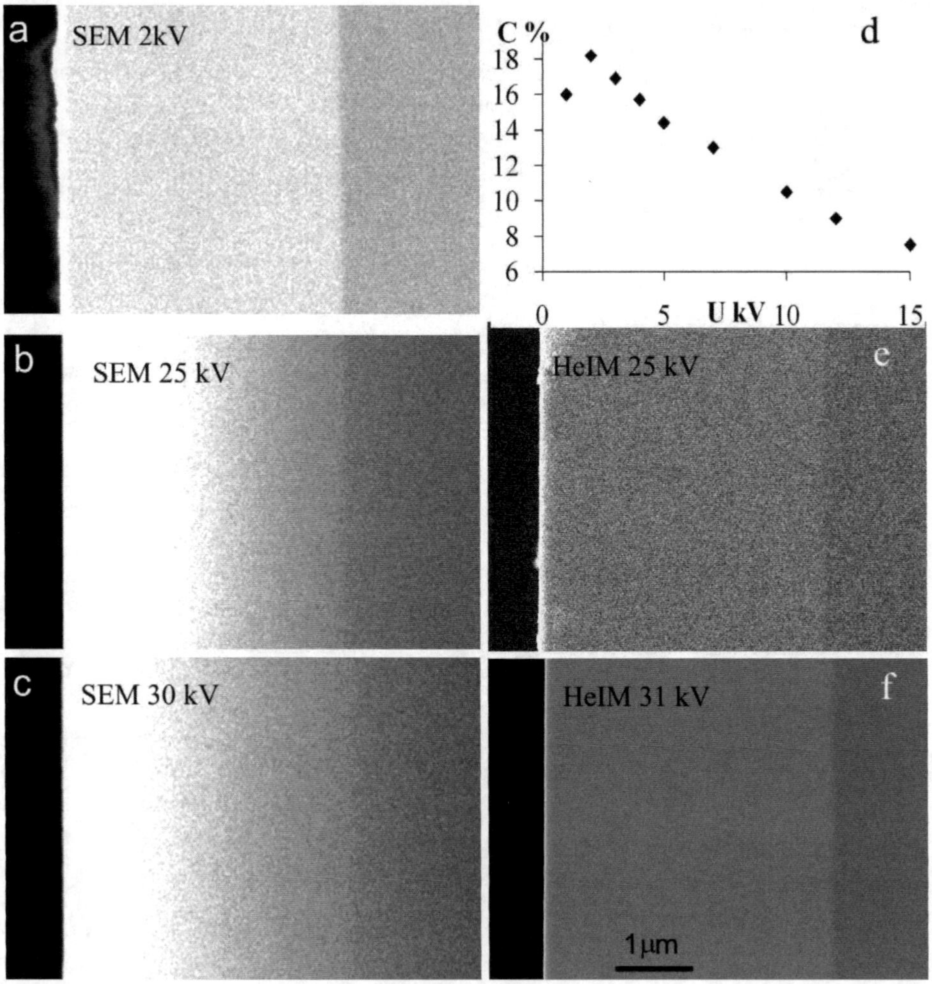

Figure 1 a-c) SEM images of p-n junction with indicated accelerating voltages, d) p-n contrast as a function of U in SEM, e-f) HeIM images with indicated accelerating voltages U. Changes in noise levels are due to changes in incident beam current, I_i.

Figure 1 d summarises the dopant contrast as a function of U in the SEM when using an Everhart-Thornley detector. Maximum contrast is obtained at $U = 2$ kV (for silicon) as was previously found in [4]. The decline in contrast with larger U could be explained by increased incident beam currents, I_i, obtained for larger U (see [9] for explanation). For example in the SEM used we measured

$I_i(U{=}1\text{kV}) = 0.07\text{nA}$ compared to $I_i(U{=}10\text{kV}) = 0.34\text{nA}$). However, the maximum in the SEM contrast which was also reported in [5] cannot be explained purely as a result of changes in I_i, instead it is explained in [4] as a result of dynamic charging. In the SEM the escaping SEs lead to a positive charge near the surface, whereas trapped primary electrons can cause a negative charge deeper within the specimen. Through this process, an internal field is established that can result in acceleration of electrons towards the surface of the specimen [4]. The magnitude of this field in intrinsic silicon depends on the accelerating voltage [4] which affects the SE yield for a material of a given dopant type and concentration and is responsible for the observed dependence of dopant contrast on accelerating voltage. According to [4] the field strength is also influenced by the dopant atom type, with the positive surface charge present in p-type silicon increasing the field, and the negative surface charges in the n-type silicon reducing the field. Increased field strength accelerates electrons towards the surface of the specimen and results in higher yields for p-type silicon when compared to that of intrinsic silicon, likewise the reduced field strength in n-type silicon results in a smaller yield compared to intrinsic silicon at constant accelerating voltage. This is suggested as the cause for SEM dopant contrast [4], with a contrast magnitude that depends on accelerating voltage for a given difference in dopant concentrations.

There is a fundamental difference in charging in the SEM and HeIM, as in the latter, the charge implanted due to the primary beam is positive, as is the charge deposited close to the surface due to the escape of SEs. However, as the SE yield for He ion impact is in the order of several SEs per incident He ion [10], more positive charge is deposited close to the surface through the escape of SEs than is deposited by the primary He ions. As a result, there will be an internal field present in HeIM similar to that suggested to form in the SEM and one may expect dopant contrast of a similar magnitude. In addition, as the SE yield for He ion impact in the range between U = 10-40 kV is reported to increase with increased U [11], we would expect HeIM contrast to increase with increasing U (an opposite trend to that in the SEM (shown in figure 1 d)).

Figure 1 e and f are images taken in an Orion Plus HeIM, where SEs are generated by a primary beam consisting of He ions accelerated to U = 31 kV (figure 1 f) and 25 kV (figure 1 e). As can be seen in figures 1 e and f, the contrast in the HeIM is of similar magnitude to that of 25 kV SEM (figure 1 b). In addition, the contrast in the HeIM at the higher accelerating voltage of U = 31 kV is increased compared to that obtained at U = 25 kV. This result supports the theory of dynamic charging as a contributor to dopant contrast in images generated by SEs.

2.2. Influence of charging

The importance of charging in relation to dopant contrast has been discussed above. In SEM, the influence of charging cannot be investigated independently from the influence of the primary beam energy (charge neutrality can only be achieved by adjusting the primary beam energy).

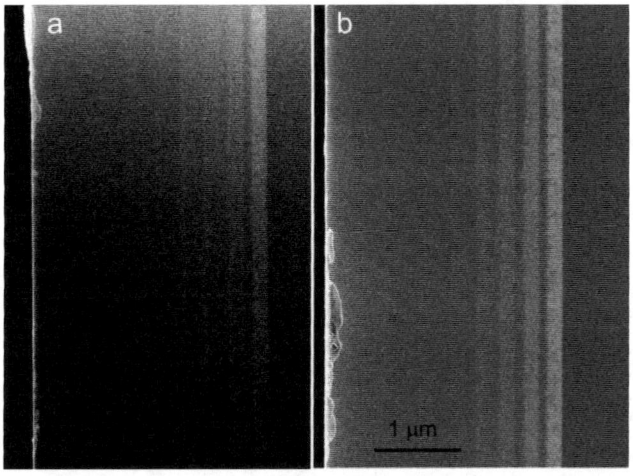

Figure 2 HeIM images obtained of an insulated staircase test structure with an accelerating voltage of 31kV: a) no flood gun, b) with electron flood gun.

However, HeIM allows the investigation of the effect of charging by neutralising the build-up of positive charge due to the impacting He ion beam through the use of an electron flood gun. The SE image obtained in the HeIM without the use of a flood gun is presented in figure 2 a. It is a cross-section of silicon containing seven boron doped layers (with dopant concentrations of 4×10^{15}, 6.5×10^{16}, 8×10^{17}, 4×10^{18}, 9×10^{18}, 2×10^{19},5×10^{19} cm^{-3} from left to right) on an n-type substrate (arsenic doped 1×10^{19} atoms cm^{-3}). The specimen was completely insulated from the specimen holder through the use of 1mm thick glass slides. At the top of the image in figure 2 a, the four highest doped layers can be distinguished. The contrast for the highest doped layer relative to the n-type substrate is ~30 %. As scanning progresses (towards bottom part of the image) the intensity is considerably reduced as a result of the build up of positive charge at the surface. The accumulation of positive charge prevents the escape of low energy SEs leading to the observed darkening effect. Nevertheless, the dopant contrast for the p-type layer with the highest dopant concentration is still visible at the bottom of the image and as a result of the charge build up, the contrast for this layer relative to the n-type substrate has increased to ~50%. This increase in dopant contrast as a result of charge accumulation during progressive scanning is consistent with the increased contrast as a result of increased acceleration voltage in the HeIM, as the latter is also related to an increase in surface charge (due to a yield increase). Both of these results demonstrate that charging does contribute to dopant contrast and should not be ignored. To collect dopant contrast images with minimum influence of charging, an electron flood gun was used to collect the image shown in figure 2 b. The flood gun was pulsed at the end of each line during image acquisition and then turned off before commencing the next line. The electron flood gun energy was 400 eV with a pulse time of 1000 μs. The fact that there is no notable reduction in intensity with progressing scanning in figure 2 b demonstrates successful charge compensation to some degree. Analysis of the dopant contrast of the highest doped layer reveals that some charging still remains as the contrast increases from 23% at top of the image to 27% at the bottom of the image.

3. Conclusion

We have investigated the influence of primary beam energy on dopant contrast obtained in HeIM and SEM, as well as the influence of charging in the HeIM. Our results indicate that charging can alter dopant contrast significantly. Notably, charging in the HeIM can be controlled in the HeIM by the use of an electron flood gun.

Acknowledgments

C. Rodenburg would like to thank the Royal Society for their support in the form of a Dorothy Hodgkin Fellowship. The authors would also like to thank the EPSRC for supporting this project (EP/E030602/1) and Ian Belding of Carl Zeiss for facilitating the use of the HeIM.

References

[1] Castell M R, Muller D A and Voyles P M 2003 *Nature Mat.* **2**, 129
[2] Hill R, Notte J and Ward B 2008 *Phys. Procedia.* **1**, 135
[3] Sealy C P, Castell M R and Wilshaw P R 2000 *J. Electron. Microsc.* **49**, 311
[4] Chakk Y and Horvitz D 2006 *J. Mater. Sci.* **41** 4554
[5] Elliott S L 2001 *PhD Thesis*, University of Cambridge p.42
[6] Jepson M A E, Inkson B J, Rodenburg C and Bell D C, 2009 *EPL* **85** 46001
[7] Jepson M A E, Inkson B J, Liu X, Scipioni L and Rodenburg C 2009 *EPL* **86** 26005
[8] Dapor M, Jepson M A E, Inkson B J and Rodenburg C 2009 *Microsc. Microanal.* **15** 237
[9] Kazemian K, Rodenburg C, Humphreys C J 2004 *Microelectronic Engineering* **73–74** 948
[10] Ramachandra R, Griffin B and Joy D 2009 *Ultramicroscopy* **109** 748
[11] Yamanaka T, Inai K and Ohva K 2009 *Proc. SPIE* **7272** 72722L

The effect of osmium staining on lamellar spacing in thin polystyrene-polyisoprene diblock copolymer films

Lech Staniewicz*[1,3], Athene M Donald[1], Debbie J Stokes[2]

[1] Biological and Soft Systems, Cavendish Laboratory, University of Cambridge, JJ Thomson Avenue, Cambridge, CB3 0HE, UK

[2] FEI Company, PO Box 80066, 5600KA, Eindhoven, The Netherlands

[3] To whom any correspondence should be addressed

e-mail: ltls2@cam.ac.uk

Abstract. Thin films of lamellar-structured polystyrene-block-polyisoprene were vapour stained with osmium tetroxide and compared using transmission imaging in an environmental scanning electron microscope to unstained films of the same polymer. Staining was found to swell the affected phase by a factor of 1.15 in-plane, compressing the unstained phase to a factor of 0.71 of its original size as it did so. Additionally, images exhibiting contrast between phases in the unstained sample were successfully taken with conventional transmission electron microscopy, showing that staining is not necessary for imaging such samples.

1. Introduction

Electron microscopy is a routine technique for directly examining the small-scale structure of polymers, but is hampered by the fact that many polymers are chemically similar and electrically insulating. To get around such difficulties, polymer samples are typically prepared by coating with conductive material and by staining one phase of a multi-component material to increase the material contrast - a review of such procedures can be found in Sawyer and Grubb [1].

The development of the environmental scanning electron microscope (ESEM) [2] enabled the imaging of insulating or wet samples in their native state, without any artefacts from sample preparation and without fear of sample charging or other such image distortions. Samples imaged "naked"' in the ESEM can look significantly different to those which have undergone preparation for traditional SEM [3] and so it is reasonable to assume that staining, being another preparation method, may cause its own share of artefacts.

This work will examine a polystyrene-polyisoprene diblock copolymer film (block copolymers such as the triblock ABS are common structural materials and other chemistries hold great promise as photonic crystals, for instance) with a lamellar structure and the microscale effects caused by using the common stain osmium tetroxide. No other preparation (eg. sample coating) has been carried out. Imaging of both the untreated and the stained samples was carried out using a new transmission detector for the ESEM [4], using the low acceleration voltage and hence stronger beam-sample interaction of (E)SEM to provide strong contrast.

2. Experimental Methods

The block copolymer used in this work was supplied by Dr. Patrick Fairclough and his group from the Department of Chemistry, University of Sheffield. It consisted of equal volume fractions polystyrene (PS, C_8H_8) and polyisoprene (PI, C_5H_8), which should result in a lamellar microstructure when annealed [5]. It has M_w= 578,000 g mol^{-1}, M_n = 426,000 g mol^{-1} and a polydispersity of 1.38.

90nm thick films of block copolymer were prepared by spin-coating a solution in toluene on to glass slides, which were then vapour annealed [6] for 16.5h to bring out the lamellar structure. Annealed films were then attached to copper TEM grids, some of which were stained by exposure to osmium tetroxide vapour for one hour.

The film-coated grids were then examined in an XL30 field-emission environmental scanning electron microscope (ESEM), produced by FEI Company, Hillsboro, Oregon, US, using a new accessory enabling transmission imaging to be conducted in the ESEM on samples which can be temperature-controlled by a Peltier element. 1 Torr (133 Pa) of water vapour was required to eliminate sample charging, but temperature control was not necessary. Such apparatus has been used by Bogner et al previously [4] on hydrated samples. This procedure will be referred to as ESEM-STEM.

Additionally, films were examined in a Tecnai 20 TEM (FEI Company, Hillsboro, Oregon, US) to compare the image quality of stained and unstained samples using conventional TEM. The accelerating voltage used was 120kV. Charging of the sample had only a minimal effect when imaging with the TEM.

3. Results and Analysis

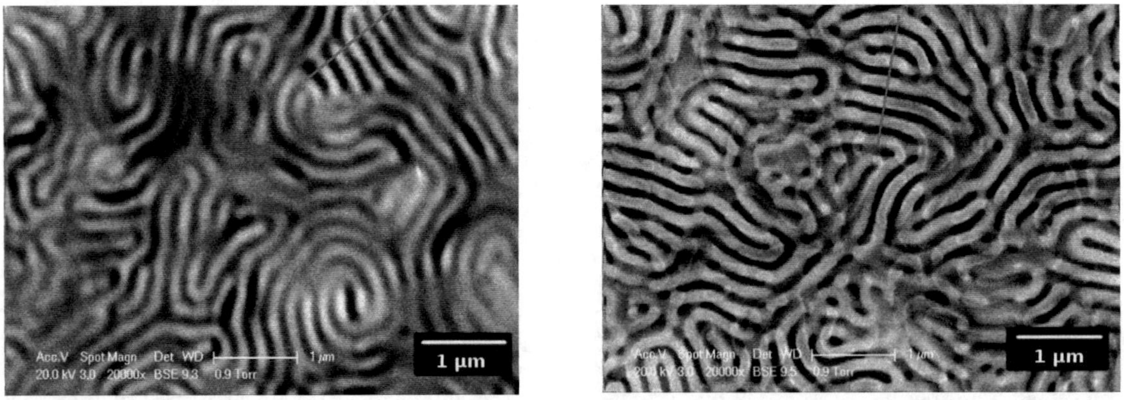

Figure 1. Dark-field ESEM-STEM images of both unstained (left) and stained (right) annealed PS-block-PI. The lines drawn at the top of each image are examples of where traces are taken for analysis. A clear difference in morphology can be seen between the two images.

Multiple images, each of a different part of the sample, were taken with ESEM-STEM in dark-field mode of stained and unstained samples, examples of which are shown in figure 1. Strong contrast can be seen from the unstained sample and as expected, the thicknesses of the two types of lamellae are essentially equal as judged by eye. To make the analysis more quantitative, line traces perpendicular to stacks of lamellae for each image were taken with the ImageJ software [7]

A custom-written C++ program using the FFTW library [8] was used to perform a discrete cosine transform on each line trace and select out the positions of either the largest peak in the power spectrum for the unstained sample (for which there was only one periodicity) or the two largest peaks in the case of the stained sample, rejecting any wavelengths over 300nm to reduce errors due to uneven film thickness causing intensity fluctuations. The most significant components in the power spectrum should correspond to the overall envelope of the lamellar shape and so a sample with two separate lamellar sizes should exhibit two large peaks - one each for the characteristic dimension of the larger and smaller spacings.

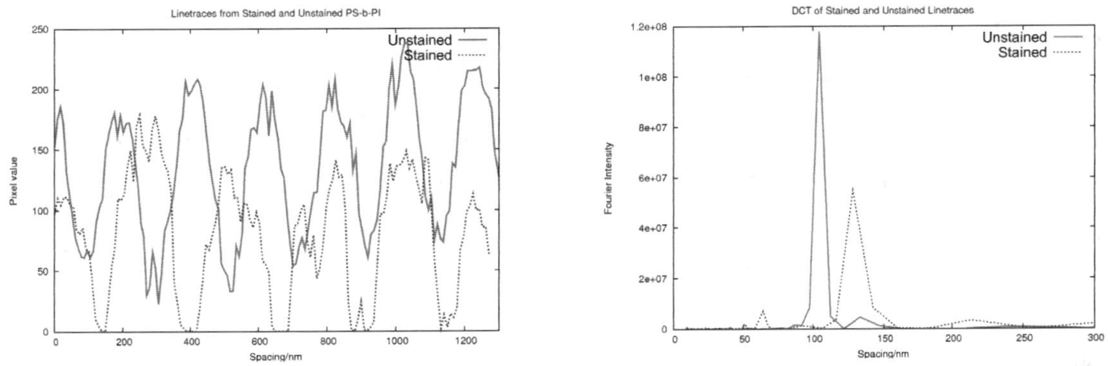

Figure 2. Examples of line traces from annealed PS-block-PI images (left), along with their discrete cosine transforms (DCT, right). Data from unstained images is shown by a red continuous line, while data from stained samples is shown by a blue dashed line. A third harmonic

Examples of such line traces, along with the power spectrum of their discrete cosine transforms, are shown in figure 2. The unstained sample has only one main peak in its power spectrum, while the stained sample has multiple peaks. The overall results are in table 1, along with an attempt to extract the same data directly from line traces.

Table 1. Lamellar spacings obtained from the discrete cosine transform of line trace data for unstained and stained samples (single distance for unstained, lower and upper distances for stained) as well as results obtained by attempting to obtain the peak and trough widths from direct analysis of the line traces (peak and trough widths). Due to the uneven shape of line traces, attempting to extract information directly from them will not be particularly accurate – this data is included for completeness only and is more to indicate which phase has swollen.

Sample	Single Distance	Lower Distance	Upper Distance	Trough Width	Peak Width
Unstained	(113±6) nm			(110±23) nm	(114±27) nm
Stained		(81±26) nm	(130±19) nm	(98±21) nm	(131±26) nm

Staining the PS-block-PI film with OsO_4 causes the stained phase to swell by a factor of approximately 1.15, compressing the other phase to a scale of approximately 0.71 of its previous size as it does so. A visual inspection of the images (figure 1) suggests a similar conclusion – the stained phase (which would be the brighter component when viewed with dark-field STEM) clearly appears larger than the other, whereas the two were comparable before staining.

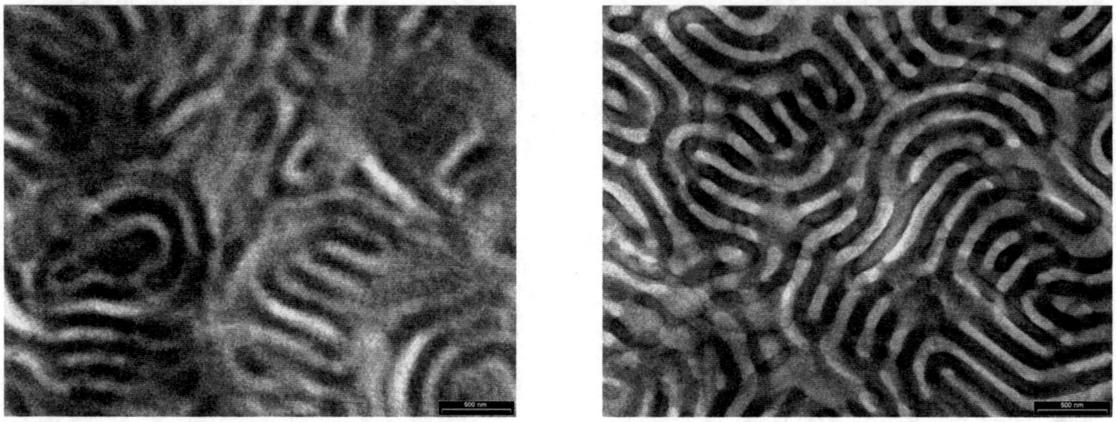

Figure 3. Bright-field TEM images of the unstained sample (left) and the stained sample (right) taken in the Tecnai 20 TEM at 120kV. Scale bar = 500nm.

Having demonstrated that staining runs the risk of altering samples such as these, we return to examine its necessity in standard TEM. Such work has been previously done by Handlin and Thomas [9], who defocussed the electron beam to enhance phase contrast within the sample. As figure 3 shows, phase contrast TEM is a viable technique for indicating the morphology of unstained polystyrene-block-polyisoprene samples. This should open up new studies via (S)TEM - rather than the more recent trend towards AFM - without the dangers of swelling artefacts highlighted here.

4. Conclusion

We have investigated the effect that vapour staining of osmium tetroxide has on the characteristic dimensions of lamellae in thin polystyrene-block-polyisoprene films. Incorporation of a heavy metal stain into the polyisoprene component of the polymer causes it to swell. Hence staining is not recommended if precise measurements are required of polymer samples on the microscale or below. Also, we demonstrate that staining is not required for imaging samples even as closely related as polystyrene ($[C_8H_8]_n$) and polyisoprene($[C_5H_8]_n$) when imaged at low energies (STEM at 20kV in the ESEM) or with a slight loss of contrast at high energies (TEM at 120kV).

Acknowledgements
This work was funded by the EPSRC (grant number EP/P50385X/1) and by a CASE studentship from FEI Company. The aid of Patrick Fairclough, Andrew Parnell, Nadejda Tzokova and Andrew Pryke from the University of Sheffield is greatly appreciated in providing the block copolymers used in this study, as well as that of FEI Company for providing the ESEM-STEM detector.

References

[1] Sawyer L C and Grubb D T 1987 Polymer Microscopy (London: Chapman and Hall)
[2] Danilatos G D 1993 *Microsc. Res. and Technique* **25** 354–61
[3] Uwins P J R, Murray M, and Gould R J 1993 *Microsc. Res. and Technique* **25** 412–18
[4] Bogner A, Jouneau P H, Thollet G, Basset D, and Gauthier C 2007 *Micron* **38** 390–401
[5] Matsen M W and Bates F S 1996 *Macromolecules* **29** 1091–8
[6] Fukunaga K, Elbs H, Magerle R, and Krausch G 2000 *Macromolecules* **33** 947–53
[7] Rasband W S 1997-2008 ImageJ US National Institutes of Health, Bethesda, Maryland, USA, http://rsb.info.nih.gov/ij/
[8] Frigo M and Johnson S G 2005 *Proc. IEEE* **93** 216–31
[9] Handlin D L and Thomes E L 1983 *Macromolecules* **16** 1514—1525

The Effect of Oxidation and Carbon Contamination on SEM Dopant Contrast

M A E Jepson, K Khan, T J Hayward, B J Inkson and C Rodenburg

Department of Engineering Materials, The University of Sheffield, Mappin Street, Sheffield, S1 3JD, United Kingdom

m.jepson@sheffield.ac.uk

Abstract. Dopant contrast in the scanning electron microscope (SEM) has the potential to provide rapid and quantitative information about the distribution of dopant atoms in semiconductors providing that a robust quantification procedure can be developed. When silicon is exposed to air, it rapidly forms a native oxide on its surface which inevitably affects dopant contrast. In this paper we compare the dopant contrast observed on an in-situ cleaved specimen and an air-cleaved specimen and further examine the influence of prolonged electron exposure on dopant contrast.

1. Introduction

Dopant contrast in the scanning electron microscope (SEM) has the potential to provide a rapid, high resolution and quantitative dopant profiling tool for the characterisation of semiconductor devices. Dopant contrast in the SEM has been known for many years [1] but only received increased attention after the widespread introduction of field emission gun SEMs (FEGSEM) which led to an increase in signal to noise ratio [2]. Even though it is widely agreed that there is an approximately linear relationship between contrast and doping level [3], the mechanism which leads to the dopant contrast is still debated [4].

Regardless of the mechanism for dopant contrast, the fact that the condition of the surface of the material can affect dopant contrast is undisputed. It has been reported that altering the scan rate of image collection (hence altering the electron dose) changes the dopant contrast [5] and that carbon contamination on the surface can even go so far as to reverse the dopant contrast [6]. Although it is clear that surface layers undoubtedly play a role in dopant contrast, its effects have not yet been quantified.

In this paper, the effect of natural oxidation and carbon contamination (produced by cleaving in air) on dopant contrast is examined, as well as the effect of surface layer build-up by prolonged electron beam exposure such as in repetitive scanning of a p-n junction.

2. Experimental

A 3.3 µm thick p-type (boron doped: 5×10^{18} atoms cm^{-3}) silicon layer was grown on an n-type (arsenic doped: 1×10^{19} atoms cm^{-3}) silicon substrate using chemical vapour deposition (CVD). In order to collect Auger spectra, part of the wafer was cleaved in air and placed into vacuum with a time between cleaving and insertion into the specimen chamber of less than 3 minutes. The Auger electron spectrometer (AES) used was a Thermo Microlab 350. Auger electron (AE) spectra were collected using a primary beam accelerating voltage of 10 kV, an energy step size of 1 eV and a dwell time of 100 mS per energy step. For in-situ cleaving experiments, an additional piece of the wafer was clamped into a cross-section specimen holder and scored using a diamond scribe part-way across its width. The specimen holder was then inserted into the pre-vacuum chamber (pressure: 6×10^{-6} Pa) and cleaved using a specimen manipulator. For in-situ cleaving, the investigated surface was not exposed to ambient air at any point.

To assess the thickness of surface layers formed by electron beam exposure, atomic force microscope (AFM) measurements were carried out using a Nanoscope III instrument with a tapping mode tip (330 kHz). For these experiments, a part of the silicon wafer was cleaved as previously described, before insertion into the vacuum chamber of the SEM again within 3 minutes of cleaving. Exposure was carried out in an FEI Sirion FEGSEM operating at an accelerating voltage of 1 kV, beam current of 78.5 pA and a working distance of 5 mm. A single scan consisted of scanning an area measuring 6 μm x 4 μm in a time of 3.25 s (a single scan delivers an electron dose of ~1.1 mC.cm^{-2}). The electron beam exposed silicon was examined using the AFM.

3. Results and Discussion

Figure 1 shows the result of imaging an ex-situ cleaved (in ambient air with ~3 minutes exposure) and an in-situ cleaved (cleaved under 6×10^{-6} Pa vacuum) silicon test structure within the AES. The measured contrast (defined as the difference in intensity between p and n-type silicon divided by the average intensity) gives values of ~5% and ~10% for the ex-situ and in-situ cleaved silicon respectively. The AE spectra presented in figure 1 c give an explanation for this difference in dopant contrast. For ex-situ cleaving, the spectra exhibit peaks at ~265 eV and ~510 eV, which correspond to elemental carbon and oxygen respectively. There is no indication of the presence of these peaks in spectra obtained from the in-situ cleaved specimen indicating that the surface is less contaminated than the ex-situ cleaved specimen.

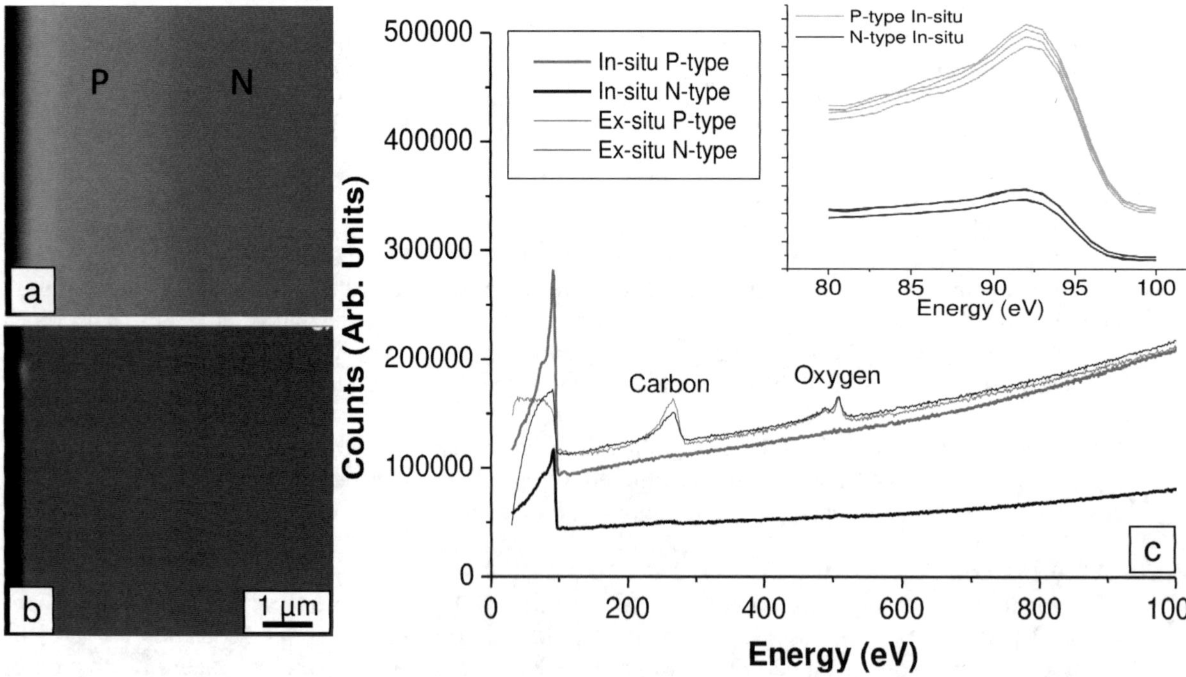

Figure 1. Secondary electron image of a silicon p-n junction cleaved (a) in-situ and (b) ex-situ and (c) Auger electron spectra showing the contamination present on p and n-type silicon after cleaving both in-situ and ex-situ with inset expansion of 80 – 100 eV.

It is remarkable that a surface layer adsorbed after only 3 minutes of exposure to ambient air, alters the dopant contrast, reducing it to half of that observed on an in-situ cleaved surface. A mechanism for this reduction was suggested as result of variation in oxide layer thickness in n-and p-doped regions [7,8] but there is no clear evidence for this in the ex-situ cleave AE spectra in figure 1 c, as there is no major difference in the oxygen peaks for p and n-doped regions respectively. Closer examination of the in-situ spectra between 30 eV and 100 eV, in particular the oxygen peak at 90 - 95 eV (figure 1 c inset) reveals that in-situ cleaved n-type silicon has a reduced peak intensity when compared to the in-situ cleaved p-type silicon. According to [9] this result suggests that a thicker natural surface layer has formed on the n-type silicon obscuring some signal from the underlying silicon, even after in-situ cleaving in a relatively high vacuum. In the ex-situ cleave, the 91 eV peak shows no consistent differences between n and p regions for the obtained data set (6 spectra).

As well as consideration of the formation of surface layers as a result of cleaving, it is also necessary to consider the dynamic effects of contamination caused by the deposition of contaminants by the scanning electron beam in an SEM, since dopant contrast has been found to strongly depend on electron dose [6].

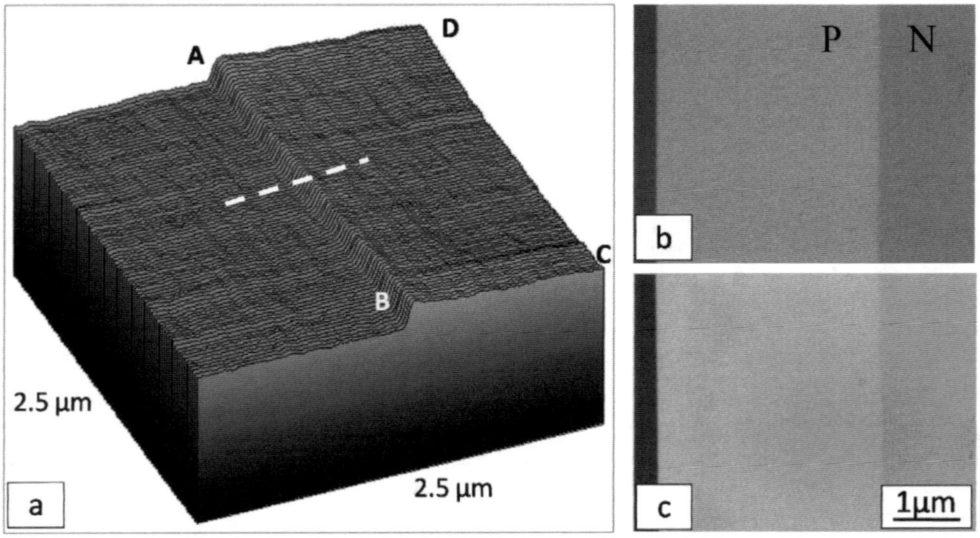

Figure 2. (a) Atomic force microscope image of contamination present on silicon after 180 scans (area ABCD) with a 1 kV, 78.5 pA electron beam (221 $mC.cm^{-2}$) (step height ~3 nm at the dashed marker) and SE dopant contrast images of a p-n junction after (b) 1 scan and (c) 180 scans.

Figure 2 a is an AFM image of the surface topography after extended exposure of an ex-situ cleaved surface to a 1 kV electron beam with a total of 180 single scans (overall electron dose: 221 $mC.cm^{-2}$) in SEM. Here it can be seen that the area exposed to the electron beam (ABCD) is raised, indicating that the electron beam has led to deposition of material from residual gases in the chamber. This effect is more pronounced at the perimeter of the scanned area than in the center. The effect of the build-up of this surface layer on dopant contrast is demonstrated in figure 2 b and c. Images collected after 1 scan (figure 2 b) and after prolonged exposure (figure 2 c) show a reduction in dopant contrast from 14 % to 5.5 %. Figure 3 shows the relationship between the number of SEM scans and the dopant contrast observed together with the approximate thickness of the SEM deposited contamination (as measured by AFM). The thickness measurements were taken from positions similar to that indicated on figure 2 a. The dopant contrast is reduced dramatically between 1 and 20 scans which corresponds to less than 1 nm of surface layer growth. A similar dramatic reduction in dopant contrast is also observed when preparing silicon surfaces using focused ion beam (FIB) for site-specific dopant contrast imaging, where a damaged layer measuring only 5 nm can cause a ~50 %

reduction in dopant contrast [10]. Both the results presented here and the result of FIB damage reported in [10] indicate the importance of the consideration of the condition of the silicon surface for the use of dopant contrast.

When examining ideal laboratory structures (as used in this study), dynamic build-up of surface layers with electron beam exposure has little impact as it is always possible to examine an adjacent unscanned area. However, when using real semiconductor devices, imaging a new area may not be an option. Therefore, this dynamic alteration of the surface during scanning adds an extra level of complexity to the the challenge of accurate quantification of dopants using SEM, and must be included in any numerical calculations which attempt to quantify dopant contrast.

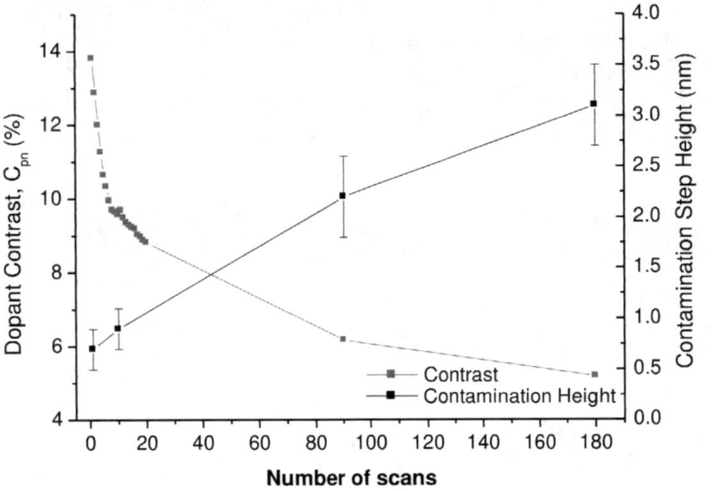

Figure 3. The effect of SEM scanning on the measured dopant contrast from a p-n junction together with the approximate step heights of the electron beam deposited contamination as measured by AFM.

4. Conclusions

A surface layer which forms naturally on cleaving silicon in air has a significant effect on the dopant contrast. We observed a reduction in dopant contrast of 50 % when comparing ex-situ cleaved silicon to an in-situ cleaved sample (when imaged in an AES). In addition, dopant contrast can be even further reduced by the build-up of deposited surface layers as a result of electron beam exposure. In order for SE dopant contrast to be considered as a quantitative dopant mapping tool, the effects of the surface contamination must be taken into account.

Acknowledgments

The authors would like to thank Professor John Watts and Rossana Grilli of the University of Surrey for their assistance with the Auger electron spectroscopy and the EPSRC for financial support (Grant no.: EP/E030602/1). C. Rodenburg would also like to thank the Royal Society for additional financial support in the form of a Dorothy Hodgkin Fellowship.

References

[1] Chang T H P and Nixon W C 1967 *Solid-State Electronics* **10** 701
[2] Castell M R, Muller D A and Voyles P M 2003 *Nature Materials* **2** 129
[3] Venables D, Jain H and Collins D C 1977 *J. Vac. Sci. Tech. b* **16** 362
[4] Walker C G H, Zaggout F and El-Gomati M M 2008 *J. Appl. Phys.* **104** 123713
[5] Kazemian P, Rodenburg C and Humphreys C J 2004 *Microelectronic Engineering,* **73** 948
[6] Mika F and Frank L 2008 *Journal of Microscopy,* **230** 76
[7] Dapor M, Jepson M A E, Inkson B J and Rodenburg C 2009 *Microsc. Microanal.* **15** 237
[8] Morita M, Ohmi T, Hasegawa E, Kawakami E and Ohwada M 1990 *J. Appl. Phys.* **68** 1272
[9] Carriére B and Deville J P 1979 *Surface Science* **80** 278
[10] Jepson M A E, Inkson B J, Beanland R, Humphrey C J and Rodenburg C 2009 *Proceedings of Microscopy of Semiconductor Materials XVI*, Oxford, 2009 (In press)

Journal of Physics: Conference Series **241** (2010) 012079 doi:10.1088/1742-6596/241/1/012079

Deep UV imaging by spectrometric full-colour cathodoluminescence microscopy

Hiroshi Saijo[1], Yasutomo Suzuki[2] and Makoto Shiojiri[3]
[1]Faculty of Biology oriented Science and Technology, Kinki University,
930 Nishimitani, Kinokawa, Wakayama, 649-6493 Japan
[2]Graduate School of Biology oriented Science and Technology, Kinki University,
930 Nishimitani, Kinokawa, Wakayama, 649-6493 Japan
[3]Prof. Emeritus, Kyoto Institute of Technology,
1-297 Wakiyama, Ohyamazaki, Kyoto 618-0091, Japan

E-mail: hirsaijo@waka.kindai.ac.jp

Abstract. A new deep-UV cathodoluminescence (CL) microscopy is tried on studying design and fabrication of recent solid-state light-emitting (LE) devices that emit near 200 nm light. CL microscopy has enough energy for electronic excitation of deep-UV emitting devices and visualizes distribution map of their electronic structures. The deep-UV imaging system is constructed on our own spectrometric full-colour CL microscope that collects CL spectra at all pixel points of the specimen during regular SEM observation, and at the end of one frame scan, a set of CL spectra is accumulated in the control computer. One frame scan of 512x512 pixels needs 8 to 80 sec, which is unbeatable by any other commercial CL microscopes. Full-colour CL micrograph is constructed using the spectra. The detector, a 32-channel photoelectron multiplier array, has higher sensitivity in UV region than any solidstate linear sensors, and a grating of 300-nm blaze wavelength instead of one at 500-nm blaze wavelength for visible lights, provide CL micrographs of 180-700 nm lights for recent deep-UV LE devices. Some unforeseen light emissions between 200-350 nm regions are observed on various materials and they are discussed correlating with the structure observed in SEM micrographs.

1. Introduction

Recent solid-state light-emitting (LE) devices go into deep-UV emission, and some emit near 200 nm light that is almost limit wavelength to transmit in air. Such devices ask for fine designing and precise fabrication in elemental, chemical and crystal structure, density and distribution of defects and impurities, etc. All above must be controlled at nano-accuracy, and the fabrication process should receive both structural and chemical verifications. Elemental distribution and crystal structure can be studied by electron microscopy with EDX, but the electronic energy levels, a key structure as LE device, are hardly examined. The wavelength of absorbing or emitting lights from a LE device is a direct measure between two energy levels among which electrons transfer during excitation and/or relaxation. The excitation of deep UV emitting devices is easily made by electric currents, *i.e.* electroluminescence, and the light emitting sites and the wavelengths are optically imaged, which means the spatial resolution is in the level of optical microscopy. Higher spatial resolution is achieved by scanning probe microscopy such as near-field microscopy, but the excitation energy is limited to near UV region. Raman

[1] To whom any correspondence should be addressed

spectroscopy, X-ray or synchrotron orbital radiation (SOR) lights excite deep energy states, but they lack microscopic imaging ability. Cathodoluminescence (CL) microscopy is the sole analysing tool that can excite deep-UV states with a narrow electron beam to draw micrograph.
Commercially available CL microscopes are designed to measure spectra of visual light to near UV in high resolution with slow imaging rate *e.g.* several hours. Among the various properties of materials, electronic structure is most irradiation sensitive and generally receives damage during one frame observation. In the present study, we tried to measure CL images of 180 nm to 700 nm lights in four sequential observations in five minutes.

2. Experimental.

2.1 Design of Deep-UV CL imaging system [1]
Figure 1 shows the block diagram of our spectrometric CL microscope, which detects CL spectra of the point where the scanning electron beam is placed and stores both secondary electron micrograph (SEM) and CL spectra signals simultaneously in the control computer. One frame scan of 512x512 pixels ends in 8s to 80s, and signal accumulation is possible by repeating frame scans. For deep-UV imaging, we replaced the grating of Jovin-Yvon spectrometer to 300-nm blaze wavelength (600 line/mm grating) instead of one at 500-nm blaze wavelength (300 line/mm grating for visible lights). Multi-channel linear array sensor is placed at the focus of the grating. The number of channels of sensing device determines spectral resolution. Most commercial microscopes use linear CCD array of 1024 channels or more, and proudly declare the spectral resolution be 0.1 nm on the data sheet. However, by the quantum chemical reasons, any luminescence from solid materials is wide spread in wavelength and the FWHM (full-width at half maximum) is generally 10 nm or more. Avoiding CCD sensors of low sensitivity to deep-UV lights and slow data reading rate due to too many channels, we employed Hamamatsu Photonics, H7260-04 32-channel photoelectron multiplier (PMT) array as a detector for the wavelengths from 200 nm to 700 nm lights. As a result, our spectral resolution is 5 nm and the highest spatial resolution in the CL micrograph is 20 nm. CL spectrum is stored at the address of the memory corresponding to the scanning electron beam position (pixel position), and is used to display full-colour CL micrographs.

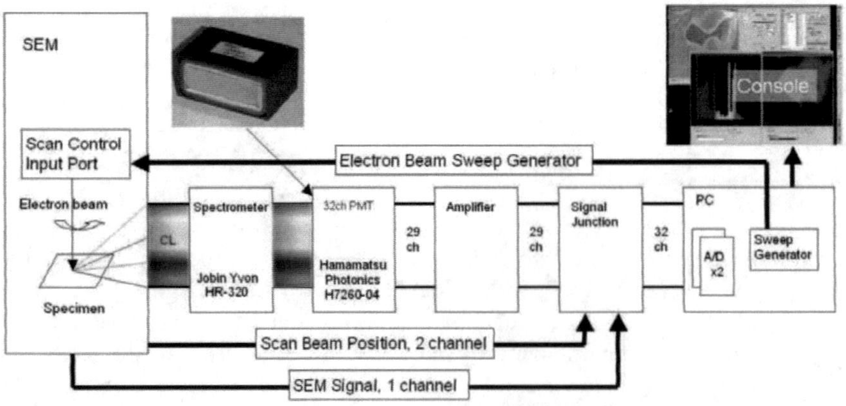

Figure 1. System block diagram of spectrometric full-colour cathodoluminescnece microscope.

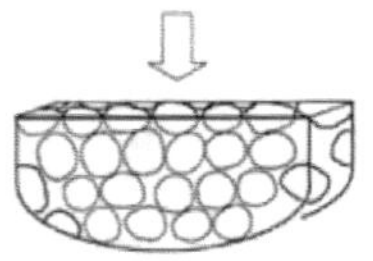

Figure 2. A schematic diagram of the grain structure of ZnO varistor. The surface is cut and polished to show the cross section of grains. Arrow is a direction of observation.

2.2. Specimen
The deep UV imaging is tested with ZnO varistors, #1 to #4, of four different SnO_2 concentrations, 0.1, 0.5, 2 and 10 % respectively. All of them contain 1% Bi_2O_3. The specimen has a grain structure and ZnO is inside the grains. SnO_2 controls grain growth and Sn-Zn spinel particles are formed inside grain. At the grain boundaries, Bi_2O_3 is believed to segregate and control the electronic properties [2] [3]. The shell and core of grains are well illustrated in the CL micrographs of visible light region [4].

3. Results and Discussion

Figure 3 is an example of CL micrograph taken using a grating of blaze wavelength 500 nm. With this grating, we record spectra of continuous 300 nm wavelengths in one scan, and for recording 200-700 nm, two measurements of 200-500 nm and 400-700 nm, are performed. In the figure, we can see strong luminescence peak at about 380 nm inside grain (see spectrum in subfigure 5) that is eminent in electroluminescence of ZnO materials. Luminescence over 400 nm is used to draw CL micrograph where wide green colour regions are grain surfaces and curved narrow green lines are grain boundaries being cross-cut on the specimen preparation. The CL micrograph composed with luminescence of 200-400 nm shows similar bright area distribution as that of 400-700 nm. However, several areas in the upper part of the figure are dark as compared to those in 400-700 nm CL micrograph.

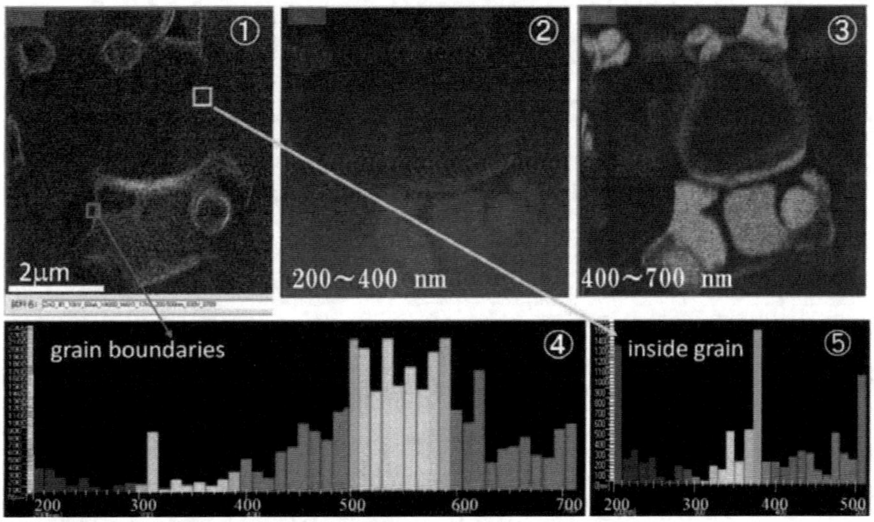

Figure 3. CL spectra and micrographs of varistor #1.

Figure 4 shows CL spectra of various points of the specimen, ZnO #1. Strong and sharp CL peaks from 200 to 250 nm are from the edges or surfaces of grain boundaries as identified in the SEM image. This understanding can be enhanced with higher incident energy.

Figure 4. SEM, CL micrographs and CL spectra of varistor #1 taken with 300 nm blaze wavelength grating. CL micrograph is composed using spectra from 160-310 nm.

Figure 5 is the case of 30 keV beam of ZnO #2, where high-energy incident electrons penetrate inside grains and excite grain boundaries hidden from the specimen surface. Spectra taken at the grain boundary edges (red square) and at the middle of the grains (blue square) show similar distribution of CL spectra. It is well known that bulk ZnO emits luminescence at about 380 nm. This specimen contains both Bi_2O_3 and SnO_2, but we can ignore Sn-Zn spinel particles because of low SnO_2 concentration in Specimens #1 and #2. Bi_2O_3 segregates on the grain boundaries and control electronic properties [3]. Bi_2O_3 has several absorptions of lights between 3.60-4.20 eV *i.e.* 344-295 nm, and with increasing thickness, it reaches 6.20 eV or 200 nm [5]. The CL peaks in Fig. 5 are, therefore, from thick Bi_2O_3 accumulated on the grain boundaries.

Figure 5. SEM, CL micrographs and CL spectra of varistor #2 taken with 300 nm blaze wavelength grating. CL micrograph is composed using spectra from 160-310 nm.

With increasing SnO_2 concentration, X-ray analysis suggests formation of Sn-Zn spinel particles inside grain [3]. Figure 6 is the case of high SnO_2 concentration. At the grain boundaries identified in the SEM image, strong 210 nm emission is again seen. Some new peaks around 255-260 nm appear in the spectra 2, 3 and 8 of Fig. 6. They seem not from the boundaries but from inside the grain. As these peaks are hardly seen in the spectra of Specimen #1 and #2, we consider them from Sn-Zn spinel or related materials.

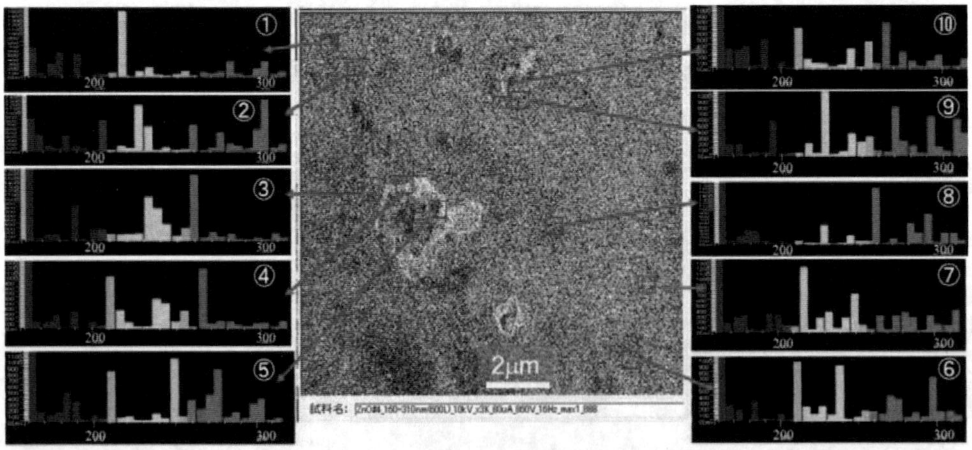

Figure 6. SEM and CL spectra of varistor #4 taken with 300 nm blaze wavelength grating.

4. Conclusion

Deep-UV luminescence is successfully imaged in CL microscopy with CL spectrum in our spectrometric full-colour electron microscope. Thick segregation of Bi_2O_3 and Sn-Zn spinel formation is clearly seen by the spectra.

References

[1] Saijo H and Shiojiri M, 2006 *16th Proc. Intern. Micros Cong., (Sapporo; Japan)* **2** 883
[2] Kindly prepared by Daneu N and Recnik A, Jozef Stefan Institute, Ljubljana, Slovenia
[3] Daneu N, Recnik A, Bernik S and Kolar D 2000 *J. Am. Ceram. Soc.* **83** 3165
[4] Saijo H, Daneu N, Recnik A and Shiojiri M, 2008 *14th Proc. European Micros Cong., (Aachen)* (Berlin: Springer) **2** 361
[5] Leonite L, Caramman M, Visinoju A and Rusu G J 2005 *Thin Solid Films*, **473** 230

Journal of Physics: Conference Series **241** (2010) 012080 doi:10.1088/1742-6596/241/1/012080

Comparison of multilayered nanowire imaging by SEM and Helium Ion Microscopy

B J Inkson[1, 3], X Liu[2], Y Peng[1], M A E Jepson[1] and C Rodenburg[1]
[1]Dept of Engineering Materials, University of Sheffield, Mappin Street, Sheffield, S1 3JD, UK
[2]Carl Zeiss NTS GmbH, Carl-Zeiss-Straße 56, 73447, Oberkochen, Germany

E-mail: Beverley.inkson@shef.ac.uk

Abstract. The helium ion microscope (HeIM) is capable of probe sizes smaller than SEM and, with intrinsically small ion/sample interaction volumes, may therefore potentially offer higher spatial resolution secondary electron (SE) imaging of nanostructures. Here 55 nm diameter CoPt/Pt multilayered nanowires have been imaged by HeIM, SEM and TEM. It is found that there is an increased resolution of nanowire surface topography in HeIM SE images compared to SEM, however there is a reduction of materials contrast of the alternating Pt and CoPt layers. This can be attributed to the increased contribution of surface contamination layers to the ion-induced SE signal, and carbon is also observed to grow on the nanowires under prolonged HeIM scanning.

1. Introduction

Scanning electron microscopy (SEM) has long been a technique of choice to image nanostructures on substrates. The SEM is able to evaluate nanowire morphology with spatial resolutions << 5 nm, depending on factors including the probe size and accelerating voltage. The recently commercialised helium ion microscope (HeIM) is capable of probe sizes of 0.25 nm [1]; a significant improvement over the probe sizes of SEMs. The incident, highly focused He ion beam generates ion induced secondary electrons (ISE) within the sample, which can be collected from the surface to form a SE image [1-4]. In principle, the smaller probe size and the intrinsically small ion/sample interaction volumes of the HeIM compared to a typical SEM, should improve the spatial resolution when imaging the smallest nanostructures [2,3], however the mechanisms of ISE generation are quite different from ESE and the benefits of HeIM over SEM are still to be proven.

Here a Zeiss Orion Plus™ HeIM is used to obtain ISE images of 55 nm diameter multilayered CoPt/Pt nanowires consisting of alternating CoPt and Pt layers each <100 nm length. The HeIM ISE images of the CoPt/Pt nanowires are compared to SEM ESE images and the known multilayer microstructure as evaluated by TEM.

2. Experimental

CoPt/Pt multilayered nanowires were fabricated by pulsed signal electrodeposition into anodic aluminium oxide (AAO) templates which had 55 nm pore diameters [5]. The nanowires consist of

[3] To whom any correspondence should be addressed.

alternating nanocrystalline layers of Pt (55 nm in diameter) and $Co_{72}Pt_{28}$ (variable diameter), whose microstructure has been extensively characterised elsewhere [5].

SEM and HeIM samples were prepared both by fracture of the templates, and by dissolving away the AAO templates in 0.2 M sodium hydroxide solution and rinsing with distilled water. Nanowires completely dissolved out of the templates were dispersed on SiO_2(100 nm)/Si substrates for HeIM analysis. The morphology of the electrodeposited multilayer nanowires were analysed using a JEOL JSM 6500F field emission gun (FEG)-SEM, a JEOL 2010F TEM, and a Zeiss Orion Plus™ HeIM. Individual nanowire samples for HeIM analysis were pre-cleaned for 5 minutes by an *in-situ* plasma cleaner (Evactron® 45 De-Contaminator, XEI Scientific, Inc.) inside the load lock of the HeIM to remove contamination. The plasma power was 10 watts with a radio frequency of 13.56 MHz. The samples were flushed with nitrogen before transfer into the HeIM chamber, vacuum <1.1×10^{-4}Pa. The helium ion beam was operated at 30-31 kV, beam current <0.3pA, and ion-induced SEs were collected using an Everhart-Thornley SE detector.

3. SEM and TEM imaging of CoPt/Pt nanowires

Fracturing the AAO templates resulted in the exposure of many aligned nanowires protruding from the AAO pores at the fracture surface. Figures 1(a,b) show typical 15kV SEM images of the CoPt/Pt multilayered nanowires, where the bright layers are Pt, and the lighter grey layers with lower secondary electron yield are $Co_{72}Pt_{28}$. Nanowires which retain a coating of alumina appear as a continuous bright column in SEM due to electron-induced charging (Figure 1(b)). Nanowires that have been dissolved out of the AAO templates also clearly demonstrate alternating contrast of the Pt and CoPt layers in both SEM (Figure 1(c)) and in TEM (Figure 1(d)). The as-grown nanowires typically exhibit a reduced diameter of the CoPt layers compared to the Pt layers, generating a modulated morphology along their length [5].

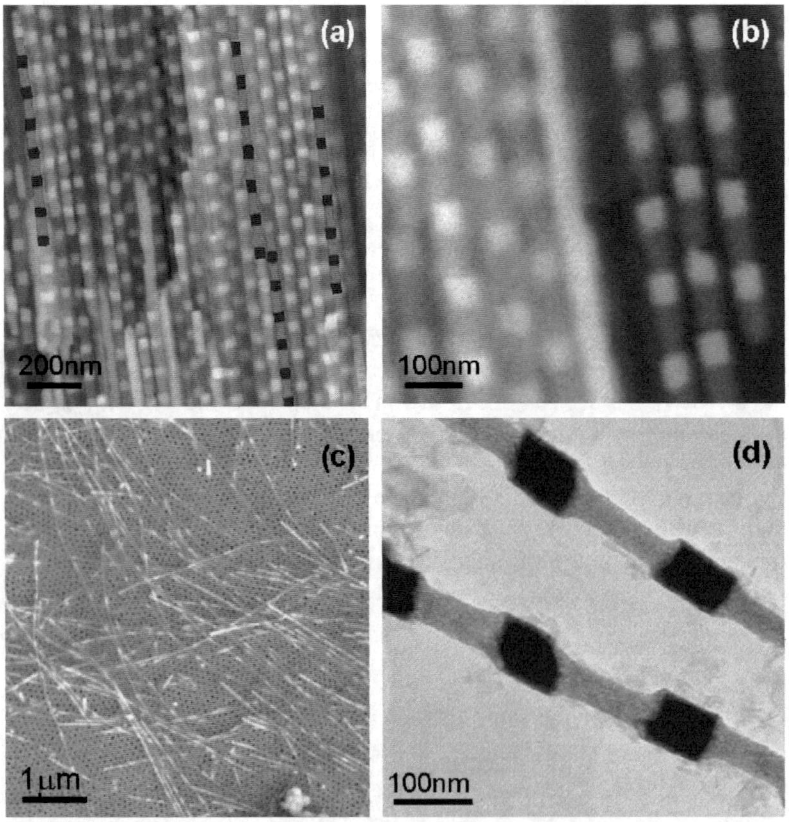

Figure 1. (a,b) SEM images of 55 nm diameter CoPt/Pt multilayered nanowires, in-situ, protruding from a fractured AAO template. The bright layers are Pt (marked in blue), the lighter grey layers are $Co_{72}Pt_{28}$ (marked in pink), and the white 'wires' are still embedded in a thin alumina coating.
(c) SEM image of individual nanowires dissolved out of the AAO template, and lying on a gold replica of the AAO template.
(d) TEM image of multilayered CoPt/Pt nanowires on a holey carbon film, showing alternating Pt (dark contrast) and $Co_{72}Pt_{28}$ (grey contrast) layers which grow with variable diameter.

4. HeIM imaging of CoPt/Pt nanowires

Figure 2 shows HeIM images of arrays of aligned CoPt/Pt multilayered nanowires, partially dissolved out of their AAO template. Clear modulated contrast can be observed along the lengths of some of the individual nanowires. The ion-induced secondary electron images (ISE) are similar to the electron induced secondary electron images (ESE), with the Pt layers generating more SE than the $Co_{72}Pt_{28}$. In these clumps of nanowires, which are not yet fully released from the AAO template, the surface and shape of the nanowires are not clearly defined, probably mainly due to a degree of residual un-dissolved alumina and contamination from the NaOH solution. For the HeIM imaging conditions used, no significant time-resolved sputtering of the nanowires was detected.

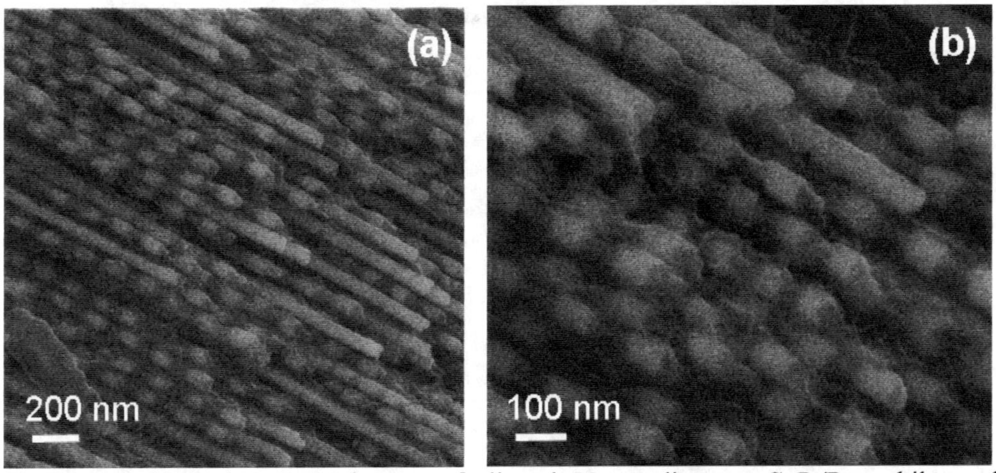

Figure 2. (a,b) 31 kV HeIM images of aligned 55 nm diameter CoPt/Pt multilayered nanowires, partially dissolved out of their AAO template. The bright layers are Pt, the lighter grey layers are $Co_{72}Pt_{28}$, and the white 'wires' are still embedded in a very thin alumina coating.

Figure 3 shows 31kV HeIM images of individual CoPt/Pt multilayered nanowires which have been fully released from their AAO template and dispersed on a SiO_2(100 nm)/Si substrate. Two key features of the HeIM images of the individual nanowires are:

1. *Increase in apparent roughness of the nanowires* (Figures 3(a,b)), with few nanowires exhibiting the regular modulated morphology apparent in SEM and TEM images (Figure 1).
2. *Loss of materials contrast of the alternating Pt and CoPt layers* (Figure 3(a)). No alternating contrast from the multilayers was detected, which is clearly visible in the SEM (and TEM) images of individual nanowires prepared by the same method (Figures 1(c,d)), and also in HeIM images of partially dissolved nanowire clumps (Figure 2).

Both of these features appear to be related to:

(a) The HeIM ISE images have an increased resolution of the outer surface topography compared to the SEM ESE images. The HeIM has a much smaller spot size and smaller interaction volume than the SEM [2-4], and clearly images very fine detail in the outer contamination on the individual nanowire and substrate surface from the dissolution process (Figures 3(a,b)). Contamination from the dissolution is also present in the SEM and TEM samples, but has a very weak contrast contribution, making the nanowires appear more smooth and regular (Figures 1(c,d)).

(b) The increased sensitivity of the HeIM ISE images to surface contamination obscures the underlying materials contrast from the Pt and CoPt layers in the individual nanowires (Figure 3(a)). There appears to be much more contamination on the individual nanowires than the partially dissolved nanowire clumps (Figure 2), possibly due to attraction of contamination to the surface localised individual wires as the specimens are dried on the substrate.

(c) Notable ion beam induced deposition (IBID) of carbon also occurs during examination in the HeIM. Figure 3(c) shows the growth of contamination, and associated reduction in nanowire contrast and resolution, as a result of exposure to the He ion beam for 35 minutes. Nanowire samples exposed to an additional 30 minutes plasma cleaning to reduce carbon contamination (Figure 3(d)) did not appear significantly different.

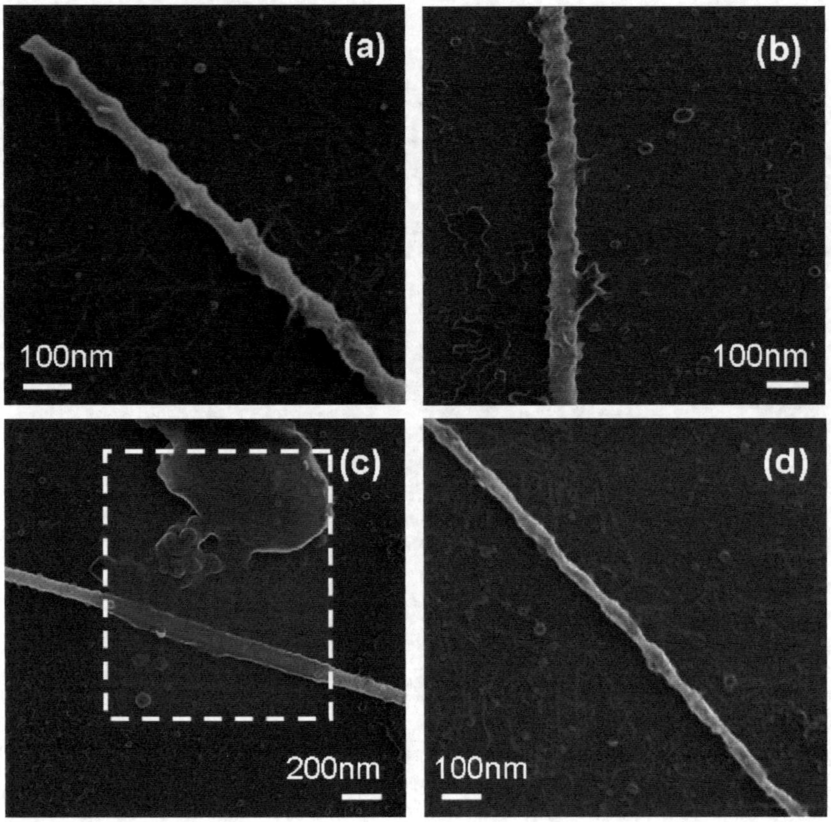

Figure 3. Role of contamination on HeIM imaging of the CoPt/Pt multilayered nanowires.
(a) 31kV HeIM image of a single multi-layered CoPt/Pt nanowire dispersed on a SiO_2 (100 nm) / Si substrate.
(b) Nanowire with very irregular external morphology.
(c) A nanowire scanned for 35 minutes in the marked box with the He ion beam accumulates contamination.
(d) A nanowire plasma cleaned for 30 minutes within the load lock of the HeIM microscope.

5. Conclusions

It is observed that HeIM ion-induced SE (ISE) images and SEM electron-induced SE (ESE) images of the same CoPt/Pt nanowires are different in character. The ISE images reveal improved detail of the outer surface topography, but are therefore highly sensitive to surface contamination layers. For the individual nanowires examined here, the dominant contribution of surface contamination to the ISE images masks material contrast from the alternating Co and Pt layers, which can be seen in the less surface-sensitive SEM ESE images. The build-up of contamination in-situ in the HeIM during prolonged scanning is also detrimental to nanowire imaging.

Acknowledgments

This work was partially funded under the EPSRC Grants GR/S85689 and EP/E030602. The authors thank Ian Belding, Carl Zeiss SMT, for access to the Zeiss Orion Plus^{TM} helium ion microscope.

6. References

[1] Hill R, Notte R and Ward B 2008 *Phys. Procedia.* **1** 701
[2] Inai K, Ohya K and Ishitani T 2007 *J. Elec. Microsc.* **56** 163
[3] Ramachandra R, Griffin B and Joy D 2009 *Ultramicroscopy* **109** 748
[4] Bell D C 2009 Microscopy and Microanalysis **15** 147
[5] Peng Y, Cullis A G, Möbus G, Xu X and B J Inkson 2007 *Nanotechnology* **18** 485704

A practical approach to test the scope of FIB-SEM 3D reconstruction

M Ritter[1] and P A Midgley
Department of Materials Science & Metallurgy, University of Cambridge, Pembroke Street, Cambridge CB2 3QZ, UK

E-mail: mwr32@cam.ac.uk

Abstract. State-of-the-art focused ion beam (FIB) instruments have an ion column for sample modification and an electron column for scanning electron microscopy (SEM). 3D reconstruction of a sample volume can be achieved by serial sectioning using the FIB in combination with high-resolution SEM imaging of each cross-section. Usually, the resolution in the direction in which the sections are milled (z-direction) is much lower than in the plane of the cross-section (xy-direction) itself. Increased sampling in the z-direction can only be achieved by decreasing the distance between single sections. For a constant volume this is equivalent to increasing the number of sections, i.e. time and effort. To perform efficient 3D reconstructions the effect of the reduced sampling in the z-direction to the overall accuracy of the 3D reconstruction has to be known. We tested this approach with FIB conical test structures that were slice-and-view processed and subsequently reconstructed. Using a reference data set with a slice thickness (z-resolution) of 22 nm, data with z-resolutions ranging from 44 nm to 440 nm were created and reconstructed with commercial software. The calculated volumes for the simulated z-resolutions and their deviations from the reference volume are shown. Deviations of up to 35% occur and reach about 10% once the z-resolution was one fifth of the upper diameter of the conical structures.

1. Introduction

3D reconstructions by serial sectioning with focused ion beam (FIB) instruments, introduced some years ago [1-3], is now a standard feature or option of most available FIB-SEM instruments. Using the SEM for imaging, instead of the ion beam, allows non-destructive quantitative analysis of features smaller than 100 nm [3]. However, the process of serial sectioning with FIB is used not only for volumetric reconstruction of the microstructure, but is also the basis for quantitative 3D analytical methods, such as 3D electron backscatter diffraction (EBSD) [4] and 3D energy dispersive X-ray spectroscopy (EDS) [5] and a combination thereof [6]. In this paper we test the ability of serial FIB sectioning, using standard features of the FIB-SEM instrument, to enable efficient 3D reconstructions using SEM image data. The resolution in the direction in which the sections are milled (z-direction) is usually much lower than that in the plane of the cross-section (xy-direction) and so it is helpful to have an estimate of the effect of z-resolution on the overall accuracy of the 3D reconstruction, e.g. for reducing FIB time and effort. If the size of a structure of interest is known approximately, this information could be used to estimate the FIB parameters of an experimental setup for 3D reconstruction.

[1] To whom any correspondence should be addressed.

Journal of Physics: Conference Series **241** (2010) 012081 doi:10.1088/1742-6596/241/1/012081

2. Methods

A Helios NanoLab 600 (FEI Company, Hillsboro, USA) FIB-SEM was used to fabricate and section the microstructures for the 3D reconstruction test. 3D reconstruction was performed with ResolveRT 4.0.1 (Visage Imaging GmbH, Berlin, Germany).

2.1. Test structure fabrication

Conical structures with diameters of 1000 nm and 500 nm were generated on a silicon wafer sample by focused ion beam milling an area of approx. 7 μm x 6 μm (figure 1). The nominal milling depth was 1000 nm but the actual depth measured was 2700 nm (figure 2). Subsequently, the cavities were exposed to gas-assisted platinum deposition and manually observed until completely covered. We chose platinum, because it provides high contrast with respect to the silicon in the SEM secondary electron (SE) images, which makes it easier to threshold the data and perform a semi-automatic segmentation of the test features.

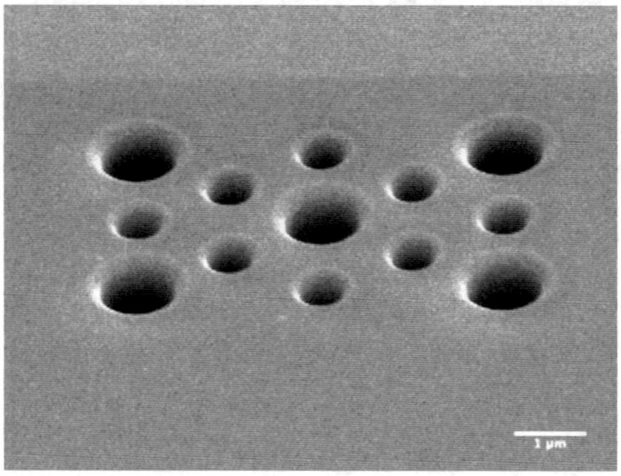

Figure 1. SEM SE image of the test structure: five large circular cavities with a diameter of 1000 nm and nine small cavities with a diameter of 500 nm.

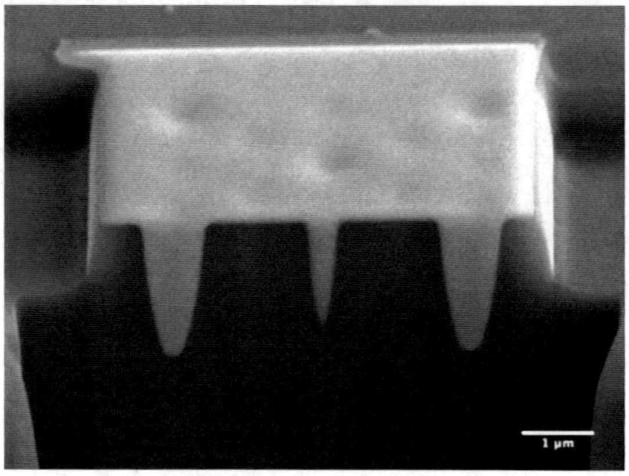

Figure 2. SEM SE image of FIB cross-section number 37 out of a total of 316 sections.

2.2. Serial FIB tomography

In total, 316 serial FIB cross-sections were made and imaged in SEM secondary electron (SE) mode at a sample tilt of 52 degrees. SEM images were recorded at 10 keV beam energy, a probe current of 0.17 nA and with a pixel size of 10.4 nm. After tilt correction, the effective pixel size in the y-direction (slow scan direction) was calculated to be 13.2 nm. Each slice that had been milled to create the serial sections had a thickness of 22 nm, determined by the spot size of the ion beam, the relative beam overlap at 30kV ion energy and the beam aperture chosen to give 0.92 nA ion current. Here, the thickness of a single slice is defined as the z-resolution.

2.3. 3D reconstruction with different virtual slice thicknesses

With the SEM image data imported into ResolveRT, the images were aligned manually so that in the reconstructed coordinate system the slicing direction became normal to the image plane. To simulate different z-resolutions, further 3D reconstructions were made with down-sampled data sets using only every second, third, fourth (and so on) SEM image (see table 1). Segmentation was performed with the wand tool and a lower threshold of 78 (black) and an upper threshold of 248 (white).

3. Results

3.1. Volume reconstruction

Table 1 shows the calculated average volumes and standard deviations of the five large and nine small cavities at various z-resolutions, ranging from 22 nm to 440 nm. At the highest z-resolution the average volume of the larger structures is 1.253 μm^3 and 0.276 μm^3 for the smaller structures. A visualization of two of the 3D reconstructions can be seen in figures 3 and 4, using z-resolutions of 22 nm and 330 nm, respectively.

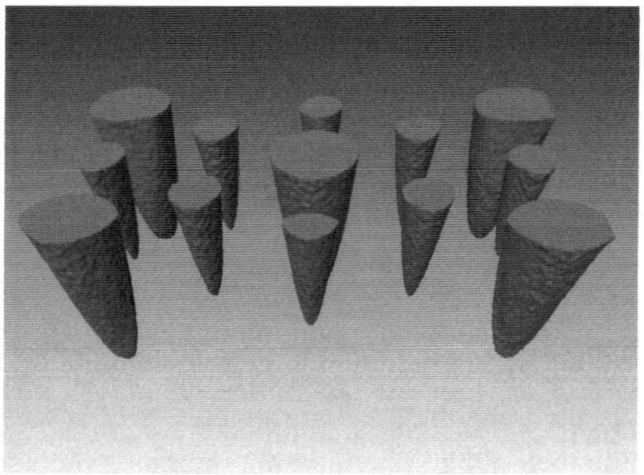

Figure 3. Reconstructed cavity volumes of the test structures (316 serial FIB cross-sections, i.e. 22 nm per slice).

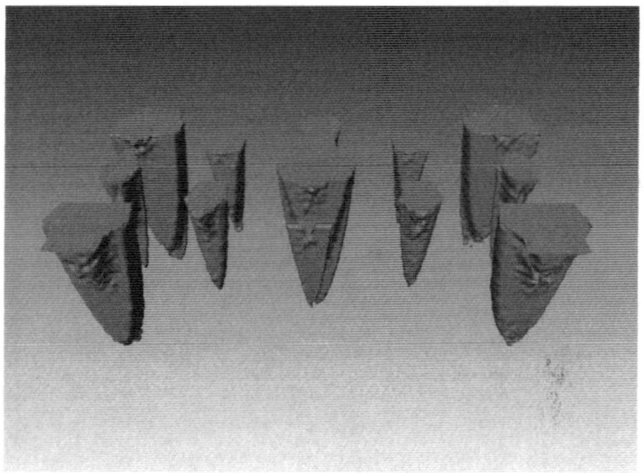

Figure 4. Reconstructed cavity volumes of the test structures (21 serial FIB cross-sections, i.e. 330 nm per slice).

Geometric distortions of the 3D reconstruction of the large cavities are already evident at a virtual z-resolution of 110 nm, and very strong geometrical errors are present at a z-resolution of 176 nm and lower. These values correspond to a z-resolution to diameter ratio of 0.11 and 0.18, respectively. At the small cavities, distortions start appearing at a z-resolution of 66 nm, and strong errors occur at 110 nm or lower. These values correspond to a z-resolution to diameter ratio of 0.13 and 0.22, respectively.

Table 1. The reconstructed average volumes and standard deviations of the five large and nine small cavities according to the chosen z-resolution.

Slice thickness (nm)	Average volume large structures (μm^3)	Standard deviation volume large structures (μm^3)	Average volume small structures (μm^3)	Standard deviation volume small structures (μm^3)
22	**1.253**	0.206	**0.276**	0.039
44	**1.282**	0.214	**0.286**	0.041
66	**1.272**	0.189	**0.284**	0.041
88	**1.259**	0.169	**0.287**	0.047
110	**1.353**	0.236	**0.310**	0.046
132	**1.310**	0.207	**0.296**	0.045
176	**1.366**	0.273	**0.294**	0.044
220	**1.405**	0.276	**0.331**	0.061
264	**1.370**	0.223	**0.310**	0.050
330	**1.481**	0.392	**0.324**	0.060
440	**1.783**	0.303	**0.284**	0.015

3.2. Deviation in reconstructed cavity volumes with respect to z-resolution

In general the results shown in table 1 indicate that the reconstructed average volumes for both types of structure in this test tend to become larger for lower z-resolutions. The deviations of the average volume of the cavities reconstructed at the various z-resolutions with respect to the reconstructed reference volume at highest z-resolution oscillate strongly, both for the large and the small cavities (see table 2). Deviations from the reference volume larger than 10% occur once the z-resolution is lower than approximately one fifth of the diameter of the test structures, i.e. at a z-resolution of 220 nm for the larger and 110 nm for the smaller cavities. Whereas the maximum deviation from the reference volume is at the same magnitude for both kinds of structures, the minimum volume deviation remains below 10% at all z-resolutions, except for the case of the large structures at 440 nm. The low deviation for the small structures at that resolution cannot be explained at present.

Table 2. Deviation of the average, maximum and minimum volumes of the reconstructed test structures at various z-resolutions with respect to the reconstructed reference volume using all 316 serial FIB cross-sections.

Slice thickness (nm)	Average volume deviation of large structures (%)	Max./min. volume deviation of large structures (%)	Average volume deviation of small structures (%)	Max./min. volume deviation of small structures (%)
44	**2.25**	3.18/1.69	**3.50**	5.68/1.62
66	**1.88**	4.00/0.21	**3.28**	8.10/0.51
88	**5.44**	6.92/2.56	**8.11**	13.68/4.33
110	**7.79**	9.82/5.87	**12.25**	16.76/7.83
132	**4.64**	5.78/3.28	**7.06**	12.11/0.47
176	**8.47**	16.59/2.69	**6.44**	9.58/0.24
220	**11.61**	15.63/6.42	**19.99**	28.30/3.47
264	**9.58**	11.52/7.73	**15.40**	29.60/4.53
330	**16.86**	33.90/2.09	**17.43**	27.90/5.58
440	**34.06**	45.69/20.70	**1.31**	2.37/0.72

4. Conclusions

Deviations with respect to the reference volume larger than 10 % were observed at a z-resolution of approximately one fifth of the diameter of the test structures. However, qualitative analysis of the 3D reconstruction shows visible geometric distortions already at a z-resolution of one tenth of the diameter of the test structures. The oscillations of the deviations itself could be due to the experimental setup of this test or due to the interpolation mechanism of the ResolveRT software. Therefore, the volume reconstruction seems only to be an indicator of the quality of the 3D reconstruction of the FIB cross-sections. Additionally, the geometric similarity of the reconstructions could be taken into account to quantify what can be seen as geometric distortions in the 3D reconstruction at lower z-resolution [7].

Acknowledgments

We thank Dr. Caterina Ducati and Dr. Jonathan Barnard for discussions, helpful suggestions and reviews. The work was supported financially by the EPSRC and the EC FP6 I3 project ESTEEM (Contract Number 026019).

References

[1] Sakamoto T, Cheng Z, Takahashi M, Owari M and Nihei Y 1998 Jpn. J. Appl. Phys. 37 2051
[2] Inkson B J, Mulvihill M and Möbus G 2001 Scripta Mat. 45 753
[3] Holzer L, Indutnyi F, Gasser P, Münch B and Wegmann M 2004 J. Microsc. 216 Pt1 84
[4] Groeber M A, Haley B K, Uchic M D, Dimiduk D M and Ghosh S 2006 Materials Characterization 57 259
[5] Kotula P G, Keenan M R and Michael J R 2006 Micros. Microanal. 12 36
[6] West G D and Thomson R C 2009 J. Microsc. 233 Pt3 442
[7] Maple C and Wang Y 2004 8th Int. Conf. on Information Visualisation (IV'04), 2004 363-369

Journal of Physics: Conference Series **241** (2010) 012082 doi:10.1088/1742-6596/241/1/012082

Device fabrication with precisely placed carbon nanotubes of known chiral vector

C S Allen[1,4], M D Elkin[1], C Zhang[2], S Hofmann[2], G Burnell[1], J Robertson[2] and B J Hickey[1]

[1] School of Physics and Astronomy, University of Leeds LS2 9JT, UK
[2] Department of Electrical Engineering, Cambridge University CB3 0FA, UK

E-mail: c.s.allen@leeds.ac.uk

Abstract. The electrical properties of single walled carbon nanotubes are strongly dependent on their precise atomic configuration and as such it is of great importance to be able to fabricate devices containing individual carbon nanotubes of known structure. We have developed a novel technique that allows for the determination of the chiral indices of an individual carbon nanotube followed by the precise placement of the tube onto pre-patterned electrodes. Carbon nanotubes are grown by chemical vapour deposition onto perforated TEM grids. Electron diffraction is performed on individual carbon nanotubes, the analysis of which reveals their structure. The desired carbon nanotube is then manipulated inside a combined STM, SEM system and placed with high precision onto pre-patterned electrodes.

1. Introduction

A single walled carbon nanotube (SWCNT) is uniquely identified by reference to its chiral indices (n, m). These describe the translation vector around the circumference of the tube in terms of the unit vectors of the graphene sheet, which it can be considered to be made of. It is well known that of all possible tube structures one third satisfy the condition $(n - m) = 3i$, where i is an integer, necessary for band structure calculations to describe them as metallic [1]. The other possible tube morphologies are semiconducting with band gaps approximately inversely proportional to the radius of the tube. In fact, due to the curvature of the tube there exists a small band gap in most "metallic" tubes, save those with $i = 0$ (i.e. $m = n$), which are commonly called armchair tubes [2].

Here we describe a technique that allows for the determination of the structure of carbon nanotubes via electron diffraction followed by the fabrication of electronic devices consisting of the desired tubes.

2. Carbon nanotube growth and characterisation

Carbon nanotubes were grown by chemical vapor deposition (CVD) on commercially available perforated silicon oxide transmission electron microscope (TEM) grids using a calcinated ferrotin catalyst. The TEM grid was placed inside a Phillips CM200 TEM operating at 80 kV, below the threshold for knock-on damage from carbon nanotubes [3]. Individual SWCNTs, individual double walled carbon nanotubes (DWCNTs) and ropes containing multiple tubes were located and their

[4] To whom any correspondence should be addressed.

positions on the grid noted. High-resolution images were captured on a CCD camera and selected area diffraction patterns recorded onto imaging plates from sections of approximately 200 nm length of selected tubes. Once developed the diffraction patterns were digitally scanned at 2400 dpi resolution to allow for subsequent analysis.

Figure 1 shows a diffraction pattern taken from an individual SWCNT spanning a hole in the TEM grid. The diffraction pattern from a SWCNT takes the form of a series of layer lines corresponding to the hexagonal reciprocal lattice of two graphene sheets, smeared in the radial direction due to curvature of the tube.

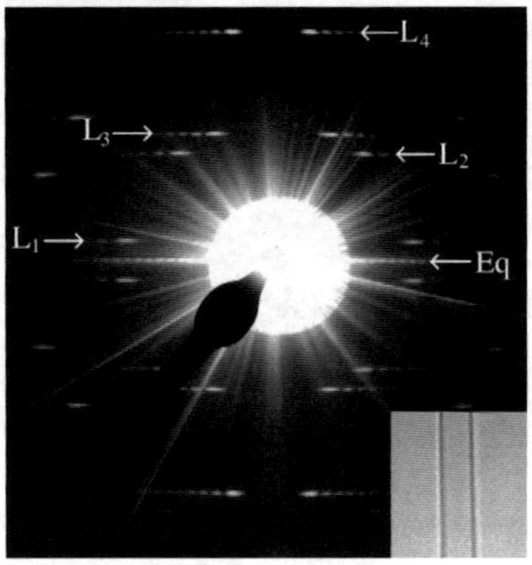

Figure 1. An electron diffraction pattern and real space CCD image (inset) of an individual SWCNT. The equatorial line, Eq and the first four layer lines, L_1-L_4 are marked.

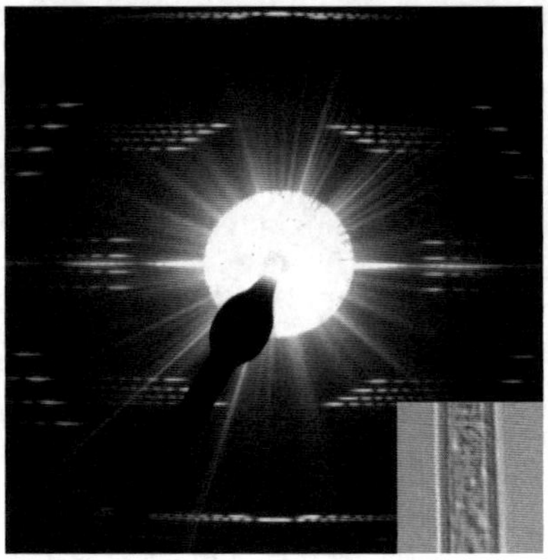

Figure 2. An electron diffraction pattern and real space CCD image (inset) of an individual DWCNT.

To determine the chiral indices of the tube the axial distances of the layer lines (L_1-L_4) from the equatorial line (Eq) are measured. The ratios of a combination of these distances are used to determine the chiral angle and ratio of chiral indices [4][5]. This analysis is independent of the calibration of the diffraction pattern and tilt of the tube with respect to the incident electron beam. In addition the intensity distribution along each layer line can be described by the square of a Bessel function, the order of which is directly related to the chiral indices of the tube [6]. Once the structure of the tube is confidently obtained from these complimentary analysis techniques, the diffraction pattern can be accurately calibrated. The diameter of the tube can be obtained from a calibrated diffraction pattern as it is related to the period of the oscillations of the Bessel function that describes each of the layer lines and the equatorial line [7]. The SWCNT in figure 1 was determined to be a (25,15) tube with a diameter of 2.7 nm and, as $(n - m)=10$, is a semiconducting tube.

The diffraction pattern from a DWCNT can be considered as the superposition of the diffraction pattern from two SWCNTs. The diffraction pattern shown in figure 2 is from an individual DWCNT and hence there are twice as many visible reflections in the same region of reciprocal space as in figure 1. The inner tube was determined to have chiral indices (28,16) and the outer tube (30,25), henceforth written (28,16)@(30,25), with diameters of 3.0 and 3.7 nm respectively and an inter tube separation of 0.35 nm, in agreement with other TEM studies of DWCNTs [7][8]. The inner tube, with $(n - m)=12$ is a small band gap metal and the outer tube with $(n - m)=5$ is a semiconducting tube.

3. Device fabrication

With the analysis of the diffraction patterns complete the TEM grid was transferred to an Omicron Nanoprobe system. This consists of four independently controllable tungsten scanning tunneling microscope (STM) tips situated inside a high-resolution scanning electron microscope (SEM) all under ultra high vacuum (~10^{-10} mbar). The desired tube was located under the SEM (figure 3 (a)) and two opposing STM tips brought into close proximity. The tips were brought into contact with the tube at either side of the hole and the length of tube lifted away from the TEM grid (figure 3(b)). The tips were then moved away from each other until the tube became detached from one of the tips (figure 3 (c)).

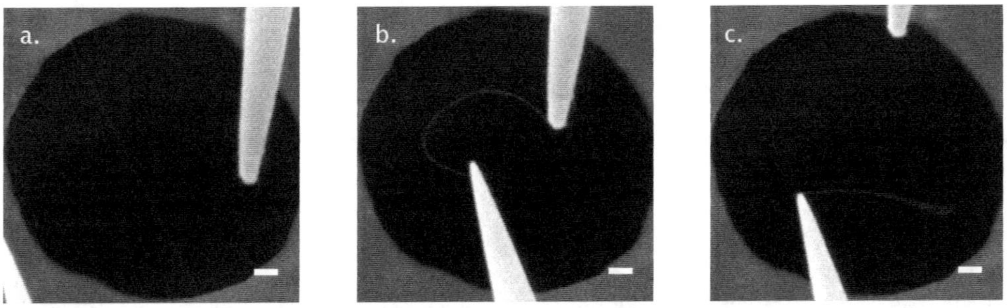

Figure 3. Removing the tube from the TEM grid. a) The tungsten tips are brought into close proximity to the desired tube. b) The tube is picked up, suspended between the tips and moved away from the surface of the TEM grid. c) The tips are moved apart leaving the tube dangling from one tip. The scale bars correspond to 200 nm.

The TEM grid was then removed and replaced with electrodes patterned using standard electron beam lithography techniques. Using the STM feedback electronics the tube was brought into tunneling contact with one electrode. The tube was then moved away from the surface to the limits of the piezomotors, the STM electronics disconnected and the tips connected to external source sense electronics. A bias of 2V was applied between the tube and the electrode and, using the piezomotors and monitoring the current flow, the tube was brought into physical contact with the electrode. Using the SEM for guidance, the tube was then laid in the desired location (figure (4)). When the tip is retracted, the tube generally remains on the substrate due to the larger van der Waals force between the tube and the substrate, compared to that between the tube and the tip.

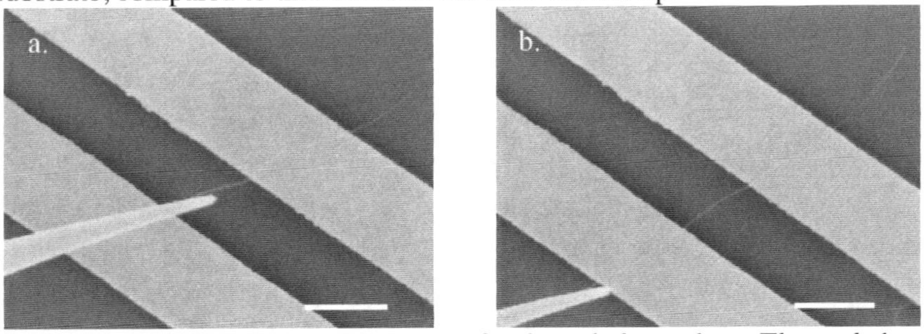

Figure 4. Placing of a tube across pre fabricated electrodes. The scale bar represents one micron.

4. Transport measurements

Figure 5 shows the current-voltage (I-V) characteristics through a SWCNT device measured immediately after fabrication, within the nanoprobe under UHV conditions. The device was fabricated using the technique described above from the semiconducting (25,15) SWCNT whose diffraction pattern is shown in figure 1. The tube was placed across pre-patterned palladium electrodes with electrode spacing of approximately 350 nm. The non-linear asymmetric I-V characteristics suggest barriers at the contacts dominate transport through the device. The choice of palladium, with its high work function, as the electrode material should limit the height of any Schottky barrier formed between the SWCNT and electrodes [9]. We therefore ascribe the asymmetric transport behaviour to the presence of poor quality tunnel barriers at the tube-electrode interfaces. We believe these barriers are formed by amorphous carbon buildup during SEM imaging, with the asymmetry due to different effective tunnel barrier thicknesses at the tube-source and tube-drain electrodes.

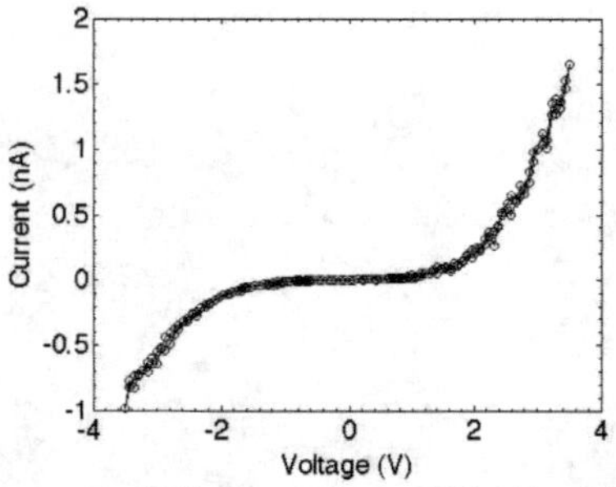

Figure 5. I-V through a (25,15) SWCNT performed at room temperature and under UHV.

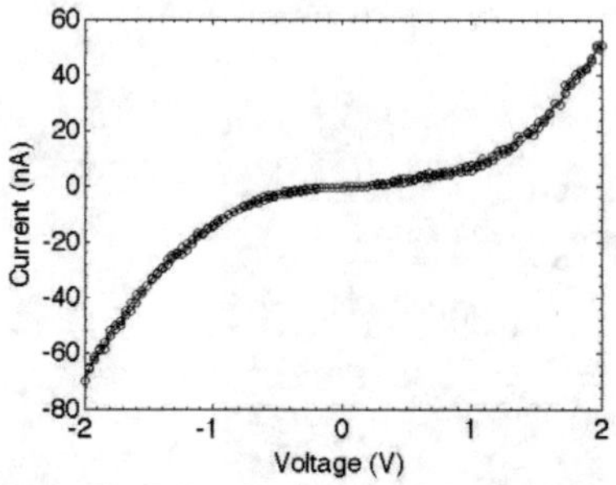

Figure 6. I-V through a (28,16)@(30,25) DWCNT performed at room temperature and under UHV.

Figure 6 shows the I-V characteristics of a device fabricated using the (28,16)@(30,25) DWCNT whose diffraction pattern is shown in figure 2. The tube was placed across pre-patterned Pd electrodes with electrode spacing of approximately 1 micron. The I-V is again highly non-linear and asymmetric about 0 V. Despite the channel length being almost three times larger than in the case of the SWCNT the current through the DWCNT device at ± 2 V is over two orders of magnitude higher than the SWCNT device (note the different voltage ranges in figure 5 and figure 6). As barriers at the tube-electrode interfaces again dominate this device the increased conductance is due to the increase in contact area between tube and electrode as a result of larger diameter of the DWCNT.

We believe that patterning of electrodes on top of the carbon nanotubes will enable Ohmic contact to be achieved.

5. Summary

We have demonstrated a technique to allow for the controlled placement of carbon nanotubes with known structures across electrodes for device fabrication. The chiral indices of carbon nanotubes is determined via electron diffraction analysis and the desired tube is then manipulated with remote controlled STM tips and placed across pre-fabricated electrodes. Provided Ohmic contact can be achieved, this technique will allow for the investigation of the electrical properties of carbon nanotubes of various morphologies and also open up the possibility of fabrication of complex devices comprising multiple tubes of differing structures.

Acknowledgements

We would like to thank Dr. A. Brown and Dr. M. Ward, School of Process, Environmental and Materials Engineering, University of Leeds, for many helpful discussions.

References

[1] Saito R, Fujita M, Dresselhaus G and Dresselhaus M S 1992 *Appl. Phys. Lett.* **60** (18)
[2] Kane C L and Mele E J 1997 *Phys. Rev. Lett.* **78** (10) 1932-1935
[3] Smith B W and Luzzi D E 2001 *J. Appl. Phys.* **90** 3509
[4] Gao M, Zuo J M, Twesten R D, Petrov I, Nagahara L A and Zhang R 2003 *Appl. Phys. Lett.* **82** (16) 2703
[5] Liu Z, Zhang Q and Qin L C 2005 *Appl. Phys. Lett* **86**(19) 191903
[6] Qin L 1994 *J. Mater. Res.* **9**(9) 2450
[7] Kociak M, Hirahara K, Suenaga K and Iijima S 2003 *Eur. Phys. J. B.* **32** 457-459
[8] Hirahara K, Koicak M, Shunji B, Nakahira T, Itoh K, Saito Y and Iijima S 2006 *Phys. Rev. B* 73 195420
[9] Javey A, Guo J, Wang Q, Lundstrom M, and Dai H 2003 *Nature* **424** 654

Journal of Physics: Conference Series **241** (2010) 012083 doi:10.1088/1742-6596/241/1/012083

Nanoconstruction by welding individual metallic nanowires together using nanoscale solder

Y Peng[1,2], A G Cullis[2] and B J Inkson[1]
[1] Department of Engineering Materials
[2] Department of Electronic and Electrical Engineering, Sheffield University, Sheffield S1 3JD, UK

E-mail: Y.Peng@sheffield.ac.uk

Abstract. This work presents a new bottom-up nanowelding technique enabling building blocks to be assembled and welded together into complex 3D nanostructures using nanovolumes of metal solder. The building blocks of gold nanowires, $(Co_{72}Pt_{28}/Pt)_n$ multilayer nanowires, and nanosolder $Sn_{99}Au_1$ alloy nanowires were successfully fabricated by a template technique. Individual metallic nanowires were picked up and assembled together. Conductive nanocircuits were then welded together using similar or dissimilar nanosolder material. At the weld sites, nanoscale volumes of a chosen metal are deposited using nanosolder of a sacrificial nanowire, which ensures that the nanoobjects to be bonded retain their structural integrity. The whole nanowelding process is clean, controllable and reliable, and ensures both mechanically strong and electrically conductive contacts.

1. Introduction

Individual building blocks used in bottom-up techniques, such as nanoparticles, nanowires and nanotubes, can now be routinely fabricated. Scanning probe technologies, such as the latest robotics nanomanipulators, make the manipulation and assembly of nanoscale building blocks into nanocircuits available to researchers around the world [1-6]. However, there is a major difference between simply bringing nano-objects into contact (formation of an interface), and integrating two nano-objects together (interfacing through a functional junction). The practical applications of nanocircuits and nanodevices require secure, functional and long-lasting bonds between their individual building blocks. Here we describe a new nanoscale electrical welding technique which radically improves the spatial resolution, flexibility and controllability of welds between individual bottom-up building blocks.

2. Fabrication of nanowires

Anodic aluminium oxide (AAO) templates with about 55nm pore diameter were fabricated by anodising 99.999% Al sheets (Goodfellow) in 0.6M oxalic acid electrolyte for three hours at room temperature with continuous stirring. The detailed fabrication method of gold nanowires and $(Co_{72}Pt_{28}/Pt)_n$ multilayer nanowires has been detailed in our previous work [1-3,7]. The nanosolder $Sn_{99}Au_1$ alloy nanowires were electrodeposited into a polycarbonate (PC) template with 80 nm pore diameter using an electrolyte composed of 0.5 M $SnCl_2$ (Alfa Aesar, 98.0-103.0%), 5g/l $AuCl_4$ (Aldrich, 99.99%), 0.01 M HCl (Aldrich, 99.5+%) and 5g/l L-Ascorbic (Alfa Aesar, 99+%). A -0.08V DC signal was applied between the PC template and a 5mm diameter tin rod anode (Goodfellow,

99.75%). The typical total deposition time for each specimen was set up to 30 minutes, and the electrolyte was continuously agitated throughout by a magnetic stirrer.

The morphology and chemistry of the electrodeposited multilayer nanowires were analysed using a scanning electron microscope (SEM) (JEOL 6500F, Japan), and a high-resolution transmission electron microscope (TEM) (JEOL 2010F, Japan) equipped with energy-dispersive X-ray analysis (EDX, Oxford Instrument INCA). SEM samples were prepared by dissolving the AAO templates using 0.2M sodium hydroxide solution and rinsing with distilled water, and by dissolving the PC template and rising with chloroform (Fisher Scientific, 99.9+%). Nanowires completely dissolved out of the templates were dispersed on holey carbon film coated grids for TEM and EDX analysis and on a SiO_2(100nm)/Si wafer for nanomanipulation, nanoassembly, electrical nanowelding and testing. The electrical nanowelding and testing was carried out using two in-situ SEM nanomanipulators (Kleindiek Nanotechnik, Germany) equipped with a Keithley 6487 picoampermeter. The data acquisition was automated using a dedicated multi-functional programme.

3. Nanowire morphology

Figures 1(a) – (c) show three representative TEM images of gold nanowires, $(Co_{72}Pt_{28}/Pt)_n$ multilayer nanowires, and nanosolder $Sn_{99}Au_1$ alloy nanowires from which the AAO and PC templates have been completely dissolved. The gold nanowires are uniform in diameter and length, indicating a good quality deposition. The average size of these gold nanowires is about 55nm in diameter and 10 μm in length. Selected-area electron diffraction (SAED) analysis confirms that the gold nanowires are polycrystalline with the fcc structure.

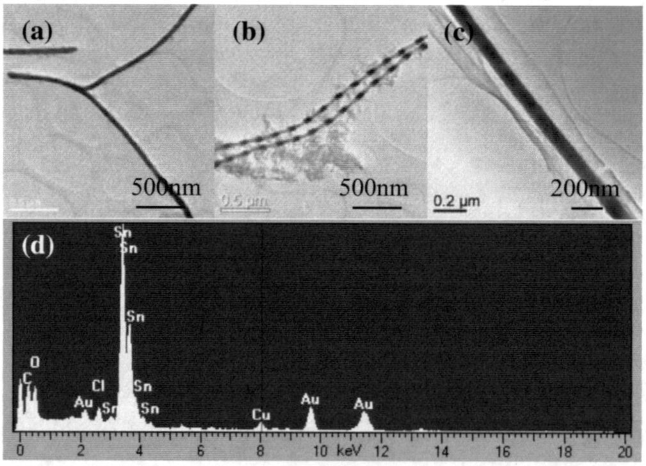

Figure 1. TEM images of (a) gold nanowires, (b) $(Co_{72}Pt_{28}/Pt)_n$ multilayer nanowires and (c) nanosolder $Sn_{99}Au_1$ alloy nanowires. (d) EDX spectrum of $Sn_{99}Au_1$ alloy nanowires.

Figure 1(b) clearly shows that each individual multilayer nanowire has a regular periodic structure like bamboo. The dark contrast layers are Pt (significant electron scattering) and the grey contrast layers mainly cobalt (lower electron scattering).The Pt layers grew to the full width of the AAO template, however the cobalt-rich layers in each nanowire did not fully fill the template pores during growth, resulting in a repeatably oscillating nanowire diameter. The chemical composition of multilayer nanowires was measured to be $(Co_{72}Pt_{28}/Pt)_n$ by EDX. Both the $Co_{72}Pt_{28}$ and Pt layers were polycrystalline, consisting of fcc nanocrystals <5nm in size. More detailed microstructure and chemical characterisation of both building block nanowires have been reported in our previous work [2, 3, 7].

The nanosolder Sn_xAu_y alloy nanowires are found to be uniform in length. The top ends of all Sn_xAu_y alloy nanowires are about 80nm, consistent with the commercially quoted PC template pore diameter, whilst the bottom diameter is about 150nm. This reveals that the channels of the commercial

PC templates have a cone shape rather than cylinder. Figure 1(d) shows a typical EDX spectrum of the $Sn_{99}Au_1$ nanowires, revealing that Sn and Au have been successfully co-electrodeposited. Cu and C (and minor Cr and Fe peaks) come from carbon coated copper grids, which was confirmed by EDX of an empty carbon coated copper grid. The O peak indicates that the surface of Sn_xAu_y alloy nanowires has been partly oxidised after they have been released from PC templates for two weeks. Quantitative analysis of EDX spectra indicates a 99:1 atomic ratio of Sn:Au, inferring a $Sn_{99}Au_1$ composition. SAED and TEM characterisation shows that individual 80nm $Sn_{99}Au_1$ alloy nanowires are polycrystalline with tetragonal βSn structure.

4. The bottom-up assembly of nanostructures

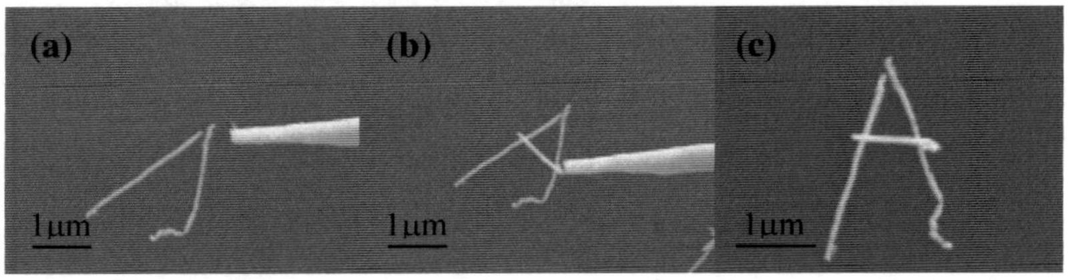

Figure 2. Bottom-up assembly of nanostructures using gold nanowires building blocks. (a) placement of gold nanowires; (b) assembly of the third nanowire; (c) complete assembly of letter "A".

The gold nanowires, $(Co_{72}Pt_{28}/Pt)_n$ multilayer nanowires, and nanosolder nanowires were dispersed on a SiO_2(100nm)/Si wafer for the bottom-up assembly of nanoconstruction. Figure 2 shows an example in which individual gold nanowires were assembled into a complex pattern by mechanical manipulation using nanomanipulators within a SEM. First the nanowires were released from the strong bonding to the substrate wafer by a mild mechanical push from a SEM nanomanipulator. The chosen nanowire is then lifted from the substrate using Van der Waals force, followed by precise release at the desired location adjacent to another nanowire (Figure 2(b)). The whole SEM mechanical manipulation process is clean, simple, quick and reliable. It also avoids the increased time, expense and contamination of methods involving nano-object pick-up via e-beam and ion-beam gas deposition [3, 8]. By means of this clean manipulation method, complex structures, like letter "A" (Figure 2(c)) and heterojunctions composed of dissimilar building blocks, can be easily assembled.

5. Electrical nanowelding of the assembled nanostructure

Once nanowire building blocks have been positioned into a required design, the goal is to securely join them together ensuring mechanically strong and electrically conducting junctions. Electrical currents can be directly and controllably applied to individual nanosolder nanowires, and through sections of nanowire circuits, using two SEM nanomanipulators (Figure 3(b) and Figure 3(e)). High quality welded nanojunctions can be achieved using nanosolder material. The solder can be the same, or different to the materials being welded. Figure 3 shows examples using similar nanosolder (gold nanowires) and dissimilar nanosolder ($Sn_{99}Au_1$ alloy nanowires). It is found that Joule heating of a sacrificial nanowire can dramatically improve the controllability, reproducibility and reliability of the nanowelding procedure. Explicitly, a ~ 70% maximum current density J_{max} ($4.94\times10^8 A\cdot cm^{-2}$) is applied to the sacrificial nanosolder for 1~2 minutes before a weld process. The high current density (typically about $3.5\times10^8 A\cdot cm^{-2}$ or 8mA) causes a softening of the solder nanowire due to Joule-heating, and induces the first join to the patterned nanowires (Figure 3(b) and 3(e)). In some cases, if the solder nanowire remains attached to the junction after welding (Figure 3(f)), mechanical force can be applied to separate them by pulling the probe away. Furthermore, if any unwanted residues of solder nanoparticles and rods exist, which are conductive and therefore dangerous and potential short-circuit sites for real nanoelectronics devices, the nanomanipulators can easily and precisely mechanically remove the welding residues.

Once formed the welds joining two nanowires are strong and very conductive. An immediate check of nanoweld quality can be done *in-situ* directly after welding which is essential for practical nanoconstruction. Nanoscale weld resistances of just 20Ω are achieved by using $Sn_{99}Au_1$ alloy solder [1].

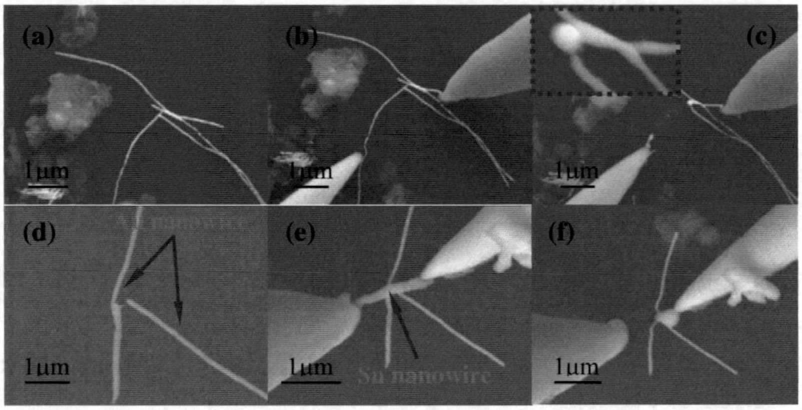

Figure 3. Electrical nanowelding of the assembled nanostructures. (a) - (c) Nanowelding nanowire structures using similar nanosolder, using 55nm Au nanowires as an example: (a) Placement of nanowires; (b) Pre-weld softening of the solder nanowire. The current flows from the right tip to the left tip via the nanowire; (c) Nanowelding nanowires together by a rectangular voltage pulse through the solder nanowire, inset showing a magnified view of the welded nanojuction. (d) – (f) Welding Chinese character "人" using a sacrificial 200nm diameter $Sn_{99}Au_1$ solder nanowire.

6. Summary

In conclusion, a high spatial resolution nanowelding method has been presented in this work, which provides a controllable and reliable approach for manipulating, assembling and welding <100nm diameter metallic nanowire building blocks into complex structures by means of SEM nanomanipulators. The key advance is to avoid detrimental current flow though the nanoobjects to be joined, and instead to locally deposit nanoscale volumes of a chosen metal at the weld site by Joule heating a sacrificial nanowire. Precise engineering of nanowelds by this technique, including the chemical flexibility of the nanowire solder, and high spatial resolution of the nanowelding method, should result in research applications including fabrication of nanosensors and nanoelectronics constructed from a small number of nanoobjects, and repair of interconnects and failed nanoscale electronics. This technique provides a promising way to conquer the challenge of the integration of nanocomponents for bottom-up nanotechnology.

Acknowledgement

This work was supported by grant GR/585689 and from the EPSRC, UK.

References

[1] Peng Y, Cullis A G and Inkson B J 2009 *Nano Lett.* **9** 91
[2] Peng Y, Cullis A G and Inkson B J 2008 *Appl. Phys. Lett.* **93** 183112
[3] Peng Y, Luxmoore I, Forster M D, Cullis A G and Inkson B J 2007 J. Phys. Conf. Ser. **126** 012031
[4] Cui Y and Lieber C 2001 *Science* **291** 851
[5] Dong L, Tao X, Zhang L, Zhang X and Nelson B 2007 *Nano Lett.***7** 58
[6] Cumings J and Zettl A 2000 *Science* **289** 602
[7] Peng Y, Cullis A G, Moebus G, Xu X J and Inkson B J 2007 *Nanotechnology* **18** 485704
[8] Matsui S, Kaito T, Fujita J, Komuro M, Kanda K, Haruyama Y 2000 *J. Vac. Sci. Technol. B* **18**, 3181

Dark field transmission electron microscopy techniques for structural characterization of nanowire heterostructures

Antonius T. J. van Helvoort§*, Dasa L. Dheeraj†, Hailong Zhou†, Sondre Grønsberg§, Gilles Patriarche‡, Bjørn-Ove Fimland† and Helge Weman†

§Department of Physics, Norwegian University of Science and Technology, NO-7491 Trondheim, Norway

†Department of Electronics and Telecommunications, Norwegian University of Science and Technology, NO-7491 Trondheim, Norway

‡CNRS-Laboratoire de Photonique et Nanostructures (LPN), Route de Nozay, F-91460 Marcoussis, France

* E-mail: a.helvoort@ntnu.no

Abstract. Synthesis and application of semiconductor nanowires (NWs) require detailed understanding of their structures. Transmission electron microscopy (TEM) has proven to be the most used structural characterisation tool for these nanostructures. For the characterisation of more complex heterostructured NWs, in this study axial (i.e. with "insert") or radial (i.e. "core-shell") III-V NWs, we use dark field TEM and high angle annular dark field scanning TEM (HAADF-STEM) as imaging modes. For the GaAs/GaAsSb/GaAs axial heterostructured NWs, dark field TEM exploits the structural difference between wurtzite GaAs NW and zinc blende GaAsSb insert, resulting in an excellent contrast between the GaAs part and the GaAsSb insert of the NW. In addition, twins, stacking faults and polytypes of the GaAs phase can easily be identified. For the radial GaAs/AlGaAs core-shell NWs where both core and shell have the wurtzite structure, HAADF-STEM is found to be a very useful technique.

1. Introduction

There has been significant progress in fabricating semiconductor nanowire (NW) structures for nano-photonic and nano-electronic devices. III-V NWs have attracted much attention as these have very interesting opto-electric properties. Recently, it has been shown that the band-alignment within a NW can be tailored by varying its crystal phase [1] as the bandgap energy and band offsets are different for different crystal phases [2].

III-V NWs can be grown by a range of different chemical and physical deposition techniques. Molecular beam epitaxy (MBE) has an advantage over other techniques due to the ability to control growth of materials with near-atomic precision and the flexibility for the growth of complex device structures like quantum cascade lasers. Using MBE, NW heterostructures can be synthesised in which the material composition is changed either in axial or radial direction. In addition, the crystal phase in these NWs can be changed from wurtzite (WZ) to zinc blende (ZB) via partially substituting the group V-element. This design opportunity raises the need for routine techniques to characterise the NW heterostructures to optimise the growth and improve the understanding of the measured properties.

Conventional transmission electron microscopy (TEM) techniques, mainly bright field (BF), high resolution TEM (HRTEM) and electron diffraction, are the most applied structural characterisation tools for these NWs. These techniques offer the required resolution to determine the NW-orientation and lattice defects such as stacking faults and twins. However, BF TEM has a low contrast and

HRTEM of a small area is not necessarily representative for the whole NW. Dark field (DF) techniques are less often applied, despite having an inherent good contrast. DF TEM would directly reflect changes in the crystal structure at relatively low magnification. DF scanning TEM, here high angle annular dark field (HAADF-STEM), would be attractive if there is an atomic number difference between different elements of a heterostructures (i.e. Z-contrast).

To determine for which NW heterostructures DF techniques would be advantageous, we compared DF TEM and HAADF-STEM to conventional TEM techniques. Here, two different III-V NW heterostructures are studied: axial GaAs(WZ)/GaAsSb(ZB)/GaAs(WZ) (NWs with "insert") and radial GaAs(WZ)/AlGaAs(WZ) ("core-shell" NWs).

2. Experimental

The GaAs/GaAsSb/GaAs axial NWs were grown in a Varian Gen II Modular MBE system equipped with a Veeco Ga dual filament cell, and Veeco Sb and As valved cracker cells. An electron-beam evaporation system was used for depositing an approximately 1 nm thick Au film on a GaAs buffer layer previously grown on a GaAs(111)B substrate. The sample was then re-loaded into the MBE system for the NW growth. An As_4 flux of 6×10^{-6} Torr and a substrate temperature of 540 °C were used for the GaAs NW growth. The GaAsSb insert was grown after 20 min of GaAs NW growth, followed by 5 min of GaAs NW growth.

The GaAs/AlGaAs core-shell NWs were grown on GaAs(111)B substrates by a Riber 32 MBE system, equipped with solid sources for Ga and Al and an As cracker source. A ~1 nm thick Au film was deposited under As_4 flux of 6×10^{-6} Torr at 550 °C using a Au effusion cell installed in the MBE growth chamber. The GaAs NW core was grown for 20 min, directly followed by the AlGaAs shell growth by opening the Al shutter. More growth details can be found elsewhere [3,4].

The NWs were scraped off from the substrate, dispersed in ethanol and transferred to a Cu grid with a lacey or full carbon film. The NWs were analyzed in a Philips CM30 TEM (BF/DF TEM, recorded on film or a 1K Tietz CCD), a JEOL 2010F (S)TEM (HAADF-STEM, 0.2 nm or 1 nm probe size, Cs =1 mm) and a JEOL 2200FS with Cs probe corrector (HAADF-STEM, 0.1 nm probe size), all operating at 200 kV. HRTEM and electron diffraction results are given elsewhere [3,4].

3. Results and discussion

Fig. 1 shows BF and DF TEM images of a GaAs/GaAsSb/GaAs axial heterostructured NW. The structure consists of five different regions: a defect-free WZ GaAs at the base, a ZB GaAsSb insert, a small ZB region twinned relative to the insert, a region with GaAs 4H polytype, and WZ GaAs with stacking faults (see Fig. 1(e)). These regions were observed in all studied NWs and the phases were confirmed by HRTEM and electron diffraction. In the BF (Fig. 1(a)) these regions are not distinguishable, whereas in the DF the WZ and ZB regions can be visualised selectively. From the DF series, the structural variation along the NW can be deduced easily (Fig. 1(b)-(e)). After growth of the GaAsSb insert, the GaAs WZ is not stabilised directly as it is for the growth of the GaAs WZ base [3]. This results in the formation of GaAs 4H polytype region and GaAs WZ with stacking faults. In all these NWs studied by DF TEM, a twin at the top of the insert was found (Fig. 1). Several NWs had also a twin at the lower boundary between GaAs WZ barrier and GaAsSb ZB insert [3]. In addition, DF TEM is an interesting technique to observe multiple inserts in NWs.

The good contrast of DF TEM was used to study the effect of Sb flux on the formation of twins in GaAsSb ZB phase. The tendency for twinning of the ZB structure is reduced with increasing Sb flux. The average twin length for the GaAsSb phase with a ~80 nm diameter is 12 nm for an Sb flux of 2×10^{-7} Torr and 38 nm for 4×10^{-7} Torr. GaAsSb phase grown with an Sb flux of 8×10^{-7} Torr had only a few twin boundaries along their total length (1-1.5 μm). The insert depicted in Fig. 1 is grown with an Sb flux of 6×10^{-7} Torr. Slightly off-zone BF TEM images can have a clear contrast between the two different ZB orientations, but still less than in DF TEM. An uneven illumination and especially NW bending can cause contrast variation not associated with the twinning. This last problem can be avoided by using a full carbon film instead of a holey or lacey carbon film, as this reduces the bending of NWs.

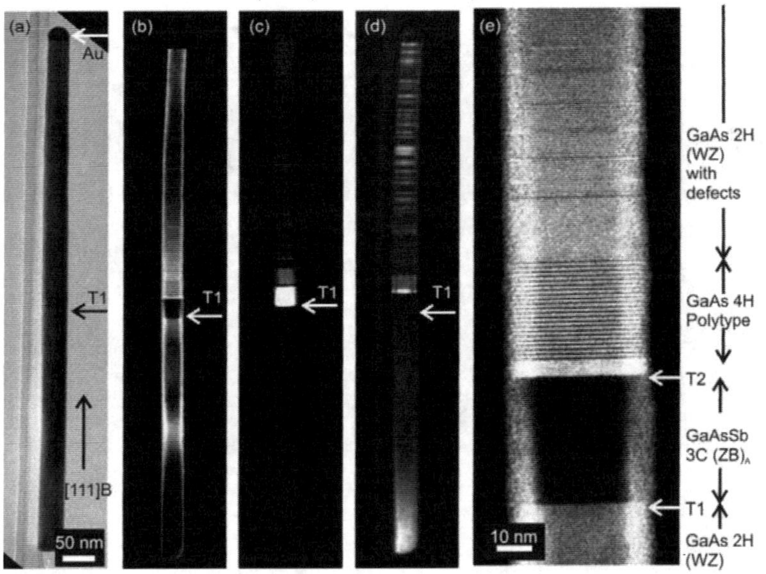

Fig. 1 GaAs NW (WZ) with GaAsSb insert (ZB): (a) BF, (b) DF-$\bar{1}100_{WZ}$, (c) DF-$11\bar{1}$ and (d) DF-$\bar{1}1\bar{1}_{ZB,B}$ and (e) detail of (b) depicting WZ(2H), ZB_A(3C), ZB_B (3C, twin), 4H polytype and stacking faults in the upper part of the WZ(2H) NW. T1 and T2 marks the lower and upper GaAs-GaAsSb interface, respectively. Images recorded on film and scanned-in at 600 dpi (a-d) or 2400 dpi (e).

These axial NW heterostructures are also studied by HAADF-STEM. The change from WZ to ZB is induced by incorporating Sb on some As lattice positions. However, Z-contrast is not the dominant contrast between the GaAs NW barrier and the GaAsSb NW insert (Fig. 2(a)). The main contrast is caused by the difference in the crystal lattice. The channeling conditions of the accelerated electrons are different for GaAs WZ and GaAsSb ZB regions. The specimen thickness is constant and the contrast is not the result of a misorientation or strain effects at the lower interface. Z-contrast can be present within one crystal phase. For example in Fig. 2(a) a strong variation in Z-contrast can be seen. The Sb content is increasing from the lower GaAs/GaAsSb boundary over several nm until it reaches a fixed value. Towards the top GaAs/GaAsSb boundary, the Sb concentration seems to be decreasing, which could cause the GaAsSb ZB crystal phase to form a twin (similar as at T2 in Fig. 1(e)). In the HAADF-STEM of Fig. 2(a) a twin in the insert is distinguishable due to a slight misorientation at T2.

The HAADF contrast and resolution can be improved by using a smaller probe in a Cs-corrected STEM. An example from an other batch of axially heterostructured NWs is shown in Fig. 2(b). These wires were grown under a lower Sb flux (2×10^{-7} Torr), hence the twinning frequency is higher. In the lattice image the atomically flat interface between WZ GaAs NW barrier and ZB GaAsSb NW insert, and the twin boundaries in the GaAsSb insert, can easily be spotted. In addition, variations in intensity of the atom columns are present due to Sb incorporation, but these need to be further analysed. Cs-corrected HAADF-STEM might be the most powerful imaging and analytical TEM technique to visualise the variation in composition and crystal phase of NWs. However, as in HRTEM, the field of view is limited, the NWs have to be oriented precisely on zone, and due to costs, Cs-corrected STEM is not a viable routine characterisation technique for optimising heterostructure NW growth.

In radial GaAs/AlGaAs core-shell heterostructures, the AlGaAs shell copies the wurtzite crystal phase of the GaAs core, hence conventional DF TEM will not distinguish core from shell. BF TEM of randomly oriented NWs (Fig. 3(a)), depicts strong diffraction contrast effects, for example caused by variation in NW orientation or stacking faults. HRTEM depicts a weak and diffuse contrast between core and shell [4]. However, in HAADF-STEM (Fig. 3(b,c)) the core-shell structure can easily be recognised, for example in Fig. 3(b) comparing wire #1 (no core-shell in this part of the NW) and wire #2 (core-shell). The core-shell structure can also be seen in wires #3 and #4, even when their relative orientation is different. This is an important remark, as tilting very thin NWs on zone for BF, DF and HRTEM imaging is very tedious. The here measured NW core diameter is consistent with the diameter of NWs before shell growth as measured by scanning electron microscopy (SEM) and TEM.

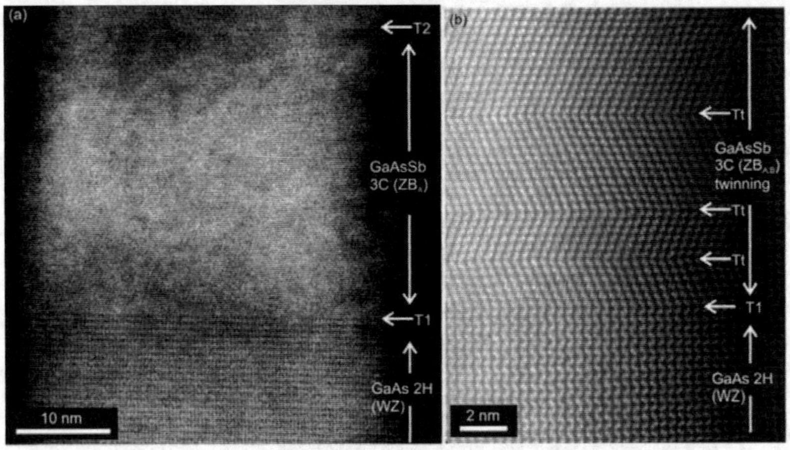

Fig. 2 HAADF-STEM lattice image of the lower WZ GaAs / ZB GaAsSb NW boundary (T1): (a) non Cs-corrected (0.2 nm probe) and (b) Cs probe-corrected STEM at higher magnification than in (a). Tt marks a twin boundary. The Sb flux was in (a) 6×10^{-7} Torr, in (b) 2×10^{-7} Torr.

SEM studies showed that during the growth of the AlGaAs shell, there is also substantial axial growth beside radial shell growth [4]. Hence, the GaAs NW core ends somewhere within the whole core-shell NW. From these HAADF images, these ends can directly be found (e.g. in wire #4 just left of the crossing with wire #3 no core-shell contrast is observed, whereas to the right of this crossing, a core-shell contrast is clearly seen). At this GaAs NW core end point, a band of stacking faults is formed, which is observed best in BF TEM and HRTEM. In HAADF, diffraction contrast from this band with stacking faults is weakly visible and depends on the relative orientation of the NW.

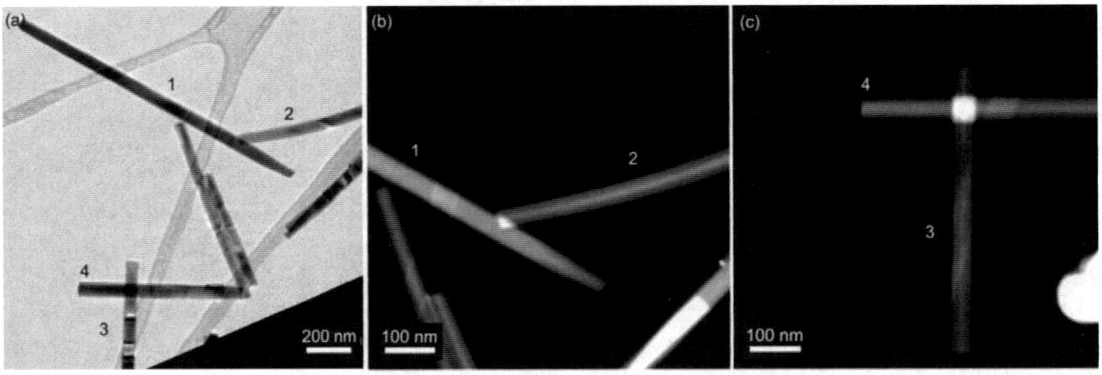

Fig. 3 Randomly oriented GaAs/AlGaAs core-shell heterostructured NWs (a) BF TEM and (b-c) HAADF STEM (1 nm probe, Cs = 1 mm). Numbers refers to specific NWs discussed in the text.

4. Conclusions

Heterostructured, axial (with "insert") or radial ("core-shell"), III-V semiconductor NWs have been studied by dark field TEM and HAADF-STEM. Dark field imaging proves to be a very powerful technique to enlighten the structural difference between the elements of the heterostructured NW. This is demonstrated for GaAs(WZ)/ GaAsSb(ZB)/ GaAs(WZ) axial heterostructured NWs grown by Au-assisted MBE. In addition, different crystal configurations in NWs, such as twins, stacking faults and GaAs 4H polytype regions, could be visualised by dark field TEM imaging, even at relatively low magnification. For radial GaAs/AlGaAs core-shell NWs, both core and shell have the WZ crystal phase. For these NWs, HAADF-STEM has proven to be a more practical characterisation technique than bright or dark field TEM.

References

[1] Mårtensson T, Trägårdh J and Samuelson L 2008 *Nano Lett.* **8** 836
[2] Hoang T B, Moses A F, Zhou H L, Dheeraj D L, Fimland B O and Weman H 2009 *Appl. Phys. Lett.* **94** 133105
[3] Dheeraj D L, Patriarche G, Largeau L, Zhou H L, Hoang T B, Moses A F, Grønsberg S, van Helvoort A T J, Fimland B O and Weman H 2008 *Nano Lett.* **8** 4459
[4] Zhou H L, Hoang T B, Dheeraj D L, van Helvoort A T J, Liu L, Harmand J C, Fimland B O and Weman H, 2009 *Nanotechnology* **20** 415701

Journal of Physics: Conference Series **241** (2010) 012085 doi:10.1088/1742-6596/241/1/012085

Low-temperature cathodoluminescence studies of GaAs nanowires in the SEM

Anders Gustafsson[1], Niklas Sköld[1,2], Jessica Bolinsson[1], Johanna Trägårdh[1,3] and Lars Samuelson[1]

[1]Solid State Physics and the Nanometer Structure Consortium (nmC), Lund University, Box 118, S-221 00 Lund, Sweden.

[2]Toshiba Research Europe Ltd, 208 Cambridge Science Park, Cambridge CB4 0GZ, UK

[3]H.H. Wills physics laboratory, University of Bristol, Tyndall Avenue, Bristol BS8 1TL, UK

Anders.Gustafsson@ftf.lth.se

Abstract. We present data from cathodoluminescence (CL) studies of nanowires (NWs) grown from size-selected gold particles using metal-organic chemical vapour deposition. The NWs in this study all have a GaAs core and are surrounded by a shell of either AlGaAs or GaInP. The spatially and spectrally resolved CL studies were performed in a scanning electron microscope, in a temperature range of 4-300 K. The NWs were studied as-grown on their substrates in side view or transferred to Si substrates. We present data from single NWs, showing that the homogeneity of the emission is influenced by the growth conditions. A general observation is that the emission is much stronger, and more well-defined from the upper part of the NW. The shell emission varies in emission energy and spatial origin in an irregular fashion, probably linked to compositional variations in the shell.

1. Introduction

Over the past few decades, there has been a quest for the perfect system for fabricating high-quality quantum wires, or nanowires (NWs). Many approaches have been tried with varying successes, for examples see e.g. reference [1]. Semiconductor NWs seeded by metal particles have attracted much attention lately. In this study we present cathodoluminescence (CL) data from NWs grown by a technique introduced by Wagner [2] that has seen a revival over the past few years, due to improvements in the fabrication steps of the techniques [3].

The NWs in this study were grown with gold as the seed particles. The growth takes place at the interface between the metal particle and the semiconductor, taking its epitaxial information from the substrate. The growth conditions were chosen so that growth on the substrate surfaces is kinetically hindered and very limited, whereas material is precipitated at the interface between the seed particles and the semiconductor. The size of the seed particle determines the area of growth of the semiconductor. This results in the growth of pillars with a very high aspect ratio. With a random size of the particles, the diameter of the pillars is also random. However, with a well-controlled size of the particles, the resulting pillars have an identical diameter and identical height.

2. Experimental details

The gold particles used as seeds in this study were produced by an aerosol technique and size-selected, as described in detail elsewhere [4]. In this study, we concentrate on NWs grown from gold particles with a diameter of either 40 or 50 nm. Under the present growth conditions in our metal-organic chemical vapour deposition (MOCVD) chamber, the preferred NW growth direction is <111> [5] and we used (111)B-oriented substrates. The NWs were grown to a length of 1 - 6 μm. To avoid leaving the GaAs side-surfaces exposed, the NWs were covered by a layer of AlGaAs (or GaInP). This shell was grown at higher temperatures, where growth takes place without the aid of metal particles. However, there is still particle-mediated growth, giving an AlGaAs core at the top of the NW. This type of core-shell structure improves the light emission from the NWs by 2-3 orders of magnitude in resonant (excitation below the bandgap of the shell) photoluminescence studies [6]. During CL experiments, the emission has a tendency to fade permanently with time unless the shell is present. Depending on the sample, the shell thickness ranges from 20 nm to 55 nm in the present study.

The CL studies were performed in a dedicated scanning electron microscope (SEM) with a lHe cold stage. The NWs were studied as grown, still attached to the substrates, in side view, or broken off and transferred to Si substrates to avoid background emission from the GaAs substrate. The temperature was typically 6-10 K. The studies were performed at an acceleration voltage of 2.5-10 keV and a probe current of 10-100 pA. The choice of the acceleration voltage was based on two criteria, (1) The acceleration voltage must be high enough to penetrate the shell to the core and (2) The acceleration voltage must be low enough to keep most of the primary excitation in the NW.

3. Results

Figure 1 (a) shows an SEM image of a typical sample in the present study. There is a slight tapering towards the tip of the NWs, and at the bottom there is a "foot" caused by different growth facets developing around the base of the wire. Figure 1 (b) shows a typical low-temperature CL spectrum of several GaAs NWs. The 1.52 eV peak is related to excitonic emission in the GaAs core and the emission at 1.49 eV is related to carbon impurities in the core. In addition, emission in the range 1.42-1.50 eV can be caused by the mixed nature of the crystal structure, zincblende (ZB) and wurtzite (WZ) in combination with a type II band alignment, similar to what has been observed in InP NWs [7]. At the interface between the two crystal structures, spatially indirect transitions will result in an emission well below the energy of the lower bandgap. This is the ZB GaAs, as the WZ bandgap is higher [8]. If the segments are short, states might be quantized in the growth direction [9] and the energy of the transition is therefore higher than the minimum. It is worth pointing out that the cores of the present NWs are too large to exhibit any quantum confinement perpendicular to the growth direction.

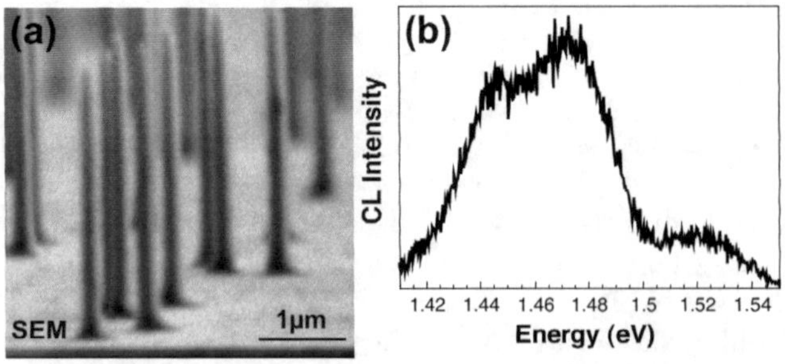

Figure 1. a) SEM image of typical NWs in side view. The dark line at the bottom of the image is the edge of the cleaved substrate. b) Typical spectrum of a bunch of NWs showing excitonic emission around 1.52 eV and defect related emission at lower energies.

Figure 2 shows a series of images from the same area of a typical sample, dominated by carbon-related emission. The CL images are recorded at different energies, as indicated on each image. The images have the intensity expanded to show the full intensity variation. For all images presented here, black means no emission. The main emission is much more prominent from the top of the NWs than the base, as shown in (b). The same pattern can be found for most of the NWs. Emission at lower energies, (c), is weaker and appears to originate further down the wire. Both these emissions come from the GaAs core. The emission from AlGaAs shell, (d), exhibits a spotty emission pattern without any trends. When varying the detection energy, the bright spots move around on the NWs in a random fashion, reflecting a local variation in the Al content. All the CL images of the GaAs emission lack emission from the top part of the NWs as the core in this part is AlGaAs, grown during the shell growth. The emission patterns are similar when the NWs are removed and transferred to non-emitting Si substrates.

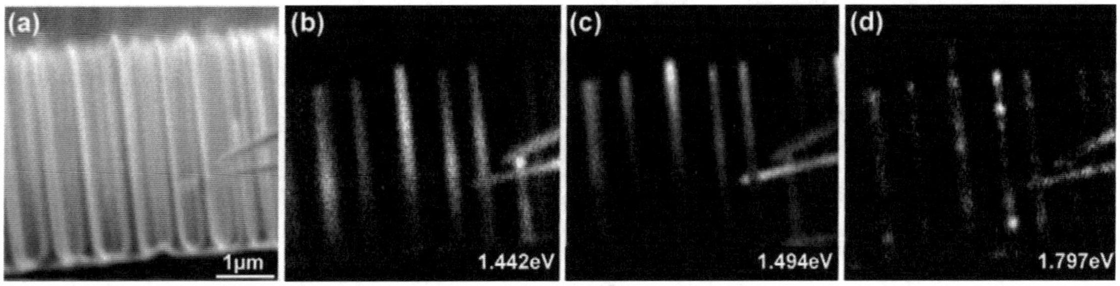

Figure 2. a) SEM image and the corresponding monochromatic CL images (b - d) of NWs with AlGaAs shells. The emission in (b) and (c) are from the GaAs core and (d) is from the AlGaAs shell. Especially (c) shows an increase in the intensity along the NWs. (d) shows a number of random emission spots along the NWs.

The increasing emission intensity towards the top is intriguing, and to understand this we have studied NWs of varying lengths in the same growth series. The emission from these NWs is dominated by excitonic emission. CL images from two samples in this series are shown in figure 3. These two sets of NWs are 3.5 μm and 5.5 μm long. They show similar trends in the emission pattern, where the first part is significantly weaker than the rest of the NW. In the case of the longer NWs, the upper micron even appears to be weaker in the image presented here. This effect can be reduced or enhanced by altering the scan speed and direction. One reason is that the long wires tend to vibrate under the electron beam leading to a lower excitation density. As the vibrations are more severe towards the tip, this leads to an apparent reduction in the emission intensity. Though not shown here, the emission from short NWs, 1-1.5 μm long, is very weak. This fits to the pattern of weak emission from the lower part of the NWs, irrespective of length.

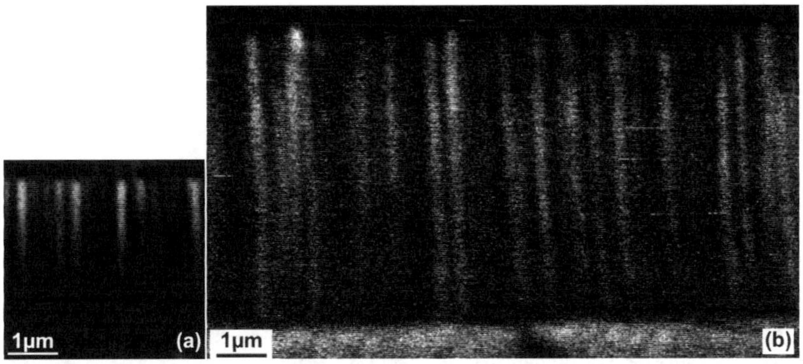

Figure 3. CL images of a) 3.5 μm-long NWs and b) 5.5 μm-long NWs recoded at 1.51 eV. Both images show the same increase in the intensity from the substrate over the first 2-3 μm of the NWs. The longer NWs reach a constant intensity from about 3 μm.

Studying the increasing emission intensity along the NWs is an ongoing project. GaAs NWs can also be grown on other substrates than GaAs, like low-cost Si substrates. We have also used GaP substrates, usually in connection with GaInP shells. The GaInP shells are used as the strain induced by the shell can tune the emission energy over a large energy range. They are grown in a similar way, with the difference that the growth is initiated with a short GaP segment. These NWs exhibit a different intensity pattern, figure 4. The main emission shows a decreasing intensity from base to tip, apart from the first 1-200 nm, which is GaP and therefore dark in the CL image of (b). In this case, the NWs are more tapered than the ones grown on GaAs substrates. The decreasing intensity is simply a volume effect where the diameter of the NW decreases and therefore the intensity decreases.

Figure 4. a) SEM image of NWs grown on a GaP substrate, and b) the corresponding CL image recorded at 1.51 eV. The dashed line indicates the position of the substrate. The first part of each NW is dark in the CL image as this part is a GaP seeding segment. The intensity of these NWs is more homogeneous than those grown on GaAs substrates.

4. Discussion

The images presented here show that it is possible to study the emission from single NWs and to study variations along them as well as using spatially and spectrally resolved CL. Though not shown here, it is possible to record spot-mode spectra along the NWs in order to study variations in the emission. Line scans of NWs with AlGaAs shells confirm that the emission from the GaAs increases from base to tip. It also confirms the trend in figure 2 (b) and (c) that the peak position of the emission shifts towards higher energy along the NW. The intensity trend is the opposite of the growth on GaP substrates, where the intensity decreases along the NW. At the same time, the peak position is unchanged. The fact that the emission is strong already from the start shows that the effect is not intrinsic to the GaAs NW growth from gold particles but rather an effect of the substrate. We also have some preliminary indications that the surface preparation prior to growth can play a significant role in the emission intensity from the lower part of the NWs.

Acknowledgements

The work has been performed with funding from, mainly, the Nanometer Structure Consortium in Lund, with support from national (e.g. VR, TFR, NFR and NUTEK), KAW and EU programs.

References

[1] Gustafsson A, Pistol M-E, Samuelson L and Montelius L 1998 *J. Applied Physics* **84** 1715.
[2] Wagner RS 1970 Whisker Technology (New York, Wiley) p 47.
[3] Ohlsson BJ, Björk MT, Magnusson MH, Deppert K and Samuelson L and Wallenberg LR 2001 *Appl. Phys. Lett.* **79**, 3335.
[4] Magnusson MH, Deppert K, Malm JO, Bovin JO and Samuelson L 1999 *J. Nanopart. Res.* **1** 243.
[5] Seifert W, Borgström M, Deppert K, Dick KA, Johansson J, Larsson MW, Mårtensson T, Sköld N, Svensson CPT, Wacaser BA, Wallenberg LR and Samuelson L 2004 J. *Crystal Growth* **272** 211.
[6] Sköld N, Karlsson LS, Larsson MW, Pistol M-E, Seifert W, Trägårdh J and Samuelson L 2005 *Nanoletters* **5** 1943.
[7] Bao J, Bell DC, Capasso F, Wagner JB, Mårtensson T, Trägårdh J and Samuelson L 2008 *Nanoletters* **8** 836.
[8] Hoang TB, Moses AF, Zhou HL, Dheeraj, DL, Fimland B and Weman H 2009 *Appl. Phys. Lett.* **94** 133105.
[9] Spirkoska D, Arbiol J, Gustafsson A, Conesa-Boj S, Glas F, Zardo I, Heigoldt M, Gass MH, Bleloch AL, Estrade S, Kaniber M, Rossler J, Peiro F, Morante JR, Abstreiter G, Samuelson L and Fontcuberta i Morral A 2009 *Phys. Rev.* B **80** 245325.

Truncated-octahedral copper-gold nanoparticles

Dung T Tran[1], Ian P Jones[1], Roy L Johnston[2], Jon A Preece[2] and Coenraad R van den Brom[2]

[1]School of Metallurgy and Materials, University of Birmingham, B15 2TT, Birmingham, UK. [2]School of Chemistry, University of Birmingham, B15 2TT, Birmingham, UK

E-mail: dtt711@bham.ac.uk

Abstract. Observations of the morphology and behaviour of chemically synthesized Cu-Au nanoparticles using high-resolution transmission electron microscopy are described. Truncated-octahedral (TO) particles, which are the result of a morphological evolution induced by the electron irradiation, are found to be the most stable. They have an fcc structure on a large size range of around 1-12 nm. Aggregation of two TO particles, to form a new TO particle, is observed directly under the electron beam. Modelling performed on 38-atom Cu-Au particles using an empirical-potential genetic algorithm also predicts the existence of fcc single-crystal TO structures.

1. Introduction

Among the various bimetallic nanoparticles which have potential applications in optics, catalysis and nano-electronics [1], Cu-Au nanoparticles have been studied by many research groups since the early 1990s [1-5].

Alloying between Cu and Au is not only well known for bulk systems [6] but also for nano-scale systems [2-5]. The dissolution of copper atoms in gold nanoparticles supported on an amorphous carbon film was observed using Transmission Electron Microscopy (TEM) [2]. Another study indicated fcc-like packing, similar to the bulk alloy phase, for 4 nm-sized stoichiometric $(Cu_3Au)_M$ (M is the number of Cu_3Au formula units) nanoparticles prepared by dual-source electron beam vaporization [3].

Different morphologies of Cu-Au nanoparticles have been observed. Decahedral and icosahedral Cu-Au alloy particles were produced by an electrochemical method and investigated by TEM [4]. In other work, decahedral and truncated octahedral morphologies were found for 1-4.5 nm-sized Cu-Au nanoparticles produced by laser vaporization of the bulk stoichiometric alloys CuAu, Cu_3Au, and $CuAu_3$ and then deposition on amorphous carbon and MgO substrates [5]. These nanoparticles all have fcc structures.

In this paper, we describe TEM observations of the morphology and the thermodynamic behaviour of Cu-Au nanoparticles, which are prepared using wet-chemical synthesis.

2. Experiment

Cu-Au nanoparticles were chemically synthesized by a two-phase reaction and galvanic exchange. This chemical method has previously been described in detail [7]. First, Cu nanoparticles capped with dodecanethiol ($C_{12}H_{25}SH$) are synthesized by $NaBH_4$ reduction of copper (II) chlorate in toluene/water and then mixed with an Au(I)-4-*tert*-butylbenzylthiolate solution. The bimetallic particles are formed by a galvanic exchange reaction between Cu and Au(I).

The particles are dispersed in toluene and then deposited on a carbon-coated copper grid for TEM investigation. Conventional TEM and High Resolution TEM (HRTEM) observations are performed on a Philips Tecnai F20 microscope operating at 200 kV. A large condenser aperture and an extraction voltage of 4200 V are used to observe the behaviour of the particles under an adequately high-density electron beam. The image processing software ImageJ 1.40g [8] is used to produce the size-distribution histogram from TEM images and to generate the Fast Fourier Transform patterns (FFT) from HRTEM images.

To study theoretically the structures of Cu-Au particles, modelling of 38-atom Cu-Au particles has been performed. A genetic algorithm code [9], incorporating the many-body Gupta empirical potential [10], is used to find the most stable structures, corresponding to global energy minima, of the particles.

3. Results and discussion

Figure 1 shows a conventional TEM image and the corresponding size-distribution histogram for the Cu-Au particles. The sizes are quite diverse, ranging from around 1 nm to around 12 nm. 2-6 nm sized particles are predominant. Although the capping thiol molecular layers prevent the particles from coalescence in solution [7], particle aggregation may still be observed in during electron microscopy, because the particle concentration is quite high and the capping layers may be vulnerable to the electron beam.

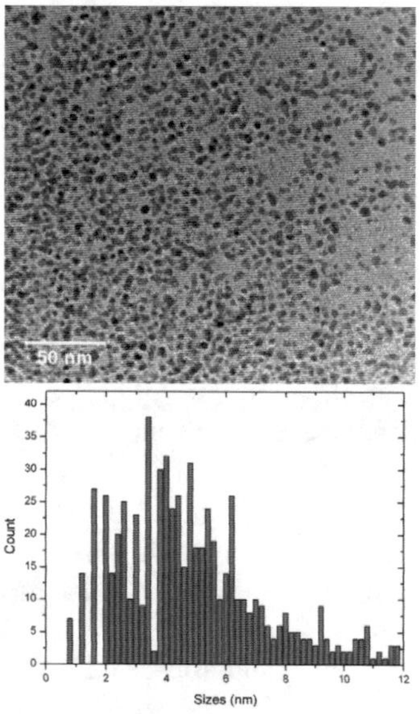

Figure 1. Conventional TEM image (upper part) and the corresponding size-distribution histogram (lower part) of the as-prepared CuAu-NPs. The sizes of the particles are their (nominal) diameters approximately measured by using the ImageJ software.

Figure 2. HRTEM images (upper part) in the [011] orientation and the corresponding FFT patterns (lower part): (a) a single-crystal particle which has the morphology of a truncated-octahedron; (b) a twinned particle which is still evolving under the electron beam, the diffraction spots of the twin are discriminated by white and grey indices.

Individual particles were investigated by HRTEM. Their morphologies changed under exposure to the high-density electron beam. Initially, the particles adopted some random indefinable shapes. Some of them seemed to exhibit circular-like or elliptic-like 2D-projections. After irradiating an individual particle with the electron beam for about 2 minutes (with the largest condenser aperture, spot size of 2, extraction voltage of 4200 V), the particle morphology begins to change. It takes approximately 15 minutes for a new stable morphology to be established. As observed, after the morphological evolution induced by the electron irradiation, the particles exhibit hexagonal 2D-projections (see Figure 2a, upper part). These hexagons are unchanged even when irradiated further with the high-density electron beam. In fact, structural rearrangements in gold nanoparticles induced by an electron beam have been reported previously [11]. In another previous work, Cu-Au nanoparticles, which were produced by laser vaporization and deposited on an amorphous carbon film, also exhibited hexagonal 2D-projections under HRTEM [5]. The stable 3D morphology of the particles is identified as a truncated-octahedron (TO) which is geometrically constructed by the uniform removal of six square pyramids from the six vertices of a regular octahedron. Figure 2b (the upper part) shows a larger particle which exhibits a distorted hexagon. This particle is twinned as it is the result of the aggregation of two smaller particles. The shape development of this particle is still in progress.

The morphological evolution of the Cu-Au particles can be interpreted thermodynamically. The initial particles, which are synthesized by chemistry, can be regarded as meta-stable aggregates, corresponding to local energy minima. When stimulated by the electron beam, these aggregates re-crystallize, which causes a transition from the meta-stable state to the thermodynamically preferred structure corresponding to a global energy minimum. Only the TO morphology is observed as the most stable morphology for the Cu-Au particles in this work. The smallest TO particles observed are around 2 nm and the largest TO particles observed are 20 nm. These TO particles are single-crystal fcc. The FFT patterns (see Figure 2a and 2b, lower parts) indicate an [011] orientation. The measurement of the d-spacing of the (111) family of planes gives an estimate of 3.78 Å for the lattice parameter of the two particles in Figure 2. This value is close to the lattice parameter of 3.76 Å for the stoichiometric $(Cu_3Au)_M$ nanoparticles in the previous work [5].

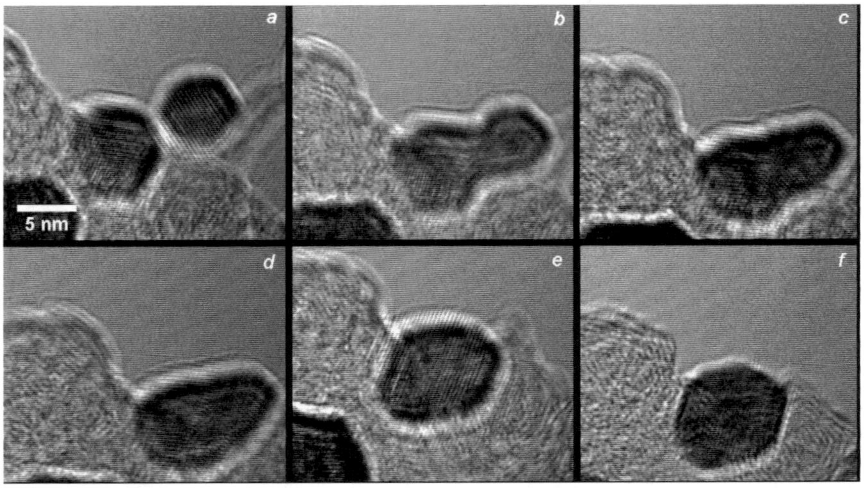

Figure 3. The aggregation of two TO particles and the formation of a new larger TO particle under the electron beam; the alphabetic order (a-f) corresponds to the time sequence; this process occurs for around 20 minutes.

Figure 3 shows an aggregation of two TO particles. Under the high-density electron beam, the particles behave like liquid drops. They unite with each other to minimize the surface energy of the system. The initial result is a twinned particle but the thermodynamic equilibrium state is eventually reached and a new single-crystal TO particle is formed. The overall process takes around 20 minutes. This confirms that the observed large TO particles (> 10 nm) can be the result of the aggregation of smaller particles when they are irradiated by the electron beam. Previously, the sintering of gold nanoparticles under electron irradiation has been observed [12].

As found from the modelling of $Cu_{38-n}Au_n$ nanoparticles, fcc-like TO structures are the most stable structures (global energy minima) for Cu-rich compositions. This has been partially verified (for n = 0-

8) by higher level Density Functional Theory (DFT) calculations [13]. Some of these structures are shown in Figure 4. Au atoms are found to occupy preferentially the centroids of the 8 hexagonal (111) facets of the TO and the TO is slightly distorted when the number of Au atoms is larger than 8. This is also supported by DFT calculations [13]. Relying on the above observed aggregation of two small TO particles (of around 5 nm) forming a larger TO particle, we would like to argue for a possible extrapolation so that the smaller particles, which are below 1 nm in size, might also have the TO morphology. This makes the modelled 38-atom Cu-Au nanoparticles, which are around 0.8 nm in size, relevant to our experiments.

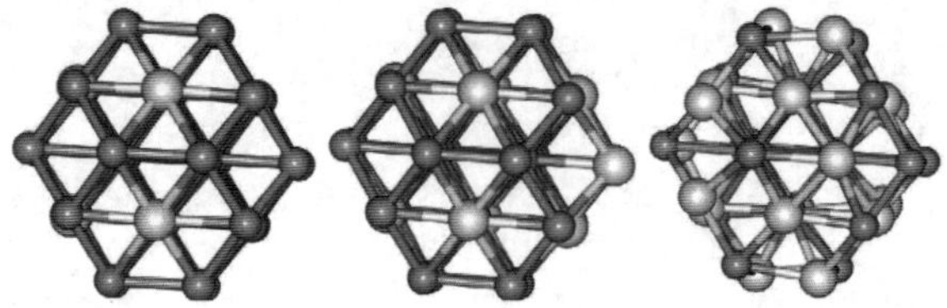

Figure 4. Some TO structures of $Cu_{38-n}Au_n$ nanoparticles found by the modelling; from left to right: $Cu_{34}Au_4$, $Cu_{29}Au_9$, and $Cu_{19}Au_{19}$; Au atoms are light grey, Cu atoms are dark grey.

4. Conclusions

Chemically synthesized Cu-Au nanoparticles have been investigated under TEM. The particle sizes are mainly around 2-6 nm. The initial particles are not stable under a high-density electron beam and morphological evolution is observed. The truncated-octahedron is found to be the morphology corresponding to the thermodynamic equilibrium state (global energy minimum) of the particles. TO particles have a single-crystal fcc structure. The aggregation of two TO particles, forming a larger TO particle, is observed under the electron beam. As observed, the TO morphology is predominant over a large range of sizes (approximately 1-12 nm) as the most stable for the Cu-Au nanoparticles. Modelling of 38-atom Cu-Au nanoparticles using a genetic algorithm also indicates a high stability of the fcc-like structures.

Acknowledgments

D.T.T's PhD studies are funded by the School of Metallurgy and Materials (University of Birmingham) and the Overseas Research Students Awards Scheme (UK).

References

[1] Ferrando R, Jellinek J and Johnston R L 2008 *Chem. Rev.* **108** 845.
[2] Yasuda H and Mori H 1994 *Z. Phys. D* **31** 131.
[3] Yasuda H and Mori H 1996 *Z. Phys. D* **37** 181.
[4] Lu D and Tanaka K 1997 *Phys. Rev. B* **55** 13865.
[5] Pauwels B, Tendeloo G V, Zhurkin E, Hou M, Verschoren G, Kuhn L T, Bouwen W and Lievens P 2001 *Phys. Rev. B* **63** 165406.
[6] Pearson W B (Ed) 1972 *Crystal Chemistry and Physics of Metals and Alloys* (Wiley: New York)
[7] Shon Y S, Dawson G B, Porter M and Murray R W 2002 *Langmuir* **18** 3880.
[8] Abramoff M D, Magelhaes P J and Ram S J 2004 *Biophotonics International* **11** 36.
[9] Johnston R L 2003 *Dalton Trans.* 4193.
[10] Cleri F and Rosato V 1993 *Phys. Rev. B* **48** 22.
[11] Smith D J, Petford-Long A K, Wallenberg L R, Bovin J O 1986 *Science* **233** 872.
[12] Plaza J L, Mendes P M, Diegoli S, Chen Y, Preece J A, Palmer R E 2005 *Nanoscience and Nanotechnology* **5** 1826.
[13] Tran D T and Johnston R L 2009 *Phys. Chem. Chem. Phys.* **11** 10340

Complementary RHEED and XRD investigation of β-FeOOH and α-Fe_2O_3 nanoparticles

Trevor Almeida*, **Yanqiu Zhu and Paul D Brown**

Division of Materials, Mechanics and Structures, Faculty of Engineering,
Department of Mechanical, Materials and Manufacturing Engineering,
University of Nottingham, University Park, Nottingham, NG7 2RD, UK

Abstract. Hydrothermal synthesis has been used systematically to produce α-Fe_2O_3 nanoparticles, via an intermediate β-FeOOH phase, as a function of temperature, time and surfactant concentration. The nanoparticles were examined using the complementary techniques of X-ray diffractometry (XRD) and reflection high-energy electron diffraction (RHEED). The relative merits of XRD and RHEED are reported on, supported by conventional transmission electron microscopy, to appraise the reaction products.

1. Introduction

Nanostructured ferromagnetic α-Fe_2O_3 (hematite) has received considerable attention as a multifunctional material, having been investigated extensively for a variety of applications including photo-catalysis, magnetic recording, magnetic resonance imaging, spin electronic devices, lithium-ion batteries and pigments; owing to its environmentally friendly properties, low processing costs and high resistance to corrosion [1]. Hydrothermal synthesis (HS) has been identified as an ideal method for the production of α-Fe_2O_3 nanoparticles, offering effective control over the size and shape of nanoparticles at relatively low reaction temperatures and short reaction times, along with high homogeneity and well-crystallized reaction products with definite composition. The use of appropriate structural characterisation techniques allows an HS 'process map' to be developed, providing insight into the growth mechanisms of α-Fe_2O_3 nanoparticles. In this context, reflection high energy electron diffraction (RHEED) and X-ray diffractometry (XRD) may be considered as complementary techniques for structural characterisation. RHEED is sensitive to the top few atomic layers of a surface and utilizes the diffraction of electrons to provide information on near surface crystal structure, preferred orientation and roughness [2]. XRD is used more commonly to characterize bulk crystallographic structures, providing information on crystalline phase, grain size and preferred orientation [3]. Here, the applicability of RHEED and XRD as techniques for the characterisation of nanostructured iron oxide are compared and their relative merits highlighted.

2. Experimental

β-FeOOH and α-Fe_2O_3 nanoparticles were prepared through the hydrothermal synthesis of 0.20 ml 45% pure $FeCl_3$ aqueous solution, further diluted in 40 ml distilled water and mixed with and without an ammonium dihydrogen-phosphate (99.999% $NH_4H_2PO_4$) surfactant. The reactant solution was mechanically stirred in a 125 ml Teflon-lined steel autoclave, sealed and inserted into a temperature

* Author to whom correspondence should be addressed: emxta1@nottingham.ac.uk

Table 1. Hydrothermal processing conditions.

Sample	$NH_4H_2PO_4$ / mg	Temperature / °C
S1	0	100
S2	0	120
S3	0	140
S4	0	160
S5	1	160
S6	2	160
S7	3	160
S8	4	160

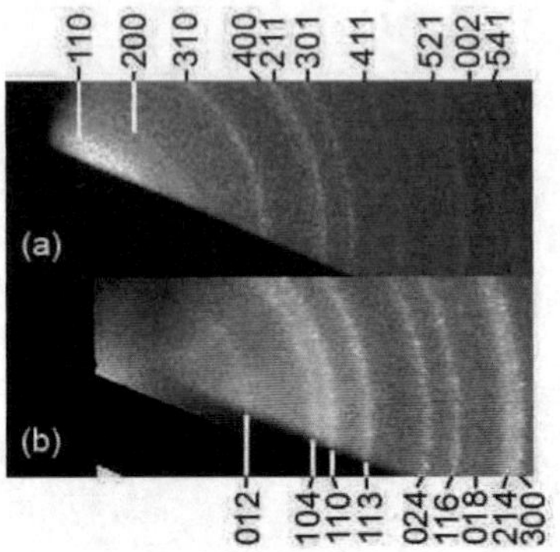

Figure 1. Example RHEED patterns indexed to (a) β-FeOOH and (b) α-Fe_2O_3.

controlled furnace for 2 hours at the reaction temperatures specified in Table 1. The autoclave, once removed from the furnace, was allowed to cool down naturally to room temperature. The synthesised reaction products were deposited straight from solution onto single crystal silicon substrates for the purpose of bulk and surface structural characterisation using a Siemens D500 X-ray diffractometer (Cu Kα radiation; θ/2θ diffraction geometry) and a dedicated RHEED stage mounted immediately beneath the projector lens of a Jeol 2000fx operated at 200 kV, respectively. For the purpose of TEM investigation, the HS product suspension was deposited straight onto lacey carbon / copper mesh support grids for conventional diffraction contrast imaging.

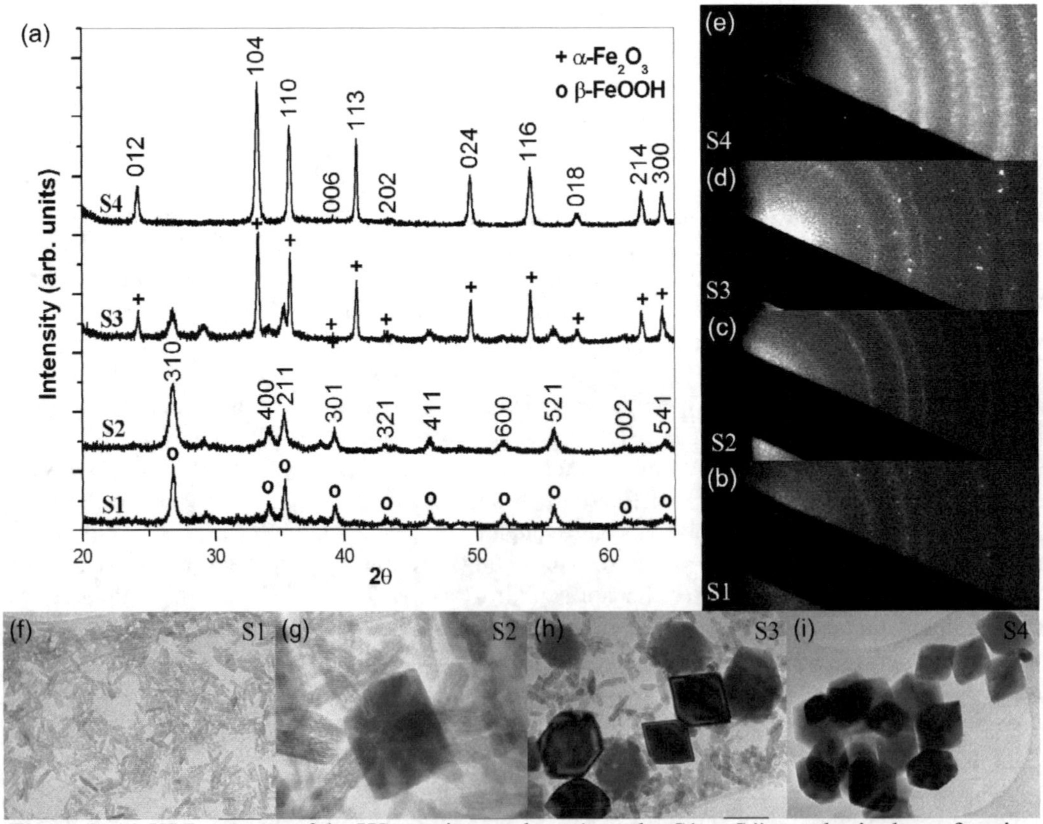

Figure 2. (a) XRD patterns of the HS reaction products (samples S1 to S4), synthesised as a function of increasing temperature in the absence of a surfactant, with indices attributed to β-FeOOH and α-Fe_2O_3. Associated RHEED patterns and bright field, diffraction contrast TEM images corresponding to samples (b,f) S1; (c,g) S2; (d,h) S3 and (e,i) S4.

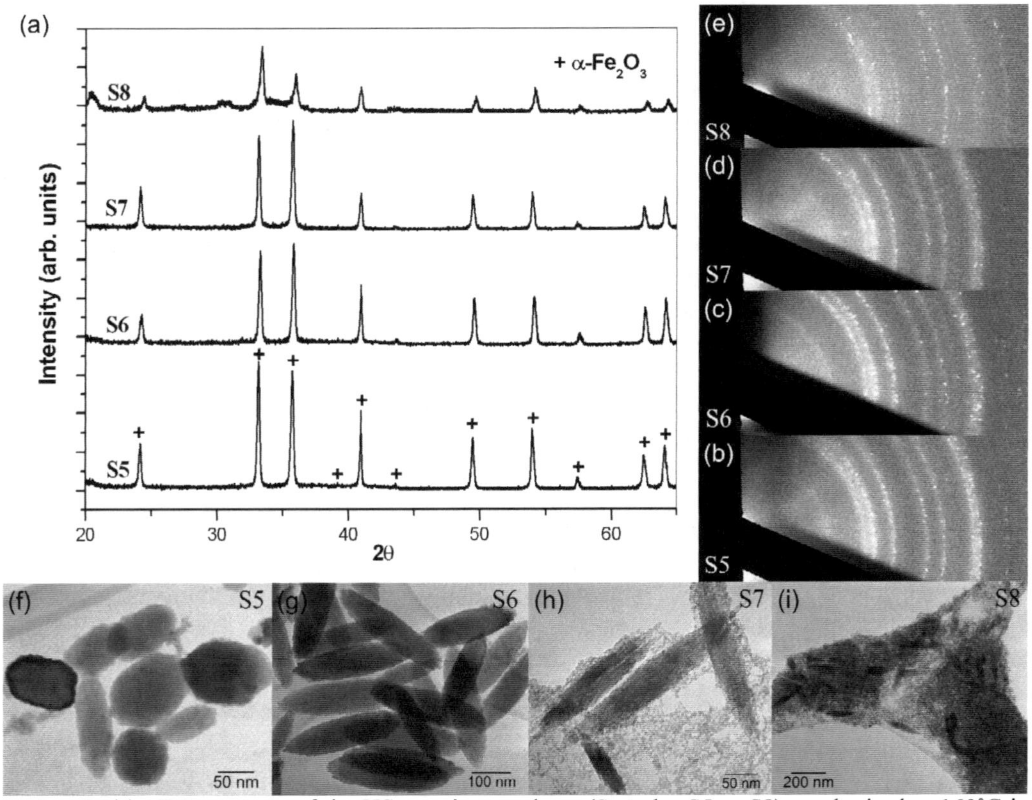

Figure 3. (a) XRD patterns of the HS reaction products (Samples S5 to S8) synthesised at 160°C in the presence of increasing phosphate concentration. Associated RHEED patterns and bright field TEM images corresponding to samples (b,f) S5; (c,g) S6; (d,h) S7 and (e,i) S8.

3. Results

Example RHEED patterns indexed to the phases of β-FeOOH and α-Fe_2O_3 are presented in Figures 1a and 1b, respectively, to assist with the interpretation of RHEED patterns recorded as a function of the HS processing conditions. Subtle differences in the spacings and intensities of the diffraction rings enabled the two distinct phases to be distinguished.

Figure 2 presents XRD and RHEED patterns of the hydrothermal products synthesised at reaction temperatures of 100°C to 160°C, in the absence of a surfactant, accompanied by conventional bright field TEM images showing the product morphologies. The XRD patterns corresponding to samples S1 to S4 (Figure 2a) chart a transformation from an intermediate phase of β-FeOOH to a well defined phase of α-Fe_2O_3 with increasing HS temperature. Figures 2b & 2c show the associated RHEED patterns from samples S1 and S2, respectively, with diffraction spacings corresponding to β-FeOOH. Conversely, the diffraction spacings from sample S4 (Figure 2e) matched accurately with α-Fe_2O_3, whilst the RHEED pattern from sample S3 exhibited rings attributable to β-FeOOH and additional, discrete diffraction spots attributable to α-Fe_2O_3 (Figure 2d). The associated TEM image of Figure 2f (sample S1) shows the small rod-like β-FeOOH particles, with varying aspect ratios, synthesised at 100°C, as compared with Figure 2i (sample S4) which shows the larger rhombohedral / hexagonal α-Fe_2O_3 nanoparticles synthesised at 160°C. The intermediate stages of α-Fe_2O_3 nanoparticle growth are seen at the reaction temperatures of 120°C and 140°C (Figures 1g,h).

Figure 3 presents associated XRD and RHEED patterns of the hydrothermal products synthesised at a reaction temperature of 160°C with surfactant additions of 1 to 4 mg phosphate, along with bright field TEM images showing the reaction product morphologies (samples S5 to S8). The XRD patterns of Figure 3a exhibit narrow, high intensity peaks characteristic of α-Fe_2O_3 at high Fe^{3+}: PO_4^{3-} molar ratios, but with an ensuing loss in intensity, and by implication extent of crystallinity, with increasing PO_4^{3-} anion concentration. The associated RHEED patterns of Figures 3b-d (samples S5 to S7) exhibit distinct, high intensity rings which reinforce the concept of superior crystallinity, as compared with sample S8, corresponding to the highest phosphate concentration, which exhibits less distinct diffraction rings (Figure 3e). The TEM image of Figure 3f illustrates the mixture of 'pseudocubes' and rounded and elliptical α-Fe_2O_3 nanoparticles comprising sample S5. Conversely, well defined lenticular α-Fe_2O_3 nanorods with an aspect ratio of 3.7 were evident within sample S6 (Figure 3g). Sample S7 (Figure 3h) comprised partial lenticular α-Fe_2O_3 nanorods exhibiting the highest aspect ratio of 6 for this sample set. A further decrease in the extent of crystallinity and the lenticular α-Fe_2O_3 nanorod aspect ratio was associated with sample S8.

4. Discussion

The data shows that both the XRD and RHEED techniques are capable of identifying and distinguishing between different β-FeOOH and α-Fe_2O_3 HS reaction products. For example, both the XRD and RHEED patterns for samples S1 and S2 identify the β-FeOOH phase, synthesised at the relatively low reaction temperatures of 100 and 120°C. However, TEM investigation of sample S2 revealed the additional presence of some small α-Fe_2O_3 nanoparticles, suggesting that the volume fraction of α-Fe_2O_3 in this instance was beneath the detection limit of either diffraction technique.

The XRD pattern of sample S3 demonstrates that α-Fe_2O_3 was the dominant phase at the intermediate temperature of 140°C, with a reduced β-FeOOH peak intensity compared to samples S1 and S2. TEM investigation reinforced this perception, with the presence of well developed α-Fe_2O_3 nanoparticles up to 150 nm in size. However, the RHEED pattern of sample S3 still showed continuous diffraction rings attributable to β-FeOOH with a few intense diffraction spots characteristic of α-Fe_2O_3. This reflects the surface sensitivity of the RHEED technique as compared with XRD. One possibility is that α-Fe_2O_3 nanoparticles are terminated with a surface structure of β-FeOOH at this intermediate temperature of 140°C. Alternatively, there may be an issue for the separation of mixed phases from suspension. It is suggested that the denser and less soluble hexagonally closed-packed α-Fe_2O_3 crystals would settle comparatively quickly compared to the smaller body-centred tetragonal β-FeOOH nanorods, during aqueous solution evaporation, leaving a higher β-FeOOH surface proportion and would dominate the RHEED in this instance. At the highest reaction temperature of 160°C (sample S4), all three characterisation techniques of XRD, RHEED and TEM confirmed the sole presence of the α-Fe_2O_3 phase.

The XRD and RHEED data for samples S5 to S8 similarly confirmed that α-Fe_2O_3 was the dominant phase at the reaction temperature of 160°C in the presence of a phosphate surfactant. TEM investigation demonstrated a variation in α-Fe_2O_3 nanoparticle aspect ratio with increasing concentration of PO_4^{3-} anions. This aspect ratio dependence is reflected in the relative peak intensities of the XRD patterns. Further, the suppression of α-Fe_2O_3 formation associated with sample S8 was attributed to the hindrance of β-FeOOH dissolution, symptomatic of higher PO_4^{3-} anion concentration and hence, absorption. This observation is echoed in the both the reduced intensities of the XRD data (Figure 3a) and less distinct diffraction rings associated with the RHEED pattern of Figure 3e.

In summary, XRD and RHEED techniques have both identified and distinguished between the different β-FeOOH and α-Fe_2O_3 HS reaction products. Subtle differences in the data arise from the surface sensitivity of the RHEED technique as compared with XRD.

References

[1] Zhao Y, Dunnill C W *et al* 2007 *Chem. Mater.* **19** 916-921
[2] Russell G J 1982 *Prog. Crystal Growth and Charact.* **5** 291-321
[3] Cullity B D 2001 *Elements of X-ray Diffraction* (Upper Saddle River: Prentice Hall)

Acknowledgements

The authors would like to thank the EPSRC for scholarship support.

Journal of Physics: Conference Series **241** (2010) 012088 doi:10.1088/1742-6596/241/1/012088

Imaging and Analysis of Axial Heterostructured Silicon Nanowires

F Li, P D Nellist, C Lang and D J H Cockayne

Department of Materials, University of Oxford, Parks Road, Oxford, OX1 3PH, UK

E-mail: fang.li@materials.ox.ac.uk

Abstract. Silicon nanowires with axially varying *n*- and *p*-doping were grown by the vapor-liquid-solid approach using gold as the catalyst. The microstructures of the axial heterostructured silicon nanowires were studied using electron microscopy techniques. The nanowire sidewalls exhibit periodic nanofaceting, which is dopant-dependent. The nanofaceting occurs during the enhanced sidewall growth that arises when the diborane dopant gas is introduced. It is also found that the silicon nanowire can crystallize in a hexagonal phase, which is thermodynamically unstable in bulk silicon at ambient and may be due to the large surface stress present during the nucleation of SiNWs.

1. Introduction

Semiconductor nanostructures have recently stimulated great interest due to their scientific importance and potential applications as the building blocks for future nanotechnologies as functional devices [1]. For example, silicon nanowires (SiNWs) with axially varying doping to generate junctions could be used as nanoscale field-effect transistors (FET) [2]. Good control of the wire surface and phase purity is an important issue for functional devices based on nanowire structures, because the physical behavior of SiNWs is strongly influenced by their microstructures. Being one of the most powerful tools in nanotechnology, the electron microscope (EM) has played an important role in revealing the microstructures of nanomaterials. In this article, using EM techniques, we characterized the microstructural properties of the axial heterostructured SiNWs synthesized using a vapor-liquid-solid (VLS) method. Here we report the observation of doping-dependent nanofaceting on SiNW surfaces and SiNWs with hexagonal crystalline structure. The possible causes of the observed phenomena are also discussed.

2. Experimental Details

Axial heterostructured SiNWs were synthesized on Si(111) substrates via VLS growth in a FirstNano EasyTube™3000 low pressure chemical vapor deposition (CVD) system with silane as the precursor gas and Au nanoclusters to catalyze the nanowire growth. The growth was carried out at ~600 °C and the SiNWs were either *n*- or *p*-doped by introducing phosphine or diborane gas, respectively. Two different axially doped nanowire heterostructures were prepared: *p*-doped, *n*-doped and intrinsic SiNWs (*p-n-i* SiNWs); and intrinsic, *p*-doped, *n*-doped and intrinsic SiNWs (*i-p-n-i* SiNWs).

After growth, the SiNWs were separated from the growth substrates by sonication in acetone. The suspension was then drop cast onto either a glass substrate or a TEM finder grid with a 20 nm thin holey carbon film, for the Raman spectroscopy and transmission electron microscopy (TEM)

characterization, respectively. TEM and high resolution transmission electron microscopy (HRTEM) imaging was made using JEOL 2000FX and JEOL 4000EX TEMs operating at 200 kV and 400 kV, respectively. Raman spectra of all samples were measured using a JY Horiba Labram Aramis imaging confocal Raman microscope and a coherent 532 nm laser.

3. Results and Discussion

3.1. Doping-dependent nanofaceting on SiNW surfaces

Any dependence of the surface properties of SiNWs on the impurity doping is crucially important to their applications as functional devices. For example, in an FET geometry, it is necessary to achieve a uniform cross section of the gate dielectric and to minimize the carrier scattering at rough interfaces.

In our previous study [3], we observed that the nanowire sidewalls exhibit periodic nanofaceting, which is found to be dopant-dependent, *i.e.*, sidewall facets occur in the boron-doped (*p*-type) region (Figure 1(a)), while both the phosphorus-doped (*n*-type) and the intrinsic SiNW regions exhibit smooth surface morphology. HRTEM imaging shows that the nanowires are axially oriented along <111>, and the crystallographic planes on which the nanofaceting occurs are determined to be {111} and {100} (Figure 1(b)). It is shown that the nanofaceting occurs during the enhanced sidewall growth that arises when the diborane dopant gas is introduced [3].

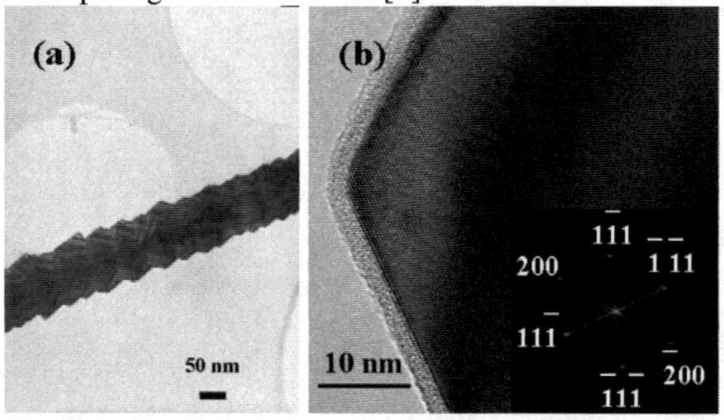

Figure 1 (a) TEM image of the sidewall facets in the B-doped region of an axially doped SiNW, viewed along the <110> direction; and (b) HRTEM image of the facet, inset: Fourier transform of the image.

3.2. SiNWs with hexagonal crystalline structure

The phase purity of SiNWs is important for the physical performance of the SiNW-based devices, since the physical properties of semiconductors are highly sensitive to their crystalline structures. It is known that the stable crystalline form of Si at ambient is a diamond cubic structure (Si I), however, Fontcuberta i Morral *et al.* [4] demonstrated the formation of hexagonal (sometimes referred to as wurtzite Si, Si IV) SiNWs by a standard CVD approach on oxidized Si substrates. Arbiol *et al.* [5] reported the switching of the Si structure from diamond to hexagonal along the wire growth direction in SiNWs synthesized via a Cu-catalyzed vapor-solid-solid mechanism. Here we report the observation of hexagonal SiNWs grown from the Au-catalyzed VLS reactions on an oxide-removed Si(111) substrate.

Both diamond cubic and hexagonal Si have a tetrahedral coordination, but the stacking of the tetrahedra is different. Two slightly different Si hexagonal structures have been reported: Si IV_A (a = b= 3.80 Å, c = 6.27 Å, c/a = 1.65) [6] and Si IV_B (a = b = 4.04 Å, c = 6.60 Å, c/a = 1.63) [7].

HRTEM characterization was carried out on the *i-p-n-i* axial heterostructured SiNWs for the determination of the crystalline structures. Figure 2(a) shows an overview image of a 150° kinked SiNW. Numbers 'I-III' in Figure 2(a) are assigned to the different positions on the kink. The corresponding HRTEM images and their Fourier transform are shown in Figure 2(b)-(d). By measuring the interplanar distances and angles, the crystalline structure of the SiNW can be determined as Si IV_B. The wire growth direction is perpendicular to the $\{10\bar{1}\bar{1}\}$ plane before the kink (part III in Figure 2(a)) and perpendicular to the $\{10\bar{1}\bar{3}\}$ plane after the kink (part II in Figure 2(a)). To further confirm the proposed structure, the same SiNW was tilted so that other zone axes were parallel to the incident beam. Figure 2(e)-(f) show the indexed selected area electron diffraction (SAED) patterns for the same specimen region for the $[\bar{2}4\bar{2}3]$ and $[\bar{1}2\bar{1}3]$ beam direction, respectively.

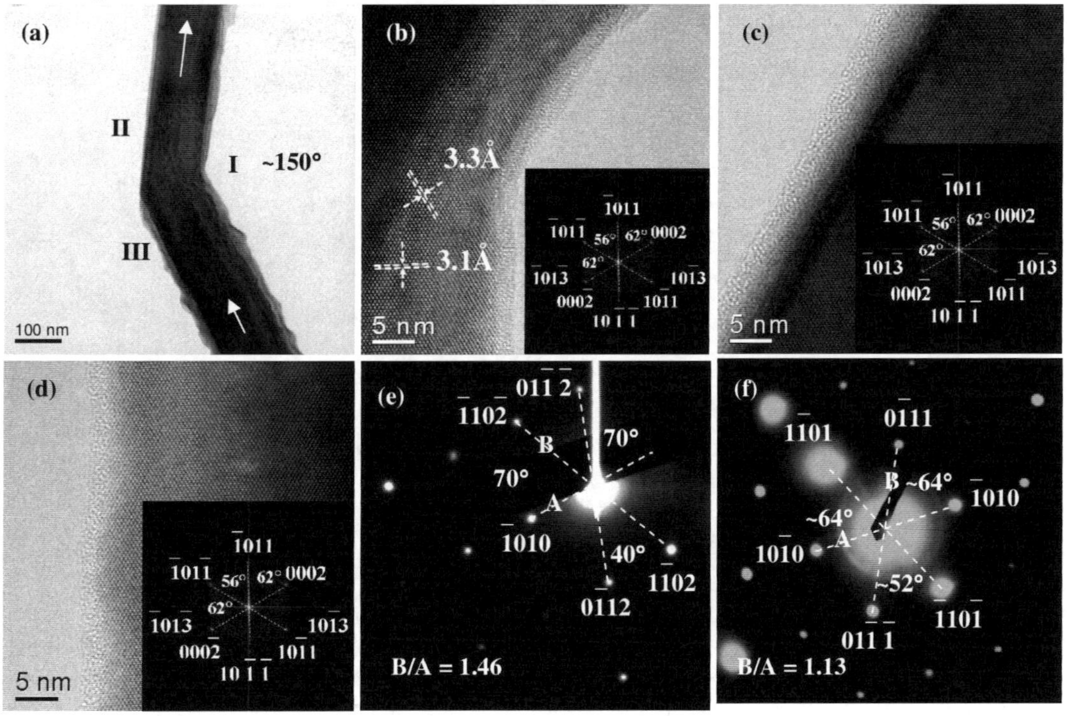

Figure 2(a) TEM image of a ~150° kink in the axial heterostructured *i-p-n-i* SiNWs, numbers 'I-III' are assigned to label different positions; HRTEM image of (b) part I, (c) part II and (d) part III in (a), insets in (b)-(d) are their corresponding Fourier transform; and indexed SAED patterns for the same kink for the (e) $[\bar{2}4\bar{2}3]$ and (f) $[\bar{1}2\bar{1}3]$ beam direction. (In (e)-(f), ratios of the spot spacings are shown as well as the angles between the plane normals).

Raman spectroscopy was also employed to characterize the structures of SiNWs. As seen from Figure 3, the first order Raman peak of the bulk Si is at ~520 cm^{-1} with a full-width at half-maximum of ~4 cm^{-1}, in good agreement with the literature [8]. To avoid the heating by the excitation laser, the laser power was kept at a relatively low value, where the Raman peak of the diamond cubic SiNWs is also ~520 cm^{-1}, suggesting that the samples are at room temperature and the laser heating effect has not yet occurred. At this low laser power, the hexagonal SiNW exhibits a sharp peak centered at 507 cm^{-1}, as well as a broad band between 223 and 288 cm^{-1}, in excellent agreement with the predicted transverse optical and the 2-fold transverse acoustical modes of hexagonal Si [9, 10].

Both the TEM and Raman analyses confirm that SiNWs with hexagonal crystalline structure can be formed during the Au-catalyzed VLS growth on a Si(111) substrate. This is different from the work done by Fontcuberta i Morral *et al.* [4], where the hexagonal SiNWs were grown on an oxidized Si substrate with an oxide thickness of 1 μm and the crystallographic alignment with the substrate was lost. It has been suggested previously that the formation of hexagonal crystalline structure in SiNWs can be caused by defects like twins [11] and stacking faults [12], or the phase transformation from diamond to hexagonal Si under high pressure [10]. Here the defect-induced hexagonal Si formation can be ruled out in this case, as such mechanisms result in the coexistence of both hexagonal and diamond phases, however, only the hexagonal phase is observed from HRTEM observations. Therefore, we suggest that the large surface stress during the nucleation of SiNWs could be a possible cause for the formation of hexagonal SiNWs observed here.

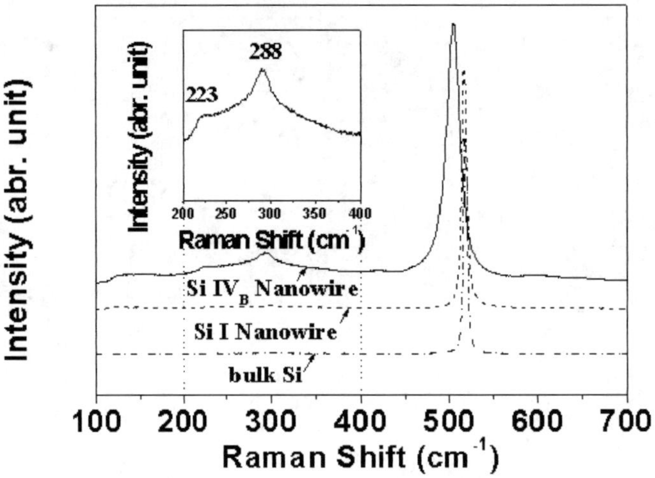

Figure 3 Stokes Raman spectra of the Si IV_B nanowire, bulk Si and Si I nanowire under the same laser power.

4. Conclusions

In summary, we have studied the microstructural properties of axially doped SiNWs using EM techniques. It is found that the diborane dopant gas alters the nanowire growth mechanism and behavior by enhancing the sidewall growth and creating a faceted sidewall surface. SiNWs with hexagonal crystalline structure can be formed during the Au-catalyzed VLS growth on the Si(111) substrate, which can be due to the large surface stress present during the nucleation.

Acknowledgment

The authors acknowledge the University of Oxford and the U.K. Engineering and Physical Sciences Research Council for a scholarship (F.L.).

References

[1] Cui Y, Zhong Z, Wang D, Wang W U and Lieber C M 2003 *Nano Lett.* **3** 149
[2] Gudiksen M S, Lauhon L J, Wang J, Smith D C and Lieber C M 2002 *Nature* **415** 617
[3] Li F, Nellist P D and Cockayne D J H 2009 *Appl. Phys. Lett.* **94** 263111
[4] Fontcuberta i Morral A, Arbiol J, Prades J D, Cirera A and Morante J R 2007 *Adv. Mater.* **19** 1347
[5] Arbiol J, Kalache B, Roca i Cabarrocas P, Morante J R and Fontcuberta i Morral A 2007 *Nanotechnology* **18** 305606
[6] Yeh C, Lu Z W, Froyen S and Zunger A 1992 *Phys. Rev. B* **46** 10086
[7] Jennings H M and Richman M H 1976 *Science* **193** 1242
[8] Richter H, Wang Z P and Ley L 1981 *Solid State Commun.* **39** 625
[9] Gogotsi Y, Baek C and Kirscht F 1999 *Semicond. Sci. Technol.* **14** 936
[10] Kailer A, Gogotsi Y and Nickel K G 1997 *J. Appl. Phys.* **81** 3057
[11] Dahmen U, Hetherington C J Priouz P and Westmacott K H 1989 *Scr. Metall.* **23** 269
[12] Liu X and Wang D 2009 *Nano Res.* **2** 575

Nanoporous structures from anodisation of non-planar aluminium surfaces

X Mu, Y Peng, T Gnanavel, B J Inkson, G Möbus

Dept Engineering Materials, University of Sheffield, Sheffield S13JD, UK

y.peng@sheffield.ac.uk

Abstract. We report about a novel combination of two etching techniques for generating nanostructures, which individually are well established: (i) metal tip etching in NaOH, as e.g. used for STM or atom probe tip making, and (ii) anodic oxidation of aluminium for the generation of self-organised ordered pores. The non-planar geometry in both these etching processes allowed us to follow the electric field mediated orientation of pattern growth observable in the SEM. Surface-near porosity is found from both etching processes, however only anodisation leads to continued growth of deep channel-like pores. FIB cross-sections of the cone and tip regions are found suitable to study the oxide-metal interface distribution and their dependence with anodisation time and current, even grain boundaries between the two opposite regions of deep pores are observed.

1. Introduction

Anodisation of aluminium foils is one of the most successful methods to generate a self-organised ordered nanochannel membrane, resulting in mostly amorphous anodic aluminium oxide (AAO) [1-2]. Most studies concentrate on perfectly planar surfaces and the mechanism of self-ordering is believed to be due to strain-equilibration induced during volume change upon oxidation [3]. Advantages of AAO over other nanopore membranes include the tunable pore diameter (via acid choice and concentration and voltage), e.g. from 3-500 nm, the homogeneity of pores, their high degree of order and the inertness of Al_2O_3 towards most fill-in materials. Applications of AAO are found for electrodeposition of metallic nanowires [4-5], growth of ordered CNTs for field emission, or membranes and masks for PVD deposition or ion sputtering.

The motivation for studying anodisation of highly non-planar starting materials, such as ultra-sharp Al-nanotips, is therefore based on two prospects: (i) by following the pore formation in cross-section with continuously varying surface curvature, improved understanding of the general self-organisation mechanism could be obtained. (ii) There are prospective applications for nanoporous materials, including their composites after filling of the pores, for which a sharply pointed tip-shape of the support is a more useful functional nanostructure, rather than a slab-shaped membrane.

2. Aluminium nanotips

2.1. Fabrication via NaOH etching

NaOH solution is traditionally used to fabricate W nanotips via an electrochemical etching method [6], e.g. for STM, nanoindentation, or nanomanipulation applications. The electrochemical etching technique also can be used to fabricate Ni nanotips via a NaCl solution [7]. We adapted this electrochemical etching method to fabricate Al nanotips. At the anode, an Al wire with 0.125 mm diameter was dipped into a 1 M NaOH solution. The etching voltage was 8 V dc. In this method the processes can be separated into two stages ("two-stage etching method").

First stage: An Al wire was immersed into the NaOH solution with a depth of 2 mm followed by turning on the etching power. The initial current was relatively high, around 2.9 mA. The current rapidly dropped to 1.0 mA within several seconds then slowly decreased. When the current dropped to about 1.0 μA, the etching circuit was switched off. During this process Al in the immersed wire dissolved into the solution and the wire diameter decreased.

Second stage: The etched tip was then immersed 150 ~ 200 μm deeper and further into the etching electrolyte. When the etching circuit was switched on again, the typical etching current was about 70 – 80 μA in the beginning. When the current has dropped to 50 – 55 μA, the etching process was completed and a sharp tip fabricated successfully. The tip was then taken out from the electrolyte and washed thoroughly by distilled water.

Sharp Al nanotips with conical shape were gained via the method described above. However, some other shapes can also be achieved by electrochemical etching. When we used a cutter to cut the Al wire, the tip was roughly formed in a roof-like shape with a sharp edge. Although the cutting-sharpness remained within the tens of micrometer range, it can be refined to achieve submicrometer radius via electrochemical etching. This etching is similar to the first stage in the method mentioned above. The immersed wires shaped roughly like a roof were etched until the etching current reached 0.4 mA. Then the wire was taken out from the NaOH solution. Roof-like tips with fine edges were obtained. In this method just one stage of etching was used, we refer to it as "one-stage etching method".

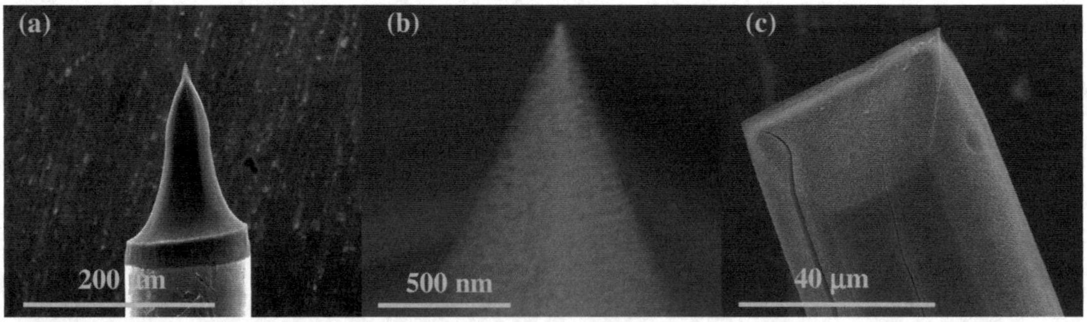

Figure 1. SEM images (secondary electrons) for two Al nanotips, FEI Inspect F FEGSEM, 15 kV: (a), (b) conical nanotip at 2 magnifications; (c) roof-shaped nanotip

2.2. SEM characterization

SEM was used to observe the fabricated tips. Figure 1 shows a typical conical nanotip with different magnified images (a), (b). The apex sharpness of this conical tip can achieve values below 50 nm. There is significant porosity in the surface layer due to amorphous oxide, which needs to be carefully distinguished from the anodization induced oxide in the next section. Figure 1(c) shows a roof-like tip obtained by the one-stage etching method.

3. Nanopore fabrication by anodization of non-planar structures: aluminium nanotips

A variety of nanotips with different geometry were anodized in 0.6 M oxalic acid with 33 V anodizing voltage for various anodizing time (15 min, 20 min, 25 min, 30 min, 50 min). Then these anodized Al nanotips were cut by focused ion beam (FIB) for in situ observation of their cross-sections by SEM (using a dual-beam JEOL/Orsay FIB). A central section can clearly show the growth behaviour of the pores into the depth of the structure under the boundary conditions of opposite surfaces of the nanotip.

During the FIB milling process very small milling current with short milling times were applied to decrease re-deposition effects which can remarkably degrade observation of the cross-sections. Figure 2 (left) shows an originally conical tip after anodization for 15 min and (right) its cross-section prepared by FIB milling imaged by SEM.

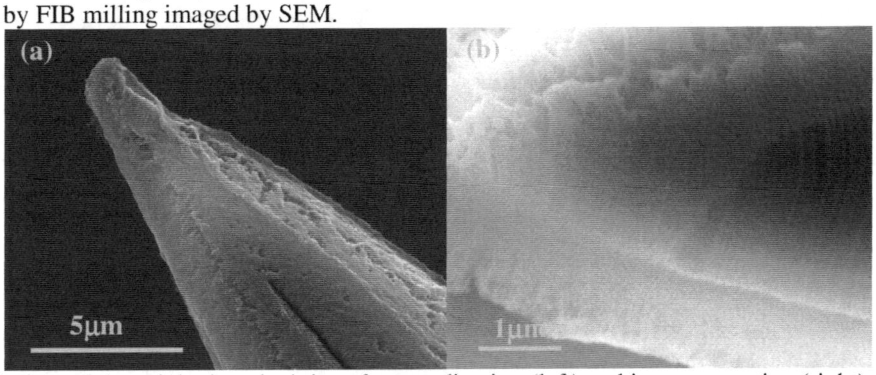

Figure 2. One of the original conical tips after anodization (left) and its cross-section (right) prepared via FIB milling. FEGSEM SE images.

During FIB milling, it is difficult to exactly cut the conical tips along their central axis. The pore channels therefore vary in contrast and visibility, and the pore direction is not always parallel to the cutting plane. Thus the roof-like tip shape was considered as preferred option. Figure 3(a) shows an SEM overview of the same tip as of figure 1(c) after anodization for 30 min, emphasizing the roughness of the roof-prism planes in contrast to the fine pore structure within the FIB cross-section. The geometry of the FIB milling setup, using two trenches, is shown as inset. Figure 3(a) also clearly separates FIB artefacts, seen as vertical tracks, from true pore structure, which is curved independent of the ion milling direction.

Figure 3. FIB cross-sections drilled into the roof-shaped Al-nanotip of figure 1(c), (a) overview with FIB cross-section placement as inset. (b) central Al core and curved Al_2O_3 pores. FEGSEM-SE images, 10 kV. Field of view of inset in (a) of 100 μm.

Figure 3(b) is a magnified image of the central area, where the two opposite oxide layers for the cross-section of this roof-like tip are seen to merge at a grain boundary (top left). Towards the bottom right, the oxide layers remain separated by the non-anodized central area of residual pure aluminium.

4. Discussion

The SEM images of the FIB cross-sections above prove that nanopores were generated in both conical and roof-like nanotips. Both shapes of tips show a 3D-disordered porous surface-near region of around 500 nm thickness, which is surprisingly sharply separated from the ordered nanochannel array below.

In the intermediate layer, where an ordered pore array phase can be found, the channels of pores are not as straight as they would be generated in a planar substrate situation. This could be a result from the generation-speed of these pores being different with different positions. The pores near the apex of the tip generate slower than further in the body of the tip. This different generation-speed can also be proven by the channel length of pores near the apex of the Al metal phase being shorter than in the area far away from the apex of the Al phase towards the wire body. One conclusion can be gained that the generation speed of anodic alumina nanopores can be affected by the surface shape (the geometry) of the substrate, hence to influence the channel geometry of pores.

Figure 3 also shows pore channels are mostly perpendicular to the interface between the phase of regular pores and the Al metal phase. And these channels are also perpendicular to the interface near the tip surface and between the disordered pore phase and the regular pore phase. This phenomenon can be explained by the anodizing electric field which is perpendicular to the surface of Al metal during the anodization process. Therefore it can be an evidence for proving that the pores grow along the direction of the anodizing electric field. In other words, via variation of the electric field direction, the pattern of pore formation can be significantly altered.

From Figure 3 it is also evident that at the centre of the regular pore phase on the thin side of the tip, no aluminium remained. Instead, a very fine bright layer is seen, formed by the combined bottom barriers below the pores. This remaining barrier layer was not dissolved into oxalic acid solution during the anodization even though the aluminium metal has already been totally consumed. This phenomenon can be explained due to the electric field vanishing suddenly as a consequence of the entire aluminium metal being converted into the insulating phase of anodic alumina. The anodic alumina cannot be dissolved by oxalic acid without assistance by an electric field.

5. Conclusions

Sharp Al nanotips with different shapes were obtained via electrochemical etching in NaOH solution. Nanoporous structures were generated in nanotips via anodization. Cross-sections of two tips with different shape were prepared via FIB milling and observed by SEM. The preparation of the cross-section of a roof-like anodic tip was more successful, as the inner cross-sections become a 2D geometry, rather than a 3D pore distribution. There is evidence to prove that the pore-generation as well as the pore-structure can be affected by the choice of substrate geometry and that the local electric field direction plays a significant role during the anodization.

Acknowledgement: We acknowledge funding contributions by EPSRC, UK.

References

[1] Masuda H 2005 *Ordered porous nanostructures and applications* (Eds. R. B. Wehrspohn), Springer Science publishing, pp. 37-55
[2] Bwana N N 2008 *J. Nanopart Res*, **10**, 313–319
[3] Li A P, Müller F, Birner A, Nielsch K, Gösele U 1998 *J. applied physics,* **84**, 6023-6026
[4] Peng Y, Cullis AG, Inkson B J 2009 *Nanoletters*, **9 (1)**, 91-96
[5] Peng Y, Cullis AG, Inkson B J 2008 *Appl. Phys. Let.*, **93 (18)**, 183112
[6] Hobara R, Yoshimoto S, Hasegawa S, Sakamoto K 2007 *e-J. Surf. Sci. Nanotech*, **5**, 94-98
[7] Gnanavel T, Peng Y, Xu X, Inkson B J, Gibbs M R J, Möbus G 2008 *Journal of Physics: Conf. Ser.*, **126**, 012071

UV-Vis plasmon studies of metal nanoparticles

J W L Eccles[1], U Bangert[1], M Bromfield[2], P Christian[2], A J Harvey[1], P Thomas[3]

[1]Materials Science Centre, The University of Manchester, Grosvenor Street, Manchester, M1 7HS, UK
[2]School of Chemistry, The University of Manchester, Oxford Road, Manchester, M13 9PL, UK
[3]Gatan UK, 25 Nuffield Way, Abingdon, OX14 1RL, UK

James.Eccles@postgrad.manchester.ac.uk

Abstract. Previous studies of the surface plasmon resonance (SPR) from alloy nanoparticles of gold and silver have focused on bulk studies of the nanoparticles in solution. Here, we present a selection of data observed from individual alloy nanoparticles using highly spatially resolved electron energy-loss spectroscopy (EELS). In order to resolve the SPR in both the pure gold and Au-Ag alloy nanoparticles, we successfully applied the Richardson-Lucy deconvolution routine. We found that resonances observed from different sized alloy nanoparticles do not follow the same trends in frequency variation as their pure metal equivalent. Additionally, we do not consistently observe a single SPR peak, as one might expect from an alloy nanoparticle. This indicates that alloy nanoparticles, produced by the simultaneous reduction of metal salts, cannot be considered truly homogeneous.

1. Introduction

Surface plasmons are created on the boundary of a metal. These represent quantised oscillations of surface charge produced by an external electric field [1]. The SPR excited on metal nanoparticles creates interesting optical properties. In solution, gold nanoparticles appear red in colour, whilst silver nanoparticles appear yellow. This is caused by the differing frequency of the SPRs in the two metals. For surface plasmons on metal nanoparticles there are a number of resonance modes (l). The most important of these is the $l = 1$ resonance (Mie mode) which has an optically active dipole field configuration [2]. For nanoparticles of gold, silver and copper, this dipole resonance occurs in the UV-Vis region making the nanoparticles useful for optical applications [3][4]. As well as the dielectric properties of the material, the frequency of the dipole resonance depends upon the size and shape of

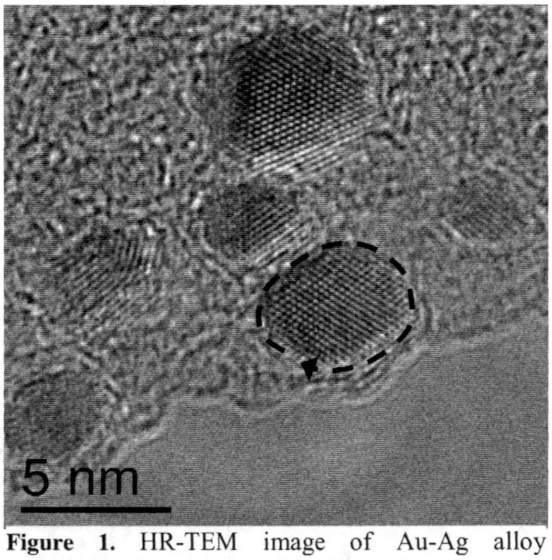

Figure 1. HR-TEM image of Au-Ag alloy nanoparticles on a lacy carbon film substrate.

nanoparticles [5]. This enables nanoparticles to be 'tuned' to different frequencies by varying these properties. For spherical nanoparticles, the greatest frequency range can be achieved by alloying the metals of gold and silver at different relative concentrations. These resonances can be scrutinised in bulk solution using UV-Vis Spectroscopy [6]. However, these resonances are a result of the superposition of resonances from individual nanoparticles, which may have a range of frequencies due to variations in size. Using a field emission gun (FEG) scanning transmission electron microscope (STEM) system, and the EELS technique, we have been able to observe the resonances of individual nanoparticles with a high degree of spatial resolution.

2. Methodology

The alloy nanoparticles were produced using equal molar amounts of the metal salts, silver nitrate and gold chloride. These salts are added to a solution of sodium citrate [7]. Sodium borohydride (SBH) is then used to reduce the metal ions to the metal. The use of sodium borohydride is advantageous since the reaction is rapid and the solution does not require heating. The citrate particles do not contribute to this reduction, but rather act as a stabiliser for the particles. The resulting nanoparticles were deposited on to a lacy carbon film substrate for analysis. Initial examination of the nanoparticles using high resolution transmission electron microscopy (HRTEM) analysis was carried out using a Tecnai F30 TEM operating at 300 kV. This revealed the nanoparticles ranged in size from ~ 3 nm up to 30 nm in diameter. Figure 1 represents a HRTEM image of several of the smaller-sized particles. The EEL spectra were acquired using a VG HB-601 STEM operated at 100 keV and equipped with a Gatan Enfina spectrometer. The STEM provides an energy resolution of 0.35 eV and is capable of a high dispersion (0.01 eV per channel). These factors enable the partial resolution of features close to the high energy zero loss peak ZLP tail. An objective aperture semiangle of 21.3 mrads was applied to limit the effect of spherical aberration, and to provide good imaging conditions. A collector aperture semiangle of 1.34 mrads was used to prevent saturation of the detector.

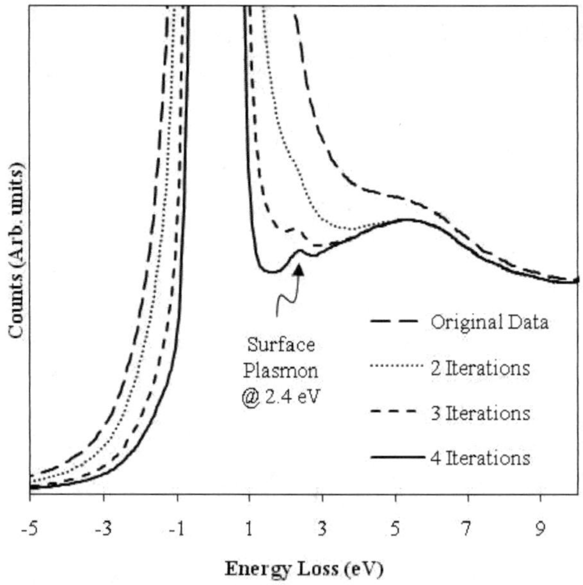

Figure 2. Deconvolved EEL spectrum from a gold nanoparticle (12nm diameter) up to 4 iterations. The surface plasmon is detectable a 2.4 eV.

The FEG, with an achievable energy resolution of around 0.35eV, has in the past enabled the direct observation of the SPR in silver nanoparticles at around 3eV. However, for nanoparticles of gold and the Au-Ag alloy, the SPR in both cases, is obscured by the high energy tail of the ZLP. We therefore applied the Richardson-Lucy (R-L) deconvolution algorithm to 'improve' the energy resolution of the spectra and to enable the SPR frequency to be observed. The RL deconvolution algorithm [8][9], whilst not necessarily being an alternative to monochromation, has been shown to be effective at improving the energy resolution of both valence-band and core-loss spectra [10][11]. It has also been successfully applied to silver nanoprisms in order to map the SPR [12]. The method assumes that the degraded image H is of the form;

$$H = W \otimes S \tag{1}$$

where W is the original image and S is the point spread function (PSF). The algorithm estimates the original image by deconvolving the degraded image with a PSF. The advantage of this routine is that it is able to conserve the total spectrum intensity and as such can cope with EEL spectra with a high dynamic range, such as the intense ZLP and weak plasmon features. The main disadvantage is that for statistically noisy data strong noise amplification arises. However, since the algorithm is iterative, it is possible to cease the process before any noise components are significantly amplified. As will be demonstrated, the use of 3-4 iterations is successful in resolving the required features without any artefacts being introduced. Figure 2 shows the R-L deconvolution routine applied to low loss spectra obtained from the surface of a gold nanoparticle. Following deconvolution up to 4 iterations, the SPR can be observed at 2.4 eV.

3. Results and Discussion

Initial studies were conducted to examine SPR behaviour in pure gold and silver nanoparticles. EEL spectra were obtained for individual nanoparticles by summing multiple spectra from the surface edge of each particle, as indicated in Figure 2 (----). Figure 3 (a) shows the variation in energy loss (frequency) as the size of the nanoparticles is varied. For silver particles with a diameter greater than 15 nm, the SPR frequency is governed largely by the dimension of the particle with respect to the

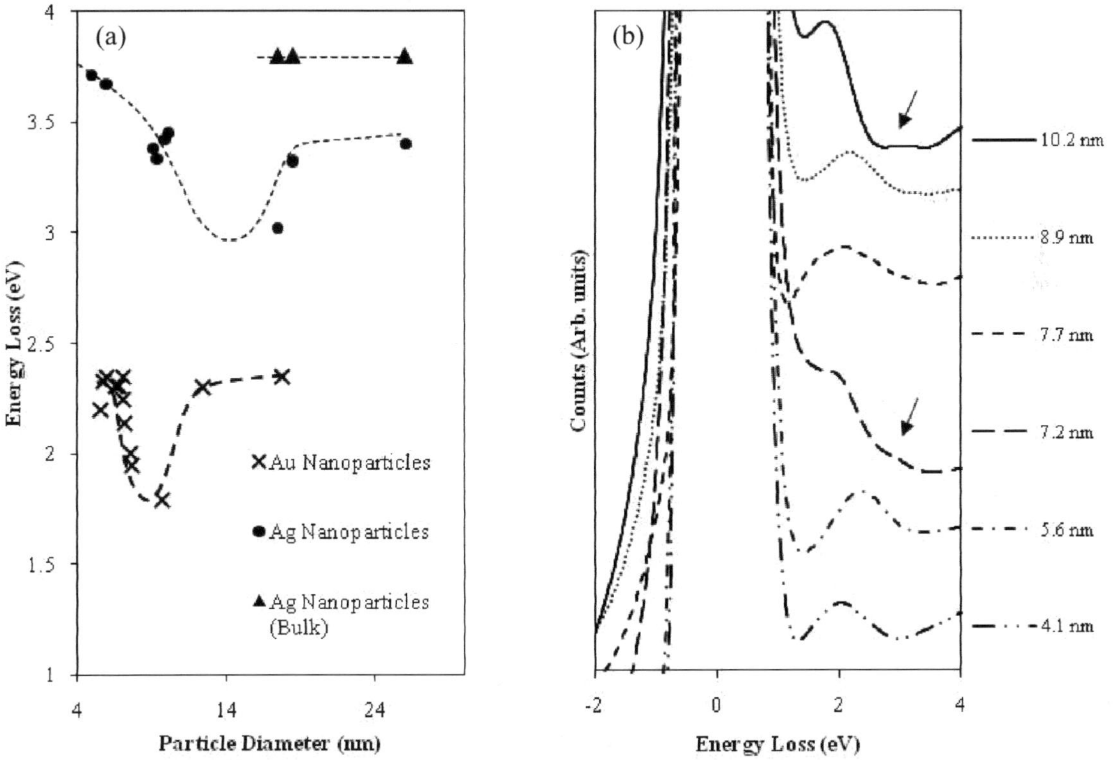

Figure 3 (a). Surface plasmon excitation energies for pure gold and pure silver nanoparticles of different particle diameter. Both pure gold and silver nanoparticles have a single SPR in the low energy regime. Silver nanoparticles have an additional bulk resonance at 3.8 eV, which only becomes dominant for larger sized particles. **(b)** EEL spectra from Au-Ag alloy nanoparticles deconvolved up to 3 iterations where arrows indicate a resolvable second resonance. The spectra are displaced vertically to make the comparison easier.

wavelength of the incoming radiation. This relationship results in a red shift in SPR frequency. For silver nanoparticles within this size regime we consider the dielectric function to be constant as a function of the radius of the particle $\varepsilon(\omega)$. However, as the size of the particle decreases, the relative size and importance of the surface layer increases. For silver nanoparticles with a diameter < 15 nm, the effective surface layer can no longer be considered a constant and the dielectric function of the material varies as a function of particle size $\varepsilon(\omega,r)$ [13]. This is known as the quantum size effect, and is a well documented nanoparticle behavior [14]. As the particle size is reduced, the quasi-continuous band structure begins to separate into discrete energy levels. This causes a blue-shift in plasmon energies. This size associated red and blue frequency shift is also observed for the gold nanoparticles, however this shift occurs at different energies due to the differing dielectric properties.

Figure 3 (b) shows the EEL spectra acquired from single Au-Ag Alloy nanoparticles deconvolved up to 3 iterations. At this level of deconvolution, the SPR can clearly be identified. One would expect, as in the case of the pure metal particles, to observe a relative shift in the nanoparticles towards a greater energy loss as particle diameter is decreased. However, this is not the case, as the energy loss of the SPRs appears to be random and show no specific trend. This could indicate the effect of other influencing factors on the energy loss of the SPR, such as the dielectric properties of the material. This suggests the dielectric function could differ from particle to particle, due to in homogeneity in the alloy. Additionally, we observe a distinct broadening of the SPR in the case of the 7.7 nm particle, and a small higher energy secondary peak indicated for 7.2 nm and 10 nm particles. This could represent segregation of the two metals within the nanoparticle. Specifically, a concentration of silver on the surface of the nanoparticle could be responsible for this second feature.

4. Conclusion

In conclusion we show that the nanoparticles cannot be considered truly a homogeneous alloy. Unlike the quantum size effects displayed by the pure metal nanoparticles, the random behaviour of alloy SPR must indicate variations in the dielectric function of each particle. It is possible this is caused by different relative concentrations of silver and gold within the particle. There is also evidence of segregation within the particles, particularly owing to excitation of a weak resonance at energies associated with pure silver SPR excitations. This indicates a possible core/shell type structure, with a thin outer shell of silver. This would support UV-Vis investigations where specifically produced core/shell particles display a twin peak structure in the extinction spectra [15].

References

[1] Raether H 1977 New York Surface Plasma Oscillations and Their Applications. In: Physics of Thin Films Vol. 9 145-171
[2] Mie G 1908 *Ann. Phys.* **25** 377
[3] Xu H, Bjerneld EJ, Kall M and Borjesson L 1999 Phys. Rev. Lett. **83** 4357-4360
[4] Ebbesen TW, Lezec HJ, Ghaemi HF, Thio T and Wolff PA 1998 *Nature* **391** 667
[5] Lance Kelly K, Coronado E, Zhao LL and Schatz GC 2003 *J. Phys. Chem. B* **107** 668-677
[6] Papavassiliou GC 1976 *J Phys. F: Metal Phys.* **6** 103-105
[7] Henglein A and Giersig M 1999 *J. Phys. Chem. B* **103** 9533-9539
[8] Richardson WH 1972 *J. Opt. Soc. Am* **62** 55
[9] Lucy LB 1974 *Astro. Journal* **79** 745-754
[10] Gloter A, Douiri A, Tencé M and Colliex C 2003 *Ultramicroscopy* **96** 385-400
[11] Kimoto K, Ishizuka K, Mizoguchi T, Tanaka I and Matsui Y 2003 *J. Elec. Micro.* **53** 299-303
[12] Nelayah J et al 2007 *Nature Physics.* **3** 348 - 353
[13] Kreibig U and Vollmer M 1995 Berlin Optical Properties of Metal Clusters 13-21
[14] Ouyang F, Batson PE and Isaacson M 1992 *Phys. Rev. B* **46** 15421-15425
[15] Morriss RH and Collins LF 1964 J. Chem. Phys. 41 3357-3363

Journal of Physics: Conference Series **241** (2010) 012091 doi:10.1088/1742-6596/241/1/012091

Controllable Growth of (001) Surface-Oriented Colloidal Crystals by Edge Meniscus Effect

Ting Zhang[1, 2], Jun Yuan[2, 3,*], Xinlin Tuo[4]

[1.] Laboratory of Advanced Materials and Department of Materials Science and Engineering, Tsinghua University, Beijing 100084, China
[2.] Beijing National Centre for Electron Microscopy, Tsinghua University, Beijing, 100084, China
[3.] Department of Physics, the University of York, Heslington, York, YO10 5DD, UK
[4.] Laboratory of Advanced Materials and Department of Chemical Engineering, Tsinghua University, 100084, Beijing, China
*Corresponding author E-mail address: jy518@york.ac.uk

Abstract. Colloidal crystal films have been fabricated on silicon substrates using a flow-controlled vertical deposition (FCVD) method. Scanning electron microscopy shows that large areas of the colloidal crystal films can exhibit regular square arrangement or related 1×n superstructure induced by the surface relaxation of colloidal spheres. They are shown to be the (100) facing crystals of face centred cubic (FCC) structure. Their formation is related to meniscus bending at the edge of the wafer, which can be controlled by varying the width of the substrate or the wetting properties of the backing substrate. This (100) surface-oriented colloidal crystal could be used as a template to fabricate square arrangements of nanopore and nanodot arrays.

1. Introduction

Nanostructured surfaces with long range periodicity are of growing importance because of their interesting physical properties such as optical responses. They are useful as templates for fabricating of highly oriented colloidal crystals, biologically active substrates, they are also useful avenue to production of high density optical devices [1], chemical and bio-sensors [2], and high-density magnetic and optical data storage [3-4]. Colloidal nanoparticles, because of the ease with which their sizes can be controlled, are favourite building blocks for low cost fabrication of such nanostructured surfaces. Monodispersed colloidal nanoparticles can easily self-assemble to form 2D or 3D lattice structure on other substrates when suitably dried of its solvent [5]. However, such self-assembly alone usually only produces close-packing ordering. In applications where the arrangement of the nanoparticles in the exposing surface is important, such as photonic crystals [6], this can be restrictive and has prompted research into methods that can produce other forms of regularly arranged nanoparticles.

To date, a popular approach is colloidal epitaxy, i.e. directed colloidal crystal growth on patterned substrates, such as on surfaces with imprinted relief [7], square pyramidal pits [8] or V-shaped grooves [9]. In addition, applications of an external electrical, magnetic, or optical field have also been examined as a promising strategy to control the growth direction [10]. Stretching of substrate in lift-

Journal of Physics: Conference Series **241** (2010) 012091 doi:10.1088/1742-6596/241/1/012091

up lithography is another alternative to produce non-close-packing ordering on surfaces [11]. Those methods are usually too complicated and expensive for large-scale applications. Relative large-scale, template-free growth of colloidal crystal domains with square surface patterns has been reported [12] using a spin coating method. However, the effects of the experimental factors such as the dispersion media, particle size and liquid concentration are not fully understood.

Here, we demonstrate template-free formation of mm^2-scale colloidal crystals with their (100) planes parallel to the surface. More importantly, they are shown to be related to the curved meniscus due to the edge effect of the substrate. Our result opens a new avenue of meniscus engineering for low cost and large-scale fabrication of nanostructured surfaces of a non-close-packing order. Additionally, we also report the use of this colloidal temperate to fabricate metal dots which can be used in plasmonics and information storage.

2. Experimental Section

The crystals in our experiment were fabricated using 290 nm polystyrene spheres purchased from the Institute of Physics and Chemistry, Chinese Academy of Sciences. Colloidal suspensions with volume fractions of 1.0% in deionised water were prepared. The method used to grow the crystals was FCVD as described elsewhere [13]. The dropping speed of the suspension level was set at 0.002ml/min and the temperature of the container was maintained at 20±2°C. Strips of silicon wafers used as supporting surfaces were treated in a mixture of H_2SO_4/H_2O_2 (98%H_2SO_4:30%H_2O_2=3:2, v/v) for 8 hours, then ultrasonicated in acetone, ethanol and deionised water respectively. This was repeated three times to make the surfaces hydrophilic. The treated silicon wafers were immerged, with a tilt angle of 10°, in the colloidal suspension. Scanning electron microscopy (SEM) (JEOL JSM6301f), Atomic Force Microscopy (AFM) (Nanoscope IIIa) operating in the tapping mode and reflection optical microscopy (NEOPHOT32) were used to characterize the morphology of the samples. For SEM inspection, the samples were coated with a thin gold film to stop charging effect as customary.

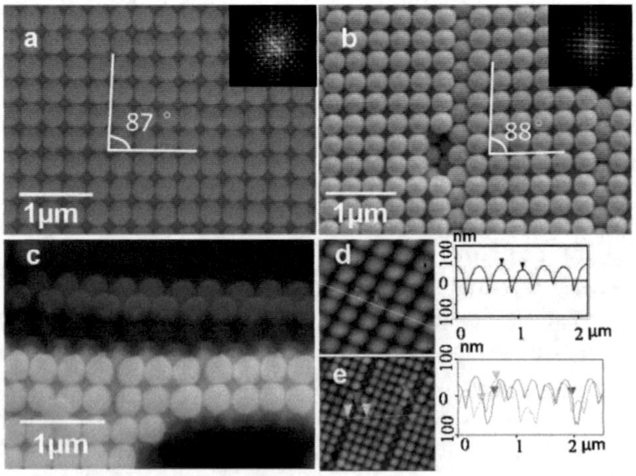

Figure 1. (Color online)
(a) An SEM top view of square arrangement of 290 nm colloidal spheres (the inset shows the corresponding Fourier transform).
(b) An SEM top view of square arrangement of 290 nm colloidal spheres with 1×6 periodicity (the inset shows the corresponding Fourier transforms).
(c) An SEM top-view of square arrangement of colloidal spheres near the edge of a cleaved colloidal film.
(d) An AFM image of square arrangement of 290 nm colloidal spheres.
(e) An AFM image of square arrangement of 290 nm colloidal spheres with 1×5 periodicity.

3. Result and discussion

Figure 1a shows a typical SEM image of colloidal crystals with simple square arrangement and the inset at the top right corner is its Fourier transformed image. Obviously the non-close-packing nature of the surface colloidal spheres can be observed. Figure1d is its corresponding AFM image. We can get the average spacing between centres of two neighbouring spheres which is 332±26 nm and the angle between the unit cell vectors is about 87°. The distorted pattern shows that it is not really a regular square arrangement. In figure 1b, a square domain with a 1×6 periodicity is also observed and this is just a typical example of many such 1×n domains observed where n can vary from 2 to 7. The angle between the unit cell vectors is about 88 °, which also varies from places to places. The inset at the top right corner is the corresponding Fourier transform image, confirming the square ordering of the spheres. According to the AFM analysis for a 1×5 structure (figure 1e), the average spacing between centres of two neighbouring spheres in the top layer is 295±16 nm and that in the trench is 312 nm, which all are very close to the nominal size of the starting nanosphere dispersed in liquid.

Figure 1c is a SEM top-view image of the colloidal films near its broken edge. Obviously, spheres in the bottom layer have also formed a regular square array and the upper spheres rest on the interstitial hollow sites formed by four neighboring spheres in a square arrangement. The side-view reveals a hexagonal stacking pattern down to the bottom layer. We therefore confirm that the square arrangement is consistent with the (100) face of the FCC colloidal crystal. From surface defects such as point vacancy (fig. 1b), we could find that the surface nanoparticles also sit on the hollow sites of the subsurface layer and the colloidal spheres in sub-layer also have a regular square arrangement. Therefore we can speculate that the 1 × n structure forms owing to the surface reconstruction of a regular square structure. However, until now, the parameters influencing the number n are not completely clear and further systematic work is required to understand their formation in details [15].

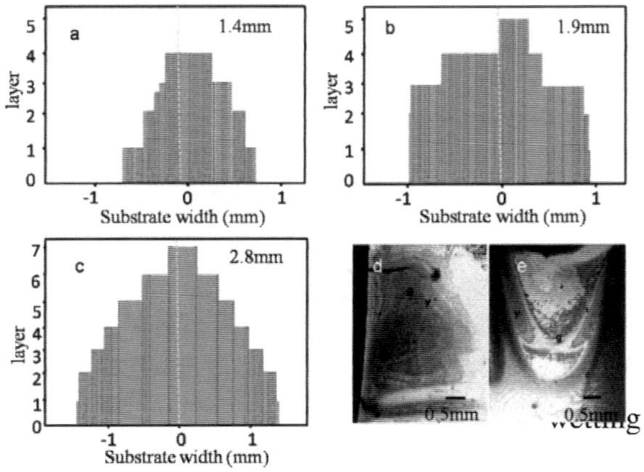

Figure 2. (Colour online)
(a-c) The layer of a cleaved colloidal crystal along the direction parallel to the suspension level, for Si strips of different widths. (pink for hexagonal arrangement and green for square arrangement, including 1× structure).

(d-e) Optical microscopy images of the colloidal films on the substrates. The growth patterns in two films are related to the shape of the receding contact lines that are varied by changing the wetting properties of the backing materials [15].

The colloidal films with either the simple square arrangement or the 1 × n structure are all multilayer stacking of nanospheres. The number of deposited colloid layers was measured, using STM on cleaved Si strip samples, from left to right along the direction parallel to the suspension level. The results for samples with different Si strip widths are summarized in figure 2. This reveals that, on average, the number of layers decreases, with the Si stripe width, down to the limit of one monolayer.

In addition, we can see clearly those square domains are mostly found near the edge of the wafers. As the wafer width reduces, the square domains at the edge usually merge to produce a large patch of square domains in the middle. For example, for the wafer with a 1.4 mm width (fig. 2a), the square domains dominate the substrate with the hexagonal arrangement decorating the step edges, i.e. the hexagonally arranged surface is now a transitory phenomenon. As for the substrate with 2.8mm width (fig. 2c), the square domains reduce to linear regions and acquire an intermediate thickness for n to n+1 hexagonal layers. To correlate the shape of the meniscus, which is an mm-scale phenomenon, with the formation of the square domains, we have exploited the different optical effect of the square and hexagonal domains [14]. The greenish patch can be associated with the Van Hover-like high density of photonic states in the (001) projections of FCC structure of 290 nm spheres. The yellow patch is associated with the high density of photonic states of the FCC crystal in its (111) projection. Film thicknesses variation only causes a slight variation within the color bands. The shape and location of the green patch representing square domain is a strong function of the curvature variation of the contact lines, later achieved by varying the wetting properties of the backing materials [15]. For the silicon substrate with a convex upward bow meniscus shape (fig. 2d), a large green patch of square domains decorated with the yellow lines corresponding to hexagonal arrangement is formed at the edge of the wafer. As for the silicon substrates whose meniscus has a concave downward bow (fig. 2e), the green region with small areas now appears at the interior. This suggests that the square domains opt to locate at regions where the meniscus is heavily curved and the contact line pinned [15].

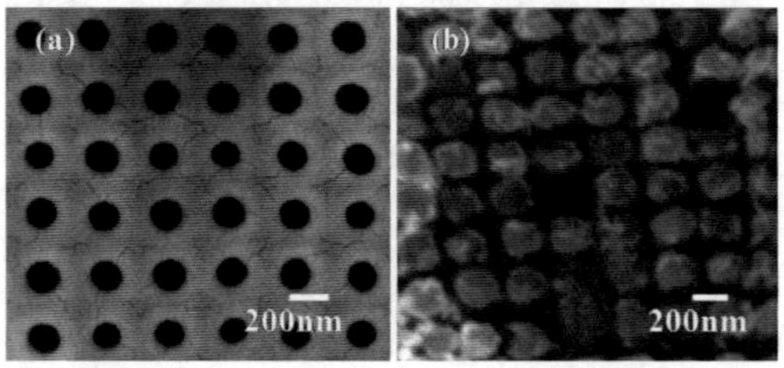

Figure 3.
(a) A SEM image of square pore arrays.
(b) A SEM image of square Ni dot arrays.

The ordered square colloidal crystal, prepared by this method, can be used as templates in diversified areas. Figure 3a shows a typical square pore array by selectively dissolving the colloidal crystal with its (100) plane parallel to the substrate [16]. Such a pore array can be used as a template to prepare an array of metallic squares. To achieve that, the porous template was first treated with reactive ion etching to remove the remaining thin layer at the bottom of the pores. Ni was then deposited by vacuum evaporation through the ordered pores. After ultrasonically washed in toluene, the polymeric template was removed, leaving an array of Ni squares on the substrate. Figure 3b shows the SEM image of the Ni squares; obviously the surface coverage of the array is quite high. The bright features around some Ni nanodisks are polymer residuals not completely removed. Such metal square arrays are important for applications such as information storage, where the readout signal can be maximized.

4. Conclusions

We have demonstrated the growth of (100)-orientated FCC colloidal crystal films on template-free silicon substrate by FCVD method. The largest square domains could exceed mm^2 magnitude. The shape of the meniscus on the substrate is found to be critical in controlling square domain size and location. We have used this colloidal template to fabricate ordered metal dot arrays. The square domains may exhibit different properties that are valuable for many practical applications.

5. Acknowledgment

We thank the Ministry of Science and Technology of China (Basic Science Research Grant No. 2002CB613501) for financial support.

References

[1] Tarhan I I, Zinkin M P and Watson G H 1995 *Opt. Lett* **20** 1571
[2] Holtz H J and Asher S A 1997 *Nature* **389** 829
[3] Sun S H, Murray C B, Weller D, *et al.* 2000 *Science* **287** 1989
[4] Siwick B J, Kalinina O, Kumacheva E, *et al.* 2001 *J. Appl. Phys* **90** 5328
[5] Dushkin C D, Nagayama K, Miwa T and Kralchevsky P A 1993 *Langmuir* **9** 3695
[6] Tarhan I I, Zinkin M P and Watson G H 1995 *Optics Letters* **20** 1571
[7] Jin C J, McLachlan M A, McComb D W, *et al.*2005 *Nano Lett* **5** 2646
[8] Yin Y, Li Z Y and Xia Y 2003 *Langmuir* **19** 622
[9] Ozin G A and Yang S M 2001 *Adv. Funct. Mater* **11** 95
[10] Holgado M, Garc′ıa-Santamar′ıa F, Blanco A, *et al.*1999 *Langmuir* **15** 4701
[11] Yan X, Yao J M, Lu G, *et al.* 2005 *J Am.Chem. Soc* **127** 7688
[12] Mihi A, Ocana M and Miguez H 2006 *AdV. Mater* **18** 2244
[13] Zhou Z and Zhao X S 2004 *Langmuir* **20** 1524
[14] Garcia P D, Galisteo-Lopez J F, Lopez C 2005 *Appl. Phys. Lett* **87** 201109
[15] Zhang T, Tuo XL and Yuan J 2009 *Langmuir* **25** 820
[16] Zhang T *et al.* 2009 *Colloids and Surfaces A: Physicochem. Eng. Aspects* **335** 202

Electron Microscopy and Analysis Group Conference 2009 (EMAG 2009) IOP Publishing
Journal of Physics: Conference Series **241** (2010) 012092 doi:10.1088/1742-6596/241/1/012092

CVD growth of boron-rich nanostructures

Zhiyang Yu[1,2], Steffan Lea[3], Jun Yuan[2,3]

[1] *Laboratory of Advanced Materials, Department of Materials Science and Engineering, Tsinghua University, Beijing 100084, China*
[2] *Beijing National Center for Electron Microscopy, Tsinghua University, Beijing 100084, China*
[3] *Department of Physics, the University of York, Heslington, York, YO10 5DD, UK*

E-mail: jy518@york.ac.uk

Abstract. Boron-rich nanostructures have been produced using a chemical vapour deposition method with iron particles as catalysts. Cyclic twinning nanowires and parallel twinning nanoplatelets have been identified using electron microscopy techniques such as selected area diffraction (SAD) for structural determination and electron energy-loss spectroscopy (EELS) for determining the chemical composition. This leads to the possibility of optimising the growth conditions of various boron-rich nanostructures, useful for further exploration of their physical properties.

1. Introduction

Boron is next to carbon in the periodic table and the structures of boron and its compounds can display as many structure variations as carbon, if not more. However, research into boron nanostructures is not as extensive as carbon nanostructures, partially because their growth conditions are poorly understood. For example, we have recently discovered a number of boron-rich nanostructures such as cyclic twinned nanowires [1-2] and parallel twinning nanoplatelets [3] as a result of a solid state reaction. However, the yield is small and the scope of optimisation in such a reaction is limited. Analysis of the growth conditions as well as the morphology of the nanostructures suggests that the actual growth may be driven by a chemical vapor deposition (CVD) process. Since CVD processes have been routinely used to produce carbon nanostructures on a large scale, we expect that a similar procedure for boron-rich nanostructures would allow us to improve the growth yield. The interesting physical properties of boron-rich materials, such as semiconductivity, thermoelectricity [4] and super-hard mechanical properties may then be investigated on the nanoscale [1]. We will provide direct evidence for the iron-assisted CVD growth of boron-rich nanostructures including cyclic twinning nanowires and parallel twinning nanoplatelets.

2. Experiments

Powders of BaO, B and Fe_3O_4 are mixed, as used in solid state reactions [1-3], and cold pressed into a pellet at a pressure of ~ 4 MPa. The pellet is placed at one end of an alumina boat, closely followed by two silicon chips of 20 mm by 5 mm in the downstream direction. One of the silicon chips was sputter-deposited with an iron film of nominally 30nm thickness, as measured by quartz thickness monitor, while the other is kept free from iron for comparison. A protective argon flow at 60 cm^3/min is maintained over the alumina boat that is heated to 1300 °C at a rate of 6 °C per minute inside a high temperature tube furnace and maintained at this temperature for one hour. The cooling is by natural convection.

3. The Results

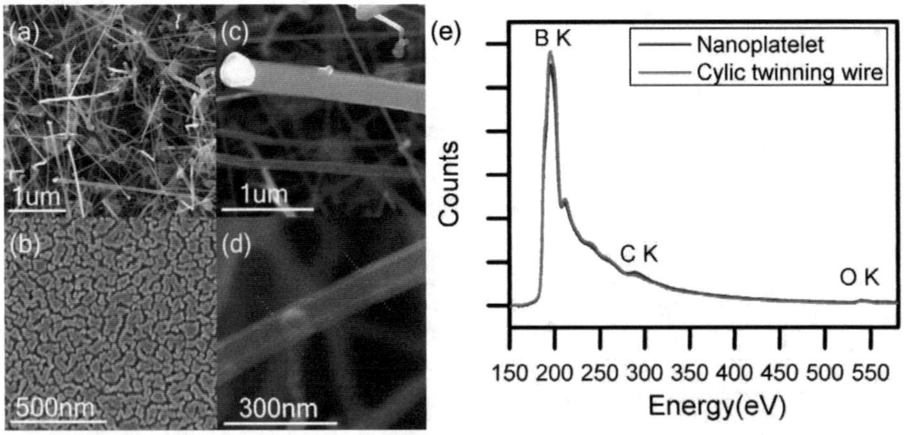

Figure 1. Morphology of nanostructures grown on the silicon chip coated with iron (a). Catalytic particles can be observed at the ends of the wires (indicated by the red dotted circles). In the absence of an iron film on the silicon chip, no boron nanostructures can be found (b). Scanning electron microscopy (SEM) image of a nanoplatelet (c) and cyclic twinning nanowire with linear contrast due to the re-entrant side facets (indicated by arrows) (d). Typical EELS spectra recorded from platelets and wires (e). The spectra have been deconvoluted using the Fourier Ratio method to remove the multiple scattering effects. (Colour online)

SEM observations show that a high density of nanostructures is found on the silicon chip pre-coated with the iron film (figure 1a) while the one without iron is quite clean (figure 1b). Moreover, small catalytic particles can be observed at the ends of the nanostructures in (figure 1a). This demonstrates the catalytic role that iron plays, consistent with the observation of boron nanostructures grown via a solid-state reaction [1-3]. Detailed observation shows that our CVD synthesis route yields both nanoplatelets and nanowires as shown in (figure 1c, d) correspondingly. The nanoplatelets display no contrast (figure 1c) while the nanowires have distinct line contrast (figure 1d). Quantitative EELS characterization reveals that the chemical composition of the platelets and wires can be determined to be $BC_{0.18\pm0.10}O_{0.15\pm0.09}$ (figure 1e), consistent with boron-rich nanostructures. In the following, we provide a summary of their characterization by transmission electron microscopy (TEM).

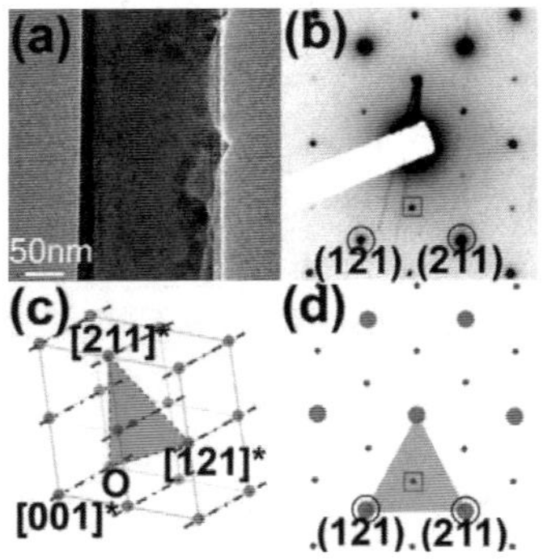

Figure 2. Bright field image of a typical nanoplatelet (a) and its diffraction pattern with one of the fractional spots indicated by a square box (b). The proposed three-dimensional reciprocal lattice with diffraction rod elongation in $[001]^*$ direction (c) and the corresponding diffraction simulation (d). (Color online)

Figure 2 shows the result of SAD characterization in nanoplatelets. The diffraction pattern shown in (figure 2b) can be indexed to the [113] zone axis of boron-rich rhombohedra crystals, apart from a set of fractional diffraction spots as indicated by a box in (figure 2b). Such anomalous spots are also characteristic of nanoplatelets grown in solid-state reaction [3, 5] and can be attributed to abundant planar defect along (001) [5]. This is characterized by elongation of three-dimensional reciprocal lattice along the $[001]^*$ direction (figure 2c). Under such circumstance fractional diffraction spots are introduced as shown in (figure 2c). The simulated pattern (figure 2d) agrees well with the experiment result (figure 2b).

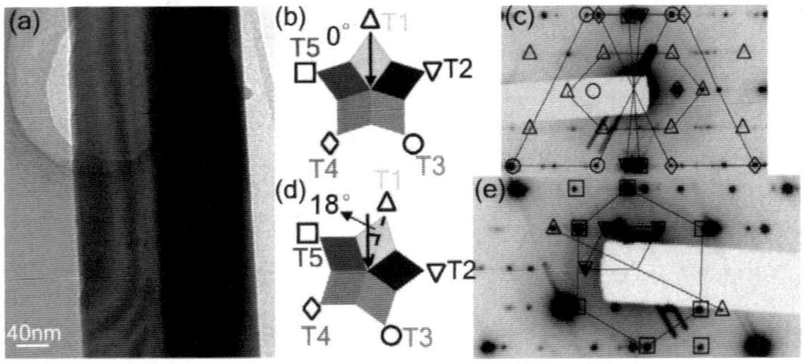

Figure 3. A plan-view TEM image of a cyclic twinning nanowire (a). A schematic cross-section view (b and d) of a cyclic twinning wire composed of T1-T5 single crystalline segments at 0° and 18° orientations respectively. The electron beam travels along the direction denoted by the vertical black arrow in (b and d). The corresponding diffraction pattern simulations are superimposed on the experimental data (c and e) for the two orientations in (b and d), with contributions from different single crystal units distinguished by shaped symbols. (Color online)

Figure 3 shows TEM analysis of a typical nanowire. The plan-view image of the nanowire in question (figure 3a), with the contrast indicating the presence of a twinning plane along the wire axis. Analysis of systematic diffraction patterns taken along two or more specific beam directions confirms that the nanowires have a cyclic twinning structure similar to that shown first for boron suboxide nanowires [1], with five crystalline segments joined along (100) twinning planes as shown in (figure 3b, d). In (figure 3c, e) are shown two 'high symmetry' diffraction patterns from this nanowire, taken with the nanowire rotated 18° clockwise in (e) from that in (c). To index these complex diffraction patterns, we have calculated simulated electron diffraction patterns from such nanowires whose cross-sections are oriented as shown in (figure 3b, d) for two orientations, 18° apart, chosen for demonstrating the twinned symmetry inherent in such a cyclic structure. Simulated electron diffraction patterns from individual twin variants, as denoted by different shaped symbols, are superimposed on top of the experimental diffraction pattern in (figure 3c, e) respectively. The simulated diffractions agree with the experimental observation except for some weaker spots in (c). The star-shaped cross-section assumed, with the re-entrant side facets, is consistent with the observation of the linear contrast in the SEM images of such nanowires (figure 1d).

4. Discussions

Electron microscopy characterization explicitly shows that the cyclic twinning wires and parallel twinning nanoplatelets grown in the solid-state reaction can be reproduced on a silicon chip in our simple CVD set-up. During this process, boron reacts with oxidants such as B_2O_3, BaO or C and produces boron rich gaseous species (for example, B_2O_2 [7]). At the growth temperature, we expect the iron film to have broken-up into droplets which can absorb boron-rich gases to produce boron-rich nanostructures. This is consistent with the observation of spatial dependence of nanostructure density on the silicon chip, with the highest density nearest to the pellet. Our demonstration of the CVD growth mechanism confirms that the same process is present in the original reaction. Products include boron-rich nanostructures, such as elongated boron carbide platelets [8], boron suboxide platelets [5] and wires [1]. The CVD growth on silicon wafer can be controlled by patterning the iron layer.

5. Summary

We have verified that the growth mechanism of boron-rich nanostructures produced in the solid-state reaction is through catalyst assisted chemical vapour deposition, by replicating the growth process on a silicon substrate. This simple synthesis approach opens new ways to produce boron-rich nanostructures through controllable methods and over a large area.

Acknowledgement

This research is supported by the Vice Chancellor's Start-up fund of University of York. Y.Z.Y would like to thank the University of York for the exchange visit that resulted in this study.

References

[1] Jiang J, Cao M, Sun Y, Wu P and Yuan J 2006 *Appl. Phys. Lett.* **88** 163107.
[2] Fu F, Jiang J, Zhang W Z and Yuan J 2008 *Appl. Phys. Lett.* **93** 043101-3.
[3] Cao M H, Jiang J, Liu H X and Yuan J 2006 *J. Electroceram.* **17** 817-20.
[4] Emin D 1987 *Physics Today* **40** 55-62.
[5] Yu Z Y, Jiang J and Yuan J (submitted).
[6] Millan A, Dewit N and Van Roost 2000 *Phys. Chem. Chem. Phys.* **2** 171-175.
[7] Inghram M G, Porter R F and Chupka W A 1956 *J. Chem. Phys.* **25** 498-501.
[8] Carlsson M, Garcia-Garcia F.J. andJohnsson M 2002 *J. Cryst. Growth* **236** 466-76.

Carbon nanotube to SiC nanorod conversion in molten salt studied by EELS and aberration corrected HRTEM

Wei Xie, Günter Möbus, Shaowei Zhang
Department of Engineering Materials, University of Sheffield, Sheffield, S1 3JD, U.K.

E-mail: mtq06wx@sheffield.ac.uk

Abstract. Silicon carbide (SiC) nanorods were prepared *via* reacting silicon nanopowders and multi-walled carbon nanotubes (MWCNTs) for 4h at 1100-1200°C in a molten NaCl-NaF binary salt in Ar atmosphere. The synthesised SiC nanorods were characterised by X-ray diffraction (XRD), high-resolution transmission electron microscopy (HRTEM), electron energy loss spectroscopy (EELS), and spherical aberration-corrected FEGTEM. XRD results revealed that MWCNTs had converted into SiC after firing at 1200°C. HRTEM observation showed a complex stacking sequence aligned along the SiC-rod direction. Combination of EELS and aberration-corrected FEGTEM illustrated the resulting nanorod composed of 3C-SiC and some stacking faults. The synthesised SiC nanorods to a large extent retained the morphology of CNT reactants, indicating that the "template-growth" mechanism had dominated the synthesis process. The conversion of MWCNTs into SiC nanorods was strongly dependent on reaction temperature and salt composition. It was found that molten NaCl-NaF binary salts could effectively accelerate the SiC formation.

1. Introduction

Silicon carbide (SiC) is one of the most important ceramic materials for abrasives, refractories, and numerous high temperature applications due to its excellent mechanical and thermal properties, such as high melting point, hardness, and chemical stability. Recent results have revealed that one-dimensional nanostructured silicon carbide (nanowires or nanorods) not only exhibits promising electronic performance but also shows outstanding elasticity and strength properties (measured Young's modulus of 600GPa for SiC nanowires) [1, 2]. The first study to synthesise SiC nanotubes could be tracked back to 1990s, from the reaction between carbon nanotubes and SiO (or SiI_2) vapour [3]. Since then, many approaches have been carried out to prepare carbide nanorods/nanotubes. Quah *et al.* reported that SiC nanotubes could be formed by reacting single-walled carbon nanotubes and SiO_2 nanopowders at 1350°C [4]. Wu and co-workers fabricated SiC nanorods through gas-solid reaction between silicon and multi-walled carbon nanotube by a direct heating method at a temperature over 1450°C [5]. These methods, however, have some drawbacks, such as the need of high temperatures, complex experimental equipment, and/or pump system to obtain a high quality vacuum. In order to overcome these drawbacks, and to simplify the synthesis procedure, a molten salt synthesis (MSS) technique was applied to fabricate SiC nanorods in this work. After heat treatment of the sample in molten salts, the reacted mass was characterised by HRTEM, EELS, and spherical aberration (Cs)-corrected FEGTEM to clarify their composition and structures.

2. Experimental procedures

Multi-walled carbon nanotubes (Aldrich, O.D. 7-20nm) were mixed with Si nanopowder (Aldrich, < 100nm) in an agate mortar, and the mixture was then dispersed in isopropyl alcohol (IPA) and sonicated in an ultrasonic bath for 60 min, followed by drying in an oven at 80°C. The mixed NaCl and NaF salts with the weight ratio of 4:1 were pre-dried overnight in an oven at 120°C. The starting material mixture was then combined with appropriate amounts of the salts. The powder mixture was placed in a graphite crucible with a graphite lid and heated for 4h at 1100-1200°C in a flowing Ar atmosphere. In order to avoid contaminations from the crucible, the graphite crucible used was pre-treated with Si powder at elevated temperature to form a stable SiC film on the inner walls. After cooling the furnace to room temperature, the reacted mass was rinsed with distilled water to leach out any residual salts. Subsequently, the resulting powders were oven-dried overnight prior to further characterisation. The phases present were identified by X-ray diffraction (XRD) (Siemens D500), and the morphology and microstructure of the resulting materials characterised by HRTEM (JEOL 2010F), EELS (Gatan GIF200), and aberration-corrected FEGTEM (JEOL 2200FS).

3. Results and discussion

X-ray diffraction analysis (not shown) of samples heated in NaCl-NaF salts for 4h at 1100-1200°C reveals that SiC starts to form at 1100°C in molten salts. On increasing temperature from 1100°C to 1150°C, SiC peak heights increase whereas CNT and Si peak heights decrease. On further increasing temperature to 1200°C, carbon peaks disappear and only SiC peaks are observed.

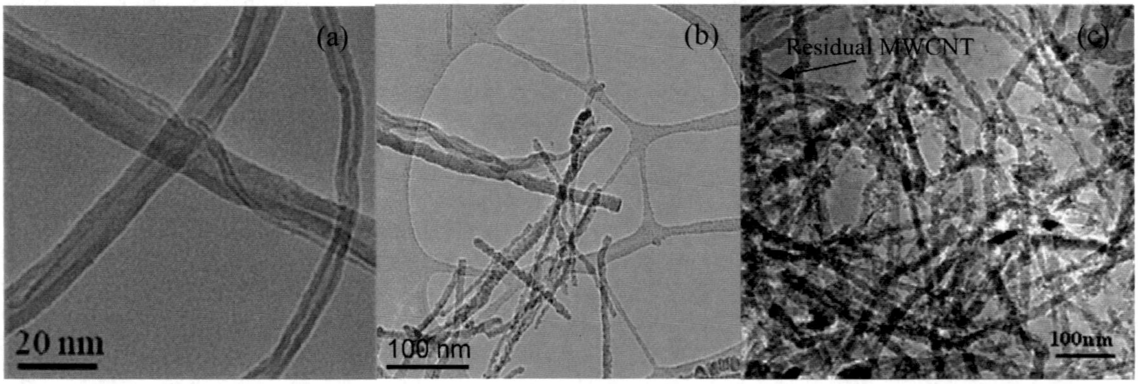

Figure 1. TEM images of the as-received MWCNTs (a) and resulting products (b) and (c).

Shown in Figure 1 are TEM images of the as-received MWCNTs (Figure 1a) and the reacted products (Figure 1b, c) obtained after heating for 4h at 1200°C in NaCl-NaF. Apart from SiC nanorods, a few residual MWCNTs were occasionally seen in the microstructure, although they were not detected by XRD. The SiC nanorods to a large extent retained the morphology of MWCNT reactants but with much rougher surfaces. To further characterise the structure of the SiC nanorods, atomic resolution TEM images of the samples were taken (Figure 2). The inset magnified images in Figure 2 illustrate that there are mainly two types of typical SiC nanorod products: a rod with complex SiC stacking sequences aligned along the rod direction (Figure 2a), slightly curved indicating possible deformation faults and weak contrast caused by the slight off-orientation of the rod; and a rod consisting of regular arrangement of lattice fringe ordering (Figure 2b). In Figure 2b, the inset magnified image reveals that the crystal lattice fringes have a d-spacing of 0.25nm which corresponds to (111) of 3C-SiC, in agreement with the XRD pattern. In addition, it can be seen that in this case the arrangement of lattice fringes is regular and uniform throughout the rod, indicating that this straight SiC nanorod has high perfection without any defects or dislocations. Moreover, some amorphous phases are also observed on the surface of the rod, and this amorphous part (mostly graphitic) might be attributed to the contamination during TEM session.

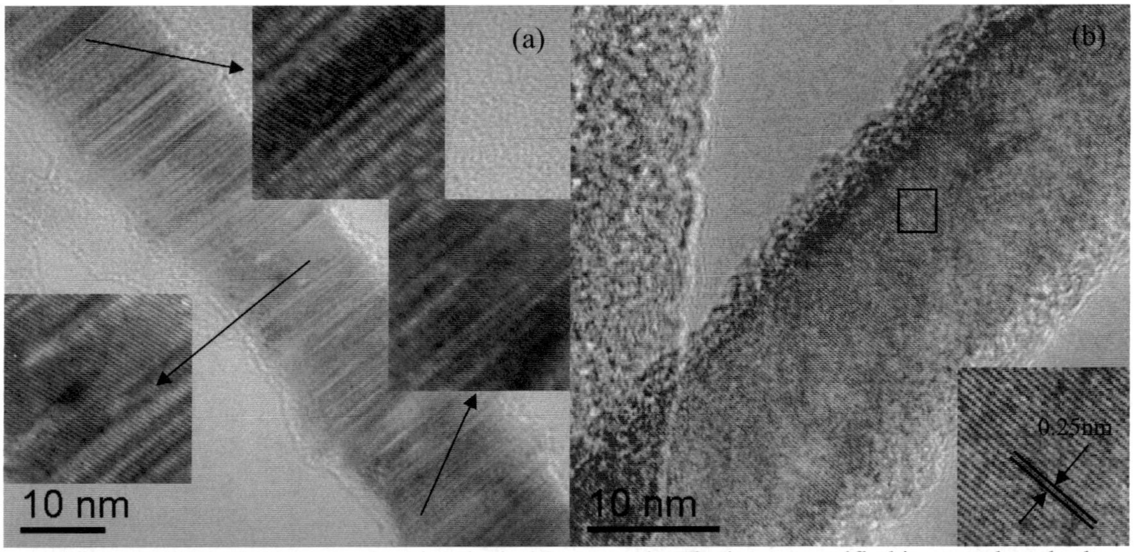

Figure 2. HRTEM micrographs of typical SiC nanorods. The inset magnified images show both complex (a) and regular (b) stacking sequences aligned along the SiC-rod direction.

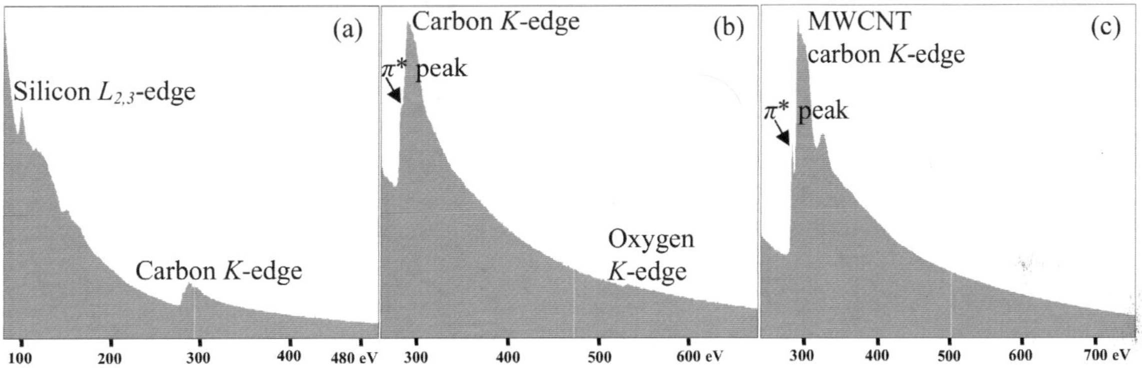

Figure 3. (a) and (b) EEL spectrum recorded from the as-produced nanorod (Figure 2b). (c) EEL spectrum of the as-received MWCNT.

An EEL spectrum of the SiC nanorod is shown in Figure 3a and b, which reveals the strong silicon L_{23}-edges (onset ~97 eV) and carbon *K*-edge (onset ~284 eV), indicating a pure silicon carbide phase is formed. In contrast with the EEL spectrum of the as-received MWCNT (Figure 3c), the carbon π* peak nearly disappears, indicating a structure change of carbon towards sp^3-bonding. In terms of peak positions, the L_{23} edges of silicon shown in Figure 3a agree well with theoretical Si L_{23} edges of 3C-SiC [6]. Also, the carbon *K*-edges of the as-produced nanorod are comparable with those of carbide materials (e.g. SiC and TiC) reported in previous researches [7, 8]. Interestingly, the energy of the maximum intensity and the shape of C *K*-edge of the sample is very similar to that of 3C-SiC as reported by Fu *et al.* [8], though there is a slight difference in carbon π* peaks which could be attributed to carbon contamination. In addition, a tiny oxygen *K*-edge (onset ~531 eV) signal (Figure 3b) is detected, suggesting there is contamination of the products. Figure 4 exhibits aberration-corrected TEM images (at $|Cs| < 0.5\mu m$) of resulting SiC-rods with mostly regular stacking sequences of cubic 3C-SiC seen along [110], however intermitted with many twins and stacking faults. Internal strain in Figure 4b leads to a variation of the viewing angle around the perfect zone axis, and in small selected areas some polar dumbbell contrast seems to appear with either positive or negative contrast, depending on focus. However, at estimated 10nm thickness, this is not a linear "structure image" and does not reveal true atomic spacing, as confirmed by HREM image simulations. The synthesised SiC nanorods to a large extent retained the morphology of CNT reactants, indicating that the "template-growth" mechanism had dominated the synthesis process. For this mechanism to function, Si has to dissolve in the molten salts during the reaction, diffuses onto the surface of MWCNTs, and then reacts with MWCNT to form SiC. MWCNT acts as a "template" during this process. It is also possible that Si diffuses together with the molten salt into the tube through the open areas, such as the tube ends. However, due to the low diffusion rate of Si in SiC, this phenomenon could be ignored since the pre-formed SiC would block further free diffusion of Si into the tube cavity. The dissolution behaviour of Si in molten salts has not been fully understood, although some researchers observed the dissolution of other metals in molten salts [9]. Based on the above discussion, it can be concluded that the MSS technique provides a convenient medium and enables the acceleration of carbide formation at relatively low temperatures. Further studies on exploring the dissolution of Si in molten salts and the exact growth mechanisms of the carbide materials are expected to be carried out in the future.

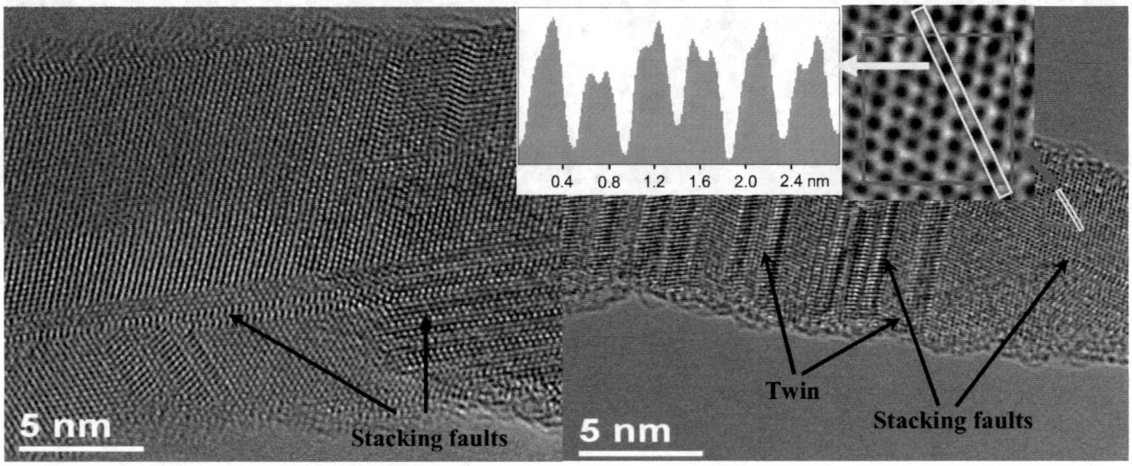

Figure 4. Aberration-corrected HRTEM images of two typical resulting SiC nanorods with inset images of perfect regions to study stacking sequence and dumbbell contrast.

4. Summary

Silicon carbide nanorods were prepared *via* reacting silicon nanopowders and MWCNTs in molten NaCl-NaF salts for 4 h at 1200ºC. This synthesis temperature is much lower than those used by other routes previously reported, indicating that the molten NaCl-NaF salt can effectively accelerate the SiC formation. The "template growth" mechanism played a dominant role in the synthesis process. The MWCNT not only acted as a carbon source but also provided a template for SiC nanorod growth.

Acknowledgement

The authors would like to acknowledge the financial support of EPSRC and Sheffield University and also thank Ian Ross and Zineb Saghi for help with the JEM 2200FS.

References

[1] Wong E W, Sheehan P E and Lieber C M 1997 *Science* **277** 1971-1975
[2] Yang W, Araki H, Tang C, Thaveethavorn S, Kohyama A, Suzuki H and Noda T 2005 *Adv. Mater.* **17** 1519-1523
[3] Dai H J, Wong E, Lu Y Z, Fan S S and Lieber C M 1995 *Nature* **375** 796
[4] Quah H, Cheong K Y and Lockman Z 2008 *J. Alloys Compd.* **475** 565-568
[5] Wu R B, Yang G Y, Pan Y and Chen J J 2007 *J. Mater. Sci.* **42** 3800-3804
[6] Egerton R F 1996 *Electron Energy-Loss Spectroscopy in the Electron Microscope* Plenum Press New York 364-365
[7] Scott A J, Brydson R, MacKenzie M and Craven A J 2001 *Phys. Rev. B* **63** 245105
[8] Fu X, Jiang J, Hu X, Yuan J, Zhang Y, Han X and Zhang Z 2008 *J. Phys.: Conference Series* **126** 012072
[9] Straumanis M E, Shih S T and Schlechten A W 1957 *J. Electrochem. Soc.* **104** 17–20

Nanostructural development of non-graphitising carbons probed using TEM/EELS: the importance of fullerenes ?

Z Zhili,[1 2] R Brydson,[1] S Reddy[1], A Westwood,[1] A Brown[1] and B Rand[1]

[1] Institute for Materials Research, University of Leeds, Leeds, LS2 9JT, UK
[2] Institute of Materials, Beijing Jiaotong University, Beijing 100044, PR China

mtlrmdb@leeds.ac.uk

Abstract. This work presents a comprehensive study of a set of non-graphitising carbons as a function of heat treatment temperature (HTT) up to 3000 °C using a combination of density measurements, X-ray and electron diffraction, high resolution TEM and low loss and core loss EELS. Results are compared to those obtained from graphitising carbons. Of interest is the variation in plasmon energy with HTT and its relationship to density as well as the ratio of sp^2 to sp^3 bonded carbon species derived from the C K-edge using a revised analytical procedure. Results are reviewed in light of the fairly recent suggestion that the intrinsic structure of non-graphitising carbons is based on the inclusion of fullerene-like units. We speculate as to whether this might have a broader significance for carbonaceous materials in general.

1. Introduction

Carbon materials are highly versatile owing to the strong dependence of their physical properties on the ratio of sp^2 (graphitic) to sp^3 (adamantine) bonds. Graphitic carbons have a variety of forms with various degrees of ordering ranging from microcrystalline graphite to glassy carbons. Amorphous carbons can have any mixture of sp^3, sp^2 and, in some cases, sp^1 bonded carbon sites with the additional possible presence of hydrogen, nitrogen or boron. The structure of amorphous or poorly crystalline carbon materials are difficult to study using conventional characterisation tools, although neutron scattering, nuclear magnetic resonance, (*uv*) Raman spectroscopy and EELS all show promise.

Non-graphitising (NG) carbon materials are hard, low density materials which fail to pass through a liquid stage upon heat treatment and do not ultimately form crystalline graphite at temperatures in the region of 3000 °C. Instead they largely retain a hard, entangled ribbon-like structure containing micropores < 2 nm in diameter. This microstructure is fixed by the cross-linking of the polymer chains within the precursor and traditionally this high surface area microporous structure has been viewed as consisting of twisted networks of carbon layer planes cross-linked by bridging groups. However, more recently a fullerene-like structure has also been proposed from high resolution electron microscopy data [1].

As a typical non-graphitisable carbon, phenolic resin is commonly used in carbon/carbon composite materials or as a refractory binder owing to its homogeneous, isotropic properties. During heating, phenolic resin is gradually transformed from a polymer to a hard, glass-like isotropic carbon containing some microporosity. The purpose of this study is to shed light on the evolution of the nanostructure and chemical composition of a phenolic resin, during the heat treatment process, ultimately in order to try and explain the observed macroscopic properties.

2. Experimental

A resol-type phenolic resin (SC 1008 - Borden Chemicals UK Ltd.) was subjected to different heat treatment temperatures (HTT) between 600 °C and 2000 °C for 2 hours under a N_2 atmosphere; the heating rate was either 1 °C /min for temperatures < 1300 °C or 500 °C /h for temperatures ≥1600 °C . Two additional samples were also high temperature heat treated using different heating profiles up to a maximum temperature of 2500 °C and 3000 °C respectively. The samples heated to temperatures ≤ 1000 °C were initially cured at 260 °C in order to remove volatile species, while those heated to temperatures ≥ 1300 °C were correspondingly cured at 1000 °C. After each heat treatment, the percentage mass loss and density (both Archimedes and Helium) of samples were determined, as well as the sample surface area and pore size distribution using nitrogen gas adsorption.

TEM analysis was conducted on an FEI CM200 FEG TEM running at 197 kV equipped with an Oxford Instruments EDX spectrometer and a Gatan Imaging Filter. Samples were prepared by dispersing in methanol, with a drop placed on a holey carbon coated copper grid. Powder XRD was also performed using a Philips APD 1700 diffractometer and Cu Kα radiation.

3. Results and Discussion

The volume plasmon in the EELS spectrum arises from a collective oscillation of the entire valence electron population and, in the free electron model the plasmon peak energy is proportional to the square root of the valence electron density. Figure 1a shows the variation of the valence plasmon peak energy, E_p, (extracted from the first derivative of the spectra) with different HTT of the phenolic resin samples, the error bars represent the standard deviation of 10 measurements from each sample. It is apparent that the plasmon peak position remains unchanged or even slightly decreases with increasing HTT, this behaviour tracks the sample density which remains relatively constant (figure 1b) or actually falls during heat treatment; as a result of the cross-linked structure any densification due to ordering is presumably counteracted by the development of microporosity. Note this data are in marked contrast to that from graphitizing carbons [2], where the plasmon peak position of a graphitizing pitch (ca. 23 eV - 26 eV) correlated extremely well with the monotonically increasing density of the material (1.5 g/cm^3 to 2.25 g/cm^3) as function of HTT.

Figure 2(a)-(e) shows a selection of high resolution electron micrographs deemed, after considerable sampling, to be representative of the microstructures observed during HTT. At low HTT there is no obvious long range order (Figure 2a). At 800 °C there is a suggestion of embryonic ordering, with the presence of 2-3 aligned basal (002) graphitic layers each containing several aromatic rings (basic structural units) in some sample regions. This ordering persists and is more obvious at 1000 °C in images underfocused below Scherzer defocus (figure 2b) where there is the suggestion of the formation of closed fullerene-like loop structures. By 1300 °C, clear (002) type lattice fringes are apparent with layer packets of thickness 3-4 layers and 3-4 nm in lateral dimension being visible. In addition pores a few nanometres in diameter are evident. These layer packets order and become straighter at 1600 °C (figure 2c) , whilst at 2000 °C and above (Figure 2d-e) circular and elongated closed loop structures some 5-6 nm in diameter become apparent with clearly visible (002) fringes and layer packets > 4-5 layers begin to form. The images at temperatures > 2000 °C are somewhat reminiscent of the structures seen by Harris and Tsang [1] in chars derived from a polyvinylidene chloride (PVDC)- polyvinylchloride (PVC) copolymer and also sucrose after pyrolysis at 2100 - 2600 °C .

The TEM images are in reasonable quantitative agreement with the XRD data at these temperatures; with increasing HTT, the (002) layer spacings are observed to progressively decrease and the layer stack coherence length L_a progressively increase up to ca. 8 nm. Furthermore, N_2 gas adsorption shows a maximum in pore volume for the 800 °C sample corresponding to a minimum in the pore width. Above this HTT, the pore width was observed to increase up to a maximum value at 1300 °C and then decrease again, with the corresponding pore volume going through a minimum at this temperature presumably arising as a result of increasing closed porosity from the formation of fullerene-like units.

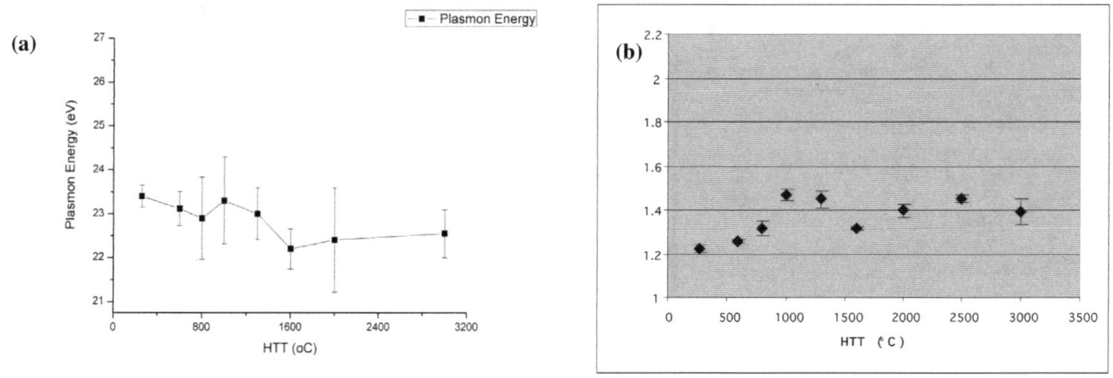

Figure 1. (a) Plasmon peak position (b) density of phenolic resin as a function of HTT.

The EELS carbon K-edges from the phenolic resin samples heat treated at different temperatures, show two main features: a π^* peak at 285 eV followed by a σ^* peak at ca. 290 eV and above. With increasing HTT, the relative π^* peak intensity is observed to increase and the σ^* structure becomes more well defined. Quantitative analysis of carbon bonding using EELS may be achieved using procedures outlined in reference [3]. Briefly the 1s to π^* peak intensity at the C K-edge (background stripped and deconvoluted to remove plural scattering), I_{π^*}, is measured using a Gaussian fitting procedure and ratioed to the total C K-edge intensity in a window of up to 20 eV extending from the edge onset. This intensity ratio is normalized to a reference spectrum containing 100% sp^2 bonding (highly oriented pyrolytic graphite). The normalized intensity ratio can then be used to obtain the sp^2 fraction, provided the different carbon bonding states are known, e.g. just sp^2 and sp^3. An example fit is shown in figure 3a and the variation in sp^2 content during the heat treatment of the non-graphitising carbon derived using this procedure is shown in figure 3b. These results suggest that, although the overall density is reasonably constant with increasing HTT presumably due to the presence of crosslinks, the sp^2 content does still increase in a similar fashion to that seen in graphitising carbons [2], albeit with the development of many very small graphitic pockets rather than an increasingly larger densified graphitic structure, as is the case with a graphitising carbon [2]. In addition, there is evidence for a reduction in sp^2 carbon content at intermediate temperatures > 1000 °C, which could be due to puffing and disordering of layers as a result of the loss of volatiles, mass loss and a corresponding density decrease from the emergent fullerene-like structure in the microporous material. Similar layer puffing was suggested to occur in graphitising carbons due to the loss of heteroatoms at somewhat higher temperatures [2].

The fit to the C K-edge in figure 3a has employed unconstrained Gaussian fits to the following features: C=C π^* (at ca. 285 eV), C-C σ^* (ca. 292 eV) and C=C σ^* (ca. 300 eV) (assignments nominally based on gas phase spectra); note that the residuals of the fit clearly show the existence of a peak at ca. 287 eV which persists even at high HTT [3]. There are a number of possible origins for this residual peak such as C-H σ^* states or even aldehyde-type bonds. It is unlikely hydrogen persists at high temperatures, however one distinct possibility based on the high resolution TEM images is the presence of fullerenes which, on the basis of high resolution X-ray absorption spectra, would be expected to give rise to a double peaked π^* peak with additional intensity at 287 eV [4].

In a broader context, we speculate as to whether such fullerene-like features are also present in other poorly crystalline carbon systems such as amorphous diamond-like carbon (DLC) films, glassy carbons, carbon nanostructures and even graphitizing carbons which may complicate a simplistic analysis of carbon bonding and possibly explain any discrepancies (some observed by the authors) between EELS analysis and other techniques for determining sp^2/sp^3 ratios such as *uv* Raman.

Journal of Physics: Conference Series **241** (2010) 012094 doi:10.1088/1742-6596/241/1/012094

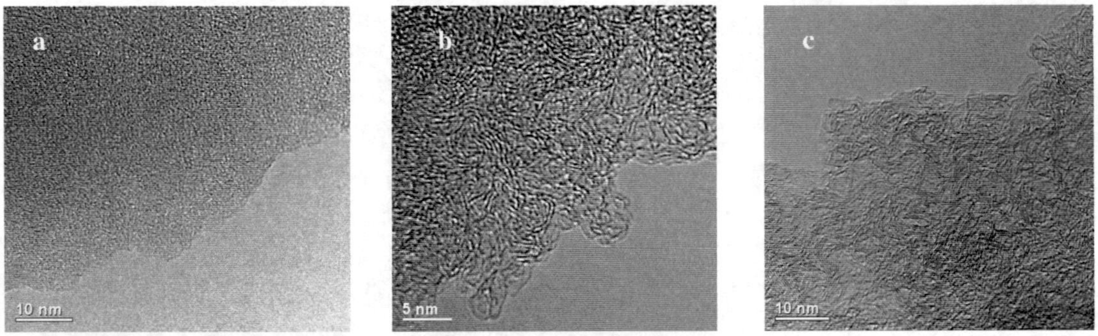

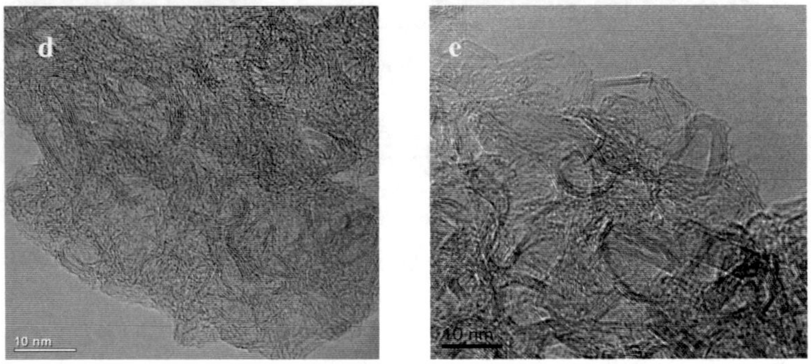

Figure 2. High resolution phase contrast images of the phenolic resin following HTT at (a) 600 °C, (b) 1000 °C, (c) 1600 °C, (d) 2000 °C and (e) 3000 °C

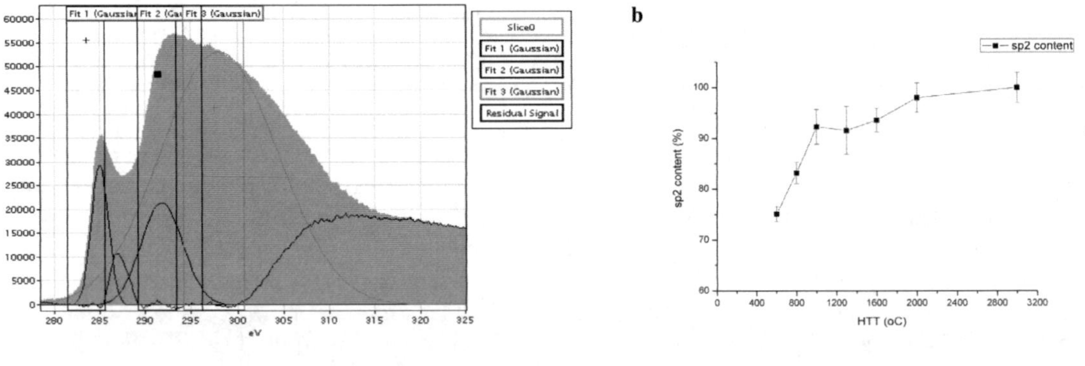

Figure 3.Analysis of C K-edge: (a) Example of Gaussian fitting on a low HTT phenolic resin (b) sp^2 content as a function of HTT for the set of phenolic resin samples.

4. Conclusions

There appears to be evidence from HRTEM and also analysis of the EELS carbon K-edge for the presence of fullerene-like structures in non-graphitising carbons heat treated above 800-1000 °C. Fullerene-like units are expected to result in additional intensity on the high energy side of the C=C π* peak at the C K-edge and may complicate a simplistic analysis of the sp^2-content in carbon materials.

References

[1] Harris PFJ and Tsang SC 1997 *Phil. Mag. A.* **76** 667-677.
[2] Daniels H et al. 2007 *Phil. Mag.* A**87** 4073-4092.
[3] Brydson R et al. 2008 Electron Microscopy EMC2008. **1** 357-358.
[4] Nyberg M et al. 1999 Phys. Rev B **60** 7956-7960.

Journal of Physics: Conference Series **241** (2010) 012095 doi:10.1088/1742-6596/241/1/012095

Variable Temperature Investigation of the Atomic Structure of Gold Nanoparticles

Neil. P. Young[1], M.A. van Huis[2], H.W. Zandbergen[2], H. Xu[3], A.I. Kirkland[1]

[1] Department of Materials, University of Oxford, Parks Road, Oxford OX1 3PH, UK

[2] Kavli Insitute of Nanoscience, Delft University of Technolgy, Lorentzweg 1, NL-2628CJ, Delft, The Netherlands

[3] Department of Geology and Geophysics, and Materials Science Program, University of Wisconsin-Madison, Madison, Wisconsin, USA.

E-mail: neil.young@materials.ox.ac.uk

Abstract. The characterisation of nanoparticle structures is the first step towards understanding and optimising their utility in important technological applications such as catalysis. Using newly developed *in-situ* transmission electron microscopy (TEM) specimen holders, the temperature dependent atomic structure of gold nanoparticles in the size range 5-12 nm has been investigated. In this size interval, the decahedral morphology has been identified as the most favourable structure at or above room temperature, while particle surface roughening becomes evident above 600°C. An icosahedral transition has also been identified at low temperature in particles under 9 nm in diameter. These experimental results are consistent with recently published temperature dependent equilibrium phase maps for gold nanoparticles.

1. Introduction

The structure of metallic nanoparticles continues to attract considerable interest due to its fundamental importance in relation their physical and chemical properties. These properties are crucial to the commercial exploitation of nanomaterial based technologies in heterogeneous catalysis. However a rigorous understanding of the structure of metallic nanoparticles, particularly as a function of particle size and temperature proves elusive. This situation has arisen through a number of factors pertinent to the physics of nanoparticles; primarily the sensitivity of small particles to the experimental conditions under which they are produced, often yielding non-equilibrium structures, and also the possibility of multiple stable structural configurations for any given size and temperature scenario. Such structural isomers may be closely related in terms of total energy, yielding structurally poly-dispersed populations, making interpretation difficult. Finally, the size of such particles makes experimental verification a challenge in terms of high-resolution *in-situ* imaging. In the present study we report results of variable temperature experiments on gold nanoparticles using newly designed *in-situ* TEM holders and compare our findings with contemporary theoretical calculations.

2. Experimental Method

Our methodology employs gold colloids prepared by the reduction of dilute solutions of tetrachloroauric acid, using maltodextrin to prevent aggregation. Particles are formed in a size range of 5 nm to 12 nm in diameter and were drop-cast from aqueous solution onto suitable TEM supports. For the heating experiments an advanced MEMS micro-hotplate type holder with SiN support membrane was used [1] in combination with a Philips CM300-UT field-emission transmission electron microscope. The low heat capacity of the micro-hotplate allows the temperature of the chip to be varied quickly and with high precision, while producing minimal specimen drift within the microscope. Temperature changes can be controlled with an accuracy of 0.1°C through feedback electronics, while the absolute uncertainty in temperature is estimated at ±30°C due to possible local temperature variations of the substrate and induced beam heating. To investigate particle cooling, a newly designed TEM cryo-holder [2] was used in combination with a JEOL JEM-2200MCO aberration-corrected microscope.

3. Results and Discussion

The first question we address is the equilibrium structure of nanoparticles at room temperature. This was investigated by selecting particles in clearly identifiable crystallographic orientations and steadily heating to ca. 400°C in the CM300-UT microscope whilst recording images at regular temperature increments. The temperature range used was expected to provide sufficient energy for the particles to thermodynamically equilibrate, regardless of their initial structure that may have resulted from kinetic trapping. An example of such a transition is shown in figure 1, where a 7.2 nm diameter particle, with an initial icosahedral structure (figure 1(a)) is heated to 400°C.

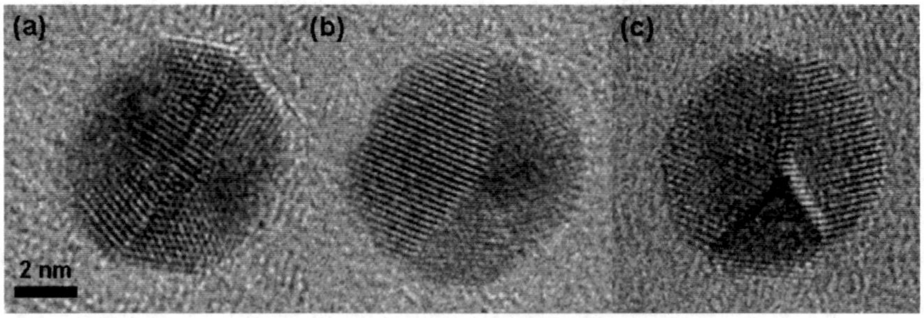

Figure 1: *In-situ* heating of an individual 7.2nm gold nanoparticle in the CM300-UT. (a) The initial icosahedral structure, close to a <112> projection, at room temperature. (b) The structure observed at 400°C, which corresponds to a decahedron close to a <110> projection. (c) Cooling the particle back to room temperature the decahedral structure is observed to be stable.

At this temperature the structure changes to the decahedral form shown in figure 1(b). No further structural transitions were observed and following gradual cooling back to room temperature the decahedral structure persists. The transition to the decahedral phase was the overwhelming trend in all of our heating experiments, whether this occurred from either an icosahedral, single crystalline or twinned starting structure. Statistically (from 33 individual, isolated particles tracked over a range of temperatures), 88% of the equilibrated nanoparticles adopted a decahedral morphology after heating to 350 - 400°C, while for 12% the as grown structure (almost exclusively icosahedral) persisted. These results indicate that the decahedral phase is energetically preferred at intermediate temperatures around 400°C and also as the room temperature equilibrium structure for gold particles in the size range investigated.

The structure of particles at high temperatures is of particular relevance for catalytic applications where changes in the surface periodicity have a strong correlation with catalytic performance. In order to investigate this, a range of pre-equilibrated particles were heated to around 650°C, with a representative heating series shown in figure 2. Evidence for the formation of re-entrant {111} facets at the particle vertices is shown in 2(c). The onset of this change in surface structure was routinely observed in our study in the temperature range 350°C – 400°C. We propose that the formation of these facets lowers the overall surface energy, offsetting contributions from internal strain and allowing the ordered decahedral motif to persist to higher temperatures. Heating the particle further reveals the onset of additional changes to the surface structure. Figures 2(e) and (f) show the particle at 600°C and 650°C respectively. At these temperatures the particle interior retains discernable decahedral features, while the particle perimeter becomes increasingly smoothed. Since the terminal layers of gold atoms have fewer neighbours, the surface is the likely location for the first signs of melting. While a full liquid melt is not suggested by the current data, the surface is certainly less ordered and resembles an amorphous layer in some locations. On cooling to room temperature, the particle did not recover a pristine decahedral structure; rather it retained the highly distorted shape presumably since insufficient kinetic energy was then available for a further transformation.

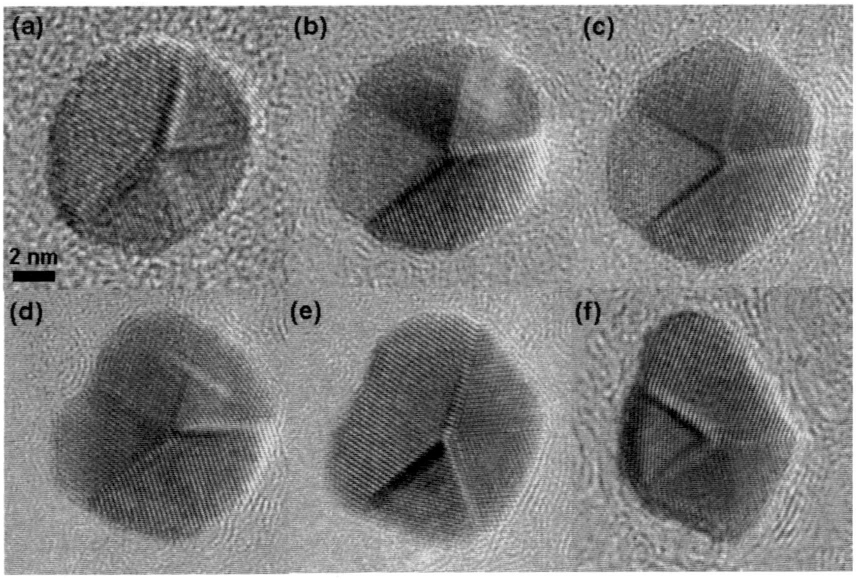

Figure 2: Controlled heating of a 10.2 nm decahedral particle in the CM300-UT. (a) Room temperature, (b) 300°C, (c) 400°C, (d) 500°C, (e) 600°C, (f) 650°C.

A quantitative phase map for gold nanoparticles, based upon relativistic *ab-initio* thermodynamics has been constructed [3], and provides an ideal comparison for the experimental data presented here. The key predictions of the phase diagram in terms of structure of gold particles as a function of temperature are summarised in table 1 below. Particle sizes studied in our experiments are predicted to adopt a decahedral morphology between room temperature and 400°C. The phase diagram predicts surface roughening of particles at 600°C, which is defined as a general deterioration of the surface structure; including melting, defect formation and growth of solid amorphous gold shells. Figure 2(f) shows such effects and is the topic of our continued research.

Having investigated particles at elevated temperature and found good agreement with the above published calculations, we have chosen to investigate the low temperature behavior of the gold particles. Figure 3(a) shows a decahedral nanoparticle imaged at room temperature. The specimen is then cooled *in-situ*, and the same particle imaged at the holder base temperature of -173°C. At this temperature this 9 nm particle is seen to revert to an icosahedral morphology. We have reproduced this for similarly sized small particles, however we have also observed that larger decahedra are structurally stable against transitions at low temperature in agreement with the predictions of the phase diagram.

Table 1: Summary of results from a recent quantitative phase diagram for gold particles [3] *(Ic: Icosahedral, Dh: Decahedral, FCC: single crystalline FCC or twinned, M: Molten, QM: Quasi-molten, SR: Surface roughening)*

Temperature / °C	**Size / nm**	**Structure**	**Temperature / °C**	**Size / nm**	**Structure**
-200	0 – 9	Ic	**200**	0 – 2	Ic
	9 – 14	Dh		2 – 18	Dh
	14 – 30	FCC		18 – 30	FCC
Room	0 – 4	Ic	**400**	0 - 1.5	Ic
Temperature	4 – 15	Dh		1.5 – 16.5	Dh
	15 – 30	FCC		16.5 – 30	FCC
			600	0 – 2.5	M / QM
				2.5 – 30	SR

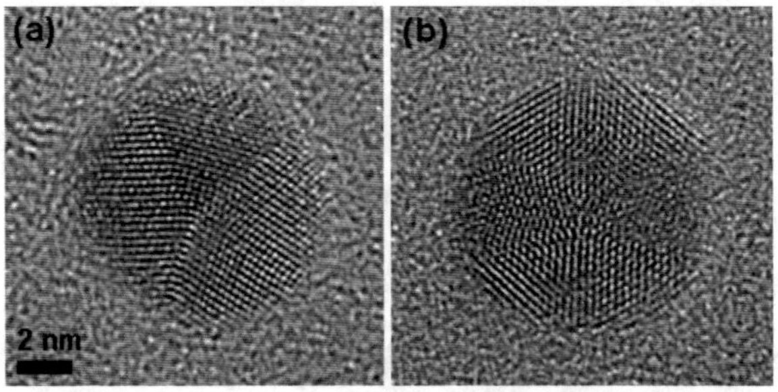

Figure 3: (a) 9 nm decahedral gold particle, (b) the same particle imaged at -173°C where the structure has changed to an icosahedral structure.

Conclusions

The structure of gold nanoparticles in the size range 5 nm – 12 nm has been investigated using variable temperature HRTEM. Heating of particles revealed the decahedral morphology to be preferred between room temperature and 400°C independent of the initial structural configuration. Heating to higher temperatures revealed the onset of surface roughening. Cooling of small particles showed an icosahedral transition. These results provide an experimental verification of a recently published theoretical phase diagram, with the methodologies employed being easily extendable to other systems of catalytic importance.

References

[1] M.A. van Huis, L.T. Kunneman, K. Overgaag, Q. Xu, G. Pandraud, H.W. Zandbergen, D. Vanmaekelbergh, Nano Letters, 8 (2008) 3959
[2] Newly developed ultra-low drfit cryo-holder for TEM, Fischione Instruments Inc.
[3] A.S. Barnard, N.P. Young, A.I. Kirkland, M.A. van Huis, H. Xu, ACS Nano, DOI: 10.1021/nn900220k

Electron Microscopy and Analysis Group Conference 2009 (EMAG 2009) IOP Publishing
Journal of Physics: Conference Series **241** (2010) 012096 doi:10.1088/1742-6596/241/1/012096

An investigation of commercial gamma-Al_2O_3 nanoparticles

Y Rozita [1,2], R Brydson[1] and A J Scott[1]
[1] Institute for Materials Research, School of Process, Environmental and Materials Engineering, University of Leeds, Leeds LS2 9JT, UK
[2] Universiti Pendidikan Sultan Idris, Tanjung Malim 39500, Malaysia

E-mail: pmry@leeds.ac.uk

Abstract. We present an analysis of the morphology and crystal structure of commercial gamma-alumina nanoparticles. Characterisation was carried out with a combination of several techniques such as transmission electron microscopy (TEM), scanning electron microscopy (SEM), energy dispersive X-ray analysis (EDX), selected area electron diffraction (SAED), in parallel with X-ray powder diffraction (XRD). TEM analysis of crystallite size and structure is in good agreement with XRD. HRTEM showed evidence for surface reconstruction on facetted cubeoctahedral nanoparticles. This is similar to results observed for cubeoctahedral nanoparticles of magnetite Fe_3O_4.

1. Introduction

Gamma-Al_2O_3 is isostructural with gamma-Fe_2O_3 and perhaps the most important nanomaterial used as a support for metal catalysts. It has been considered as one of the most promising advanced materials for a variety of applications due to its distinctive chemical, mechanical and thermal properties. The high surface area to volume ratio in nanoparticles can lead to modifications in structure and properties. There is great interest in the study of the surfaces of gamma-alumina nanoparticles as their catalytic activity is closely related to the structure of exposed crystalline faces. Inasmuch as gamma-alumina is widely used as a catalyst support, its structure and properties had been the subject of numerous studies. Gamma-alumina has been conventionally described as a defect spinel ($Fd\bar{3}m$) with the idealized formula $Al_{21+1/3}\square_{2+2/3}O_{32}$, where $\square$ denotes a vacancy [1]. Extensive reviews have reported that when gamma-Al_2O_3 is derived from amorphous aluminas, it always has a cubic lattice [2]. Both a cubic lattice [3] and a tetragonal distortion [4] are found for boehmite [AlO(OH)]- or gibbsite [$Al(OH)_3$]- derived gamma-Al_2O_3. Other studies, however, have proposed the existence of only a tetragonal structure [5]. Using TEM, Jefferson [6] found evidence for enhanced surface contrast of exposed crystalline (111) surface facets of gamma-Al_2O_3 nanoparticles and similar results were observed for cubeoctahedral nanoparticles of magnetite Fe_3O_4 [7]. The principal difficulty in analysing the gamma-alumina structure is because it does not yield single crystals suitable for a standard structural analysis [8]. This study examines two commercial gamma-aluminas using a combination of XRD, SEM and TEM.

2. Methodology and Materials

2.1. Materials

Commercial gamma-alumina nanopowders were obtained from Alfa Aesar (99.97% purity -product code 39812: batch B17U053 labelled as AA1 and batch K16104 labelled as AA2) and from Johnson Matthey (99.99% purity - product code 37212 labelled as JM) with surface areas nominally between approximately 80-150 m^2/g and were used without further treatment.

2.2 Methodology

The powder samples were stored in a desiccator until they were needed. Samples were analyzed using a Panalytical X'Pert ProMPD X-ray diffractometer with a Cu Kα source and an X'cellerator detector. The diffractometer was set up to measure in step-scan mode over the 2θ range (10 °- 90°), the range was chosen because it covered all the important diffraction peaks for the crystal structures to be examined. A step size or sample interval of 0.05 degrees and a scan time of 0.01°/scan were used.

The characterisation of gamma-alumina nanoparticles with transmission electron microscopy (TEM) was carried out using a Philips CM200 thermally assisted field emission gun TEM operated at 197 kV (point resolution of 0.24nm) equipped with a UTW Oxford Instruments EDX detector running ISIS software and Gatan (GIF200) imaging filter. All TEM samples were examined using standard holey carbon films supported on 400 mesh copper grids. The sizes of the nanoparticles were determined by measuring the length and width (aspect ratio) of the crystals for at least 100 particles in each sample.

3.0 Results and Discussion

XRD analysis was carried out in order to characterize the phase and crystal structure of the bulk samples. The polycrystalline diffractograms of the JM, AA1 and AA2 samples (Figure 1) were compared to an XRD standard for gamma-alumina structure JCPDS reference no. 00-010-0425 in the International Centre for Diffraction Data (ICDD) database. The diffractograms for all samples in Figure 1 display three distinct reflections at $2\theta = 37.2^\circ \pm 0.1^\circ$ (311 reflection), $45.6^\circ \pm 1^\circ$ (400) and $66.9^\circ \pm 1^\circ$ (440) which is in agreement with the database standard and also the powder XRD studies of Lippens [8]. Generally, the XRD peak positions matched with the gamma-alumina standard apart from the absence of a (111) reflection in samples JM and AA1, while only a very diffuse (111) reflection was found at $2\theta = 19.4^\circ$ in AA2. Additional peaks are located at $2\theta = 14.3^\circ$ in JM and at $2\theta = 44.7^\circ$ in AA2 which correspond to delta-alumina peaks. There are also four sharp peaks in all samples located at $2\theta = 38.5^\circ$, 44.5°, 65.0° and 78.0° (labelled with an asterix) which were assigned to aluminium peaks (reference: JCPDS no. 00-004-0787) from the sample holder and provides an internal calibration. The gamma-alumina XRD relative peak intensities are similar for the three samples but do not agree with the standard which may due to differences in site occupancies within the unit cell. The widths of the XRD peaks allow us to calculate the average crystallite size by using the Debye-Scherrer equation: the results derived from the (400) and (440) reflections being averaged. This analysis shows the average crystallite size of JM, AA1 and AA2 to be approximately 10 nm, 5 nm and 16 nm respectively. The formation of gamma-alumina is expected, because it is reported to be the most thermodynamically stable phase when the specific surface area of alumina is larger than $125 m^2/g$, or when the crystallite size (particle diameter) is less than about 13 nm. However, the fact that different peaks give different crystallite sizes suggests that there may be more than one alumina phase present or crystal shape anisotropy (e.g. preferential growth in certain directions).

Low magnification TEM images shown in Figure 2 were used to assess any agglomeration of the nanoparticles and to allow the measurement of the crystallite size. All of the nanoparticles were found to be agglomerated, although this was least for the AA2 sample. SEM and TEM image analysis revealed that the average agglomerate size was 550 nm ± 110 nm (JM), 535 ±140 nm (AA1) and 345 nm ± 120 nm (AA2). Both samples JM and AA1 clearly revealed a mixture of spherical and facetted particles while the AA2 sample possessed a mixture of 'long needlelike', 'cubic' and also facetted particles. The average projected particle diameter was 16 nm ± 6 nm (JM), 15 nm ±10 nm (AA1) and 31 nm ± 14 nm (AA2). The crystallite sizes from the TEM analysis are in agreement with those derived from XRD traces. EDX analysis of the samples demonstrates their purity with only aluminium and oxygen being detected. The average atom% of O and Al is (62.04 and 37.96) ± 0.08 respectively in JM, (62.72 and 37.28) ± 0.05 in AA1 and (64.28 and 35.72) ± 0.04 in AA2.

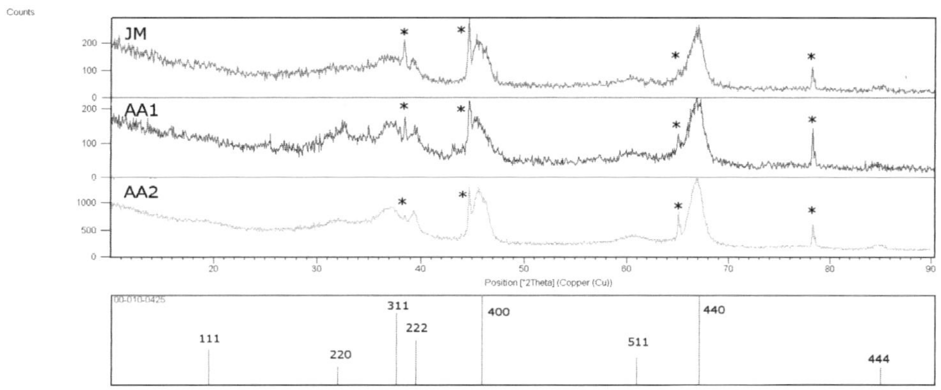

Figure 1. XRD diffractograms for JM, AA1 and AA2 together with ICDD database pattern for gamma-alumina JCPDS reference no. 00-010-0425

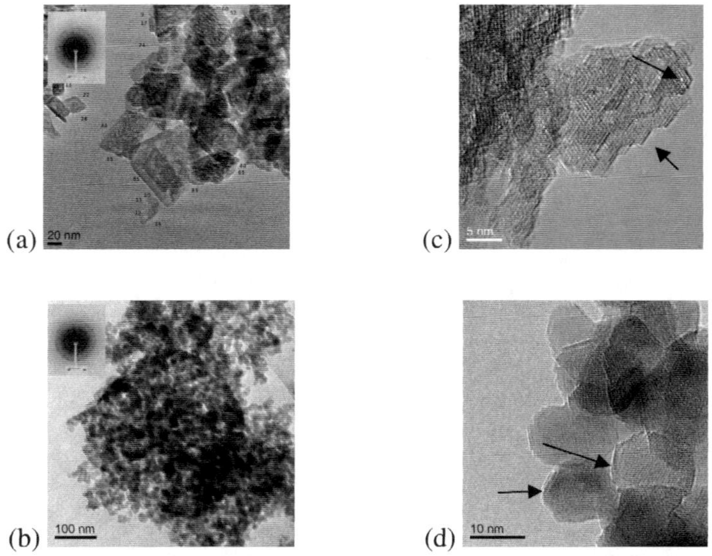

Figure 2. Bright Field TEM images of (a) AA2 and (b) JM

Black arrows showing the facetted particles in TEM micrographs of (c) AA2 (d) JM

The crystallinity of the nanoparticles was additionally confirmed by selected-area electron diffraction (SAED), which for all three samples revealed three broad diffraction rings corresponding to the (311), (400) and (440) reflections (Figure 2) and these d-spacings are consistent with the gamma-alumina structure. Additional proof for the crystallinity of the nanoparticles was obtained by high-resolution TEM investigations. The nanoparticles were found to be agglomerated on both the carbon film and also over the edges of holes within the film, complicating the process of acquiring high resolution images as the particles of interest are often obscured by overlapping particles. However, HRTEM imaging of all samples found evidence for surface reconstruction on single crystal facetted cubeoctahedral nanoparticles. The enhanced contrast is shown in figure 3 for the AA2 sample and was found to occur on facets that terminate with (111) planes {d_{111}= 4.55A} and (400) planes {d_{400}= 1.99A}, the latter being a (100) surface. The angle between the (111) and (100) faces is ≈55°. This enhanced contrast was similar to enhanced contrast observed at cation-terminated (111) surfaces of gamma-Al_2O_3 [6] and magnetite Fe_3O_4 [6, 7]. The difference in the scattering powers of oxygen and aluminium implies that the observed enhanced contrast at the (111) surface facets could be explained if the surface is cation-terminated with few or no surface oxygen atoms. Assuming that the enhanced surface contrast at the surfaces of the nanoparticles within the micrographs is a result of additional cations, in this case Al^{3+} ions from within the bulk of the nanoparticles, a disordered vacancy-containing layer should be apparent in the images just back from the edge. This defect structure is observed in TEM micrographs taken in [6, 7] and is also perhaps visible in Figure 3. The presence of excess cations at the nanoparticle's surface is a modification of the structure from the bulk and a reconstructed surface may have significant implications for the role of gamma-alumina in enhancing the dispersion of fine metal particles in catalysts [6].

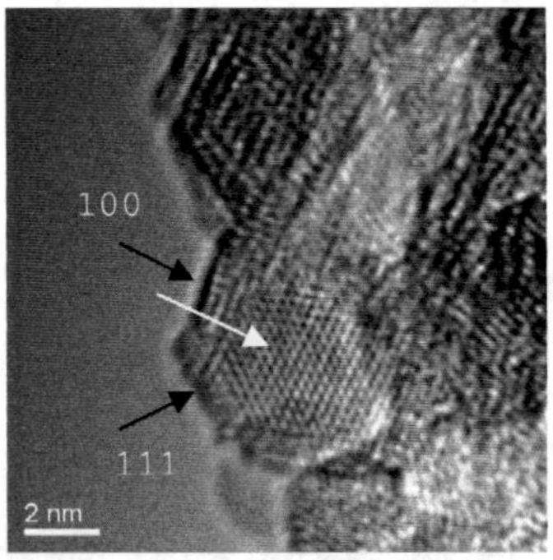

Figure 3. HRTEM image of AA2 with black arrows showing the enhanced surface contrast on facets that terminate with (111) planes and (100) planes at the end of the edges and a disordered sub surface apparent in the image just back from the edges. White arrow showing ordered lattice fringes.

References

[1] Verwey E J W 1935 *Z. Krystallogr.* **91** 317
[2] Levin L A, Bendersky D G and Brandon M R 1997 *Acta Mater.* **45** 3659
[3] Zhou S and Snyder R. L. 1991 *Acta Crystallogr.* **47** 617
[4] Wilson S J 1979 *J. Solid State Chem.* **30** 247
[5] Paglia G, Buckley C E, Rohl A L, Hart R D , Winter K, Studer A J, Hunter B A and Hanna J V 2004 *Chem. Mater.* **16** 220
[6] Jefferson D.A 2000 *Phil.Trans.R. Soc. Lond. A* **358** 2683
[7] Lovely G R, Brown A P, Brydson R, Kirkland A I, Meyer R R, Chang L, Jefferson D A, Falke M and Bleloch A 2006 *Applied Physics Lett.* **88** 93124
[8] Lippens B C and Boer J H 1964 *Acta Cryst.* **17** 1312

Study of erbium-doped silicon nanocrystals in silica

R J Kashtiban[1], U Bangert[1], I F Crowe[2], M Halsall[2], A J Harvey[1] and M Gass[3]

[1] Nanostructured Materials Research Group, School of Materials, The University of Manchester, P.O. Box 88, Manchester, M1 7HS, UK
[2] School of Electrical and Electronic Engineering, University of Manchester, Manchester M60 1QD, United Kingdom
[3] SuperSTEM, STFC Daresbury Laboratory, Warrington WA4 4AD, UK

E-mail: Reza.jalilikashtiban@postgrad.manchester.ac.uk

Abstract: Er-doped SiO_2 and Er-doped Si-NCs embedded in a SiO_2 matrix were produced by Er and/or Si ion beam implantation of a Si (100) substrate. The composition and distribution of implanted Er varies in samples either with or without Si implants. HAADF and EELS detail in samples with Si implants, the Si and Er distribution is identical, and within a band of ~110 nm width at ~75 nm below the SiO_2 surface. Intense PL emission at 1.54 μm confirms formation of $ErSi_2$, for the majority of aggregates, is unlikely. The present investigation details most Si-NCs are surrounded by Er_2O_3, or possess this phase within.

1. Introduction

The incorporation of the Er^{+3} ion in silica makes optical amplification through sharp and temperature-stable radiative transitions at ~1.54 μm possible [1]. SiO_2, doped with erbium, is currently used in silicon-based optical amplifiers [2-3]. The low absorption cross section of erbium ions in this structure leads to low gain and the necessity of pumping with expensive lasers. To improve the optical gain of intra band transitions originating from the 4f shell, erbium-doped silica, containing silicon nanocrystals (Si-NCs), have been suggested as a promising candidate [4-5]. The internalisation of Si-NCs increases the absorption cross section $\approx 10^4$ times. The larger excitation cross section of erbium sensitised Si-NCs compared to pure silica [6], the broad continuous absorption band of the Si-NCs, the enhanced emission cross section of the sensitized erbium ions [7], as well as the possibility of using cheap LEDs as pumping source, represent numerous advantages of this structure compared with erbium-doped SiO_2 in absence of Si-NCs.

In previous work, erbium-implanted Si-NCs embedded in to a silica matrix, resulted in an amorphous silicon structure evidenced by energy-filtered transmission electron microscopy (EFTEM), nevertheless, these amorphous silicon clusters were able to sensitise erbium ions [8]. In a recent study, authors employed EFTEM and high-angle-annular-dark-field (HAADF) imaging to characterise samples with erbium concentrations greater than 10^{19} cm^{-3}, observing erbium-rich clusters ≈5-10 nm in size, separated from the Si-NCs [9]. In our more recent study [10], we investigated highly erbium-doped silica and Si-NCs, within a dedicated aberration-corrected scanning transmission electron microscope (STEM). We demonstrated co-implantation of erbium and silicon leads to a band of nanostructures in a silica matrix, whereas implantation of erbium without silicon leads to large amorphous erbium-rich clusters, presumably Er_2O_3; owing to the tendency of erbium migrate to towards the surface of the silica layer. We further detailed that silicon ions might act as agglomeration centres for erbium ions. Present electron energy loss spectroscopy (EELS) elemental maps confirm that strong correlation between the position of silicon and erbium exists. We also address the composition of erbium in the structure.

2. Experimental

Two types of samples, (1) exclusively erbium-doped SiO_2, or; (2) erbium-containing Si-NCs in SiO_2, were grown on a silicon substrate, with a similar SiO_2 thickness and composition, in order to investigate erbium-implanted microstructure and composition. In the latter, a beam of 85 keV silicon ions providing a dose of 8×10^{16} at/cm^2 was targeted at a 700 nm thick SiO_2 layer, grown on the top of a Si (1 0 0) substrate. Erbium ion implantation was then performed using a beam of 320 keV and a dose of $3x10^{15}$ at/cm^2. Post implantation, the samples were annealed at 1100 °C in a N_2 atmosphere for 1 hour, in order to form Si-NCs. The erbium and silicon concentration peaks in the samples were $3x10^{20}$ cm^{-3} and $8x10^{21}$ cm^{-3}, respectively. The growth conditions and implantation parameters for the sample type (1) were similar to those for sample type (2), except there was no silicon implantation.

An aberration corrected STEM, at the *SuperSTEM* Labratory Daresbury, fitted with a *Nion* Mark II quadrupole-octupole corrector[11] operating at 100keV was employed to acquire HRTEM, Bright Field (BF) and High Angle Annular Dark Field (HAADF) images as well as EEL spectrum images (SIs) [12] with 0.2s/pixel exposure time, 0.7 eV /channel energy dispersion, the convergence angle of 24mrad and the collection semi-angle of <5mrad . The instrument allows Z-contrast lattice images with 1Å resolution to be obtained. An Ultra High Vacuum (UHV) *Gatan Enfina* system was used to obtain spectrum images.

3. Results and discussion

Figure 1(a) represents typical EEL spectra of a NC (red line) and of the oxide layer (green line), in which the Si $L_{2,3}$-, Er $N_{4,5}$- and O K-edges can be identified . The erbium peak is clear in the red line spectrum in (a), whereas there are no clear features consistent with the presence of erbium in the green line spectrum, which represents an area off the NCs in the oxide layer. The C K-edge was employed to calibrate EEL spectra. Figure 1(b) represents an HREM image of a NC of size ≈3.5 nm (clear lattice fringes are existed). Figure 2(b-c) details elemental maps of the Er-$N_{4,5}$- and Si-$L_{2,3}$- edges, following a power law background fitting. There is a clear and strong correlation between the silicon and erbium distribution in the maps. No substantial erbium clusters within the SiO_2 exist apart from the Si-NCs, detailing no sign of formation of separate Er_2O_3 clusters in between the Si-NCs. Figure 1(d ,e) represents the FWHM and energy of a Gaussian peak fitted to the onset peak of O-K edge. Rare earth oxides exhibit a split OK-edge structure and earlier onset, compared to a later onset and single peak characteristics of SiO_2 [13].

Fitting a single Gaussian envelope leads to a larger FWHM, than for a Gaussian peak fitted to the single OK-edge peak of SiO_2. The fact that a larger FWHM and lower peak energy are spatially correlated with the Si-NCs indicates the presence of erbium oxide in/near the Si-NCs. HREM inspection shows the NCs are crystalline (not shown here); hence due to the strong evidence of co-location of Er, Si and O and the well known low solubility of Er in Si and high quality silica [14] the erbium oxide presumably forms a shell surrounding the Si-NCs. In contrast, the narrower Gaussian peak and blue-shifted energy, indicates the presence of SiO_2.

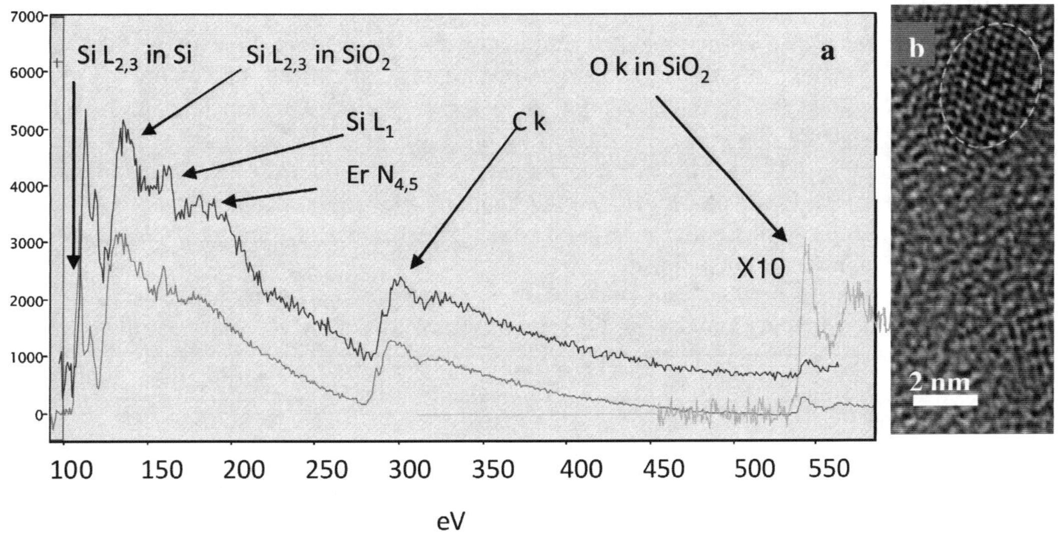

Figure 1 (a) Typical EEL spectrum on a NC(red) shown on (b), and on the SiO_2 layer(green) together with OK edge intensity multiplied by 10 (light green) within a SI with edges used to extract compositional maps of Si $L_{2,3}$, Er $N_{4,5}$ and OK intensities at ~99 eV , ~167-177 eV and ~532-542 eV respectively. (b) HREM image of a NC

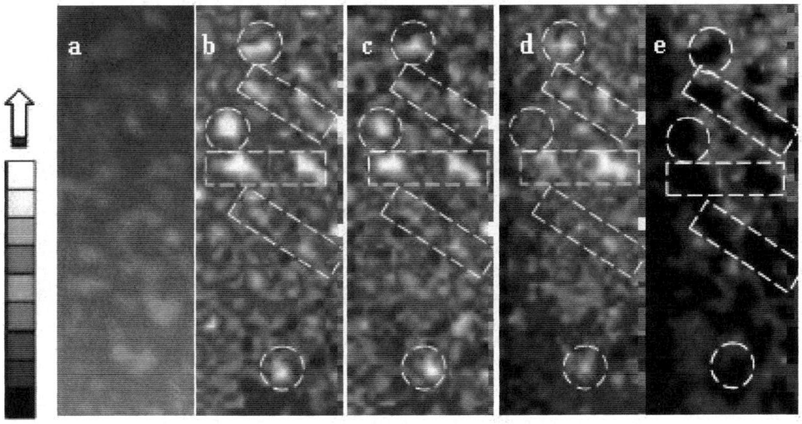

Figure 2 (a) HAADF image of erbium doped Si-NCs of the mapped area, (b) Si $L_{2,3}$ edge map (98-101 eV), (c) Er $N_{4,5}$ edge map(167-177 eV), following a suitable power low background fitting. (d) FWHM and, (e) energy of the Gaussian peak fitted to the O K-edge. Dashed circles and rectangles indicate correlations in quantities of the images. The contrast of the maps is presented on the colour-scale with white/bright signifying high and blue/dark low values of the respective quantity

4. Conclusion

EELS and HAADF measurements detail erbium and silicon co-implantation leads to identical distributions of erbium and silicon. The dominant presence of SiO_2 within the matrix results in direct oxygen elemental mapping, by employing the O K-edge of the expected Er_2O_3 phase, extremely difficult. Indirect investigations, using modifications of the O K-edge characteristics such as width and position arising from the presence of the rare earth oxides, demonstrate that the majority of erbium is in the form of Er_2O_3. This erbium is distributed identically with Si-NCs in the structure, but as no structural phases other than silicon can be detected in HREM. It must be concluded that the erbium oxide surrounds the Si-NCs.

4. Acknowledgment

We gratefully acknowledge Dr. Andrew P. Knights at the Centre for Emerging Device Technologies, McMaster University, Canada and Prof. Russell M. Gwilliam at the Surrey Ion Beam Centre, University of Surrey for producing the samples.

5. References

[1] Miniscalco W J, Journal of Lightwave Technology **9** 234
[2] Desurvire E, Simpson J R and Becker P C, Optics Letters **12** 888
[3] Kik P G, Brongersma M L and Polman A, Appl. Phys. Lett. **76** 2325
[4] Fujii M, Masato Y, Kansawa Y, Hayashi S and Yamamoto K Appl. Phys. Lett. **71** 1198
[5] Kenyon A J, Chryssou C. E, Pitt C W, Shimizu-Iwayama T, Hole D E, Sharma N and Humphreys C J, J. Appl. Phys. **91** 367
[6] Kik P G and Polman A, J. Appl. Phys. **91**(1) 534
[7] Pacifici D, Franzò G, Iacona F, Boninelli S, Irrera A, Miritello M and Priolo F, Materials Science and Engineering: B **105** 197
[8] Kik P G and Polman A, J. Appl. Phys., **88**(4) 1992
[9] Pellegrino P, Garrido B, Arbiol J, Garcia C, Lebour Yand Morante J R, Appl. Phys. Lett. **88** 121915
[10] Kashtiban R J, Bangert U, Crowe I, Halsall M P, Sherliker B, Harvey A J, Eccles J, Knights A P, Gwilliam R and Gass M, to be published in IOP
[11] O. L. Krivanek, N. Dellby, A. R. Lupini, Ultramicroscopy 78 (1999) 1-11.
[12] Hunt, J A and Williams D B, Ultramicroscopy 38 47
[13] Schamm S, Scarel G and Fanciulli M, Rare Earth Oxide Thin Films in Topics in Applied Physics **107** 153
[14] Horak P, Loh W H, Kenyon A J, Opt. Express **17** 906

Atomic scale graphene landscapes - natural, dosed and doped

U Bangert[1], M Gass[2], A L Bleloch[2] and R R Nair[1,3]

[1] School of Materials, The University of Manchester, P.O. Box 88, Manchester, M1 7HS, UK
[2] SuperSTEM, STFC Daresbury Laboratory, Warrington WA4 4AD, UK
[3] School of Physics, The University of Manchester, P.O. Box 88, Manchester, M1 7HS, UK

E-mail: ursel.bangert@manchester.ac.uk

Abstract: Graphene surfaces are scrutinised for topographic peculiarities, which occur naturally or have been introduced. High-angle-annular-dark-field (HAADF) and bright-field lattice images of graphene acquired in an aberration-corrected scanning transmission electron microscope (STEM), in conjunction with electron energy-loss spectroscopy (EELS), show vacancy- and ad-atom-related point defects, as well as the presence of hydrogen. In boron implanted graphene, the edges of graphene sheets appear to be capture centres for boron atoms. Furthermore, rippling effects in the graphene sheets can be revealed by fast Fourier transform (FFT) procedures; these help visualise changes in the bond length projection arising from inclinations of the sheet.

1. Introduction

In the last five years, graphene has become one of the most famous and well researched materials. We are interested in atomic detail of these 2-D crystals: high-angle-annular-dark-field (HAADF) lattice images of graphene acquired in a scanning transmission electron microscope (STEM), with the electron beam focussed onto the sheet, are a direct representation of the ball-and-stick model of the atomic lattice: bright contrast corresponds to atoms, and dark contrast to the empty space in between. In addition, variations in brightness can be attributed to ad-atoms. The stability of freely suspended, extended two-dimensional structures represents a long-standing theoretical debate, and it has previously been suggested that graphene is crinkled, or that defects might act as stabilisers. Here, we investigate phenomena related to the stability issue, such as the structure of point defects - vacancy related, as well as arising from carbon ad-atoms – and also sheet undulations, via atomic resolution imaging in HAADF (and simultaneously in BF), in an aberration-corrected 100 keV STEM. Doping and dosing of graphene has become an important issue for electronic functionalisation; in this context the capability of hydrogen uptake and incorporation of functional impurities (i.e. boron) is of interest, both of which we will address.

2. Experimental

The experiments reported here were conducted on an aberration-corrected dedicated scanning transmission electron microscope [1-3] with probe size 1Å (the Daresbury SuperSTEM), equipped with an Enfina electron energy-loss (EEL) spectrometer. The operating voltage was mostly 100 keV; some spectrum images were obtained at 80 keV. The angle of the HAADF detector annulus is 70-210 mrad; the instrumental resolution for EEL spectroscopy 0.28 eV.

Journal of Physics: Conference Series **241** (2010) 012098 doi:10.1088/1742-6596/241/1/012098

Graphene samples were obtained by micromechanical cleavage, a method which has been extensively reported [4]. Freely suspended graphene sheets detailed in figure 2 were exposed to a hydrogen plasma, and those detailed in figure 3 were subjected to boron implantation in a low energy ion implanter at 100 eV and a dose of 10^{14} cm^{-2}, corresponding to one boron atom nm^{-2}.

3. Results and discussion

Figure 1(a) details an HAADF STEM image of single layer graphene, following application of the low-pass filter shown in the top left inset, in order to remove noise from the images. The chicken-wire contrast arises from carbon atom rings, with the atoms positioned at the junctions. However, the graphene lattice is not of uniform contrast; some of the bridges between atoms are significantly brighter. The green spots mark such 'bridges : they have been obtained by setting an intensity saturation threshold at twice the intensity of that of the surrounding atoms. The blue intensity traces (top curves) in figure 1(c) demonstrate this effect further: these traces are acquired along C-C bonds across locations of the highlighted (green) patches (1a), where the HAADF intensity is twice that of the neighbouring carbon atoms. This indicates existence of carbon ad-atoms, bridging C-C bonds in the 6-rings. The red intensity trace (bootom curves) in figure 1(c) is acquired in vacuum, and is of a similar level as the intensity in the centre of the 6-rings. The arrow in figure 1(a) points to a vacancy. The contrast is straight-forwardly interpretable: black areas represent empty space, and hence missing atoms can be directly seen. Vacancies, or aggregates thereof, are commonly observed in atomic resolution HAADF images [5] and thought to be introduced by the electron beam in the microscope.

The image in figure 1(b) is obtained of the same area, after applying the ring-filter in the inset (top left; the procedure and results are detailed in [6]). By imposing a narrow ring mask on the Fourier transform (FFT) of the raw images (with band width corresponding to ~0.04 Å), spatial frequencies within ~2% of the frequency, corresponding to the real space graphitic a-plane distance are passed. The inverse transform (IFFT) reveals locations of atoms with correct lattice spacings, although with possible rotations, through intense lattice fringes. The colour (intensity) coding in figure 1(b) is chosen such that atoms possessing the correct bond length appear yellow (bright). Should the projected bond length change by as little as 2% owing to out-of-plane bending of the atoms, the lattice periodicity will become less visible in the IFFT. Such regions are made to appear blue (dark). Atomic scale detail can be seen as a fine raster superimposed on the colours. Thus undulations of up to 0.5 nm in amplitude were observed everywhere in freely suspended mono-layer graphene patches, in the absence of visible topological defects. Figure 1(d) provides a larger area view of this scenario. However, the ripple patterns can be influenced by point defects, as shown in figure 1(b): here, the vacancy (arrowed in figure 1(a) causes deformation in an extensive surrounding area,

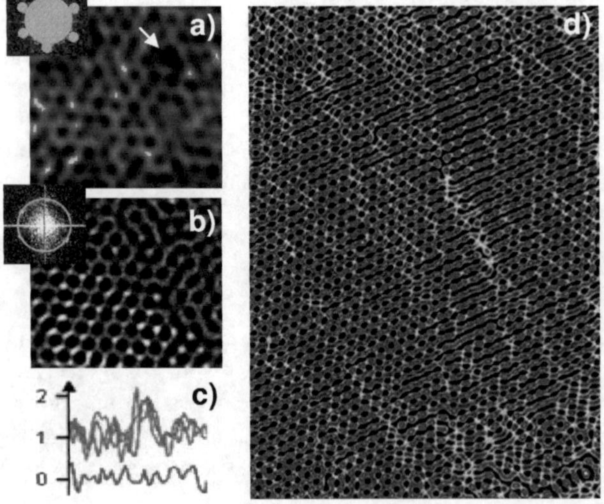

Figure 1 **(a)** HAADF image of pristine graphene; the inset shows the low frequency bandpass filter applied to the FFT of the original raw image to obtain image (a); green spots mark locations of enhanced intensity, **(b)** IFFT obtained from same area as in (a), with the ring filter overlaid on the FFT of the original image; the procedure is explained in detail in [6] **(c)** intensity traces (blue lines, top) along sides of hexagons across locations of enhanced intensity (marked green); the red, lower trace is of vacuum, **(d)** IFFT as in (b), but of a larger area.

manifest in the IFFT (in figure 1(b) by the blue (dark) shadesr, which relate to projected bond shortening of >5%. Here the sheet is bent; the ripple pattern appears to be the result of local changes in stiffness, which is reduced at vacancies (occurring in flanks of ripples) and enhanced in regions with ad-atoms (which are flatter).

Carbon and oxygen ad-atoms have previously been discussed [7,8]. Furthermore, the possibility for atomic hydrogen chemi- or physisorbed onto graphite surfaces [9] and graphene [10] has been researched theoretically and experimentally [11], and direct experimental observation has been attempted via transmission electron microscopy imaging [7]. In order to assess hydrogen adsorption onto graphene, we have carried out electron energy-loss spectroscopy (EELS). We applied the spectrum imaging method [12], and scrutinised the EEL spectrum images for occurrence of the core-level excitation signal of hydrogen (around 13 eV). Figures 2(c) and (f) show low loss spectra extracted from spectrum images (SIs) obtained of the graphene patches shown in figures 2(a) and (d), typically a few 100 nm^2 in size. The pixel area in the SIs was ~1 nm^2. Spectrum (g) in figure 2 is from a pixel in a hydrogen-free graphene patch. All spectra are power-law background subtracted, and start with the rise of the $\sigma+\pi$ plasmon at 14.7 eV [13], the signature of mono-layer

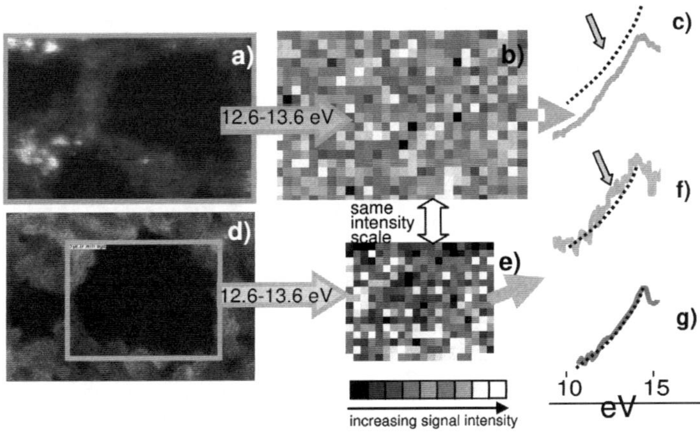

Figure 2 (**a**) HAADF image of spectrum image region in hydrogen-dosed graphene. The whitish contrast in HAADF images arises from hydro-carbon deposits. (**b**) Integrated intensity of the residual signal after π- and $\pi+\sigma$-plasmon removal in the energy window 12.6-13.6 eV. (**c**) Power-law background-fitted spectrum from pixel in (b). **d**) HAADF ~~image of~~ pristine graphene with the spectrum image region framed. (**e**) Intensity of the residual signal following the same procedures as in (b). (**f**) Power-law background-fitted spectrum from pixel in (e) and (**g**) from hydrogen-free graphene. The frame width of the HAADF images is 30 nm

graphene. Principal component analysis (PCA) was performed on the spectrum images [e.g., 14], in order to reduce noise. To obtain hydrogen-free areas, few layer graphene regions were repeatedly scanned. The scan raster was successively decreased in total size, and HAADF images in combination with EEL measurements revealed areas being peeled off layer-by-layer, until finally a hole appeared. Spectra of such freshly revealed areas do not show a bump at 13 eV as can be witnessed in figure 2(g). The dashed line shows a Lorentzian fit to the rise of the $\sigma+\pi$ plasmon; this curve presents a perfect fit for hydrogen-free spectra and was used as plasmon 'background' for spectra detailed in figure 2(c) and (f). Curve (f) shows a spectrum from a pixel in the SI-raster detailed in figure 2(e) of pristine graphene, exhibiting a bump on the $\sigma+\pi$ plasmon, and hence revealing presence of hydrogen. The hydrogen seems to be omni-present and possibly marginally accumulated in the vicinity of hydro-carbon contamination (white 'clouds' in the image in figure 2(d). Spectrum (c) is from a pixel in the SI-raster detailed in figure 2(b) of single layer graphene, which has been exposed to a hydrogen plasma, i.e. intentionally *dosed*. The bump at 13 eV is more pronounced here. This is further illustrated in the hydrogen distribution maps in figure 2(b) and (e), which show relative intensities of the signals in each pixel at ~13eV, following subtraction of the Lorentzian background curve. The signals were integrated and normalised to the total spectrum intensity. These hydrogen distribution maps reveal that the hydrogen-related intensity in the hydrogen -dosed film is about twice that of the pristine film.

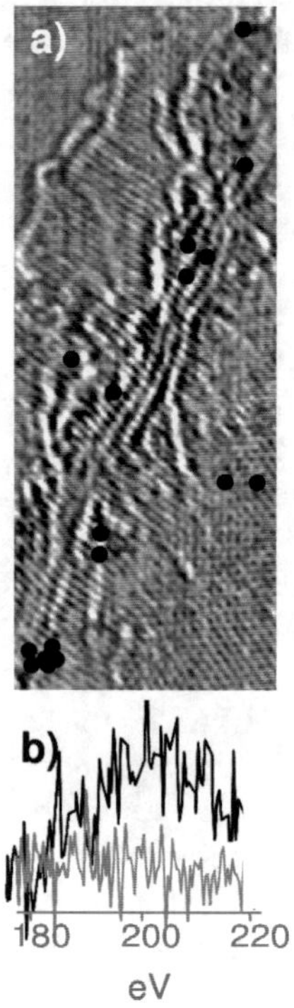

Figure 3 (a) HREM BF STEM image of boron implanted (~1 at nm^{-2}) staggered graphene sheets (leftmost edge is of a single sheet; edges of subsequent sheets occur as frayed and/or parallel lines); black dots are locations of pixels with B K-edge signals in the corresponding SI, **(b)** examples of signals extracted from pixels of black (B-K edge; black line) and unmarked (no signal; grey line) locations.

Intentional *doping* with boron via low energy ion implantation revealed a boron retention of less than 1% at implantation energies of 100 eV. Furthermore, it occurs that the boron preferably attaches to edges of graphene sheets: figure 3(a) shows a BF HREM STEM image of the area from which a SI was acquired. The edge of a mono-layer graphene sheet can be seen to run diagonally (right top, to left middle) across the image with edges of a further four staggered sheets located inwards from the single sheet edge. The black dots are positions of SI pixels, where single atom boron K-edge signals (188 eV losses) were detected via EELS. (Details of this study are presented elsewhere [15]). Most of these positions line up with the staggered edges. Furthermore, the boron edge shape (figure 3(b) is not that of substitutional B, but rather resembles that of amorphous boron or of B_4C. This leads us to conclude that ion implantation doping of graphene at energies <100 eV does not lead to effective substitutional incorporation for electronic modification.

4. Conclusion

We have demonstrated that free-standing monolayer graphene is remote from the ideal structural model, but possesses an intricate surface topology owing to undulations and point defects (ad-atoms and point defects). The former always occur, but also specifically arise from the latter. Hydrogen is present on all graphene surfaces, but the coverage is increased by dosing. Intentional boron doping via low-energy ion implantation has so far predominantly led to boron ad-atom incorporation at graphene sheet edges, rather than substitution.

5. References

[1] Krivanek O L, Delby N and Lupini A R 1999 *Ultramicroscopy* **78** 1
[2] Howie A 1979 *J. Microscopy* **117** 11
[3] Pennycook S and Boatner L A 1988 *Nature* **336** 565
[4] Booth T J, Blake P, Nair R R, Jiang D, Hill E W, Bangert U, Bleloch A, Gass M, Novoselov K S, Katsnelson M I and Geim A K 2008 *Nano Lett.* 2442
[5] Gass M H, Bangert U, Bleloch A L, Wang P, Nair R R and Geim A K 2008 *Nature Nanotech.* **3** 676
[6] Bangert U, Gass M H, Bleloch A L, Nair R R and Geim A K 2009 *Phys Stat Sol A* **206** 1117
[7] Meyer J C, Girit C O, Crommie M F and Zettl A 2008 *Nature* **454** 319
[8] Nordlund K, Keinonen J and Mattila T 1996 *Phys. Rev. Lett.* **77** 699
[9] Jeloaica L and Sidis V 1999 *Chem. Phys. Lett.* **300** 157
[10] Ito A, Nakamura H and Takayama A, preprint at http://arxiv.org/abs/condmat/0703377
[11] Elias D C et al. 2009 *Science* **323** 610
[12] Jeanguillaume C and Colliex C 1989 *Ultramicroscopy* **28** 252
[13] Eberlein T et al. 2008 *Phys. Rev. B* **77** 233406
[14] Borglund N, Astrand P-G and Csillag S 2005 *Microsc. Microanal.* **11** 88
[15] Bangert U, Seepujak A, Gass M H and Bleloch A L, accepted y *Phys Rev B.*

Journal of Physics: Conference Series **241** (2010) 012099 doi:10.1088/1742-6596/241/1/012099

Radiative collective modes in ion-implanted multiwalled carbon nanotubes

A Seepujak[1], U Bangert[1], J W Eccles[1], A J Harvey[1], J van den Berg[2] and M Reading[2]
[1] Department of Materials Science, University of Manchester, Oxford Road, Manchester M13 9PL, UK
[2] School of Computing, Science and Engineering, University of Salford, Salford, Manchester M5 4WT, UK

E-mail: A.Seepujak@manchester.ac.uk

Abstract. Surface mode fields are strongly confined to a MWCNT surface, representing a conceivable solution for realization of quasi-2D optics. In the present Contribution, MWCNTs are ion implanted with guest ions. Implantation enables engineering of the local dielectric tensor $\varepsilon(\mathbf{k},\omega)$, allowing specific selection of the γ emission eigenfrequency. C_s-corrected STEM is used to acquire EEL spectra with atomic resolution. Lines are subsequently characterized using EEL spectra simulated using first-principles DFT calculations, utilising the full-potential linearised augmented plane wave formalism. In addition to presenting virtual collective modes, implanted guest ions are seen to introduce waveguide eigensolutions which provide a possible mechanism for light transmission in MWCNTs.

1. Introduction

With the metallic or semiconducting character of MWCNTs (multi-walled carbon nanotubes) dictated by chirality, MWCNTs present a potentially versatile component for use in plasmonic circuits. Doping modifies the dielectric response. The intra-layer sp^2-hybidization, combined with the p_z orbitals orthogonal to the basal planes, results in MWCNTs being particularly receptive to introduction of foreign species, in typically substitutional or intercalated configurations. In the present Contribution, doping is performed using ion implantation, offering an intrisically clean nature and precise dose control. EEL (electron energy-loss) spectroscopy, in combination with scanning transmission electron microscopy (STEM), represents an ideal characterization technique capable of providing information concerning chemical bonding and electronic structure of dopants in the nanometre and sub-nanometre regimes, from volumes not merely restricted to the surface. Lines and features arising in low energy loss EEL spectra are then characterised using optical loss spectra and dielectric functions calculated within the DFT (density functional theory) framework.

Graphite possesses a characteristic band structure near E_F, manifesting in a distinctive optical anisotropy, and polarization-dependent semimetallic behaviour. Using a simple ridged band filling model, one can view the graphite π-collective modes as originating from the $\pi\rightarrow\pi^*$ electric dipole transition. Whilst a zero DOS exists at E_F, crucially, degeneracy of the π and π^* bands allows excitations at small **k** close to the Γ point, presenting the possibility of collective modes in the UV to IR frequency regimes. Whilst few studies exist concerning radiative modes and waveguide modes propagating on graphitic materials [8], the radiative or waveguided nature of certain graphite collective eigensolutions is utilised in the present study to suggest a possible mechanism for γ emission from MWCNTs. Literature details electrically induced radiation from a single s-SWCNT FET [1]; light emission through electroluminescence and photoluminescence in InP nanowires [2] and field-emission-induced luminescence [3]. However, the ion implantation method for modifying eigenfrequency is particularly suitable for mass production methods, which avoids the need for device structures or field emission.

2. Experimental

EEL spectra were acquired using a C_3-aberration-corrected STEM operated at 100 kV at the SuperSTEM Laboratory, UK. The 0.1 nm probe enables EEL spectra of projected features the size of individual atomic columns to be obtained. In nano-structured specimens consisting of only a few atoms, the probability of single impurity atom detection amongst matrix atoms is thereby dramatically improved. The EEL spectrometer consists of a CCD detector-based Gatan UHV Enfina instrument. The energy resolution of the entire system, at high dispersion (0.02 eV/channel), is ~0.3 eV. All spectra presented herein represent raw data, from which a power-law fitted background was subtracted. The acquisition time per spectrum was 1 s. Commercially available pristine MWCNTs, from the same manufactured batch, were dispersed using an identical method for each sample, onto standard, lacy carbon film-coated TEM copper grids. The MWCNTs were of high purity and graphitisation, and possessed few (<10) walls. All glassware were vigorously washed in ethanol in order to prevent silicon contamination. Ion implantation was carried out into the TEM-ready specimens; the Salford University low energy implanter was used to deliver ion beams at the following energies and doses: B – 100 eV / $1\text{x}10^{14}$ cm^{-2}; BF_2 – 200 eV / $2.5\text{x}10^{15}$ cm^{-2}; Ag – 200 eV / $20\text{x}10^{14}$ cm^{-2}.

3. Discussion

The linear response of a medium to an external field is characterized by its complex dielectric function. Consider an optically anisotropic medium (such as graphite); the complex dielectric constant assumes the form of a 3 x 3 tensor. The dielectric functions can be described by,

$$\varepsilon_{||c}=\varepsilon_{1,||c}+i\varepsilon_{2,||c} \tag{1a}$$

$$\varepsilon_{\perp c}=\varepsilon_{1,\perp c}+i\varepsilon_{2,\perp c} \tag{1b}$$

$(\varepsilon_{1,||c})\cdot\omega^2/c^2$ defines the light line in the medium. The light line is of fundamental importance, since it defines regions in the $(\omega,|\mathbf{k}|)$ space, where virtual, waveguide and surface modes exist [5], as detailed in figure 1(a). One can immediately see the effect of $\varepsilon_{1,||c}>1$ in a CW rotation the light line. For a given $|\mathbf{k}|$, $\varepsilon_{1,||c}>1$ enables a mode to be within the virtual domain, in comparison to $\varepsilon_{1,||c}=1$. $\varepsilon_{1,||c}$ thus represents the term dictating transformation of surface mode characteristics to virtual mode characteristics. Should $0\leq\sqrt{\varepsilon_{1,||c}}<1$, then the medium will possess a smaller refractive index than vacuum, largely resulting in total internal reflection of γ.

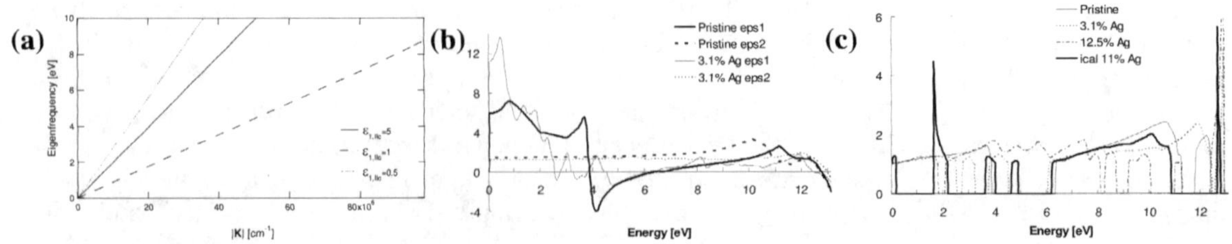

Figure 1. (a) Definition of the virtual mode and surface mode domains by the light line, for varying values of the dielectric constant. (b) DFT calculations of $\varepsilon_{1,||c,\perp c}$ of pristine graphite and graphite doped with Ag adopting a substitutional configuration. (c) The calculated number of waveguide modes for various media.

Furthermore, when $\varepsilon_{2,\|c,\perp c}$ are small, the more pronounced an eigensolution (i.e. the narrower a peak), in a loss spectrum. Modification of the graphite dielectric function is detailed in figure 1(b), consisting of a calculation using the Wien2K DFT code [4], demonstrating Ag atoms adopting a simple substitutional configuration in planar graphite can result in $\varepsilon_{1,\|c}>>1$. For DFT calculations detailed herein, the muffin-tin radii were set at R=1.34 atomic units, 250 k-points were utilized, and self-consistent iterations were performed until convergence (0.1 eV) on the total energy. An Rk_{max} value of 5 was used. The number of waveguide modes [5] calculated [8] for various media is detailed in figure 1(c): when $\varepsilon_{1,\|c}<1$, waveguide modes cannot be excited. EEL spectra acquired from various MWNCTs doped with various species are detailed in figure 2; crucially, the dimensions and number of walls were nearly identical for each MWCNT.

Figure 2(a) and 2(b) detail the change in eigenfrequency and linewidth of a single Gaussian peak fitted to the EEL spectra of doped MWCNTs. For each spectrum, the zero-loss peak was extracted using and identical model and energy window, and the Gaussian peak was fitted to an identical energy window between 3.0 eV – 8.4 eV. The pristine MWCNT exhibits the characteristic redshift and increase in linewidth from b=0 to grazing incidence. These trends are ascribed in to optical anisotropy and cylindrical anisotropy [5].

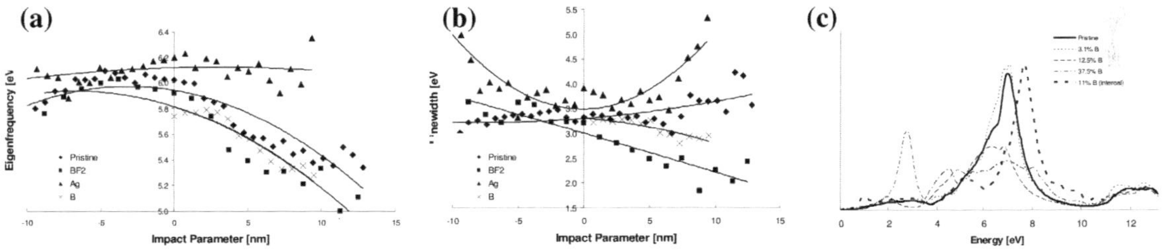

Figure 2. (a) Measured eigenfrequency, and (b) linewidth, of a single Gaussian peak fitted to an identical energy window for MWCNTs doped with various species. (c) DFT simulated dielectric functions of planar graphite doped with B.

For the BF_2 and boron doped MWCNTs, the pristine MWCNT redshift trend from b=0 to grazing incidence is preserved. B is predisposed to assuming substitutional configurations in MWCNTs. The consistently lower energies of the BF_2 and boron doped MWCNTs, compared to those of the pristine MWCNT, is attributed to transitions lower in energy than the pristine graphite π-plasmon, as detailed in the Wien2K [4] DFT calculated spectra (figure 2(c)). The similarity between observed trends for the B and BF_2 doped MWCNTs can be ascribed to the dissociation of the BF_2 radical, solely resulting in B, and not F incorporation. Figure 2(c) details a lifting of degeneracy or near-degeneracy of the graphite π-plasmon with B doping, constituting another possible cause of the Gaussian peak redshift. Figure 2 details the introduction of a Drude plasmon for very high (>25% at. %) substitutional levels. For the Ag implanted MWCNT, the largely invariant eigenfrequency and clear increase in linewidth can be attributed to substitutional or intercalated configurations, which are predicted [5] to introduce (higher energy) states at around 8 eV. It is feasible that very low level Ag intercalation may also contribute to the observed trends, in such an instance, a Drude plasmon obscured by the ZLP fails to weight the fitted Gaussian peak. Individual spectra are detailed in figure 3, which reproduce the trend in eigenfrequency identified in figure 2: the uppermost spectra correspond to zero impact parameter, with the lowermost spectra representing grazing incidence. For equivalent impact parameter, the BF_2 doped MWCNT always has a lower plasmon frequency than that of the pristine MWCNT. Furthermore, for the BF_2 doped MWCNT, splitting of the graphite π-plasmon is evident, consistent with the simulations detailed in figure 2(c).

BF and HAADF (high-angle-annular-dark-field) images (figure 4) confirm that structural integrity has been preserved following the ion implantation, but in the case of Ag, HAADF imaging (not shown) reveals the Ag has clustered on the outermost surface of the MWCNT, thus accounting for an unexpectedly low level of Ag doping observed during single atom EELS of present specimens [6].

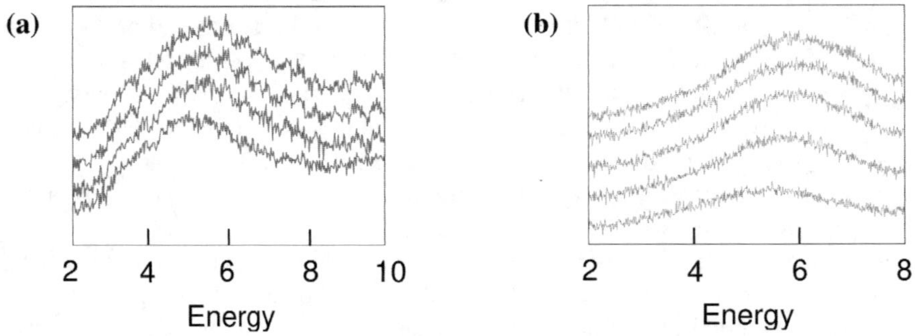

Figure 3. (a) Experimentally measured spectra from a BF_2 doped and, (b) pristine MWCNT. (c) Simulated dielectric functions of B or Ag atoms, adopting various configurations in planar graphite.

However, single atom EELS reveals the retention of B dopant atoms is much greater [6]. The strength of the present technique is now clearly evident: by performing first-principles DFT calculations, one can ascertain the effect of dopant atoms on modifying the dielectric response, with ion implantation parameters adjusted accordingly.

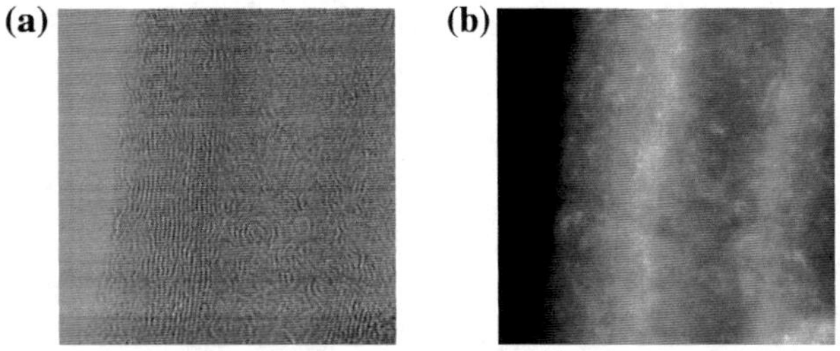

Figure 4. BF and HAADF images of (a) a pristine MWCNT, and (b) a BF_2 implanted MWCNT. Visible in the HAADF image are bright spots corresponding to high scattering centres. Since B is of lower Z than C, we attribute this effect to the possible presence of B-O bonding.

4. Summary

Careful modification of the dielectric function can transform surface modes into virtual radiative modes or waveguide modes. Such modes would conceivably allow γ emission, or γ transmission over long distances, respectively. The relative cleanliness of ion implantation doping is of fundamental importance to the present technique, largely avoiding the inadvertent doping associated with other methods. Use of C_s-corrected STEM presents the unique possibility of deducing the bonding state of individual atoms within a matrix [6].

5. References

[1] Misewich J A , Martel R, Avouris P, Tsang J C, Heinze S and Tersoff J 2003 *Science* **300** 783
[2] Duan X, Huang Y, Cui Y, Wang J and Lieber C M 2001 *Nature* **409** 66
[3] Bonard J-M, Stockli T, Maier F, de Heer W A, Chatelain A, Salvetat J P and Forro L 1998 *Phys. Rev. Lett.* **81** 1441
[4] Nelhiebel M, Louf P H, Schattschneider P, Blaha P, Schwarz K and Jouffrey B 1999 *Phys. Rev.* B **59** 12807
[5] Seepujak A, Bangert U, Eccles J W, Harvey A J, van den Berg J and Reading M. To be submitted to *Phys. Rev.* B
[6] Bangert U, Bleloch A, Gass M H, Seepujak A and van den Berg J. Submitted to *Nature Nano.*
[7] Seepujak A, Bangert U, Harvey A J, Costa P M F J and Green M L H 2006 *Phys. Rev.* B **74** 075402
[8] Wessjohann H G 1976 *Phys. Stat. Sol.* (b) **77** 535

Journal of Physics: Conference Series **241** (2010) 012100 doi:10.1088/1742-6596/241/1/012100

Aberration-corrected electron microscopy processing and imaging of novel organic and inorganic nanostructures

V. Nicolosi[1], E C Cosgriff[1], J N Coleman[2] and P D Nellist[1]

1. University of Oxford, Department of Materials, Parks Road OX1 3PH, Oxford, UK

2. University of Dublin, Trinity College, School of Physics, D2, Dublin, Ireland

E-mail: valeria.nicolosi@materials.ox.ac.uk

Abstract. One dimensional nanostructures such as organic and inorganic nanotubes, nanowires and graphene have generated much excitement among materials scientists in recent years. However difficulties associated with their lack of processability have seriously hampered the number of possible applications. Poor dispersability in most common solvents and tendency to bundle forming larger aggregates has been a hurdle in the way of real exploitation. The recent development of techniques able to disperse and exfoliate these 1D objects have captured much attention, strongly re-launching these materials towards the applicative scene. It is mandatory to understand at what extent these techniques are efficient and whether the atomic structures of these nanomaterials are damaged or remain pristine after dispersion and exfoliation in liquid media. Aberration-corrected electron microscopy becomes in this context a very attractive and unique tool to investigate these structures at atomic scale.

1.Introduction

One-dimensional nanostructured materials such as organic and inorganic nanotubes[1] and nanowires[2] are potentially useful in a number of areas of nanoscience and nanotechnology due to their remarkable mechanical, electrical and thermal properties[3]. However, isolated and pure objects are rarely available to experimentalists. They tend to be produced in presence of large quantities of impurities and tend to aggregate into larger objects whose properties are generally inferior to those of isolated ones. In the last few years thermodynamic dispersion and exfoliation theories have been developed and showed to apply universally to 1D nanostructures of very diverse nature[4,5]. The first material which has benefitted from these newly developed processing methods are the inorganic nanowires made up from molybdenum, sulfur and iodine (MoSI nanowires). Purification and dispersion in liquid phase media and exfoliation of the as-synthesised nanowire ropes down to single, sub-nanometer wide wires or very thin nano-sized bundles have allowed to study their previously unknown atomic structure[6-7]. Here we present an aberration-corrected scanning transmission electron microscopy (STEM) that was essential and extraordinarily suitable to pin down the extremely complicated structure of both individual nanowires and nano sized bundles of them. This was essential for unlocking the whole range of unexplored potential of these materials and render them applicable for innovative technologies. Another nanomaterial which has been at the centre of the most advanced

[1] valeria.nicolosi@materials.ox.ac.uk

nanotechnology research and that has most benefitted from a liquid phase dispersion method has been graphene. Until a few years ago the standard procedure used to make graphene was micromechanical cleavage[8], which is a very low yield production method. In order to fully exploit graphene's outstanding properties, a mass production method was necessary and the development of a method to exfoliate cheap, commercial graphite in organic solvents down to large area single graphene flakes with high yield was one major achievement[5]. Before applying this method for real high-tech purposes it is imperative to see whether the graphene's structure endures the exfoliation procedure. In order to do so, flakes were characterised by aberration-corrected high resolution transmission electron microscopy (HRTEM), revealing extremely high quality graphene structures.

Experimental method

MoSI nanowires have been obtained in the stoichiometry $Mo_6S_3I_6$ by direct synthesis from elemental material[9]. The as grown material was purified as previously reported[6], dispersed in acetone at a concentration of 0.01 mg/ml and deposited by drop casting on holey carbon grids (400 mesh). Scanning transmission electron microscopy (STEM) studies were performed using the Oxford Jeol 2200MCO instrument fitted with a Ceos Cs corrector. STEM characterisation on the nanowires was performed at 200kV. STEM simulations were performed using the Melbourne STEM program developed by L J Allen, S D Findlay and M P Oxley[10].

The commercial graphite purchased from Sigma-Aldrich (Product Number 332461) was dispersed and exfoliated in N-methylpyrrolidone (NMP) at concentration of 0.1 mg/ml as previously described (ref). The dispersion obtained was drop casted onto holey carbon grids (400 mesh size) and HRTEM performed using the Oxford Jeol 2200MCO instrument fitted with a Ceos Cs corrector, operated at 80kV.

Results and Discussion

1.1 Mo6S3I6 nanowires

Crystallographic libraries on molybdenum chalcogenides and a combined X-ray powder diffraction (XRD) and X-ray absorption fine structure (XAFS) study on $Mo_6S_3I_6$ carried out by A. Meden and co-workers[11] were used to get preliminary structural information as a starting point for our systematic electron microscopy study. The predicted structure is shown in Figure 1a. In this model the hollow striped spheres represent Mo atoms, arranged in octahedral symmetry; the grey and black spheres represent instead two structural positions, peripheral (P) and linkage (L) sites respectively. XRD and XAFS were able to accurately determine the positions of the molybdenum octahedra and of both P and L sites but could not determine their occupancy in terms of sulfur or iodine atoms. Space groups of hexagonal-lattice P63 or trigonal lattice R-3 were suggested to describe the packing. Occupation of the L sites by iodine atoms was suggested.

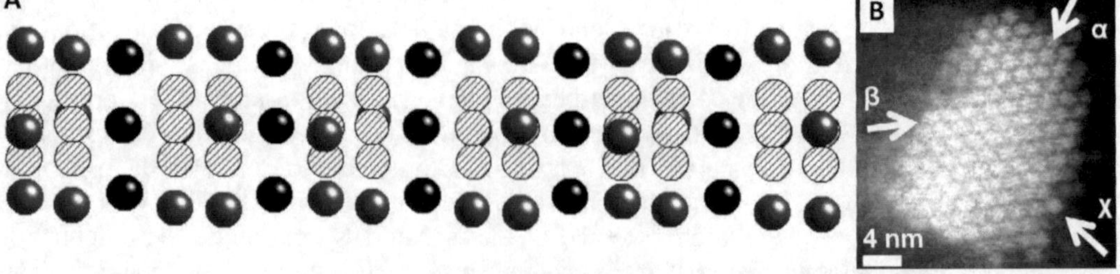

Figure 1. a)The nanowire atomic structure as determined by XRD and XAFS [11]. Hollow striped, grey and black spheres represent Mo atoms, and Peripheral and Linkage sites respectively; b) ADF STEM image of a nanowire bundle cross section

Aberration corrected HAADF STEM was used to make detailed refinements to the suggested atomic structure. Making use of high angles of electron scattering, HAADF STEM is strongly dependent on atomic number (Z-contrast) and becomes an attractive methodology to detect the different atomic species in the structure of the nanowires. The heavy iodine atoms become with this technique particularly detectable. Figure 1b shows a cross section of a nanowire bundle confirming an hexagonal close packing with a nanowire-to-nanowire distance of 0.96nm. Representative images of the three different high symmetry projections are shown in Figure 2a,b,c. The fact that three different image types were observed for the three high symmetry directions immediately suggests deviations from the previously predicted structure (both R-3 and P63 possess rotational symmetries and that would lead to identical projections for the three high symmetry directions). These three patterns can only be reproduced if iodine is allocated to the P sites and the sulfur atoms in the L sites respectively. ADF simulations, obtained using allocation of the I in the P sites and S atoms in the L sites, are shown for a nanowire bundle from the three high-symmetry directions, α, β, and *X*. The out-of-phase zig-zag pattern appearing in figure 2a can only be obtained for a nanowire bundle configuration P-1

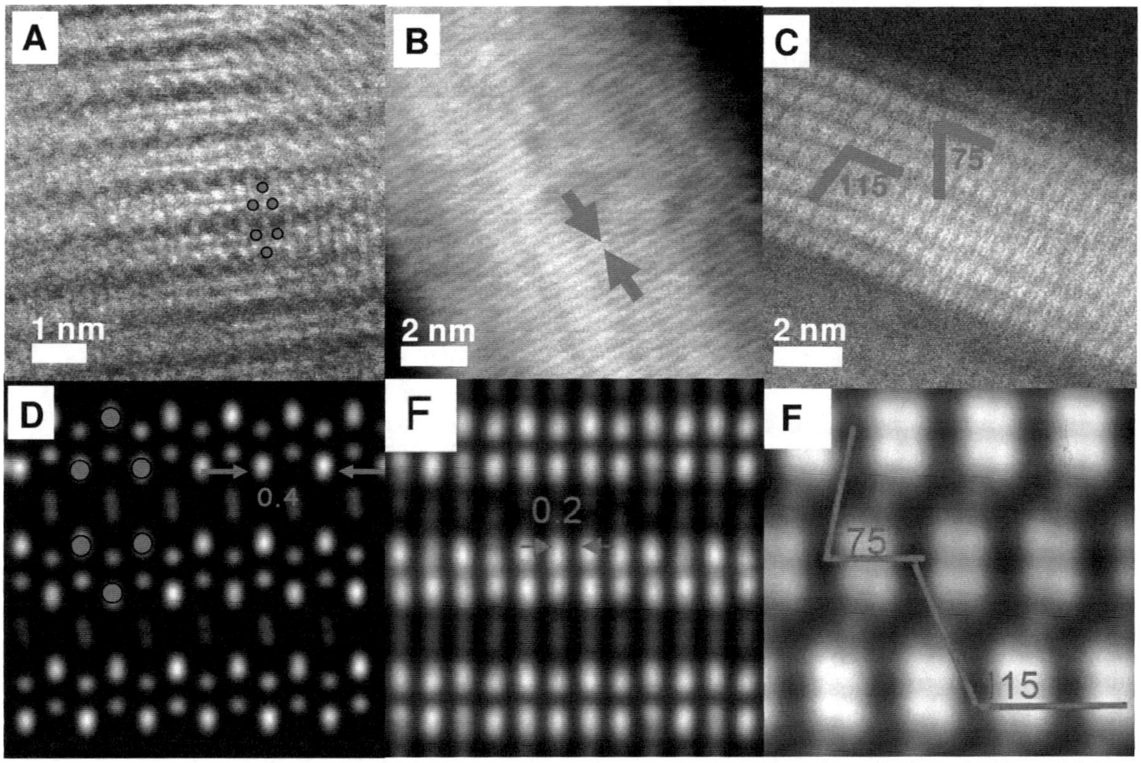

Figure 2. a-b-c) Experimental ADF STEM images of nanowire bundles viewed along the three high-symmetry directions shown as α (A), β (B), and *X*(C); **d-e-f)** Simulated ADF images obtained using a $P1^{-}$ structure and sulfur in the L sites.

1.2 Graphene

Methods recently developed to obtain individual carbon nanotubes suspended in liquids[5] were used to disperse and exfoliate commercial graphite into single graphene flakes. According to these studies, 1D nanomaterials disperse well into solvents whose surface energy matches theirs. The key question was to understand whether this dispersion method was giving dispersed graphite or suspended graphene. Shown in figure 3a is a bright field TEM images of an object typically observed. Statistical analysis show that Figure 3a is representative of about 80% of the material deposited on the TEM grid. Electron diffraction was used to confirm the presence of monolayers The two insets show in fact selected area diffraction patterns acquired from the two different areas indicated by the two arrows respectively. The main difference between the two diffraction patterns is that for the multi-layer (lower inset), the {2110} spots appear to be more intense relative to the {1100} spots. The large flake over the grid's hole is therefore certainly a monolayer. Aberration-corrected HRTEM was carried out to investigate whether the exfoliation technique was introducing structural defects. Figure 3a shows a typical, unfiltered HRTEM image of a flake over a grid's hole, that elegantly proves the non-destructive nature of this exfoliation technique.

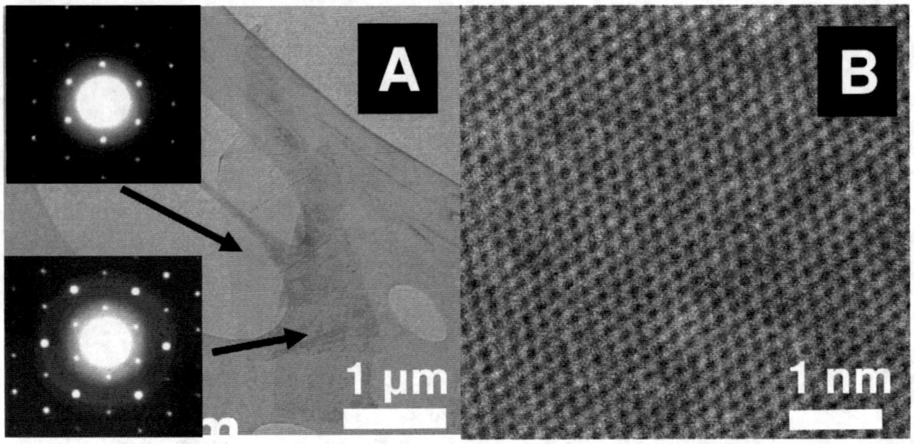

Figure 3. a) a Bright field TEM image of a typical graphene flakes found on the TEM grid after liquid phase dispersion. The insets show electron diffraction patterns relative to the two selected area. The upper diffraction pattern corresponds to a monolayer, while the lower one to a multi-layer; **b)** Aberration-corrected HRTEM image of an area over a grid's hole.

Results and Discussion

We have developed a method to disperse, exfoliate and process organic and inorganic 1D-nanostructures, such as inorganic nanowires made of Molybdenum, Sulfur and Iodine and graphene. To make real applications feasible it is crucial to fully characterize their structure down to atomic resolution and correlate this to their actual physical and chemical properties. Advances in aberration-corrected optics in electron microscopy have revolutionised methods of characterising nano-materials, and have opened new frontiers for materials science. Here we have presented an overview of a variety of low-dimensional materials issues, showing what aberration-corrected electron microscopy can do for material scientists.

References

[1] Iijima S., 1991 *Nature* **354**, 56-58
[2] Tenne R, Margulis L, Genut M et al., 1992 *Nature* **360**, 444-446
[3] Zaric S, Ostojic G N, Kono J et al., 2004 *Science* **304**, 1129
[4] Bergin S et al. 2008 *Advanced Materials,* 20, 10, 1876-1881
[5] Hernandez Y et al. 2008 *Nature Nanotechnology,* 3, 9, 563-568
[6] Nicolosi V et al. 2005 *J. Phys. Chem. B* **109**, 7124
[7] Nicolosi V et al. 2007 *Advanced Materials,* **l19**, 4, 543-547
[8] Novoselov K S et al.2005 *Proceedings of the National Academy of Sciences of the United States of America* **102**, 30, 10451
[9] Vrbanic D et al 2004 *Nanotechnology* **15**, 635
[10] Allen L J, Findlay S D, Oxley M P and Rossouw C J 2003 *Ultramicroscopy* 96 47-63
[11] Meden A, Kodre A, Gomilsek J P, Arcon I, Vilfan I, Vrbanic D, Mrzel A, Mihailovic D 2005 *Nanotechnology* **16**, 1578

Journal of Physics: Conference Series **241** (2010) 012101 doi:10.1088/1742-6596/241/1/012101

Effect of ECAP on microstructure and mechanical properties of cast AZ91 magnesium alloy

C W Chung[1, 2], R G Ding[2], Y L Chiu[2] and W Gao[1]

[1] Department of Chemical & Materials Engineering, University of Auckland, New Zealand

[2] School of Metallurgy and Materials, University of Birmingham, United Kingdom

Email: y.chiu@bham.ac.uk

Abstract. An as-cast AZ91 magnesium alloy was processed by Equal Channel Angular Pressing (ECAP) at 320°C. The microstructure and mechanical properties were studied. It has been found that ECAP refines both the grains and precipitates, thus modifies the strength and ductility of the processed alloy. After the first pass of ECAP, the yield stress improves significantly from 71 MPa to 140 MPa.

1. Introduction

Magnesium casting alloys such as Mg-9Al-1Zn (AZ91) have attracted increasing attention from the automotive, aircraft and electronic industries because of its low density, high damping properties, excellent machining abilities and good recycling potentials [1]. However, at room temperature, as-cast AZ91 exhibits poor formability and limited ductility due to the high fraction of coarse $Mg_{17}Al_{12}$ intermetallic phase [2]. Conventionally, grain refinement is usually used in improving the strength and ductility of alloys. Studies have shown that severe plastic deformation, such as ECAP, can improve the strength and ductility of processed materials by significantly reducing the grain size [3-6]. The ECAP processing of the AZ91 alloy has been extensively investigated at high temperature, and with solution treatment prior to ECAP to dissolve the coarse intermetallic particles. In this study, ECAP was carried out directly on an as-cast AZ91. The mechanical properties and microstructure before and after the ECAP have been investigated.

2. Experimental Procedures

A sand-cast ingot of AZ91 magnesium alloy with nominal chemical composition of 9 wt% Al, 1 wt% Zn and 0.5 wt% Mn was used in this study. ECAP experiments were performed in a die with 90° internal angle at 320°C and 4 mm/min pressing speed via the B_C route [7]. Tensile specimens with gauge dimensions of 2×2×7.5 mm^3 were machined from the ECAP processed sample, with the tensile axis parallel to the extrusion direction. Tensile tests were carried out at room temperature at a nominal strain rate of 5×10^{-4} s^{-1}. The microstructure observations were carried out on a JEOL 7000F FEGSEM and an FEI Tecnai F20 TEM equipped with an Oxford Instrument EDS system.

3. Results

3.1. Microstructures

Typical SEM micrographs of the as-cast and ECAP processed AZ91 are respectively shown in figure 1a-c. Primary intermetallic phase ($Mg_{17}Al_{12}$) and eutectic product consisting of the intermetallic phase and α-Mg are located at the interdendritic regions (figure 1a). The intermetallic particles with sizes of up to 80 μm were observed. After 1 pass of ECAP, some of the large intermetallic phases and dendrites were broken, and the grain size shows a bimodal distribution with the average size of about 8 μm for the fine grains (figure 1b). The grain size remains at about 6 μm with more than 2 passes of ECAP (figure 2). An EDS linescan across the matrix and part of particle (figure 3) indicates that the particle is Al and Zn rich. The diffraction patterns in figures 3e and 3f confirmed the cubic nature of the particles, which could be indexed as $Mg_{17}Al_{12}$. In addition, a high density of dislocations was often observed in the matrix, as shown in figure 4. Diffraction contrast analysis (not shown here) has indicated that most dislocations are <a> type dislocations.

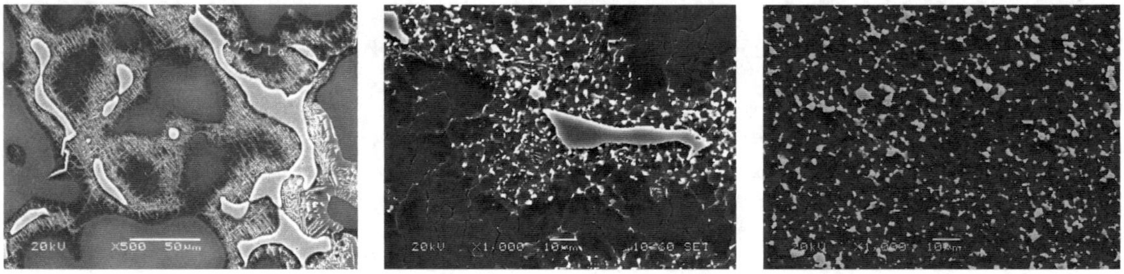

Figure 1. Typical SEM micrographs of (a) as-cast, (b) 1 ECAP pass and (c) 6 ECAP processed samples.

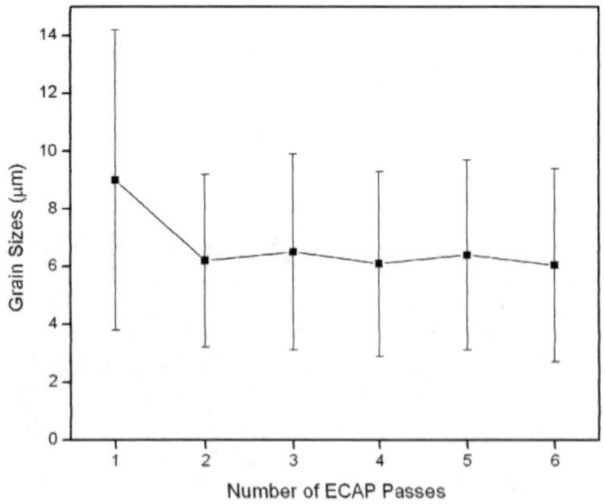

Figure 2. Variation in grain size with the number of ECAP passes. Note: In 1 passed sample, the size of large grains were not included.

3.2. Mechanical Properties

Figure 5 shows the mechanical properties of the alloy as a function of number of ECAP passes. After one pass of ECAP, the strength improves significantly (i.e. the yield stress increased from 71 MPa to 140 MPa), while no obvious increase in ductility was observed. After the second pass of ECAP, the ductility of the alloy improved. Further passes of ECAP shows no significant improvements in both the strength and elongation of the processed alloys.

4. Discussions

The tensile tests indicate that both the strength and ductility of as-cast AZ91 have been improved via ECAP processing. The strengthening mechanisms in the ECAP processed alloy mainly include solid solution strengthening, precipitate strengthening and grain boundary strengthening. The lack of peak shifts in X-ray diffraction experiments (not shown here) suggested that ECAP processing has no effect on the lattice parameters of both the matrix and the precipitates, indicating that the ECAP has not modified the solid solution strengthening. On the other hand, precipitate (beta phase) strengthening depends on the chemistry, volume fraction, size, morphology and distribution of the precipitates. The constant lattice parameters upon the ECAP processing suggest that the chemistry of the precipitates has not been changed. After 6 passes of ECAP, the precipitate size has been refined to about 1 μm, which is not expected to provide an obvious strengthening. Therefore, precipitate strengthening is not expected to be the major strengthening. Grain boundary strengthening is usually described by Hall-Petch equation. The grain size of AZ91 alloy has been significantly refined by ECAP (i.e. 300 μm for the as-cast sample vs. 6 μm for the 6-pass processed sample). The Hall-Petch coefficient for Mg was reported to be about 0.7 $MPa{\cdot}m^{-1/2}$ for coarse grains (30-87 μm) and 0.13 $MPa{\cdot}m^{1/2}$ for fine grains (17-30 μm) [8]. If the coefficient is regarded to be 0.13 $MPa{\cdot}m^{1/2}$, the contribution of grain boundary strengthening to the improvement of strength in the 6-pass sample is about 44 MPa. It seems likely that grain boundary strengthening is the major strengthening factor in these ECAP processed alloys. Consistently, beyond 2 passes of ECAP, the strength of AZ91 alloy has not been improved by the ECAP processing, as no further grain size refinement was observed.

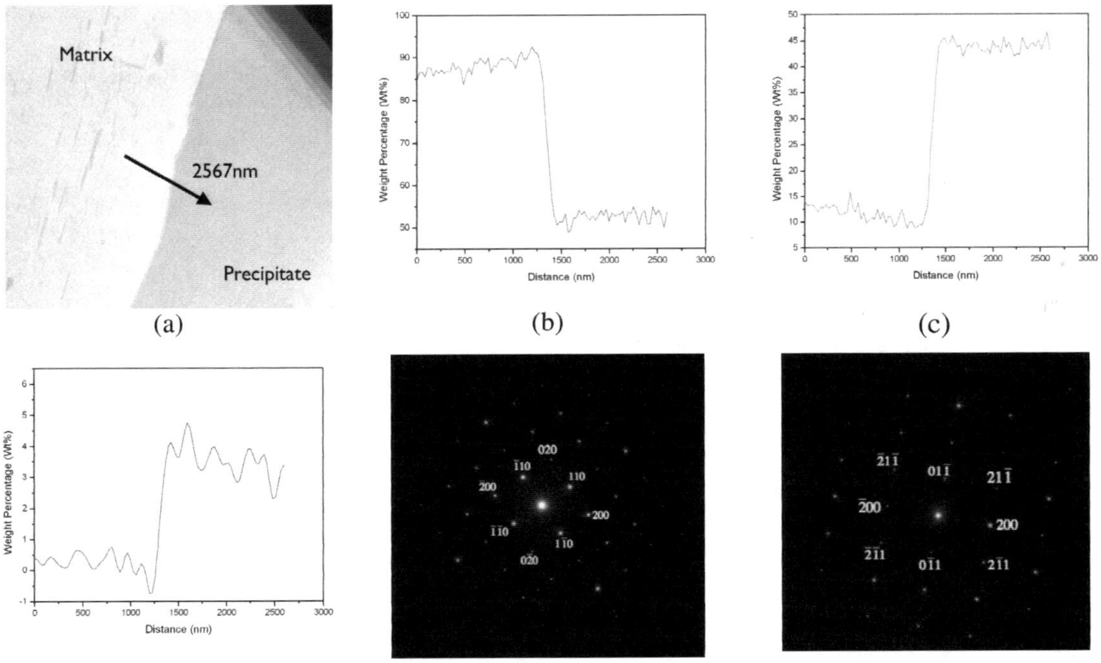

Figure 3. A BF STEM image of the intermetallic particle in the as-cast alloy (a); EDS linescan across matrix and the particle illustrate the contraction changes of Mg (b), Al (c) and Zn (d); Diffraction patterns obtained along [001] (e) and [011] (f) of the particle.

Figure 4. A BF TEM image taken from 6-passed sample, showing a high density of dislocations.

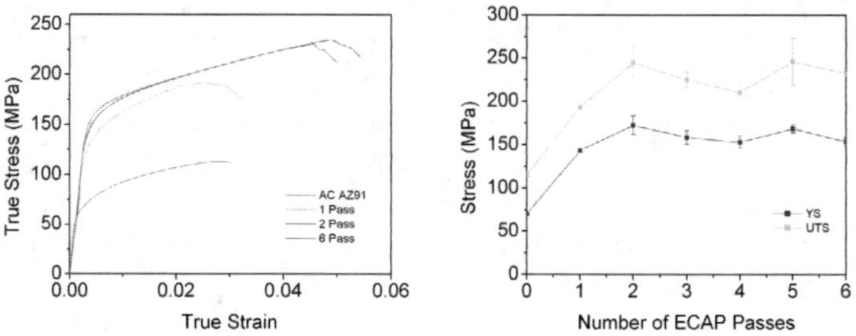

(a) (b)

Figure 5. (a) True stress - true strain curves measured at room temperature from the as-cast and ECAP processed samples with 1, 2 and 6 passes; (b) Variation in yield strength (YS) and ultimate tensile strength (UTS) with the number of pass.

5. Summary

The ECAP effectively refines the grains and beta phase of AZ91 alloy, thus improves the strength and ductility of AZ91 alloy. Grain boundary strengthening is the major strengthening mechanism the ECAP processed alloy.

References

[1] Kainer U 2003 *Magnesium – Alloys and Technologies* (Weinheim, Wiley-VCH)

[2] Chen B, Lin D L, Li J, Zeng X Q and Lu C 2007 *Mater. Sci. Eng. A* **483-484** 113-116

[3] Kim J K, Jeong H G, Hong S I, Kim Y S and Kim W J 2001 *Scr. Mater.* **45** 901-907

[4] Kim W J, Chung C S, Ma D S, Hong S I and Kim H K 2003 *Scr. Mater.* **49** 333-338

[5] Kocich R, Greger M and Machackova A 2006 *J. Achieve. Mater. Manufacturing Eng.* **18** 295-298

[6] Valiev R Z and Langdon T G 2006 *Prog. Mater. Sci.* **51** 881-981

[7] Langdon T G 2007 *Mater. Sci. Eng. A* **462** 3-11

[8] Koike J 2005 *Metall. Mater. Trans. A* **36** 1689-96

EELS and STEM analysis of metal nitride/substrate interfaces

I M Ross[1], Z Zhou[1], W M Rainforth[1], A Bleloch[2], A P Ehiasarian[3] and P Eh Hovespian[3]
[1]Department of Engineering Materials, University of Sheffield, Mappin Street, Sheffield S1 3JD, United Kingdom
[2]SuperSTEM, Daresbury Laboratory, Daresbury WA4 4AD, United Kingdom
[3]Nanotechnology Centre for PVD Research, Sheffield Hallam University, Sheffield, S1 1WB, United Kingdom

i.ross@shef.ac.uk

Abstract. Nitride based protective coatings produced by physical vapour deposition offer a promising means to extend the oxidation and wear performance of γ-TiAl. Metal ion pre-etching of the substrate surface has been shown to significantly enhance the adhesion between the coating and substrate. In this study we have applied spectroscopic imaging and advanced STEM analysis to study such interfaces. For a conventional Cr cathodic arc discharge Cr ion implantation of ~6nm into the substrate was observed. For an etch using high power impulse magnetron sputtering the metal ion implantation depth was reduced to ~2-3nm. In both cases a corresponding depletion zone of Ti and Al was observed. In addition, nitrogen diffusion into the substrate during coating deposition was also revealed to a similar depth respectively.

1. Introduction

A requirement of protective surface coatings, designed for high temperature and wear resistance applications is an excellent adhesion between the substrate and coating. For coatings produced by physical vapour deposition (PVD) the principal technique to enhance adhesion usually involves the application of a substrate pre-treatment using an Ar glow discharge plasma. Previous studies have shown that the Ar glow discharge can sometimes result in poor adhesion due to insufficient removal of surface contaminants or coalescence of implanted Ar at the interface. On the other hand, etching with a metal arc discharge has resulted in an improvement in coating adhesion compared to glow discharge [1]. Such a pre-treatment process utilizes a cathodic arc discharge to generate the metal ion plasma. Cathodic arc discharges produce a high flux of metal ions effectively removing surface contamination such as oxide films while metal ion implantation helps to reduce the lattice mismatch between the substrate and coating, reducing interface stresses [2]. One drawback however of cathodic arc etches is the production of droplets on the coating surface which can lead to the formation of detrimental growth defects. A recent solution to this problem is the development of High Power Impulse Magnetron Sputtering (HIPIMS) which significantly reduces the formation of such growth defects resulting in dense coatings with a high adhesion to the substrate [3]. Potential alloys for aerospace and automotive applications such as γ-TiAl required protective coatings to extend component life times particularly at elevated temperatures. Knowledge of the structure of the interface between the applied coating and γ-TiAl substrate is therefore important in the evaluation of the coating/component performance. Hence, in this work, we present the application of electron energy-loss spectroscopy

(EELS), energy filtered TEM (EFTEM) and conventional and aberration corrected scanning transmission electron microscopy (STEM) to investigate the interface between two PVD coatings utilizing different etching regimes deposited on γ-TiAl.

2. Experimental methods

Deposition of the nitride coating was performed in a Hauzer HTC 1000-4 PVD coater at the Nanotechnology Centre for PVD Research using unbalanced magnetron sputtering (UBM). Polished samples of γ-TiAl of composition Ti-45Al-8Nb-0.17C (at%) were subjected to either a Cr cathodic arc discharge or a Cr based HIPIMS etch prior to the UBM deposition of a TiAlN [2] or CrAlN [4] base layer respectively. The cathodic arc etching was performed with a substrate bias voltage of -1200V [1,2] whereas the HIPIMS etch utilized a substrate bias of -600V [5]. In both cases etching was performed in an Ar atmosphere (0.1 Pa) with a substrate temperature of 400°C.

Analysis of the coating and interface was performed using a JEOL 2010F TEM/STEM operating at 200kV equipped with an Oxford Instruments ISIS EDS system and a Gatan image filter (GIF). EDS analysis was performed in STEM mode across the interface between the base-layer and γ-TiAl substrate using a ~1nm probe at probe intervals of ~1.9nm and a spectrum acquisition time of 15 seconds per probe position. HAADF images of the interface were recorded before and after the line scan to assess specimen drift. Energy filtered images of the TiAlN/γ-TiAl interface where acquired at the Al $L_{2,3}$ (73eV), N K (401eV), Ti $L_{,2,3}$, (456eV) and Cr $L_{2,3}$ (575eV) edges. A slit width of 10eV was used for the Al $L_{2,3}$ and 20eV for the N K, Ti $L_{,2,3}$ and Cr $L_{2,3}$ each with an exposure time of 3x5 seconds.

Aberration corrected STEM was performed on the CrAlN/γ-TiAl interface in a modified Vacuum Generators STEM equipped with a Nion spherical aberration corrector and a Gatan Enfina EELS spectrometer. The EELS data was acquired at 100kV with a beam convergence half-angle of 24 mrad and a spectrometer collection angle of 19 mrad. High angle annular dark-field (HAADF) images were recorded with a detector acceptance angle of 70-210 mrad. Spectrum image EELS line scans were obtained across the interface with a nominal 0.15nm probe, a step size of 0.5nm and a spectrometer dispersion of 0.5eV/channel.

3. Results and Discussion

3.1. TiAlN/γ-TiAl with cathodic arc etch

A conventional STEM HAADF image of the lower region of the nitride coating and substrate is given in figure 1a revealing a band of bright contrast at the interface boundary. The results of EDS probe analysis across the interface (defined in figure 1a) is shown in figure 1b. A number of key features are evident, firstly an obvious Cr peak is observed within the substrate surface clearly demonstrating the effective ion implantation process during cathodic arc etching. Secondly a corresponding decrease in the Al signal and slight decrease in the Ti signal are also detected at the interface. The presence of significant Cr and reduction in the Al and Ti signal are therefore consistent with the observed bright contrast in the HAADF STEM image (figure 1a) since the average atomic number at the interface has clearly increased. However, due to the limitations with the detector efficiency it was not possible to record the relative nitrogen signal using EDS. Therefore additional energy filtered imaging was performed. The results of this investigation are presented in figure 2. A clear Cr signal is observed with a sharp onset in the intensity profile, tailing off with an implantation depth of 5-8nm into the substrate. Likewise the regions of Al and Ti depletion appear to correlate well with the region of Cr implantation. Interestingly, the N map exhibits the distinct presence of nitrogen diffusion into the substrate to a similar depth to that observed for the Cr implantation. Titanium alumindes are known to have a high affinity for nitrogen, forming nitrides preferentially during the initial stages of oxidation in air. Since the etch process was performed in Ar atmosphere the observed N diffusion region must have developed during the deposition stage of the nitride base-layer most likely during the very initial stages when significant nitrogen species would be present in the reactive atmosphere [1,2].

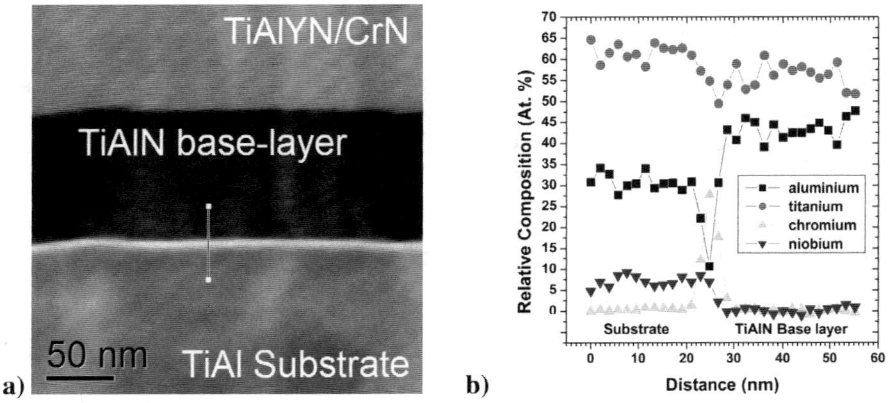

Figure. 1. a) ADF-STEM image of the interface between the substrate and nitride coating, b) EDS line profile across the interface between the substrate (left hand side) and base-layer (right hand side).

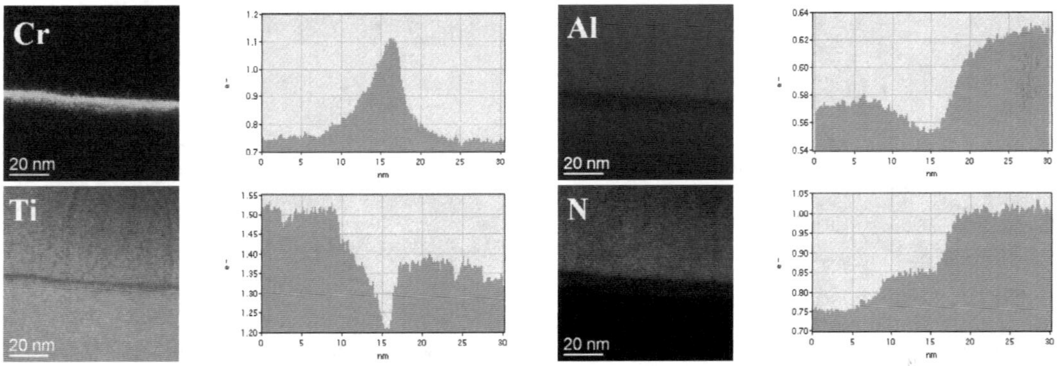

Figure. 2. EFTEM maps and corresponding intensity profiles across the interface between the TiAlN base-layer and Cr ion cathodic arc etched γ-TiAl substrate.

3.2. CrAlN/γ-TiAl with HIPIMS etch

A bright field aberration corrected STEM image of the interface between the substrate and base-layer of the HIPIMS etched sample is shown in figure 3a. The image illustrates the presence of coherent lattice fringes across the interface supporting previous reports of localized epitaxy between the coating and substrate after HIPIMS treatment [5]. A HAADF STEM image of the same region is shown in figure 3b indicating the position of the EELS and EDS line scans. The interface imaged in dark contrast in the BF-STEM image (figure 3a) is clearly observed in bright contrast in figure 3b suggesting a significant increase in the average atomic number at the interface as previously observed for the cathodic arc etched substrate. The result of EDS STEM analysis, shown in figure 4a, reveals a significant decrease in the Al content at the interface and a slight segregation of Nb (from the bulk substrate towards the substrate surface) both consistent with the HAADF image contrast. However, in this instance there is little evidence for any major Cr implantation compared to the cathodic arc treatment. The EELS spectrum image profile shown in figure 4b however confirms the presence of shallow Cr implantation with a slight increase in the Cr concentration at the substrate surface. As observed for the TiAlN cathodic arc etched sample, N diffusion is also revealed to a depth of 4-5nm into the substrate surface, most likely during the initial deposition of the nitride base-layer. It was not possible to record the Al EELS signal during the acquisition of the data shown in figure 4b (due to limitations with the coupling optics between the microscope and spectrometer). However, subsequent studies of the Al K edge (1560eV) across the interface boundary likewise showed a similar decrease at the interface supporting the previous EDS data.

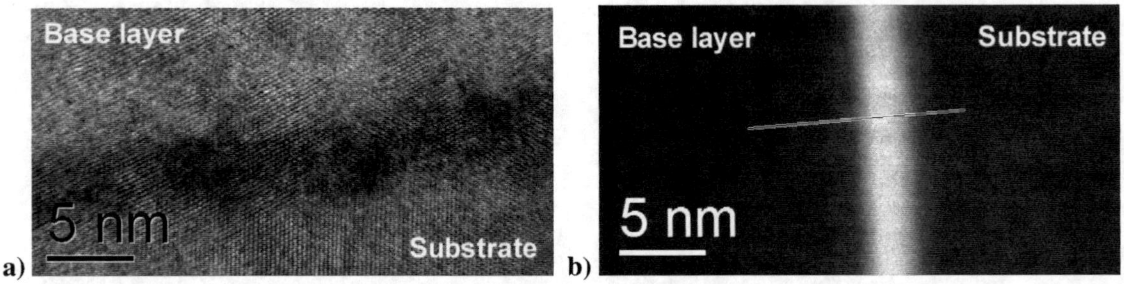

Figure. 3. (a) Bright-Field C_s corrected STEM image showing the coherent interface between a CrAlN base-layer and a HIPIMS etched γ-TiAl substrate, (b) corresponding HAADF image (interface rotated 90 degrees) showing the position of the EELS spectrum image line scan.

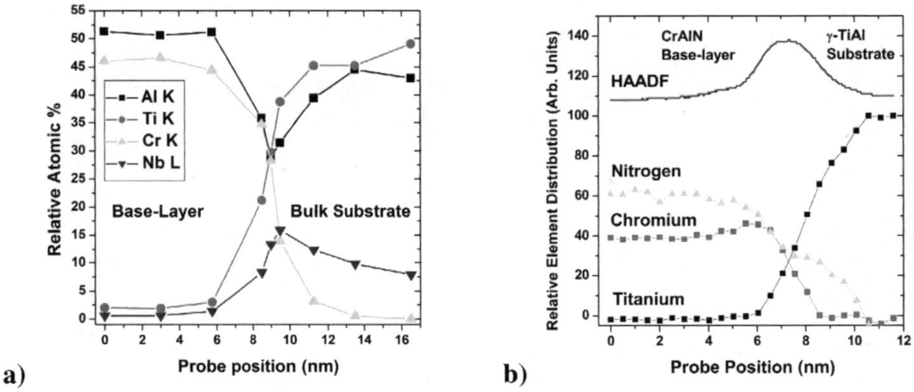

Figure. 4. Interface analysis between a CrAlN baselayer and a HIPIMS etched γ-TiAl substrate: (a) conventional FEG-STEM EDS analysis and (b) corresponding C_s corrected HAADF intensity profile and relative elemental distribution from the EELS spectrum image scan as defined in figure 3b.

4. Conclusions

We have successfully performed EELS, EFTEM and advanced STEM imaging of the interface between two nitride coatings and a γ-TiAl substrate. Regions of localised epitaxy were observed between the coating and substrate in each case. For a conventional Cr ion cathodic arc discharge and TiAlN coating a Cr ion implantation of ~6nm was observed into the substrate surface. For a Cr based HIPIMS etch and CrAlN coating the metal ion implantation depth was reduced to ~2-3nm. In both samples a corresponding depletion zone of Ti and Al was observed within the substrate surface. The diffusion of nitrogen into the substrate during coating deposition was also revealed to a similar depth to that corresponding to the ion implanted region.

References

[1] Ehiasarian A P, Anders A and Petrov I 2007 *J. Vac. Sci. Technol. A* **25** 543
[2] Schonjahn C, Ehiasarian A P, Lewis D B, New R, Münz W D Twesten R D and Petrov I 2001 *J.Vac. Sci. Technol. A* **19** 1415
[3] Ehiasarian A P, Munz W D, Hultman L, Helmersson U and Petrov I 2003 *Surf. Coat. Technol.* **163-164** 277
[4] Hovsepian P Eh, Reinhard C and Ehiasarian A P 2006 *Surf. Coat. Technol* **201** 4105
[5] Ehiasarian A P, Wen J G and Petrov I 2007 *J. Appl. Phys.* **101** 054301

Acknowledgements

The Authors gratefully acknowledge the European Commission (INNOVATIAL NMP3-CT-2005-515844).

Journal of Physics: Conference Series **241** (2010) 012103 doi:10.1088/1742-6596/241/1/012103

Microtextures and grain boundary misorientation distributions in controlled heat input titanium alloy fusion welds

R Leary[1], **E Merson**[2] **and R Brydson**[1]

[1]Institute for Materials Research, University of Leeds, Leeds, LS2 9JT, UK
[2]National Metals Technology Centre, Swinden House, Moorgate Road, Rotherham, South Yorkshire, S60 3AR, UK

E-mail: pm06rkl@leeds.ac.uk

Abstract. Microstructures, macrotextures and microtextures in commercial purity titanium and Ti-6Al-4V fusion welds produced by the InterPulse gas tungsten constricted arc welding (GTCAW) technique have been characterised. At the cooling rates associated with the InterPulse technique, α variants sharing a common $(1\bar{1}20)$ pole are found to cluster together into groups within prior β grains, leading to large areas where all variants are separated by a misorientation of 60º. These present potential easy slip paths, hence increasing the "effective structural unit size." Characterisation of these microtextures may provide new insight into microtexture-properties relations and the mechanisms of microtextural evolution.

1. Introduction

Titanium alloys are one of the most widely used groups of materials in structural engineering applications. However, joining them by fusion welding is still seen as difficult. As such, there is an ongoing need for development of improved welding techniques and characterization of the welds which they produce. The work here shows that textural and microtextural characterization is an essential part of this, and may provide insight into properties which remain to be fully understood.

One area of development in the search for improved titanium weld properties has been towards constricted arc processes. Gas tungsten constricted arc welding (GTCAW), and specifically the InterPulse technique of the Vacuum Brazing Consultants (VBC) group [1] has shown the potential for high quality results. The InterPulse technique uses magnetic constriction and high frequency modulation of the arc current (20,000Hz) to greatly reduce the overall heat input compared to conventional gas tungsten arc welding.

It has long been known that when characterizing titanium in general, texture should not be ignored. More recently, with the advent of techniques such as X-ray diffraction (XRD) for texture analysis, this importance has been increasingly demonstrated. Moreover, with analysis by electron backscatter diffraction (EBSD) it has now become possible to link microstructure and texture together - so-called "microtexture" - to provide new insight into materials processing-structure-properties relationships. Recent application of these techniques to titanium alloy welds [2] has highlighted the importance of microtextural characterization of fusion welds. In titanium alloys, microtextural characterisation takes on particular importance because the α to β phase transformation, which occurs according to the Burgers orientation relationship $\{0001\}_\alpha // \{110\}_\beta$ and $<1\bar{1}20>_\alpha // <\bar{1}11>_\beta$ [3], is often biased towards the preferential selection of certain α variants (so called "variant selection"). Of particular significance are the findings that the formation of titanium weld microtextures may be related to the cooling rate of the fusion zone [2].

2. Experimental

Welds were produced in 1.2mm thick Ti-6Al-4V and Cp-Ti sheet using an InterPulse 150 narrow bead manual TIG precision welding system. Welding parameters typically consisted of an arc voltage of 9V, average current of 20-40A and a linear travel speed of $50 mm \cdot min^{-1}$. These resulted in heat inputs of $\sim 240 J \cdot mm^{-1}$ and consequent cooling rates of $\sim 5°C \cdot s^{-1}$ at the β-transus temperature. Some welds were produced using low frequency current pulsing in addition to the 20,000Hz InterPulse waveform. Characterization was performed by polarized light microscopy, electron backscatter diffraction (EBSD), X-ray diffraction (XRD) and microindentation hardness measurements.

3. Results

The use of different welding parameters had a significant effect on the overall microstructures and textures in different welds. Figure 1 shows polarized light micrographs of two welds produced with an identical heat input, but using different welding parameters, intended for (a) high penetration and (b) structure build up. The use of polarized light gives textural contrast between different basal plane orientations because of the anisotropy of the unit cell of the hexagonal α phase.

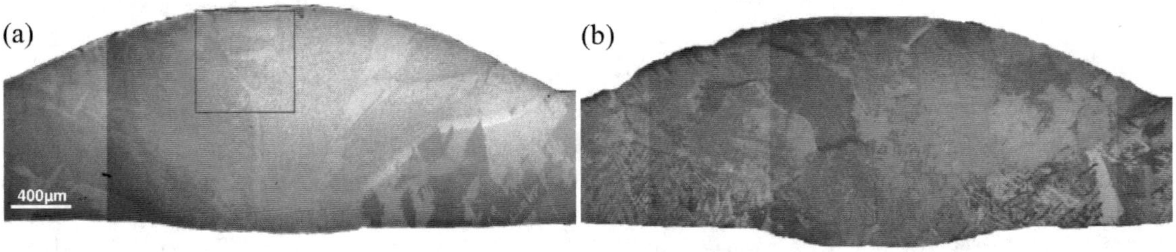

Figure 1. (Colour online) Polarized light micrographs of the fusion zone in InterPulse welds produced using Cp-Ti base metal and Ti-6Al-4V filler wire, with welding parameters intended for: (a) high penetration, (b) structure build up. The area outlined by the box in (a) corresponds to the polarized light micrograph and orientation map in figure 2. (NB The images are montages of more than one micrograph).

While the overall microstructures and textures vary significantly between these and other welds examined, there are important microtextures which were found to be present in all of the welds (although to differing extents). An example is highlighted in figure 2, showing an orientation map from EBSD and a corresponding polarized light micrograph. In the orientation map, prior β grain boundaries are outlined by solid black lines. These boundaries have been determined by isolating an area in the orientation map suspected to be a single prior β grain and examining the (0001) pole figure for that region. For a single prior β grain, the pole figure should, because of the Burgers relationship [3], show six strong peaks. Within the prior β grain in the centre of the map, certain α variants have clustered into "sub-grains," which are divided by dashed white lines. For two of the sub-grains, the α texture components present have been identified (using the INCA Crystal software "autotexture" routine), and their respective (0001) and $(1\bar{1}20)$ pole figures are displayed. These show that in each sub-grain, there are three α variants, with (0001) poles which lie 60° apart on a great circle. Hence these variants share a common $(1\bar{1}20)$ pole (circled) and therefore a common $[1\bar{1}20]$ direction.

4. Discussion

The finding that α variants within sub-grains are separated by a grain boundary misorientation of 60° and share a common $[11\bar{2}0]$ direction is significant. This is because $\langle 11\bar{2}0 \rangle$ is the slip direction in α titanium. Hence if slip were to occur within the sub-grain, it would be expected to proceed relatively more easily than in a region with more mixed grain boundary misorientations. Where the grain size is small, such as in the fusion zone, these grain boundaries may take on great importance in determining the mechanical properties of the weld.

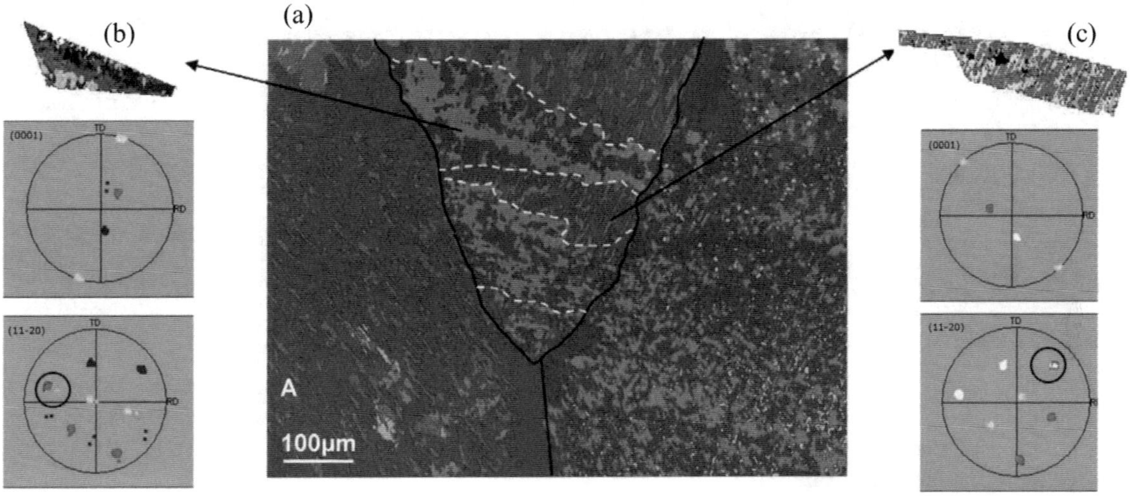

(d)

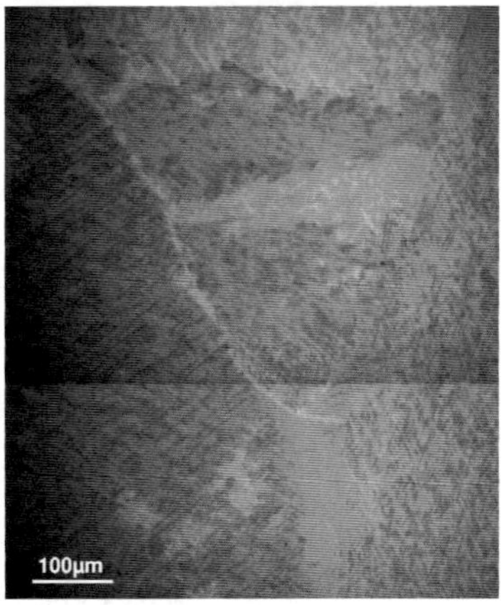

Figure 2. (Colour online) (a) Orientation map from EBSD showing sub-grain clustering. Prior β grain boundaries are outlined by black solid lines and sub-grain clusters are divided by dashed white lines. (b) and (c) show texture components within the two arrowed sub-grains. The associated pole figures show that each sub-grain consists of three texture variants, whose (0001) planes lie 60° apart on a great circle and hence the variants share a common $(11\bar{2}0)$ pole (circled). (d) Polarized light micrograph (montage of two images) corresponding to the central area in (a).

The "clustering" into sub-grains has been observed in the past in plasma welds [2] and was proposed as an explanation of higher than expected impact toughness in those welds. In this work, the significance in practice has been observed directly by performing microindentation hardness measurements into sub-grains. These generally showed a lower hardness compared to other more "random" microtextural regions, although further work is required in this aspect.

As proposed by Merson [2], the sub-grain clustering effect appears to be related to the cooling rate of the welding process, with it being able to predominate at intermediate cooling rates, such as those associated with plasma welding (investigated by Merson [2]) or the InterPulse welding examined here. A number of authors have supposed that the high occurrence of the 60° grain boundary misorientation must be related to some energetic favourability of this boundary type e.g. [2,4,5], although a full explanation of why has not yet been made. 60° is one of the five allowed grain boundary misorientations between α laths within a single prior β grain according to the Burgers relationship [6], but the proportions seen here are much higher than random and are associated with the sub-grain clustering effect.

In this work, a feature which frequently occurred in the microstructures/microtextures was the formation of laths into a distinct "triangular" arrangement, some examples of which are shown in figure 3. Analysis of the grain boundaries between these laths suggests that they are separated by the favourable 60° misorientation. The fact that the three laths in the triangle have grown thicker than others could suggest that the boundary is favourable in growth. On the other hand, the fact that many laths have formed together in the sub-grains could suggest it is related to nucleation. Further investigation is required here.

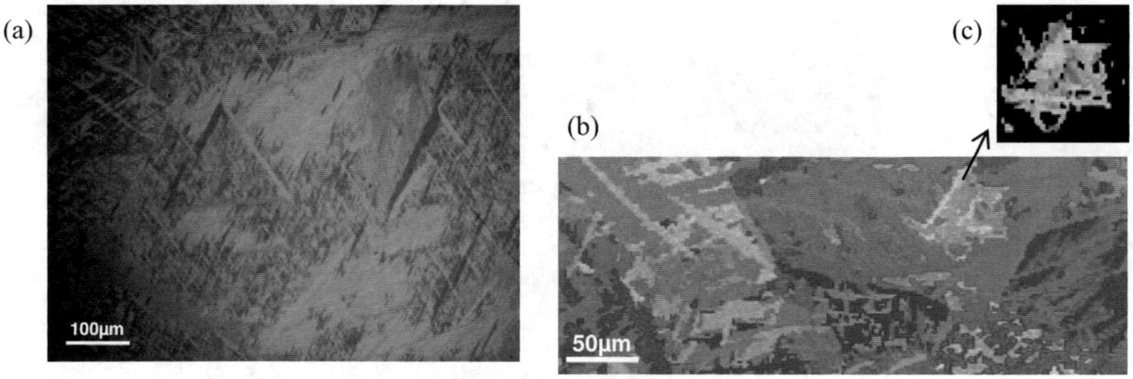

Figure 3. (Colour online) (a) Polarized light micrograph from the fusion zone of an InterPulse weld showing the formation of α laths into clear triangular morphologies. (b) Fusion zone orientation map from EBSD showing a "triangle" cluster (not corresponding to (a)). (c) Isolated texture component map for the sub-grain cluster in (b).

Importantly, it was found in this work that the clustering effect need not be isolated to small areas. Indeed, when the prior β grain labelled A in figure 2 is run through the INCA Crystal autotexture routine, it is found to possess the "sub-grain microtexture." Correlating the orientation map in figure 2 with the polarized light micrograph in figure 1(a) shows that this prior β grain spans the entire width of the fusion zone, and hence is highly significant in terms of the overall weld macrostructure. A strong macrotexture was indeed confirmed by XRD, although these results are not shown here.

5. Conclusions

This work has shown that consideration of macrostructure, microstructure, texture and microtexture are necessary in order to characterize titanium alloy fusion welds. Important microtextures which may explain as yet poorly understood mechanical properties have been found and characterized, and it is proposed that their occurrence and nature are strongly related to the cooling rate associated with the welding process. The findings warrant further work in this area, which may yield greater understanding of both phase transformations and mechanical properties in titanium alloy fusion welds and in titanium alloys in general.

Acknowledgements
We would like to thank the VBC group for their involvement in this work.

References

[1] VBC group http://www.vbcgroup.com/
[2] Merson E 2008 Institute for Materials Research, University of Leeds. PhD thesis.
[3] Burgers W G 1934 *Physica* **1** 561-86
[4] Randle V, Rohrer G S and Hu Y 2008 *Scripta Materialia* **58** 183-6
[5] van Bohemen S M C, Kamp A, Petrov R H, Kestens L A I and Sietsma J 2008 *Acta Materialia* **56** 5907-14
[6] Gey N and Humbert M 2002 *Acta Materialia* **50** 277-87

Journal of Physics: Conference Series **241** (2010) 012104 doi:10.1088/1742-6596/241/1/012104

Accurate analysis of EBSD data for phase identification

Y Palizdar[1], R C Cochrane[1], R Brydson[1], R Leary[1] and A J Scott[1]
[1] Institute for Materials Research, University of Leeds, Leeds LS2 9JT UK

E-mail: preyp@leeds.ac.uk

Abstract. This paper aims to investigate the reliability of software default settings in the analysis of EBSD results. To study the effect of software settings on the EBSD results, the presence of different phases in high Al steel has been investigated by EBSD. The results show the importance of appropriate automated analysis parameters for valid and reliable phase discrimination. Specifically, the importance of the minimum number of indexed bands and the maximum solution error have been investigated with values of 7-9 and 1.0-1.5° respectively, found to be needed for accurate analysis.

1. Introduction

Electron backscatter diffraction (EBSD) in the scanning electron microscope (SEM) is now a well established technique for examination of the crystallographic microstructure of single and multiphase materials [1]. The ability to rapidly analyse large areas of samples and the ease of specimen preparation when compared to transmission electron microscopy (TEM) is a major advantage promoting its use. However, in certain instances it can be difficult to detect and differentiate second phases which may be present in small amounts, especially when present as small or finely dispersed particles.

As part of on-going research to investigate the effect of aluminium on the microstructure and mechanical properties of low carbon, low nitrogen steels, EBSD has been employed as a technique to perform quantitative metallography. The results of this study have been discussed elsewhere [2, 3]. However, during the processing of the EBSD results, the authors have realized that by consideration of some practical points the EBSD results will be more reliable.

It should be noted that EBSD is becoming an increasingly popular characterisation technique with many researchers inexperienced in this area. It is well known that optimisation of hardware and the microscope set up such as accelerating voltage, specimen tilt and probe size together with good sample preparation are essential to obtain good quality results [1, 4]. Since EBSD data analysis is computer based to obtain reliable and accurate results we must pay particular attention to certain processing parameters such as the minimum number of indexed bands and the maximum solution error, questioning the given default values. The maximum solution error is a measure of the allowed angular error between the collected and predicted patterns i.e. the pixel will be unindexed if the angle between the pattern and solution is greater than the selected maximum solution error). In this paper the effect of these parameters on the EBSD results will be discussed.

2. Experimental Methods

EBSD investigations were carried out on a steel sample (0.02 wt% C, 1.40 wt% Mn, 0.48 wt% Al, 0.001 wt% N) which was prepared in a laboratory vacuum melt furnace as a 50kg cast and hot rolled to 13 mm thick plate, the finish rolling temperature being in the range of 1000-950°C.

The samples were prepared for EBSD by polishing on a Buehler Vibromat vibratory polisher using non-crystallising colloidal silica. EBSD was then carried out using a FEI Quanta FEGSEM from the centre of specimens with patterns detected using an Oxford Instruments camera. Patterns were analysed in real-time using the INCA software from Oxford

Journal of Physics: Conference Series **241** (2010) 012104 doi:10.1088/1742-6596/241/1/012104

Instruments. Maps were collected using different step sizes from approximately 2 μm down to 100 nm. To investigate the size of AlN particles, TEM has been employed. AlN particles were extracted by the carbon replica technique and examined using a Philips CM20 TEM operated at 200 kV.

3. Results and discussion

EBSD pattern quality and phase maps were obtained at different levels of magnification using step sizes of approximate 2, 0.5 and 0.1μm, with analysis for the expected phases of ferrite, cementite and AlN; these are shown in figures 1a-c respectively. The default settings of the INCA crystal software were used with the minimum number of bands for indexing being 5 and the maximum solution error 2 degrees. The microstructure is indexed as being fully ferritic with some AlN and cementite particles. The corresponding pattern quality maps are shown to be of high quality. The AlN particle distributions look feasible since thermodynamic calculations performed using either the equation proposed by Seiverts [5] or the Thermocalc software suggest that the solvus temperature of aluminium nitride for this specific composition would not be lower than 1250 °C. This suggests that relatively large AlN particles should be expected within the bulk. In addition, TEM shows the AlN particle size to be approximately 500 nm (figure 2). Therefore, in theory, EBSD results obtained using a 100 nm step size can be expected to be reliable since at this step size there would be more than 4 pixels covering one particle which facilitate the phase identification process. The other issue which needs to be considered is spot size. However, since EBSD has been performed using a FEGSEM, the spot size and consequently special resolution is small.

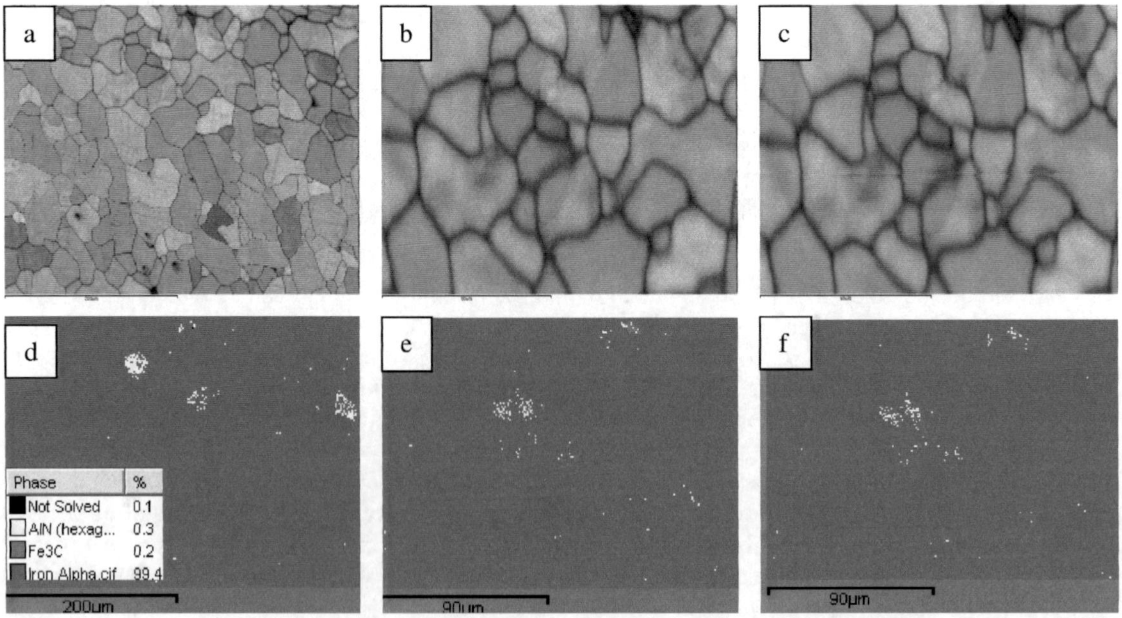

Figure 1. Pattern quality maps with expected phases of ferrite, cementite and AlN. The minimum number of bands and maximum error are the default settings of 5 and 2 respectively. Approximate 2 μm step size applied for "a" and "d" and 0.5 and 0.1 performed for "b" and "e" and "c" and "f" respectively.

Following the EBSD analysis of the steel which indicated the presence of AlN, both EDX mapping and backscattered imaging within the SEM were employed to add further confirmation. However, no contrast indicative of the presence of different phases within the ferrite matrix was found by BS-SEM and also no aluminium was detected using EDX. This could be attributed to the poor spatial resolution for both backscattered imaging and EDX analysis but since the attempt to show the presence of AlN by back scatter imaging and EDX technique did not succeed it was thought that more investigation was required to confirm the presence of AlN.

To examine the reliability of the EBSD results, reprocessing of the data was performed with analysis solely for the phases, ferrite and cementite. However, the pixels previously indexed as AlN do not now appear as unindexed points but are indexed as ferrite. This could be the consequences of overlapping two patterns. On the other hand, it could be an indication of using inappropriate tolerances when processing the EBSD data. To examine the suitability of the default tolerances, the raw data were reprocessed with the expected phases being Al_3Ni_2, Al_3Zr and the silicate mineral, Augite; these "nonsense" phases possess a range of different crystal structures and are most definitely not present in the sample. Figure 3a shows the subsequent phase map. The majority of the sampled area has been indexed as one of these three phases. Therefore, it was believed that the applied tolerances used to process the EBSD data were not tight enough to avoid mis-indexing of the data.

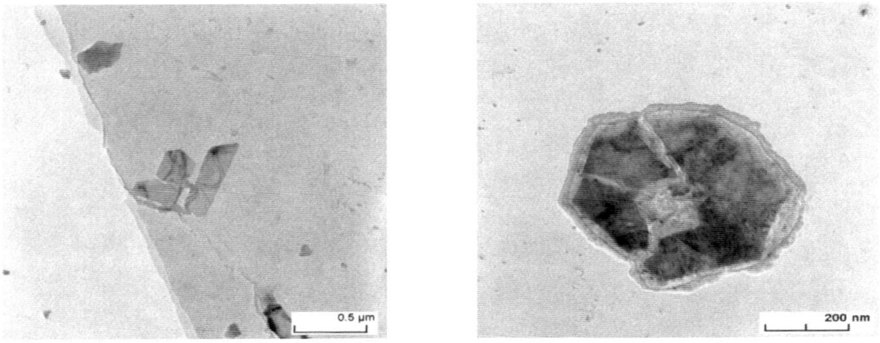

Figure 2. TEM micrograph of AlN particles

We believe that the two key parameters which can affect the accuracy of the pattern indexing are the minimum number of indexed bands and the maximum solution error. Figure 3b shows the effect of varying these parameters on pattern indexing. As can be seen, with the minimum number of bands set to 7 and a maximum allowable error of 2°, a significant number of data points are still mis-indexed, implying that the values of these parameters are insufficient to give reliable results. Increasing the number of matching bands to 9 reduces the number of mis-indexed pixels, and a further reduction is made by lowering the allowed angular error to 1.5°; the number of indexed points with these inappropriate phases now falls to almost zero. However, some pixels (0.3%) are still being indexed as one of these clearly incorrect phases.

Figures 4a and 4b show the examination of the same parameters to reprocess the initial sample. Based on given results the optimum minimum indexed bands can be a range between 7 and 9 and the optimum value for maximum error solution could be around 1.5. It should be noted that after reprocessing the initial data with new obtained criteria (9 minimum indexed band and 1.5 maximum solution error) negligible number of pixels indexed as AlN particles and also the indexed ratio fall to 95 %.

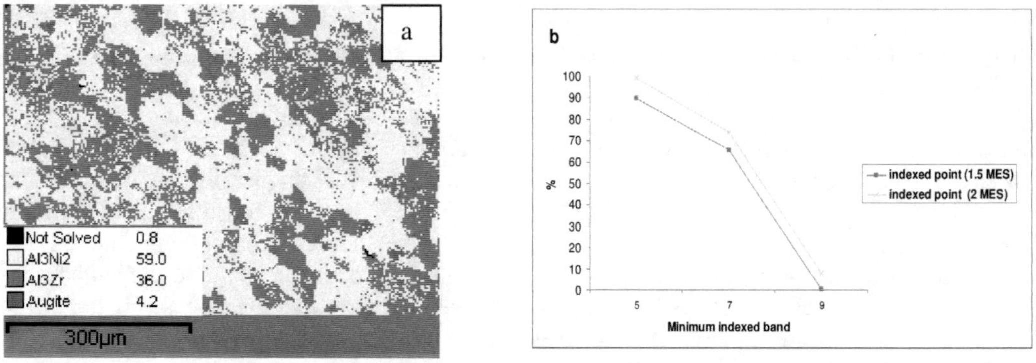

Figure 3. EBSD data reprocessed with the expected phases being Al_3Ni, Al_3Zr and Augite. Fig. 3a shows reprocessed EBSD data with 5 minimum indexed bands and a maximum solution error of 2° together with the relative assignments (% of the indexed pixels). Fig. 3b shows the effect of the minimum indexed bands and maximum error solution on the percentage of indexed pixels.

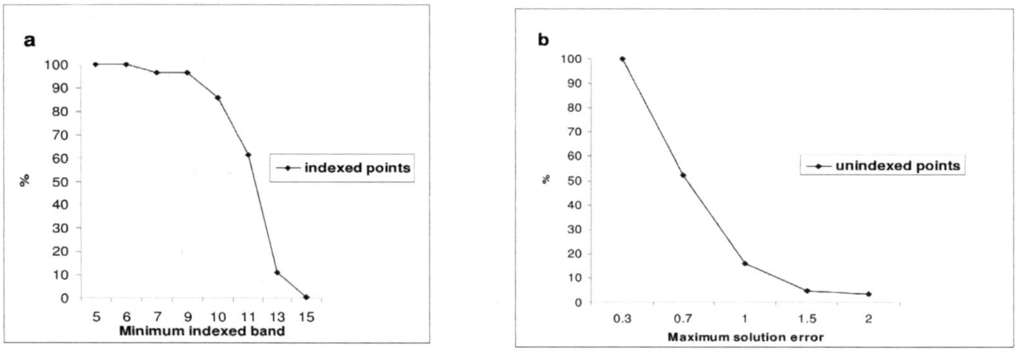

Figure 4. The effect of the minimum number of indexed bands (a) and the maximum solution error (b) on the percentage of indexed and unindexed pixels respectively as either ferrite, AlN and cementite as the expected phases.

4. Conclusions

The results obtained from this study show that although the importance of hardware set up can not be ignored, the used data processing criteria plays a remarkable role in making the EBSD results reliable and reproducible. In particular, we have shown the importance of the minimum indexed band and maximum error solution. Thus, treating automated ESBD as a standard "package" which will provide quick and easy data is not on its own adequate to produce good quality results. However, by employing the modified criteria the authors still failed to observe a reliable and reproducible AlN distribution by ESBD.

4. Acknowledgments

The authors would like to thank the EPSRC, grant EP/C512839/1 for funding this work.

References

[1] Gourgues-Lorenzon A F, *Int. Mater. Rev.* **52** 65
[2] Palizdar Y, Scott A J., Cochrane R C and Brydson R, *Mater. Sci. Tech.* **25** 1243
[3] Palizdar Y, Cochrane R C, Brydson R, Crowther D, San Martin D and Scott A *Mater. Charact.* (In press)
[4] Humphreys F J, *J Mater. Sci.* **36** 3833
[5] Gladman T 1997, *The Physical Metallurgy of Microalloyed Steels* (London: Institute of Materials)

Microstructure and Mechanical Property of ECAPed ZE41 Magnesium Alloy

R G Ding[1], C W Chung[1, 2], and Y L Chiu[1]
[1] School of Metallurgy and Materials, University of Birmingham, United Kingdom
[2] Department of Chemical & Materials Engineering, University of Auckland, New Zealand

Email: y.chiu@bham.ac.uk

Abstract. A ZE41 alloy was processed by equal channel angular pressing (ECAP) at 320°C. The microstructure and tensile properties were investigated. It has been observed that ECAP refines both the grains and precipitates, thus modifies the strength and ductility of the alloy. The sample after 6-passes of ECAP processing has a yield stress of 230MPa and elongation of 20%, compared with 160 MPa and 8% prior to the ECAP. The deformation of the alloy is predominated via dislocation slip with mechanical twining.

1. Introduction

Magnesium alloys have a unique combination of properties such as low density and high strength-to-weight ratio. This brings magnesium alloys great potential for applications in automobile and aeronautic industries. However, the use of the alloys has been limited by the poor formability and limited ductility at room temperature, ascribed to their hexagonal close-packed crystal structure. Grain refinement is an important means to improve the mechanical properties of magnesium alloys, for example their strength and ductility can be changed via different severe plastic deformation techniques, such as ECAP. Studies have shown that the strength and ductility of some Mg alloys could not be improved simultaneously via ECAP [1]. ZE41 is a popular Mg-Zn-Re alloy and has moderate strength and creep resistance [2]. The present study evaluates the microstructure and mechanical properties of a ZE41 alloy upon ECAP processed.

2. Experimental procedure

A commercial ZE41 alloy with nominal chemical composition of 4 wt% Zn, 0.7 wt% Zr and 1.4 wt% La+Ce was used in this study. Cylindrical samples of 10 mm in diameter and 55 mm in length were ECAP processed at a speed of 4 mm/min at 320°C in a die with 90° inter-channel angle. After each pass, the sample was rotated by 120°. Tensile specimens with gauge dimensions of 2×2×7.5 mm^3 were machined from the ECAP processed samples, with the tensile axis parallel to the extrusion direction. Tensile tests were carried out at room temperature at nominal strain rate of $5x10^{-4}$ s^{-1}. The microstructure examination was carried out on an FEI Quanta 3D FIB/SEM and an FEI Tecnai F20.

3. Results and analysis

Typical SEM micrographs of as-received and ECAP processed ZE41 samples are shown in Figure 1. Secondary precipitates of the as-received sample show nearly continuous distribution along grain boundaries. However, a small amount of particles was found to be located at grain interior. After 1 pass of ECAP, some particles were broken and grains were elongated showing a bi-modal distribution of grain size. The variation in grain and particle size with ECAP pass is illustrated in Figure 2, indicating that the average size of grains and particles reduces with increasing number of ECAP passes, but such refinement is not significant after 5 passes. Small grains are believed to be mainly resulted from dynamic recrystallization (the current ECAP temperature is higher than half of the melting temperature).

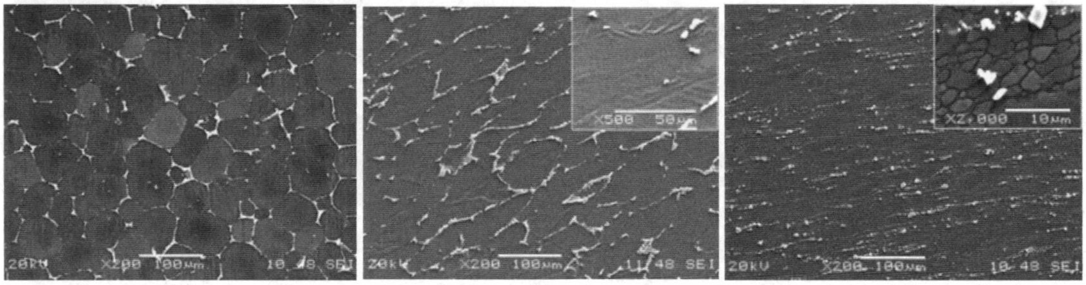

(a) (b) (c)

Figure 1 SEM images taken from as-received alloy (a), the 1 pass (b) and 6 passes of ECAP. Inset is of higher magnification.

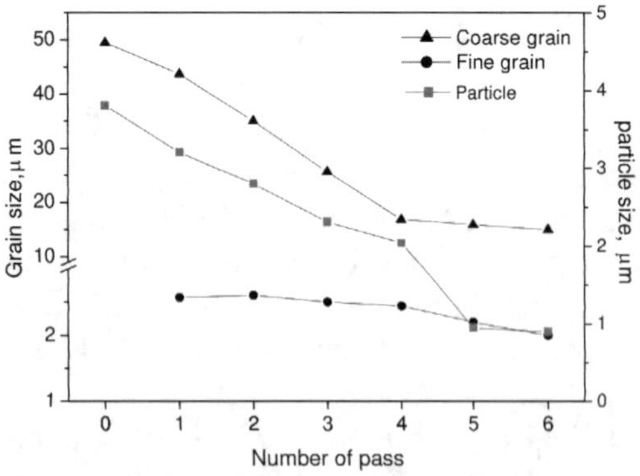

Figure 2 Variation in grain and particle size with the number of ECAP passes.

Figure 3 are TEM images obtained from the as-received and ECAP processed samples. Based on the information from EDX analysis and diffraction patterns (Figs. 3 and 4), grain boundary particle of orthorhombic T-phase (Mg_7Zn_3Re) can be identified with a = 0.99 nm, b = 1.15 nm and c = 0.98 nm, and the particle at grain interior is tetragonal Zn_2Zr_3 with a = 0.763 nm and c = 0.697 nm.

Figure 5 shows the mechanical properties of the alloy as a function of the number of ECAP passes. After the first pass of ECAP, the strength improves significantly from 160 MPa to 245 MPa, while no obvious increase in ductility has been observed. After 2nd pass, the strength of the alloy is not improved with the increasing number of passes, but the ductility shows a noticeable increase upon the 3rd pass of ECAP.

After tensile test, high density of dislocations and some mechanical twins were observed (Fig. 6) in the ECAP processed sample. It suggests that ZE41 alloy is deformed prominently via dislocation slip with mechanical twining.

(a) (b) (c)

Figure 3 TEM images taken from as-received alloy (a), the 1 pass (b) and 6 passes of ECAP. Insets in Fig.3a are EDX spectrum and diffraction patterns from grain boundary particle in the as-received sample.

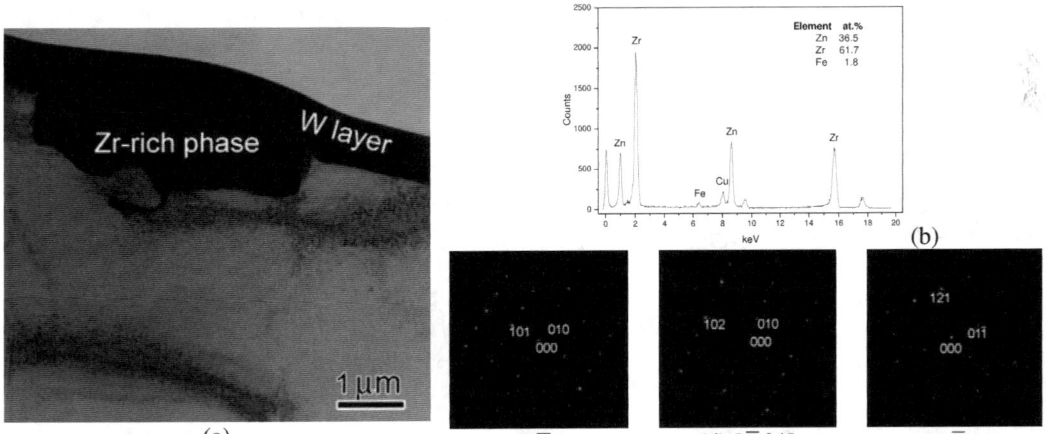

(a) (c) [$\bar{1}01$] (d) [$\bar{2}01$] (e) [$\bar{3}11$]

Figure 4 TEM image of Zr-rich phase (a); EDX spectrum from the Zr-rich phase (b) and diffraction patterns of [$\bar{1}01$] (c); [$\bar{2}01$] (d) and [$\bar{3}11$] (e) from the Zr-rich phase. Note: this TEM foil was prepared using FIB.

4. Discussion

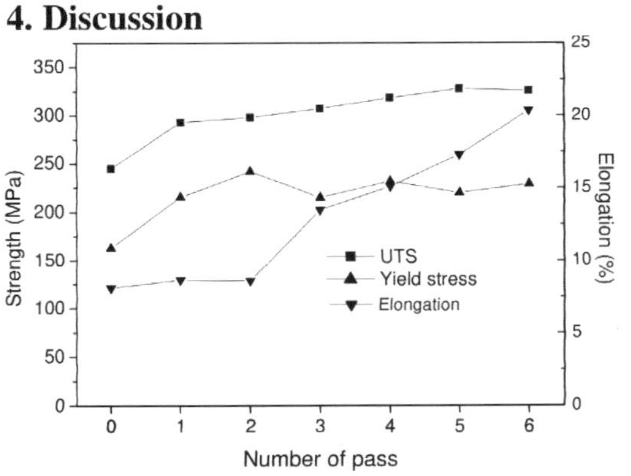

Figure 5 Strength and elongation of the alloy as a function of the number of pass

The tensile test result illustrates that the strength and ductility of ZE41 have been improved via ECAP processing. The strengthening mechanisms in the ECAP processed alloy include solid solution strengthening, dispersion strengthening and grain size strengthening. XRD results (not present here) showed that ECAP processing has no effect on the lattice parameters measured, which is consistent with the fact that the volume fraction of precipitates remains invariant with the number of ECAP passes. In other words, the ECAP is not expected to modify the solid solution strengthening. ECAP has reduced the precipitate size (Fig. 2), which could improve the strength of the alloy. However, dispersion strengthening is not expected to be the major strengthening because the particle size (of ~0.9 μm even after 6 passes) is not sufficiently small. This is consistent with the fact that, while the particle size continuously decreases with the increasing number of passes, the strength does not follow the exact trend (Fig.5). Grain boundary strengthening is usually described by the Hall-Petch equation. This strengthening could be an important one in the ECAPed ZE41 because of large Hall-Petch coefficient for Mg alloys and the obvious grain refinement of ZE41 via ECAP. Other metallurgical factors such as texture could influence the strength. XRD analysis (not shown here) indicated that a basal texture (0002) was observed in the 1- and 2-passes of ECAP processed samples, but not in the other samples. This may explain the rapid strengthening observed in the 1- and 2-passes processed samples. Fracture analysis showed that secondary cracks initiated from the grain boundary T-phase. This suggests the ductility of the alloy is dependent on the distribution and size of the T-phase. Although some grain boundary T-phases were broken after 1 pass of ECAP processing, a nearly continuous distribution of T-phase along grain boundaries is still observed clearly (Fig. 1b). Not surprisingly, after 1 and 2 passes of ECAP processing, the ductility of the alloy has not been improved. Further refinement of grain boundary precipitates and grain resulted in the tensile elongation of up to 20% (Fig.5).

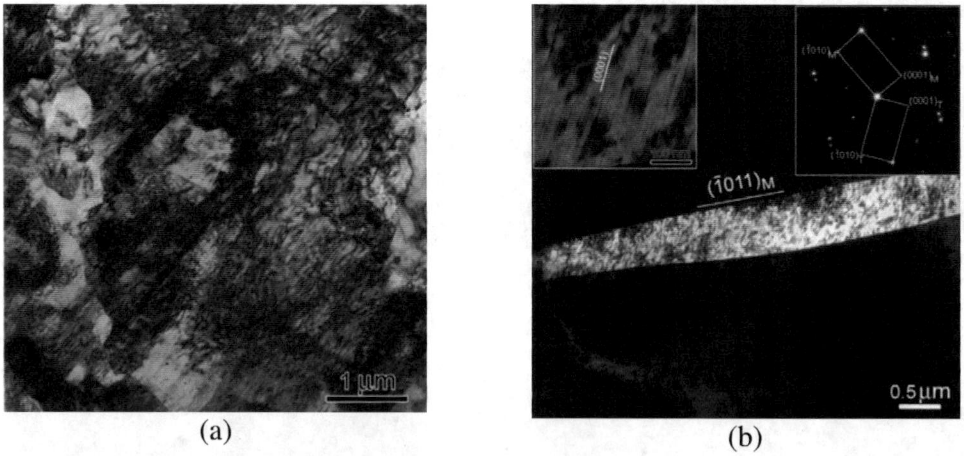

(a) (b)

Figure 6 (a) a high density of dislocations in ECAP sample and (b) a DF TEM image of twin taken from a tensile sample. Inserts in Fig.6b are higher magnification of the twin and composite SAD of the twin and the matrix.

5. Summary

ECAP brought about the refinement of grains and T-phase, thus influenced on the strength and ductility of the alloy. It is expected that the ductility is predominantly controlled by the size and distribution of grain boundary T-phase.

References

[1] Tang W, Chen R, Zhou J, and Han E 2009 *Mater. Sci. & Eng. A* **499** 404
[2] Ma A, Jiang J, Saito N, Shigematsu N, Yuan Y, Yang D, and Nishida Y 2009 *Mater.Sci. & Eng. A* **513-514** 122
[3] Mendis C, Ohishi K, Kawamura Y, and Hono K 2009 *Acta Mater.* **57** 749

Annihilation behaviour under electron irradiation of athermal ω-phase crystals formed by cooling at 131K in a β Ti-Mo alloy

E Sukedai[1], H Matsumoto[2]

[1] Okayama University of Science, Okayama 700-0005, Japan,
[2] Graduate student: Okayama University of Science, Okayama 700-0005, Japan (Present address: Mitsubishi Chemical Group Science and Technology Research Center, Inc. Yokohama 227-8502, Japan

E-mail address : sukedai@mech.ous.ac.jp

Abstract. Formation of athermal ω-phase crystals due to cooling to 131 K has been directly observed in a β-type Ti-15mass%Mo alloy. The athermal ω-phase crystals easily disappear by electron irradiation during the *in-situ* observation at 131 K. Incubation phenomenon of the annihilation is also recognized. The annihilation behaviour was investigated based on the dependence on electron irradiation conditions and incubation phenomena. It is concluded that the annihilation mechanism is concerned with interactive effects of temperature rise due to electron irradiation and collective oscillation resulted from inelastic scattering of electron beam.

Introduction

Negative temperature dependence of electrical resistivity in β-Ti and Nb alloy systems has been reported, and the origin has been postulated on the formation of athermal ω-phase crystals due to cooling [1-3]. The present authors have reported that a direct observation of the formation of postulated athermal ω-phase crystals has been achieved in a Ti-15mass%Mo alloy cooled to 131 K using electron microscope technique [4, 5]. At the same time, it has been also mentioned that (1) atomic structure of athermal ω-phase crystals consists with atomic structures of ω-phase crystals formed by aging and stressing, and (2) the athermal ω-phase crystals easily disappear by electron irradiation during the *in-situ* observation, which means that an inverse, ω–β phase transformation occurs due to electron irradiation.

The occurrence of phase transformations caused by electron irradiation under different conditions has been reported in many kinds of alloy systems, and it is considered that the phase transformations are resulted from one of factors such as radiation damage, radiation enhanced and induced diffusions, and thermal diffusion due to temperature rising, or their interactive effect(s) [6, 7]. In the previous paper, the main factor for the annihilation caused by electron irradiation has been discussed on temperature rise and radiation enhanced diffusion [5]. Radiation damage has not been considered as a main factor, since the used electron microscope was operated at 200 kV and the threshold voltage for atomic displacement for Ti and Ti alloys is over 400 kV [8]. But, a reasonable conclusion has not been taken. In the present paper, a predominant factor on the onset of annihilation mechanism of athermal ω-phase crystals due to electron irradiation at 160 kV and 200 kV has been discussed, because a clarification of the predominant factor might be important to understand one of the factors of the onset of ω–β phase transformation, and the relationship between electron irradiation and phase transformations. The discussion has been done based on experimental results about (1) a dependence of onset times of annihilation of athermal ω-phase crystals upon electron irradiation conditions, and (2) the existence of incubation time to start the annihilation.

2. Experimental procedures

A Ti-15mass%Mo alloy was used for the specimen, which was solution-treated at 1073 K for 3.6 ks in vacuum, followed by water quenching. Thin foil specimens for electron microscope observations were prepared using twin-jet electropolishing method. A specimen holder, EM-31100 was used for *in-situ* observation at 131 K, and for controlling specimen temperatures. A JEM 2010F electron microscope operated at 160 kV and 200 kV was used. Dark field images and high resolution electron microscope (HREM) image observations were carried out at 131 K. Results of *in-situ* dark field image observations during cooling and heating processes were recorded using a TV-VTR system. Electron irradiation conditions were adjusted by changing acceleration voltages to 160 and 200 kV, and electron beam currents of approximately 20 and 5 pA/cm^2 were also adjusted by changing condenser apertures and condenser lens.

3. Experimental Results

In order to consider a predominant factor for the annihilation behaviour due to electron irradiation, the annihilation behaviour under four different irradiation conditions has been investigated. Specimens were cooled down to 131 K and held in the temperature for 10.8 ks without electron irradiation, and athermal ω-particles of approximately 40 nm in diameter were irradiated (*in-situ* observed). Their annihilation behaviour was recorded using a TV-VTR system, and their onset times of the annihilation (incubation times) and life times (times until disappearing) were measured by playing back the tapes.
When *in-situ* observation was carried out while a specimen was cooled down from room temperature to 131 K, the formation of athermal ω-particles was not observed [5]. Figure 1 shows an example of an athermal ω-particle and the process of annihilation due to electron irradiation with the condition of 160 kV and 5.7 pA/cm^2 [5]. It is noted that until irradiating for 255 sec, the size change of athermal ω-particles is hardly recognized. After that, the annihilation due to electron irradiation started. In this case, the irradiation time for the onset of annihilation was decided as 255 sec. This result indicates that an incubation phenomenon occurs during the annihilation behaviour.
Similar observations have been performed under four different irradiation conditions. On each condition, annihilation behaviour has been recorded several times and each onset time was decided.

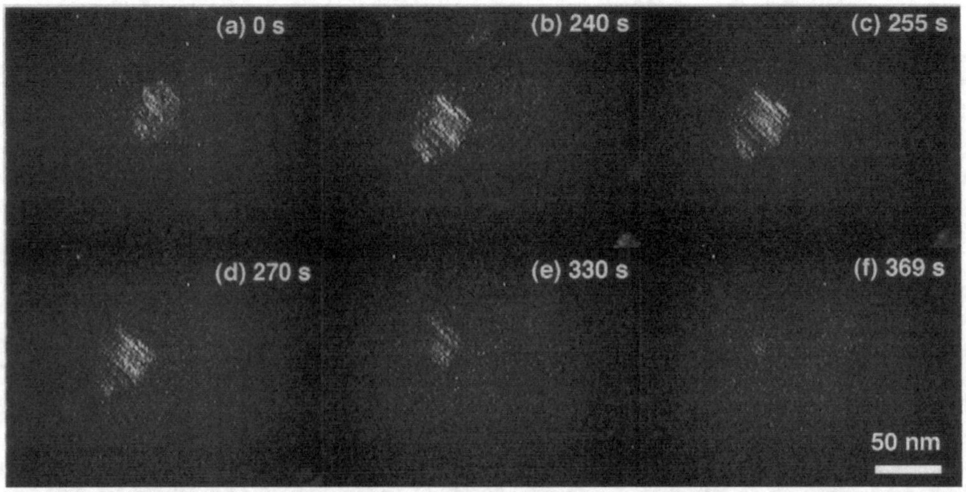

Figure 1 Annihilation behaviour of an athermal ω-particle formed by cooling at 131 K for 10.8 ks due to the irradiation of 160 kV and 5.7 pA/cm^2.

Those results are summarized in Figure 2.

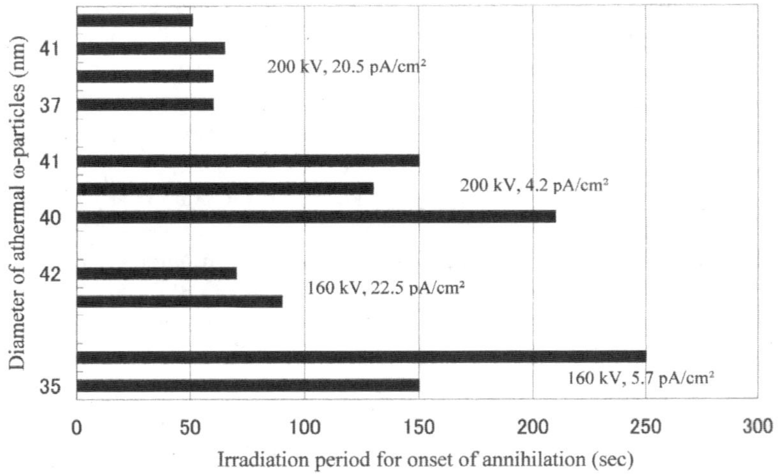

Figure 2 Electron irradiation condition dependence of the onset time for annihilation of athermal ω-particles formed by cooling at 131 K for 10.8 ks.

It is found that the incubation times depend on irradiation conditions. The following two results are revealed: (1) In the case of the same acceleration voltage, the onset times of annihilation under the irradiation with lower electron beam current density, 5 pA/cm^2 is more than two times longer than those under the irradiation with 20 pA/cm^2. (2) In the case of the same electron beam current density, the onset times of annihilation under the irradiation at 200 kV is slightly shorter than those under the irradiation at 160 kV. In the latter case, a significant difference is hardly recognized since the difference of acceleration voltages is smaller compared with the difference between beam currents. It is recognized that both of acceleration voltage and beam current affect the annihilation behaviour.

4. Discussion

The annihilation mechanism of athermal ω-particles due to electron irradiation has been considered. The present study has been carried out under the following conditions: (1) under acceleration voltages lower than the threshold voltages of atomic displacement in Ti and its alloys, and (2) at a lower temperature, 131 K. Also the atomic structure of athermal ω-phase is confirmed to consist with de Fontaine's model [9] using electron diffraction technique and HREM method [4]. Therefore, it is thought that the origin of the annihilation is not due to knock-on phenomenon but due to the effect of energy transfer from electron beam; that is, inelastic scattering between electron bean and atoms. Temperature rise of metallic and semiconductor materials due to electron irradiation has been investigated theoretically and experimentally [10, 11]. Theoretical consideration indicates that the amount of temperature rise (ΔT) is proportional to $I_o\rho/\kappa$, where I_o is the beam current at the specimen, ρ atomic density and κ thermal conductivity of the specimen [10]. A temperature rise in the present study was estimated using the experimental conditions. Thermal conductivity of some Ti-Mo system alloys is approximately 7 W/mK [12] (5 W/mK in Ti-6Al-4V at 131 K), and the beam current was 2 x 10^{17} e/cm^2s. The amount of temperature rise was 1.46 K. This result leads us to conclude that the temperature rise due to electron irradiation in the present work is not the predominant origin, because it is known that athermal ω-particles formed at 131 K were disappeared at temperature over 149 K by heating-up experiments [13].
It is also found that the annihilation of athermal ω-particles occurs only in a local area irradiated by electron beam [4]. Therefore the annihilation is strongly affected by encountering of irradiated electrons with atoms of crystal lattice. Therefore, this annihilation might be able to divided into two

primitive processes: (1) the first process occurs the moment electrons encounter atoms, and (2) the second process occurs after the first one.

The conduction of heat, which is treated in theoretically corresponds to the second process. The first process is thought a phenomenon in tremendously short period. However, irradiated electrons interact with atoms inelastically and these electrons transfer their energy to atoms of crystal lattice continuously. As the result, free energy of the irradiated area might be gradually increased.

Two collective oscillation mechanisms are described as the energy transfer mechanisms in a textbook: one is plasmon oscillation between electron cloud around atoms and electron beam, and the other one is phonon oscillation between atoms of crystal lattice and electron beam [14]. The latter one corresponds to heat, treated in the theoretical estimation. By means of these mechanisms, the continuously transferred energy from electron beam is accumulated in the irradiated area. However, energy dispersion due to the second process occurs simultaneously.

Experimental results indicated that the incubation periods exist in the annihilation process and depend on the irradiation conditions. In the present study, it is considered that the supplied energy by electron irradiation is lager than the dispersed energy due to the second process. And the moment the free energy of athermal ω-particles increased by the accumulated energy reaches to a critical value, the annihilation starts. Therefore, the origin of annihilation of athermal ω-particles due to electron irradiation is considered as not only thermal activation due to temperature rise but also the energy transferred by inelastic scattering between electron beam and atoms of crystal lattice.

5. Summary

Formation of athermal ω-phase crystals due to cooling to 131 K is recognized in a Ti-15Mo alloy. The athermal ω-phase crystals easily disappear during the *in-situ* observation even at 131 K. The origin of the annihilation was investigated based on temperature rise due to electron irradiation and the incubation time for the annihilation. It is considered that the annihilation mechanism is concerned with interactive effects of temperature rise due to electron irradiation and collective oscillations resulted from inelastic scattering of electron beam to atoms of crystal lattice.

Acknowledgement

The authors would like to express their sincere thanks to Emeritus Professor H. Takahashi and Dr N Sakaguchi of Hokkaido University for fruitful suggestions to the discussion about temperature rise and to Professors M. Ikeda of Kansai University for preparing a Ti-15Mo alloy used in the present study.

References

[1] Ames S L and McQuillan A D: 1954 *Acta Metall.* **2** 831.
[2] Vohra Y K, Menon E S K, Sikka S K and Krishnan R: 1981 *Acta Metall.* **29** 457.
[3] Yoshida S and Tsuya Y: 1956 *J. Phys. Soc. Jpn.* **11** 1206.
[4] Matsumoto H, Sukedai E and Hashimoto H 1999 *Mater. Trans. JIM* **40** 1436.
[5] Matsumoto H, Sukedai E and Hashimoto H 2000 *Microsc. Microanal.* **6** 362.
[6] Kiritani M and Takata H 1978 *J. Nucl. Mater.* **69-70** 277.
[7] Ballesteros C, Ortiz M I, Morana B, Rodriguez A and Rodriguez T 2008 *Proceedings of EMC* pp. 467-468.
[8] Makin M J, Bucley S N and Walters G P 1977 *J. Nucl. Mater.* **68** 161-167.
[9] de Fontaine D and Buck O 1973 *Philos.Mag.* **27** 967.
[10] Fisher S B 1970 *Radiat. Eff. Defect S.* **5** 239.
[11] Jenec I, Bench M W, Robertson I M and Kirk M A 1995 *J. Appl. Phys.*, **78** 974.
[12] Boyer R, Welsh G and Collings E W 1994 *Materials Properties Handbook: Titanium Alloys*, (ASM International), pp. 515, 903, 949.
[13] Sukedai E, Matsumoto H and Hashimoto H J. 2002 *Electron Microsc.* **51** S143.
[14] Williams D B and Carter C B: 1996 *Transmission electron microscopy* (Plenum Press, New York/London) pp. 51-65.

Stability of Nanoparticles in an Oxide Dispersion Strengthened Alloy

V. de Castro, S. Lozano-Perez and M. L. Jenkins
Department of Materials, University of Oxford, Parks Road, OX1 3PH, Oxford, UK

E-mail: vanessa.decastro@materials.ox.ac.uk

Abstract. Oxide Dispersion-Strengthened (ODS) reduced-activation ferritic/martensitic steels are strong candidates for structural applications in the first wall and blanket of future fusion reactors. In the present work, a model ODS alloy based on a Fe-Cr matrix has been produced in order to gain insight into the processes which occur during processing and after high-energy ion irradiation. The model ODS alloy has been characterized by Energy Dispersive X-Ray and Electron Energy Loss Spectroscopy and Energy Filtered Transmission Electron Microscopy, focusing on the evolution of the nanoparticle dispersion and the chemical composition of individual nanoparticles. The chemical analyses reveal that there are three different kinds of nanoparticles in the alloys. Particle sizes, distributions and morphologies are not changed significantly by irradiation up to doses of ~ 7 dpa.

1. Introduction

Oxide dispersion-strengthened reduced-activation ferritic/martensitic steels (ODS-RAFMS) are being considered as structural materials for the first wall and blanket of future fusion reactors. An homogeneous dispersion of hard oxide nanoparticles in the steel could block the motion of dislocations and allow an increase in their maximum service temperature of more than 100°C compared with conventional RAFMS [1]. It has been demonstrated that a dispersion of ~ 0.3wt% of Y_2O_3 particles results in better tensile and creep properties than those of the base material [2]. A major concern with these materials has been their poor impact behaviour, but this could be improved by applying suitable thermo-mechanical treatments [3]. The nanoparticles could also help to trap helium atoms and point defects, which would reduce the swelling and hardening effects which cause the deterioration of the steels under irradiation [4]. ODS steels are normally produced by pulvimetallurgical routes that include mechanical alloying (MA) of the initial powders and consolidation by hot isostatic pressing (HIP) or hot extrusion. Recent studies have shown that the chemistry of the oxide nanoparticles in ODS steels can evolve with respect to the starting powders [5, 6] and their crystallographic phase may also change. For example, the milling conditions may lead to dissolution of the initial particles. Re-precipitation of particles with different crystallography or chemical composition than the starting particles could significantly affect their dissolution behaviour under irradiation and the mechanical performance of the material.

In order to gain insight into these processes, we focus here on the study of a simplified Fe12%Cr ODS alloy. This work reports the evolution of the nanoparticle dispersion after ion irradiation at 500° C up to doses of ~ 7 dpa.

2. Experimental methods

The model ODS alloy had a nominal composition Fe-12Cr-0.4 Y_2O_3 (wt%). The Y_2O_3 starting powders were spherical and had sizes ranging from ~ 9 to 100 nm. The lattice parameters obtained from X-ray diffraction of the powders were consistent with a monoclinic cell [7]. The alloy was produced by MA followed by HIP as described in [8]. After consolidation, the alloy was tempered at 1023 K for 4h. Disks of diameter 3 mm and thickness ~ 60 μm were then cut and prepared for irradiation by flash electropolishing for ~ 1 s to remove possible superficial damage produced during the grinding process. The disks were irradiated at 500ºC to doses of 2.5 and 7 dpa. Both 0.5 and 2 MeV Fe+ ions were used in order to obtain a flat damage profile within about 1 μm of the ion-entry surface. The irradiated disks were subsequently prepared for TEM by flash electropolishing for ~ 2 s to remove about ½ μm from the irradiated surface, followed by back thinning by jet electropolishing in a TENUPOL 5 twinjet polisher using 5 % $HClO_4$ + 95 %CH_3OH as electrolyte.

TEM characterisation of the particles before and after irradiation was carried out by high angle annular dark field (HAADF) scanning TEM (STEM), energy dispersive spectroscopy (XEDS) and electron energy loss spectroscopy (EELS) using a JEOL 3000F TEM operated at 300kV. Energy filtered TEM (EFTEM) maps were obtained with a JEOL 2200MCO prototype microscope equipped with two spherical aberration (Cs) correctors.

3. Results and discussion

EFTEM maps in figure 1 show the Y-rich nanoparticle dispersion before and after irradiation. Details of the extraction of Y maps can be found in [9].

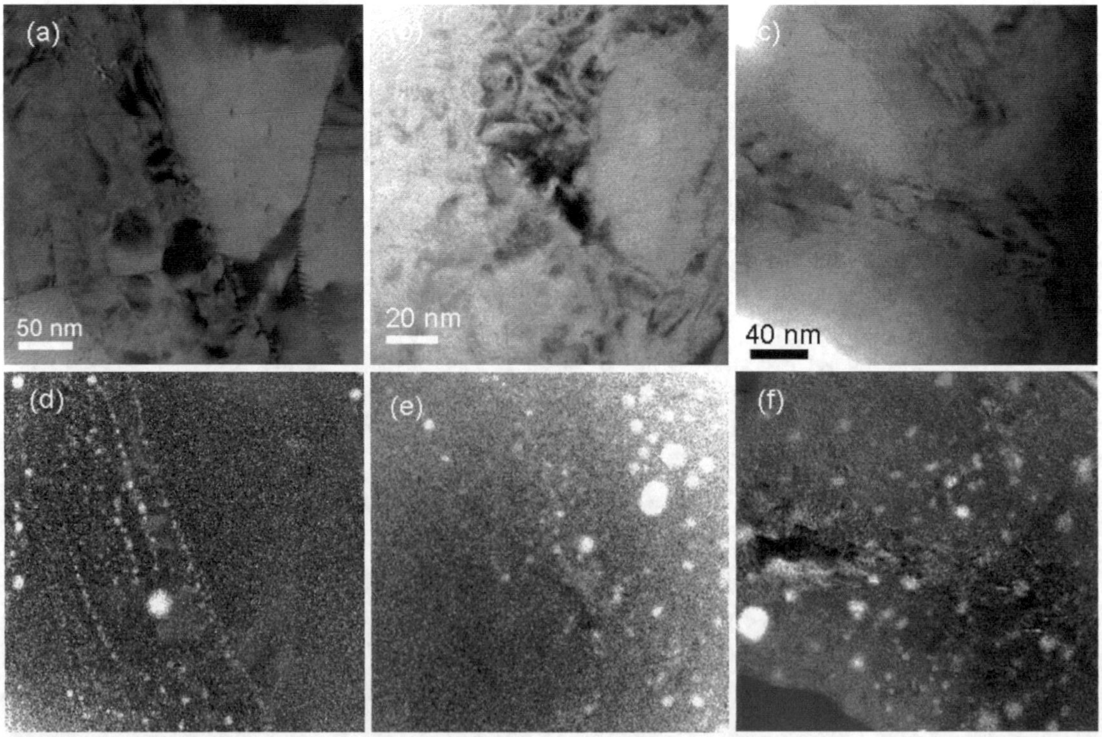

Figure 1. BF images and corresponding Y $N_{2,3}$ EFTEM maps of the ODS/Fe12Cr showing the distribution of Y-rich ODS particles before (a), (d) and after irradiation at 2.5 dpa (b), (e) and 7dpa (c), (f).

Particle sizes, distribution and densities were similar in the three conditions, suggesting that they had not undergone dissolution under irradiation. The particle size distribution varied from one region to another, with individual ODS particle sizes in the range 1-50 nm. The particles were not homogeneously distributed. Particle densities were calculated for different regions with average volumes of 250x250x100 nm^3. The foil thickness was measured using the log-ratio technique where the Malis formula was used to estimate the inelastic mean free path [10]. Densities were found to vary between 0 and 1.1×10^{23} m^{-3}. The particles were frequently found within the grains, but they can also be aligned along grain boundaries as shown in Figure 1 (a) and (d).

The chemistry of the particles was studied before irradiation in order to analyze their evolution during processing and after irradiation, with the aim of looking for possible changes caused for example by irradiation-enhanced diffusion. Nanoparticles with different chemical compositions were found. Figure 2 (a) is a STEM-HAADF image showing nanoparticles in the non-irradiated alloy. The chemical compositions of the particles marked as A, B and C were analyzed by EELS and XEDS, see figure2 (b). Particle B has a higher contrast than A and C. Particles A and B are attached, appearing as halves of a bigger particle. Contrast changes in HAADF images can be due to differences in atomic number, density or thickness. The linescans reveal that the change in contrast found for particle B is due to a difference in chemical composition, as shown in figure 2 (b). This particle contains Y, Cr and O, while particles A and C, with a lower contrast, contain Cr and O but no Y. The high O content found in the matrix is attributed to the formation of an oxide film on the sample. The increment in the O concentration in the matrix in the direction of the linescan is probably an artefact arising from a decrease in thickness, which is suggested by the decrease in the HAADF signal of the image. Thinner regions would have less Fe and Cr content, but the same surface oxide thickness, forcing the quantitative O signal to increase.

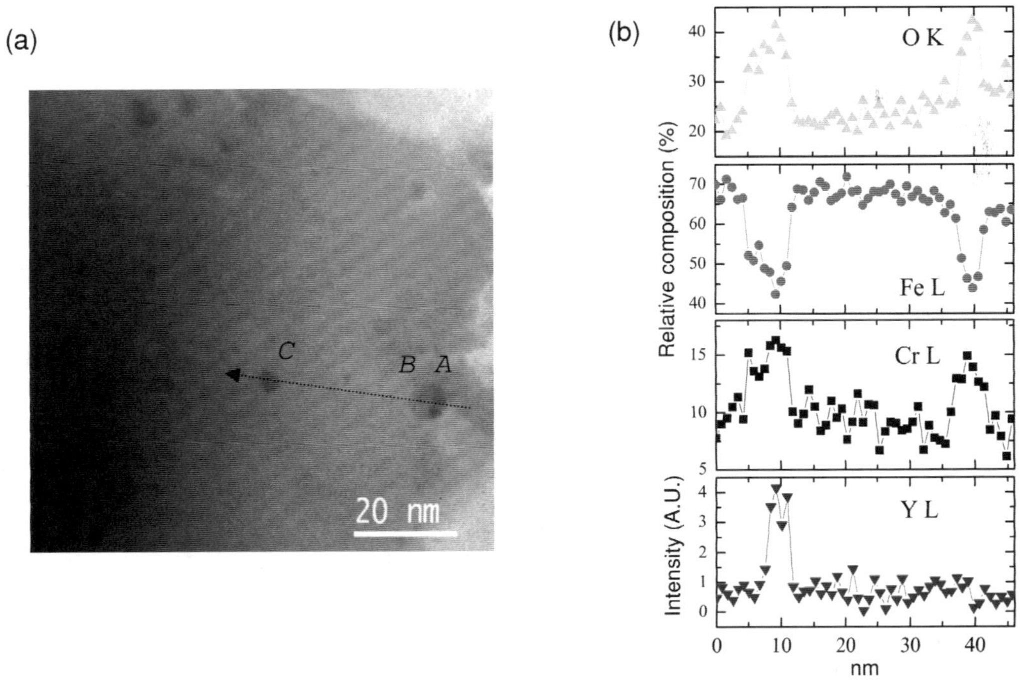

Figure 2. (a) STEM-HAADF image of a region in the non-irradiated ODS alloy, and (b) EELS elemental profile of O, Fe and Cr and EDS elemental profile of Y across the line marked in (a).

Other nanoparticles present in the alloy consist of an Y-O rich core and a Cr-rich shell. This core-shell structure is shown in figure 3. The chemical compositions of the particles do not seem to change after irradiation, at least up to a dose of 2.5 dpa, and the Cr shell can be still observed [9]. Analyses of the particle composition, especially the core-shell structure at 7 dpa, are ongoing.

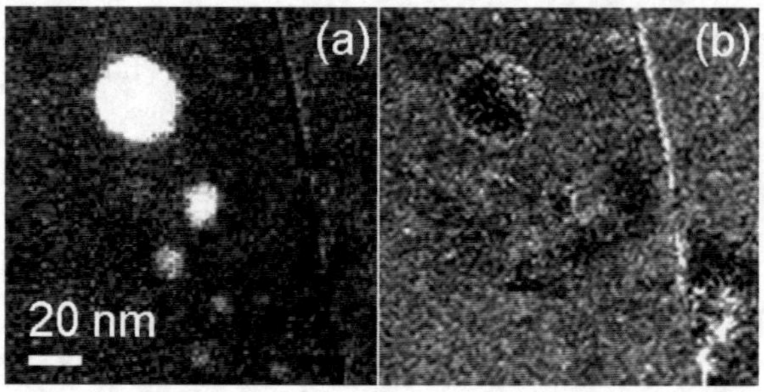

Figure 3. (a) Y $N_{2,3}$ EFTEM map showing Y-rich nanoparticles in the non-irradiated alloy and (b) Cr $M_{2,3}$ EFTEM map showing that a Cr shell forms around some of the particles.

4. Conclusions

This work describes the nanoparticle dispersion present in an oxide dispersion-strengthened ODS/Fe12wt%Cr alloy and its evolution after Fe^{+} irradiation at 500ºC at doses of 2.5 and 7 dpa. An inhomogeneous dispersion of nanoparticles with sizes < 50 nm was found. Particle sizes, distributions and morphologies appear unchanged up to an implantation dose of 7 dpa, suggesting that they have not undergone dissolution under irradiation. The chemical composition of the particles was studied by STEM-EELS-EDS and EFTEM. Some Y-rich particles consist of an Y-O rich core and a Cr rich shell, while others contain Y-Cr-O. Particles containing only Cr and O were also detected. The chemistry of the particles appears to be unaltered at least up to a dose of 2.5 dpa, confirming their stability under these irradiation conditions. Compositional analyses at 7 dpa are in progress.

5. References

[1] A. Ramar, N. Baluc and R. Schaüblin 2007 *J. Nucl. Mater.* **367-370** 217.
[2] R. Schäublin, T. Leguey, P. Spätig, N. Baluc, M. Victoria 2002 *J. Nucl. Mater.* **307-311** 778.
[3] R. Lindau, A. Möslang, M. Rieth, M. Klimiankou, E. Materna-Morris, A. Alamo, A. F. Tavassoli, C. Cayron , A. M. Lancha , P. Fernandez , N. Baluc, R. Schäublin, E. Diegele, G. Filacchioni, J.W. Rensman, B.v.d. Schaaf , E. Lucon , W. Dietz 2005 *Fus.Eng.Des.* **75–79** 989.
[4] G. R. Odette, M. J. Alinger ad B. D. Wirth 2008 *Anu. Rev. Mater. Res.* **38** 471.
[5] M. K. Miller, D. T. Hoelzer, E. A. Kenik and K. F. Russel 2004 *J. Nucl. Mater.* **329-333** 338.
[6] E. A. Marquis 2008 *Appl. Phys. Lett.* **93** 181904.
[7] V. de Castro, T. Leguey, M.A. Monge, A. Munoz, R. Pareja, D.R. Amador, J.M. Torralba and M. Victoria 2003 *J. Nucl. Mater.* **322** 228.
[8] V. de Castro, T. Leguey, A. Muñoz, M.A. Monge, R. Pareja, E. A. Marquis, S. Lozano-Perez and M.L. Jenkins 2009 *J. Nucl. Mater.* **386–388** 449.
[9] S. Lozano-Perez, V. de Castro and R. Nicholls 2009 *Ultramicroscopy* **109** 1217.
[10] T. Malis, S. C. Chend and R. F. Egerton 1988 J. Electron Microscopy Technique **8** 193.

Acknowledgments

The authors thank Prof. R. Pareja's group at the Universidad Carlos III de Madrid for providing the ODS alloy. V. de Castro acknowledges the FP6 Euratom Research and Training Programme on Nuclear Energy for funding this research. S Lozano-Perez is grateful to INSS (Japan) for sponsoring his research.

Tracing C changes in a C/CrC PVD coating using Raman spectroscopy and EELS

Z Zhou[a]*, IM Ross[a], WM Rainforth[a], and P Eh Hovsepian[b]
[a]Department of Engineering Materials, University of Sheffield, Sheffield, S1 3JD, UK
[b]Materials Engineering Research Institute, Sheffield Hallam University, Sheffield, S1 1WB, UK

E-mail: z.zhou@sheffield.ac.uk

Abstract. A C/CrC coating was deposited by unbalanced magnetron sputtering of graphite and Cr metal targets in argon atmosphere for low friction tribological applications. The coating possessed a nanocomposite structure with amorphous carbon embedded in a metastable fcc CrC matrix. The nanocomposite structure was annealed at 750°C in Ar+5%H_2 atmosphere for 30 minutes to evaluate its thermal stability. Microstructures of the as-deposited and annealed coating were characterised using Raman spectroscopy and cross sectional transmission electron microscopy coupled with electron energy loss spectroscopy (TEM/EELS). Raman spectroscopy suggested the presence of graphitic carbon in the coating after annealing. EELS (ELNES of C K) combined with HRTEM investigation of the cross sections revealed graphitic C at the very top (~200nm) of the coating and amorphous C at the remaining part of the coating after annealing at 750°C. The advantages and limitations of Raman spectroscopy and TEM techniques in studying C/CrC coating are also discussed.

1 Introduction

Nanoscale hydrogen free C/Cr coatings have been produced by unbalanced magnetron sputtering of graphite and Cr metal targets with microstructures and properties strongly dependent on the substrate bias voltage [1]. The coatings evolve from an amorphous (-65V to -95V substrate bias voltage) through an 'onion-like' carbon structure (-120V) to nanoscale multilayer structure (-350V and -450V) and finally a uniform fine grain structure (-550V). The multilayer coating deposited at a bias voltage of -350V has attracted considerable attention. The coating possesses a nanocomposite structure with amorphous carbon embedded in a CrC matrix. The nanoscale amorphous carbon clusters self-assemble into layers alternated by CrC, giving a multilayer appearance. The fcc CrC is a thermodynamically non-equilibrium phase which can be stabilized by high ion energy magnetron sputtering. The thermal stability of the coating is therefore of great concern.

Raman spectroscopy is a popular non-destructive technique for characterization of carbon in C based coatings especially diamond like carbon coatings with minimum sample preparation. The main features in the Raman spectra of carbons are the D (disorder) and G (graphite) peaks, which lie at ~1360 and ~1560cm^{-1} for visible excitation. Both D and G peaks are due to sp2 vibrations only. The G peak is due to the band stretching of all pairs of sp2 atoms in both rings and chains. The D peak is due to the breathing modes of sp2 atoms in rings. For diamond or samples containing a significant fraction of diamond, the sp3 peak is seen at 1332cm^{-1} [2].

In the current work, the thermal stability of the multilayer structure (-350V substrate bias) was investigated by annealing heat treatment at 750°C in an Ar+H_2 atmosphere for 30 minutes. Detailed Raman spectroscopy and cross sectional TEM study was conducted to understand the microstructure changes of the coating.

2 Experimental procedures

C/Cr coatings were deposited by the combined steered cathodic arc/UBM sputtering (ABS™: Arc-Bond sputtering) technique using a Hauzer HTC 1000-4 PVD coater. The coating chamber had three graphite and one chromium metal targets, with AISI304 stainless steel substrates. The coating process comprises three steps: (i) Cr^+ etching in cathodic arc mode by applying substrate bias of -1200V; (ii) CrN base layer deposition using reactive sputtering of Cr in $Ar+N_2$ atmosphere with substrate bias -75V; (iii) C/Cr multilayer formation using UBM sputtering in a non-reactive Ar atmosphere with substrate bias at -350V. More details of the coating process have been reported elsewhere [1].

The Raman spectroscopy analysis was conducted at room temperature by using a Renishaw inVia Raman microscope with 20mW 514 nm laser. The spectra were acquired with 10% power filter, which was used to avoid heating of the samples. The objective lenses (x50) produced a ~1μm laser spot. Cross sections for TEM were prepared by conventional methods with ion milling to electron transparency by a Gatan precision ion polishing system (PIPS). Bright field TEM micrographs and selected area diffraction patterns were recorded using a Phillips EM 420 TEM and a JEOL 2010 field emission gun (FEG) TEM. Electron energy loss spectra were acquired on the JEOL 2010 FEGTEM equipped with a Gatan GIF2000 energy filter.

3 Results and discussion

Figure 1 compares the Raman spectra of the plan-view surfaces of the as-deposited coating and the coating after annealing at 750°C (intensity normalized to maximum). The as-deposited coating shows a featureless spectrum, whereas the annealed coating has Raman bands at 1356, 1591, 2950cm^{-1} assigned to carbon D, G and D'' [2,3]. The additional band at 2710cm^{-1}, was historically named G', since it is the second most prominent band always observed in graphite samples (the first is G), now designated as 2D as it is actually 2nd order of D peak [2]. Thus, the spectrum of the annealed surface is believed to contain polycrystalline graphitic carbon. Raman studies of similar C/Cr coatings suggested a transformation from graphitic (sp2 C-C) bond to metal carbide (Cr-C) bond owing to increased ion energy bombardment from -65V to -350V bias voltage [4]. The as-deposited coating at -350V was then believed to be essentially carbide with minimum sp2 bonded carbon. After annealing at 750°C, more ordered carbon was present in the coating. Discovering where the ordered carbon is located is the first step to understand the stability of the carbide coating. Cross section TEM coupled with EELS was then conducted to trace the change of C changes after annealing.

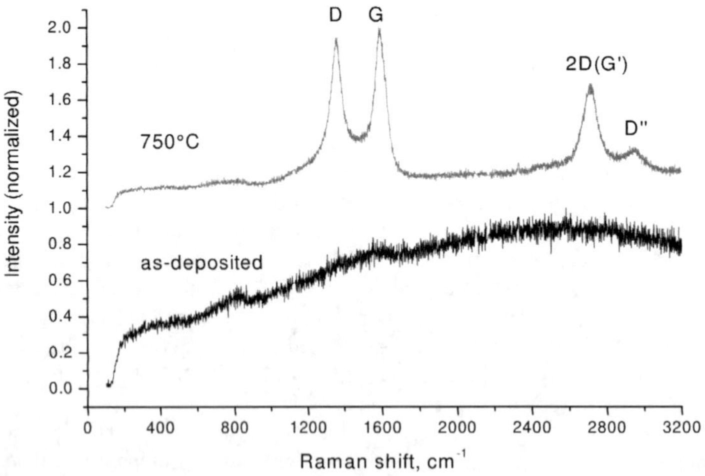

Figure 1. Raman spectra of the plan-view surface of as-deposited coating and coating annealed at 750°C.

Figure 2 gives the TEM cross sections of the as-deposited coating (1.29 μm thick) and 750°C annealed coating. The as-deposited coating is composed of a CrN base layer (~250nm) and the C/CrC coating (~1040nm). The C/CrC coating is composed of C clusters (bright contrast) and CrC matrix (dark contrast), as shown in Figure 2a. The C clusters segregated into non-continuous layers, giving the coating multilayer appearance. Three major zones are shown in Figure 2b after annealing at 750°C. (1) 1/6 from the very top of the coating lost its multilayer and became nanocrystalline. (2) ½ of the coating in the middle zone was still multilayer C/CrC, as identified by electron diffraction (inset). (3) 1/3 from the bottom of the coating had lost its multilayer structure, but the columnar grain structure could be clearly seen. C clusters were still visible with larger size than the as-deposited condition. Carbon cluster coalescence occurred at the very bottom of the coating. Selected area electron diffraction revealed that the columnar grains at the bottom of the coating were probably Cr_3C_2 and the top nanocrystallites contained Cr_2O_3.

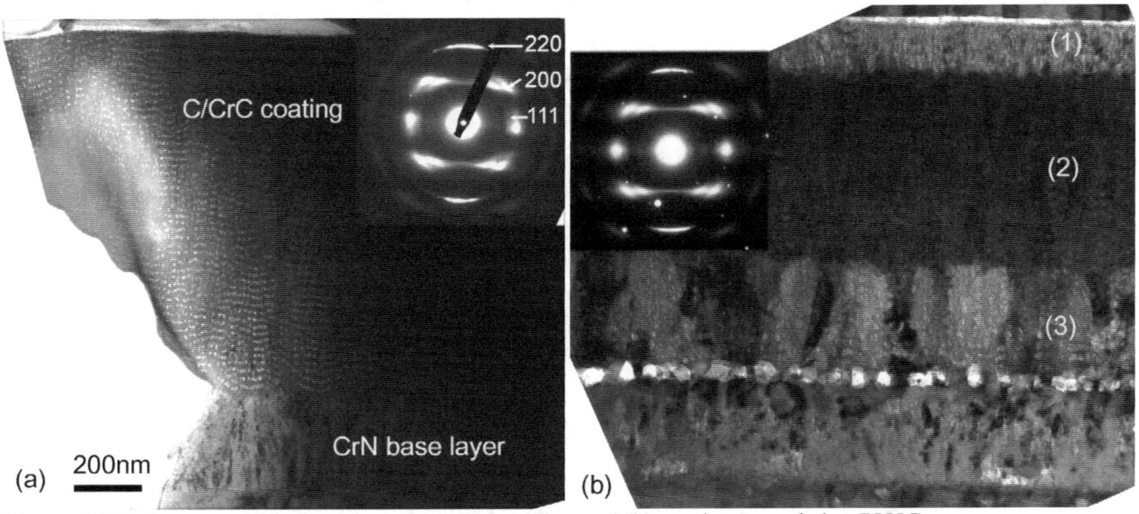

Figure 2. TEM cross sections of (a) as-deposited coating and (b) coating annealed at 750°C.

Figure 3 shows EEL spectra from all these three zones compared to the multilayer as-deposited coating. Reference spectra of CrC[5], Cr_3C_2[6] and natural graphite[7] were also included. Note (1) no significant difference in the carbon core loss between CrC and Cr_3C_2 reference spectra, (2) CrC and Cr_3C_2 both have no C sp2 bond but show sharp peaks at 285.0eV, the same energy as π peak in graphite. All the spectra are aligned and normalised to the peak at 285.0eV (i.e. the shift of C K edge is not considered here). The EEL spectra were acquired using a 200nm selected area aperture with microscope in diffraction mode (i.e. image-coupling to spectrometer). The microscope condition yielded a 1.1eV energy resolution. With a 0.5ev/channel dispersion, both C-K and Cr $L_{2,3}$ edges were observed in all these three parts of the coating. The spectrum of the as-deposited coating resembles the CrC reference with a higher σ peak than the reference, probably due to large contribution of amorphous C clusters included in the selected area. The C K ELNES of the columnar grains at the bottom of the annealed coating (zone 3) is consistent with Cr_3C_2. The mid zone 2 had reduced σ intensity compared to the as-deposited condition, suggesting it may have less amorphous C than the as-deposited condition. It is evident that the size of the C clusters became larger in the 750°C annealed sample than the as-deposited coating. The spectrum of the top nanocrystalline zone 1 resembles the features of typical hydrogenated diamond like carbon coatings [8], having a sharp π peak followed by a pronounced σ peak. However, the graphite σ peak at 292.5eV followed by a plateau was not observed, even with a 0.1eV/channel dispersion. Since the spectra intensity was normalized to π peak, it allows comparison of the Cr concentration, suggesting a depletion of Cr towards the coating surface.

Figure 4 compares the morphology of the C clusters in the as-deposited condition to those at zones 2 and 1 of the annealed coating. It is evident that the C retained its amorphous nature in the C/CrC nanocomposite region (zone 2), as shown in Figures 4a and b, although the population of C clusters reduced after annealing. In contrast to the EELS data, there was clear evidence of graphitic C at zone 1 (Figure 4c). The EEL spectrum from the top zone 1 was more likely to be collected from an area of random orientated graphite crystallites.

The lateral spatial resolution of the laser Raman microscopy is ~1 μm, which is similar to the thickness of the entire C/CrC coating and therefore Raman was not performed on cross sectional samples. The penetration depth of a 632nm laser Raman beam was estimated to be in the range of 50-100nm with x100 objective [9], so the Raman spectra acquired from the plan-view surfaces should have only included the top zone featuring graphitic carbon detected by HREM. Thus, the cross section TEM coupled with selected area EELS and HREM suggested that the only at the top zone 1 was graphitic carbon found, C remained amorphous in the middle C/CrC and bottom C/Cr_3C_2 zones.

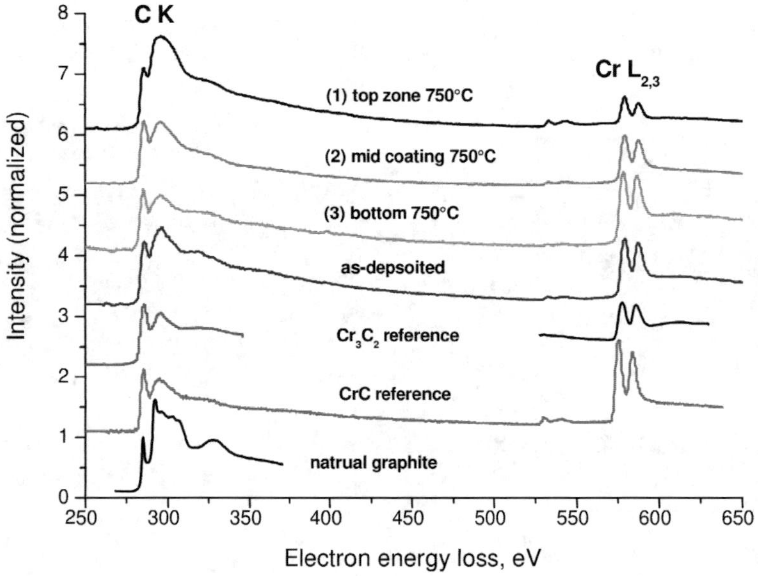

Figure 3. EEL spectra from the three zones in the coating annealed at 750°C compared to the as-deposited coating and references.

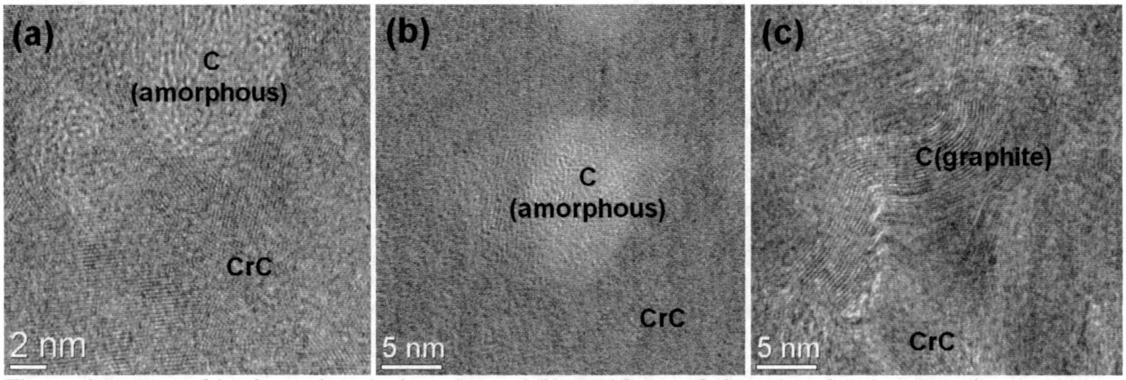

Figure 4. HREM of (a) the as-deposited coating and (b) 750°C annealed coating showing amorphous carbon and defect containing CrC matrix. (c) HREM evidence of graphitic C on the top zone in the 750°C annealed coating.

4 Conclusions

The C/CrC nanocomposite has partially changed its multilayer structure after annealing at 750°C into three zones: top nano-crystallites, middle C/CrC composite (with larger size of C clusters than as-deposited condition) and bottom columnar Cr_3C_2 with significant C clusters coalescence. Raman spectroscopy on the plan-view surfaces detected graphitic carbon. EELS and HREM study revealed graphitic carbon at the top 200nm layer whereas C clusters remains largely unchanged amorphous at the middle C/CrC and bottom Cr_3C_2 zones.

References:

[1] Hovsepian P Eh, Kok Y N, Ehiasarian A P, Haasch R, Wen J G, Petrov I 2005 *Surf. Coat. Technol.* **200** 1572
[2] Ferrari A C 2007 *Solid State Commun.* **143** 47
[3] Vidano R P, Fishbach D B, Willis L J, Loehr T M 1981 *Solid State Commun.* **39** 341
[4] Kok Y N, Hovsepian P Eh, Haasch R, Petrov I 2005 *Surf. Coat. Tech.* **200** 1117.
[5] Gassner G, Patscheider J, Mayrhofer P H, Hegedus E, Toth L, Kovacs I, Pecz B, Srot V, Scheu Ch, Mitterer C 2006 *J. Vac. Sci. Technol.* **24B** 1837
[6] Hug G, Leapman R, and Jaouen M. *Microbeam Analysis-1995*, Breckenridge: VCH Publishers, Inc. p303-304.
[7] Abidate O. http://www.cemes.fr/~eelsdb/.
[8] Fan X, Dickey E C, Pennycook S J, Sunkara M K 1999 *Appl.Phys.lett* **75** 2740.
[9] Vallerot J-M, Bourrat X, Mouchon A, Chollon G 2006 *Carbon* **44** 1833.

AUTHOR INDEX

AUTHOR INDEX

AUTHOR INDEX